# Carboxylic Ortho Acid Derivatives

*Preparation and Synthetic Applications*

### ROBERT H. DeWOLFE
UNIVERSITY OF CALIFORNIA
SANTA BARBARA, CALIFORNIA

 1970

ACADEMIC PRESS  New York and London

COPYRIGHT © 1970, BY ACADEMIC PRESS, INC.
ALL RIGHTS RESERVED
NO PART OF THIS BOOK MAY BE REPRODUCED IN ANY FORM,
BY PHOTOSTAT, MICROFILM, RETRIEVAL SYSTEM, OR ANY
OTHER MEANS, WITHOUT WRITTEN PERMISSION FROM
THE PUBLISHERS.

ACADEMIC PRESS, INC.
111 Fifth Avenue, New York, New York 10003

*United Kingdom Edition published by*
ACADEMIC PRESS, INC. (LONDON) LTD.
Berkeley Square House, London W1X 6BA

LIBRARY OF CONGRESS CATALOG CARD NUMBER: 70-84226

PRINTED IN THE UNITED STATES OF AMERICA

# Preface

This book is a critical survey of the preparation, properties, and reactions of the principal classes of ortho acid derivatives. It grew out of a literature survey on the chemistry of ortho esters, undertaken in connection with a research program. As the survey progressed, it became clear that the voluminous literature on reactions of ortho acid derivatives, poorly indexed by the abstract journals, is not readily accessible to one unfamiliar with the field. Yet these reactions are synthetically useful and mechanistically interesting, and merit the attention of chemists interested in a broad spectrum of problems.

Ortho acid derivatives are substances which have three or four oxygen, nitrogen, sulfur, or halogen atoms bonded to the same carbon atom. The more important classes of ortho acid derivatives are the ortho esters, the ortho thioesters, the amide acetals, the ester aminals, and the ortho amides. These substances are extremely versatile intermediates in synthetic organic chemistry, and are the reagents of choice for the preparation of a large number of acyclic and heterocyclic compounds. In spite of their importance, no recent review of the chemistry of ortho acid derivatives is available. Post's book on ortho esters, published twenty-five years ago, covers only about a fourth of the literature on these compounds. The ortho amides, amide acetals, and ester aminals were discovered relatively recently, and their chemistry has not been summarized.

Where justified, mechanistic interpretations are provided for the various reactions discussed. Since the long-term value of this type of treatise is as a key to the literature, I have made every effort to make the bibliography as complete as possible through mid-1968.

Of the various classes of ortho acid derivatives, two are discussed only as intermediates in the preparation of other ortho acid derivatives. These are the halides derived from ortho acids (dihaloethers and haloacetals), and carbonium salts having two oxygen, nitrogen, or sulfur functions bonded to the carbonium carbon. Trinitromethane and its derivatives, which differ fundamentally in their chemical properties from other ortho acid derivatives, are not considered.

I am indebted to Professor Stanley J. Cristol and his colleagues in the Department of Chemistry at the University of Colorado for their hospitality during the fall and winter of 1965, when this project was begun, and to my wife Barbara, whose helpful suggestions contributed much to its completion.

*Santa Barbara, California*  ROBERT H. DEWOLFE
*May, 1969*

# Contents

*Preface* . . . . . . . . . . . . . . . . . . . . . v

## Chapter 1. Synthesis and Properties of Carboxylic Ortho Esters and Related Compounds

I. Introduction . . . . . . . . . . . . . . . . . . 1
II. Synthesis of Ortho Esters from Nitriles and Imidic Esters . . . . . . 2
III. Synthesis of Ortho Esters from Trihalomethyl Compounds and $\alpha$-Halo Ethers 12
IV. Synthesis of Ortho Esters from Other Ortho Esters . . . . . . . 18
V. Formation of Ortho Esters by Reactions involving addition of Alcohols to Carbon–Carbon Multiple Bonds. . . . . . . . . . . . . 32
VI. Synthesis of Ortho Esters from Carbonium Salts . . . . . . . . 36
VII. Synthesis of Cyclic and Spirocyclic Ortho Esters from Epoxides and Formate Esters or Lactones . . . . . . . . . . . . . . . . 38
VIII. Formation of Ortho Esters by Cyclizations of Acyloxy Compounds. . . . 40
IX. Synthesis of Carboxylic Ortho Esters from Tetraalkyl Orthocarbonates . . 44
X. Formation of Carboxylic Ortho Esters by Reactions of Dialkoxycarbenes with Alcohols . . . . . . . . . . . . . . . . . . 45
XI. Synthesis of Ortho Esters by Addition of Ketene Acetals to $\alpha,\beta$-Unsaturated Carbonyl Compounds . . . . . . . . . . . . . . . 46
XII. Miscellaneous Reactions Which Yield Ortho Esters . . . . . . . 47
XIII. Synthesis of Peroxy and Hydroperoxy Ortho Esters . . . . . . . 50

| | | |
|---|---|---|
| XIV. | Synthesis of Acyloxy Ortho Esters | 52 |
| XV. | Physical Properties of Ortho Esters | 54 |
| | References | 122 |

## Chapter 2. Reactions of Ortho Esters Which Result in Carbon–Oxygen and Carbon–Halogen Bond Formation

| | | |
|---|---|---|
| I. | Hydrolysis of Carboxylic Ortho Esters | 134 |
| II. | Reactions of Carboxylic Ortho Esters Which Yield Ethers | 146 |
| III. | Reactions of Carboxylic Ortho Esters Which Yield Carboxylate Esters | 152 |
| IV. | Conversion of Carbonyl Compounds to Hemiacetals, Acetals, and Ketals | 154 |
| V. | Reactions of Carboxylic Ortho Esters Which Yield Esters of Acids Other than Carboxylic and Carbonic Acids | 164 |
| VI. | Reactions of Carboxylic Ortho Esters Which Yield Alkyl Halides | 167 |
| | References | 168 |

## Chapter 3. Reactions of Ortho Esters Which Result in Carbon–Nitrogen or Carbon–Phosphorus Bond Formation

| | | |
|---|---|---|
| I. | Introduction | 178 |
| II. | Reactions Which Yield Acyclic Products | 179 |
| III. | Reactions Which Yield Nitrogen Heterocycles | 195 |
| IV. | Reactions Resulting in Carbon–Phosphorus Bond Formation | 212 |
| | References | 215 |

## Chapter 4. Reactions of Ortho Esters Which Involve Carbon–Carbon or Carbon–Hydrogen Bond Formation

| | | |
|---|---|---|
| I. | Introduction | 223 |
| II. | Reactions of Ortho Esters with Organometallic Compounds | 224 |
| III. | Reactions of Ortho Esters with Metal Hydrides | 231 |
| IV. | Reactions of Ortho Esters with Compounds Which have Activated Carbon–Hydrogen Bonds | 231 |
| V. | Electrophilic Additions of Orthocarboxylates to Carbon–Carbon Double Bonds | 266 |
| VI. | Aromatic Substitution Reactions of Ortho Esters | 269 |
| VII. | Reactions of Ortho Esters with Ethyl Diazoacetate and with Diazo Ketones | 271 |
| VIII. | Miscellaneous Ortho Ester Addition Reactions Which Form Carbon–Carbon Bonds | 272 |
| IX. | Elimination Reactions of Ortho Esters–Synthesis of Ketene Acetals | 274 |
| X. | Other Elimination Reactions of Ortho Esters | 277 |
| | References | 278 |

## Chapter 5. Carbohydrate Ortho Esters

| | | |
|---|---|---|
| I. | Introduction | 298 |
| II. | Synthesis and Properties of Carbohydrate Ortho Esters | 299 |
| III. | Reactions of Carbohydrate Ortho Esters Which Yield Non-Ortho Ester Products | 331 |
| | References | 342 |

## Chapter 6. Thioorthocarboxylates, Thioorthocarbonates, and Related Compounds

I. Introduction . . . . . . . . . . . . . . . . . . . . 348
II. Syntheses of Trithioorthocarboxylates, Tetra-Thioorthocarbonates, and Related Compounds . . . . . . . . . . . . . . . 349
III. Properties and Sources of Thioorthocarboxylic and Thioorthocarbonic Acid Derivatives . . . . . . . . . . . . . . . . . . . . . . 373
IV. Reactions of Thioorthocarboxylates, Thioortho-Carbonates, and Related Compounds . . . . . . . . . . . . . . . . . . . 373
References . . . . . . . . . . . . . . . . . . . . . . 414

## Chapter 7. Amide Acetals, Ester Aminals, and Ortho Amides

I. Introduction . . . . . . . . . . . . . . . . . . . . 420
II. Syntheses of Amide Acetals, Ester Aminals, and Ortho Amides . . . . . 421
III. Properties and Sources of Amide Acetals, Ester Aminals, and Ortho Amides 450
IV. Reactions of Amide Acetals, Ester Aminals, and Ortho Amides . . . . . 475
References . . . . . . . . . . . . . . . . . . . . . . 501

Author Index . . . . . . . . . . . . . . . . . . . . . . 507

Subject Index . . . . . . . . . . . . . . . . . . . . . . 547

CHAPTER I

# Synthesis and Properties of Carboxylic Ortho Esters and Related Compounds

## I. INTRODUCTION

Carboxylic ortho acids, $RC(OH)_3$, are hydrates of ordinary carboxylic acids. They are thermodynamically so unstable relative to the carboxylic acid and water that the equilibrium concentration of ortho acid in aqueous solutions of carboxylic acids is ordinarily too small to be detectable (63b).

Although ortho acids themselves have not been isolated, a number of their derivatives are reasonably stable substances. This chapter discusses the preparation and properties of carboxylic ortho acid derivatives in which three oxygen atoms are bonded to an acyl carbon by single bonds—that is, compounds having the grouping

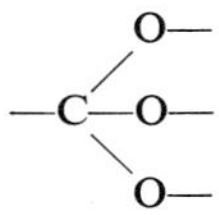

$$-C(-O-)(-O-)(-O-)$$

in their structures.

Substances possessing this structural feature include the carboxylic ortho esters, $RC(OR')_3$, and heterocyclic compounds related to them. Closely related to the ortho esters are peroxy ortho esters, which have an –OOR or –OOH group bonded to the acyl carbon; hydrogen ortho esters, which have an –OH group bonded to the acyl carbon; and acyloxy ortho esters, which have an –OCOR group bonded to the acyl carbon.

Several of the reactions used for preparing ortho esters and related compounds are widely applicable; their scope, limitations, and mechanisms are discussed in detail. Other reactions, which were used to synthesize one or a few ortho esters, or which yield an ortho ester as a by-product, are described briefly.

The last section of the chapter is a compilation of physical properties of ortho esters and related compounds, together with methods which have been used to synthesize each compound and references to the original literature. This should provide a convenient source of information on known ortho esters and suggest methods of synthesizing related esters which have not yet been described in the literature.

## II. SYNTHESIS OF ORTHO ESTERS FROM NITRILES AND IMIDIC ESTERS

### A. Introduction

The most generally applicable synthesis of ortho esters involves alcoholysis of imidic ester hydrochlorides, which are usually prepared by the addition of an alcohol to a nitrile in the presence of anhydrous hydrogen chloride.

$$RCN + R'OH + HCl \rightarrow RC(=NH_2)OR'^+Cl^- \tag{1}$$

$$RC(=NH_2)OR'^+Cl^- + 2 R'OH \rightarrow RC(OR')_3 + NH_4^+Cl^- \tag{2}$$

This reaction, sometimes called the Pinner synthesis, was first used in 1883 for the preparation of a series of trialkyl orthoformates from hydrogen cyanide (*275, 276*). Pinner, whose main interest was the preparation and properties of imidic esters and their salts (*278*), alcoholyzed only alkyl formimidate salts.

A quarter of a century elapsed before Pinner's reaction was used to synthesize ortho esters from nitriles. In 1907 Reitter and Hess described the preparation of triethyl orthoacetate and triethyl orthopropionate [Eq. (2), R = $CH_3$, $C_2H_5$; R' = $C_2H_5$] by this method (*286*). Fifteen years later Staudinger and Rathsam prepared triethyl phenylorthoacetate from phenylacetonitrile (*333*), and shortly after this Sah reported the synthesis of a series of orthoacetates and phenylorthoacetates from the corresponding nitriles (*309, 311, 313*). In 1935 Brooker and White described improved procedures for the conversion of nitriles to ortho esters, and reported the synthesis of trimethyl esters of orthopropionic, orthobutyric, orthovaleric, orthocaproic, and orthoisocaproic acids, as well as triethyl phenoxyorthoacetate and triethyl orthobenzoate (*50*).

Most of our knowledge of the mechanism of the Pinner synthesis and the side reactions which interfere with it is due to McElvain and co-workers, who used the reaction to prepare ortho esters as intermediates in the synthesis of ketene acetals (*35, 202, 203, 206, 208, 213, 214, 217, 221–223, 225–230*).

**B. Synthesis of Imidic Ester Hydrochlorides**

The synthesis of ortho esters from nitriles is usually carried out as a two-step process, the first step of which is preparation of an imidic ester hydrochloride [Eq. (1)].

Preparation of lower aliphatic imidic ester hydrochlorides usually involves addition of a slight excess of anhydrous hydrogen chloride to a chilled solution of 1 equivalent of the nitrile in about 1.1 equivalents of the alcohol. The resulting solution is allowed to stand in the cold until the imidate hydrochloride begins to precipitate. Anhydrous diethyl ether is then added to the reaction mixture, the suspension is allowed to stand in the cold, and the imidic ester hydrochloride is collected by suction filtration and freed from solvent and hydrogen chloride. In a variant of this procedure the ether is added before the hydrogen chloride. Yields of imidic ester hydrochlorides usually exceed 70% of theoretical and may be nearly quantitative. The dry salts are reasonably stable if protected from moisture. Detailed procedures are described in the literature (*127, 132, 222, 236, 309*).

The purpose of the ether is to prevent the imidate salt from crystallizing in a hard cake containing occluded solvent and reactants. It does not improve yields. Other inert solvents, including benzene, chloroform, nitrobenzene, and 1,4-dioxane have also been used. Dioxane is particularly useful as a solvent and diluent if the nitrile is only slightly soluble in ether (*46, 167, 223*).

1. Factors Influencing Rates and Yields in Imidate Syntheses

Although satisfactory yields of imidic ester hydrochlorides are obtainable from most nitriles, both the yield of the imidate salt and its rate of formation are influenced by the structure of the nitrile.

Yields seem to be determined mainly by steric factors. For example, ethyl isovalerimidate hydrochloride was obtained in only 35–40% yields under conditions which gave yields of imidic ester hydrochlorides of 70% or higher from most aliphatic nitriles (*222*). Some *o*-substituted benzonitriles and some α-naphthonitriles cannot be converted to imidic ester hydrochlorides by Pinner's method (*191, 277*), and 2,2-diphenyl-4-chlorobutyronitrile did not react with alcoholic hydrogen chloride in 44 days (*166*). Yields reported in the literature for syntheses of a number of imidic ester hydrochlorides from the corresponding nitriles are given in Table I.

TABLE I

YIELDS OBTAINED IN THE SYNTHESIS OF IMIDIC ESTER HYDROCHLORIDES AND ORTHO ESTERS

| R | R' | Yield of $RC(=NH_2)OR'^+Cl^-$ from RCN | Yield of $RC(OR')_3$ from $RC(=NH_2)OR'^+Cl^-$ | References |
|---|---|---|---|---|
| H | $CH_3$ | 90 | 74 | *236* |
| H | $C_2H_5$ | 90 | 70 | *236* |
| $CH_3$ | $C_2H_5$ | 90 | 77 | *222* |
| $C_2H_5$ | $CH_3$ | 88 | 69 | *50* |
| $C_2H_5$ | $C_2H_5$ | 90 | 77 | *222* |
| $C_3H_7$ | $CH_3$ | 77 | 79 | *208, 214* |
| $C_3H_7$ | $C_2H_5$ | 67 | 62 | *222* |
| $CH(CH_3)_2$ | $CH_3$ | 99 | 70 | *202* |
| $CH(CH_3)_2$ | $C_2H_5$ | 80 | 28 | *222* |
| $C_4H_9$ | $CH_3$ | 79 | 79 | *214* |
| $C_4H_9$ | $C_2H_5$ | 75 | 60 | *222* |
| $CH_2CH(CH_3)_2$ | $C_2H_5$ | 37 | 22 | *222* |
| $C_5H_{11}$ | $CH_3$ | 75 | 40 | *50* |
| $(CH_2)_2CH(CH_3)_2$ | $CH_3$ | 71 | 9 | *50* |
| Cyclopentyl | $CH_3$ | 97 | 84 | *225* |
| Cyclohexyl | $CH_3$ | 100 | 58 | *225* |
| $C_7H_{15}$ | $CH_3$ | — | 68 | *214* |
| $C_8H_{17}$ | $CH_3$ | 79 | 78 | *214* |
| $C_6H_5$ | $C_2H_5$ | 90 | 30 | *50, 168* |
| $C_6H_5CH(CH_3)$ | $CH_3$ | 91 | 21 | *227* |
| $C_6H_5CH_2$ | $CH_3$ | 95 | 67 | *226* |
| $C_6H_5CH_2$ | $C_2H_5$ | 75 | 66 | *226* |
| $C_6H_5CH_2CH_2$ | $C_2H_5$ | 38 | 30 | *217* |
| $ClCH_2$ | $C_2H_5$ | 85 | 72 | *222* |
| $NCCH_2$ | $CH_3$ | 87 | 65 | *223* |
| $NCCH_2$ | $C_2H_5$ | 99 | 62 | *223* |
| $CH_3SCH_2$ | $C_2H_5$ | — | 25 | *169* |
| $C_2H_5OCH_2$ | $C_2H_5$ | 88 | 47 | *230* |
| $(C_2H_5O)_2CH$ | $C_2H_5$ | 71 | 12 | *206* |
| $C_6H_5OCH_2$ | $C_2H_5$ | 82 | 30 | *50* |
| $C_2H_5O_2CCH_2$ | $C_2H_5$ | 93 | 82 | *223* |
| $NCCH_2CH_2$ | $CH_3$ | 80 | 77 | *223* |
| $Cl_3CH_2CH_2$ | $CH_3$ | 92 | 84 | *223* |
| $CH_3O_2CCH_2CH_2$ | $CH_3$ | 93 | 63 | *223* |

The rate of reaction of nitriles with alcoholic hydrogen chloride is influenced by both steric and electronic factors. Rates of formation of ethyl imidate hydrochlorides decrease as the size of R in RCN increases. In the case of hydrogen cyanide, the exothermic addition reaction is so rapid that special care must be taken to provide for adequate cooling (*59, 236*). With acetonitrile, the reaction is complete in 2 hours, and with propionitrile about 6 hours are required (*222*). One to 2 days is an adequate reaction period for most of the higher aliphatic nitriles.

No systematic study of electronic substituent effects on rate of imidic ester formation has been reported. The limited data available indicate that, as expected for a nucleophilic addition, electron-withdrawing substituents in the nitrile facilitate formation of the imidate hydrochloride. For example, bromo- and chloroacetonitrile react with phenol and hydrogen chloride much more rapidly than does acetonitrile (*213*).

2. Alcohols Used in the Pinner Imidic Ester Synthesis

The alcohols most often used in the preparation of imidic ester hydrochlorides are methanol and ethanol. Other primary alcohols such as 1-propanol (*122*), 1-butanol (*271*), 1-decanol (*271*), 2-methyl-1-propanol (*279*), and benzyl alcohol (*148*) have also been used, as have the secondary alcohols 2-propanol (*67*) and 2-butanol (*336*). Phenols have also been used (*147, 213*). Most of these imidic ester hydrochlorides were not converted to ortho esters. A more detailed survey of the literature on imidic ester syntheses is given by Roger and Nielson (*302*).

3. Side Reactions in the Synthesis of Imidic Ester Hydrochlorides

The most serious competing reaction in the conversion of nitriles to imidic ester hydrochlorides is decomposition of the imidate salt to an amide and an alkyl chloride [Eq. (3)]. This reaction is minimized by carrying out the addition reaction at low temperatures.

$$RC(=NH_2^+)OR'Cl^- \rightarrow RCONH_2 + R'Cl \qquad (3)$$

C. Alcoholysis of Imidic Ester Hydrochlorides

Imidic ester hydrochlorides are converted to ortho esters by reaction with alcohols [Eq. (2)]. In the simplest alcoholysis procedure a solution of the imidic ester hydrochloride in an excess of the alcohol is allowed to stand at room temperature until precipitation of ammonium chloride is complete (*50, 309, 311, 313*). This process is time-consuming (up to 6 weeks is required for complete reaction in some cases), and frequently gives poor yields of the ortho ester.

McElvain and Nelson (*222*) improved the alcoholysis step by refluxing the imidic ester hydrochloride with a five- to tenfold excess of the alcohol in ether solution. This reduced the time required for complete reaction to 2 days or less, and resulted in significantly improved yields of several aliphatic ortho esters. The temperature of the refluxing ether solution is below that at which the competing decomposition of the imidic ester hydrochloride to a carboxamide and an alkyl chloride [Eq. (3)] becomes serious. Even higher yields of ortho esters are obtained if a suspension of the imidic ester hydrochloride in a mixture of the alcohol and petroleum ether is stirred at room temperature (*202, 208, 225*).

Ortho esters are destroyed by strong acids such as hydrogen chloride, and are rapidly hydrolyzed by moisture in the presence of acids.

$$RC(OR')_3 + HCl \longrightarrow RCO_2R' + R'OH + R'Cl \tag{4}$$

$$RC(OR')_3 + H_2O \xrightarrow{H^+} RCO_2R' + 2\,R'OH \tag{5}$$

It is therefore essential that the imidic ester hydrochloride be completely free of hydrogen chloride, and that moisture be carefully excluded during the reaction.

Detailed procedures for syntheses of representative ortho esters by alcoholysis of imidic ester hydrochlorides are given in the literature (*202, 214, 222, 223, 225, 236, 255a, 325*). Table I records yields reported in the synthesis of ortho esters by this reaction.

1. Competing Reactions in the Alcoholysis of Imidic Ester Hydrochlorides

Even when moisture and hydrogen chloride are excluded from the reaction mixtures, carboxamides and alkyl chlorides are always formed as by-products during the alcoholysis of imidic ester hydrochlorides. In some cases a carboxylate ester and a dialkyl ether are also formed. The reactions leading to these by-products may compete with the desired reaction to such an extent that little or no ortho ester is formed.

The nature and extent of the competition between amide, ester, and ortho ester formation is determined mainly by the structure of the imidic ester hydrochloride, and to a lesser extent by the medium in which the alcoholysis reaction occurs.

Alcoholysis of imidic ester hydrochlorides derived from straight-chain nitriles gives the corresponding ortho esters in yields of 60–80%. A single α-substituent such as methyl, phenyl, ethoxy, carbethoxy, cyano, or chloro in the imidic ester hydrochloride does not appreciably change these yields; however, two α- or β-substituents usually lower the yield of ortho ester to 20–30% or less (*222, 228*).

Amide formation is a serious competing reaction in the alcoholysis of $\alpha,\alpha$- and $\alpha,\beta$-disubstituted imidic ester hydrochlorides, where the yield of this side product may be of the order of 50%. Some normal ester is formed when there is an $\alpha$-phenyl substituent, and much larger amounts are formed when the imidic ester hydrochloride has two $\alpha$- or $\beta$-substituents. The amount of normal ester formed increases as the size of the $\alpha$-substituent increases.

McElvain and Tate studied the decomposition of imidic ester hydrochlorides to the corresponding carboxamides and alkyl chlorides [Eq. (3)] in chloroform and *tert*-butyl alcohol solutions, in which alcoholysis does not occur (*228*). It was found that disappearance of the imidate hydrochloride is a first-order process which is much faster in the more polar solvent *tert*-butyl alcohol than in chloroform. The rate of decomposition of $RC(=NH_2)OR'^+Cl^-$ is very sensitive to the structure of R' [relative rates: $R' = CH_3$, 1.00; $C_2H_5$, 0.060; $(CH_3)_2CH$, 0.019], but is less sensitive to variation of the structure of R [relative rates: $R = CH_3$, 1.00; $(CH_3)_2CH$, 1.58; $C_6H_5CH_2$, 2.68; $C_2H_5O_2CCH_2$, 9.32]. An imidic ester hydrobromide decomposes more rapidly than the corresponding hydrochloride under the same conditions.

Normal esters are formed by reaction of ortho esters with imidic ester hydrochlorides. For reactions of a series of ethyl imidate hydrochlorides with ortho esters it was found that the rate of formation of normal ester is insensitive to the structure of the acyl substituent of the ortho ester, but is determined mainly by the acidity of the imidic ester hydrochloride (*228*). The observation that this side reaction is an acid-catalyzed decomposition of the ortho ester has practical implications. In some cases normal ester formation may be minimized by neutralizing the alcoholysis reaction mixture with an alkoxide prior to distillation of the product mixture (*206*).

It should be possible to minimize amide formation during imidate alcoholysis by using an imidate salt with a nonnucleophilic anion. No investigation of this idea has been reported. Use of boron trifluoride instead of hydrogen chloride to promote the alcoholysis of nitriles and imidic esters avoids amide formation, but the boron trifluoride catalyzes the decomposition of the ortho ester to normal ester and ether. McElvain and Stevens found that methanolysis of the boron trifluoride adduct of methyl 2-phenylpropionimidate yielded only dimethyl ether and methyl 2-phenylpropionate (*227*), and McKenna and Sowa reported that the reaction of acetonitrile with boron trifluoride in propanol formed only propyl acetate and dipropyl ether (*232*).

2. Synthesis of Mixed Ortho Esters from Imidic Ester Hydrochlorides

Alcoholysis of imidic ester hydrochlorides has been used to synthesize ortho esters having two different alkoxy groups.

$$RC(=NH_2)OR'^+Cl^- + 2\ R''OH \rightarrow RC(OR'')_2OR' + NH_4^+Cl^- \qquad (6)$$

# 1. CARBOXYLIC ORTHO ESTERS AND RELATED COMPOUNDS

Pinner prepared a number of mixed orthoformates this way (*275, 276*), and Sah reported the synthesis of several mixed orthoacetates (*311*). Because of the extreme ease with which mixed ortho esters undergo acid-catalyzed disproportionation [Eq. (7)], this reaction has little preparative significance.

$$RC(OR'')_2OR' \rightleftharpoons RC(OR')_3 + RC(OR'')_3 + RC(OR')_2OR'' \qquad (7)$$

### 3. FORMATION OF HETEROCYCLIC ORTHO ESTERS FROM IMIDIC ESTER HYDROCHLORIDES

Heterocyclic ortho esters are formed by the reaction of acyclic imidic ester hydrochlorides with diols and triols, and by the reaction of heterocyclic imidate hydrochlorides with alcohols.

2-Alkoxydioxolanes are the major products of reactions of imidic ester hydrochlorides with ethylene glycol. Some diortho ester is formed simultaneously (*202, 225*).

$$RC(=NH_2)OCH_3{}^+Cl^- + HOCH_2CH_2OH \longrightarrow$$

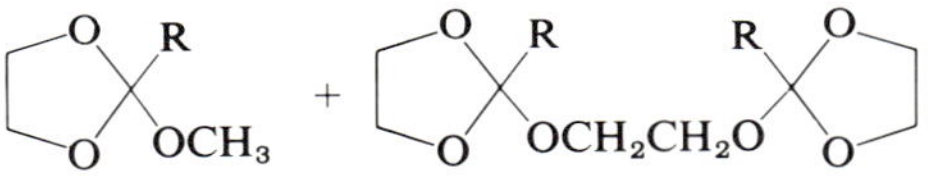

$$\text{(dioxolane with OCH}_3\text{)} + \text{(dioxolane with OCH}_2\text{CH}_2\text{O-dioxolane)} \qquad (8)$$

A related reaction was reported by Pauer, who treated glycerol with hydrogen cyanide and hydrogen chloride and claimed to obtain "monoglyceryl orthoformate," $C_4H_6O_3$, in 70% yield (*270*). The observed boiling point of his product, 126°C at 12 mm, is much higher than would be expected for the bicyclic ortho ester **1**, and suggests that the substance was a dimer, possibly the tricyclic diortho ester **2**.

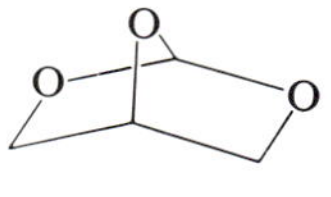

    **1** 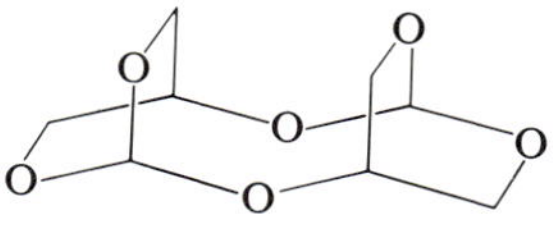        **2**

Other heterocyclic ortho esters have been prepared by the alcoholysis of the hydrochlorides of 2-iminofurans and 2-iminopyrans (*179, 182, 374*).

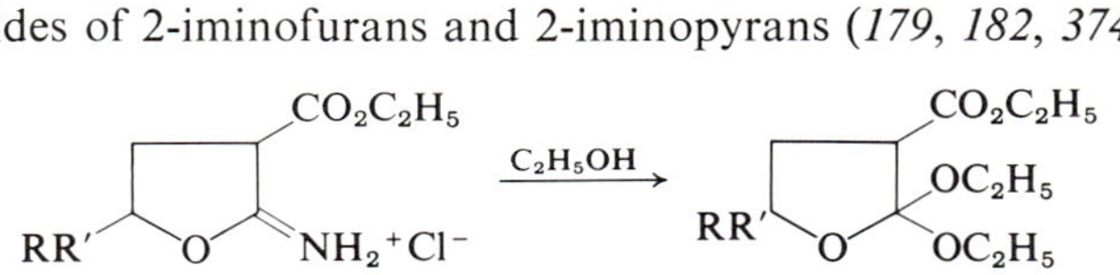

$$\text{RR'-iminofuran-CO}_2C_2H_5 \cdot NH_2{}^+Cl^- \xrightarrow{C_2H_5OH} \text{RR'-furan-CO}_2C_2H_5, OC_2H_5, OC_2H_5 \qquad (9)$$

$$\text{2-iminochromene-C}_6H_5 \cdot NH_2{}^+Cl^- \xrightarrow{CH_3OH} \text{chromane-C}_6H_5, OCH_3, OCH_3 \qquad (10)$$

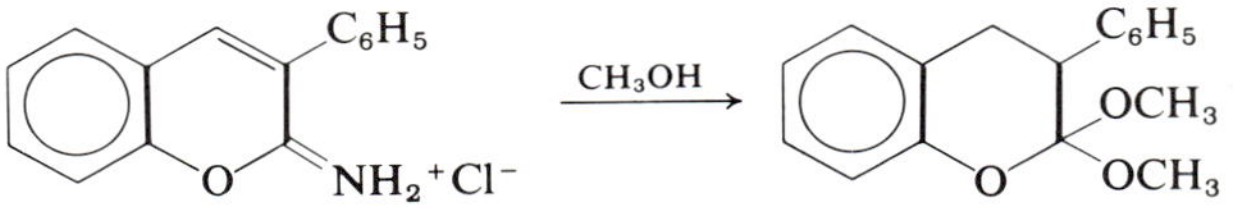

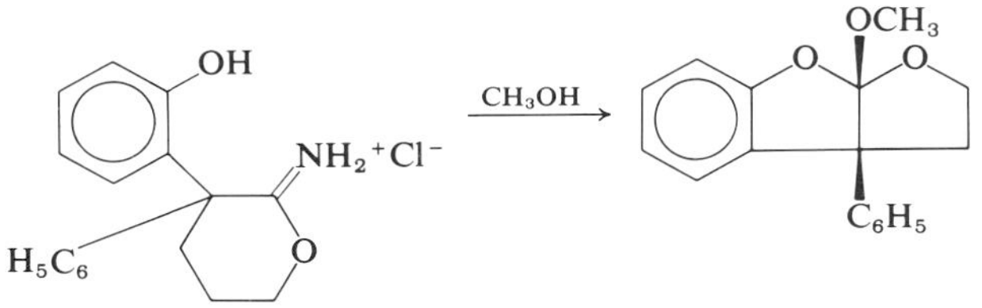

<img: structure showing conversion of a cyclic imidate hydrochloride with phenyl and 2-hydroxyphenyl substituents to a bicyclic orthoester with OCH₃ and C₆H₅ groups via CH₃OH> (11)

### 4. Formation of Ortho Esters from N-Substituted Imidic Esters

Cations derived from N-substituted and N,N-disubstituted imidic esters are alcoholyzed to ortho esters. For example, Knott (*175*) found that the conjugate acid of ethyl *N*-phenylformimidate reacts with ethanol to form triethyl orthoformate as one product.

$$2\ C_6H_5\overset{+}{N}H{=}CHOC_2H_5 + C_2H_5OH \xrightarrow{-H^+}$$
$$HC(OC_2H_5)_3 + C_6H_5NH\cdots\overset{+}{C}H\cdots NHC_6H_5 \quad (12)$$

The reaction is of no practical importance, since N-arylformimidic esters are most conveniently prepared from aromatic amines and orthoformic esters (*296*).

A potentially useful synthesis of ortho esters from the corresponding carboxylic acids has been described by Eilingsfeld, Seefelder, and Weidinger (*93*), who alcoholyzed the imide chlorides obtained from propionyl and benzoyl piperidides. These reactions, which involve piperidinoethoxycarbonium chlorides as intermediates, give triethyl orthopropionate and triethyl orthobenzoate in yields of 62% and 45%, respectively.

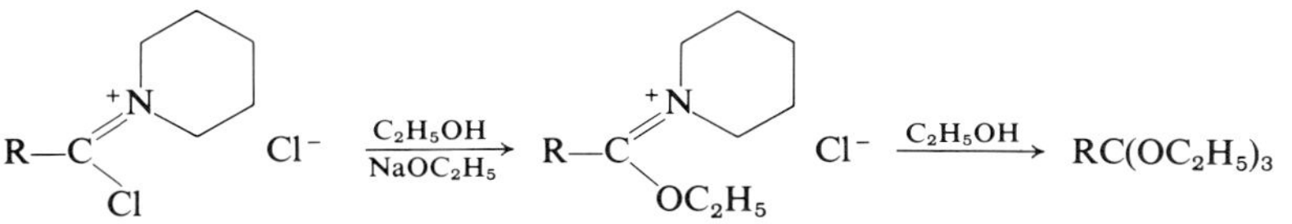

$$R-C\overset{\overset{+N(\text{piperidino})}{|}}{\underset{Cl}{|}} Cl^- \xrightarrow[NaOC_2H_5]{C_2H_5OH} R-C\overset{\overset{+N(\text{piperidino})}{|}}{\underset{OC_2H_5}{|}} Cl^- \xrightarrow{C_2H_5OH} RC(OC_2H_5)_3 \quad (13)$$

### D. Mechanism of the Pinner Ortho Ester Synthesis

Observations on imidic ester hydrochloride alcoholyses and accompanying side reactions suggest that decomposition of imidic ester hydrohalides to carboxamides is the result of $S_N2$ displacements by halide on the *O*-alkyl groups of the imidate salts. The effects of both alkoxy and acyl substituents on reaction rate are compatible with this conclusion. If amide formation occurs only from the dissociated salt or a solvent-separated ion pair, this hypothesis also accounts for observed solvent effects on amide formation.

The formation of carboxylate ester apparently is an acid-catalyzed reaction of the ortho ester with chloride ion or an alcohol. Since chloride ion is more nucleophilic than alcohol, much more alkyl chloride than dialkyl ether is formed as a by-product of imidate alcoholysis.

An acceptable mechanism for the conversion of nitriles to ortho esters must account for observed substituent effects on rates and yields in formation

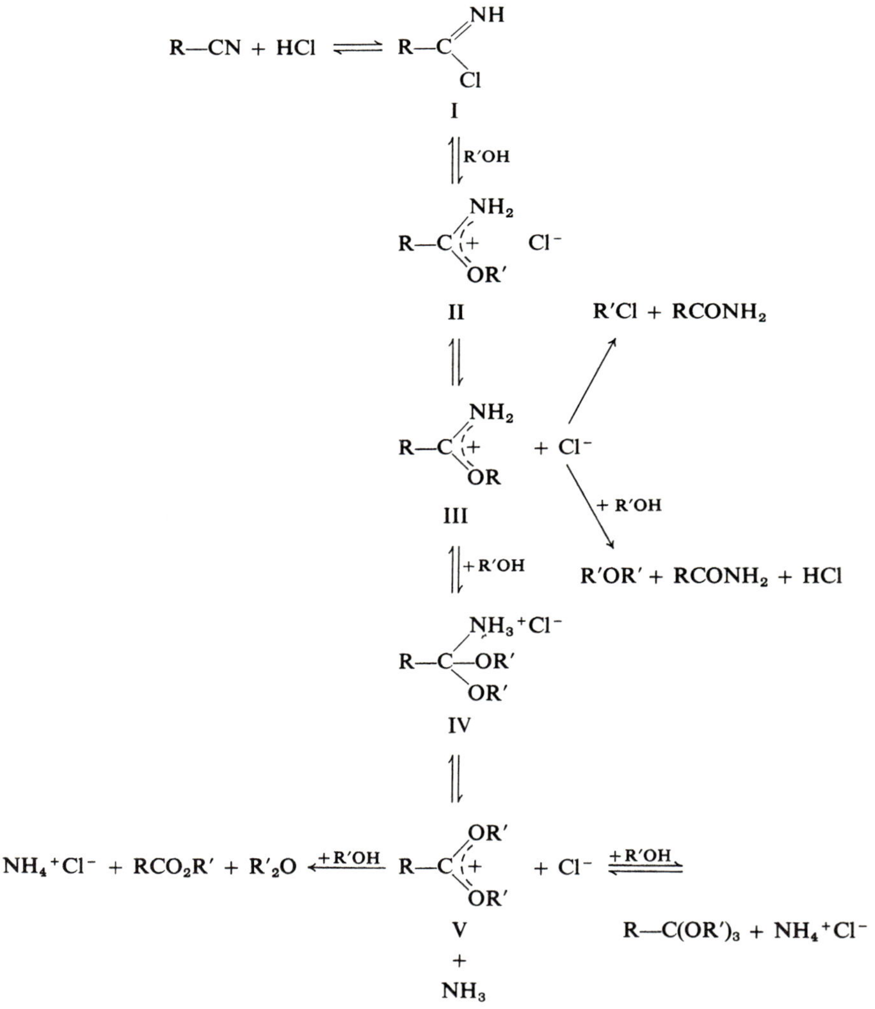

$$R-CN + HCl \rightleftharpoons R-C\begin{matrix}NH\\Cl\end{matrix}$$
$$\text{I}$$

$$\Updownarrow R'OH$$

$$R-C\begin{matrix}NH_2\\(+\\OR'\end{matrix} \quad Cl^-$$
$$\text{II}$$

$$\Updownarrow$$

$$R-C\begin{matrix}NH_2\\(+\\OR\end{matrix} + Cl^- \quad \longrightarrow R'Cl + RCONH_2$$
$$\text{III}$$

$$\downarrow +R'OH \qquad \searrow +R'OH$$

$$\qquad\qquad\qquad R'OR' + RCONH_2 + HCl$$

$$R-C\begin{matrix}NH_3{}^+Cl^-\\-OR'\\OR'\end{matrix}$$
$$\text{IV}$$

$$\Updownarrow$$

$$NH_4{}^+Cl^- + RCO_2R' + R'_2O \xleftarrow{+R'OH} R-C\begin{matrix}OR'\\(+\\OR'\end{matrix} + Cl^- \xrightleftharpoons{+R'OH}$$
$$\text{V}$$
$$+$$
$$NH_3$$

$$\qquad\qquad\qquad\qquad R-C(OR')_3 + NH_4{}^+Cl^-$$

(14)

of imidic ester hydrochlorides from nitriles and formation of ortho esters from the imidic ester hydrochlorides. The mechanism should also rationalize substituent effects on the rates and extents of the various competing reactions. A reaction scheme which meets these requirements is outlined in Eq. (14).

Intermediates II and III, which may be regarded as an ion pair and a dissociated salt, are invoked to account for the fact that decomposition of imidic ester hydrochlorides to amides and alkyl chlorides is facilitated by increasing the polarity of the solvent (228). Although amide chlorides, I, have not been definitely identified as intermediates in the Pinner synthesis, they are known compounds (92) and are products of addition of hydrogen chloride to nitriles (347b).

### E. One-Step Syntheses of Ortho Esters from Nitriles or Amides

Ortho esters are usually synthesized from hydrogen cyanide or nitriles by first preparing an imidic ester hydrochloride and then alcoholyzing the imidate salt. In this way the side reactions caused by unreacted hydrogen chloride can be avoided. It would obviously be convenient to prepare ortho esters without isolating the imidic ester hydrochlorides, and this has been accomplished in a few instances. Erickson (97) synthesized trimethyl, triethyl, and tributyl orthoformates in 23–55% yields in a one-step process which consists of introducing hydrogen chloride into a large excess of both the alcohol and hydrogen cyanide. One disadvantage of this method is the problem of disposing of large quantities of hydrogen cyanide. A different one-step procedure for preparing trimethyl orthoformate involves carrying the reaction out in chlorobenzene or *o*-dichlorobenzene and neutralizing the unreacted hydrogen chloride with ammonia prior to isolation of the ortho ester (199a, 255). The yield of trimethyl orthoformate, based on hydrogen cyanide, exceeds 70%.

Another one-step synthesis of orthoformate uses formamide rather than hydrogen cyanide as the starting material. A solution of formamide and benzoyl chloride or ethyl chlorocarbonate in a petroleum ether–alcohol mixture is allowed to stand at room temperature until precipitation of ammonium chloride is complete, and then the product is isolated. Triethyl, triisopropyl, and tributyl orthoformates were prepared in 40–50% yields in this manner (265a). *O*-Acyl formimidate salts are probably intermediates in these reactions.

Triethyl orthoacetate and triethyl orthopropionate were obtained in yields of 80% and 50%, respectively, by preparing and alcoholyzing the imidate salts in chloroform solution without isolating them (101). A similar process is described in the Russian patent literature (372).

## III. SYNTHESIS OF ORTHO ESTERS FROM TRIHALOMETHYL COMPOUNDS AND α-HALO ETHERS

### A. Introduction

In 1854 Williamson reported that chloroform reacts with sodium ethylate and sodium amylate to form products now known to be triethyl and triamyl orthoformates (*178, 362, 363*).

$$CHCl_3 + 3\ NaOR \rightarrow HC(OR)_3 + 3\ NaCl \qquad (15)$$
$$(R = C_2H_5, C_5H_{11})$$

This extension of the Williamson ether synthesis resulted in the discovery of an important new class of compounds, the carboxylic ortho esters.

The reaction of alkoxides and phenoxides with trihalomethyl compounds is a useful procedure for preparing trialkyl and triaryl orthoformates, trialkyl orthobenzoates, and tetraalkyl orthocarbonates. This reaction is limited to trihalomethyl compounds such as chloroform, benzotrichloride, chloropicrin, and a few others which lack α-hydrogen atoms and hence cannot undergo base-promoted elimination reactions.

### B. Synthesis of Orthoformates

1. Trialkyl Orthoformates

Chloroform undergoes a complex series of reactions in alcoholic solutions of alkoxides. One of the reaction products is a trialkyl orthoformate. Competing reactions produce carbon monoxide (*28*) and alkenes derived from the alcohol (*28, 136*). Dichlorocarbene is almost certainly an intermediate in this reaction (*134, 136, 170*), and the chief factor limiting yields of orthoformate is the competing reactions which dichlorocarbene undergoes in alkaline alcoholic solutions. An oversimplified representation of the reactions which occur when chloroform is added to an ethanolic solution of sodium ethoxide is as follows:

$$HCCl_3 + C_2H_5O^- \rightarrow\ :CCl_3^- + C_2H_5OH$$
$$:CCl_3^- \rightarrow\ :CCl_2 + Cl^-$$
$$:CCl_2 + C_2H_5O^- \rightarrow C_2H_5O\ddot{C}Cl + Cl^- \qquad (16)$$
$$C_2H_5O\ddot{C}Cl + C_2H_5O^- \rightarrow C_2H_5OH + CH_2{=}CH_2 + CO + Cl^-$$
$$C_2H_5O\ddot{C}Cl + C_2H_5O^- \rightarrow (C_2H_5O)_2C: + Cl^-$$
$$(C_2H_5O)_2C: + C_2H_5OH \rightarrow HC(OC_2H_5)_3$$

A number of trialkyl orthoformates have been prepared from chloroform and sodium alkoxides. These include trimethyl orthoformate (*73, 99, 195a*,

*312*), triethyl orthoformate (*1, 11, 24, 27a, 53, 58, 73, 164, 281, 312, 354, 360, 362, 363, 367*), tripropyl orthoformate (*73, 197, 281, 312*), triisopropyl orthoformate (*312*), tributyl orthoformate (*58, 197, 312*), triisobutyl orthoformate (*73, 312*), tri-2-butyl orthoformate (*274*), triamyl orthoformate (*362, 363*), triisoamyl orthoformate (*73, 281, 312*), trihexyl orthoformate (*58*), and triallyl orthoformate (*29*).

The reaction of sodium alkoxides with chloroform has been carried out by adding chloroform to the alkoxide (either dry or in alcohol solution), by adding sodium metal to a mixture of the alcohol and chloroform, and by adding chloroform to a suspension of the alkoxide in an inert solvent such as ether or benzene. Yields of trialkyl orthoformates are usually poor (30–40%) regardless of the procedure used, but seem to be increased by raising the reaction temperature. Two of the more successful procedures for preparing trimethyl and triethyl orthoformates in yields exceeding 50% are described in the patent literature (*99, 360*). Detailed procedures for preparing these ortho esters have been published (*164, 237*).

Post and Erickson (*281*) prepared several mixed orthoformates (diethylpropyl orthoformate, dipropylisoamyl orthoformate, and propyldiisoamyl orthoformate) by reaction of chloroform with mixtures of the alkoxides and alcohols, followed by fractional distillation of the product mixtures.

Trialkyl orthoformates have also been prepared by reaction of alkoxides with fluorodichloromethane and difluorochloromethane (*135, 137*). Triisopropyl orthoformate was obtained in 71% yield from the latter of these haloforms.

An interesting variation of the usual synthesis of orthoformates from chloroform is the ferric chloride-catalyzed reaction of fluoroalkanols with chloroform. $HC[OCH_2(CF_2)_3CHF_2]_3$ and $HC[OCH_2(CF_2)_5CHF_2]_3$ were synthesized in 66% and 80% yields by refluxing the fluoroalkanols, chloroform, and ferric chloride for 90 hours (*133*). Prolonged heating with ferric chloride, which would destroy ordinary ortho esters, appears to have little effect on the fluoroalkyl orthoformates.

Tetrahalomethanes react with alkoxides to form trialkyl orthoformates rather than the expected orthocarbonates. Thus, Nef (*262*) and Ingold and Powell (*154*) prepared triethyl orthoformate by allowing carbon tetrachloride to react with ethanolic sodium ethoxide, and Cleaver isolated trialkyl orthoformates as the main products of reaction of alkoxides with dichlorodifluoromethane (*62*). This anomalous result is understandable in view of the fact that tetrahalomethanes are converted to the corresponding haloforms by hydroxide and alkoxides (*22, 177, 262*), presumably via trihalomethylcarbanion intermediates.

$$CX_4 + RO^- \rightarrow ROX + :CX_3^- \qquad (17)$$

The haloform and trialkyl orthoformate could both be formed from the trihalomethylcarbanion [Eq. (16)].

### 2. Triaryl Orthoformates

The Reimer-Tiemann synthesis of phenolic aldehydes involves electrophilic attack by dichlorocarbene (generated from chloroform under the alkaline reaction conditions) on the ortho and para positions of the phenoxide ions (*170*). Electrophilic attack by the carbene on phenoxide oxygen should compete with the reaction leading to hydroxy aldehydes. It is therefore not surprising that triaryl orthoformates, formed by a reaction sequence analogous to Eq. (16), are occasionally isolated as minor products of the formylation reaction.

These reactions have yielded triphenyl orthoformate (*14, 21, 345*), tri-*o*-, *m*-, and *p*-cresyl orthoformates (*15, 81*), tri-α-tetralyl orthoformate (*369*), tri-*o*-, *m*-, and *p*-chlorophenyl orthoformates (*108a, 162*), tri-*p*-benzylphenyl orthoformate (*181*), and tri-*o*- and *p*-nitrophenyl orthoformates (*358*). Phenoxides whose rings are too deactivated to undergo the Reimer-Tiemann reaction undergo attack by the carbene intermediate on oxygen to form ortho esters. Yields are usually poor.

### C. Synthesis of Higher Ortho Esters

### 1. Trialkyl Orthobenzoates

Benzotrichloride and substituted benzotrichlorides react with alcoholic alkoxides to form trialkyl orthobenzoates. Orthobenzoates which have been prepared in this way include trimethyl orthobenzoate (*189, 201, 229*), triethyl orthobenzoate (*198, 204, 310, 351*), trimethyl *p*-chloroorthobenzoate (*189*), trimethyl *p*-fluoroorthobenzoate (*283*), and hexamethyl *m*- and *p*-diorthophthalates (*193*).

The only reported attempt to prepare a triaryl orthobenzoate (from benzotrichloride and sodium phenoxide) was unsuccessful (*76*).

### 2. Other Orthocarboxylate Esters

Other trihalomethyl compounds having no hydrogen α to the trihalomethyl group may yield ortho esters on reaction with alcoholic alkoxides. Hexachloropropene has been converted to triethyl 2,3,3-trichloroorthoacrylate in this manner (*102, 300*), and trimethyl and triethyl 2,3,4-trifluoro-4,4-dichloroorthocrotonates have been prepared similarly (*176*).

Ethyl trichloroacetate was reported to react with alcoholic sodium triiodophenoxide to give ethyl tri-2,4,6-triiodophenoxy acetate (*72*). 6-Trichloromethylpurine reacts with methanolic potassium hydroxide to form

6-trichloromethylpurine (63), and with methanolic sodium p-cresylate to form 6-tri(4-methylphenoxy)methylpurine (63a).

2-Trimethoxymethylbenzimidazole, 2-triphenoxymethylbenzimidazole, and 2-tri-p-chlorophenoxymethylbenzimidazole were prepared by reaction of 2-trichloromethylbenzimidazole with methanol or the sodium phenoxides under alkaline conditions (143a). The trimethoxymethylbenzimidazole was also obtained in 30% yield by reaction of N-chloro-N'-phenyltrichloroacetamidine with methanolic sodium hydroxide (93a).

Heiber (125) claimed to have prepared triphenyl orthoacetate and other triaryl orthoacetates by reaction of aqueous sodium phenoxides with 1,1,1-trichloroethane. However, Cope later showed that the compounds described by Heiber were actually 1,2-diphenoxyethanes, formed from 1,2-dichloroethane present in the trichloroethane as an impurity (65).

## D. Synthesis of Tetraalkyl Orthocarbonates

Although tetraalkyl orthocarbonates, $C(OR)_4$, cannot be obtained by reaction of carbon tetrachloride with alkoxides, they may be prepared in satisfactory yields from chloropicrin or trichloromethanesulfenyl chloride.

$$CCl_3NO_2 + 4\ NaOR \rightarrow C(OR)_4 + 3\ NaCl + NaNO_2 \qquad (18)$$

$$ClSCCl_3 + 4\ NaOR \rightarrow C(OR)_4 + 4\ NaCl + S \qquad (19)$$

The preparation of tetraethyl orthocarbonate by reaction of sodium ethoxide with chloropicrin was described by Bassett in 1864 (28). Subsequently chloropicrin was used in the synthesis of tetramethyl orthocarbonate (344, 353a), tetraethyl orthocarbonate (295, 303, 344), tetrapropyl orthocarbonate (344), tetrabutyl orthocarbonate (344), and tetraisobutyl orthocarbonate (303). Yields range from 30% to 60%. The detailed course and mechanism of this reaction have not been established.

Tetraalkyl orthocarbonates are formed in higher yields from trichloromethanesulfenyl chloride [Eq. (19)] than from chloropicrin [Eq. (18)]. Tetramethyl orthocarbonate has been prepared by reaction of trichloromethanesulfenyl chloride with an alcoholic solution of the alkoxide (329, 344), as have tetraethyl orthocarbonate (64, 126, 341, 344), tetrapropyl orthocarbonate (344), tetrabutyl orthocarbonate (344), and tetraisobutyl orthocarbonate (64). Trichloromethanesulfenyl chloride may be prepared by chlorination of carbon disulfide (88) and is commercially available.

Although the detailed mechanism of orthocarbonate formation from trichloromethanesulfenyl chloride has not been established, it is known that the reaction occurs in two steps, the first of which is a solvolysis (64, 146).

$$ClSCCl_3 + ROH \rightarrow ROSCCl_3 + HCl \qquad (20)$$

The alkyl trichloromethane sulfinate formed in this reaction is then converted to the orthocarbonate. The exact fate of sulfur in this reaction is not known. Since acidification of the residue remaining after extraction of the orthocarbonate from the reaction mixture results in deposition of sulfur and liberation of hydrogen sulfide, it is assumed that a polysulfide is formed (344).

An exception to the general rule that orthocarbonates cannot be prepared from carbon tetrachloride was reported by Hill and co-workers (133), who found that polyfluoroalkyl orthocarbonates are formed in 35–40% yield by the ferric chloride-catalyzed reaction of polyfluoroalkanols with carbon tetrachloride.

$$4\ CHF_2(CF_2)_nCH_2OH + CCl_4 \xrightarrow{FeCl_3} C[OCH_2(CF_2)_nCHF_2]_4 + 4\ HCl \qquad (21)$$

### E. Preparation of Ortho Esters from α-Haloalkyl Ethers

Although α-haloalkyl ethers are probably intermediates in the conversion of trichloromethyl compounds to ortho esters, they are too reactive to be isolated under the usual reaction conditions. However, 1,1-dihaloalkyl ethers can be prepared by other means, and are easily converted to ortho esters by reaction with alcoholic sodium alkoxides. Some of the resulting ortho esters have been prepared only in this way.

1,1-Dichloromethyl methyl ether was converted into a number of mixed orthoformates in 36–58% yields (43, 117).

$$CH_3OCHCl_2 + 2\ NaOR \rightarrow CH_3OCH(OR)_2 + 2\ NaCl \qquad (22)$$
$$(R = C_6H_5,\ p\text{-}O_2NC_6H_4,\ cyclo\text{-}C_6H_{11},\ CH_3,\ C_2H_5,\ C_3H_7)$$

Aryl- and alkyldiphenyl orthoformates are formed in yields exceeding 80% by reaction of chlorodiphenoxymethane with alcohols and phenols in the presence of pyridine (317, 318).

$$ClCH(OC_6H_5)_2 + ROH + C_5H_5N \rightarrow ROCH(OC_6H_5)_2 + C_5H_5NH^+Cl^- \qquad (23)$$
$$(R = C_6H_5,\ p\text{-}O_2NC_6H_4,\ \beta\text{-}C_{10}H_7,\ CH_3,\ C_2H_5)$$

Triethyl orthoacetate and triethyl α-haloorthoacetates have been prepared from ethyl haloethyl ethers and ethanolic sodium ethoxide (130, 342) [Eq. (24)]. Triethyl orthoacetate has also been prepared from ethyl 1,1-diazidoethyl ether (327) [Eq. (24), X = $N_3$].

$$C_2H_5OCX_2R + 2\ NaOC_2H_5 \rightarrow RC(OC_2H_5)_3 + 2\ NaX \qquad (24)$$
$$(X = F,\ R = CHCl_2,\ CHFCl;\ X = Cl,\ R = CH_3,\ CH_2Cl;\ X = N_3,\ R = CH_3)$$

Baganz and Kruger (19, 20) obtained hexaalkyl orthooxalates by reaction of sodium alkoxides with 1,2-dialkoxy-1,1,2,2-tetrachloroethanes.

$$ROCCl_2CCl_2OR + 4\ NaOR' \rightarrow ROC(OR')_2C(OR')_2OR + 4\ NaCl \qquad (25)$$
$$(R = R' = C_2H_5;\ R = C_4H_9,\ R' = C_2H_5,\ C_4H_9)$$

Hemi-orthooxalates (alkyl trialkoxy acetates) have been prepared similarly by alcoholysis of alkyl alkoxydichloroacetates in the presence of alkoxides, ammonia, or pyridine (*5*, *6*, *38*, *161*).

$$ROCCl_2CO_2R + R'OH \rightarrow ROC(OR')_2CO_2R \quad (26)$$

$$(R = R' = CH_3, C_2H_5, C_3H_7, CH_2CH(CH_3)_2, C_5H_{11};\ R = CH_3, R' = C_2H_5)$$

The alkyl alkoxydichloroacetates were prepared by reaction of dialkyl oxalates with phosphorus pentachloride. An extensive series of alkyl alkoxydiaryloxyacetates was prepared by reaction of the alkyl alkoxydichloroacetates with sodium phenoxides in dioxane (*156*, *157*), and a number of alkoxydiaryloxyacetamides have been synthesized similarly (*172*, *173*).

$$ROCCl_2CO_2R + 2\,NaOAr \rightarrow ROC(OAr)_2CO_2R + 2\,NaCl \quad (27)$$

$$ROCCl_2CONR_2 + 2\,NaOAr \rightarrow ROC(OAr)_2CONR_2 + 2\,NaCl \quad (28)$$

Two interesting heterocyclic ortho esters, 2-triethoxymethylimidazoline and 2-triethoxymethyltetrahydropyrimidine, are formed by the reaction of 1,2-diaminoethane and 1,3-diaminopropane with 1,2-diethoxy-1,1,2,2-tetrahaloethanes in ethanol solution (*18*).

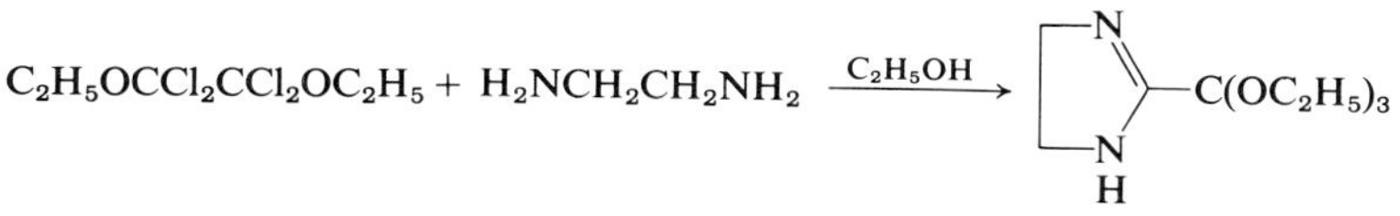

C2H5OCCl2CCl2OC2H5 + H2NCH2CH2NH2
C2H5OH
N
C(OC2H5)3
N
H

(29)

C2H5OCClBrCClBrOC2H5 + H2N(CH2)3NH2
C2H5OH
N
C(OC2H5)3
N
H

(30)

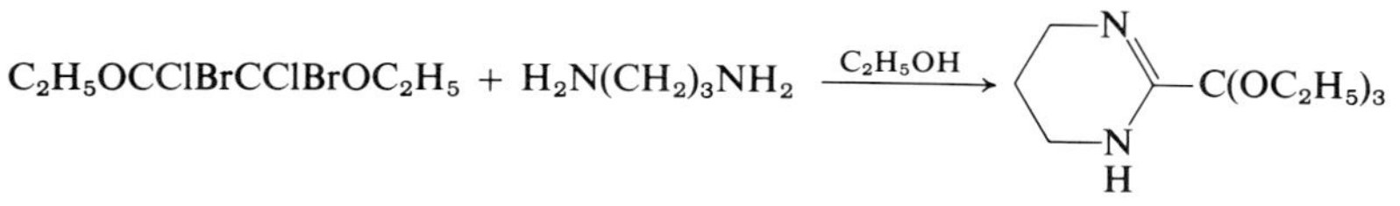

α-Halo ethers were unisolated intermediates in syntheses of 2-methoxy-2-trichloromethyl-1,3-dioxolane and 2-methoxy-2-dichloromethyl-1,3-dioxolane from 2-chloromethylene-1,3-dioxolane (*116a*) [Eq. (30a)].

O
O
=CCl2
Cl2
CH2Cl2
O
Cl
O
CCl3
CH3OH
(C2H5)3N
O
CCl3
O
OCH3
HCl
(C2H5)2O
O
Cl
O
CHCl2
CH3OH
(C2H5)3N
O
CHCl2
O
OCH3

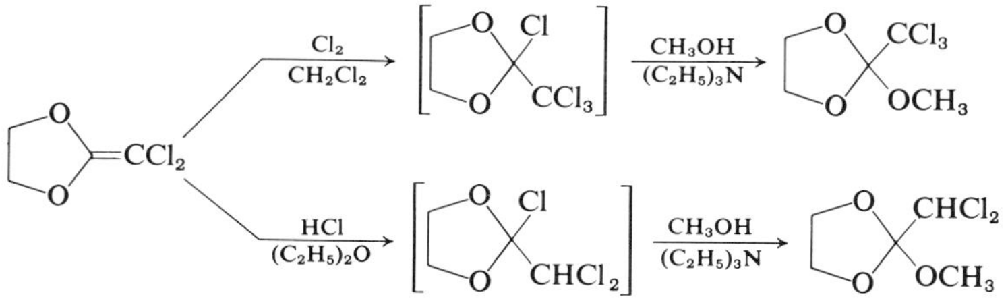

(30a)

2-Chloro-1,3-dioxolane was presumably an intermediate in the conversion of 1,3-dioxolane to 2-(2-chloroethyl)-1,3-dioxolane by low-temperature photochlorination in the presence of ethylene oxide (*160a*) [Eq. (30b)].

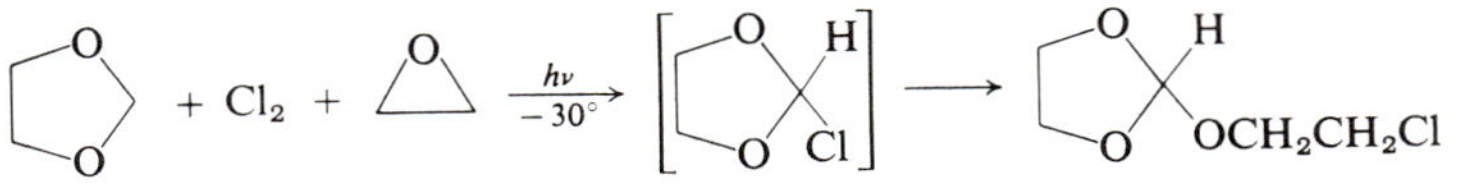

$$\underset{O}{\overset{O}{\bigcirc}} + Cl_2 + \underset{}{\overset{}{\triangle}} \xrightarrow[-30°]{h\nu} \left[ \underset{O}{\overset{O}{\bigcirc}}\underset{Cl}{\overset{H}{\bigvee}} \right] \longrightarrow \underset{O}{\overset{O}{\bigcirc}}\underset{OCH_2CH_2Cl}{\overset{H}{\bigvee}} \quad (30b)$$

A few orthocarbonates have been prepared from halomethyl ethers. Dichlorodiphenoxymethane reacts with phenol to form tetraphenyl orthocarbonate in 78% yield (*118*). 2,2-Dichlorobenzo-1,3-dioxolane, obtained from the reaction of phosphorus pentachloride with pyrocatechyl carbonate, reacts with alcohols and phenols to form 2,2-dialkoxy- and 2,2-diaryloxy-benzo-1,3-dioxolanes in high yields (*116, 118, 119*) [Eq. (31)]. Several attempts to prepare tetramethyl orthocarbonate by reaction of methyl trichloromethyl ether with methanolic sodium methoxide yielded only dimethyl carbonate (*80*).

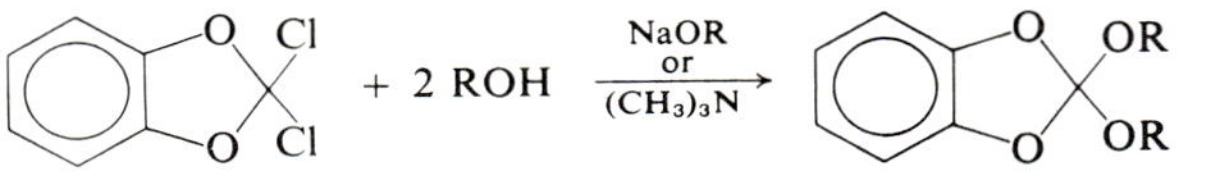

$$\underset{Cl}{\overset{O}{\underset{O}{\bigcirc}}}\underset{Cl}{\overset{}{\bigvee}} + 2\,ROH \xrightarrow[(CH_3)_3N]{NaOR \text{ or}} \underset{O}{\overset{O}{\underset{O}{\bigcirc}}}\underset{OR}{\overset{OR}{\bigvee}} \quad (31)$$

(R = $CH_3$, $C_2H_5$, $C_3H_7$, $C_6H_5$)

## IV. SYNTHESIS OF ORTHO ESTERS FROM OTHER ORTHO ESTERS

### A. Transesterification Reactions of Ortho Esters

1. INTRODUCTION

Ortho esters react with alcohols with exchange of alkoxy groups to form new ortho esters. In the case of acyclic ortho esters and monohydroxy alcohols, the result is represented by Eqs. (32)–(35).

$$RC(OR')_3 + R''OH \to RC(OR')_2OR'' + R'OH \quad (32)$$

$$RC(OR')_2OR'' + R''OH \to RC(OR'')_2OR' + R'OH \quad (33)$$

$$RC(OR'')_2OR' + R''OH \to RC(OR'')_3 + R'OH \quad (34)$$

$$RC(OR')_3 + 3\,R''OH \to RC(OR'')_3 + 3\,R'OH \quad (35)$$

If R'OH is lower boiling than R"OH, the equilibrium of Eq. (35) [which is the sum of Eqs. (32)–(34)] may often be displaced completely to the right by distilling the more volatile alcohol from the reaction mixture. In some cases $RC(OR'')_3$ is sufficiently destabilized by steric hindrance so that only mixed ortho esters [the products of Eqs. (32) and (33)] can be isolated. The composition of the product mixtures depends on reaction time, catalyst, and catalyst concentration, efficiency of removal of R′OH from the reaction mixture, and the nature (particularly the size) of R′ and R″.

Ortho ester alcoholysis, like ortho ester hydrolysis, almost certainly involves acid-catalyzed formation of dialkoxycarbonium ion intermediates [e.g., $R'—\overset{+}{C}(OR')_2$]. The formation of these intermediates is so extremely sensitive to acid catalysis that usually it is not necessary to add a catalyst to the reaction mixture—indigenous acidic impurities cause the reaction to occur at a feasible rate. However, acids such as sulfuric, hydrochloric, and toluenesulfonic acids have been used to accelerate the transesterification reaction.

Most of the acyclic ortho esters which have been prepared by transesterification reactions are orthoformates. A limited number of orthoacetates, orthopropionates, and orthocarbonates have also been synthesized in this manner.

Transesterification as a means of synthesizing acyclic ortho esters is subject to severe steric limitations. The reaction proceeds more difficultly with secondary than with primary alcohols, and with tertiary alcohols is incomplete if it occurs at all.

Perhaps the most interesting applications of ortho ester transesterification involve reactions of ortho ester with diols, triols, and polyols, including carbohydrates, steroids, and other cyclic polyhydroxy compounds, to form alkoxydioxolanes, alkoxydioxanes, and more complex bicyclic and polycyclic ortho esters. In addition to yielding some interesting heterocyclic compounds, these reactions provide a means of protecting hydroxyl groups in some kinds of multistep syntheses, and are occasionally useful in assigning relative orientations to hydroxyl groups in cyclic molecules.

## 2. Synthesis of Acyclic Ortho Esters

a. *Trialkyl Orthoformates.* i. *Alcoholysis of other trialkyl orthoformates.* Transesterification was first used for the preparation of high-boiling orthoformates by Hunter (*153*) and by Mkhitaryan (*249*). This procedure was developed into a general synthesis of trialkyl orthoformates by Alexander and Bush (*2*). By refluxing mixtures of triethyl orthoformate and an excess of the higher alcohol in an apparatus fitted with an efficient fractionating column until the theoretical quantity of ethanol distilled over, and then fractionating the residue, they obtained tripropyl, tributyl, triisobutyl, tri-2-butyl, triamyl, triisoamyl, and trihexyl orthoformates in 75–95% yields. Tribenzyl orthoformate, which decomposed during distillation, was isolated as a low-melting solid. Triisopropyl and tri-*tert*-butyl orthoformates could not be prepared.

This method of preparing orthoformates was extended and improved by Roberts and co-workers (298), who found that reaction times could be decreased and yields improved by the addition of a small amount of sulfuric acid to the reaction mixture. The propyl, butyl, amyl, hexyl, and isoamyl esters were synthesized in 80–95% yields. By using trimethyl orthoformate as the starting material, it was possible to synthesize triisopropyl orthoformate in 75% yield, although the reaction was slow. Synthesis of orthoformates derived from higher-boiling secondary alcohols was complicated by acid-catalyzed dehydration of the alcohols and subsequent hydrolysis of the ortho esters during distillation of the reaction mixture. These side reactions could be minimized by neutralizing the acid catalyst with an alkoxide prior to isolation of the products. Using this procedure, tricyclohexyl orthoformate was prepared in 44% yield, along with diethylcyclohexyl and ethyldicyclohexyl orthoformates.

Higher trialkyl orthoformates can also be prepared by passing a mixture of trimethyl orthoformate and the higher alcohol through an acidic ion-exchange resin column. The effluent contains methanol, the higher alkanol, the desired trialkyl orthoformate, and mixed orthoformates. By recycling the mixed ortho esters, yields exceeding 95% are achieved (87a).

Since the alcoholysis reaction does not involve fission of the carbon–oxygen bond of the alcohol, it is possible to prepare optically active orthoformates by transesterifying triethyl orthoformate with an optically active secondary alcohol. (−)-Trimethyl orthoformate (249), (−)-tri-2-octyl orthoformate (153), tri-(S)-2-methylbutyl orthoformate (273, 273a) and tri-(s)-3-methyl-2-pentyl orthoformate (303a, 303b) were prepared by transesterification.

Steric requirements for alcohol exchange in orthoformate transesterification have not been thoroughly explored. It is known, however, that alcoholyses using secondary alcohols occur much more slowly and are more difficult to drive to completion than those involving primary alcohols, and that transesterification using tertiary alcohols are even more difficult. All attempts to prepare tri-*tert*-butyl orthoformate by this method have failed (2, 298), although synthesis of ethyldi-*tert*-butyl and diethyl-*tert*-butyl orthoformates by partial alcoholysis of triethyl orthoformate with *tert*-butyl alcohol was recently reported (261). Trineopentyl orthoformate can be prepared by the transesterification method (192).

Hydroxy compounds having electronegative substituents undergo partial or complete exchange with triethyl orthoformate. Thus, tri-2-chloroethyl orthoformate was prepared in high yield from 2-chloroethanol (124, 234). Phenol and triethyl orthoformate give diethylphenyl orthoformate (328), and hydroxyacetonitrile forms diethylcyanomethyl orthoformate (199).

ii. *Alcoholysis of trialkyl orthothioformates.* Trialkyl orthoformates have also been prepared by zinc chloride-catalyzed alcoholysis of trialkyl orthothioformates (*121, 251*). Triethyl and tributyl orthoformates were prepared in 66% and 46% yields by reaction of the alcohols with triethyl orthothioformate.

This reaction provides a means of synthesizing trialkyl orthoformates from alkyl formates or formic acid, since both formic acid and its esters can be converted to trialkyl orthothioformates (*103, 145, 149, 165*). Acid chlorides of higher carboxylic acids can be converted to trialkyl orthothiocarboxylates (*293*), but attempts to convert these to trialkyl ortho esters have not been reported.

b. *Esters of Higher Orthocarboxylic Acids.* Transesterification as a route to esters of the higher orthocarboxylic acids has barely been explored. Tri-2-chloroethyl orthoacetate (*124*) and diethylphenyl orthoacetate (*328*) were prepared by reaction of triethyl orthoacetate with 2-chloroethanol and with phenol. Tripropyl, tributyl, trihexyl, and trioctylorthopropionates were obtained by transesterification of triethyl orthopropionate with the appropriate alcohols (*9*).

c. *Tetraalkyl Orthocarbonates by Transesterification.* Transesterification reactions of tetraalkyl orthocarbonates apparently are more difficult to drive to completion than are analogous reactions of orthoformates. Smith and Delin reported that heating tetramethyl orthocarbonate and *n*-butyl alcohol in the presence of *p*-toluenesulfonic acid produced a mixture of trimethylbutyl orthocarbonate, dimethyldibutyl orthocarbonate, and tributylmethyl orthocarbonate (*329*). No tetrabutyl orthocarbonate was isolated. The only ester they obtained in isolable amounts when *sec*-butyl alcohol was used was trimethyl-2-butyl orthocarbonate. Arbusov, however, claimed to have prepared tetra-*n*-hexyl, tetra-*n*-octyl, tetra-*n*-nonyl, and tetra-*n*-decyl orthocarbonates by transesterification reactions with tetramethyl orthocarbonate (*6a*).

3. Syntheses of Monocyclic Ortho Esters

Transesterification of acyclic ortho esters with 1,2- and 1,3-diols provides the best means of synthesizing 2-alkoxy-1,3-dioxolanes (**3**) and 2-alkoxy-1,3-dioxanes (**4**).

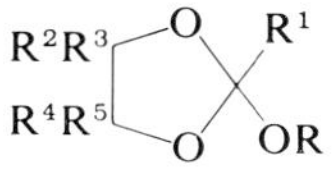

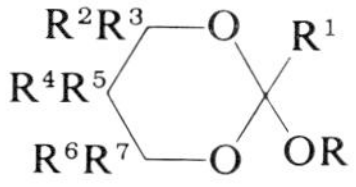

These reactions are usually carried out by heating a mixture of the glycol and ortho ester, with or without a catalyst such as sulfuric or *p*-toluenesulfonic acid, and distilling off the alcohol as it is formed. The preparation of 2-ethoxy-1,3-dioxolane from ethylene glycol and triethyl orthoformate is an example [Eq. (36)]. Yields are usually excellent.

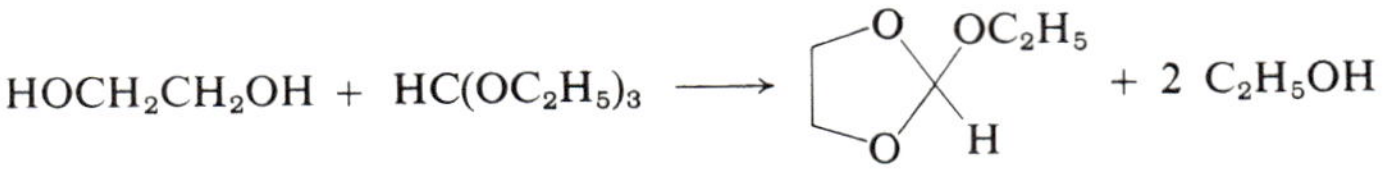

$$HOCH_2CH_2OH + HC(OC_2H_5)_3 \longrightarrow \begin{array}{c}\text{dioxolane structure}\end{array} + 2\ C_2H_5OH \quad (36)$$

a. *2-Alkoxy-1,3-dioxolanes.* Syntheses of 2-alkoxy-1,3-dioxolanes (3) by transesterification of acyclic ortho esters with 1,2-diols and by reaction of alcohols with 2-alkoxy-1,3-dioxolanes are summarized in Table II.

b. *2-Alkoxyl-1,3-dioxanes.* Several substituted 2-alkoxy-1,3-dioxanes (4) have been prepared by reaction of trimethyl and triethyl orthoformates with 1,3-diols. These include 2-ethoxy-5-hydroxy-1,3-dioxane *(68)*, 2-ethoxy-4-methyl-1,3-dioxane *(246)*, 2-ethoxy-4-chloromethyl-1,3-dioxane *(47)*, 2-ethoxy-5,5-dimethyl-1,3-dioxane *(47)*, 2-ethoxy-4,4,6-trimethyl-1,3-dioxane *(246)*, and 2-methoxy-5,5-dimethyl-4-isopropyl-1,3-dioxane *(47)*.

4. SYNTHESIS OF BICYCLIC AND POLYCYCLIC ORTHO ESTERS

a. *Bicyclic Ortho Esters.* Bicyclic ortho esters are formed by reactions of acyclic ortho esters with suitably constituted cycloalkanediols or with acyclic triols.

Examples of the first type of reaction are the preparation of *cis-* and *trans-*8-ethoxy-7,9-dioxabicyclo[4.3.0]nonanes (5, R = H) from the cyclohexanediols and triethyl orthoformate *(69)*, synthesis of *cis-* and *trans-*8-ethoxy-8-methyl-7,9-dioxabicyclo[4.3.0]nonanes (5, R = CH$_3$) from the same diols and triethyl orthoacetate *(364)*, and formation of *cis-*3-ethoxy-3-methyl-2,4-dioxabicyclo[4.4.0]decane (6) from triethyl orthoacetate and *cis-*2-hydroxymethylcyclohexanol *(79, 180a)*.

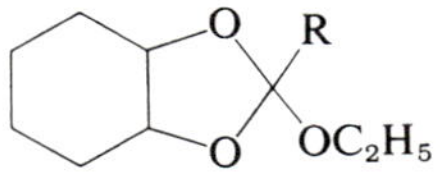

5 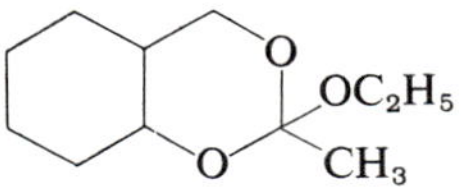    6

Reactions of acyclic triols with triethyl orthoformate were used to synthesize 4-methyl-2,6,7-trioxabicyclo[2.2.2]octane (7) *(68, 78, 265)*, 2,7,8-trioxabicyclo[3.2.1]octane (**8**) (*68*), 2,8,9-trioxabicyclo[4.2.1]nonane (**9**) (*68*), and 2,8,9-trioxabicyclo[3.3.1]nonane (**10**) (*68*).

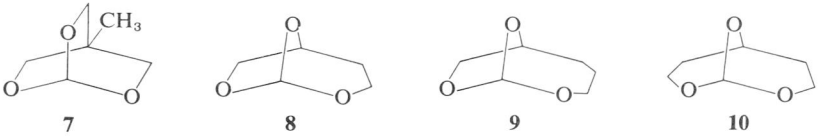

       **7**             **8**             **9**           **10**

b. *Tricyclic Ortho Esters*. Reactions of acyclic ortho esters with all-*cis*-1,3,5-cyclohexanetriol yield a group of ortho esters possessing interesting chemical and stereochemical properties. These compounds, which have the rigid tricyclic ring system of the hydrocarbon adamantane, were christened trioxaadamantanes by Stetter (*334*). They are named systematically as derivatives of 2,4,10-trioxatricyclo[3.3.1.1$^{3,7}$]decane (**11**, R = H).

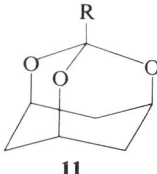

**11**

A series of 3-substituted 2,4,10-trioxaadamantanes (**11**, R = H, CH$_3$, C$_2$H$_5$, CH$_2$Cl, CH$_2$Br, C$_6$H$_5$) was synthesized by treating *cis*-1,3,5-cyclohexanetriol (phloroglucitol) with the appropriate trimethyl or triethyl orthocarboxylates in methanolic hydrogen chloride solution (*334*). The tricyclic ortho ester function in these compounds is not attacked by Grignard reagents, metal hydrides, diazomethane, acid chlorides, or hydrogen in the presence of palladium, and is relatively resistant to acid hydrolysis. These facts led Stetter to suggest the use of the trioxaadamantyl group to protect carboxyl functions in multistep syntheses (*335*). Following this suggestion, Stetter and others (*44, 266*) have synthesized a number of complex 3-substituted-2,4,10-trioxaadamantanes as intermediates in multistep syntheses.

Formation of another trioxaadamantane derivative, 6,6,9,9-tetramethyl-2,4,10-trioxatricyclo[3.3.1.1$^{3,7}$]decane, by reaction of hexahydrosyncarpic acid (2,2,6,6-tetramethyl-1,3,5-cyclohexanetriol) established the cis relationship of the three hydroxyl groups of the triol (*139*). Other examples of the use of tricyclic orthoester formation in reactions of cyclic polyhydroxy compounds with ethyl orthoformate to assign relative orientations to the hydroxyl groups are the preparation of the 1,5,19-orthoformate of an ouabagenin derivative (**12**) (*23, 346*), and formation of the 1,3,5-orthoformates of

TABLE II
2-ALKOXY-1,3-DIOXOLANES PREPARED BY TRANSESTERIFICATION REACTIONS

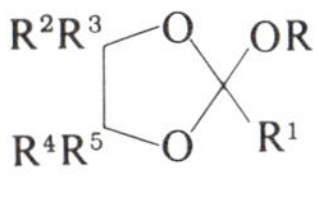

$$\begin{array}{c} R^2R^3 \diagdown O \diagdown OR \\ \diagup \diagdown R^1 \\ R^4R^5 \diagup O \end{array}$$

| R | R$^1$ | R$^2$ | R$^3$ | R$^4$ | R$^5$ | Yield[a] (%) | References |
|---|---|---|---|---|---|---|---|
| CH$_3$ | H | H | H | H | H | 70 | 17 |
| C$_2$H$_5$ | H | H | H | H | H | 90 | 12, 17, 250 |
| C$_3$H$_7$ | H | H | H | H | H | 76 | 17 |
| C$_4$H$_9$ | H | H | H | H | H | 62 | 17 |
| ClCH$_2$CH$_2$ | H | H | H | H | H | 55 | 17[b] |
| Menthyl | H | H | H | H | H | — | 250[c] |
| Bornyl | H | H | H | H | H | — | 250[d] |
| C$_2$H$_5$ | H | CH$_3$ | H | H | H | 59 | 17 |
| C$_2$H$_5$ | H | ClCH$_2$ | H | H | H | 64 | 17, 272 |
| C$_2$H$_5$ | H | HOCH$_2$ | H | H | H | 88 | 68 |
| C$_2$H$_5$ | H | C$_6$H$_5$OCH$_2$ | H | H | H | — | 47 |
| C$_2$H$_5$ | H | p-ClC$_6$H$_4$OCH$_2$ | H | H | H | — | 272, 343 |
| C$_2$H$_5$ | H | o-CH$_3$C$_6$H$_4$OCH$_2$ | H | H | H | — | 343 |
| C$_2$H$_5$ | H | o-CH$_3$OC$_6$H$_4$OCH$_2$ | H | H | H | — | 343 |

# SYNTHESIS FROM OTHER ORTHO ESTERS

| | | | | | | |
|---|---|---|---|---|---|---|
| C$_2$H$_5$ | H | o-C$_4$H$_9$OC$_6$H$_4$OCH$_2$ | H | H | — | *343* |
| C$_2$H$_5$ | H | CH$_3$ | H | CH$_3$ | — | *246* |
| C$_2$H$_5$ | H | C$_2$H$_5$O$_2$C | H | H | — | *69* |
| C$_2$H$_5$ | H | C$_6$H$_5$ | H | H | — | *69* |
| C$_2$H$_5$ | H | CH$_3$ | CH$_3$ | CH$_3$ | 95 | *69, 246* |
| CH$_3$ | CH$_3$ | H | H | H | — | *339* |
| CH$_3$ | CH$_3$ | ClCH$_2$ | H | H | 68 | *365* |
| CH$_3$ | CH$_3$ | BrCH$_2$ | H | H | 75 | *365* |
| C$_2$H$_5$ | CH$_3$ | H | H | H | 73 | *238* |
| C$_2$H$_5$ | CH$_3$ | ClCH$_2$ | H | H | 75 | *365* |
| CH$_3$ | CH$_3$ | o-CH$_3$C$_6$H$_4$OCH$_2$ | H | H | — | *343* |
| C$_2$H$_5$ | CH$_3$ | o-CH$_3$C$_6$H$_4$OCH$_2$ | H | H | — | *343* |
| C$_4$H$_9$ | CHCl$_2$ | H | H | H | — | *245$^e$* |
| C$_3$H$_7$ | CCl$_3$ | H | H | H | — | *244$^e$* |
| C$_2$H$_5$ | C$_2$H$_5$ | o-CH$_3$C$_6$H$_4$OCH$_2$ | H | H | — | *343* |
| CH$_3$ | C$_4$H$_9$ | o-CH$_3$C$_6$H$_4$OCH$_2$ | H | H | — | *343* |
| CH$_3$ | C$_6$H$_5$ | H | H | H | 60 | *289* |
| CH$_3$ | C$_6$H$_5$ | C$_6$H$_4$= | (4,5-benzo) | H | — | *75* |
| C$_2$H$_5$ | C$_6$H$_5$ | C$_6$H$_4$= | (4,5-benzo) | H | — | *75* |
| CH$_3$ | CH$_3$O | C$_6$H$_4$= | (4,5-benzo) | H | 91 | *119, 328* |

$^a$ By transesterification with R'C(OR)$_3$, unless otherwise specified.
$^b$ From 2-ethoxy-1,3-dioxolane and 2-chloroethanol.
$^c$ From 2-ethoxy-1,3-dioxolane and menthol.
$^d$ From 2-ethoxy-1,3-dioxolane and borneol.
$^e$ From 2-methoxy-2-R'-dioxolane and R'OH.

ketigenin (**13**) (*315*), convallagenin-A (*165c*), and 1,3,5,17-tetrahydroxyandrostanes (*89*).

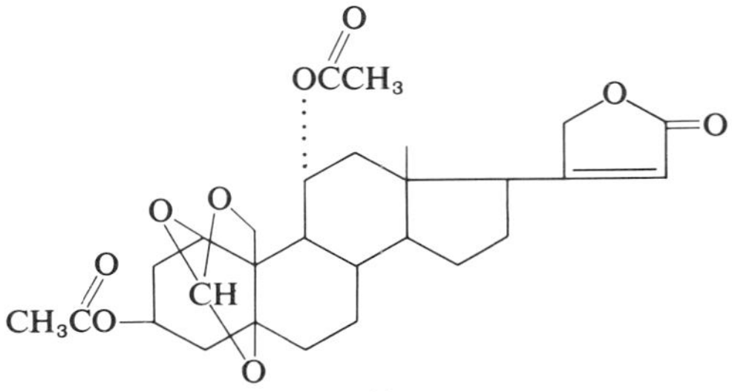

**12**

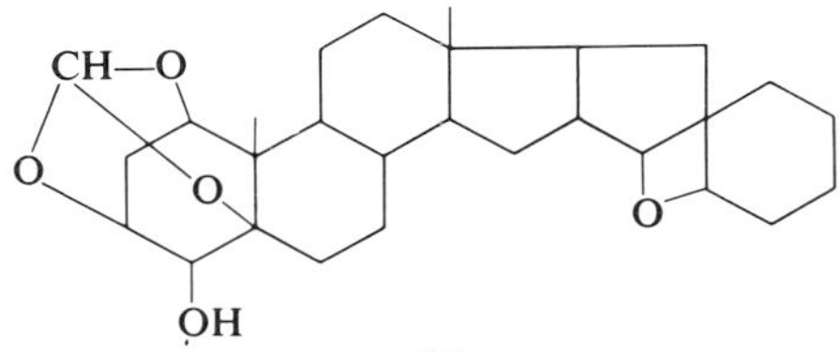

**13**

c. *A Pentacyclic Ortho Ester.* Scyllitol (*cis*-1,3,5-*trans*-2,4,6-cyclohexanehexol) reacts with triethyl orthoformate in dimethyl sulfoxide solution at 200°C to form a pentacyclic diortho ester, 3,5,8,10,13,14-hexaoxapentacyclo[7.3.1.1$^{4,12}$.0$^{2,7}$.0$^{6,11}$]tetradecane (**14**) (*353*). This substance, for which the trivial name hexaoxadiamantane was proposed, possesses unusual thermal stability for an ortho ester: it melts at 303°–305°C in a sealed tube, and is stable at 400°C.

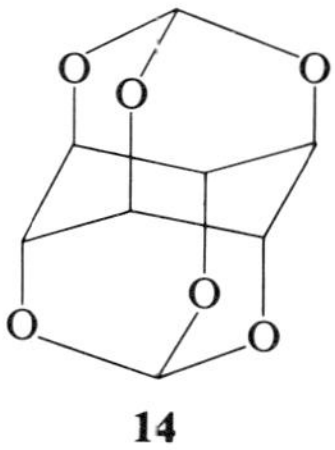

**14**

## 5. Steroid Ortho Esters

A number of cyclic and bicyclic ortho esters have been prepared by reactions of acyclic ortho esters with steroids having two or three hydroxyl groups on ring D or on the side chain at C-17. Most of these compounds were

## SYNTHESIS FROM OTHER ORTHO ESTERS

synthesized for pharmacological screening, but in several instances the ortho ester function was used to protect the esterified hydroxyl groups during chemical transformations of other parts of the molecule (74, 84, 106–108, 304). Formation of cyclic ortho esters, followed by mild acid hydrolysis to the hydroxyacyloxy derivatives, has been used as a method of selective acylation of polyhydroxy steroids (85, 95, 105, 107, 110). This sequence provides the only route to 21-hydroxy-17-acyloxy steroids (100a).

$16\alpha,17\alpha$-Dihydroxy steroids are converted to $16\alpha,17\alpha$-alkoxy-alkylenedioxy derivatives by reaction with acylic orthocarboxylates in acidic solutions. A specific example is the reaction of $16\alpha$-hydroxycortisone with trimethyl orthoformate (84).

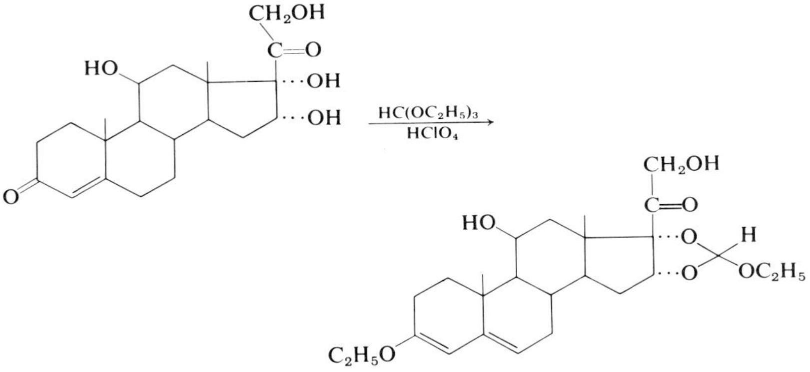

(37)

The 3-keto group is converted to an enol ether function. Other examples of formation of $16\alpha,17\alpha$-cyclic ortho esters are described in the literature (3, 32, 51, 74, 83–86, 330, 331).

Cortisone itself reacts with triethyl orthoformate to form a $17\alpha,21$-ethoxymethylenedioxy derivative (160).

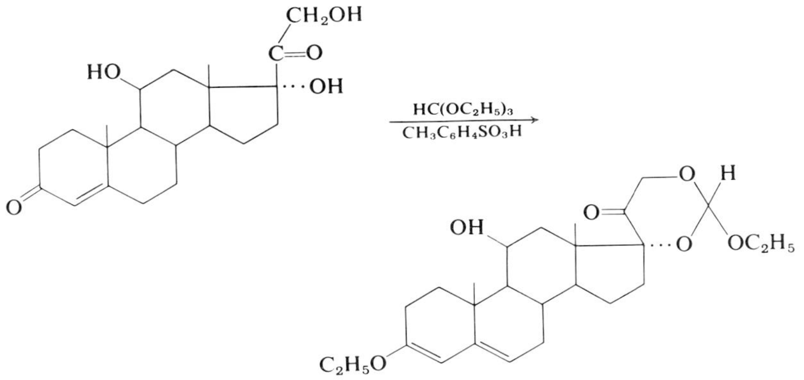

(38)

Again, the 3-keto group is converted to an enol ether grouping. Similar syntheses of a number of closely related 17α,21-cyclic ortho esters have been published (*94–96, 104–107, 109, 110, 110a, 160, 323e, 351b*).

Another 2-alkoxy-1,3-dioxane steroid derivative, **15**, is formed by the boron trifluoride-catalyzed reaction of 3β-acetoxy-5-α-pregnane-16β,20-diol with triethyl orthoformate (*254*).

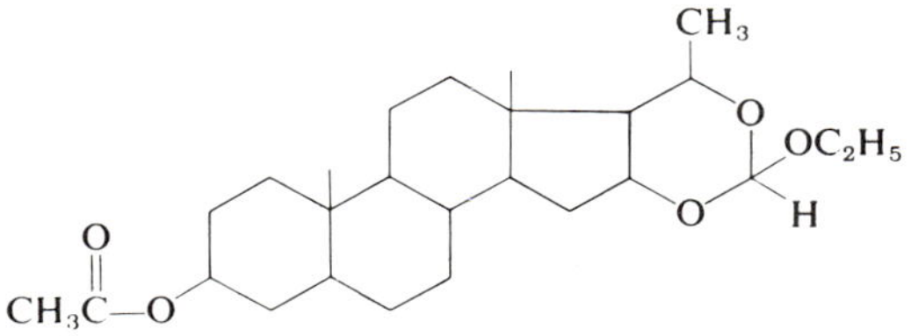

**15**

Several trioxabicyclo[2.2.2]octanes with steroidal substituents at the 4-position have been synthesized from 20,20-bis(hydroxymethyl)-21-hydroxy steroids and triethyl orthoformate (*33, 304–306*). An example is the bicyclic ortho ester prepared from 3α,21-dihydroxy-11-oxo-20,20-bis(hydroxymethyl)-5β-pregnane (**16**) (*305*).

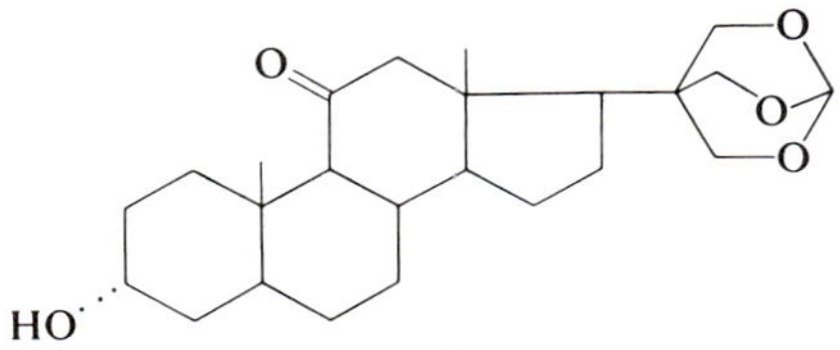

**16**

### 6. Ortho Esters Derived from Sugars, Ribonucleosides, and Ribonucleotides

A number of carbohydrates and carbohydrate derivatives have been converted to cyclic ortho esters by transesterification with acyclic ortho esters. These reactions are discussed in Chapter 5.

### B. Formation of Mixed Ortho Esters by Disproportionation Reactions

Post and Erickson (*281*) found that fractionating a mixture of tripropyl orthoformate and triisoamyl orthoformate which had stood for a month at room temperature yielded all four possible ortho esters [Eq. (39), $R = C_3H_7$, $R' = CH_2CH_2CH(CH_3)_2$].

$$HC(OR)_3 + HC(OR')_3 \rightleftharpoons HC(OR)_2OR' + HC(OR')_2OR \qquad (39)$$

This disproportionation reaction probably involves acid-catalyzed formation of dialkoxycarbonium ions, followed by transfer of alkoxy groups between ortho ester molecules and the carbonium ions. Apparently indigenous acidic impurities or acidic surfaces suffice to catalyze the reaction, since disproportionation usually occurs in the absence of added acid catalysts. Reported isolations of mixed ortho esters from reactions of imidic ester hydrochlorides or α-halo ethers with alcohols and alkoxides indicate that disproportionation may be slow in the absence of acids.

A quantitative study of the equilibration of trimethyl and triethyl orthoformates [Eq. (39), $R = CH_3$, $R' = C_2H_5$)] showed that the alkoxy groups are nearly randomly distributed at equilibrium (*252*).

A different kind of disproportionation reaction is the equilibration of methoxybis(ethylthio)methane to a mixture consisting mostly of trimethyl orthoformate and triethyl orthothioformate (*117*).

$$3\ CH_3OCH(SC_2H_5)_2 \rightleftharpoons HC(OCH_3)_3 + 2\ HC(SC_2H_5)_3 \quad (40)$$

## C. Synthesis of Ortho Esters by Alteration of Acyl or Alkoxy Substituents of Other Ortho Esters

The ortho ester function, $-C(OR)_3$, is usually quite stable under neutral and basic conditions. This fact makes possible a number of reactions which produce changes in ortho ester acyl and alkoxy substituents without affecting the ortho ester function itself. Substitution, elimination, addition, and condensation reactions have been used for this purpose. Most of these transformations involve alteration of the acyl substituent R of an ortho ester $RC(OR')_3$, but a few examples of modification of the alkoxy substituent $R'$ have also been reported.

### 1. Preparation of α-Bromo Ortho Esters

A number of α-bromo ortho esters were prepared by direct bromination of ortho esters having one or more α-hydrogens [Eq. (41)].

$$RR'CHC(OR'')_3 + Br_2 + C_5H_5N \rightarrow RR'CBrC(OR'')_3 + C_5H_5NH^+Br^- \quad (41)$$

These reactions are carried out by adding the theoretical quantity of bromine dropwise to a mixture of equimolar amounts of the ortho ester and pyridine at temperatures ranging from 0° to 35°C, depending on the reactivity of the ortho ester. In this way McElvain and co-workers prepared trimethyl and triethyl orthobromoacetate, trimethyl and triethyl 2-bromoorthopropionate, trimethyl 2-bromoorthovalerate, trimethyl 2-bromoorthononanoate, trimethyl 2-bromo-3-methylorthobutyrate, and trimethyl α-bromophenylorthoacetate in yields ranging from 67% to 80% (*35, 204, 207, 214, 226, 230, 355*).

It is possible to replace a second α-hydrogen atom by using 2 moles each of bromine and pyridine, and carrying out the reaction at a higher temperature. For example, triethyl orthodibromoacetate was prepared in 53% yield from triethyl orthoacetate (230).

## 2. Reduction Reactions of Acyl Substituents of Ortho Esters

The ortho ester function is resistant to catalytic reduction (163) and, if the acyl substituent is bulky, is also relatively inert toward metal hydrides. These properties have been used to advantage in the synthesis of a number of ortho esters by chemical and catalytic reduction of functions in the acyl groups of other ortho esters.

4 - Methyl - 1 - trichloromethyl - 2,6,7 - trioxabicyclo[2.2.2]octane (**17**, R = $CCl_3$) is converted to 1,4-dimethyl-2,6,7-trioxabicyclo[2.2.2]octane (**17**, R = $CH_3$) by catalytic hydrogenolysis (25). Tetramethyl monoorthooxalate is reduced by lithium aluminum hydride to unstable trimethyl hydroxyorthoacetate (264) and carbomethoxymethyl trioxaadamantane (**18**, R = $CH_2CO_2CH_3$) is similarly reduced to the hydroxyethyl analog (**18**, R = $CH_2CH_2OH$) (44). Lithium tri-*tert*-butoxyaluminum hydride was used to reduce an ortho ester having an acid chloride function (**18**, R = $CH_2COCl$) to the corresponding aldehyde (**18**, R = $CH_2CHO$) (44). This aldehyde was

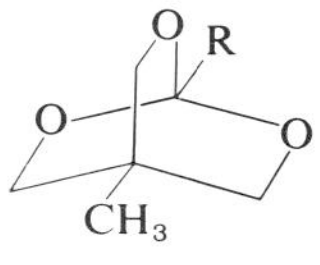

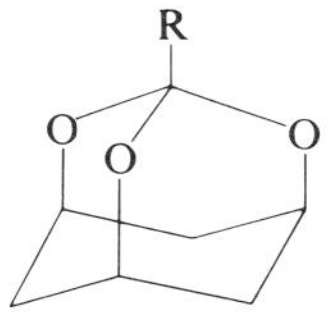

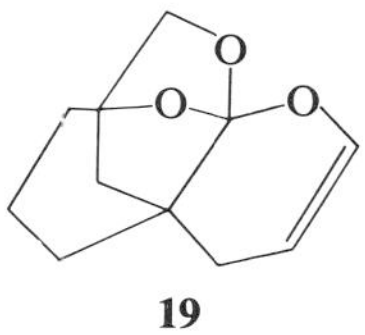

also prepared by reducing **18** [R = $CH_2CON(CH_3)C_6H_5$] with lithium aluminum hydride.

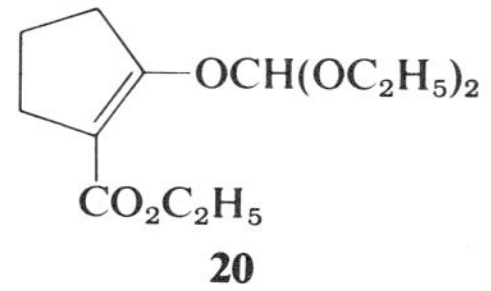

Carbon–carbon multiple bonds in ortho esters can also be partially or completely saturated without destroying the ortho ester function. Raney nickel-catalyzed hydrogenation of the tricyclic ortho ester **19** converted it to the saturated analog, 2,10,11-trioxatricyclo[4.4.4.0$^{1,6}$]tetradecane (216). Diethyl-2-carbethoxycyclopentenyl orthoformate (**20**) is hydrogenated over platinum oxide to diethyl-2-carbethoxycyclopentyl orthoformate (190). There are also examples of lithium aluminum hydride reduction of acetylenic to ethylenic linkages in ortho ester acyl substituents (45).

## 3. Substitution Reactions of Ester and Acid Chloride Functions in Ortho Ester Acyl Side Chains

Ester and acid chloride functions in ortho ester side chains undergo substitution reactions at the carboxyl carbon which do not affect the ortho ester function. Tetramethyl monoorthooxalate and tetramethyl monoorthosuccinate react with dimethylamine and ammonia, respectively, to yield carboxamido ortho esters, $R_2NCO(CH_2)_nC(OCH_3)_3$ (R = $CH_3$, $n = 0$; R = H, $n = 2$) (216, 223). Compound **18**, R = $CH_2CO_2CH_3$, reacts with ammonia to form the amide **18**, R = $CH_2CONH_2$; and **18**, R = $CH_2COCl$ is converted by N-methylbenzylamine and by hydrazine to **18**, R = $CH_2CON(CH_3)C_6H_5$, $CH_2CONHNH_2$ (44). A carboalkoxy function in the acyl side chain of an ortho ester can be saponified without hydrolyzing the ortho ester group. Examples are the alkaline hydrolysis of a steroid ortho ester derivative (106) [Eq. (42)], and of **18**, R = $CH_2CO_2C_2H_5$ (44).

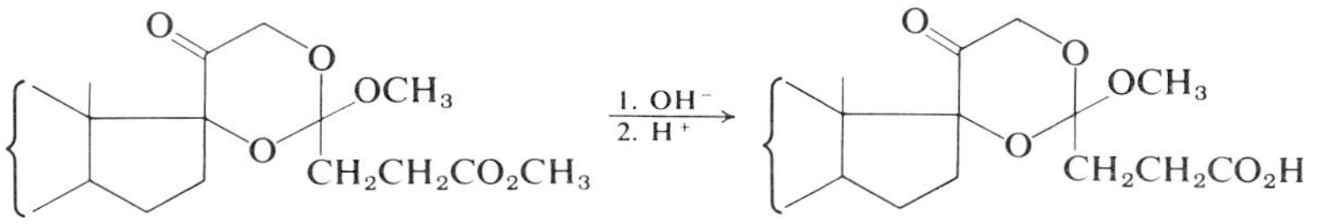

$$\text{[structure]} \xrightarrow[\text{2. H}^+]{\text{1. OH}^-} \text{[structure]} \quad (42)$$

## 4. Reactions in Which Acyl Side Chains of Ortho Esters Are Lengthened

Tricyclic ortho esters **18** having aldehyde or carbalkoxy groups in the acyl side chain R undergo the usual addition reactions with Grignard reagents, without affecting the ortho ester function (45, 334). The ortho ester organomagnesium bromide **18**, R = $CH_2CH_2CH_2C{\equiv}CMgBr$, reacted with propargyl halides to form polyynetricyclic ortho esters (266). Straight-chain ortho esters condense with chloral and other α,α-dichloroaldehydes to form esters of 2-substituted 3-hydroxy-4,4-dichloroorthocarboxylic acids. The resulting hydroxy ortho esters react with a second mole of the α-haloaldehyde, forming 6,6-dialkoxy-1,3-dioxane derivatives (128, 183, 184) [Eq. (43)].

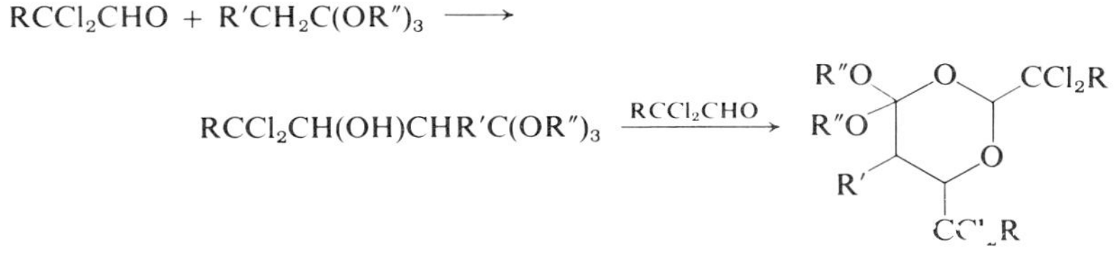

$$RCCl_2CHO + R'CH_2C(OR'')_3 \longrightarrow$$

$$RCCl_2CH(OH)CHR'C(OR'')_3 \xrightarrow{RCCl_2CHO} \text{[dioxane structure]} \quad (43)$$

Trialkyl orthoformates react with 2 moles of hexafluoroacetone at 150°C to form 4,4-dialkoxy-2,2,5,5-tetratrifluoromethyl-1,3-dioxolanes (47, 48). Dialkoxy carbenes may be intermediates in these reactions.

## 5. Elimination and Nucleophilic Substitution Reactions in Acyl Side Chains of Ortho Esters

Trimethyl ortho-4-chloro-3-butynoate was synthesized from trimethyl ortho-4,4,4-trichlorobutyrate by treatment with sodium methoxide (*223*). Triethyl orthobromoacetate reacts with ethanolic sodium iodide to form triethyl orthoiodoacetate (*35*).

## 6. Synthesis of Ortho Esters by Chemical Alteration of Alkoxy Substituents of Other Ortho Esters

A few syntheses of ortho esters have been reported which involve elimination, substitution, and reduction reactions of substituted alkoxy groups of other ortho esters. Examples are the base-promoted elimination reactions used to convert 2-chloroethyl groups to vinyl groups in the synthesis of trivinyl orthoformate (*234*) and trivinyl orthoacetate (*205*), the reactions of tri-2-chloroethyl orthoformate and orthoacetate with trimethylamine to form quaternary salts of the structure $RC[OCH_2CH_2N(CH_3)_3]_3^{+3} \cdot 3\,Cl^-$ (*124*), and the catalytic hydrogenation of diethylcyanomethyl orthoformate to diethyl-2-aminoethyl orthoformate (*199*).

## V. FORMATION OF ORTHO ESTERS BY REACTIONS INVOLVING ADDITION OF ALCOHOLS TO CARBON–CARBON MULTIPLE BONDS

### A. Introduction

Alcohols add to the carbon–carbon double bonds of 1,1-dialkoxy-1-alkenes (or 1,1-diaryloxy-1-alkenes) (ketene acetals) to yield carboxylic ortho esters [Eq. (44)]. Closely related is the addition of ethanol to ethoxyacetylene to form triethyl orthoacetate [Eq. (45)]. Less closely related is the conversion of 1,1-dihalo-1-alkenes to ortho esters by reaction with alkoxides [Eq. (46)].

$$RR^1C{=}C(OR^2)_2 + R^3OH \rightarrow RR^1CHC(OR^2)_2OR^3 \qquad (44)$$

$$HC{\equiv}COC_2H_5 + 2\,C_2H_5OH \rightarrow CH_3C(OC_2H_5)_3 \qquad (45)$$

$$RR^1C{=}CX_2 + R^2OH + 2\,R^2O^- \rightarrow RR^1CHC(OR^2)_3 + 2\,X^- \qquad (46)$$

### B. Addition of Alcohols to Ketene Acetals

In 1936 McElvain and Beyerstedt found that 1,1-diethoxyethylene (ketene diethylacetal) reacts rapidly with ethanol to form triethyl orthoacetate in almost quantitative yield (*34*). Since then a number of ortho esters have been prepared by reactions analogous to this.

Additions of alcohols to ketene acetals are accelerated by acids such as hydrogen chloride. The first step of the acid-catalyzed reaction probably is formation of a dialkoxycarbonium ion (*297*), which reacts with alcohol to yield the ortho ester and regenerate the catalyst.

$$RR^1C{=}C(OR^2)_2 + H^+ \longrightarrow RR^1CHC{\overset{OR^2}{\underset{OR^2}{+}}} \xrightarrow[-H^+]{+R^3OH} RR^1CHC(OR^2)_2OR^3 \quad (47)$$

The addition reaction occurs without added acids, however, and there are several reported examples of ortho ester formation from ketene acetals and alcohols under alkaline conditions (*34, 36, 230*).

The addition of alcohols to ketene acetals is useful for synthesizing mixed ortho esters. For example, a series of mixed trialkyl orthoacetates, $CH_3C(OR)(OC_2H_5)_2$ (R = butyl, isobutyl, *sec*-butyl, isoamyl, neopentyl, and benzyl), was prepared from ketene diethyl acetal and the appropriate alcohols (*204*). The reaction also provides a means for converting α-haloaldehydes and α-haloacetals to ortho esters. Examples illustrating the synthesis of ortho esters which would be difficult or impossible to prepare by other means are given in Eqs. (48)–(51) (*205, 212, 218, 370*).

$$CH_2{=}CHOC_6H_5 \xrightarrow{Br_2} CH_2BrCHBrOC_6H_5 \xrightarrow{C_6H_5ONa}$$
$$CH_3BrCH(OC_6H_5)_2 \xrightarrow{tert\text{-BuOK}}$$
$$CH_2{=}C(OC_6H_5)_2 \xrightarrow{C_6H_5OH} CH_3C(OC_6H_5)_3 \quad (48)$$

$$(C_6H_5)_2CClCH(OCH_3)_2 \xrightarrow{tert\text{-BuOK}} (C_6H_5)_2C{=}C(OCH_3)_2 \xrightarrow{CH_3OH}$$
$$(C_6H_5)_2CHC(OCH_3)_3 \quad (49)$$

$$ClCH_2CH(OCH_2CH_2Cl)_2 \xrightarrow{tert\text{-BuOK}} CH_2{=}C(OCH{=}CH_2)_2 \xrightarrow{HOCH_2CH_2Cl}$$
$$CH_3C(OCH{=}CH_2)_2OCH_2CH_2Cl \xrightarrow{tert\text{-BuOK}} CH_3C(OCH{=}CH_2)_3 \quad (50)$$

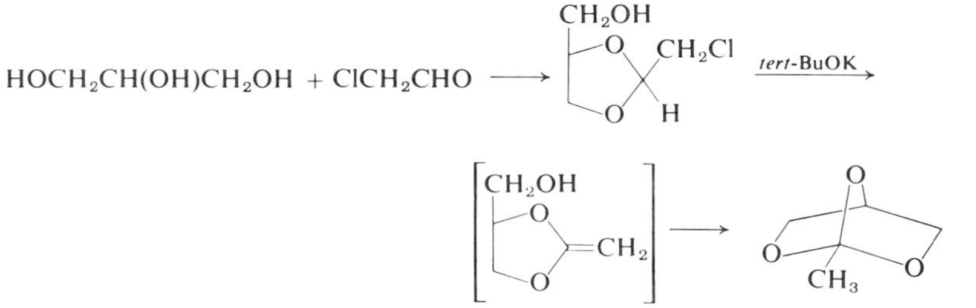

$$HOCH_2CH(OH)CH_2OH + ClCH_2CHO \longrightarrow \underset{\text{(cyclic acetal with }CH_2OH\text{ and }CH_2Cl\text{)}}{\text{intermediate}} \xrightarrow{tert\text{-BuOK}}$$

$$\left[\text{cyclic intermediate with }CH_2OH, {=}CH_2\right] \longrightarrow \text{bicyclic ortho ester with }CH_3 \quad (51)$$

The reaction of alcohols with ketene acetals is not a practical method of synthesizing most ortho esters, since the necessary ketene acetals are difficult to prepare and are frequently best prepared by dealcoholation of the desired ortho ester. In spite of these limitations a number of ortho esters have been synthesized from ketene acetals, and several of them have been prepared in no other way. These include trimethyl orthodiphenylacetate (*218*), divinylethyl orthoacetate (*205*), divinyl-2-chloroethyl orthoacetate (*205*), three divinyl-2-nitroalkyl orthoacetates (*100*), trimethyl dimethoxyorthoacetate (*143*), triethyl 3,3-diethoxyorthopropionate (*219*), triethyl ortho-3-methyl-3-butenoate (*220*), 2-methoxy- and 2-ethoxy-2-chloromethyl-1,3-dioxolanes (*98*, *209*), 2-methoxy- and 2-ethoxy-2-dichloromethyl-1,3-dioxolanes (*209*, *308*), and 2-methoxy-2-benzyl-1,3-dioxolane (*209*). Triphenyl orthoacetate, the only known triaryl orthoacetate, was synthesized by the addition of phenol to ketene diphenylacetal (*212*) [Eq. (48)].

### C. Addition of Alcohols to Alkoxyacetylenes

Ethoxyacetylene is converted to triethyl orthoacetate in 44% yield by prolonged refluxing with ethanolic sodium ethoxide (*294*) [Eq. (45)]. The generality of this reaction has not been established, and its mechanism is not known. The reaction probably involves two nucleophilic additions, with ketene diethylacetal as an intermediate. Several orthoacetic acid derivatives having structures $CH_3C(OC_2H_5)(ON=CRR')_2$ were prepared by addition of aldoximes and ketoximes to ethoxyacetylene (*345a*).

### D. Reaction of Alkoxides with 1,1-Dihalo-1-alkenes and Other 1,1-Disubstituted 1-Alkenes

Several 1,1-dihalo-1-alkenes have been converted to ortho esters by reaction with alkali metal alkoxides. The detailed course of these reactions is not known.

The first step in the conversion of 1,2-dichloro-1,2-difluoroethylene or 1-chloro-1,2,2-trifluoroethylene to triethyl fluorochloroorthoacetate by reaction with ethanolic ethoxide is probably formation of ethyl 2-chloro-1,2-difluorovinyl ether by a substitution reaction, since this substance can be isolated from the reaction mixture and subsequently converted to the ortho ester (*247*, *307*).

This substitution–addition mechanism may be general for reactions of 1,1,2-trihalo-2-substituted alkenes with alkoxides. Park and co-workers (*269*) found that ethers of the structure $CF_3CCl=CClOR$ (R = $CH_3$, $C_2H_5$, $C_3H_7$, $C_4H_9$) are the principal products of reaction of alcoholic potassium hydroxide with $CF_3CCl=CCl_2$, and isolated trimethyl 2-chloro-3,3,3-trifluoroorthopropionate as one product of the reaction with methanol. The synthesis of $CHCl_2C(OC_2H_5)=CClC(OC_2H_5)_3$ from perchloro-1,3-butadiene and sodium ethoxide (248) can also be rationalized by this mechanism if it is assumed that ortho ester formation is followed by prototropic rearrangement and substitution of ethoxide for one of the remaining vinylic chlorines.

$$CCl_2=CClCCl=CCl_2 \xrightarrow[C_2H_5OH]{C_2H_5O^-} CCl_2=CClCHClC(OC_2H_5)_3 \xrightarrow{C_2H_5O^-}$$
$$CHCl_2CCl=CClC(OC_2H_5)_3 \xrightarrow{C_2H_5O^-} CHCl_2C(OC_2H_5)=CClC(OC_2H_5)_3 \quad (52)$$

A different mechanism, involving chloroacetylene, alkoxyacetylenes, and ketene acetals as intermediates, was proposed for the reaction of β-alkoxyalkoxides with 1,1-dichloroethylene to form tri-β-alkoxyalkyl orthoacetates (188).

$$H_2C=CCl_2 \xrightarrow{NaOR} HC\equiv CCl \xrightarrow{ROH} H_2C=CClOR \xrightarrow{NaOR}$$
$$HC\equiv COR \xrightarrow{ROH} CH_2=C(OR)_2 \xrightarrow{ROH} CH_3C(OR)_3 \quad (53)$$
$$(R = CHR'CH_2OR'')$$

Triethyl chlorocyanoorthoacetate, $NCCHClC(OC_2H_5)_3$, is formed by the reaction of 2,3,3-trichloroacrylonitrile with sodium ethoxide (300a), and tetrakis(trifluoromethyl)ethylene is converted to tetramethyl hexafluoroorthoisobutyrate, $(CF_3)_2CHC(OCH_3)_3$, by methanolic sodium methoxide (121a).

1,1-Dichloro-2-nitro-1-pentene reacts with ethanolic sodium ethoxide to form the *aci*-nitro salt of triethyl 2-nitroorthopentanoate (371).

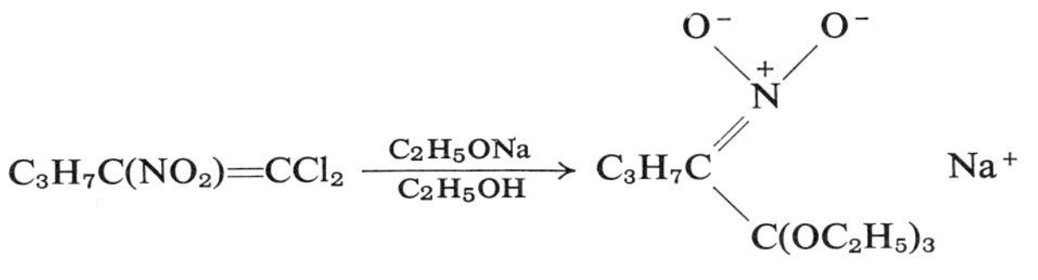

$$C_3H_7C(NO_2)=CCl_2 \xrightarrow[C_2H_5OH]{C_2H_5ONa} C_3H_7C\begin{matrix}\diagup N^+\diagdown \\ O^- \quad O^- \\ \diagdown \\ C(OC_2H_5)_3\end{matrix} Na^+ \quad (54)$$

Another addition–substitution reaction leading to a stable ortho ester salt is the reaction of tetracyanoethylene with alcoholic alkoxides (357).

$$(NC)_2C=C(CN)_2 \xrightarrow[ROH]{MOR} M^+[(NC)_2CC(OR)_3]^- \quad (55)$$
$$(M = Na, K; R = CH_3, C_2H_5)$$

There is one example of an alkyl sulfinate ion serving as the leaving group in an addition–substitution reaction yielding an ortho ester. 7,7,8,8-Tetrakis-(ethylsulfonyl)quinodimethane reacts with methanolic sodium methoxide to give a 28% yield of trimethyl *p*-bis(ethylsulfonyl)methylorthobenzoate. The reaction involves a 1,6-addition as well as replacement of the two ethylsulfinate groups (*129*).

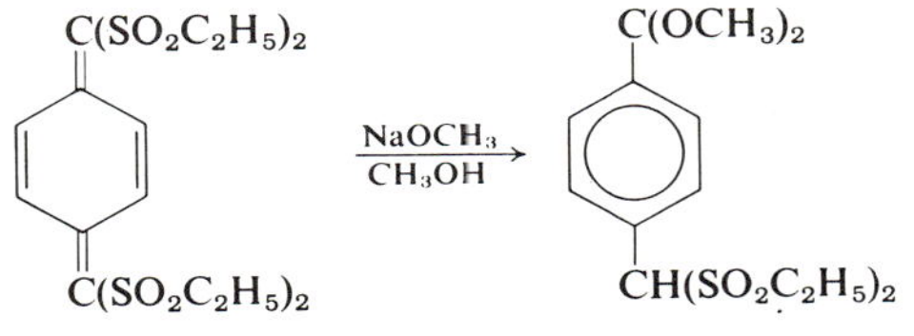

$$\underset{\underset{C(SO_2C_2H_5)_2}{\Big|}}{\overset{\overset{C(SO_2C_2H_5)_2}{\Big|}}{\bigcirc}} \xrightarrow[CH_3OH]{NaOCH_3} \underset{\underset{CH(SO_2C_2H_5)_2}{\Big|}}{\overset{\overset{C(OCH_3)_2}{\Big|}}{\bigcirc}} \qquad (56)$$

### E. Reactions of Alkoxides with 1,2-Dihaloacetylenes

1,2-Diiodoacetylene reacts with ethanolic sodium ethoxide to form triethyl iodoorthoacetate, $ICH_2C(OC_2H_5)_3$, in low yield (*261a*). The reaction probably involves two nucleophilic additions followed by alcoholysis of the resulting 1,1-diethoxy-1,2-diiodoethane. No other example of this type of ortho ester synthesis has been reported.

## VI. SYNTHESIS OF ORTHO ESTERS FROM CARBONIUM SALTS

### A. Introduction

Dioxolenium, dioxenium, and *O*-alkyllactonium salts react with alcohols under basic conditions to form cyclic ortho esters [Eqs. (57), (58), and (59)]. These salts, particularly the carbonium tetrafluoroborates and hexachloroantimonates, are reasonably stable compounds which can, in many cases, be prepared in excellent yields. There are a few instances of syntheses of acylic ortho esters from dialkoxycarbonium salts.

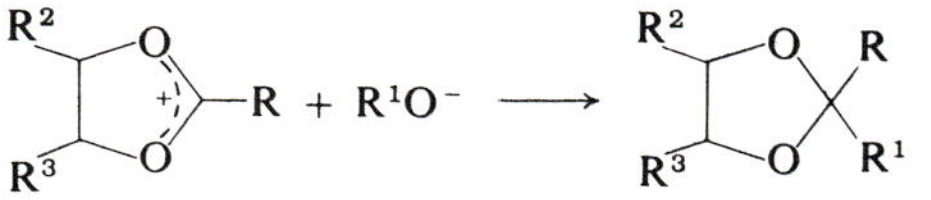

$$\underset{R^3}{\overset{R^2}{\diagdown}}\underset{O}{\overset{O}{\diagup}}\!\!\!\!\!\!\!\!\!\!>\!\!-R + R^1O^- \longrightarrow \underset{R^3}{\overset{R^2}{\diagdown}}\underset{O}{\overset{O}{\diagup}}\!\!\!\!\!\!\!\!\!\!<\!\!\!{\overset{R}{R^1}} \qquad (57)$$

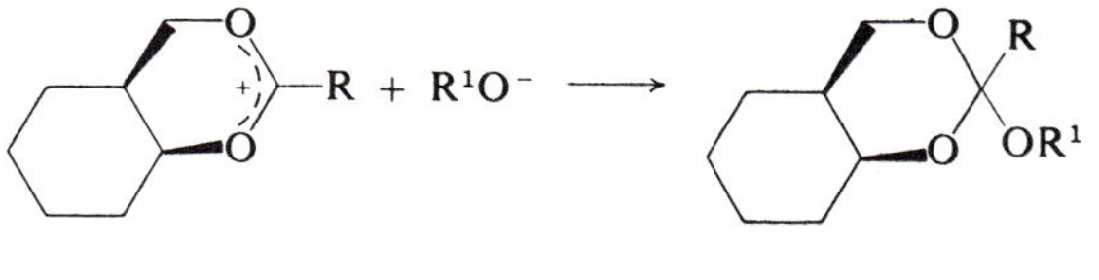

$$\text{(58)}$$

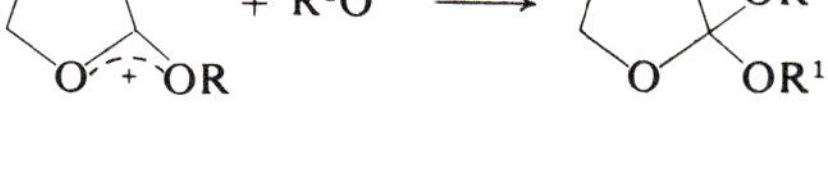

$$\text{(59)}$$

## B. Reactions of Alcohols with Dioxolenium Salts

Reactions of dioxolenium tetrafluoroborates with alcoholic alkoxides yielded 2-ethoxy-2-methyl-1,3-dioxolane [Eq. (57), R = $CH_3$, $R^1$ = $C_2H_5$, $R^2$ = $R^3$ = H] (*241*), 2-ethoxy-2-phenyl-1,3-dioxolane [Eq. (57), R = $C_6H_5$, $R^1$ = $C_2H_5$, $R^2$ = $R^3$ = H] (*241*), 2-methoxy-2-(2,6-dimethoxyphenyl)-1,3-dioxolane [Eq. (57), R = 2,6-$(CH_3O)_2C_6H_3$, $R^1$ = $CH_3$, $R^2$ = $R^3$ = H] (*31a*), 2-methoxy-2-phenyl-*cis*-4,5-tetramethylene-1,3-dioxolane [Eq. (57), R = $C_6H_5$, $R^1$ = $CH_3$, $R^2$, $R^3$ = $(-CH_2-)_4$] (*320a*) and 2-methoxy-2-methyl-*cis*-4,5-tetramethylene-1,3-dioxolane [Eq. (57), R = $R^1$ = $CH_3$, $R^2$, $R^3$ = $(-CH_2-)_4$] (*4*). The last two reactions yield mixtures of *exo* and *endo* isomers. Wilcox and Nealy converted the dioxolenium salt derived from *trans*-1,2,3,4-tetramethylcyclobutene-3,4-diol dibenzoate to the corresponding 2-methoxy-2-phenyl-1,3-dioxolane in 60% yield (*361*).

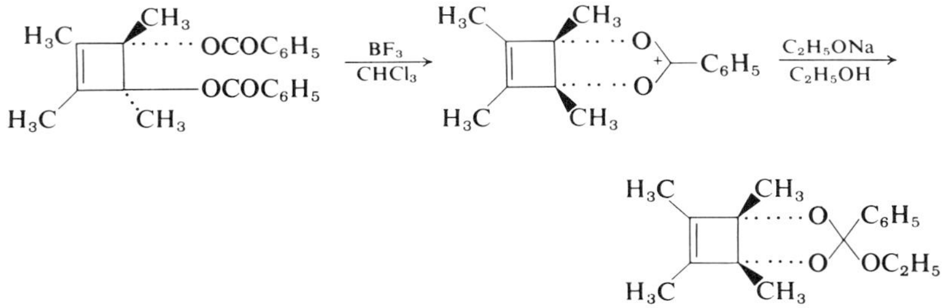

(60)

## C. Reactions of Alcohols with Dioxenium Salts

2-Ethoxy-2-methyl-*cis*-4,5-tetramethylene-1,3-dioxane and 2-methoxy-2-phenyl-*cis*-4,5-tetramethylene-1,3-dioxane were prepared from the corresponding dioxenium tetrafluoroborates and alkoxides [Eq. (58), R = $CH_3$, $R^1$ = $C_2H_5$; R = $C_6H_5$, $R^1$ = $CH_3$] (*180a, 320a*). The synthesis of other 2-substituted 2-alkoxy-1,3-dioxolanes from 1,3-dioxenium tetrafluoroborates was claimed in a recent communication which, however, gave no details of the reactions (*319a*).

## D. Reactions of Alcohols with *O*-Alkyllactonium Salts

The *O*-alkyllactonium tetrafluoroborates prepared from γ-lactones and trialkyloxonium tetrafluoroborates react with alkoxides to form 2,2-dialkoxytetrahydrofurans (*242*). 2,2-Dimethoxy- and 2,2-diethoxytetrahydrofuran were synthesized in 69% and 85% yields in this way, and 2,2-diethoxy-2,5*H*-dihydro-3,4-benzofuran (**21**) was prepared in 62% yield. 2,2-Diethoxychroman (**22**), 2,2-dimethoxy- and 2,2-diethoxybenzo-2*H*-pyran (**23**) have been prepared from the corresponding δ-lactones (3,4-dihydrocoumarin and coumarin) via the *O*-alkyl-δ-lactonium tetrafluoroborates in 80–90% yields (*239*).

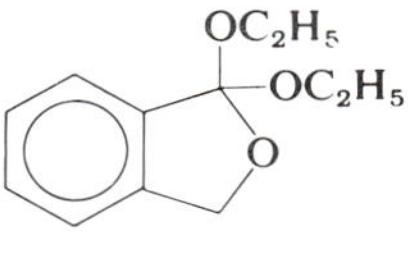

**21** 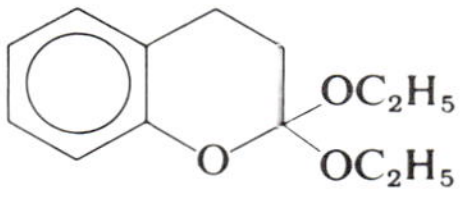    **22** 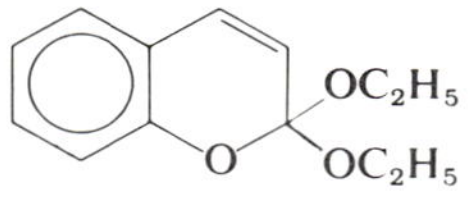    **23**

### E. Acyclic Ortho Esters from Carbonium Salts

Hexamethyl orthooxalate was prepared by reaction of the dicarbonium ion formed by oxidation of tetramethoxyethylene (with iodine) and sodium methoxide (*142a*). The same carbonium salt reacts with 9,10-phenanthraquinone to form 2,2,3,3-tetramethoxy-2,3-dihydrophenanthro-9,10*b*-1,4-dioxin (**24**) in 24% yield (*142a*).

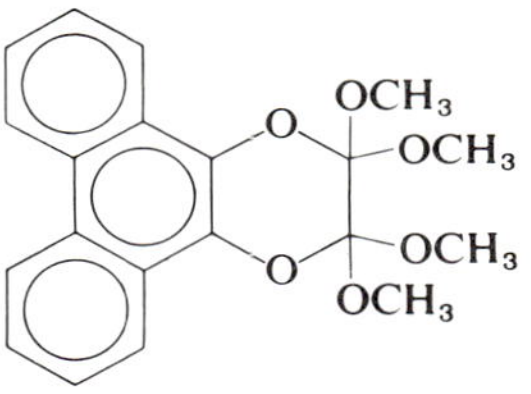

**24**

Alkyl carbamate acetals, $R_2NC(OR')_3$, are prepared by reactions of alkoxides with carbonium salts. These reactions are discussed in Chapter 7.

## VII. SYNTHESIS OF CYCLIC AND SPIROCYCLIC ORTHO ESTERS FROM EPOXIDES AND FORMATE ESTERS OR LACTONES

### A. Introduction

Epoxides add to the carbonyl groups of formate esters, γ-lactones, and δ-lactones in the presence of catalytic amounts of boron trifluoride to form 2-alkoxy-1,3-dioxolanes and spirocyclic ortho esters (*39*). Examples of these reactions are given in Eqs. (61) and (62).

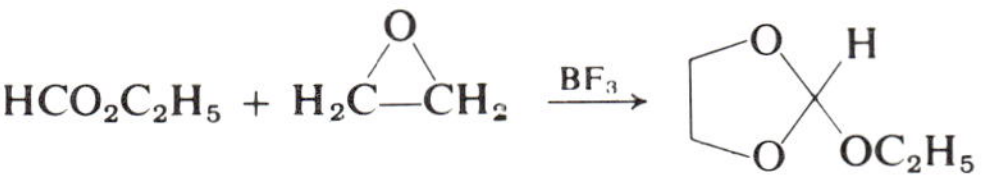

$$HCO_2C_2H_5 + H_2C\text{—}CH_2 \xrightarrow{BF_3} \tag{61}$$

$$(CH_2)_n + H_2C\text{—}CH\text{—}R \xrightarrow{BF_3} (CH_2)_n \tag{62}$$

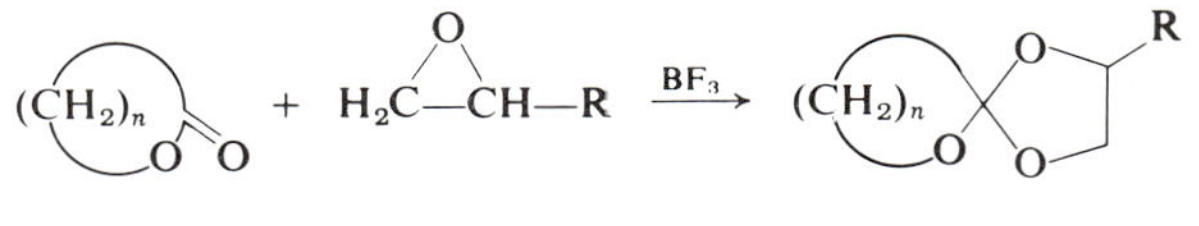

## B. 2-Alkoxy-1,3-dioxolanes from Formates and Epoxides

Ethyl formate and ethylene oxide in the presence of boron trifluoride yield 2-ethoxy-1,3-dioxolane [Eq. (61)], and 2-ethoxy-4-chloromethyl-1,3-dioxolane is formed from ethyl formate and epichlorohydrin. Epoxides do not react with acetate esters unless equivalent amounts of boron trifluoride are used, in which case 2-methyl-1,3-dioxolenium tetrafluoroborates are the principal products.

These reactions probably involve nucleophilic coordination of the epoxide with the carboxyl carbon of the ester–boron trifluoride complex, followed by nucleophilic attack by the carbonyl oxygen on epoxide carbon (240).

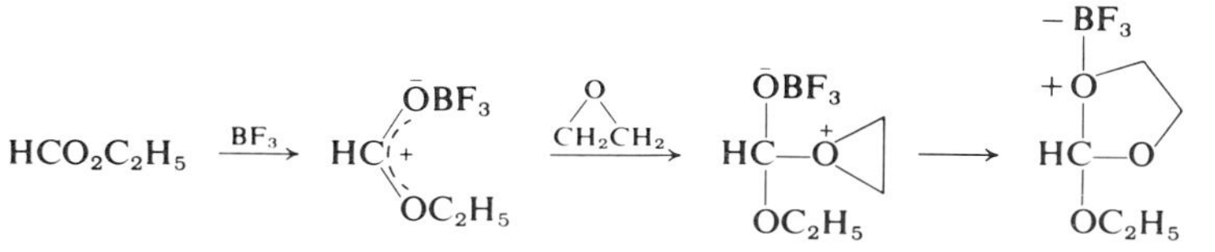

$$HCO_2C_2H_5 \xrightarrow{BF_3} HC\overset{\bar{O}BF_3}{\underset{OC_2H_5}{\diagdown}}{+} \xrightarrow{\overset{O}{\underset{CH_2CH_2}{\triangle}}} HC\overset{\bar{O}BF_3}{\underset{OC_2H_5}{\diagdown}}\overset{+}{O}\triangleleft \longrightarrow HC\overset{-BF_3}{\underset{OC_2H_5}{\diagdown}}\overset{+O}{O} \quad (63)$$

## C. Spirocyclic Ortho Esters from Lactones and Epoxides

A number of spirocyclic ortho esters have been synthesized by the addition of epoxides to γ- and δ-lactones, using boron trifluoride or stannic chloride as catalysts (40, 41, 163a). γ-Butyrolactone and α-chloro-γ-butyrolactone react with ethylene oxide, propylene oxide, epichlorohydrin, and 3-phenoxy-1,2-epoxypropane to yield 25 (R = R' = H; R = CH$_3$, R' = H; R = CH$_2$Cl, R' = H; R = CH$_2$OC$_6$H$_5$, R' = H; R = CH$_2$Cl, R' = Cl). Compound 26 is prepared from phthalide and epichlorohydrin, while epichlorohydrin and δ-valerolactone form 27. Reaction of coumarin with ethylene oxide or epichlorohydrin gives 28 (R = H, CH$_2$Cl). Compounds of structure 25 were obtained in 30–70% yields, 26 in 21% yield, 27 in 28% yield, and 28 in 26% (R = H) and 39% (R = CH$_2$Cl) yields.

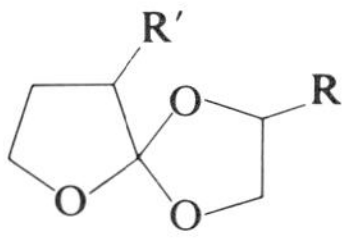

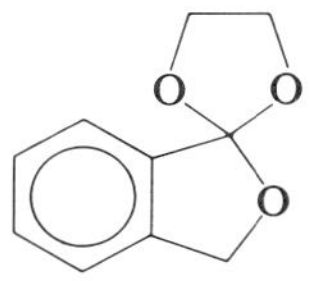

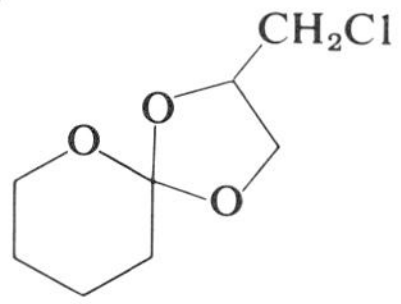

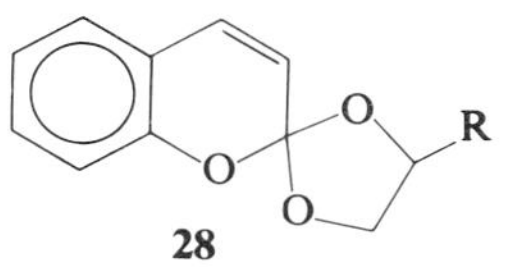

## VIII. FORMATION OF ORTHO ESTERS BY CYCLIZATIONS OF ACYLOXY COMPOUNDS

### A. Introduction

There are two classes of reactions, other than additions of epoxides to formate esters and lactones, which convert ordinary esters to cyclic ortho esters. The first of these involves intramolecular substitutions in which the acyloxy group functions as the nucleophile. The second involves intramolecular addition of a hydroxyl group of a monoester of a diol or triol to the carbonyl group of the acyloxy function.

### B. Synthesis of 2-Alkoxy-1,3-dioxolanes and 2-Alkoxy-1,3-dioxanes from Acyloxy Arenesulfonates and Halides

If an acyloxy group is properly located and oriented relative to a neighboring leaving group such as tosylate or bromide, solvolytic replacement of the leaving group is particularly facile. The enhanced reactivity of such leaving groups is due to anchimeric assistance (366) of departure of the leaving halide or arenesulfonate ion by the acyloxy group. Under most conditions the reactive dioxycarbonium ion thus formed is converted to a mixture of isomeric substituted carboxylate esters [Eq. (64)]. Anchimeric assistance is usually limited to substrates in which the cyclic dioxycarbonium ion has a five- or six-membered ring.

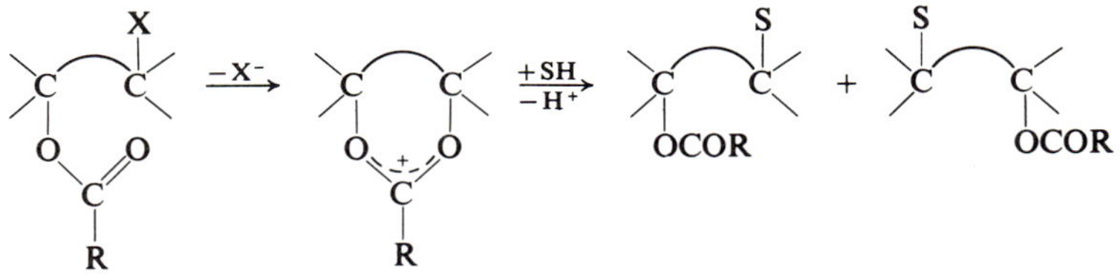

$$\underset{R}{\overset{X}{\underset{O\diagdown_{C}\diagup O}{\diagup C\diagdown C\diagdown}}} \xrightarrow{-X^-} \underset{R}{\overset{}{\underset{O\diagdown_{\overset{+}{C}}\diagup O}{\diagup C\diagdown C\diagdown}}} \xrightarrow[-H^+]{+SH} \underset{OCOR}{\overset{S}{\diagup C\diagdown C\diagdown}} + \underset{OCOR}{\overset{S}{\diagup C\diagdown C\diagdown}} \qquad (64)$$

If the acyloxy derivative is alcoholyzed under basic conditions, the carbonium ion may react with the alcohol to form a cyclic ortho ester which is sufficiently stable to be isolated. For example, solvolysis of *trans*-2-acetoxycyclohexyl tosylate in absolute ethanol containing slightly more than 1 equivalent of potassium acetate afforded 2-ethoxy-2-methyl-*cis*-4,5-tetramethylene-1,3-dioxolane in 51% yield (364).

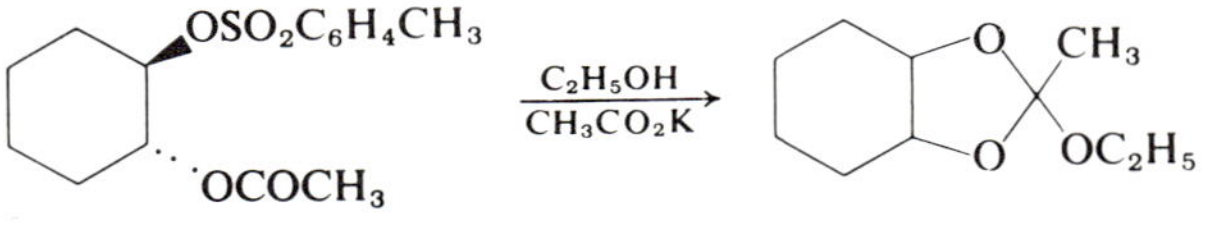

$$\text{cyclohexyl}(OSO_2C_6H_4CH_3)(OCOCH_3) \xrightarrow[CH_3CO_2K]{C_2H_5OH} \text{dioxolane}(O)(CH_3)(O)(OC_2H_5) \qquad (65)$$

Both the cis and trans isomers of 2-acetoxycyclohexylmethyl tosylate yield ortho esters when solvolyzed under the same conditions (*180, 320*).

$$\underset{\text{OCOCH}_3}{\overset{\text{CH}_2\text{OSO}_2\text{C}_6\text{H}_4\text{CH}_3}{\text{C}_6\text{H}_{10}}} \xrightarrow[\text{CH}_3\text{CO}_2\text{K}]{\text{C}_2\text{H}_5\text{OH}} \underset{\text{O}}{\overset{\text{O}}{\text{C}_6\text{H}_{10}}}\!\!\!\!\overset{\text{OC}_2\text{H}_5}{\underset{\text{CH}_3}{}} \quad (66)$$

2-Ethoxy-2-methyl-*cis*-5,6-tetramethylene-1,3-dioxane was obtained in 71% yield from the *cis*-acetoxytosylate; the *trans*-acetoxytosylate yielded 23% of the trans ortho ester.

A number of carbohydrate ortho esters have been prepared by alcoholysis of acyloxy halides (*267*). These reactions are discussed in Chapter 5.

### C. Synthesis of Polycyclic Ortho Esters and Hydrogen Ortho Esters by Acylation of Diols and Triols

Acylation of triols sometimes yields cyclic ortho esters or hydrogen ortho esters* rather than normal carboxylate esters. Formation of these ortho acid derivatives must involve several discrete steps. The most likely reaction scheme is as follows:

$$\begin{matrix}\text{OH}\\ \text{OH}\\ \text{OH}\\ \mathbf{29}\end{matrix} \xrightarrow{\text{RCY}} \begin{matrix}\text{OCR}\\ \text{OH}\\ \text{OH}\\ \mathbf{30}\end{matrix} \rightleftharpoons \begin{matrix}\text{O}\\ \text{O} \!\!\!\diagdown\!\!\! \text{R}\\ \text{OH}\\ \end{matrix} \!\!\!\!\!\!\underset{\overset{+\text{H}_2\text{O}}{-\text{H}^+}}{\overset{+\text{H}^+,-\text{H}_2\text{O}}{\rightleftharpoons}} \begin{matrix}\text{O}\\ \text{O}\!\!\!-\!\!\!\text{R}\\ \text{OH}\\ \mathbf{31}\end{matrix} \underset{+\text{H}^+}{\overset{-\text{H}^+}{\rightleftharpoons}} \begin{matrix}\text{O}\\ \text{O}\!\!\!-\!\!\!\text{R}\\ \text{O}\\ \mathbf{32}\end{matrix} \quad (67)$$

#### 1. FORMATION OF HYDROGEN ORTHO ESTERS BY CYCLIZATION OF HYDROXY ESTERS

The equilibrium between simple carboxylate esters and hydrogen ortho esters is usually greatly in favor of the normal ester, although salts of hydrogen dialkyl perfluoroorthocarboxylate esters have been prepared by addition of sodium alkoxides to the carboxylate esters (*144, 340*).

$$C_nF_{2n+1}CO_2R + RONa \rightarrow C_nF_{2n+1}C(OR)_2ONa \quad (68)$$

Hydroxyalkyl esters having electron-attracting acyl substituents exist as equilibrium mixtures which may contain significant amounts of the 2-substituted 2-hydroxyl-1,3-dioxolanes formed by intramolecular cyclization.

---
* The term hydrogen ortho ester is used to refer to substances having the system $-\text{C}(-\text{O}-\text{CH}_2-)_2\text{OH}$ in their structures.

The physical properties of the ethylene glycol ester of trichloroacetic acid suggest that it exists mostly as 2-hydroxy-2-trichloromethyl-1,3-dioxolane *(131)*, an assumption supported by the fact that it reacts with diazomethane to form 2-methoxy-2-trichloromethyl-1,3-dioxolane *(243–245)*. The analogous ester of dichloroacetic acid appears to exist as a mixture of comparable amounts of normal ester and 2-hydroxy-2-dichloromethyl-1,3-dioxolane, while chloroacetic acid and ethylene glycol give the normal ester *(245)*. It was found recently that trifluoroacetylation of *cis*-3,4-dihydroxytetrahydrofuran affords the cyclic hydrogen ortho ester **33** *(37)*.

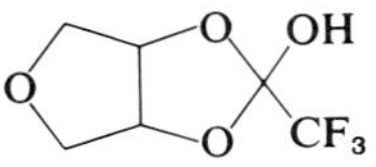

**33**

The amount of hydrogen ortho ester in equilibrium with the normal carboxylate ester is increased by electron-withdrawing acyl substituents, and, more importantly, by the presence in the hydrogen ortho ester of a dioxolane or dioxane ring. There are examples in natural product chemistry of hydrogen ortho esters which are stable relative to the normal hydroxy esters. Spectroscopic evidence supports hydrogen ortho ester structure **34** for the formyl derivative of 3α-benzylamino-5β-prognane-20β,21-diol *(268)*. Tetrodotoxin (**35**), a poison isolated from the puffer fish *Spheroides* and from the salamander *Taricha torosa*, has a tricyclic hydrogen ortho ester grouping *(112,*

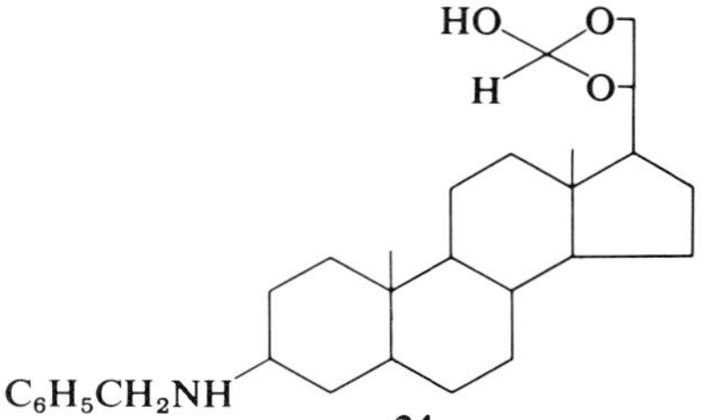

**34**

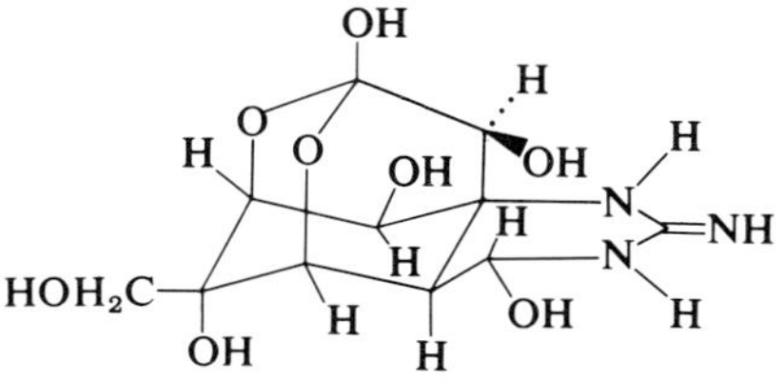

**35**

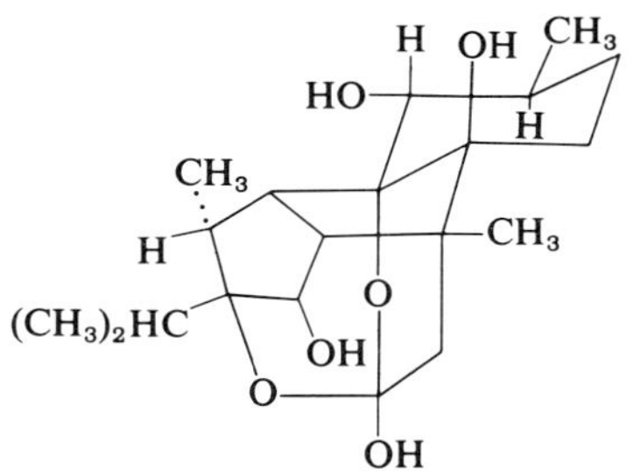

**36**

*340b, 347, 368*). The ortho ester structure originally proposed for ryanodol (**36**) (*16a, 314*) is probably incorrect (*360a*).

Another group of substances for which hydrogen ortho ester structures have been suggested are the products of photoaddition of aromatic aldehydes to *o*-quinones (*256–258, 321, 322*).

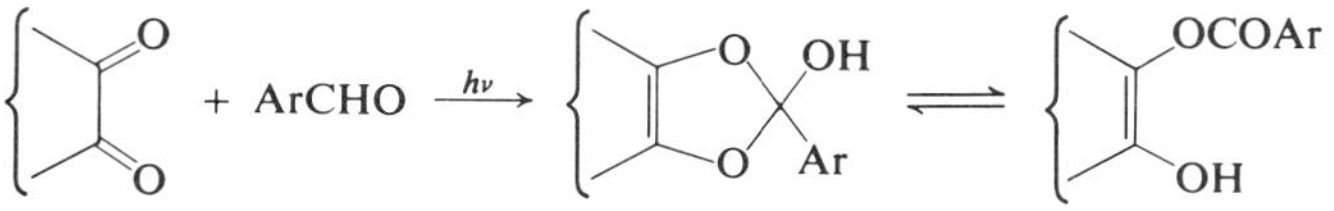

$$\text{[o-quinone]} + \text{ArCHO} \xrightarrow{h\nu} \text{I} \rightleftharpoons \text{II} \quad (69)$$

The hydrogen ortho ester, I, may be the major component of the equilibrium mixture in some instances. The photoaddition product of phenanthraquinone and *p*-chlorobenzaldehyde, for example, was converted by diazomethane to a methyl derivative which yielded methyl *p*-chlorobenzoate on hydrolysis (*321*). In most cases, however, the open-chain normal ester, II, probably predominates at equilibrium (*253, 259*).

### 2. Formation of Polycyclic Ortho Esters by Cyclizations of Hydroxy Esters

If there is a free alcoholic hydroxyl group located near the acyl carbon of a cyclic hydrogen ortho ester (as in **30**), acid-catalyzed intramolecular dehydration to a tricyclic or polycyclic ortho ester may occur. The cyclization reaction probably involves a cyclic dioxycarbonium ion (**31**) as an intermediate.

The only examples of ortho ester formation by acylation of acyclic triols are acid-catalyzed reactions of a number of carboxylic acids with 2-alkyl-2-hydroxymethyl-1,3-propanediols (*24, 54*).

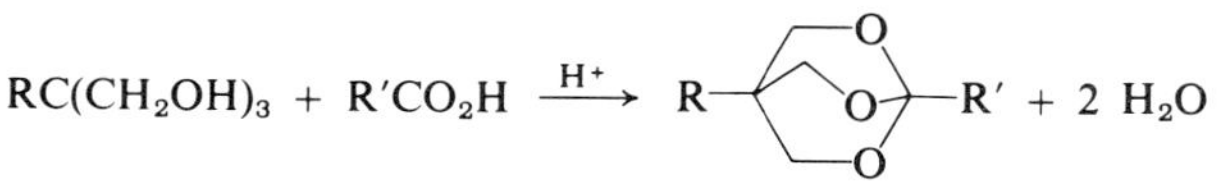

$$RC(CH_2OH)_3 + R'CO_2H \xrightarrow{H^+} R-\underset{O}{\overset{O}{\bigtriangleup}}-R' + 2\ H_2O$$

$R = CH_3,\ C_2H_5,\ C_3H_7;\ R' = H,\ CH_3,\ C_2H_5,\ C_5H_{11},\ C_6H_{13},\ C(CH_3)_3,$
$\qquad\qquad CH_2Cl,\ CCl_3,\ CHCl_2,\ CF_3,\ 3,5\text{-}(NO_2)_2C_6H_3 \quad (70)$

The reaction, which is facilitated by electron-withdrawing acyl substituents in the carboxylic acid, was forced to completion by azeotropic distillation of the water from the reaction mixture with benzene or xylene.

Acetylation of a number of polyhydroxy alkaloids and related compounds yields polycyclic orthoacetates. Examples are ortho esters derived from alkaloids of the cevine (*27, 185, 187, 338, 349*) and sabadine (*13, 186*) series, and from anhydroryanodol (*348*). Formation of these ortho esters and substances derived from them has played an important role in structure proofs and configurational assignments for these complex substances. Representative examples of this class of compounds are cevadine orthoacetate-4,16-diacetate *(185)* (**37**) neosabadine orthoacetatetriacetate *(13)* (**38**) and anhydroryanodol orthoacetatediacetate *(348)* (**39**).

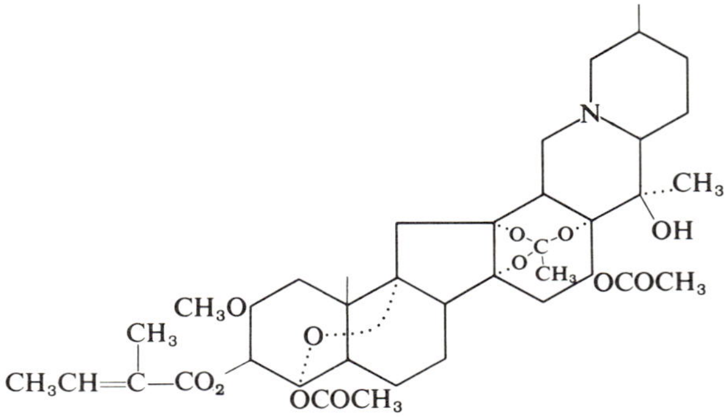

**37**

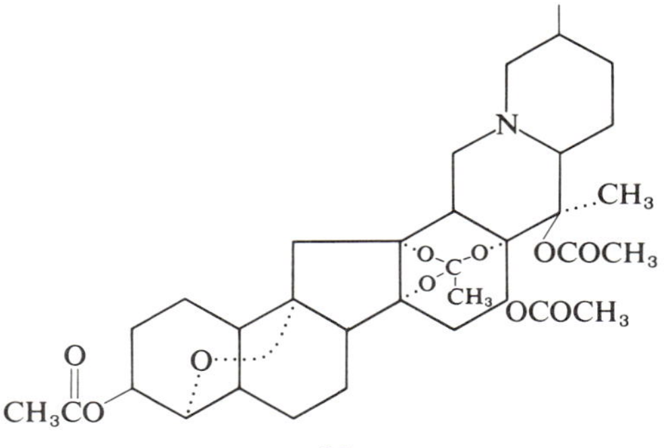

**38** 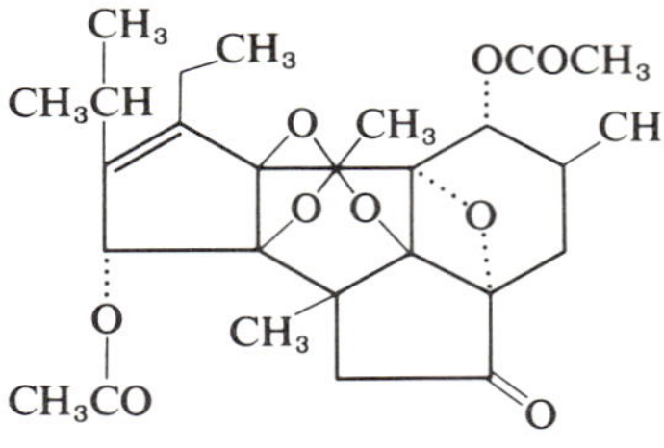    **39**

## IX. SYNTHESIS OF CARBOXYLIC ORTHO ESTERS FROM TETRAALKYL ORTHOCARBONATES

In 1908 Chichibabin *(56)* reported that triethyl orthocarboxylates are among the products of reaction of Grignard reagents with tetraethyl orthocarbonate.

$$C(OC_2H_5)_4 + RMgX \rightarrow RC(OC_2H_5)_3 + MgXOC_2H_5 \qquad (71)$$

This reaction, like the Bodroux-Chichibabin synthesis of acetals from Grignard reagents and trialkyl orthoformates *(42, 55)* (p. 224), involves nucleophilic replacement of an ortho ester alkoxy group by the carbanionoid group of the organomagnesium compound, aided by coordination between alkoxy oxygen and magnesium. Chichibabin emphasized that short reaction times and relatively low reaction temperatures are essential if the reaction is to yield ortho esters rather than acetals. He claimed to obtain triethyl orthobenzoate in about 70% yield from the reaction of phenylmagnesium bromide with tetraethyl orthocarbonate, but did not specify the relative amounts or the concentrations of the reactants. Reaction of ethylmagnesium iodide with the orthocarbonate yielded a mixture from which no pure triethyl orthopropionate was isolated. A. J. Hill reported obtaining trialkyl orthobenzoates in yields ranging from 50% to 77% by treating tetramethyl, tetraethyl, tetrabutyl, and tetraamyl orthocarbonates with phenylmagnesium bromide (*280*). Iotsitch synthesized triethyl orthophenylpropiolate by reaction of phenylethynylmagnesium iodide with tetraethyl orthocarbonate (*155*).

McElvain and McBane (*200, 222*) attempted to prepare triethyl orthocarboxylates by Chichibabin's procedure. They isolated only diethyl ketals and ethyl alkyl ethers—presumably formed by reaction of the Grignard reagents with initially formed ortho esters. Barre and Ladoucer also found that ketals are the principal products of reaction of alkylmagnesium halides with tetraethyl orthocarbonate (*26*). However, they used 2 moles of Grignard reagent per mole of orthocarbonate, and high reaction temperatures.

The synthesis of orthocarboxylate esters from orthocarbonates is obviously complicated by the reaction of the orthocarboxylates with Grignard reagents to form acetals. No systematic attempt to determine conditions which would give maximum yields of ortho esters has been reported.

Zinc chloride catalyzes the reaction of phenylacetylene with tetraethyl orthocarbonate to form triethyl phenylorthopropiolate in 14% yields (*150, 151*). The generality of this reaction has not been established.

Acetyl cyanide and tetraethyl orthocarbonate react to form the oxalic acid derivative $NCC(OC_2H_5)_3$ (*43b*).

## X. FORMATION OF CARBOXYLIC ORTHO ESTERS BY REACTIONS OF DIALKOXYCARBENES WITH ALCOHOLS

Thermal decomposition of 1,2,3,4-tetrachloro-7,7-dimethoxy-5-phenylbicyclo[2.2.1]hepta-2,5-diene (**40**) in the presence of methanol yields trimethyl orthoformate (*141, 142, 195*).

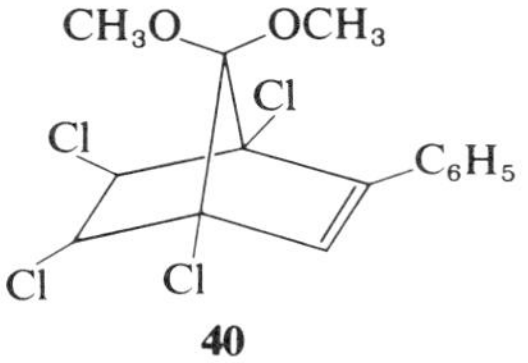

**40**

The ortho ester presumably was produced by reaction of dimethoxycarbene [$(CH_3O)_2C$:], formed by decomposition of the bicyclic ketal, with methanol. Photolysis of $N',N'$-diethoxymethylenetoluene-$p$-sulfonylhydrazide, $(C_2H_5O)_2C$=$NNHSO_2C_6H_4CH_3$-$p$, in methanol afforded a 33% yield of diethylmethyl orthoformate from the reaction of diethoxycarbene with methanol (70).

Pyrolysis of 7,7-dimethoxycycloheptatriene at 350°C results in formation of a 5% yield of trimethyl orthoformate, presumably via dimethoxycarbene (142c).

## XI. SYNTHESIS OF ORTHO ESTERS BY ADDITION OF KETENE ACETALS TO α,β-UNSATURATED CARBONYL COMPOUNDS

Ketene dimethyl acetal undergoes Diels-Alder condensations with α,β-unsaturated carbonyl compounds to form 2,2-dimethoxy-2,3-dihydropyran derivatives (222).

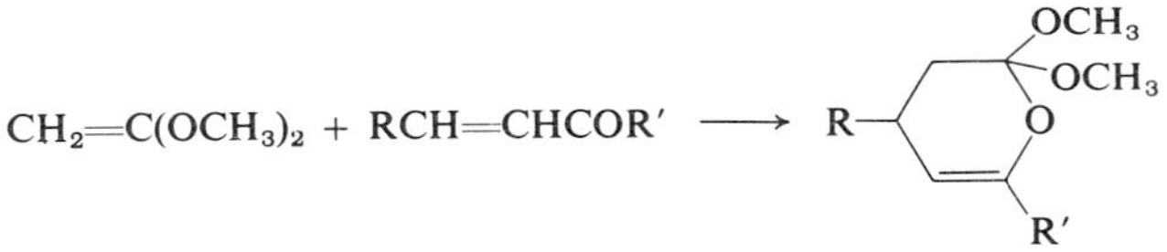

$$CH_2=C(OCH_3)_2 + RCH=CHCOR' \longrightarrow R-\underset{R'}{\overset{OCH_3,OCH_3}{\diagup}} \quad (72)$$

Yields range from 70% when R = R' = H to 30% when R = $C_6H_5$, R' = $CH_3$. Similarly, 2-methoxy-5,6-dihydropyran and acrolein yielded 1-methoxy-2,10-dioxabicyclo[4.4.0]-3-decene,

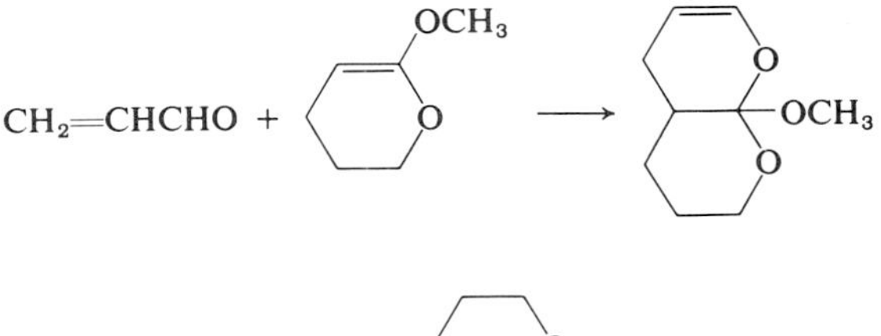

$$CH_2=CHCHO + \text{(2-methoxy-5,6-dihydropyran)} \longrightarrow \text{(bicyclic OCH}_3\text{)} \quad (73)$$

$$CH_2=CHCHO + \mathbf{41} \longrightarrow \mathbf{19}$$

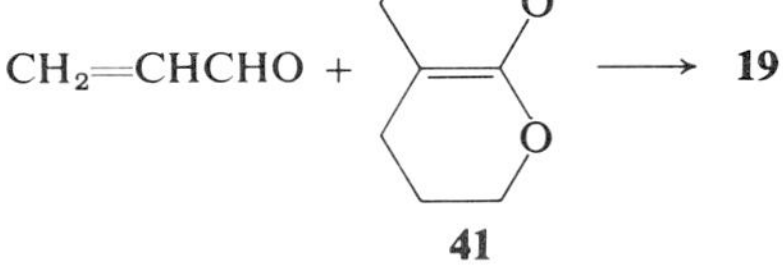

and 2,10-dioxabicyclo[4.4.0]-1-decene (**41**) and acrolein condense to form **19** (216).

## XII. MISCELLANEOUS REACTIONS WHICH YIELD ORTHO ESTERS

Several reactions whose general applicability has not been established yield ortho esters as products. Oxidation of tetramethoxyethylene with oxygen affords tetramethyl monoorthooxalate in 39% yield (*140, 142b*). Tetramethoxyethylene reacts with sulfur in chloroform solution to form tetramethyl thionmonoorthooxalate, $CH_3OC(S)C(OCH_3)_3$, in 63% yield (*142b*). Electrochemical methoxylation of 1,2-dimethoxybenzene forms hexamethyl orthomuconate, *cis,cis*-$(CH_3O)_3CCH=CHCH=CHC(OCH_3)_3$ (*30*), and electrochemical methoxylations of 2-bromofuran, 5-bromofurfuraldehyde, and 5-bromofuroic acid in acidic methanol yield 2,5,5-trimethoxy-2,5-dihydrofuran, 2-dimethoxymethyl-2,5,5-trimethoxy-2,5-dihydrofuran, and 2-carbomethoxy-2,5,5-trimethoxy-2,5-dihydrofuran, respectively (*332a*). Reaction of 2,5-dimethoxy-2,5-dihydro-2-methylfuran with methanolic hydrogen chloride produces trimethyl ortholevulinate, $CH_3COCH_2CH_2C(OCH_3)_3$, in 77% yield (*324*).

Oxidation of $\gamma,\delta$-unsaturated ketones or 2,7-dioxabicyclo[2.2.1]heptanes with *m*-chloroperbenzoic acid yields substituted 2,7,8-trioxabicyclo[3.2.1]-octanes (*103a, 103b*).

Methylation of the sodium methoxide addition products of methyl perfluoroalkanoates (*144, 199b*) with dimethyl sulfate yields trimethyl perfluoroorthocarboxylates (*144*).

Trimethyl $\alpha$-chloro-$\beta$-methylorthocrotonate is formed by reaction of trimethyl orthoisobutyrate with 2,2-dichloro-3,3-dimethylcyclopropanone dimethyl ketal [Eq. (74)] (*231*). Other 2,2-dichlorocyclopropanone ketals react with potassium *tert*-butoxide to form dialkyl-*tert*-butyl orthopropiolates (*231*) [Eq. (75)]. Another reaction which was reported to yield a substituted trialkyl orthopropiolate occurs between ethanolic sodium ethoxide and *N*,*N*-dimethyl-*N*-allyl-2,2,3,3-tetrachlorobutylammonium bromide (*16*) [Eq. (76)].

$$\text{(cyclopropane with } H_3C, H_3C, Cl, Cl, OCH_3, OCH_3) + (CH_3)_2CHC(OCH_3)_3 \longrightarrow$$
$$(CH_3)_2C=CClC(OCH_3)_3 + (CH_3)_2CHCO_2CH_3 + CH_3Cl \quad (74)$$

$$\text{(cyclopropane with } R, Cl, Cl, OR', OR') + 2\ (CH_3)_3CO^- \longrightarrow$$
$$RC\equiv CC(OR')_2OC(CH_3)_3 + 2\ Cl^- + (CH_3)_3COH$$
$$(R = C_6H_5,\ R' = CH_3;\ R = H,\ R' = C_2H_5) \quad (75)$$

$$(CH_3)_2\overset{+}{N}(C_3H_5)CH_2CCl_2CCl_2CH_3Br^- \xrightarrow[C_2H_5OH]{NaOC_2H_5}$$
$$CH_3C\equiv C-C(OC_2H_5)_3 + (CH_3)_2NC_3H_5 \quad (76)$$

Diethyl maleate reacts with triethyl orthoformate in the presence of benzoyl peroxide at 95°C to form tetraethyl monoortho-2-carbethoxysuccinate, $C_2H_5OCOCH_2CH(CO_2C_2H_5)C(OC_2H_5)_3$ (*260*).

Treatment of 2-phenyl-5,5-bis(*p*-nitrobenzoxymethyl)-1,3-dioxane with *p*-nitrophenylhydrazine and acetic acid gives an 80% yield of 1-*p*-nitrophenyl-4-*p*-nitrobenzoxy-2,6,7-trioxabicyclo[2.2.2]octane (*31*).

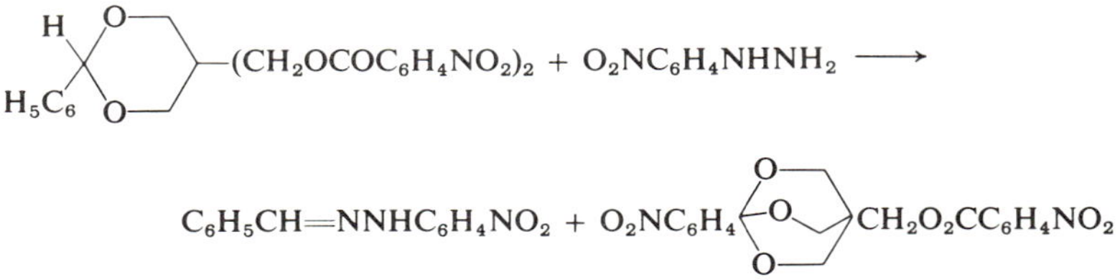

$$\underset{H_5C_6}{\overset{H}{\diagup}}\!\!\!\!\diagdown\!\!\overset{O-}{\underset{O-}{\diagup}}\!\!-(CH_2OCOC_6H_4NO_2)_2 + O_2NC_6H_4NHNH_2 \longrightarrow$$

$$C_6H_5CH\!\!=\!\!NNHC_6H_4NO_2 + O_2NC_6H_4\!\!\underset{O-}{\overset{O-}{\diagup}}\!\!\!\!\diagdown\!\!CH_2O_2CC_6H_4NO_2 \quad (77)$$

Diazomethyl ketones react with ketene acetals to form 2,2-dialkoxy-1,2-dihydropyrans (*315a*) [Eq. (78)]. Ketene acetals and nitrones yield 5,5-dialkoxyisoxazolidines (*315b*) [Eq. (79)].

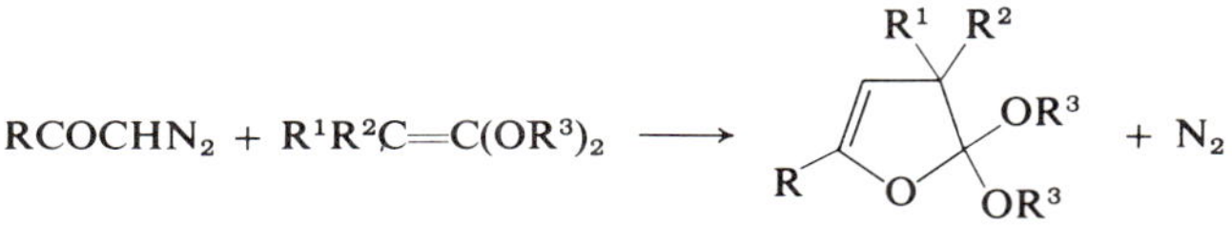

$$RCOCHN_2 + R^1R^2C\!\!=\!\!C(OR^3)_2 \longrightarrow \underset{R}{\overset{R^1\ R^2}{\diagup\!\!\!\diagdown}}\!\!\!\!\!\underset{O}{\diagdown}\!\!\overset{OR^3}{\diagup}\!\!\!\!\!\diagdown\!\!\!OR^3 + N_2 \quad (78)$$

(R = $C_6H_5$, $R^1$ = H, $R^2$ = H, $CH_3$, $R^3$ = $C_2H_5$; R = $C_6H_5$, $R^1$ = $R^2$ = $R^3$ = $CH_3$; R = *p*-$O_2NC_6H_4$, *p*-$CH_3OC_6H_4$, $CH_3$, $C_2H_5$, $R^1$ = H, $R^2$ = H, $CH_3$, $R^3$ = $C_2H_5$; R = *p*-$O_2NC_6H_4$, *p*-$CH_3OC_6H_4$, $R^1$ = $R^2$ = $R^3$ = $CH_3$)

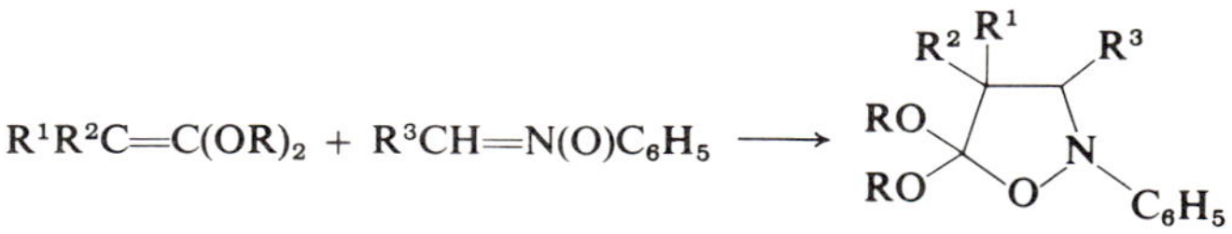

$$R^1R^2C\!\!=\!\!C(OR)_2 + R^3CH\!\!=\!\!N(O)C_6H_5 \longrightarrow \underset{RO}{\overset{RO}{\diagup}}\!\!\!\!\diagdown\!\!\overset{R^2\ R^1}{\underset{O}{\diagup\!\!\!\diagdown}}\!\!\!\!\overset{R^3}{\underset{N}{\diagdown}}\!\!\!\!\diagdown\!C_6H_5 \quad (79)$$

(R = $C_2H_5$, $R^1$ = $R^2$ = H, $R^3$ = $C_6H_5$, *p*-ClC$_6H_4$; R = $CH_3$, $R^1$ = $CH_3$, $C_2H_5$, $R^2$ = H, $CH_3$, $R^3$ = $C_6H_5$, *p*-ClC$_6H_4$; R = $C_2H_5$, $R^1$ = $CH_3$, $R^2$ = H, $R^3$ = $C_6H_5$, *p*-ClC$_6H_4$)

Ketene dimethyl acetal adds to 2,2-dimethylcyclopropane to form 1,1-dimethyl-5,5-dimethoxy-4-oxaspiro[2.3]hexene (*347a*).

α-(*N,N,N*-Trimethylhydrazonium)acetal iodides fail to undergo the Neber rearrangement when treated with isopropyl alcoholic sodium isopropoxide, but yield α-iminoorthocarboxylates instead (*127a*).

$$RC[\!\!=\!\!\overset{+}{N}N(CH_3)_3]CH(OR^1)_2I^- \xrightarrow{(CH_3)_2CHO^-Na^+} RC(\!\!=\!\!NH)C(OR^1)_2OCH(CH_3)_2 \quad (80)$$

[R = $CH_3$, $R^1$ = $C_2H_5$; R = $C_6H_5$, $(OR^1)_2$ = $OCH_2CH_2O$]

Diphenylacetylene, chloroform, and ethylene oxide react at elevated temperatures in the presence of tetraethylammonium bromide to form a mixture of products which contains $C_6H_5CH=C(C_6H_5)C(OCH_2CH_2Cl)_3$ (*264a*).

1,2-Dihydroxybenzene thioncarbonate is converted by treatment with trimethyl phosphite to the polycyclic ortho ester **42** (*66a, 152a*).

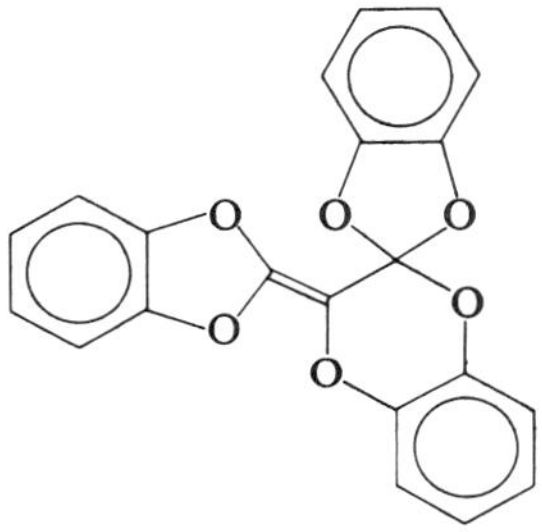

**42**

Methyl and phenyl *o*-diazoacetylbenzoates react with methanolic HCl to form 1,1-dimethoxy-4-oxoisochroman (*82*).

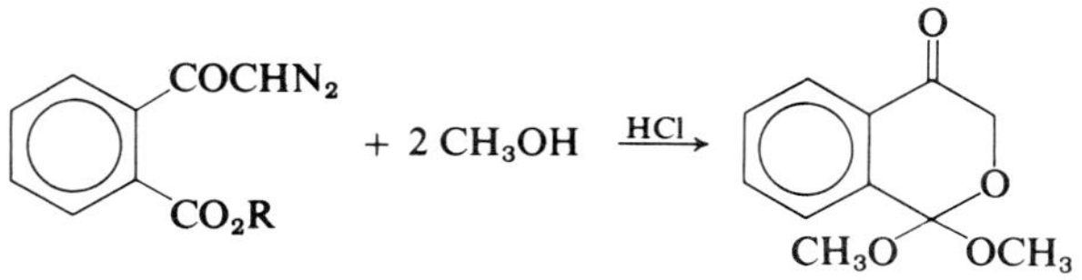

$$\text{[structure]} + 2\ CH_3OH \xrightarrow{HCl} \text{[structure]} + ROH + HCl \quad (81)$$

(R = $CH_3$, $C_6H_5$)

Phenyl cyanate reacts with alcohols in the presence of acids to form tetraalkyl orthocarbonates, $C(OR)_4$ (R = $CH_3$, $C_2H_5$) (*232a, 233*). The polycyclic orthocarbonate **43** is formed by reaction of 2,2-dihydroxybiphenyl with thiophosgene (*370a*).

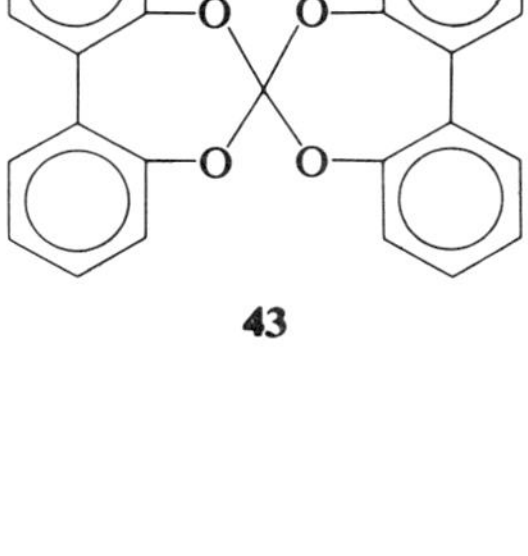

**43**

## XIII. SYNTHESIS OF PEROXY AND HYDROPEROXY ORTHO ESTERS

Syntheses of several peroxy ortho esters $RC(OR')_2OOR''$ and hydroperoxy ortho esters $RC(OR')_2OOH$ have been reported recently.

Reactions of hydroperoxides with orthocarboxylate esters yield peroxy ortho esters. Examples are reactions of trialkyl orthoformates, orthoacetates, and orthobenzoates, as well as 2-phenyl-2-methoxy-1,3-dioxolanes, with *tert*-butyl hydroperoxide and cumyl hydroperoxide to form peroxy ortho esters (*193, 287, 289, 290*).

$$RC(OR')_3 + R''OOH \rightarrow RC(OR')_2OOR'' + R'OH \qquad (82)$$
$$[R = H, C_6H_5, CH_3; \ R' = CH_3, C_2H_5; \ R'' = (CH_3)_3C, C_6H_5C(CH_3)_2]$$

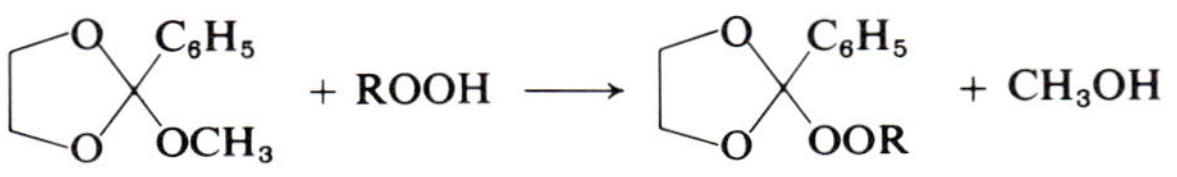

$$\text{(dioxolane-OCH}_3, C_6H_5) + ROOH \longrightarrow \text{(dioxolane-OOR, } C_6H_5) + CH_3OH \qquad (83)$$
$$R = (CH_3)_3C, C_6H_5C(CH_3)_2$$

A number of trityl, *tert*-butyl, and other peroxy ortho esters were prepared by addition of hydroperoxides to ketene acetals (*319*).

$$RCH{=}C(OR')_2 + R''OOH \rightarrow RCH_2C(OR')_2OOR'' \qquad (84)$$
$$[R = H, C_6H_5, CN; \ R' = CH_3, C_2H_5;$$
$$R'' = (C_6H_5)_3C, (CH_3)_3C, \alpha\text{-tetralyl, isochromanyl}]$$

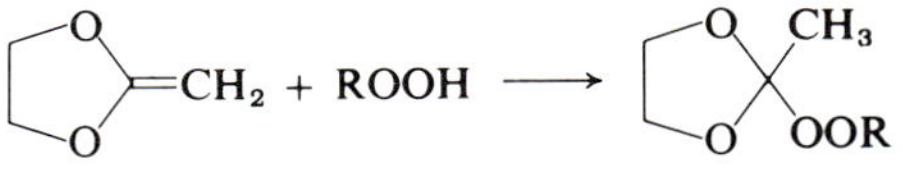

$$\text{(dioxolane)}{=}CH_2 + ROOH \longrightarrow \text{(dioxolane-CH}_3\text{, OOR)} \qquad (85)$$
$$R = (CH_3)_3C, (C_6H_5)_3C$$

Peroxy diortho esters were the principal products of reaction of ketene diethylacetal and ketene ethyleneacetal with hydrogen peroxide (*319*).

$$2\ CH_2{=}C(OC_2H_5)_2 + H_2O_2 \rightarrow (C_2H_5O)_2C(CH_3)OOC(CH_3)(OC_2H_5)_2 \qquad (86)$$

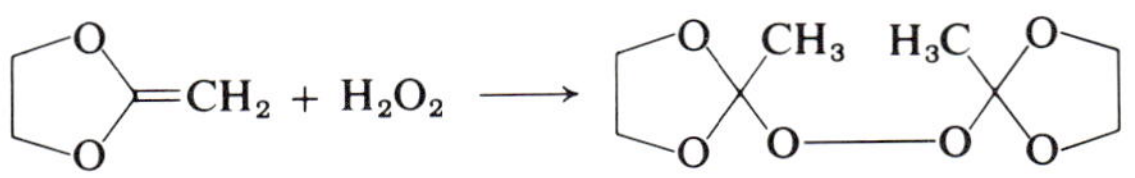

$$\text{(dioxolane)}{=}CH_2 + H_2O_2 \longrightarrow \text{(bis-dioxolane peroxide)} \qquad (87)$$

2-Phenyl-2-cumylperoxy-1,3-dioxolane is formed in low yield by the reaction of cumyl hydroperoxide with 2-phenyl-1,3-dioxolenium tetrafluoroborate (289), and 2-phenyl-2-methylperoxy-1,3-dioxolane was prepared in 43% yield by methylation of 2-phenyl-2-hydroperoxy-1,3-dioxolane with dimethyl sulfate (288).

Ethyleneacetals and other cyclic acetals react with oxygen when irradiated with ultraviolet light to form 2-hydroperoxy-1,3-dioxolanes (71, 288, 291, 292, 323b–323d) and 2-hydroperoxy-1,3,5-trioxanes (323b). Hydroperoxydioxolanes are also obtained by treatment of 2-alkoxy-1,3-dioxolanes with hydrogen peroxide (289). The hydroperoxydioxolanes may react with a second mole of ethylene acetal to form peroxy diortho esters (71, 292, 323c, 323d) [Eq. (90)]. 2-Hydroperoxy-2,4,6-trimethyl-1,3,5-trioxane undergoes a similar reaction (323b).

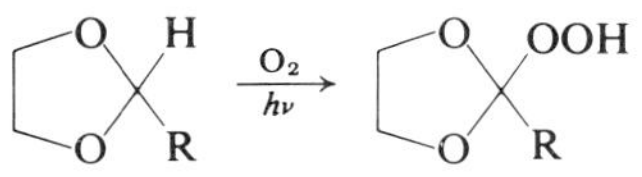

$$\text{(dioxolane-H,R)} \xrightarrow[h\nu]{O_2} \text{(dioxolane-OOH,R)} \qquad (88)$$

$$\text{(benzodioxole-H,C}_6\text{H}_5\text{)} \xrightarrow[h\nu]{O_2} \text{(benzodioxole-OOH,C}_6\text{H}_5\text{)} \qquad (89)$$

$$\text{(dioxolane-OOH,R)} + \text{(dioxolane-H,R)} \xrightarrow[h\nu]{O_2} \text{(bis-dioxolane-O-O)} \qquad (90)$$

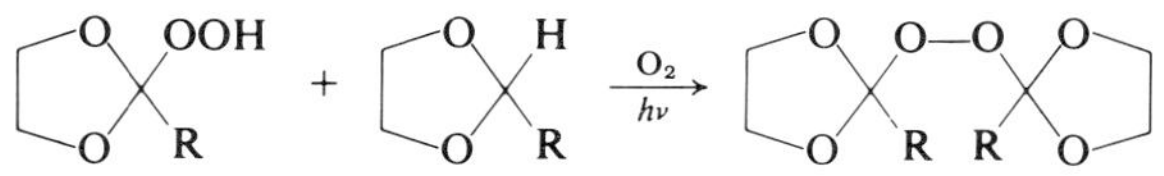

A number of 1,2,6-trioxaspiro[4.4]nonanes (**44**) were prepared by reaction of 2-vinyl-1,3-dioxolanes with oxygen (*153a, 323a*). 3,3-Di-*tert*-butylperoxyphthalide (**45**) is formed by reaction of phthalyl chloride with *tert*-butyl hydroperoxide (*247a*).

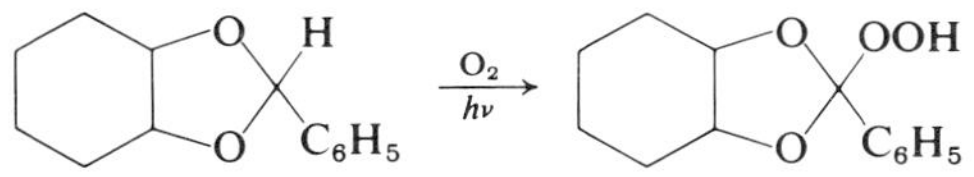

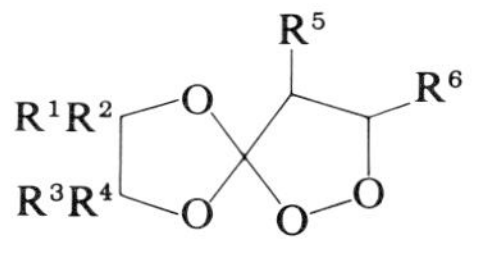

**44**: spiro structure with $R^1R^2$, $R^3R^4$, $R^5$, $R^6$ substituents

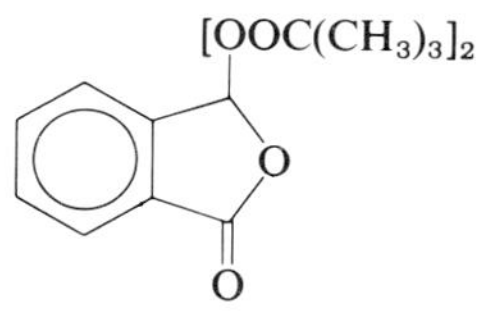

**45**: phthalide with $[OOC(CH_3)_3]_2$

## XIV. SYNTHESIS OF ACYLOXY ORTHO ESTERS

Acyclic acyloxy ortho esters are compounds having the generalized structures $RC(OCOR')(OR')_2$ and $RC(OCOR'')_2OR'$. Dialkoxymethyl and 1,1-dialkoxyethyl carboxylates, the simplest and most useful of the acyloxy ortho esters, are formed by reaction of acid anhydrides with trialkyl orthoformates and trialkyl orthoacetates (282).

$$(R'CO)_2O + RC(OC_2H_5)_3 \rightarrow RC(OC_2H_5)_2OCOR' + R'CO_2C_2H_5 \qquad (91)$$
$$(R = H, CH_3; R' = CH_3, C_2H_5)$$

Diethoxymethyl acetate and diethoxymethyl propionate can be isolated by fractional distillation at reduced pressures. 1,1-Diethoxyethyl acetate, $CH_3C(OC_2H_5)_2OCOCH_3$, however, decomposes to α-ethoxyvinyl acetate when distilled.

Reaction (**91**) is irreversible and does not occur at room temperature in the absence of catalysts. When $R = H$ and $R' = CH_3$, it proceeds slowly at 100°C and at a moderate rate at 140°C (*315d*). At elevated temperatures, however, diethoxymethyl acetate decomposes to ethyl formate and ethyl acetate. The maximum yield of diethoxymethyl acetate is obtained by halting the reaction when the rate of decomposition of the product becomes equal to its rate of formation, and then fractionating the reaction mixture under reduced pressure (*315d*).

Improved yields of diethoxymethyl acetate and diethoxymethyl propionate were obtained by substituting acetic-formic and propionic-formic anhydrides for acetic and propionic anhydrides (*315d*). Diethoxymethyl esters of propionic, butyric, isobutyric, valeric, pivalic, benzoic, and mesitoic acids were obtained in yields exceeding 75% by distilling equimolar mixtures of the carboxylic acid and diethoxymethyl acetate under reduced pressure (*315d*). Diethoxymethyl formate was prepared in 50% yield by reaction of formic acid with diethoxymethyl benzoate in pentane solution (*315d*).

Alkoxydiacyloxymethanes were formed in 53–87% yields by reaction of carboxylic acids with alkyl dichloromethyl ethers in the presence of tertiary amines (*117*).

$$2\ RCO_2H + HCCl_2OR' \xrightarrow{(C_2H_5)_3N} (RCO_2)_2CHOR' \qquad (92)$$
$$(R = CH_3, C_2H_5, C_6H_5, p\text{-}O_2NC_6H_4; R = CH_3, C_4H_9)$$

Malonic acid reacts with methoxyacetylene to form a cyclic diacyloxy ortho ester, 2-methyl-2-methoxy-4,6-dioxo-1,3-dioxane (**90**) [Eq. (93)]. Similar 1,1-diacyloxy-1-alkoxyethanes are transient intermediates in the conversion of carboxylic acids to anhydrides by reaction with alkoxyacetylenes (*90, 263, 356*).

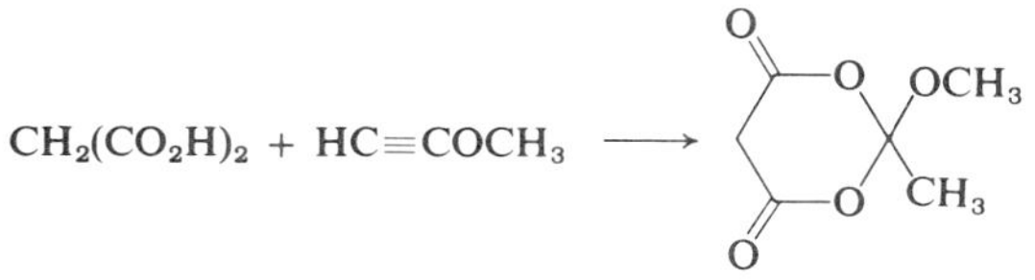

$$CH_2(CO_2H)_2 + HC\equiv COCH_3 \longrightarrow \text{[dioxanone structure]} \qquad (93)$$

Lead tetraacetate oxidation of dibutyl ether gave a mixture of products containing 1-butoxy-1,1-diacetoxybutane and 1-(1-acetoxybutoxy)-1,1-diacetoxybutane (*158*).

Three methods of preparing acyloxy ortho esters in which the acyloxy group is incorporated into a lactone ring have been reported. $\alpha$-Acyloxy acid chlorides react with alcohols in the presence of tertiary amines to form 2-alkyl-2-alkoxy-4-oxo-1,3-dioxolanes (*235*) [Eq. (94)]. Reaction of phthaloyl peroxide with alkenes yields spirocyclic acyloxy ortho esters (*113, 114*). A specific example is the reaction with *trans*-stilbene [Eq. (95)].

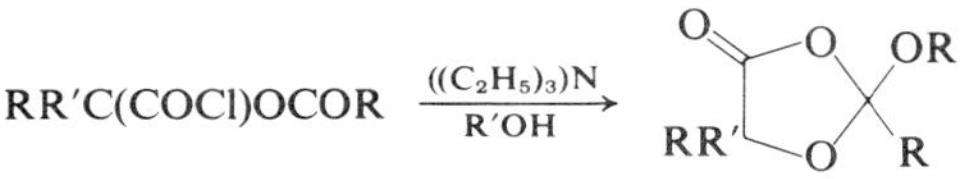

$$RR'C(COCl)OCOR \xrightarrow[R'OH]{((C_2H_5)_3)N} RR'\text{[dioxolanone]}R \qquad (94)$$

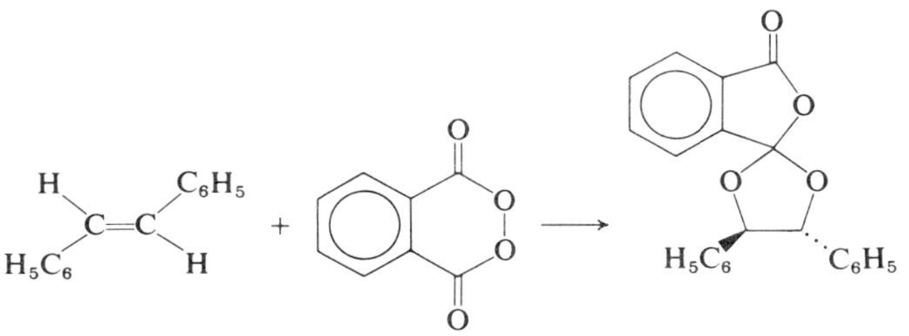

$$\underset{H_5C_6}{\overset{H}{\diagup}}C=C\underset{H}{\overset{C_6H_5}{\diagup}} + \text{[phthaloyl peroxide]} \longrightarrow \underset{H_5C_6}{\text{[spirocyclic product]}}C_6H_5 \qquad (95)$$

The product of reaction of cyclohexanone with oxygen under alkaline conditions reacts with triethyl orthoformate in the presence of hydrogen chloride to yield the cyclic acyloxy diethyl orthoformate **46** (*77*).

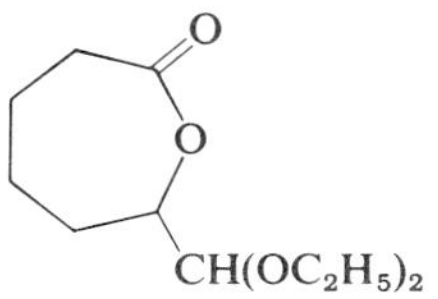

[structure of **46**: seven-membered lactone ring with CH(OC₂H₅)₂ substituent]

**46**

2-(Dichloromethylene)-1,3-dioxolane reacts with carboxylic acids to form 2-acyloxy-2-dichloromethyl-1,3-dioxolanes (*325a*).

## XV. PHYSICAL PROPERTIES OF ORTHO ESTERS

Most of the aliphatic orthocarboxylates are colorless liquids with characteristic ethereal odors. A few ortho esters—particularly bicyclic and tricyclic compounds—are colorless solids. Typically they are slightly soluble to very slightly soluble in water (in which they are stable at neutral and alkaline pH), and are soluble in or miscible with nonpolar organic solvents.

The ortho ester function does not absorb light in the ultraviolet portion of the spectrum, and pure, saturated ortho esters are transparent down to 2000 Å. The characteristic feature of the infrared absorption spectra of carboxylic ortho esters is a strong C—O stretching band near 1100 cm$^{-1}$. The NMR spectra of ortho esters correlate in a straightforward manner with their structures.

Dipole moments have been determined for trialkyl orthoformates and orthopropionates (*7, 8*), tetraalkyl orthocarbonates (*6a, 6b, 353a*), 2-alkoxydioxolanes and 2-alkoxydioxanes (*10*), and trioxaadamantane (*349a*). Surface tensions (and parachors) were reported for trialkyl orthoformates (*8a, 120, 152, 352, 352a*) and orthopropionates (*9, 345b*) and tetraalkyl orthocarbonates (*8a*). Refractive indexes at several wavelengths were measured for trialkyl orthoformates (*345b, 352*). X-Ray diffraction spectra have been determined for several trialkyl orthoformates and tetraethyl orthocarbonate (*163b*). In general, physical properties of ortho esters present no anamolous features.

Table III is a compilation of data for all of the ortho esters which have been described in the literature except those derived from carbohydrates, steroids, and other natural products. For each compound, where available, are given the boiling point (and the pressure at which it was measured), the melting point, the refractive index (measured at 25°C with sodium D light unless otherwise specified), the density (at 25°C, unless otherwise specified), the method or methods used in synthesizing the ortho ester, and references to publications reporting syntheses or physical properties. The ortho esters are arranged in the table according to the system used in *Chemical Abstracts* formula indexes.

Spectroscopic and other data on many of the compounds in Table III appear in the original literature, but are not referred to in the table. Due to thermal instability of many ortho esters—particularly in the presence of acids or acidic surfaces—boiling points of higher-molecular-weight ortho esters which were measured at ordinary pressures should be viewed with suspicion, as should other physical properties measured on the same samples.

## TABLE III
### Physical Properties of Ortho Esters

| Structure | B.p., °C/mm (M.p., °C) | Refractive index $n^a$ | Density $d^b$ | Method[c] | References |
|---|---|---|---|---|---|
| 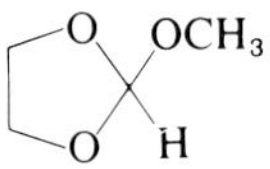 OCH₃, H (dioxolane) | 129.5°/760 | C₄H₈ 1.4070 | — | 3 | 17 |
| HC(OCH₃)₃ | 101°/760 | C₄H₁₀ 1.377 | 0.963 | 1, 2 | 7, 66, 73, 117, 142c, 165a, 174, 195a, 199a, 273, 276, 312 |
| 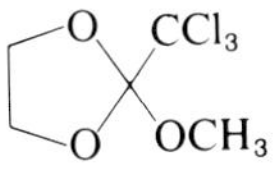 CCl₃, OCH₃ (dioxolane) | 112°–113°/10 (78°–78.5°) | C₅H₇ — | — | 8 | 116a, 245 |
| 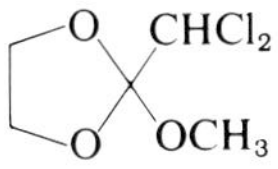 CHCl₂, OCH₃ (dioxolane) | 67°–70°/1.7 | C₅H₈ 1.4680 | 1.400 | 8 | 116a, 209 |
| 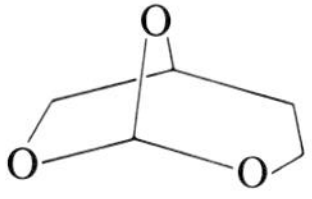 bicyclic ortho ester | (47°–49°) | 1.4625 | — | 3 | 68 |
| 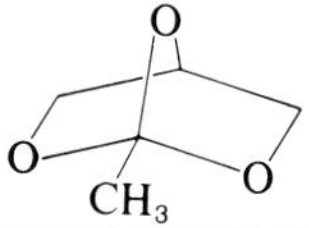 bicyclic ortho ester with CH₃ | 169°–170°/760 | 1.4530 (20°) | 1.2346 (20°) | 5 | 370 |

TABLE III—continued

| Structure | B.p., °C/mm (M.p., °C) | Refractive index $n^a$ | Density $d^b$ | Method[c] | References |
|---|---|---|---|---|---|
| $C_5H_9$ | | | | | |
| 2-oxa-spiro(tetrahydrofuran) | (22.5°–23.5°) | — | — | 11 | 323a |
| 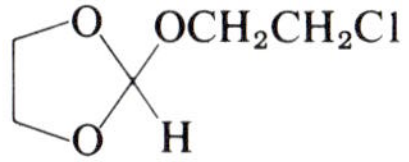 OCH$_2$CH$_2$Cl, H on dioxolane | 92°/11 | 1.4489 | — | 2, 3 | 17, 160a |
| 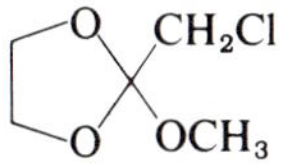 CH$_2$Cl, OCH$_3$ on dioxolane | 60°–63°/6 | 1.4449 | 1.2630 | 5 | 209 |
| CCl$_3$C(OCH$_3$)$_3$ | 63°–65°/5 | 1.4392 (20°) | 1.3302 (20°) | 1 | 324a |
| CF$_3$C(OCH$_3$)$_3$ | 106°/760 | 1.3380 | — | 11 | 144 |
| $C_5H_{10}$ | | | | | |
| 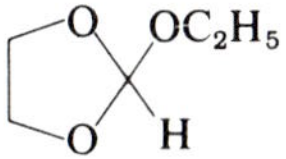 OC$_2$H$_5$, H on dioxolane | 145.5°/760 | 1.4100 | 1.053 (20°) | 3 | 12, 17, 250 |
| 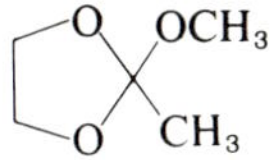 OCH$_3$, CH$_3$ on dioxolane | 132°–134°/745 | 1.4060 | 1.0424 | 3, 5 | 209, 339 |
| 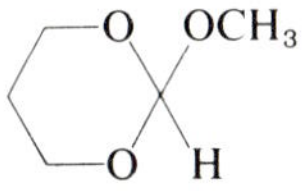 OCH$_3$, H on dioxane | 144°/745 | 1.4250 (20°) | — | 3 | 93a |

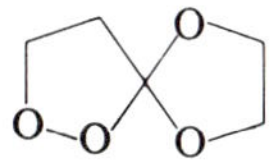

# PHYSICAL PROPERTIES OF ORTHO ESTERS

| Compound | bp/mp | Formula | $n_D$ | $d$ | | Refs |
|---|---|---|---|---|---|---|
| BrCH$_2$C(OCH$_3$)$_3$ | 74°–75°/17 | C$_5$H$_{11}$ | 1.4501 | 1.4771 | 1, 4 | 194, 204 |
| H$_2$NC(O)C(OCH$_3$)$_3$ | (123°–125°) | | — | — | 4 | 6, 173 |
| HC(OCH$_3$)$_2$OC$_2$H$_5$ | 115°–120°/760 | C$_5$H$_{12}$ | — | — | 1 | 275 |
| CH$_3$C(OCH$_3$)$_3$ | 107°–109°/760 | | 1.3859 | 0.9438 | 1 | 309, 351 |
| C(OCH$_3$)$_4$ | 114°/760 (−5.5°) | | 1.3837 | 1.0010 | 2, 5 | 6a, 8a, 42, 232a, 233, 283, 344, 353a |
| ![CHCl$_2$/OCOCH$_3$ dioxolane] | 82°/2 | C$_6$H$_8$ | 1.4677 (20°) | 1.4195 (20°) | 5 | 325a |
| 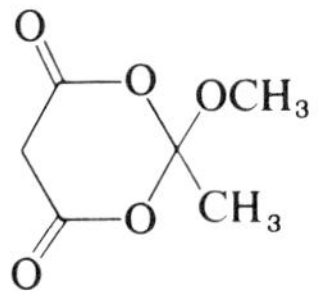 ![OCH$_3$/CH$_3$ dioxanedione] | 85°–95°/0.1 | | 1.4345 | — | 11 | 90 |
| C$_2$F$_5$C(OCH$_3$)$_3$ | 116°/760 | C$_6$H$_9$ | 1.3289 | — | 11 | 144 |
| CF$_3$CHClC(OCH$_3$)$_3$ | 150°/626 | C$_6$H$_{10}$ | 1.3842 (20°) | 1.3311 (20°) | 5 | 269 |
| 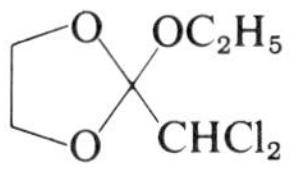 ![OC$_2$H$_5$/CHCl$_2$ dioxolane] | 100°–110°/17 | | — | — | 5 | 308 |
| ![H$_2$C= OC$_2$H$_5$/H dioxolane] | — | | — | — | 4 | 17 |

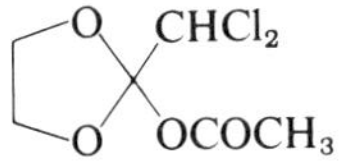

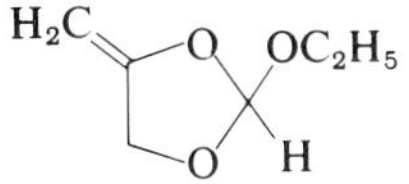

TABLE III—continued

| Structure | B.p., °C/mm (M.p., °C) | Refractive index $n^a$ | Density $d^b$ | Method$^c$ | References |
|---|---|---|---|---|---|
| 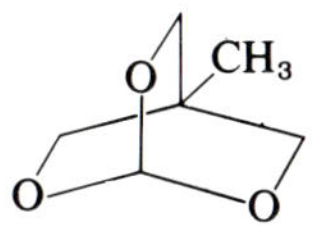 | (105°–106°) | — | — | 3 | 68, 78 |
| 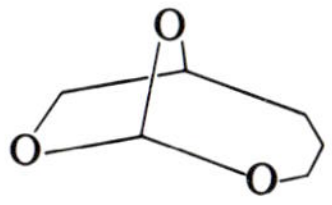 | 140°–180°/20–30 | 1.4658 | — | 3 | 68 |
| 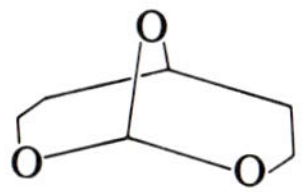 | 140°–170°/30–40 | 1.465 | — | 3 | 68 |
| 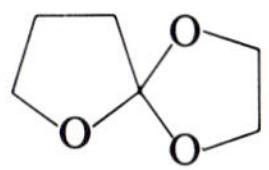 | 64°–67°/14 | 1.4460 (20°) | — | 7 | 40, 163a |
| 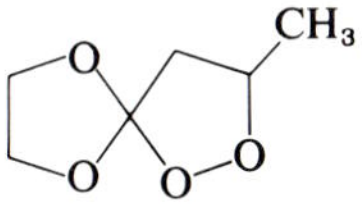 | 51°/0.4 | 1.4415 | — | 11 | 153a |
| 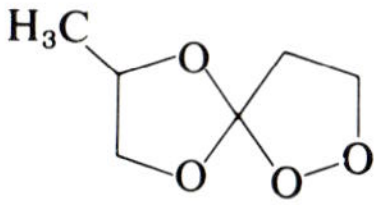 | 50°/0.7 | 1.4425 | — | 11 | 153a |

## PHYSICAL PROPERTIES OF ORTHO ESTERS

| Structure | bp (°C/mm) | $n_D$ | $d$ | — | Refs. |
|---|---|---|---|---|---|
| 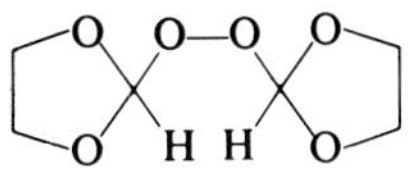 (spiro bis-dioxolane) | (73.5°–74.5°) | — | — | — | 5, 11  292, 319, 323d |
| | | $C_6H_{11}$ | | | |
| 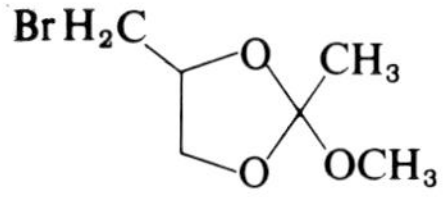 BrH$_2$C—(dioxolane, CH$_3$, OCH$_3$) | 89°–90°/15 | 1.4620 | 1.4625 | | 3  365 |
| 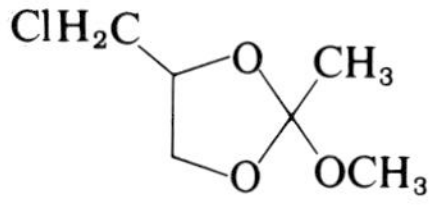 ClH$_2$C—(dioxolane, CH$_3$, OCH$_3$) | 79°–80°/15 | 1.4385 | 1.1839 | | 3  365 |
| 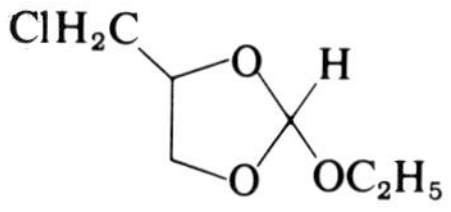 ClH$_2$C—(dioxolane, H, OC$_2$H$_5$) | 78°–81°/12 | 1.4407 | — | | 3, 7  17, 39, 272 |
| 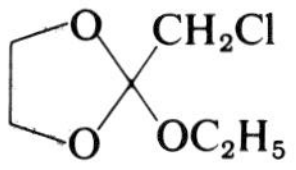 (dioxolane, CH$_2$Cl, OC$_2$H$_5$) | 198°/760 | 1.4221 | 1.1873 | | 5  98 |
| NCCH$_2$C(OCH$_3$)$_3$ | 93°/8 | 1.4215 | 1.079 | | 1, 5  223, 224 |
| | | $C_6H_{12}$ | | | |
| 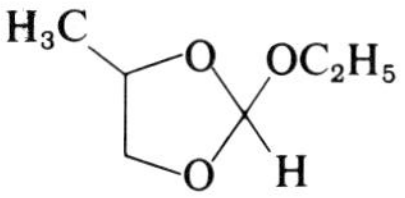 H$_3$C—(dioxolane, OC$_2$H$_5$, H) | 141°/760 | 1.4190 | — | | 3  17 |
| 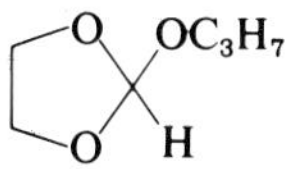 (dioxolane, OC$_3$H$_7$, H) | 165°/760 | 1.4143 | — | | 3  17 |

TABLE III–continued

| Structure | B.p., °C/mm (M.p., °C) | Refractive index $n^a$ | Density $d^b$ | Method[c] | References |
|---|---|---|---|---|---|
| 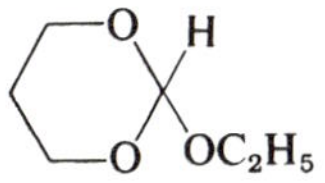 2-ethoxy-1,3-dioxane (H, OC₂H₅) | 156°/760 | 1.4267 | — | 3 | 17 |
| 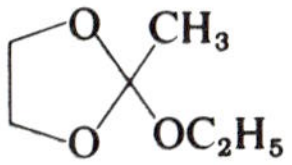 2-methyl-2-ethoxy-1,3-dioxolane (CH₃, OC₂H₅) | 144°–145°/745 | 1.4079 | 1.0038 | 3 | 209 |
| 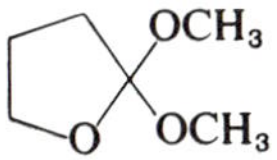 2,2-dimethoxy-tetrahydrofuran (OCH₃, OCH₃) | 42.5°–44°/10 | — | — | 6 | 242 |
| 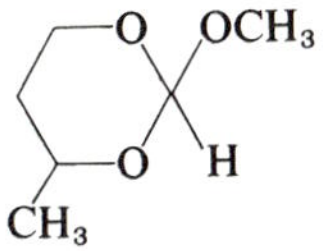 2-methyl-4-methyl-2-methoxy-1,3-dioxane | 130–143°/745 | 1.4250 (20°) (cis) 1.4178 (20°) (trans) | — | 3 | 93a |
| HCO₂CH(OC₂H₅)₂ | 60°/11 | 1.3958 (25°) | — | 3 | 315d |
| 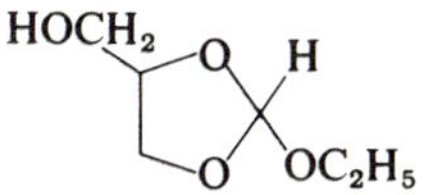 HOCH₂– dioxolane –OC₂H₅, H | — | — | — | 3 | 68 |
| 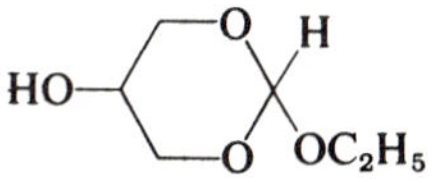 HO– dioxane –OC₂H₅, H | — | — | — | 3 | 68 |

## PHYSICAL PROPERTIES OF ORTHO ESTERS

| Compound | bp/mm (or mp) | $n_D$ | $d$ | Method | References |
|---|---|---|---|---|---|
| $CH_3OC(S)C(OCH_3)_3$ | 76°/12 | — | — | 11 | 142b |
| $CH_3O_2CC(OCH_3)_3$ | — | — | — | 2 | 5, 6, 142b |
| $C_6H_{14}$ | | | | | |
| $HC(OC_2H_5)_2OCH_3$ | 133°–134°/760 | 1.3868 (20°) | — | 2, 11 | 70, 117 |
| $CH_3C(OCH_3)_2OC_2H_5$ | 123°–126°/760 | 1.3889 | 0.9192 | 1 | 309 |
| $C_2H_5C(OCH_3)_3$ | — | — | — | 1 | 373 |
| $CH_3OCH_2CH_2OCH(OCH_3)_2$ | 174°/743 | 1.4012 | 1.001 | 11 | 72b, 72c |
| $C_6H_{15}$ | | | | | |
| $(CH_3)_2NC(OCH_3)_3$ | 130°–132°/760 | 1.4098 (20°) | — | 6 | 49, 91, 359 |
| $C_7-$ bicyclic with $CF_2$, $CF_3$, $F_3C$, $CF_3$ | — | — | — | 4 | 340a |
| $C_7H_9$ | | | | | |
| $CFCl_2CF=CFC(OCH_3)_3$ | 69.5°–72.5°/8 | 1.4182 (20°) | 1.4117 (20°) | 5 | 176 |
| 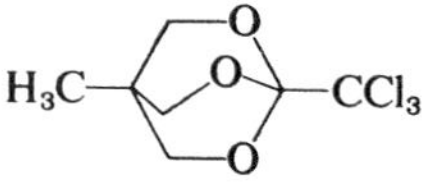 bicyclic $H_3C$–$CCl_3$ | (218°–221°) | — | — | 8 | 25 |
| 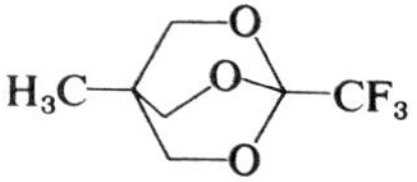 bicyclic $H_3C$–$CF_3$ | (144°–145°; 153.5°) | — | — | 8 | 25, 340a |
| $C_3F_7C(OCH_3)_3$ | 147°/760 | 1.3207 | — | 11 | 144 |
| $C_7H_{10}$ | | | | | |
| 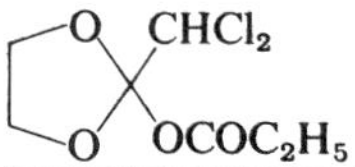 dioxolane with $CHCl_2$, $OCOC_2H_5$ | 105°/2 | 1.4630 | 1.3681 | 5 | 325a |

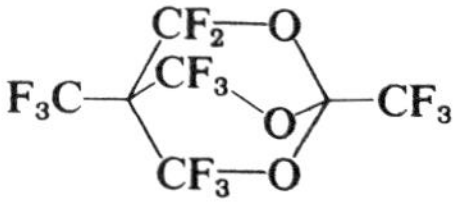

## TABLE III—continued

| Structure | B.p., °C/mm (M.p., °C) | Refractive index $n^a$ | Density $d^b$ | Method[c] | References |
|---|---|---|---|---|---|
| $(F_3CS)_2CHC(OCH_3)_2$ | 66°/9 | 1.399 (25°) | — | 5 | 121a |
| $HC(OCH=CH_2)_3$ | 46°–47°/16 | 1.4281 (20°) | 0.9547 (20°) | 4 | 234 |
| 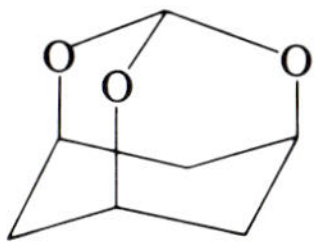 [adamantane-like structure] | (219°–220°) | — | — | 3 | 334 |
| 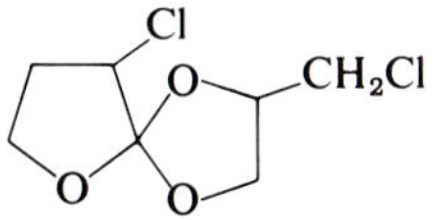 [bicyclic structure with CH₂Cl and Cl] | 82°–84°/0.3 | 1.4893 (20°) | — | 7 | 40 |

$C_7H_{11}$

| Structure | B.p., °C/mm (M.p., °C) | Refractive index $n^a$ | Density $d^b$ | Method[c] | References |
|---|---|---|---|---|---|
| 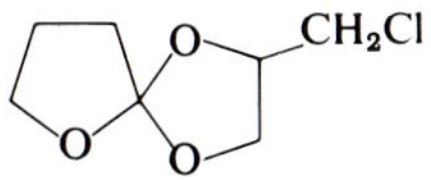 [spiro structure with CH₂Cl] | 105°–112°/13 | 1.4680 (20°) | — | 7 | 40, 41 |
| $ClC\equiv CCH_2C(OCH_3)_3$ | 62°–73°/5 | — | — | 4 | 223 |
| $HC(OCH=CH_2)_2OCH_2CH_2Cl$ | 83°–84°/12.5 | 1.4462 (20°) | 1.0998 (20°) | 4 | 234 |
| 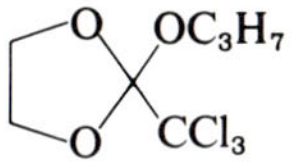 [structure with OC₃H₇ and CCl₃] | 128°–130°/12 | — | — | 3 | 244 |
| 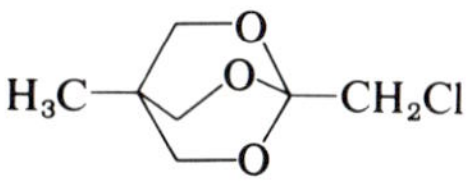 [bicyclic structure with H₃C and CH₂Cl] | (138°–139°) | — | — | 8 | 25 |

# PHYSICAL PROPERTIES OF ORTHO ESTERS

| Structure | bp (°C/mm) | $n_D$ | $d$ | | Ref. |
|---|---|---|---|---|---|
| HC(OCH₂CH₂Cl)₂OCH=CH₂ | 99°–102°/3 | 1.4629 (20°) | — | 4 | 234 |
| 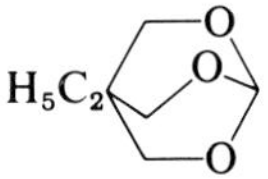 [C₇H₁₂ bicyclic ortho ester, H₅C₂] | (60°) | — | — | 8 | 54 |
| 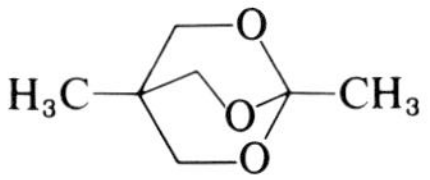 [bicyclic ortho ester, H₃C, CH₃] | (85°–90°) | — | — | 4 | 25 |
| 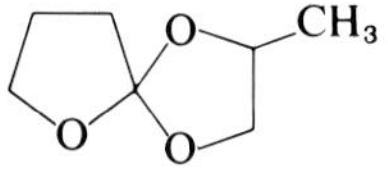 [spiro dioxolane, CH₃] | 67°–69°/15 | 1.4394 (20°) | — | 7 | 40 |
| 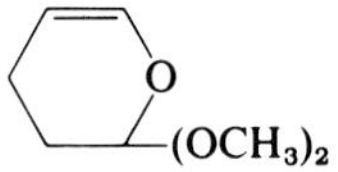 [pyran (OCH₃)₂] | 47°–48°/7 | 1.4427 | 1.055 | 10 | 211 |
| 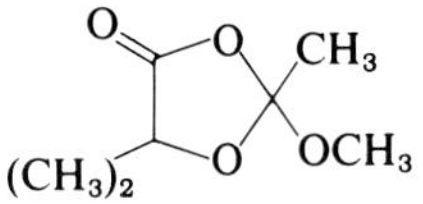 [dioxolanone, CH₃, OCH₃, (CH₃)₂] | 33°/0.2 | 1.4115 (20°) | — | 8 | 325 |
| 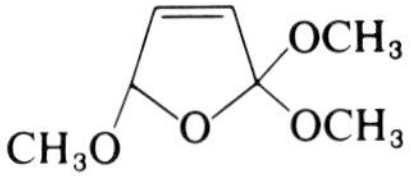 [dihydrofuran, OCH₃, OCH₃, CH₃O] | 85°/12 | 1.4360 | 1.081 | 11 | 332 |
| 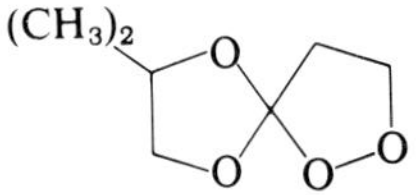 [spiro bis-dioxolane, (CH₃)₂] | 39°/0.4 | 1.4365 (25°) | — | 11 | *153a* |

TABLE III—continued

| Structure | B.p., °C/mm (M.p., °C) | Refractive index $n^a$ | Density $d^b$ | Method[c] | References |
|---|---|---|---|---|---|
| **C₇H₁₃** | | | | | |
| 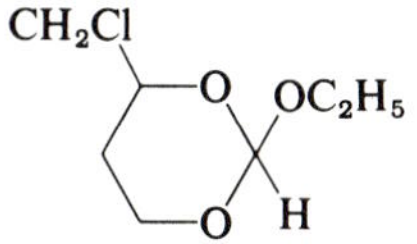 2-ethoxy-4-(chloromethyl)-1,3-dioxane | 100°/22 | — | — | 3 | 47 |
| 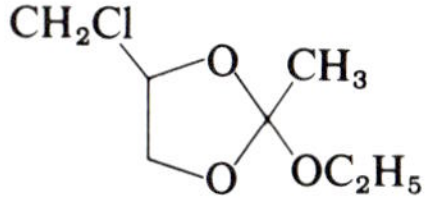 2-methyl-2-ethoxy-4-(chloromethyl)-1,3-dioxolane | 86°–87°/15 | 1.4367 | 1.1332 | 3 | 365 |
| $HC(OCH_2CH_2Cl)_3$ | 157°/14 | 1.4678 (20°) | 1.2982 (20°) | 3 | 124, 234 |
| $Cl_3CCH_2CH_2C(OCH_3)_3$ | 91°–92°/4 (87°–88°) | 1.4578 | 1.277 | 1 | 223 |
| $Cl_3CCH(OH)CH_2C(OCH_3)_3$ | | — | — | 4 | 184 |
| $HC(OCH_2CH_2F)_3$ | 128°/21 | 1.3946 (20°) | 1.2316 (20°) | — | 284 |
| $HC(OC_2H_5)_2OCH_2CN$ | 59°/1 | — | — | 3 | 199 |
| $HCCH_2CH_2C(OCH_3)_3$ | 73°–74°/0.5 104°–105°/10 | 1.4269 | 1.055 | 1 | 221, 223 |
| **C₇H₁₄** | | | | | |
| 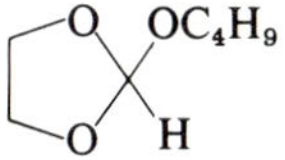 2-butoxy-1,3-dioxolane | 183°/760 | 1.4202 | — | 3 | 17 |
| 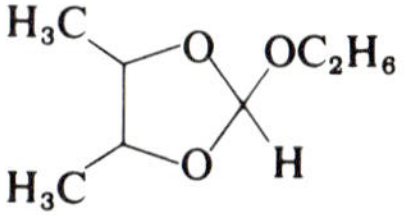 2-ethoxy-4,5-dimethyl-1,3-dioxolane | 45°/6 | — | — | 3 | 246 |

## PHYSICAL PROPERTIES OF ORTHO ESTERS

| Compound | bp | $n_D$ | $d$ | | Refs. |
|---|---|---|---|---|---|
| 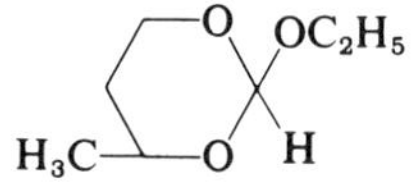 [2-ethoxy-1,3-dioxane with H3C] | 169°–171°/760; 61°/7.5 | 1.4185 | 0.9918 | 3 | 10, 246 |
| 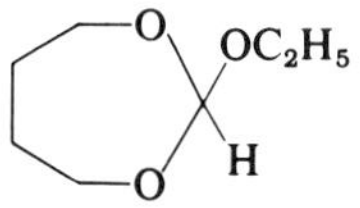 [2-ethoxy-1,3-dioxepane] | 85°/4 | — | — | 3 | 246 |
| 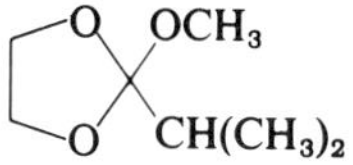 [2-methoxy-2-isopropyl-1,3-dioxolane] | 160.5°/744 | 1.4188 | 1.007 | 1 | 203 |
| 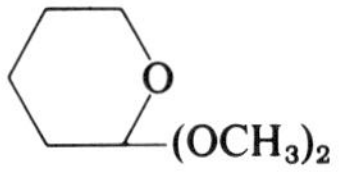 [2,2-dimethoxytetrahydropyran] | 69°–70°/20 | 1.4298 | 1.029 | 4 | 216 |
| $CH_3CO_2CH(OC_2H_5)_2$ | 70°/15 | 1.3959 | 0.9910 | 3 | 282, 315d |
| $(CH_3)_2CBrC(OCH_3)_3$ | 77°–79°/18 | $C_7H_{15}$ 1.4510 | — | 11 | 210 |
| $H_2NC(O)CH_2CH_2C(OCH_3)_3$ | (94°–97°) | — | — | 4 | 223 |
| $C_2H_5SO_2CHClC(OCH_3)_3$ | (59°–60°) | — | — | 2 | 43a |
| $HC(OC_2H_5)_3$ | 145°/760 | $C_7H_{16}$ 1.391 | 0.886 | 1, 2, 3 | 7, 52, 53, 73, 87, 87a, 255a, 265a, 273, 275, 276, 281, 352, 362, 363 |
| $C_3H_7C(OCH_3)_3$ | 140°–147°/760 | 1.4017 | 0.9621 | 1 | 208 |
| $(CH_3)_2CHC(OCH_3)_3$ | 134°–136°/760 | 1.4003 | 0.9253 | 1 | 202, 229 |
| $CH_3C(OC_2H_5)_2OCH_3$ | 135°–136°/760 | 1.3919 | 0.9009 | 1 | 309 |
| $CH_3SCH_2CH_2C(OCH_3)_3$ | 51°–52°/1 | — | — | 1 | 60, 61 |
| $(CH_3O)_2CHC(OCH_3)_3$ | 85°/25 | — | — | 5 | 143 |

TABLE III—continued

| Structure | B.p., °C/mm (M.p., °C) | Refractive index $n^a$ | Density $d^b$ | Method[c] | References |
|---|---|---|---|---|---|
| HC(OC₂H₅)₂OCH₂CH₂NH₂ | 88°/10 | C₇H₁₇ — | — | 4 | 199 |
| 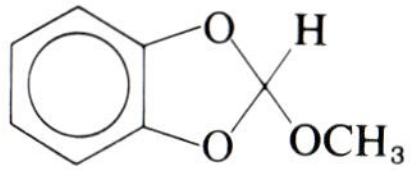 [benzodioxole with H and OCH₃] | 88°–89°/16 | C₈H₈ 1.5136 (20°) | — | 2 | 47, 118a |
| 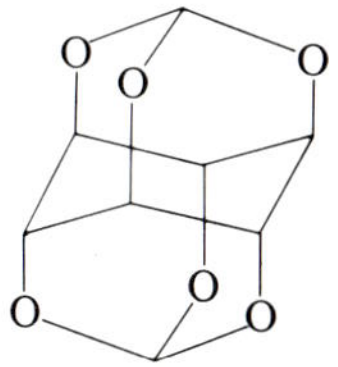 [cage structure] | (303°–305°) | — | — | 3 | 353 |
| 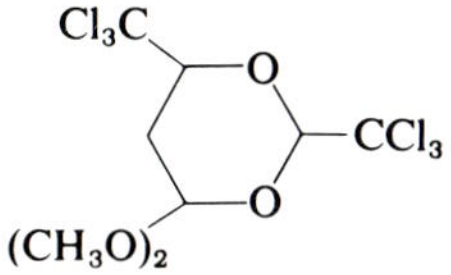 Cl₃C—[dioxane]—CCl₃ (CH₃O)₂ | (108°–109°) | — | — | 4 | 184 |
| 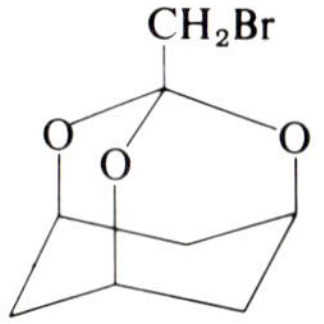 [adamantane-type structure with CH₂Br] | (95°) | C₈H₁₁ — | — | 3 | 279a, 334 |

# PHYSICAL PROPERTIES OF ORTHO ESTERS

| Structure | B.p. (°C) | $n_D$ | $d$ | Yield | Ref. |
|---|---|---|---|---|---|
| 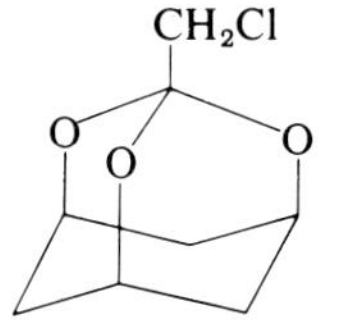 CH₂Cl-adamantane ortho ester | (107°) | — | — | 3 | 334 |
| H₃C-C(CCl₂CH₃) cage | (151°–153°) C₈H₁₂ | — | — | 8 | 25 |
| CH₃C(OCH=CH₂)₃ | 145°–147°/760 | 1.4328 | 0.941 | — | 205 |
| 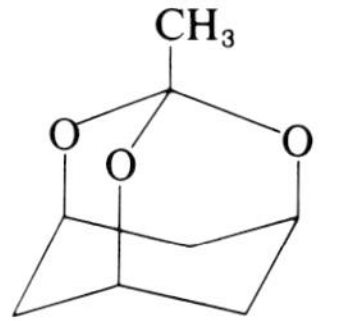 CH₃-adamantane ortho ester | (126°) | — | — | 3 | 279a, 334 |
| 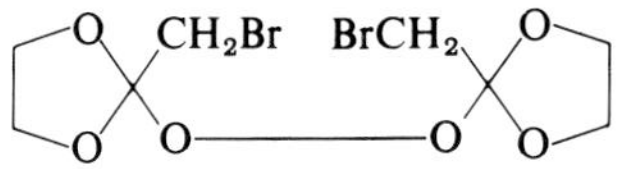 CH₂Br, BrCH₂ spiro bis-dioxolane | (99°) | — | — | 11 | 323, 323c |
| 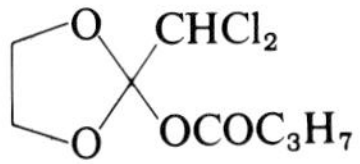 CHCl₂, OCOC₃H₇ dioxolane | 99.5°/1 | 1.4650 | 1.3186 | 5 | 325a |
| 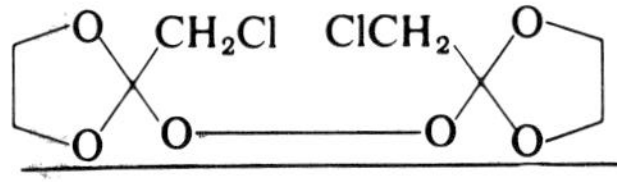 CH₂Cl, ClCH₂ spiro bis-dioxolane | (101°–102°) | — | — | 11 | 323, 323c |

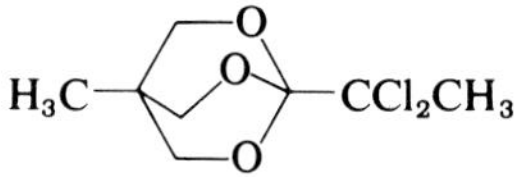

## TABLE III—continued

| Structure | B.p., °C/mm (M.p., °C) | Refractive index $n^a$ | Density $d^b$ | Method[c] | References |
|---|---|---|---|---|---|
| $C_8H_{13}$ | | | | | |
| 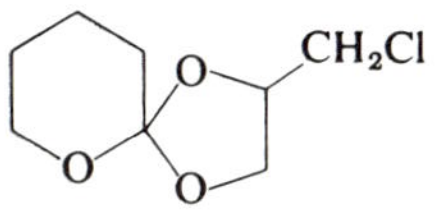 (CH₂Cl, spiro cyclohexane-dioxolane) | 116°–117°/15 | 1.4718 (20°) | — | 7 | 40 |
| 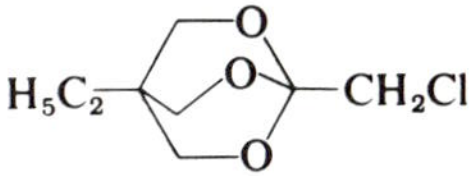 (CH₂Cl, bicyclic orthoester with H₅C₂) | (54°) | — | — | 8 | 54 |
| 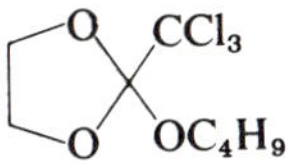 (CCl₃, OC₄H₉ dioxolane) | 117°–119°/10 | — | — | 3 | 245 |
| 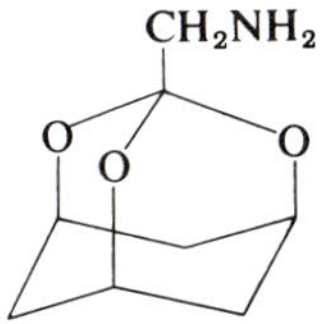 (CH₂NH₂ adamantane orthoester) | (163°) | — | — | 4 | 44 |
| CH₃C(OCH=CH₂)₂OCH₂CH₂NO₂ | 80°–85°/2 | 1.4492 | — | 5 | 100 |
| CH₃C(OCH=CH₂)₂OC₂H₅ | 144°–145°/760 | 1.4221 | 0.924 | 5 | 205 |
| $C_8H_{14}$ | | | | | |
| 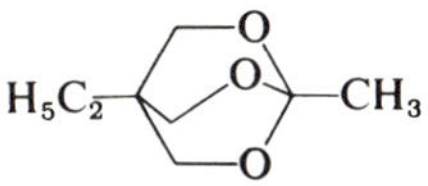 (CH₃, bicyclic orthoester with H₅C₂) | 85°/8 (40°–41°) | — | — | 8 | 54 |

# PHYSICAL PROPERTIES OF ORTHO ESTERS

| Structure | bp/mp | $n_D$ | Density | | Ref. |
|---|---|---|---|---|---|
| 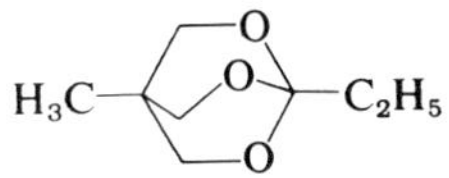 (bicyclic with $C_2H_5$, $H_3C$) | 60°–65°/3 | 1.4473 (20°) | 1.0837 (20°) | 8 | 54 |
| 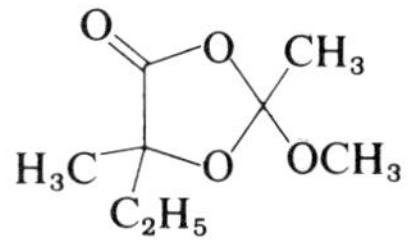 ($CH_3$, $OCH_3$, $C_2H_5$ dioxolanone) | 49°/0.3 | 1.4198 (20°) | — | 8 | 235 |
| 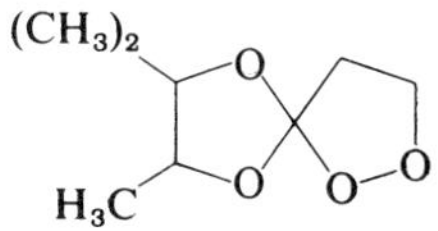 $(CH_3)_2$ spiro dioxolane | 55°/5 | 1.4412 (25°) | — | 11 | 153a |
| 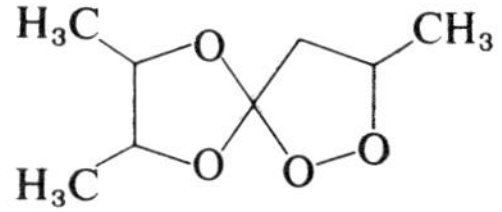 $H_3C$, $CH_3$, $H_3C$ spiro | 60°/1.1 | 1.4408 (25°) | — | 11 | 153a |
| 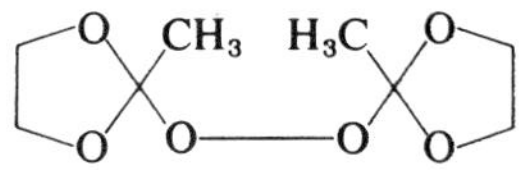 $C_8H_{15}$ spiro | (68°) | — | — | 11 | 71, 323, 323d |
| 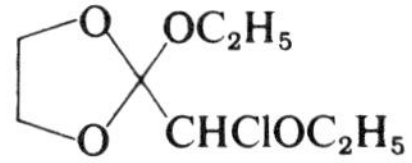 $CH_3$, $H_3C$ bis-dioxolane | 125°–130°/22 | — | — | 5 | 308 |
| $OC_2H_5$, $CHClOC_2H_5$ dioxolane | | | | | |
| $CH_3C(OCH_2CH_2Cl)_3$ | 155°–156°/13 | 1.4372 (20°) | 1.1980 (20°) | 3 | 124 |
| $CCl_3C(OC_2H_5)_3$ | 101°–102°/2 | 1.4292 | 1.035 | 1 | 324a |
| $NC(CH_2)_3C(OCH_3)_3$ | 108°–109°/8 | | | 1 | 221 |
| $NCC(OC_2H_5)_3$ | 60°/13 | 1.3995 | — | 11 | 43b |

TABLE III—continued

| Structure | B.p., °C/mm (M.p., °C) | Refractive index $n^a$ | Density $d^b$ | Method[c] | References |
|---|---|---|---|---|---|
| | | $C_8H_{16}$ | | | |
| $CH_3C(OC_2H_5)(ON=CHCH_3)_2$ | 78°/4 | 1.4448 | — | 5 | 345a |
| 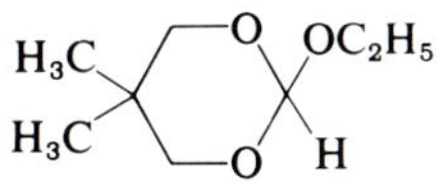 H₃C─O─C(OC₂H₅)(H)─O (with H₃C substituent) | 91°/41 | 1.4412 | — | 3 | 47, 74a |
| 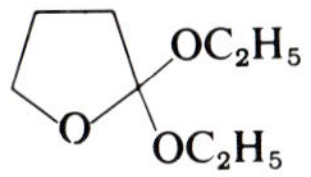 (tetrahydrofuran with OC₂H₅, OC₂H₅) | 60°–61.5°/10 | — | — | 6 | 242 |
| $CH_3C(O)CH_2CH_2C(OCH_3)_3$ | 88.5°–89.5°/10 | 1.4225 | 1.0274 | 11 | 324 |
| 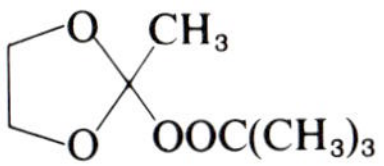 (dioxolane with CH₃ and OOC(CH₃)₃) | 68°–69°/12 | 1.4206 (20°) | — | 5 | 319 |
| $C_2H_5CO_2CH(OC_2H_5)_2$ | 85°/20 | 1.4052 | 0.9857 | 3 | 282, 315d |
| $CH_3O_2CCH_2CH_2C(OCH_3)_3$ | 64°–65°/1 | 1.4230 | 1.084 | 1 | 223 |
| $CH_3O_2CC(OC_2H_5)_2OCH_3$ | 89°/12 | — | — | 2 | 6 |
| $CHBr_2C(OC_2H_5)_3$ | 102°–104°/8 | 1.4691 | 1.5272 | 4 | 230 |
| $CHClFC(OC_2H_5)_3$ | 67.5°/10  181°/747 | 1.4059 | 1.0815 | 2, 5 | 247, 307, 342 |
| $CHCl_2C(OC_2H_5)_3$ | 124°–129°/60 | 1.4336 | 1.3145 | 2 | 342 |
| | | $C_8H_{17}$ | | | |
| $BrCH_2C(OC_2H_5)_3$ | 77°–79°/9 | 1.4393 | 1.2639 | 4, 5 | 36 |
| $C_3H_7CHBrC(OCH_3)_3$ | 93°–96°/14 | 1.4507 (24°) | 1.268 (27°) | 4 | 214 |

# PHYSICAL PROPERTIES OF ORTHO ESTERS

| Compound | bp | $n_D$ | density | formula | | refs |
|---|---|---|---|---|---|---|
| $ClCH_2C(OC_2H_5)_3$ | 81.5°–82.5°/17 | 1.4199 | — | | 1, 2 | 35, 130, 222 |
| $ICH_2C(OC_2H_5)_3$ | 96°–98°/10 | 1.4660 | — | | 4 | 36, 261a |
| | | | | $C_8H_{18}$ | | |
| $HC(OC_2H_5)_2OC_3H_7$ | 165°/747 | 1.3989 (20°) | 0.8813 (23°) | | 2 | 281 |
| $HC(OC_3H_7)_2OCH_3$ | 61.5°–62°/11 | 1.4010 (20°) | — | | 2 | 117 |
| $CH_3C(OC_2H_5)_3$ | 144°–146°/760 | 1.3949 | 0.8847 | | 1, 2, 5 | 34, 35, 130, 222, 286, 309, 326 |
| $C_4H_9C(OCH_3)_3$ | 164°–166°/760 | 1.4090 (24°) | 0.941 (27°) | | 1 | 214 |
| $C_4H_9OC(OCH_3)_3$ | 65.5°/18 | 1.4003 (20°) | 0.9553 | | 3 | 329 |
| $(CH_3)_2CHCH_2OC(OCH_3)_3$ | 139°–141°/760 | 1.394 (20°) | 0.949 (20°) | | 3 | 329 |
| $(CH_3O)_3CC(OCH_3)_3$ | 195°/760 | — | — | | 6 | 142a |
| | | | | $C_9H_4$ | | |
| 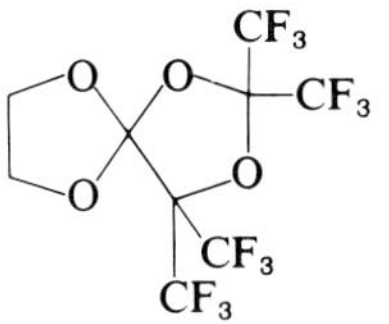 | 176°/760 | — | — | | 11 | 47 |
| | | | | $C_9H_6$ | | |
| 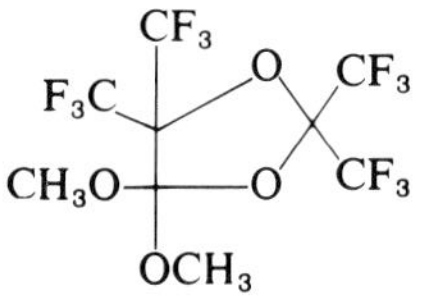 | 161°/760 | 1.3225 | — | | 11 | 47 |
| | | | | $C_9H_8$ | | |
| $C(OCH_2CF_2Cl)_4$ | 80°/2.8 | — | — | | 2 | 133 |
| | | | | $C_9H_{10}$ | | |
| 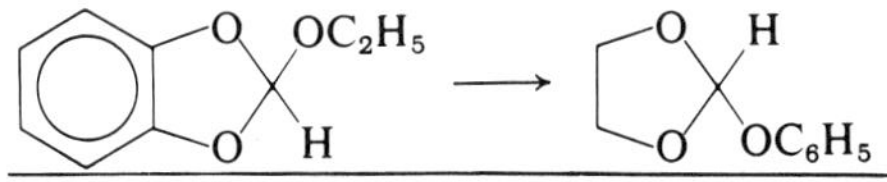 | 94°/11 | 1.5033 | — | | 3 | 17 |
| | 73°–74°/0.5 | — | — | | — | 2a |

TABLE III—continued

| Structure | B.p., °C/mm (M.p., °C) | Refractive index $n^a$ | Density $d^b$ | Method$^c$ | References |
|---|---|---|---|---|---|
| 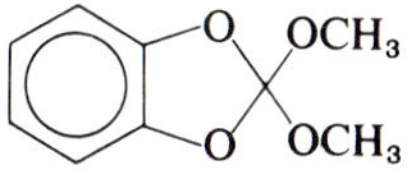 (catechol dimethyl orthoester, OCH₃/OCH₃) | 115°/19 | 1.4978 (20°) | 1.1959 (20°) | 2, 3 | *119, 328* |
| 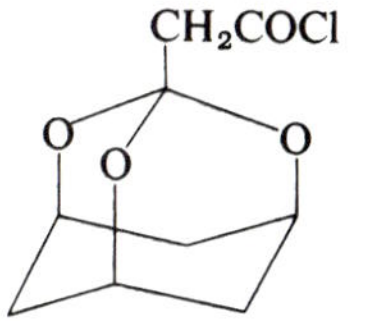 CH₂COCl adamantane-type orthoester | (50°) | — | — | 4 | 44 |
| CH₂CON₃ adamantane-type orthoester, C₉H₁₂ | — | — | — | 4 | 44 |
| 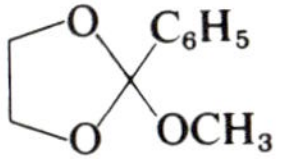 C₆H₅/OCH₃ dioxolane | 139°/20 | 1.5136 (20°) | — | 3 | 289 |
| 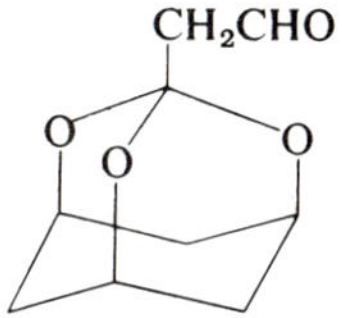 CH₂CHO adamantane-type orthoester | (87°) | — | — | 4 | 44 |

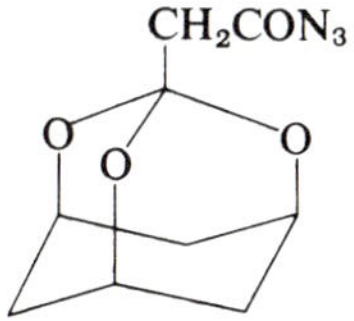

# PHYSICAL PROPERTIES OF ORTHO ESTERS

| Structure | bp/mp | $n_D$ | — | n | Ref |
|---|---|---|---|---|---|
| 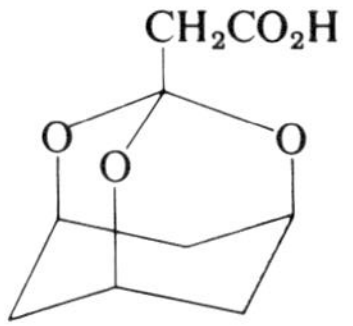 CH$_2$CO$_2$H (adamantane-like) | (137°) | — | — | 4 | *44* |
| CH$_3$OCCl=CClCCl=CClC(OCH$_3$)$_3$ | (97.5°) | — | — | 2 | *301* |
| 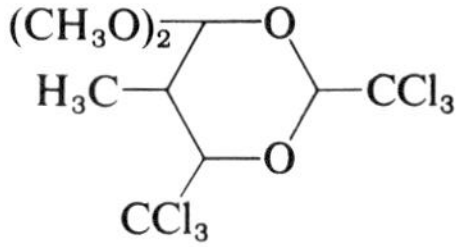 (CH$_3$O)$_2$ / H$_3$C-pyran-CCl$_3$, CCl$_3$ | 105°–107°/0.2 | 1.495 (27°) | — | 4 | *184* |
| 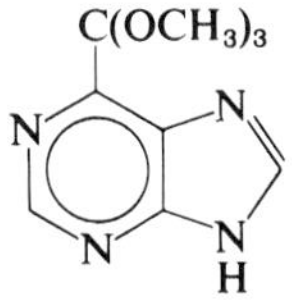 C(OCH$_3$)$_3$ purine | (179°–180°) | — | — | 2 | *63* |
| 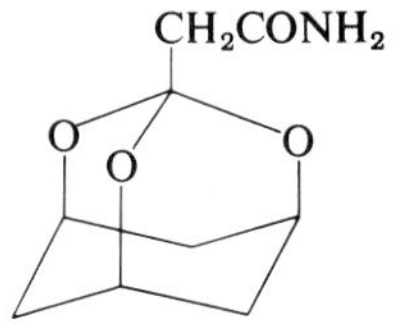 CH$_2$CONH$_2$ (adamantane-like) | (166°) | — | C$_9$H$_{13}$ | 4 | *44* |
| C$_2$H$_5$ (adamantane-like) | (38°) | — | C$_9$H$_{14}$ | 3 | *334* |

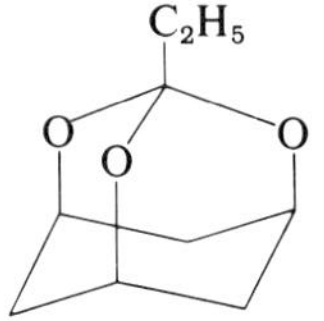

TABLE III—continued

| Structure | B.p., °C/mm (M.p., °C) | Refractive index $n^a$ | Density $d^b$ | Method[c] | References |
|---|---|---|---|---|---|
| 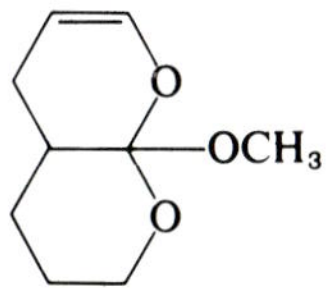 [structure: bicyclic with OCH₃] | 99°/20 | 1.4727 | 1.106 | 10 | 216 |
| 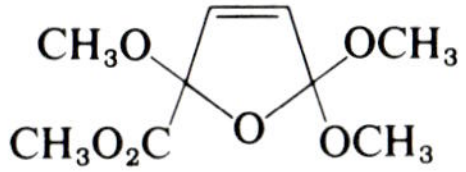 [structure: CH₃O₂C, CH₃O₂C, OCH₃, OCH₃ furan] | 134°/11 | 1.4522 | 1.1540 | 11 | 332a |
| 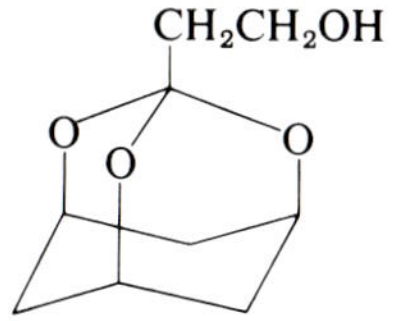 [structure: CH₂CH₂OH orthoester cage] | (59°) | — | — | 4 | 44 |
| 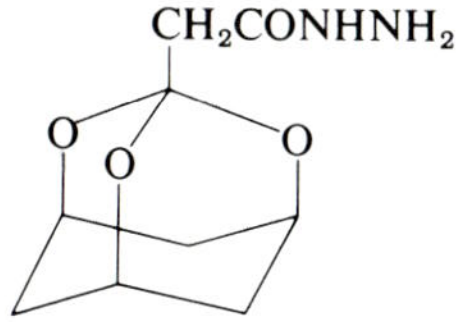 [structure: CH₂CONHNH₂ orthoester cage] | (124°) | — | — | 4 | 44 |
| $CCl_2={=}CClC(OC_2H_5)_3$ | 149°/50 | $C_9H_{15}$ 1.4694 (20°) | 1.2183 (20°) | 2 | 102 |

## PHYSICAL PROPERTIES OF ORTHO ESTERS

| Structure | bp/mmHg | $n_D$ | $d$ | | Ref. |
|---|---|---|---|---|---|
| NCCHClC(OC$_2$H$_5$)$_3$ | (63°) | C$_9$H$_{16}$ | — | 5 | 300a |
| 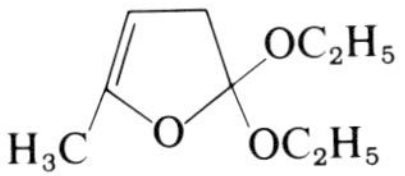 (structure with OC$_2$H$_5$, OC$_2$H$_5$, H$_3$C) | 68°/14 | — | — | 5 | 315a |
| 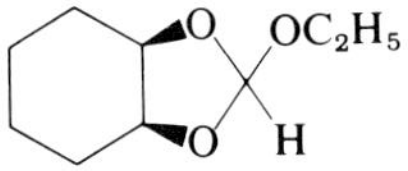 (cyclohexane fused dioxolane, OC$_2$H$_5$, H) | 66°–69°/5 | 1.4474 | — | 3 | 69 |
| 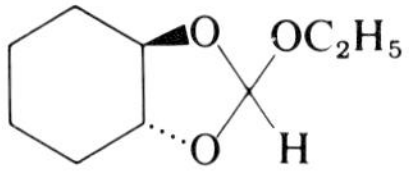 (cyclohexane fused dioxolane, OC$_2$H$_5$, H) | 98°–104°/8 | 1.4533 | — | 3 | 69 |
| 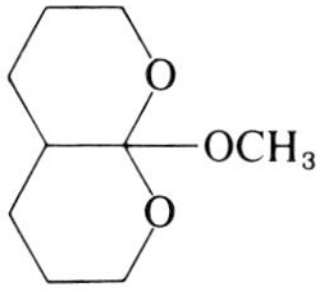 (bicyclic with OCH$_3$) | 38°/0.1 | 1.4653 | 1.087 | 4 | 216 |
| 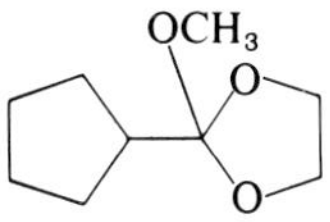 (cyclopentyl dioxolane, OCH$_3$) | 81°–82.5°/7 | 1.4545 | — | 1 | 225 |
| 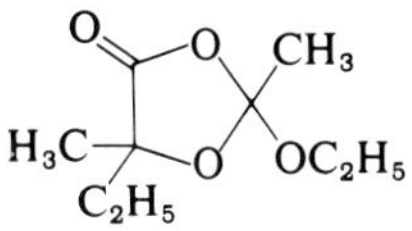 (dioxolane, CH$_3$, OC$_2$H$_5$, H$_3$C, C$_2$H$_5$) | 61°–62°/0.6 | 1.4207 (20°) | — | 8 | 235 |
| 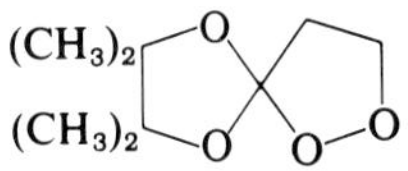 (spiro bis-dioxolane, (CH$_3$)$_2$, (CH$_3$)$_2$) | 58°/0.3 | 1.4468 | — | 11 | 153a |

## TABLE III—continued

| Structure | B.p., °C/mm (M.p., °C) | Refractive index $n^a$ | Density $d^b$ | Method[c] | References |
|---|---|---|---|---|---|
| 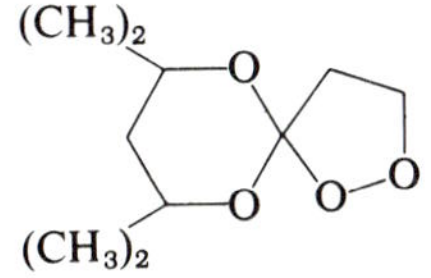 (CH₃)₂ — [bicyclic acetal structure] — (CH₃)₂ | 58°/0.6 (38°) | — | — | 11 | *153a* |
| | | $C_9H_{17}$ | | | |
| NCCH₂C(OC₂H₅)₃ | 83°–84°/2 | 1.4189 | 0.978 | 1, 5 | *223, 224* |
| NC(CH₂)₄C(OCH₃)₃ | 128°–129°/9 | 1.4344 | 1.0197 | 1 | *221* |
| | | $C_9H_{18}$ | | | |
| 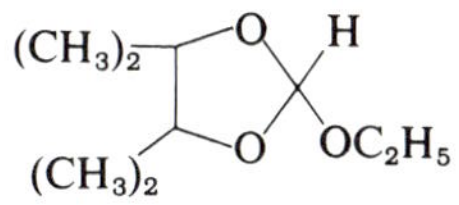 (CH₃)₂ — [dioxolane with H, OC₂H₅] — (CH₃)₂ | 72°–75°/17 | 1.4135 | 0.947 | 3 | *69, 246* |
| 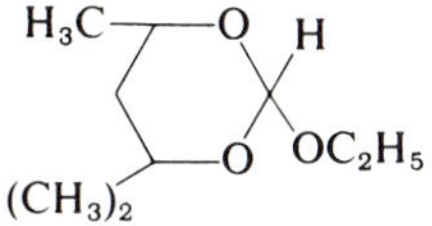 H₃C — [dioxane with H, OC₂H₅] — (CH₃)₂ | 80°/2.5 | — | — | 3 | *246* |
| 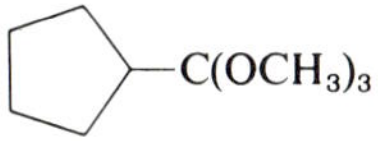 cyclopentyl-C(OCH₃)₃ | 76°/18 | 1.4366 | — | 1 | *225* |
| C₃H₇-CO₂CH(OC₂H₅)₂ | 90°/13 | 1.4047 | 0.9611 | 3 | *315d* |
| (CH₃)₂CHCO₂CH(OC₂H₅)₂ | 68°/5 | 1.4020 | 0.9524 | 3 | *315d* |
| | | $C_9H_{19}$ | | | |
| CH₃CHBrC(OC₂H₅)₃ | 73°/8 | 1.4338 | 1.181 | 4 | *355* |

## PHYSICAL PROPERTIES OF ORTHO ESTERS

| Compound | bp/mm (or mp) | $n_D$ | $d$ | | References |
|---|---|---|---|---|---|
| | | $C_9H_{20}$ | | | |
| $HC(OC_2H_5)_2OC(CH_3)_3$ | 57°–59°/14 | | | 3 | 261 |
| $HC(OC_3H_7)_2OC_2H_5$ | 184°/745 | 1.4031 (20°) | 0.8973 (23°) | 2 | 281 |
| $C_2H_5C(OC_2H_5)_3$ | 160°/760 | 1.3987 | 0.8826 | 1, 2 | 7, 222, 286, 326, 373 |
| | 60°/17 | | | 1 | 168 |
| $CH_3SCH_2C(OC_2H_5)_3$ | 78°–80°/30 | | | 1 | 168 |
| $CH_3OCH_2C(OC_2H_5)_3$ | 173°–175°/760 | | | — | 337 |
| $C(OC_2H_5)_4$ | 158°–161°/760 | 1.3905 | 0.9197 (18°) | 2, 5 | 6a, 8a, 28, 52, 64, 232a, 233, 295, 341, 344 |
| | 71°–73°/11 | | | | |
| $HC(OC_2H_5)_2OOC(CH_3)_3$ | 71°–73°/11 | 1.4070 (20°) | 0.918 | 3 | 287, 289 |
| $C_2H_5OCH_2CH_2OCH(OC_2H_5)_2$ | 112°–117°/35 | 1.4060 | 0.9254 | 11 | 72b |
| $(CH_3O)_3CCH_2C(OCH_3)_3$ | 71°/1 | | | 5 | 140a |
| | | $C_9H_{21}$ | | | |
| $(CH_3)_2NC(OC_2H_5)_3$ | 76.5°/23 | 1.4084 (20°) | — | 6 | 49, 91, 359 |
| | | $C_{10}H_5$ | | | |
| 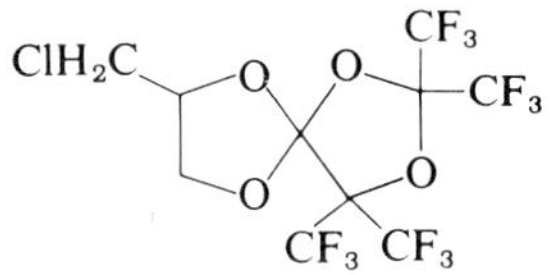 ![structure]  (CF3 dioxolane) | 94°/12 | 1.3523 | — | 11 | 47 |
| ClH$_2$C structure | | | | | |
| | | $C_{10}H_6$ | | | |
| 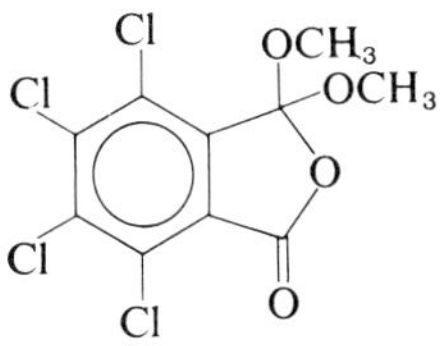 (tetrachlorophthalide dimethyl ketal) | (161.5°–162.5°) | — | — | 2 | 350 |
| | | $C_{10}H_{12}$ | | | |
| 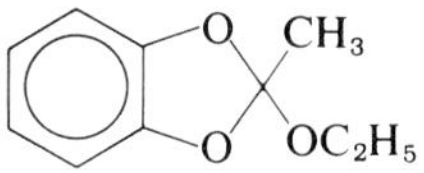 (benzodioxole derivative) | 85°–87°/11 | 1.4925 (22°) | — | 3 | 118a |

TABLE III—continued

| Structure | B.p., °C/mm (M.p., °C) | Refractive index $n^a$ | Density $d^b$ | Method$^c$ | References |
|---|---|---|---|---|---|
| 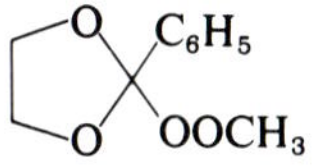 [C₆H₅ / OOCH₃ dioxolane structure] | 86°–87°/0.01 | 1.5165 (20°) | — | 11 | 288 |
| **C₁₀H₁₃** | | | | | |
| p-ClC₆H₄C(OCH₃)₃ | 83°–85°/2 | — | — | 2 | 189 |
| p-FC₆H₄C(OCH₃)₃ | 106°–107°/20 | — | — | 2 | 283 |
| p-O₂NC₆H₄C(OCH₃)₃ | (88°–90°) | — | — | 1 | 189 |
| **C₁₀H₁₄** | | | | | |
| C₆H₅C(OCH₃)₃ | 114°–115°/25 | 1.4858 | 1.0637 | 2 | 189, 201, 229, 351 |
| 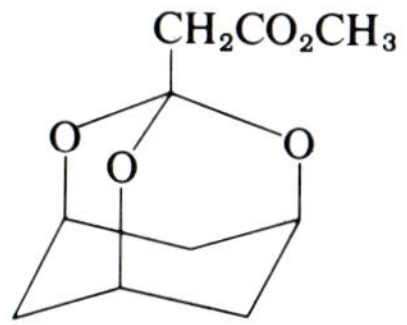 [CH₂CO₂CH₃ bicyclic ortho ester structure] | (96°) | — | — | 4 | 44 |
| **C₁₀H₁₅** | | | | | |
| CFCl₂CF=CFC(OC₂H₅)₃ | 67.5°/2 | 1.4230 (20°) | 1.2607 (20°) | 2 | 176 |
| **C₁₀H₁₆** | | | | | |
| HC(OCH₂CH=CH₂)₃ | 195°–206°/760 | — | — | 2 | 29 |
| **C₁₀H₁₇** | | | | | |
| CH₃C(OCH=CH₂)₂OCH₂CH(NO₂)C₂H₅ | 90°–93°/2 | 1.4480 | — | 5 | 100 |
| CH₃C(OCH=CH₂)₂OCH(CH₃)CH(NO₂)CH₃ | 85°–86°/2 | 1.4512 | — | 5 | 100 |

# PHYSICAL PROPERTIES OF ORTHO ESTERS 79

| Structure | Formula | bp/mm | $n_D$ | $d$ | Ref. | Ref. |
|---|---|---|---|---|---|---|
| (CH₃, OC₂H₅ on cyclohexane ketal) | C₁₀H₁₈ | 92°–93°/10 | 1.4489 | 1.0273 | 5 | 297, 364 |
| (CH₃, OC₂H₅ on cyclohexane ketal, stereoisomer) | | 95°–96°/10 | 1.4498 | 1.0244 | 3 | 364 |
| (C₂H₅ bicyclic orthoester, C₃H₇) | | 103°/5 (24°–25°) | 1.4517 (20°) | 1.0419 (20°) | 8 | 54 |
| CH₃C≡CC(OC₂H₅)₃ | | 74°–75°/9.5 | 1.4533 (20°) | 0.9780 (20°) | 11 | 16 |
| (cyclohexane spiro dioxolane, OCH₃) | | 100°–101°/17 | 1.4606 | — | 1 | 225 |
| (CH₃)₂ dihydrofuran, C₂H₅, (CH₃O)₂ | | 90°/40 | — | — | 11 | 315a |
| H₃C, (C₂H₅O)₂, CH₃ dihydrofuran | | 78°/18 | — | — | 11 | 315a |
| (CH₃)₂ spiro dioxane, (CH₃)₂ | | 63°/0.6 | 1.4501 (25°) | — | 11 | 153a |

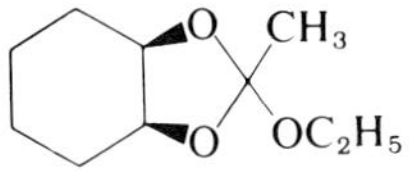

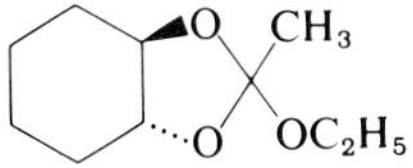

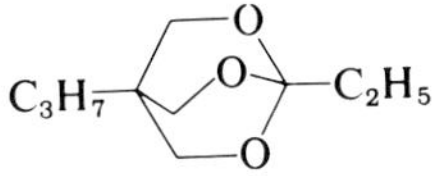

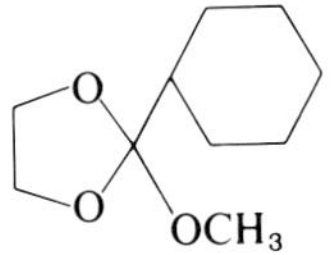

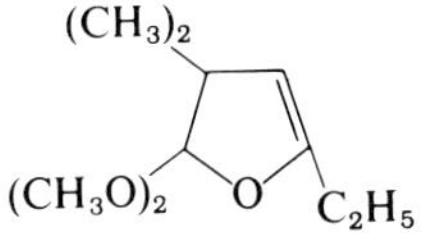

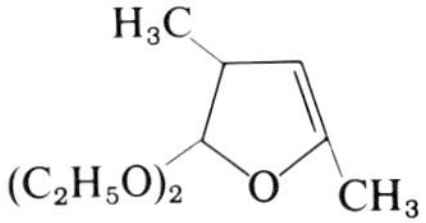

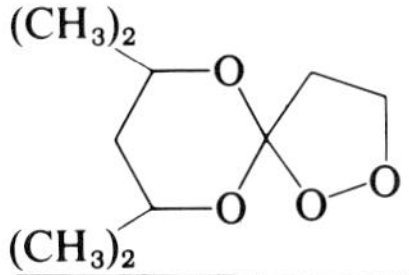

## TABLE III—continued

| Structure | B.p., °C/mm (M.p., °C) | Refractive index $n^a$ | Density $d^b$ | Method$^c$ | References |
|---|---|---|---|---|---|
| 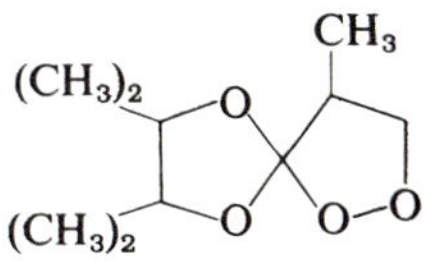 [structure 1] | 66°/0.3 | 1.4486 (25°) | — | 11 | *153a* |
| 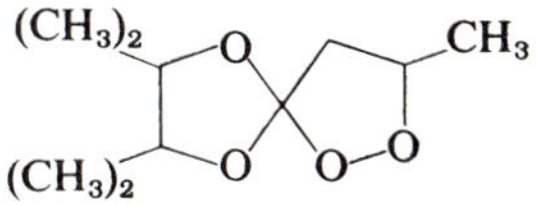 [structure 2] | 56°/0.2 | 1.4432 (25°) | — | 11 | *153a* |
| 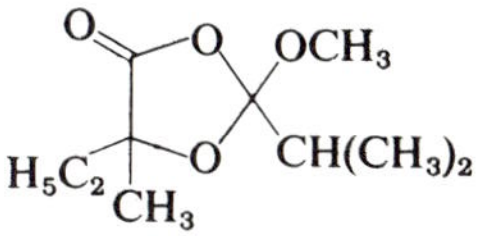 [structure 3] | 56°/0.04 | 1.4290 | — | 8 | *235* |
| 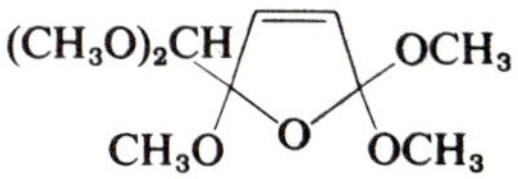 [structure 4] | 92°/2 | 1.4510 | 1.140 | 11 | *332* |
| $Cl_3CCH(OH)CH_2C(OC_2H_5)_3$ $C_{10}H_{19}$ | (91°–92°) | — | — | 4 | *183* |
| $CH_3C(OC_2H_5)[ON=CHC_2H_5]_2$ $C_{10}H_{20}$ | 90°/4 | — | — | 3 | *345a* |
| $CH_3C(OC_2H_5)[ON=C(CH_3)_2]_2$ | 78°/4 | 1.4470 | — | 5 | *345a* |

# PHYSICAL PROPERTIES OF ORTHO ESTERS

| Compound | bp (°C/mm) | $n_D$ | $d$ | | | References |
|---|---|---|---|---|---|---|
| 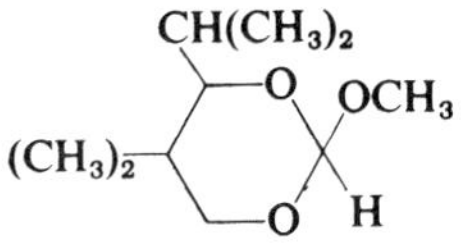 CH(CH$_3$)$_2$ — OCH$_3$ — (CH$_3$)$_2$ (cyclic with H) | 100°/17 | — | — | | 3 | 47 |
| 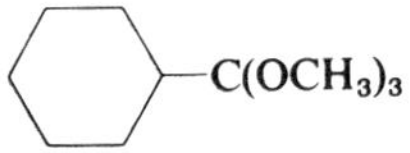 C(OCH$_3$)$_3$ (cyclohexyl) | 87°–88°/13 | 1.4436 | — | | 1 | 225 |
| 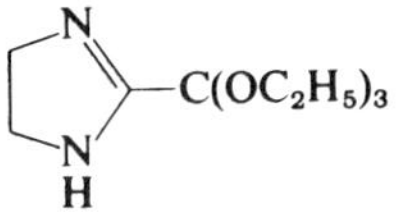 C(OC$_2$H$_5$)$_3$ (imidazoline) | (64°–65°) | — | — | | 2 | 18 |
| C$_2$H$_5$O$_2$CC(OC$_2$H$_5$)$_3$ | 84°–86°/6 | — | — | | 2 | 38, 161 |
| C$_4$H$_9$CO$_2$CH(OC$_2$H$_5$)$_2$ | 68°/2 | 1.4090 | 0.9488 | | 3 | 315d |
| (CH$_3$)$_3$CCO$_2$CH(OC$_2$H$_5$)$_2$ | 89°/18 | 1.4300 | — | | 3 | 315d |
| CH$_2$C(=NH)C(OC$_2$H$_5$)$_2$OCH(CH$_3$)$_2$ | — | — | — | $C_{10}H_{21}$ | 11 | 127a |
| HC(OC$_3$H$_7$)$_3$ | 196°–198°/760 91°–92°/15 | 1.4058 | 0.8783 | $C_{10}H_{22}$ | 1, 2, 3 | 2, 73, 87a, 197, 252, 265a, 273, 276, 281, 298, 312 |
| HC[OCH(CH$_3$)$_2$]$_3$ | 166°–168°/760 65°–66°/18 | 1.3940 | 0.8600 | | 2, 3 | 62, 87a, 135, 138, 265a, 273, 298, 312 |
| CH$_3$C(OC$_2$H$_5$)$_2$OC$_4$H$_9$ | 70°–72°/15 | 1.4057 | 0.8682 | | 5 | 204 |
| CH$_3$C(OC$_2$H$_5$)$_2$OCH$_2$CH(CH$_3$)$_2$ | 64°–66°/14 | 1.4017 | 0.8616 | | 5 | 204 |
| CH$_3$C(OC$_2$H$_5$)$_2$OCH(CH$_3$)C$_2$H$_5$ | 63°–65°/15 | 1.4016 | 0.8648 | | 5 | 205 |
| CH$_3$C(OC$_3$H$_7$)$_2$OC$_2$H$_5$ | 190°–194°/760 | 1.4064 | 0.8713 | | 1 | 309 |
| C$_3$H$_7$C(OC$_2$H$_5$)$_3$ | 58°–59°/7 | 1.4028 | 0.875 | | 1 | 222 |
| (CH$_3$)$_2$CHCH(OC$_2$H$_5$)$_3$ | 50°–51°/7 | 1.4002 | 0.871 | | 1 | 222 |
| C$_6$H$_{13}$C(OCH$_3$)$_3$ | 190°–192°/725 | — | — | | 1 | 106 |
| CH$_3$SCH$_2$CH$_2$C(OC$_2$H$_5$)$_3$ | 71°–72°/0.8 | — | — | | 1 | 60, 61 |

TABLE III—continued

| Structure | B.p., °C/mm (M.p., °C) | Refractive index $n^a$ | Density $d^b$ | Method[c] | References |
|---|---|---|---|---|---|
| $C_2H_5OCH_2C(OC_2H_5)_3$ | 69°–70°/10 | 1.4055 | 0.921 | 1 | 230 |
| $CH_3C(OC_2H_5)_2OOC(CH_3)_3$ | 47°/4 | 1.4073 (20°) | — | 3, 5 | 287, 289, 290, 319 |
| $(CH_3O)_3CCH_2CH_2C(OCH_3)_3$ | 104°–108°/5 | 1.4261 | — | 1 | 223 |
| $C_{11}H_{10}$ structure | 114°–118°/0.3 (43°) | — | — | 7 | 40 |
| $C_{11}H_{11}$ structure | 120°–122°/0.1 (53°–71°) | — | — | 7 | 40 |
| $C_{11}H_{12}$ structure | (88°–89°) | — | — | 11 | 82 |
| $C_{11}H_{14}$ structure | (202°) | — | — | 2 | 93b, 143a |

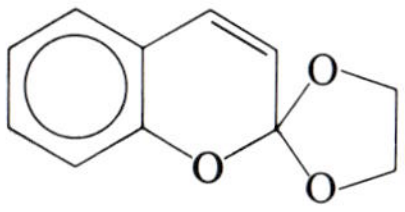

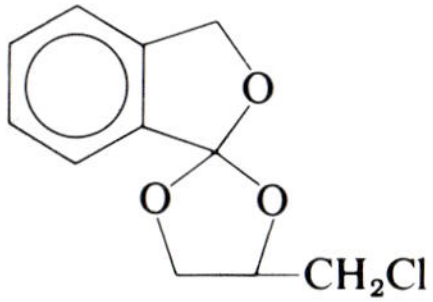

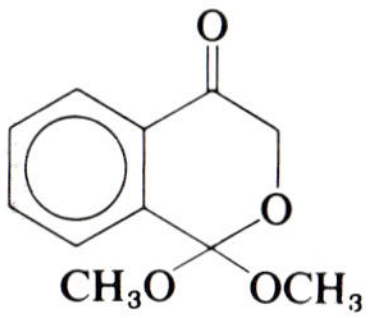

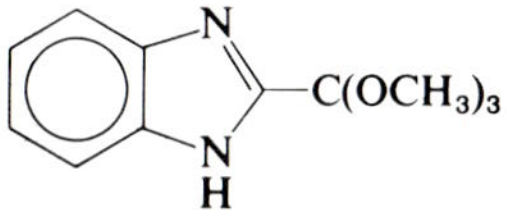

# PHYSICAL PROPERTIES OF ORTHO ESTERS 83

| Structure | bp/mm (mp) | $n_D$ | $d$ | | Ref. |
|---|---|---|---|---|---|
| 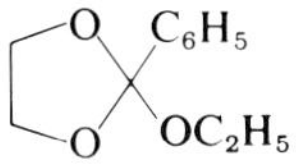 (cyclic with $C_6H_5$, $OC_2H_5$) | 121°–126°/12 | — | 1.101 (20°) | 6 | 241 |
| 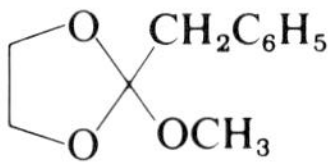 (cyclic with $CH_2C_6H_5$, $OCH_3$) | 75°–77°/0.8 | 1.5060 | 1.1140 | 5 | 209 |
| $CH_2CH=CHCHO$ | | | | | |
| 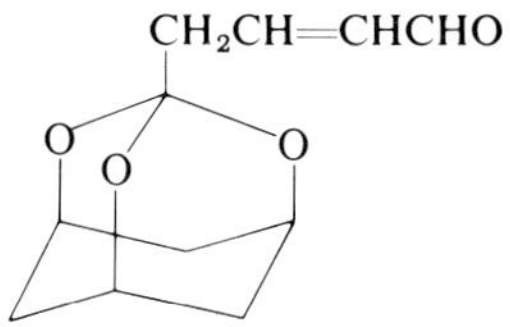 (adamantane-like structure) | (92°) | — | — | 4 | 44 |
| 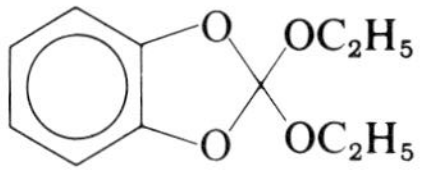 (benzodioxole with $OC_2H_5$, $OC_2H_5$) | 123°/15 | 1.4943 (20°) | — | 2, 3 | 119 |
| $C_6H_5CHBrC(OCH_3)_3$ | 105°–107°/1 $C_{11}H_{15}$ (66°–68°) | — | — | 4 | 226 |
| $C_6H_5SO_2CHClC(OCH_3)_3$ | — $C_{11}H_{16}$ | — | — | 2 | 43a |
| $HC(OC_2H_5)_2OC_6H_5$ | 111°/11 | 1.4799 (20°) | 1.0185 (20°) | 3 | 328 |
| 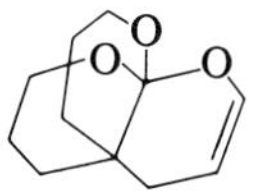 (bicyclic ether) | (82.5°–84°) | — | — | 10 | 216 |
| $p\text{-}CH_3C_6H_4C(OCH_3)_3$ | 102°–104°/10 | — | — | 1 | 189 |
| $C_6H_5CH_2C(OCH_3)_3$ | 72°–76°/0.5 | 1.4948 | 1.0644 | 1 | 226, 229, 313 |
| $p\text{-}CH_3OC_6H_4C(OCH_3)_3$ | 114°–115°/5 | — | — | 1 | 189 |
| $C_6H_5CO_2CH(OC_2H_5)_2$ | 99°/0.7 | — | 1.0658 | 1 | 315d |

TABLE III–continued

| Structure | B.p., °C/mm (M.p., °C) | Refractive index $n^a$ | Density $d^b$ | Method[c] | References |
|---|---|---|---|---|---|
| CH$_2$CO$_2$C$_2$H$_5$ (structure) | — | — | — | 3 | 44 |
| 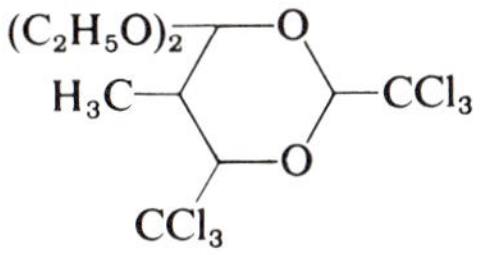 (C$_2$H$_5$O)$_2$ ... CCl$_3$ / H$_3$C ... CCl$_3$ (structure) | (81°–82°) | — | — | 4 | 184 |
| 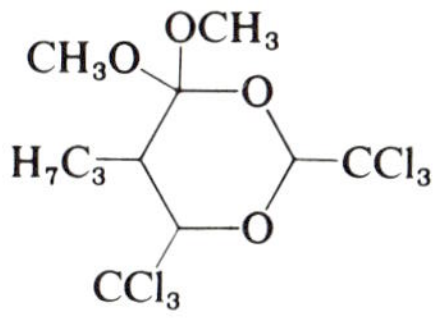 CH$_3$O, OCH$_3$ ... CCl$_3$ / H$_7$C$_3$ ... CCl$_3$ (structure) | 123°–125°/2 | 1.4950 (20°) | — | 4 | 184 |
| C$_{11}$H$_{18}$ (structure with CH$_3$ groups) | (59°–60°) | — | — | 3 | 139 |

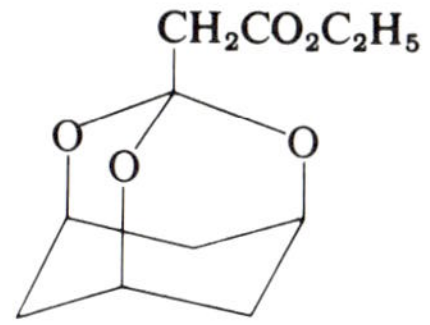

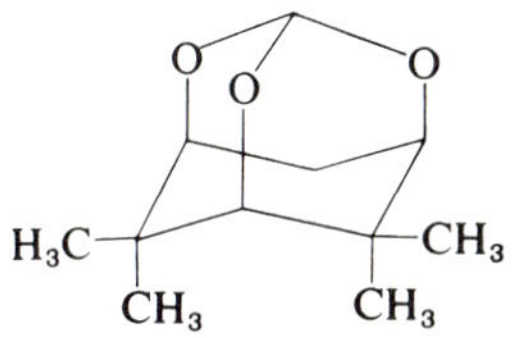

# PHYSICAL PROPERTIES OF ORTHO ESTERS

| Structure | bp/mp | $n_D$ | $d$ | | | Ref. |
|---|---|---|---|---|---|---|
| 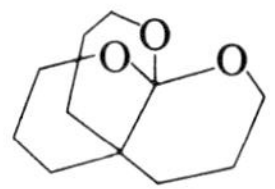 (bicyclic ortho ester) | (116°–117°) | — | — | — | 4 | 216 |
| 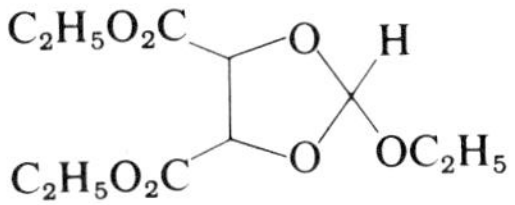 $C_2H_5O_2C$–(ring with H, $OC_2H_5$) | 150°–154°/20 | 1.4327 | 1.152 | — | 3 | 69 |
| **$C_{11}H_{20}$** | | | | | | |
| 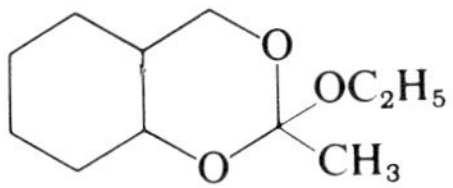 $C_2H_5O_2C$–(ring with $OC_2H_5$, $CH_3$) | 105°–106°/14 | 1.4534 | — | — | 3, 8 | 79, 180, 180a |
| 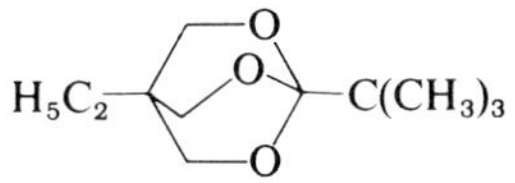 $H_5C_2$–(adamantane-like with $C(CH_3)_3$) | (94°–95°) | — | — | — | 8 | 54 |
| $HC{\equiv}CC(OC_2H_5)_2OC(CH_3)_3$ | 179°–180°/760 | 1.4170 | 0.9100 | — | 5 | 231 |
| 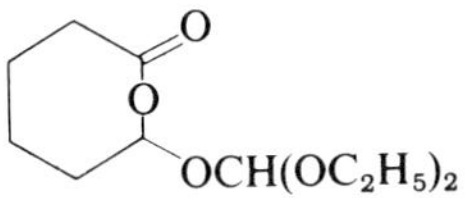 (cyclohexanone–$OCH(OC_2H_5)_2$) | — | 1.4319 (22.5°) | 1.047 (22.5°) | — | 11 | 77 |
| 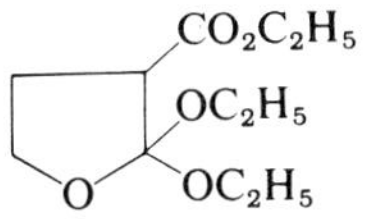 (furanone with $CO_2C_2H_5$, $OC_2H_5$, $OC_2H_5$) | 48°–52°/0.01 | — | — | — | 1 | 179 |
| **$C_{11}H_{22}$** | | | | | | |
| 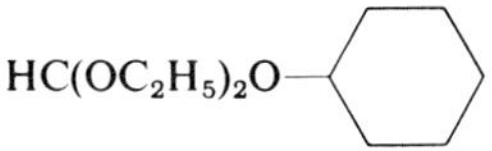 $HC(OC_2H_5)_2O$–cyclohexyl | 64°/1.1 | 1.4328 | 0.945 | — | 3 | 298 |
| $CH_2{=}C(CH_3)CH_2C(OC_2H_5)_3$ | 66°–69°/7 | 1.4520 | — | — | 5 | 220 |
| $C_2H_5O_2CCH_2C(OC_2H_5)_3$ | 120°–121°/18 | 1.4220 | 0.995 | — | 1 | 223 |

TABLE III—continued

| Structure | B.p., °C/mm (M.p., °C) | Refractive index $n^a$ | Density $d^b$ | Method[c] | References |
|---|---|---|---|---|---|
| | | $C_{11}H_{23}$ | | | |
| $C_3H_7CHBrC(OC_2H_5)_3$ | 69°–70°/2 | 1.4390 | — | 4 | 207 |
| $(CH_3)_2CHCHBrC(OC_2H_5)_3$ | 63°–64°/1.3 | 1.4408 | 1.150 | 4 | 207 |
| $NH=C(OC_2H_5)CH_2C(OC_2H_5)_3$ | 94°–95°/3 | 1.4272 | 0.979 | 1 | 223 |
| | | $C_{11}H_{24}$ | | | |
| $HC[OC(CH_3)_3]_2OC_2H_5$ | 33°–35°/10–12 | — | — | 3 | 261 |
| $CH_3C(OC_2H_5)_2OCH_2CH_2CH(CH_3)_2$ | 80°–82°/15 | 1.4077 | 0.8626 | 5 | 204 |
| $CH_3C(OC_2H_5)_2OCH_2C(CH_3)_3$ | 87°–88°/28 | 1.4037 | 0.8481 | 5 | 204 |
| $CH_3C[OCH_2CH(CH_3)_2]_2OCH_3$ | 205°–206°/760 | — | — | 1 | 311 |
| $C_4H_9C(OC_2H_5)_3$ | 49°–50°/3 | 1.4086 | 0.873 | 1 | 222 |
| $(CH_3)_2CHCH_2C(OC_2H_5)_3$ | 57°–59°/7 | 1.4056 | 0.869 | 1 | 222 |
| $(CH_3O)_2C(OC_4H_9)_2$ | 101°/20 | 1.4100 (20°) | 0.9234 (20°) | 3 | 329 |
| | | $C_{11}H_{25}$ | | | |
| $(C_2H_5)_2NC(OC_2H_5)_3$ | 48°/12 | 1.4212 (20°) | — | 6 | 49 |
| | | $C_{12}H_{10}$ | | | |
| 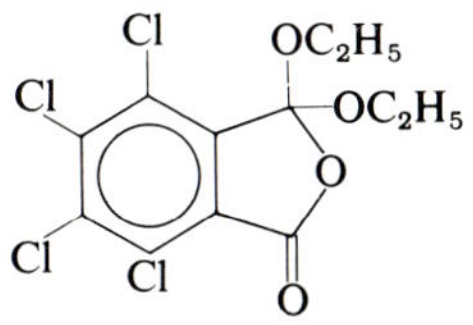 [structure: tetrachloro phthalide with $OC_2H_5$ and $OC_2H_5$ groups] | (126°) | — | — | 2 | 171 |
| 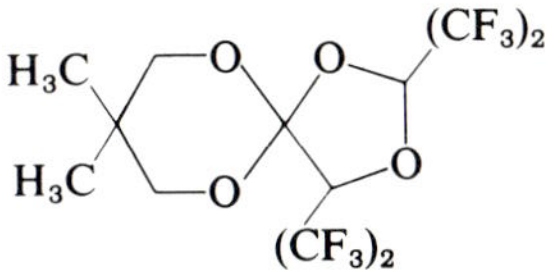 [structure: spiro dioxolane/dioxane with $(CF_3)_2$ and $H_3C$ groups] | 43°/0.18 | 1.3391 | — | 11 | 47 |

# PHYSICAL PROPERTIES OF ORTHO ESTERS

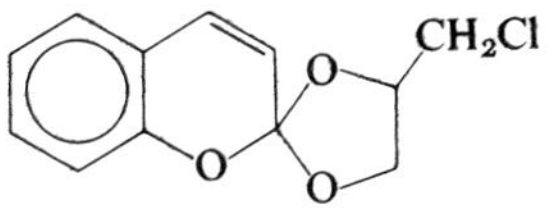

| Structure | Formula | bp/mp | $n_D$ | | Refs |
|---|---|---|---|---|---|
| (2-chloromethyl-benzodioxole spiro) | $C_{12}H_{11}$ | 134°–136°/0.4 | 1.5731 (20°) | — | 7  40 |
| (2-methyl-benzodioxole spiro) | $C_{12}H_{12}$ | 105°–107°/? | 1.5746 (20°) | — | 7  41 |
| 3,5-dinitrophenyl OC bicyclic (H₃C) | | (224°–225°) | — | — | 8  25 |
| benzodioxole with CO₂C₂H₅, OC₂H₅ | $C_{12}H_{14}$ | 144°/10 | 1.4982 (21°) | — | 2  118a |
| p-ClC₆H₄OCH₂ dioxolane OC₂H₅, H | $C_{12}H_{15}$ | 120°/4 | — | — | 3  272 |
| CH₂CH₂CH₂C≡CH adamantane-like ortho ester | $C_{12}H_{16}$ | (53°–54°) | — | — | 3  266 |

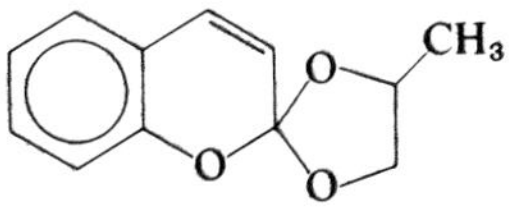

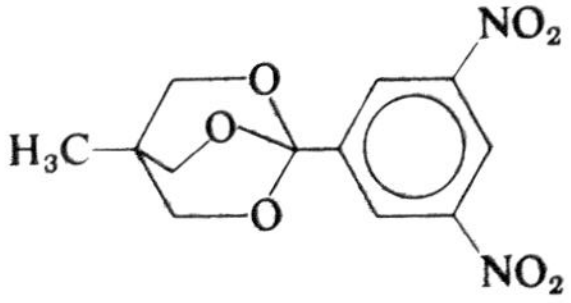

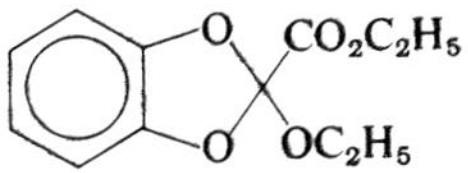

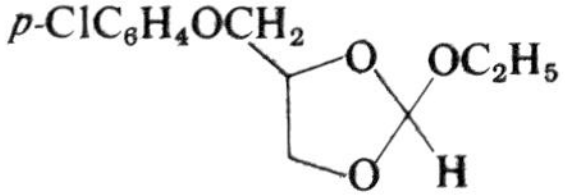

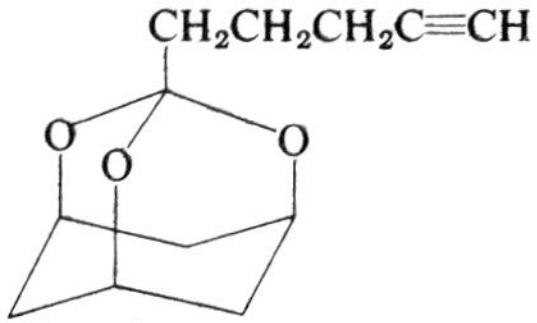

TABLE III—continued

| Structure | B.p., °C/mm (M.p., °C) | Refractive index $n^a$ | Density $d^b$ | Method[c] | References |
|---|---|---|---|---|---|
| 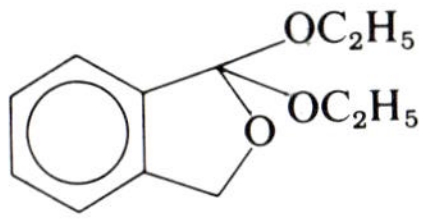 ![structure with OC₂H₅, OC₂H₅] | 122°–124°/13 | 1.4995 (20°) | 1.07 (20°) | 6 | 242 |
| 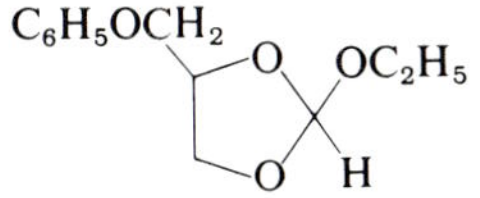 $C_6H_5OCH_2$—[dioxolane with OC₂H₅] | 173°–174°/12 | — | — | 3 | 47 |
| 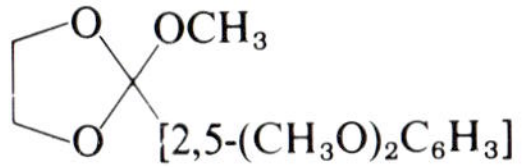 [dioxolane with OCH₃] | (45°) | — | — | 6 | 31a |
| [2,5-$(CH_3O)_2C_6H_3$] | | | | | |
| | | $C_{12}H_{17}$ | | | |
| $BrCH_2C(OC_2H_5)_2OC_6H_5$ | 84°–86°/2 | 1.5048 (20°) | 1.3192 (20°) | 1 | 213 |
| $ClCH_2C(OC_2H_5)_2OC_6H_5$ | 78°/10 | 1.4988 (20°) | 1.1498 (20°) | 1 | 213 |
| $C_6H_5CH_2SO_2CHClC(OCH_3)_3$ | (91°) | — | — | 2 | 43a |
| | | $C_{12}H_{18}$ | | | |
| $CH_3C(OC_2H_5)_2OC_6H_5$ | 103°–104°/14 | 1.4783 | 0.9904 | 1, 3, 5 | 213, 215, 328 |
| $C_6H_5CH_2C(OCH_3)_2OC_2H_5$ | 217°–219°/760 | 1.5080 (20°) | 1.0640 (20°) | 1 | 313 |
| $C_6H_5CH(CH_3)C(OCH_3)_3$ | 70°–71°/0.5 | 1.4928 | 1.0334 | 1 | 227 |
| 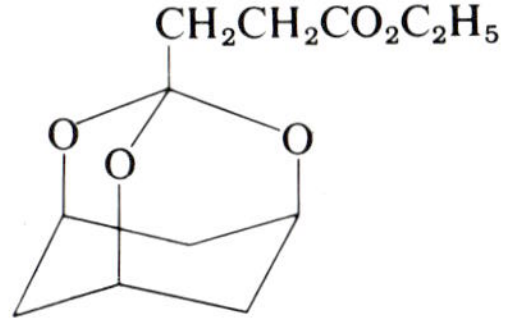 [adamantane-like cage with $CH_2CH_2CO_2C_2H_5$] | (50°) | — | — | 3 | 335 |

# PHYSICAL PROPERTIES OF ORTHO ESTERS

| Compound | bp/mmHg (mp) | $n_D$ | $d$ | | Ref. |
|---|---|---|---|---|---|
| | | $C_{12}H_{21}$ | | | |
| $CHCl_2C(OC_2H_5)=CClC(OC_2H_5)_3$ | 150°–151°/11 | 1.4685 (20°) | — | 2 | 299 |
| | | $C_{12}H_{22}$ | | | |
| 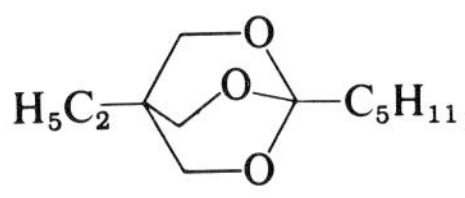 (structure with $H_5C_2$, $C_5H_{11}$ bicyclic) | (41.5°–42°) | — | — | 8 | 54 |
| $HC≡C(CH_2)_3C(OC_2H_5)_3$ | 104°–106°/14 | 1.4309 (22°) | — | 1 | 266 |
| 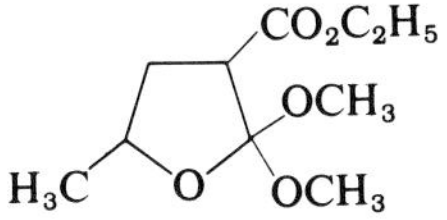 (structure with $CO_2C_2H_5$, $OCH_3$, $OCH_3$, $H_3C$) | 57°–58°/0.01 | — | — | 1 | 179 |
| $(CH_3O)_3CCH=CHCH=CHC(OCH_3)_3$ | (100°) | — | — | 11 | 30 |
| 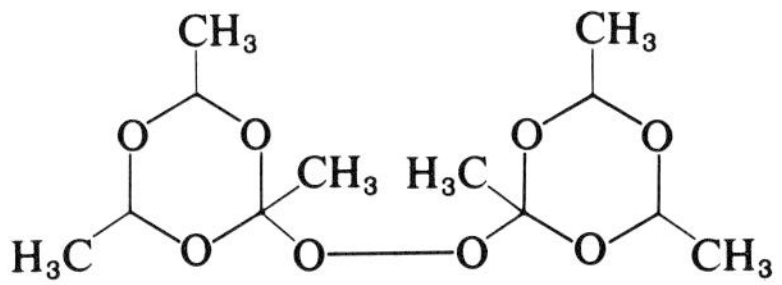 (bicyclic structure with $CH_3$ groups) | (127°) | — | — | 11 | 323b |
| $C_2H_5O_2CCH_2CH_2C(OC_2H_5)_3$ | 127°–128°/18 | 1.414 (23°) | — | 1 | 335 |
| $(CH_3O)_3CCH_2CH=CHCH_2C(OCH_3)_3$ | — | — | — | 1 | 44 |
| | | $C_{12}H_{25}$ | | | |
| $C_7H_{15}CHBrC(OCH_3)_3$ | 110°–112°/0.7 | 1.4532 (24°) | 0.883 (27°) | 4 | 214 |
| 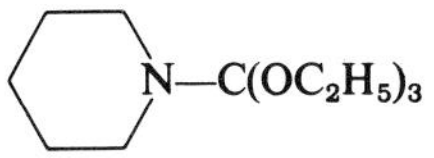 (piperidine)$N—C(OC_2H_5)_3$ | 100°–105°/13 | — | — | 6 | 91, 359 |
| | | $C_{12}H_{26}$ | | | |
| $HC(OC_3H_7)_2OCH_2CH_2CH(CH_3)_2$ | 124°–130°/24 | 1.415 (20°) | 0.8647 (23°) | 2 | 281 |
| $CH_3C(OC_4H_9)_2OC_2H_5$ | 98°–100°/13 | 1.4119 | 0.8623 | 5 | 204, 309 |

TABLE III—continued

| Structure | B.p., °C/mm (M.p., °C) | Refractive index $n^a$ | Density $d^b$ | Method$^c$ | References |
|---|---|---|---|---|---|
| CH$_3$C[OCH$_2$CH(CH$_3$)$_2$]$_2$OC$_2$H$_5$ | 208°–210°/760 | — | — | 1 | 311 |
| C$_2$H$_5$C(OC$_3$H$_7$)$_3$ | 93°/14 | 1.4100 | 0.8831 | 1, 3 | 7, 9, 373 |
| C$_8$H$_{17}$C(OCH$_3$)$_3$ | 138°–140°/45 | 1.4255 (24°) | 0.898 (27°) | 1 | 214 |
| (C$_2$H$_5$O)$_2$CHC(OC$_2$H$_5$)$_3$ | 85°–86°/8 | 1.4072 | — | 1, 11 | 206, 316 |
| (C$_2$H$_5$O)$_2$C(CH$_3$)OOC(CH$_3$)(OC$_2$H$_5$)$_2$ | (−6°) | — | — | 5 | 319 |
| [CH$_2$CH$_2$C(OCH$_3$)$_3$]$_2$ | 115°/2 | — | — | 1 | 112a |
| C$_{13}$H$_8$ structure | (109°–110°) | — | — | 2 | 118 |
| C$_{13}$H$_{10}$ structure | 67°/4.3 | 1.3532 | — | 11 | 47 |
| C$_{13}$H$_{12}$ structure | 65°/0.1 | — | — | 2 | 133 |
| C$_{13}$H$_{14}$ structure | (176°–177°) | — | — | 3 | 334 |

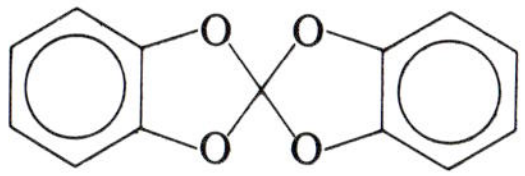

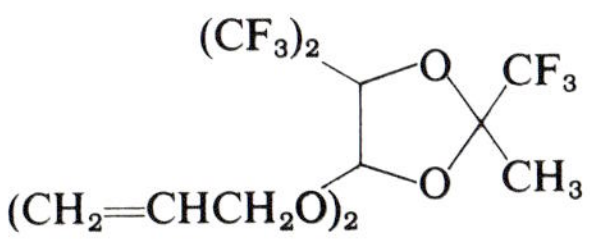

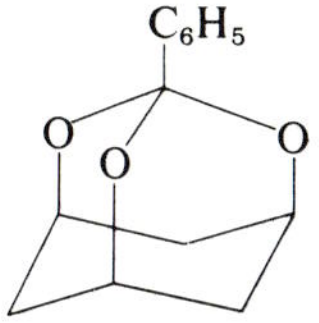

PHYSICAL PROPERTIES OF ORTHO ESTERS 91

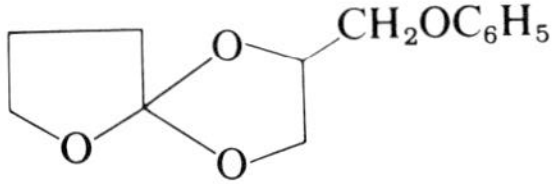

| Structure | Formula | bp/mm (mp) | $n_D$ | $d$ | Yield (%) | Ref. |
|---|---|---|---|---|---|---|
| (structure with CH₂OC₆H₅) | C₁₃H₁₆ | 135°–141°/0.3 (61°–74°) | — | — | 7 | 40, 41 |
| (structure with OCH₃, OCH₃, H₅C₆) | | 150°–154°/11 | 1.5277 | 1.083 | 10 | 211 |
| CH(CH=CH)₂CHO (bicyclic structure) | | (101°) | — | — | 4 | 44 |
| C(=NH)C₆H₅, OCH(CH₃)₂ | C₁₃H₁₇ | — | — | — | 11 | 127a |
| (CH₂)₄C≡CH (bicyclic structure) | C₁₃H₁₈ | (47°–49°) | — | — | 3 | 266 |
| o-CH₃C₆H₄OCH₂, H, OC₂H₅ | | 102°/0.5 175°–176°/15 | — | — | 3 | 272, 343 |

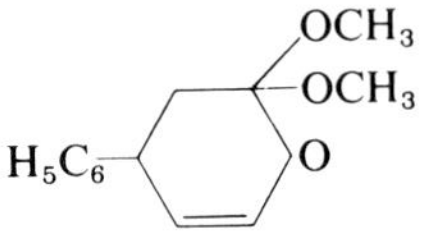

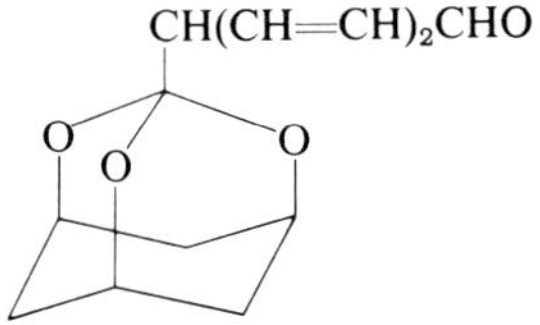

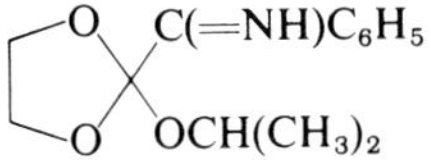

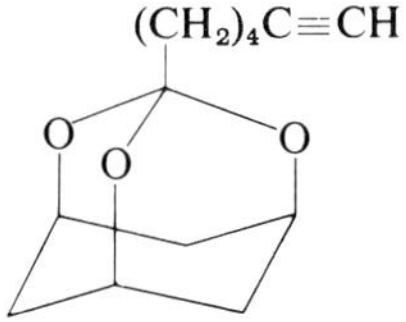

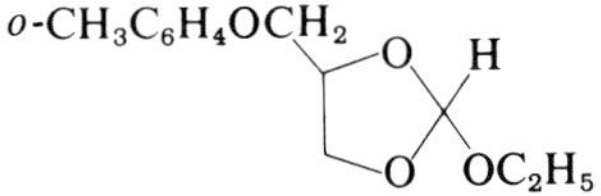

TABLE III—continued

| Structure | B.p., °C/mm (M.p., °C) | Refractive index $n^a$ | Density $d^b$ | Method[c] | References |
|---|---|---|---|---|---|
| 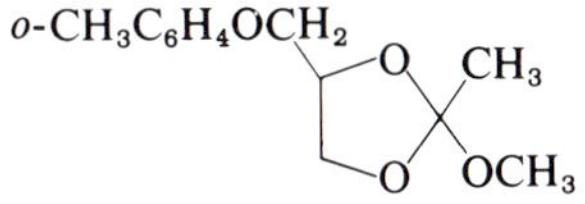 o-CH₃C₆H₄OCH₂—[dioxolane with CH₃, OCH₃] | 144°/8 | — | — | 3 | 343 |
| 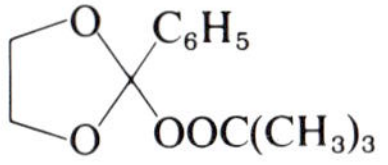 [dioxolane with C₆H₅, OOC(CH₃)₃] | 60°–70°/0.001 | 1.501 (20°) | — | 3 | 287, 289 |
| 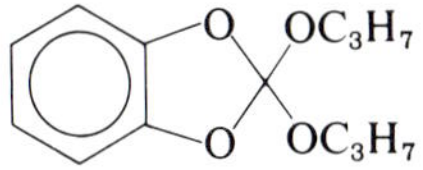 [benzodioxole with OC₃H₇, OC₃H₇] | 147°/12 | 1.4855 (20°) | — | 2 | 119 |
| CH₂CH(OH)C≡COC₂H₅ [adamantane-like structure] | — | — | — | 4 | 44 |
| CH₃C(OC₂H₅)₂OCH₂C₆H₅ | $C_{13}H_{20}$ 121°–122°/8 128°–129°/10 | 1.4778 | 0.9839 | 5 | 204 |
| C₆H₅C(OC₂H₅)₃ | 240°/760 | 1.4930 | 1.0325 | 1, 2 | 168, 198, 204, 351 |

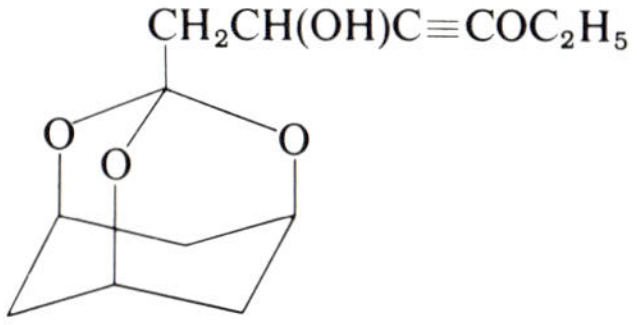

# PHYSICAL PROPERTIES OF ORTHO ESTERS

| Compound | bp/mp | $n_D$ | density | | Ref. |
|---|---|---|---|---|---|
| $C_6H_5CH_2C(OC_2H_5)_2OCH_3$ | 224°–226°/760 | 1.5000 (20°) | 1.0356 (20°) | 1 | *313* |
| $C_6H_5C(OCH_3)_2OOC(CH_3)_3$ | 55°–59°/0.09 (25.5°–26.5°) | — | — | 3 | *193* |
| 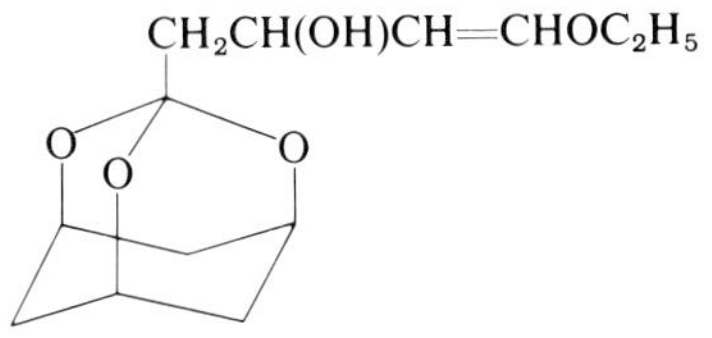 $CH_2CH(OH)CH=CHOC_2H_5$ (structure) | (81°) | — | — | 4 | *44* |
| 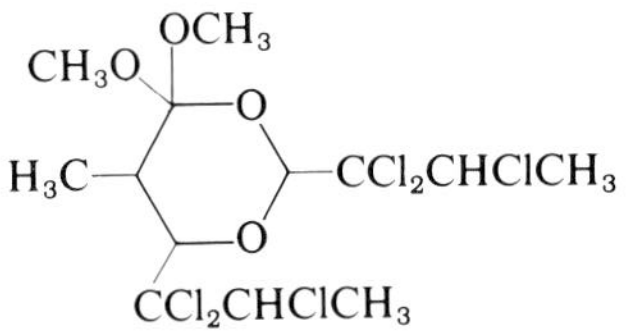 (structure with $OCH_3$, $CH_3O$, $H_3C$, $CCl_2CHClCH_3$, $CCl_2CHClCH_3$) | (112°–113°) | — | — | 4 | *184* |
| (structure) | 138°/11 | $C_{13}H_{22}$  1.4543 (20°) | — | 3 | *165b* |
| $HC(OCH_2CH=CHCH_3)_3$ | 157°–158°/18 | — | — | 2 | *190* |
| 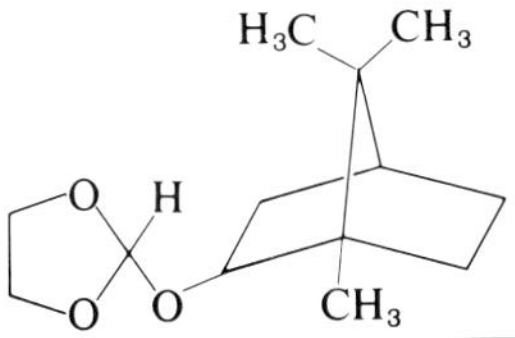 cyclopentene with $OCH(OC_2H_5)_2$ and $CO_2C_2H_5$ | 148°–152°/16 | 1.4720 (20°) | 1.0496 (20°) | 3 | *250* |
| bicyclic structure with $H_3C$, $CH_3$, $CH_3$, H, O, O | | | | | |

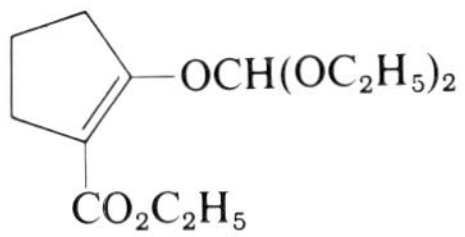

## TABLE III—continued

| Structure | B.p., °C/mm (M.p., °C) | Refractive index $n^a$ | Density $d^b$ | Method[c] | References |
|---|---|---|---|---|---|
| | | $C_{13}H_{24}$ | | | |
| 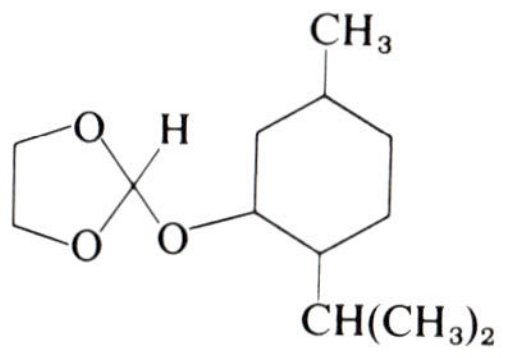 [structure with CH₃, H, O, O, CH(CH₃)₂] | (34.2°) | — | — | 3 | 250 |
| 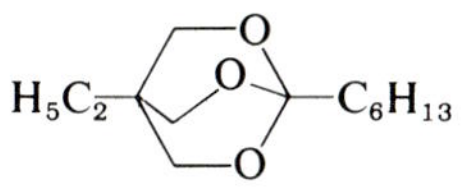 [H₅C₂ bicyclic structure with C₆H₁₃] | (29°–30°) | — | — | 8 | 54 |
| HC≡C(CH₂)₄C(OC₂H₅)₃ | 103°–118°/14 | 1.4348 (24°) | — | 1 | 266 |
| 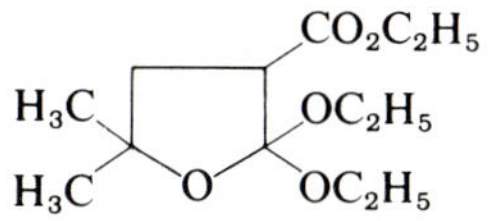 [H₃C, H₃C, CO₂C₂H₅, OC₂H₅, OC₂H₅ structure] | 53°–58°/0.01 | — | — | 1 | 179 |
| 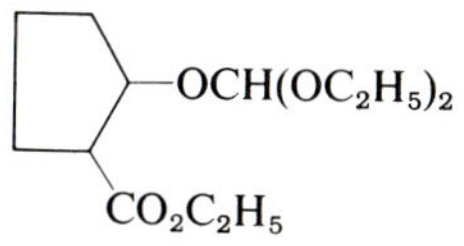 [cyclopentane with OCH(OC₂H₅)₂ and CO₂C₂H₅] | 157°–158°/14 | — | — | 4 | 190 |
| | | $C_{13}H_{28}$ | | | |
| HC(OC₄H₉)₃ | 245°–247°/760<br>132°–133°/21.5 | 1.4155 | 0.8726 | 2, 3 | 2, 7, 87a, 197, 265a, 273, 298, 312, 352 |

# PHYSICAL PROPERTIES OF ORTHO ESTERS

| Compound | bp/mm | Formula | $n_D$ | $d$ | # | Ref |
|---|---|---|---|---|---|---|
| HC[OCH$_2$CH(CH$_3$)$_2$]$_3$ | 224°–226°/760 | | 1.4100 | 0.8550 | 2, 3 | 2, 73, 312 |
| | 118°–120°/22 | | | | | |
| HC[OCH(CH$_3$)C$_2$H$_5$]$_3$ | 213°–215°/760 | | 1.4102 | 0.8638 | 2, 3 | 2, 273, 273a, 274 |
| | 115°/23 | | | | | |
| HC(OC$_5$H$_{11}$)$_2$OC$_2$H$_5$ | 255°/760 | | — | — | 1 | 275 |
| CH$_3$C[OCH$_2$CH$_2$CH(CH$_3$)$_2$]$_2$OCH$_3$ | 219°–223°/760 | | — | — | 1 | 311 |
| C(OC$_3$H$_7$)$_4$ | 224°/760 | | 1.4100 (20°) | 0.897 (20°) | 2 | 8a, 303, 344 |
| (C$_2$H$_5$O)$_2$CHCH$_2$C(OC$_2$H$_5$)$_3$ | 103°–106°/6 | | 1.4140 | 0.944 | 5 | 219 |
| ![structure: C$_6$H$_5$/OCH$_3$ benzodioxole] | 167°–168°/12 | C$_{14}$H$_{12}$ | 1.5654 (21°) | — | 3 | 75, 118a |
| | 112°–113°/0.5 | | | | | |
| HC(OC$_6$H$_5$)$_2$OCH$_3$ | 175°/12 | C$_{14}$H$_{14}$ | 1.5517 (20°) | — | 2 | 117, 318 |
| [CH$_2$C$_6$H$_5$ orthoester cage structure] | (133°) | C$_{14}$H$_{16}$ | — | — | 3 | 44 |
| (CH$_3$)$_2$ / (CH$_3$O)$_2$ — C$_6$H$_4$NO$_2$-$p$ furanone | 150°/0.25 | C$_{14}$H$_{17}$ | — | — | 11 | 315a |
| 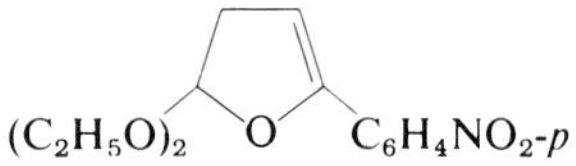 (C$_2$H$_5$O)$_2$ — C$_6$H$_4$NO$_2$-$p$ furanone | 158°/0.3 | | — | — | 11 | 315a |

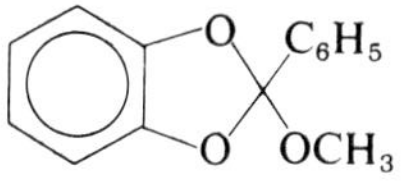

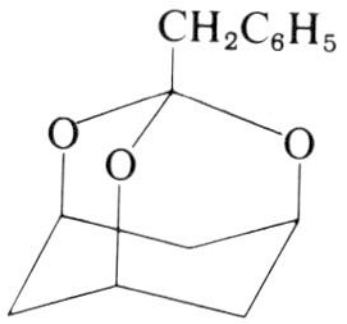

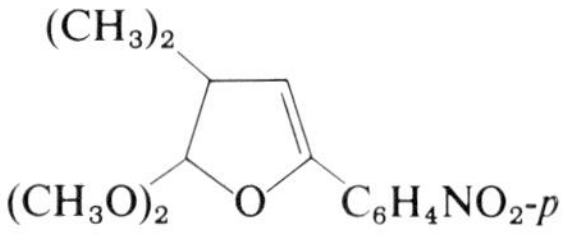

## TABLE III—continued

| Structure | B.p., °C/mm (M.p., °C) | Refractive index $n^a$ | Density $d^b$ | Method[c] | References |
|---|---|---|---|---|---|
| $C_{14}H_{18}$ | | | | | |
| 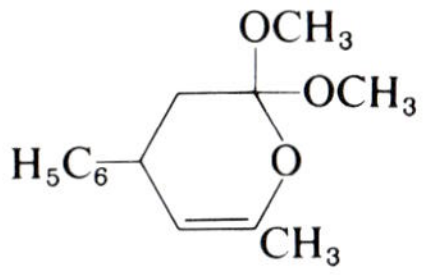 (structure with OCH₃, OCH₃, CH₃, H₅C₆) | 106°–108°/0.4 | 1.5167 | 1.062 | 10 | 211 |
| 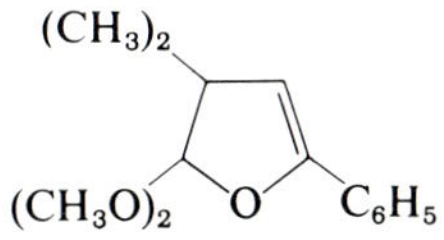 (CH₃)₂ ... C₆H₅ / (CH₃O)₂ | 100°/0.3 | — | — | 11 | 315a |
| 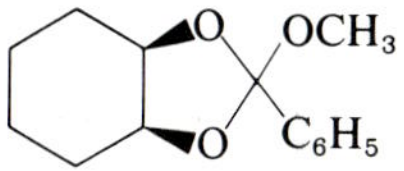 (cyclohexane-fused dioxolane with OCH₃, C₆H₅) | 104°/0.5 | — | — | 11 | 315a |
| 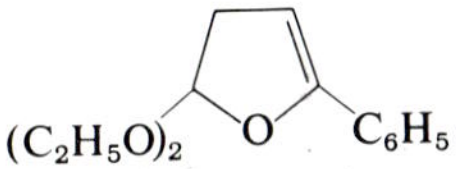 (C₂H₅O)₂ ... C₆H₅ | — | — | — | 6 | 320a |
| 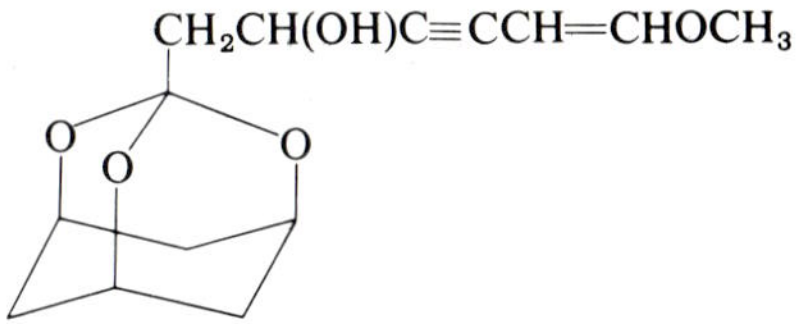 CH₂CH(OH)C≡CCH=CHOCH₃ (caged structure) | (118°) | — | — | 4 | 44 |

# PHYSICAL PROPERTIES OF ORTHO ESTERS

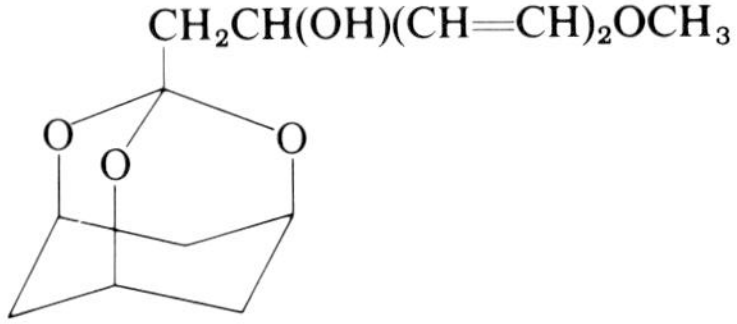

| Compound | | | | | |
|---|---|---|---|---|---|
| | $C_{14}H_{20}$ | | | | |
| 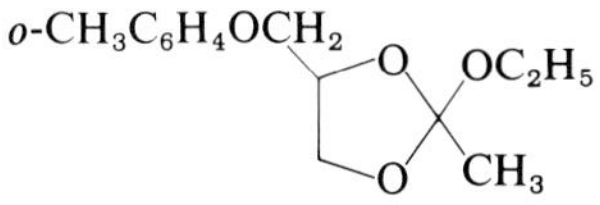 o-$CH_3C_6H_4OCH_2$ with dioxolane ($OC_2H_5$, $CH_3$) | 171°–172°/14 | — | — | 3 | 343 |
| $CH_2CH(OH)(CH=CH)_2OCH_3$ | — | — | — | 4 | 44 |
| 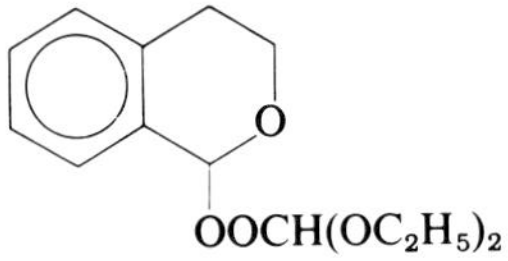 (adamantane-type structure) | (49°) | — | — | 5 | 319 |
| (isochroman-type) $OOCH(OC_2H_5)_2$ | | | | | |
| | $C_{14}H_{21}$ | | | | |
| p-$ClC_6H_4OCH_2C(OC_2H_5)_3$ | 170°/18 | — | — | 1 | 377 |
| | $C_{14}H_{22}$ | | | | |
| $C_6H_5CH_2C(OC_2H_5)_3$ | 225°–227°/760<br>88°–91°/0.1 | 1.5050 | 1.000 | 1 | 226, 313, 333 |
| 2,3,6$(CH_3)_3C_6H_2CO_2CH(OC_2H_5)_2$ | 144°/1.5 | — | — | 3 | 315d |
| $HC(OC_2H_5)_2OOC(CH_3)_2C_6H_5$ | 70°–80°/0.002 | 1.488 (20°) | — | 3 | 289 |
| p-$(CH_3O)_3CC_6H_4C(OCH_3)_3$ | (124.8°–125.2°) | — | — | 2 | 193 |
| m-$(CH_3O)_3CC_6H_4C(OCH_3)_3$ | (95.4°–96.6°) | — | — | 2 | 193 |
| | $C_{14}H_{26}$ | | | | |
| 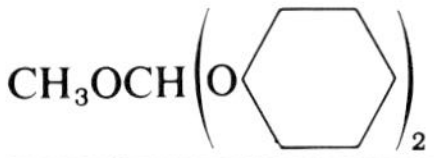 $CH_3OCH(O-cyclohexyl)_2$ | 107°–109°/0.4 | 1.4671 (20°) | — | 2 | 117 |

## TABLE III—continued

| Structure | B.p., °C/mm (M.p., °C) | Refractive index $n^a$ | Density $d^b$ | Method[c] | References |
|---|---|---|---|---|---|
| 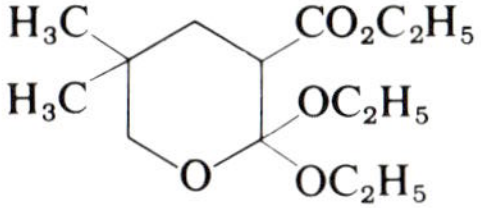 H₃C⟨CO₂C₂H₅, OC₂H₅, OC₂H₅⟩ (pyran structure) | 52°–56°/0.01 | — | — | 1, 3 | *179* |
| 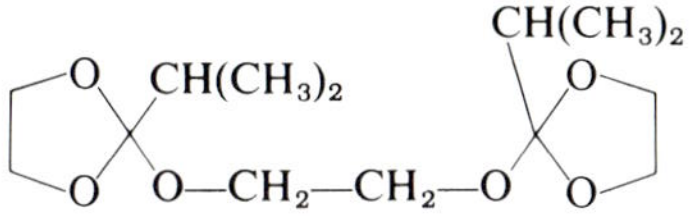 CH(CH₃)₂ dioxolane with CH(CH₃)₂ | 110°–112°/0.2 | 1.4451 | 1.0823 | 1 | *203* |
| 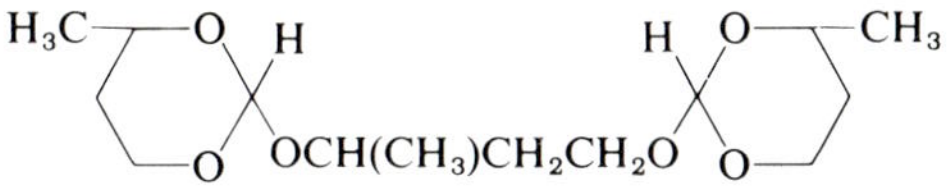 Spiro dioxane/CH₃ structure | 91°/7 | — | — | 3 | *246* |
| 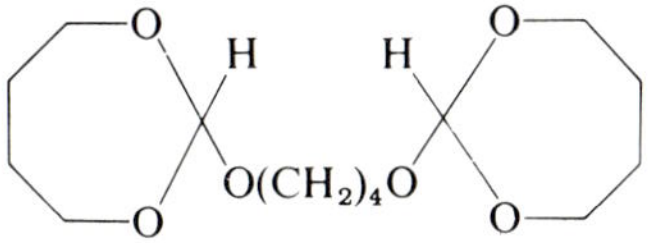 H₃C spiro OCH(CH₃)CH₂CH₂O structure | 101°/2.5 | — | — | 3 | *246* |
| 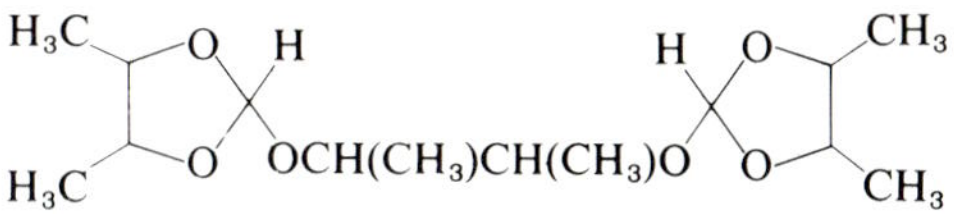 H₃C spiro OCH(CH₃)CH(CH₃)O structure | 120°/3 | — | — | 3 | *246* |
| C₃H₇O₂CC(OC₃H₇)₃  C₁₄H₂₈ | 129°–130°/12 | — | — | 2 | 5 |

# PHYSICAL PROPERTIES OF ORTHO ESTERS

| Compound | Formula | bp (°C/mm) | $n_D$ | $d$ | Refs. | |
|---|---|---|---|---|---|---|
| $HC[OCH_2CH_2CH(CH_3)_2]_2OC_3H_7$ | $C_{14}H_{30}$ | 140°–147°/30 | 1.4194 (20°) | 0.8626 (23°) | 2 | 281 |
| $CH_3C[OCH_2CH(CH_3)_2]_3$ | | 207°–208°/760 | — | — | 1 | 326 |
| $CH_3C[OCH_2CH_2CH(CH_3)_2]_2OC_2H_5$ | | 236°–238°/760 | — | — | 1 | 311 |
| $CH_3OC(OC_4H_9)_3$ | | 130°/16 | 1.4165 (20°) | 0.9018 (20°) | 3 | 329 |
| $CH_3C(OCH_2CH_2OC_2H_5)_3$ | | 98°–100°/0.5 | — | — | 5 | 188 |
| $CH_3C[OCH(CH_3)CH_2OCH_3]_3$ | | 78°–80°/0.3 | — | — | 5 | 188 |
| $(C_2H_5O)_3CC(OC_2H_5)_3$ | | 63°/0.1 | — | — | 2 | 19, 156 |
| 2-ethoxy-2-phenyl-1,3-benzodioxole (OC₂H₅, C₆H₅) | $C_{15}H_{14}$ | 115°–116°/0.5 172°–173°/13 | 1.5548 (20°) | — | 2, 3, 6 | 75, 118a |
| $H_2NC(O)C(OC_6H_5)_2OCH_3$ | $C_{15}H_{15}$ | (157°–159°) | — | — | 2 | 173 |
| $HC(OC_6H_5)_2OC_2H_5$ | $C_{15}H_{16}$ | 150°/3 | — | — | 2 | 318 |
| $CH_2CH=CHCH(OH)(C{\equiv}C)_2H$ (adamantane-like structure) | | (126°) | — | — | 4 | 45 |
| $CH(CH_3)_2 ... (CF_3)_2 ... (CF_3)_2$ spiro dioxolane with $H_3C$, $H_3C$ | | (73°–74°) | — | — | 11 | 47 |

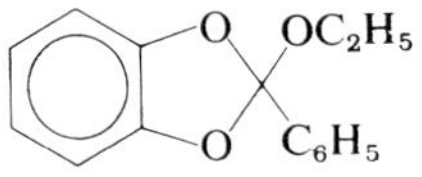

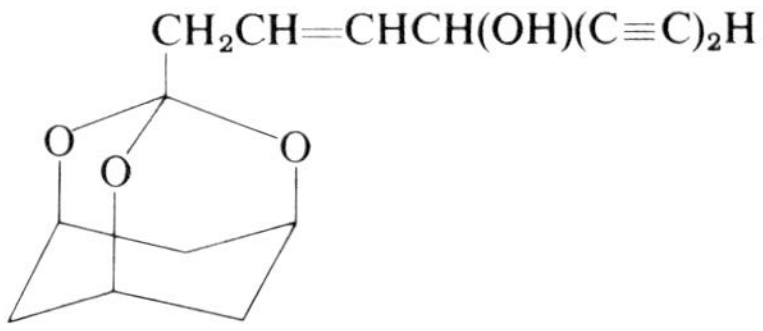

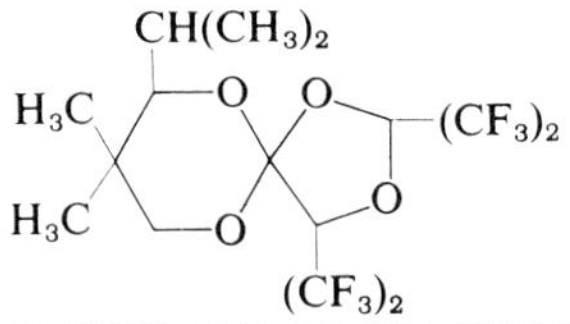

TABLE III—continued

| Structure | B.p., °C/mm (M.p., °C) | Refractive index $n^a$ | Density $d^b$ | Method[c] | References |
|---|---|---|---|---|---|
| CH₂CH=CHCH(OH)CH=CHC≡CH | | $C_{15}H_{18}$ | | | |
| 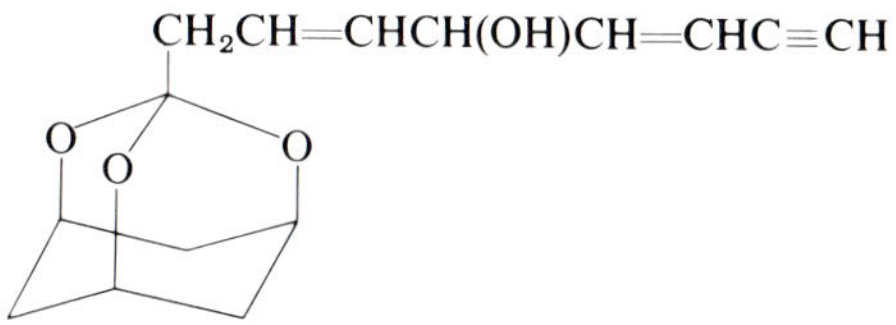 (adamantane-like bicyclic orthoester structure) | (98°) | — | — | 4 | *45* |
| NCCH₂C(OCH₃)₂OO (tetrahydronaphthalene structure) | (87°) | $C_{15}H_{19}$ — | — | 5 | *319* |
| 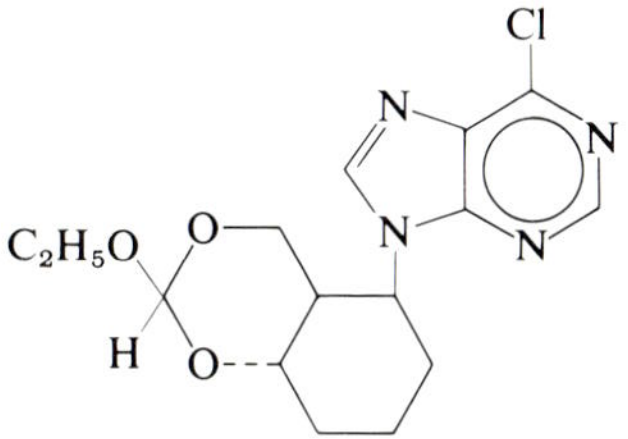 (chloro-purine/imidazole bicyclic structure with C₂H₅O, H) | — | — | — | 3 | *315c* |
| 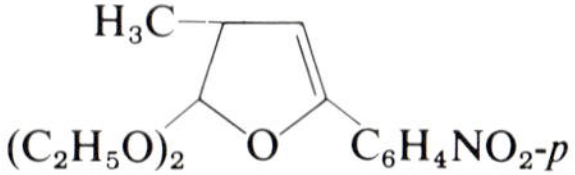 H₃C (C₂H₅O)₂ / C₆H₄NO₂-p furan structure | 155°/0.5 | — | — | 11 | *315a* |

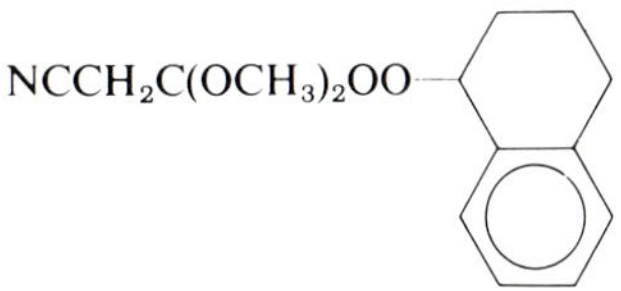

# PHYSICAL PROPERTIES OF ORTHO ESTERS

| Compound | bp | Formula | $n$ | $d$ |  | Ref. |
|---|---|---|---|---|---|---|
| $C_6H_5C{\equiv}CC(OC_2H_5)_3$ | 147°/12 | $C_{15}H_{20}$ | 1.5004 | 0.9962 | 9 | *150, 151, 155* |
| 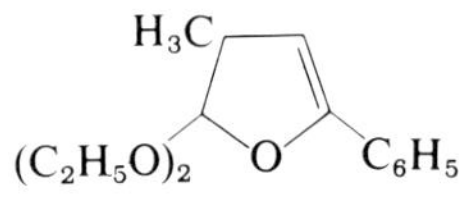 $H_3C$–[furan, $C_6H_5$, $(C_2H_5O)_2$] | 100°/0.3 | — | — | — | 11 | *315a* |
| [bicyclic, $OCH_3$, $C_6H_5$] | — | — | — | — | 6 | *320a* |
| 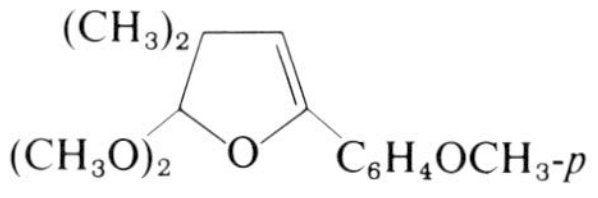 $(CH_3)_2$–[furan, $C_6H_4OCH_3{-}p$, $(CH_3O)_2$] | 140°/1.5 | — | — | — | 11 | *315a* |
| 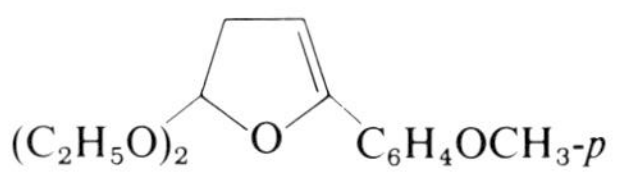 [furan, $C_6H_4OCH_3{-}p$, $(C_2H_5O)_2$] | 130°/0.4 | — | — | — | 11 | *315a* |
| $o{-}(C_2H_5OCH{=}NCO)C_6H_4OCH(OC_2H_5)_2$ | — | $C_{15}H_{21}$ | — | — | 3 | *351a* |
| $H_3C$–[furan, $C_6H_5$, $(C_2H_5O)_2$] | 94°/0.2 | $C_{15}H_{22}$ | — | — | 4 | *315a* |
| 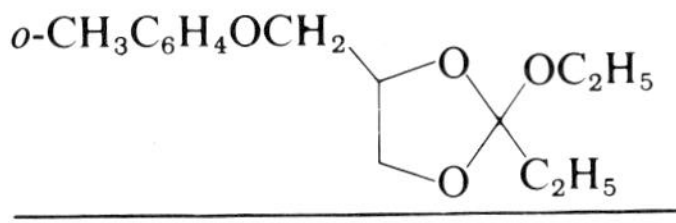 $o{-}CH_3C_6H_4OCH_2$–[dioxolane, $OC_2H_5$, $C_2H_5$] | 194°–195.5°/27 | — | — | — | 3 | *343* |

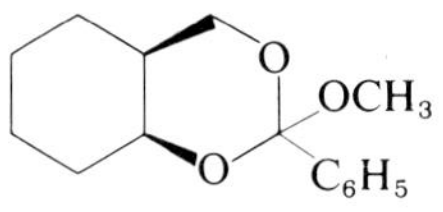

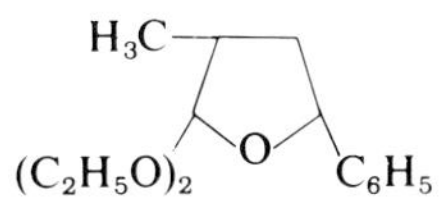

TABLE III—continued

| Structure | B.p., °C/mm (M.p., °C) | Refractive index $n^a$ | Density $d^b$ | Method[c] | References |
|---|---|---|---|---|---|
| | | $C_{15}H_{24}$ | | | |
| $C_6H_5CH_2C(OC_3H_7)_2OCH_3$ | 239°–242°/760 | 1.4950 (20°) | 1.0109 (20°) | 1 | 313 |
| $C_6H_5CH_2C[OCH(CH_3)_2]_2OCH_3$ | 227°–229°/760 | 1.4913 (20°) | 1.0079 (20°) | 1 | 313 |
| $C_6H_5CH_2CH_2C(OC_2H_5)_3$ | 138°–139°/10 | 1.4722 | — | 1 | 217, 311 |
| $p\text{-}(C_2H_5SO_2)_2CHC_6H_4C(OCH_3)_3$ | (122°–123.5°) | — | — | 5 | 129 |
| | | $C_{15}H_{28}$ | | | |
| 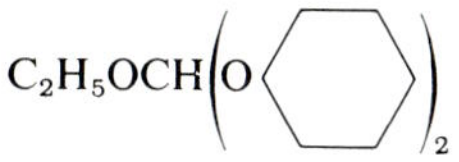 $C_2H_5OCH{\left(O\!-\!\!\bigcirc\right)}_2$ | 109°/1.1 | 1.4605 | 0.977 | 3 | 292 |
| 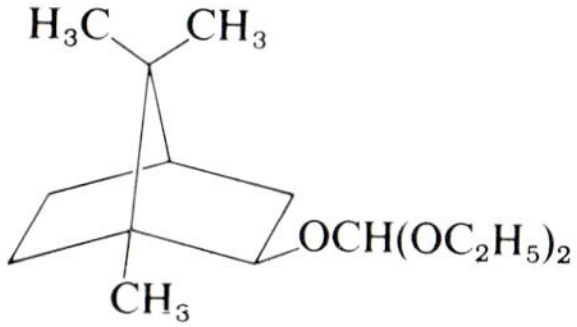 [bicyclic structure with $H_3C$, $CH_3$, $OCH(OC_2H_5)_2$, $CH_3$] | 132°–134°/13 | 1.4531 (20°) | 0.9624 (20°) | 3 | 249 |
| | | $C_{15}H_{30}$ | | | |
| 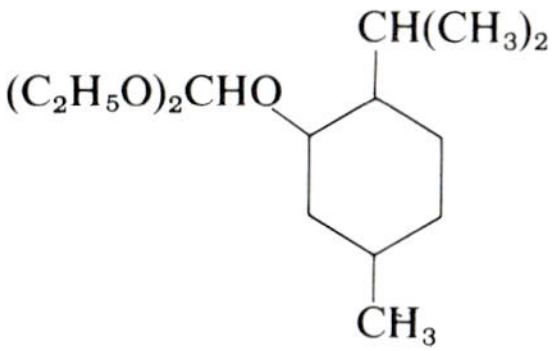 [cyclohexane with $CH(CH_3)_2$, $(C_2H_5O)_2CHO$, $CH_3$] | 135°/14 | 1.4446 (20°) | 0.9252 (20°) | 3 | 249 |
| | | $C_{15}H_{32}$ | | | |
| $C_2H_5C(OC_4H_9)_3$ | 117°/5 | 1.4208 | 0.8699 | 1, 3 | 7, 9, 373 |

# PHYSICAL PROPERTIES OF ORTHO ESTERS

| Compound | BP/MP | $n_D$ | Formula | $d$ | | Ref |
|---|---|---|---|---|---|---|
| $HC[OCH_2(CF_2)_3CHF_2]_3$ | 95°/0.08 | — | $C_{16}H_{10}$ | — | 2 | 133 |
| 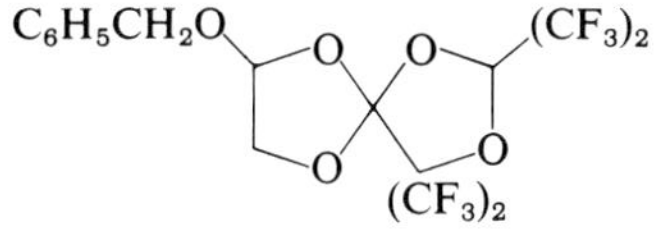 $C_6H_5CH_2O$ ⟨spiro structure with $(CF_3)_2$⟩ | 120°/1.45 | 1.4044 | | — | 11 | 47 |
| $CH_3O_2CC(OC_6H_3Cl_2\text{-}2,4)_2OCH_3$ | (95°–96°) | — | $C_{16}H_{12}$ | — | 2 | 20 |
| $CH_3O_2CC(OC_6H_4Cl\text{-}p)_2OCH_3$ | (63°–64°) | — | $C_{16}H_{14}$ | — | 2 | 20 |
| $CH_3O_2CC(OC_6H_4NO_2\text{-}p)_2OCH_3$ | (168°–169°) | — | | — | 2 | 20 |
| $CH_3O_2CC(OC_6H_4NO_2\text{-}m)_2OCH_3$ | (76°–77°) | — | | — | 2 | 20 |
| $H_2NC(O)C(OC_6H_4NO_2\text{-}p)OC_2H_5$ | (193°–194°) | — | $C_{16}H_{15}$ | — | 2 | 173 |
| $CH_3O_2CC(OC_6H_5)_2OCH_3$ | 208°–209°/19 | — | $C_{16}H_{16}$ | — | 2 | 20 |
| $H_2NC(O)C(OC_6H_5)_2OC_2H_5$ | (113°–115°) | — | $C_{16}H_{17}$ | — | 2 | 173 |
| $CH_3C(OC_6H_5)_2OC_2H_5$ | 135°–136°/4 | 1.5369 (20°) | $C_{16}H_{18}$ | 1.0934 (20°) | 3 | 328 |
| $CH_2NHCO_2CH_2C_6H_5$ (adamantane-like structure) | (120°) | — | $C_{16}H_{19}$ | — | 4 | 44 |

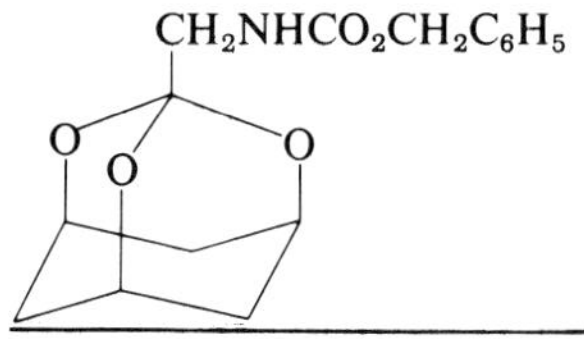

TABLE III—continued

| Structure | B.p., °C/mm (M.p., °C) | Refractive index $n^a$ | Density $d^b$ | Method[c] | References |
|---|---|---|---|---|---|
| $C_{16}H_{20}$ | | | | | |
| 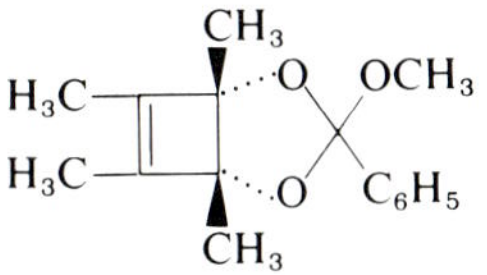 (structure with CH₃, OCH₃, C₆H₅) | (67.5°–68.5°) | — | — | 6 | 361 |
| $C_{16}H_{22}$ | | | | | |
| 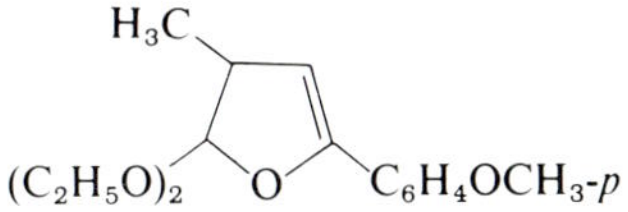 (structure with H₃C, C₆H₄OCH₃-p, (C₂H₅O)₂) | 179°/0.4 | — | — | 11 | 315a |
| 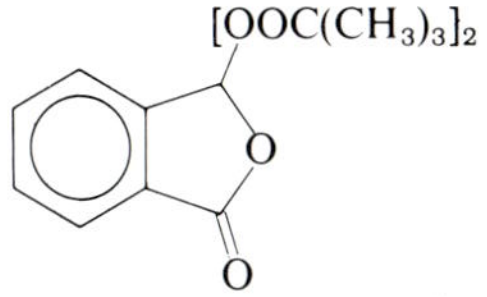 [OOC(CH₃)₃]₂ | (87°) | — | — | 2 | 247a |
| $C_{16}H_{24}$ | | | | | |
| 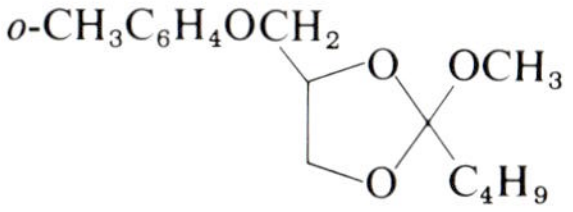 o-CH₃C₆H₄OCH₂ (dioxolane with OCH₃, C₄H₉) | 166°/5 | — | — | 3 | 343 |
| $C_{16}H_{25}$ | | | | | |
| C₆H₅CH₂CONHCH₂C(OC₂H₅)₃ | (40°) | — | — | 1 | 376 |

## PHYSICAL PROPERTIES OF ORTHO ESTERS

| Compound | Formula | BP/mm (MP) | $n_D$ | $d_4$ | Method | Ref. |
|---|---|---|---|---|---|---|
| $C_6H_5CH_2C(OC_3H_7)_2OC_2H_5$ | $C_{16}H_{26}$ | 238°–241°/760 | 1.4967 (20°) | 1.0094 (20°) | 1 | 313 |
| $C_6H_5CH_2C[OCH(CH_3)_2]_2OC_2H_5$ | | 228°–230°/760 | 1.4908 (20°) | 1.0030 (20°) | 1 | 313 |
| $HC(OC_5H_{11})_3$ | $C_{16}H_{34}$ | 135°–137°/4 | 1.4237 | — | 1, 2, 3 | 2, 276, 298, 362, 363 |
| $HC[OCH_2CH_2CH(CH_3)_2]_3$ | | 105°–106°/1.5 | 1.4205 | 0.8578 | 2, 3 | 2, 73, 298, 312 |
| $HC[OCH_2CH(CH_3)C_2H_5]_3$ | | 99°/1 | 1.4254 (20°) | 0.8660 (20°) | 3 | 273 |
| $HC[OCH_2C(CH_3)_3]_3$ | | (71.5°–73.5°) | — | — | 3 | 192 |
| (structure with $C_6H_5$, $OCH_3$, $OCH_3$) | $C_{17}H_{16}$ | (68°–69°) | — | — | 1 | 182 |
| (structure with $OCH_3$, $O$, $C_6H_5$) | | (105°–106°) | — | — | 1 | 374 |
| $(CH_3)_2NC(O)C(OC_6H_4NO_2\text{-}p)_2OCH_3$ | $C_{17}H_{17}$ | (175°–176°) | — | — | 2 | 173 |
| (dioxolane structure with $H_5C_6$, $OC_2H_5$) | $C_{17}H_{18}$ | 136°–140°/0.7 | 1.5495 | 1.098 | 3 | 69 |
| $(CH_3)_2NC(O)C(OC_6H_5)_2OCH_3$ | $C_{17}H_{19}$ | (81°–83°) | — | — | 2 | 173 |
| $(C_6H_5)_2CHC(OCH_3)_3$ | $C_{17}H_{20}$ | 136°–141°/0.8 (39°–43°) | 1.5515 | — | 5 | 218 |

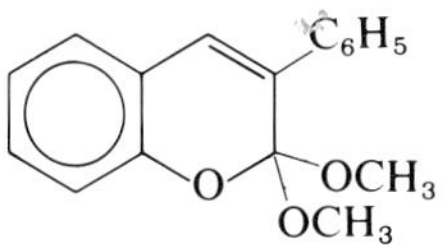

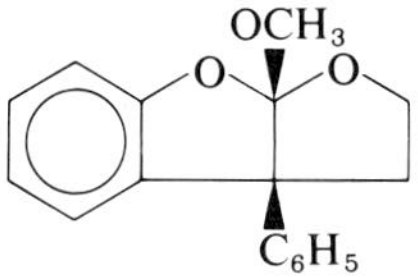

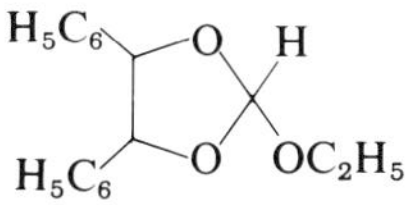

TABLE III—continued

| Structure | B.p., °C/mm (M.p., °C) | Refractive index $n^a$ | Density $d^b$ | Method[c] | References |
|---|---|---|---|---|---|
| $C_{17}H_{22}$ | | | | | |
| 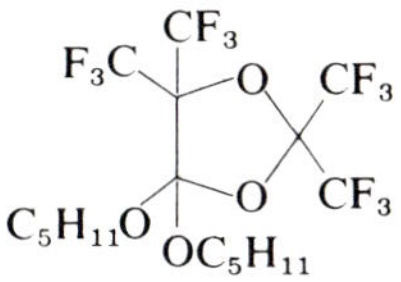 [F_3C-CF_3 / CF_3 ortho ester with $C_5H_{11}O$, $OC_5H_{11}$] | 152°/45 | 1.3717 | — | 11 | 47 |
| $C_{17}H_{26}$ | | | | | |
| 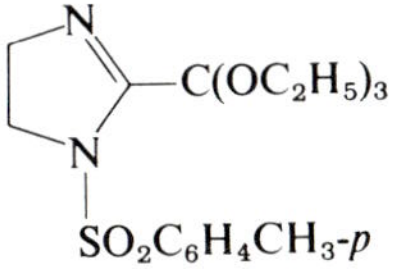 $C(OC_2H_5)_3$ imidazoline with $SO_2C_6H_4CH_3$-$p$ | (98°–99°) | — | — | 4 | 18 |
| $C_{17}H_{28}$ | | | | | |
| $C_6H_5CH_2C(OC_4H_9)_2OCH_3$ | 254°–257°/760 | 1.4911 (20°) | 0.9953 (20°) | 1 | 313 |
| $C_6H_5CH_2C[OCH_2CH(CH_3)_2]_2OCH_3$ | 245°–248°/760 | 1.4898 (20°) | 0.9929 (20°) | 1 | 313 |
| $C_{17}H_{29}$ | | | | | |
| 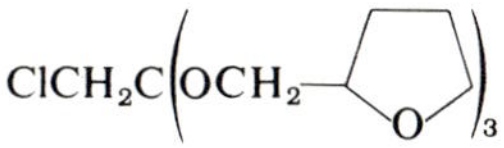 $ClCH_2C(OCH_2\text{-THF})_3$ | 145°–160°/1.5 | — | — | 5 | 188 |
| $C_{17}H_{30}$ | | | | | |
| 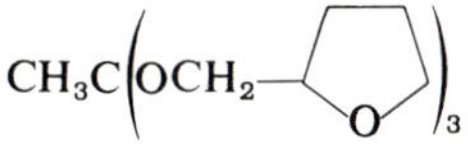 $CH_3C(OCH_2\text{-THF})_3$ | 151°–154°/0.5 | — | — | 5 | 188 |

| Compound | | BP/MP | $n_D$ | $d$ | n | Ref |
|---|---|---|---|---|---|---|
| C(OC$_4$H$_9$)$_4$ | C$_{17}$H$_{36}$ | 273°/760 | 1.4216 | 0.8850 | 2 | 6a, 8a, 344 |
| C[OCH$_2$CH(CH$_3$)$_2$]$_4$ | | 245°/760 | — | 0.900 (8°) | 2 | 64, 303 |
| CH$_3$C(OCH$_2$CH$_2$OCH$_2$CH$_2$OCH$_3$)$_3$ | | 170°–172°/0.5 | — | — | 5 | 188 |
| C$_2$H$_5$O$_2$CC(OC$_6$H$_3$Cl$_2$-2,4)$_2$OC$_2$H$_5$ | C$_{18}$H$_{16}$ | 220°–222°/4 | — | — | 2 | 156 |
| CH$_3$O$_2$CC(OC$_6$H$_3$Cl$_2$-2,4)$_2$OC$_3$H$_7$ | | 211°–212°/1 | — | — | 2 | 173 |
| C$_2$H$_5$O$_2$CC(OC$_6$H$_4$Cl-$p$)$_2$OC$_2$H$_5$ | C$_{18}$H$_{18}$ | (62°–63°) | — | — | 2 | 156 |
| C$_2$H$_5$O$_2$CC(OC$_6$H$_4$NO$_2$-$p$)$_2$OC$_2$H$_5$ | | (176°–179°) | — | — | 2 | 156 |
| C$_2$H$_5$O$_2$CC(OC$_6$H$_4$NO$_2$-$m$)$_2$OC$_2$H$_5$ | | (86°–87°) | — | — | 2 | 156 |
| CH$_3$O$_2$CC(OC$_6$H$_4$NO$_2$-$p$)$_2$OC$_3$H$_7$ | | (95°–97°) | — | — | 2 | 173 |
| CH$_3$O$_2$CC(OC$_6$H$_4$NO$_2$-$m$)$_2$OC$_3$H$_7$ | | (85°–87°) | — | — | 2 | 173 |
| C$_6$H$_5$\\OOC(CH$_3$)$_2$C$_6$H$_5$ (cyclic) | C$_{18}$H$_{20}$ | (75°–76°) | — | — | 3 | 287, 289, 290 |
| C$_2$H$_5$O$_2$CC(OC$_6$H$_5$)$_2$OC$_2$H$_5$ | | 216°–217°/25 | — | — | 2 | 156 |
| H$_2$NC(O)C(OC$_6$H$_5$)$_2$OC$_4$H$_9$ | C$_{18}$H$_{21}$ | (108°–110°) | — | — | 2 | 173 |
| H$_2$NC(O)C(OC$_6$H$_5$)$_2$OCH$_2$CH(CH$_3$)$_2$ | | (122°–124°) | — | — | 2 | 173 |
| (CH$_3$)$_2$NC(O)C(OC$_6$H$_5$)$_2$OC$_2$H$_5$ | | (71°–73°) | — | — | 2 | 173 |
| [adamantane-type diorthoester with CH$_2$CH=CHCH$_2$ bridge] | C$_{18}$H$_{24}$ | (274°) | — | — | 3 | 44 |

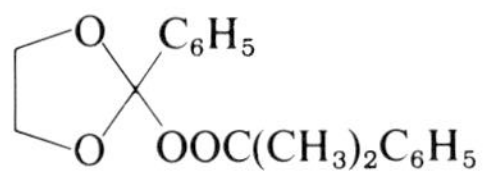

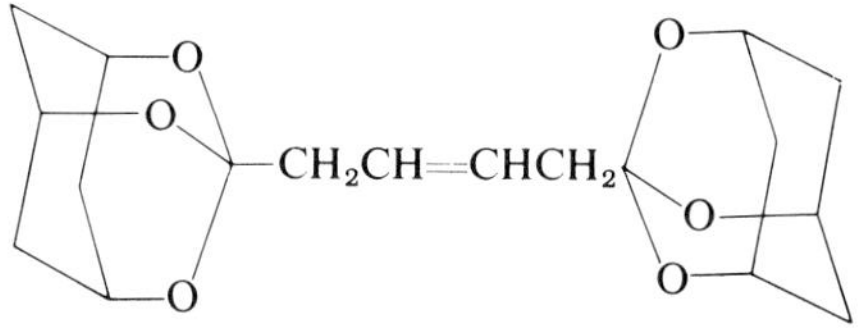

TABLE III—continued

| Structure | B.p., °C/mm (M.p., °C) | Refractive index $n^a$ | Density $d^b$ | Method[c] | References |
|---|---|---|---|---|---|
| | $C_{18}H_{30}$ | | | | |
| $C_6H_5CH_2C(OC_4H_9)_2OC_2H_5$ | 254°–257°/760 | 1.4916 (20°) | 0.9974 (20°) | 1 | 313 |
| $C_6H_5CH_2C[OCH_2CH(CH_3)_2]_2OC_2H_5$ | 248°–251°/760 | 1.4883 (20°) | 0.9867 (20°) | 1 | 313 |
| 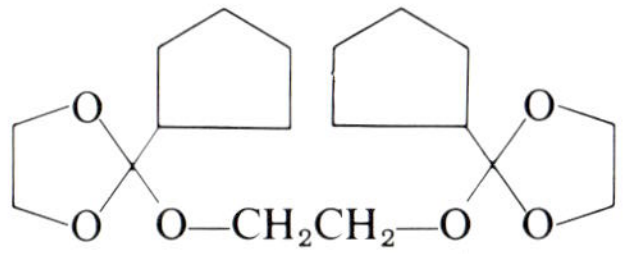 (spiro bis-dioxolane with cyclopentane rings, $-O-CH_2CH_2-O-$ bridge) | 154°–157°/0.2 (31°–33°) | — | — | 1 | 225 |
| | $C_{18}H_{34}$ | | | | |
| 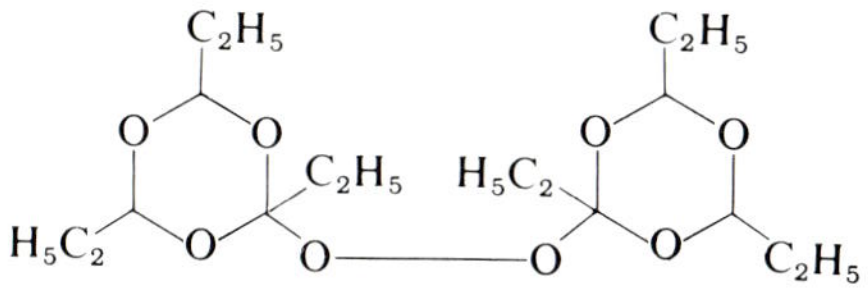 (trioxane with $C_2H_5$ substituents) | — | — | — | 11 | 323b |
| | $C_{18}H_{36}$ | | | | |
| $(CH_3)_2CHCH_2O_2CC[OCH_2CH(CH_3)_2]_3$ | 146°/8.5 | — | — | 2 | 5 |
| | $C_{18}H_{38}$ | | | | |
| $C_4H_9O(C_2H_5O)_2CC(OC_2H_5)_2OC_4H_9$ | 102°/0.1 | — | — | 2 | 19 |
| $S[CH_2CH_2C(OC_2H_5)_3]_2$ | 125°–140°/0.7 | — | — | 1 | 112a |
| | $C_{19}H_7$ | | | | |
| $(2,4,5\text{-}Cl_3C_6H_2O)_3CH$ | (230°–231°) | — | — | 2 | 162 |
| | $C_{19}H_{10}$ | | | | |
| $(2,4\text{-}Cl_2C_6H_3O)_3CH$ | (205.5°–206°) | — | — | 2 | 162, 248 |

# PHYSICAL PROPERTIES OF ORTHO ESTERS

| Compound | Formula | mp/bp | | Ref. |
|---|---|---|---|---|
| | $C_{19}H_{13}$ | | | |
| $(o\text{-ClC}_6H_4O)_3CH$ | | $(129°–130°)$ | — | 2  162 |
| $(p\text{-ClC}_6H_4O)_3CH$ | | $(106°)$ | — | 2  108a |
| $(p\text{-O}_2NC_6H_4O)_3CH$ | | $(232°)$ | — | 2  358 |
| $(o\text{-O}_2NC_6H_4O)_3CH$ | | $(182°)$ | — | 2  358 |
| (benzodioxole, $C_6H_5$, $OC_6H_5$) | $C_{19}H_{14}$ | $(76°–78°)$ | — | 2  118a |
| (benzodioxole, $OC_6H_5$, $OC_6H_5$) | $C_{19}H_{15}$ | $(100°–102°)$ | — | 2  118 |
| $HC(OC_6H_5)_2OC_6H_4NO_2\text{-}p$ | | $(98°)$ | — | 2  318 |
| $HC(OC_6H_5)_3$ | $C_{19}H_{16}$ | $(76°–77°)$ | — | 2  14, 21, 318, 345 |
| $p\text{-O}_2NC_6H_4CO_2H_2C\text{—}C_6H_4NO_2\text{-}p$ (bicyclic) | | 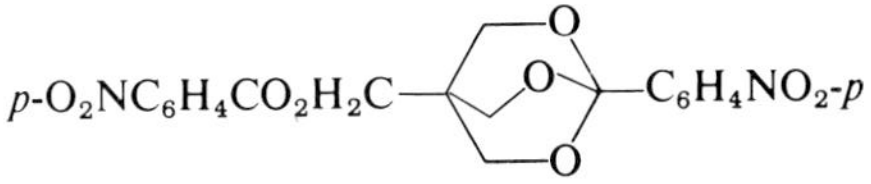 $(137°–138°)$ | — | 11  31 |
| | $C_{19}H_{18}$ | | | |
| $CH_3O_2CC(OC_6H_3Cl_2\text{-}2,4)_2OC_4H_9$ | | $215°–216°/1$ | — | 2  157 |
| $CH_3O_2CC(OC_6H_3Cl_2\text{-}2,4)_2OCH_2CH(CH_3)_2$ | | $206°–208°/1$ | — | 2  157 |
| $C_2H_5O_2CC(OC_6H_3Cl_2\text{-}2,4)_2OC_3H_7$ | | $207°–208°/1$ | — | 2  157 |
| | $C_{19}H_{20}$ | | | |
| $CH_3O_2CC(OC_6H_4NO_2\text{-}m)_2OC_4H_9$ | | $243°–244°/1$ | — | 2  157 |
| $CH_3O_2CC(OC_6H_4NO_2\text{-}p)_2OC_4H_9$ | | $(102°–104°)$ | — | 2  157 |
| $CH_3O_2CC(OC_6H_4NO_2\text{-}m)_2OCH_2CH(CH_3)_2$ | | $(78°–80°)$ | — | 2  157 |
| $CH_3O_2CC(OC_6H_4NO_2\text{-}p)_2OCH_2CH(CH_3)_2$ | | $(104°–105°)$ | — | 2  157 |
| $C_2H_5O_2CC(OC_6H_4NO_2\text{-}m)_2OC_3H_7$ | | $(71°–73°)$ | — | 2  157 |
| $C_2H_5O_2CC(OC_6H_4NO_2\text{-}p)_2OC_3H_7$ | | $(152°–154°)$ | — | 2  157 |

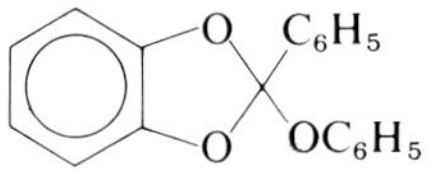

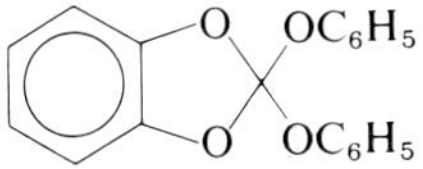

## TABLE III—continued

| Structure | B.p., °C/mm (M.p., °C) | Refractive index $n^a$ | Density $d^b$ | Method[c] | References |
|---|---|---|---|---|---|
| $C_{19}H_{22}$ | | | | | |
| 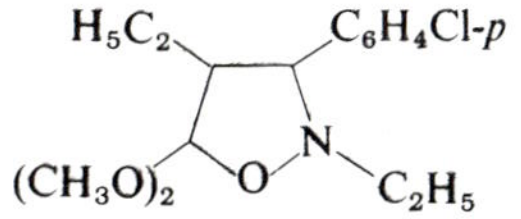 H₅C₂ ⟨ring⟩ C₆H₄Cl-p, N–C₂H₅, (CH₃O)₂ | (113°) | — | — | 11 | 315b |
| 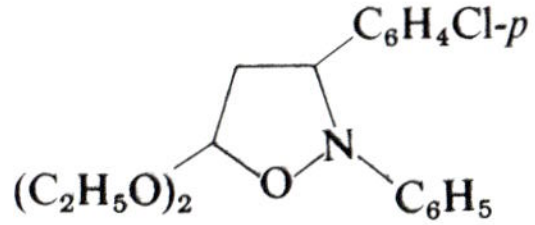 ⟨ring⟩ C₆H₄Cl-p, N–C₆H₅, (C₂H₅O)₂ | (65°) | — | — | 11 | 315b |
| CH₃O₂CC(OC₆H₅)₂OC₄H₉ | 170°–172°/1 | — | — | 2 | 157 |
| CH₃O₂CC(OC₆H₅)₂OCH₂CH(CH₃)₂ | 195°/4.5 | — | — | 2 | 157 |
| C₂H₅O₂CC(OC₆H₅)₂OC₃H₇ | 159°–161°/1 | — | — | 2 | 157 |
| $C_{19}H_{23}$ | | | | | |
| 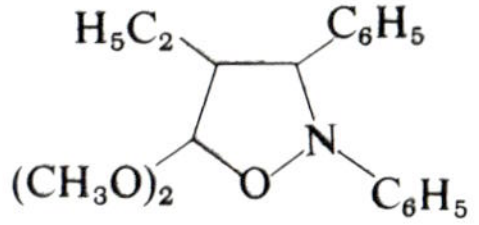 H₅C₂ ⟨ring⟩ C₆H₅, N–C₆H₅, (CH₃O)₂ | (59°) | — | — | 11 | 315b |
| 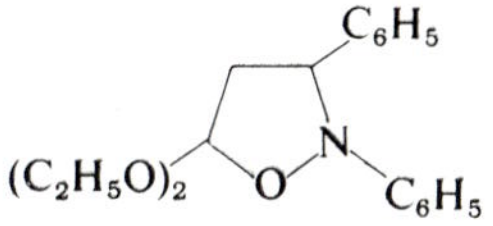 ⟨ring⟩ C₆H₅, N–C₆H₅, (C₂H₅O)₂ | 163°/0.4 | — | — | 11 | 315b |

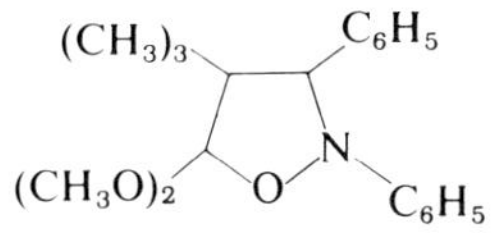

# PHYSICAL PROPERTIES OF ORTHO ESTERS

| Compound | bp/mp | Formula | $n_D$ | $d$ | | Refs |
|---|---|---|---|---|---|---|
| pyrrolidine [(CH3)3, C6H5, N-C6H5, (CH3O)2] | (84°) | — | — | — | 11 | *315b* |
| C2H5C(OCH2C6H5)2OC2H5 | — | C19H24 | — | — | 1 | *326* |
| C6H5CH2C[OCH2CH2CH(CH3)2]2OCH3 | 260°–265°/760 | C19H32 | 1.4900 (20°) | 0.9880 (20°) | 1 | *313* |
| HC(OC6H11)3 | 200°–205°/15 (72.6°–73.8°) | C19H34 | — | — | 3 | *123, 298* |
| HC(OC6H13)3 | 153°/1.8 | C19H40 | — | — | 3 | *2, 298* |
| HC[OCH2CH(C2H5)2]3 | 135°/1.5 | | 1.4329 | — | 3 | *273* |
| HC[OCH(CH3)CH(CH3)(C2H5)]3 | 108°/0.5 | | 1.4323 (25°) | 0.8684 (25°) | 3 | *303a, 303b* |
| CH3C(OC6H5)3 | 148°–153°/0.5 (61°–62°) | C20H18 | — | — | 5 | *212* |
| C2H5O2CC(OC6H3Cl2-2,4)OCH2CH(CH3)2 | 214°–215°/1 | C20H20 | — | — | 2 | *157* |
| C3H7O2CC(OC6H3Cl2-2,4)2OC3H7 | 233°–234°/4 | | — | — | 2 | *156* |
| triphenylene-dioxole with (OCH3)2 | (103°) | | — | — | 11 | *142a* |
| C2H5O2CC(OC6H3Cl2-2,4)2OC4H9 | 215°–216°/1 | | — | — | 2 | *157* |

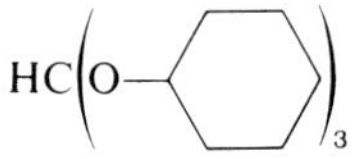

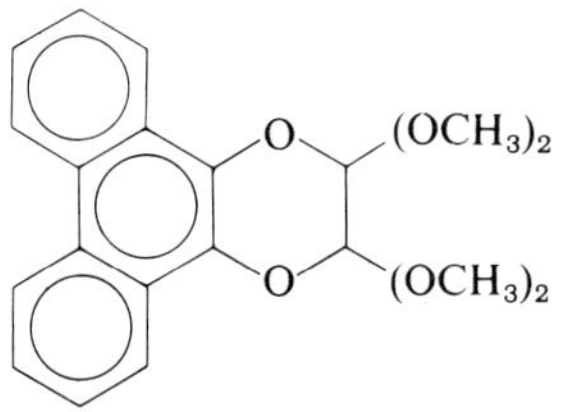

TABLE III—continued

| Structure | B.p., °C/mm (M.p., °C) | Refractive index $n^a$ | Density $d^b$ | Method[c] | References |
|---|---|---|---|---|---|
| $C_{20}H_{22}$ | | | | | |
| $C_3H_7O_2CC(OC_6H_4Cl\text{-}p)_2OC_3H_7$ | 220°–221°/5 | — | — | 2 | 156 |
| $C_2H_5O_2CC(OC_6H_4NO_2\text{-}p)_2OC_4H_9$ | (97°–99°) | — | — | 2 | 157 |
| $C_3H_7O_2CC(OC_6H_4NO_2\text{-}p)_2OC_3H_7$ | (115°–116°) | — | — | 2 | 156 |
| $C_2H_5O_2CC(OC_6H_4NO_2\text{-}m)_2OCH_2CH(CH_3)_2$ | 250°–252°/1.5 | — | — | 2 | 157 |
| $C_3H_7O_2CC(OC_6H_4NO_2\text{-}m)_2OC_3H_7$ | 240°–241°/3 | — | — | 2 | 156 |
| $C_2H_5O_2CC(OC_6H_4NO_2\text{-}p)_2OCH_2CH(CH_3)_2$ | (162°–164°) | — | — | 2 | 157 |
| $C_{20}H_{24}$ | | | | | |
| 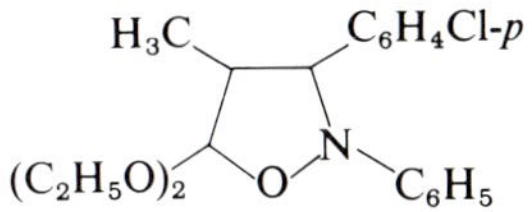 H₃C—C₆H₄Cl-p / N—C₆H₅ / (C₂H₅O)₂ | (61°) | — | — | 11 | 315b |
| $CH_3C[ON=C(CH_3)C_6H_5]_2OC_2H_5$ | (78°) | — | — | 5 | 345a |
| $C_2H_5O_2CC(OC_6H_5)_2OC_4H_9$ | 172°–174°/1 | — | — | 2 | 157 |
| $C_3H_7O_2CC(OC_6H_5)_2OC_3H_7$ | 175°–176°/2 | — | — | 2 | 156 |
| $C_2H_5O_2CC(OC_6H_5)_2OCH_2CH(CH_3)_2$ | 175°–177°/1.5 | — | — | 2 | 157 |
| $C_{20}H_{25}$ | | | | | |
| $(CH_3)_2NC(O)C(OC_6H_5)_2OC_4H_9$ | (73°–74°) | — | — | 2 | 173 |
| $(CH_3)_2NC(O)C(OC_6H_5)_2OCH_2CH(CH_3)_2$ | (70°–72°) | — | — | 2 | 173 |
| 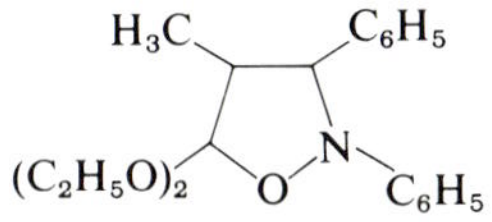 H₃C—C₆H₅ / N—C₆H₅ / (C₂H₅O)₂ | (82°) | — | — | 11 | 315b |
| $C_{20}H_{34}$ | | | | | |
| $C_6H_5CH_2C[OCH_2CH_2CH(CH_3)_2]_2OC_2H_5$ | 260°–265°/760 | 1.4887 (20°) | 0.9867 (20°) | 1 | 313 |

# PHYSICAL PROPERTIES OF ORTHO ESTERS

| Structure | bp/mp | $n_D$ | Formula | — | Count | Ref |
|---|---|---|---|---|---|---|
| 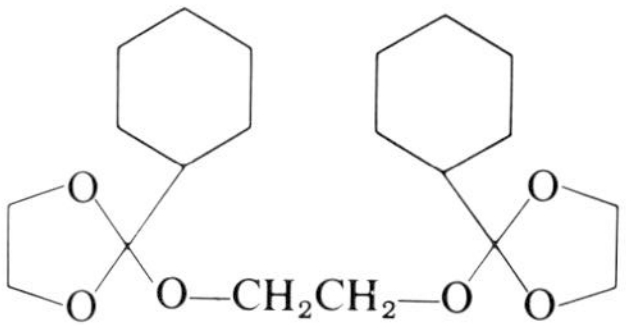 (cyclohexane-dioxolane spiro compound with -O-CH₂CH₂-O- bridge) | 166°–169°/0.2 (52°–55°) | — | — | — | 1 | 225 |
| $m$-C$_6$H$_4$[C(OCH$_3$)$_2$OOC(CH$_3$)$_3$]$_2$ | — | 1.4470 (20°) | — | — | 3 | 193 |
| $p$-C$_6$H$_4$[C(OCH$_3$)$_2$OOC(CH$_3$)$_3$]$_2$ | (90.6°–91.7°) | — | — | — | 3 | 193 |
| (dioxane with CH₃ groups) OCH(CH$_3$)CH$_2$C(CH$_3$)$_2$O | 106°/0.8 | — | C$_{20}$H$_{38}$ | — | 3 | 246 |
| 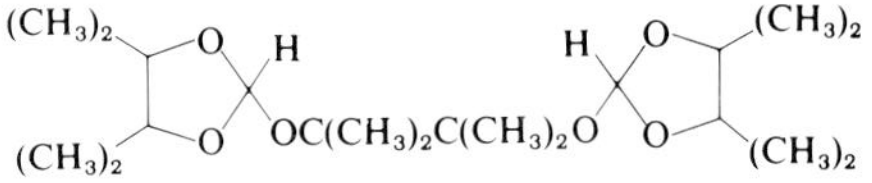 (dioxolane with isopropyl groups) OC(CH$_3$)$_2$C(CH$_3$)$_2$O | 63°/0.4 | — | — | — | 3 | 246 |
| C[OCH$_2$(CF$_2$)$_3$CHF$_2$]$_4$ | 135°/0.05 | — | C$_{21}$H$_{12}$ | — | 2 | 133 |
| 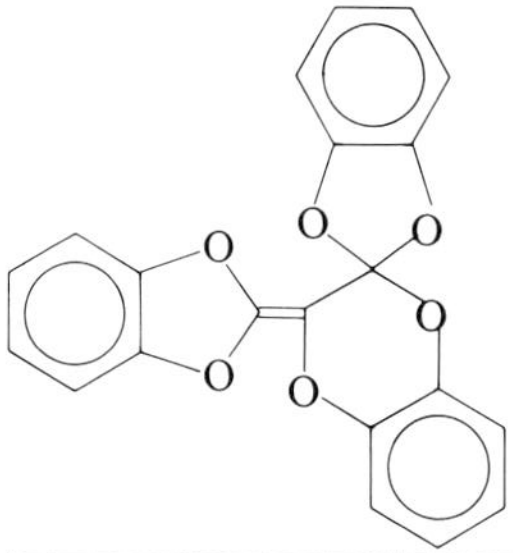 (spiro benzodioxole compound) | (178.5°) | — | — | — | 11 | 66a, 152a |

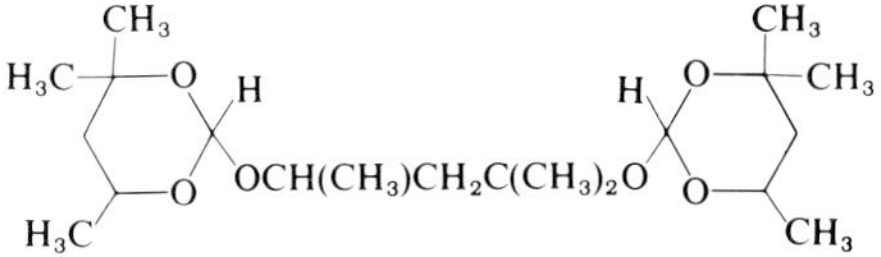

TABLE III—continued

| Structure | B.p., °C/mm (M.p., °C) | Refractive index $n^a$ | Density $d^b$ | Method[c] | References |
|---|---|---|---|---|---|
| $C_{21}H_{22}$ | | | | | |
| $C_3H_7O_2CC(OC_6H_3Cl_2-2,4)_2OC_4H_9$ | 216°–217°/1 | — | — | 2 | 157 |
| $C_3H_7O_2CC(OC_6H_3Cl_2-2,4)_2OCH_2CH(CH_3)_2$ | 223°–224°/1 | — | — | 2 | 157 |
| $(CH_3)_2CHO_2CC(OC_6H_3Cl_2-2,4)_2OC_4H_9$ | 216°–218°/1.5 | — | — | 2 | 157 |
| $(CH_3)_2CHO_2CC(OC_6H_3Cl_2-2,4)_2OCH_2CH(CH_3)_2$ | 214°–216°/1 | — | — | 2 | 157 |
| $C_{21}H_{23}$ | | | | | |
| $C_6H_5CH=C(C_6H_5)C(OCH_2CH_2Cl)_3$ | (101°) | — | — | 11 | 264a |
| $C_{21}H_{24}$ | | | | | |
| $C_3H_7O_2CC(OC_6H_4NO_2\text{-}m)_2OC_4H_9$ | 245°–246°/1 | — | — | 2 | 157 |
| $C_3H_7O_2CC(OC_6H_4NO_2\text{-}p)_2OC_4H_9$ | (106°–107°) | — | — | 2 | 157 |
| $(CH_3)_2CHO_2CC(OC_6H_4NO_2\text{-}p)_2OC_4H_9$ | (108°–110°) | — | — | 2 | 157 |
| $C_3H_7O_2CC(OC_6H_4NO_2\text{-}m)_2OCH_2CH(CH_3)_2$ | 253°–254°/2.5 | — | — | 2 | 157 |
| $(CH_3)_2CHO_2CC(OC_6H_4NO_2\text{-}p)_2OCH_2CH(CH_3)_2$ | (112°–114°) | — | — | 2 | 157 |
| $C_{21}H_{26}$ | | | | | |
| $C_3H_7O_2CC(OC_6H_5)_2OC_4H_9$ | 173°–174°/1 | — | — | 2 | 157 |
| $(CH_3)_2CHO_2CC(OC_6H_5)_2OCH_2CH(CH_3)_2$ | 170°–171°/1 | — | — | 2 | 157 |
| $C_{21}H_{44}$ | | | | | |
| $C_2H_5C(OC_6H_{13})_3$ | 173°/4 | 1.4329 | 0.8668 | 3 | 7, 9 |
| $C_{22}H_{10}$ | | | | | |
| $HC[OCH_2(CF_2)_5CHF_2]_3$ | 137°–139°/0.1 | — | — | 2 | 133 |
| $C_{22}H_{11}$ | | | | | |
| $C_2H_5O_2CC(OC_6H_{13}\text{-}2,4,6)_3$ | (208°–211° dec) | — | — | 2 | 72 |
| $C_{22}H_{18}$ | | | | | |
| 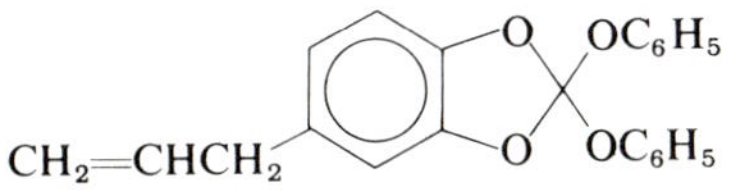 ![structure with OC₆H₅ groups]  $CH_2=CHCH_2$ | 256°–258°/17 | — | — | 2 | 72a |

## PHYSICAL PROPERTIES OF ORTHO ESTERS

| | | | | |
|---|---|---|---|---|
| | | $C_{22}H_{22}$ | | |
| HC(OCH$_2$C$_6$H$_5$)$_3$ | (8°) | 1.5621 | — | 3 | 2, 273 |
| HC(OC$_6$H$_4$CH$_3$-o)$_3$ | (96°) | — | — | 2 | 81 |
| HC(OC$_6$H$_4$CH$_3$-m)$_3$ | (50°) | — | — | 2 | 81 |
| HC(OC$_6$H$_4$CH$_3$-p)$_3$ | (112°) | — | — | 2 | 15, 81 |
| HC(OC$_6$H$_4$OCH$_3$-p)$_3$ | (50°–51°) | — | — | 2 | 248 |
| | | $C_{22}H_{24}$ | | | |
| C$_4$H$_9$O$_2$CC(OC$_6$H$_3$Cl$_2$-2,4)$_2$OC$_4$H$_9$ | 294°/4 | — | — | 2 | 156 |
| (CH$_3$)$_2$CHCH$_2$O$_2$CC(OC$_6$H$_3$Cl$_2$-2,4)$_2$OC$_4$H$_9$ | 219°–221°/1 | — | — | 2 | 157 |
| (CH$_3$)$_2$CHCH$_2$O$_2$CC(OC$_6$H$_3$Cl$_2$-2,4)$_2$-OCH$_2$CH(CH$_3$)$_2$ | 239°–240°/5 | — | — | 2 | 156 |
| 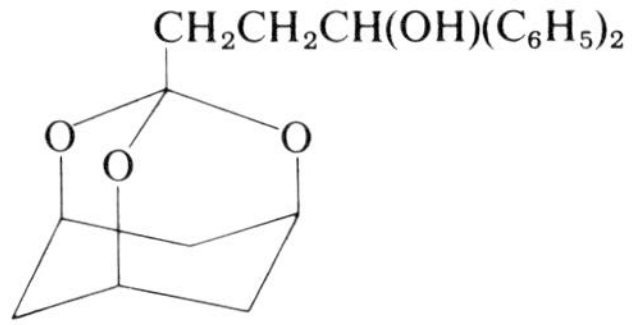 CH$_2$CH$_2$CH(OH)(C$_6$H$_5$)$_2$ | (164°) | — | — | 4 | 335 |
| | | $C_{22}H_{26}$ | | | |
| C$_4$H$_9$O$_2$CC(OC$_6$H$_4$Cl-p)$_2$OC$_4$H$_9$ | 221°/3 | — | — | 2 | 156 |
| (CH$_3$)$_2$CHCH$_2$O$_2$CC(OC$_6$H$_4$Cl-p)$_2$-OCH$_2$CH(CH$_3$)$_2$ | 200°–201°/4 | — | — | 2 | 156 |
| C$_4$H$_9$O$_2$CC(OC$_6$H$_4$NO$_2$-m)$_2$OC$_4$H$_9$ | 268°/3 | — | — | 2 | 156 |
| (CH$_3$)$_2$CHCH$_2$O$_2$CC(OC$_6$H$_4$NO$_2$-m)$_2$-OCH$_2$CH(CH$_3$)$_2$ | 253°–254°/3 | — | — | 2 | 156 |
| (CH$_3$)$_2$CHCH$_2$O$_2$CC(OC$_6$H$_4$NO$_2$-m)$_2$OC$_4$H$_9$ | 247°–248°/1 | — | — | 2 | 157 |
| C$_4$H$_9$O$_2$CC(OC$_6$H$_4$NO$_2$-p)$_2$OC$_4$H$_9$ | (101°–102°) | — | — | 2 | 156 |
| (CH$_3$)$_2$CHCH$_2$O$_2$CC(OC$_6$H$_4$NO$_2$-p)$_2$OC$_4$H$_9$ | (90°–91°) | — | — | 2 | 157 |
| (CH$_3$)$_2$CHCH$_2$O$_2$CC(OC$_6$H$_4$NO$_2$-p)$_2$-OCH$_2$CH(CH$_3$)$_2$ | (99°–100°) | — | — | 2 | 156 |

TABLE III—continued

| Structure | B.p., °C/mm (M.p., °C) | Refractive index $n^a$ | Density $d^b$ | Method[c] | References |
|---|---|---|---|---|---|
| $C_{22}H_{28}$ | | | | | |
| $C_4H_9O_2CC(OC_6H_5)_2OC_4H_9$ | 201°–202°/4 | — | — | 2 | 156 |
| $(CH_3)_2CHCH_2O_2CC(OC_6H_5)_2OC_4H_9$ | 174°–176°/1 | — | — | 2 | 157 |
| $(CH_3)_2CHCH_2O_2CC(OC_6H_5)_2OCH_2CH(CH_3)_2$ | 182°–183°/4 | — | — | 2 | 156 |
| $C_{22}H_{44}$ | | | | | |
| $C_5H_{11}O_2CC(OC_5H_{11})_3$ | 190°/14 | — | — | 2 | 156 |
| $C_{22}H_{46}$ | | | | | |
| $HC[OCH(C_2H_5)C_4H_9]_3$ | 133°/1.1 | 1.4322 | 0.861 | 3 | 298 |
| $C_{23}H_{18}$ | | | | | |
| $\beta$-$C_{10}H_7$-$OCH(OC_6H_5)_2$ | (92°) | — | — | 2 | 318 |
| $C_{23}H_{22}$ | | | | | |
| ![structure with CH3, O, OOC(C6H5)3] (139°–140°) | | — | — | 5 | 319 |
| $C_{23}H_{34}$ | | | | | |
| 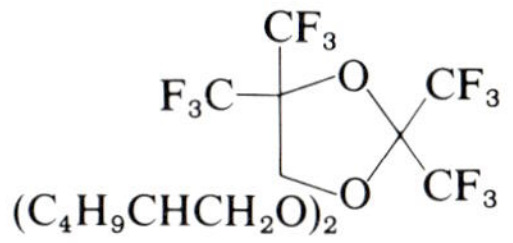 ![CF3 structure] $(C_4H_9CHCH_2O)_2$ | 130°/2.4 | 1.3870 | — | 11 | 47 |
| $C_{23}H_{40}$ | | | | | |
| 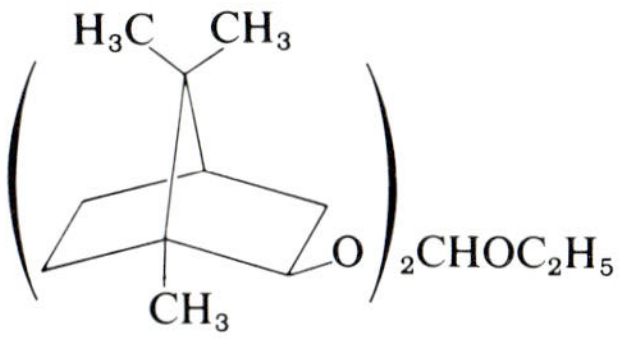 ![bicyclic structure with H3C, CH3, (O)2CHOC2H5] | 192°–194°/10 | — | — | 3 | 249 |

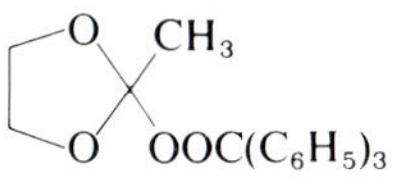

# PHYSICAL PROPERTIES OF ORTHO ESTERS

| Structure | Formula | bp/mp | $n_D$ | $d$ | n | Ref. |
|---|---|---|---|---|---|---|
| [structure: C₂H₅OCH(O-cyclohexyl with CH₃ and CH(CH₃)₂)₂] | $C_{23}H_{44}$ | 195°–197°/11 | 1.463 (20°) | 0.932 (20°) | 3 | *249* |
| $CH_2CH_2C(OH)(C_6H_4CH_3\text{-}m)_2$ | $C_{24}H_{28}$ | (131°) | — | — | 4 | *335* |
| [adamantane-like orthoester structure] | | | | | | |
| [spiro dibenzo structure] | $C_{25}H_{16}$ | (340°) | — | — | 11 | *370a* |
| $C(OC_6H_5)_4$ | $C_{25}H_{20}$ | (97°–98°) | — | — | 2 | *118* |
| $HC(OCH_2CH_2C_6H_5)_3$ | $C_{25}H_{28}$ | 83°–83.5°/6 | 1.5320 (20°) | 1.0417 (20°) | — | *111* |
| $CH_3C(OC_2H_5)_2OOC(C_6H_5)_3$ | | (49°–51°) | — | — | 5 | *319* |
| $HC[OCH(CH_3)C_6H_{13}]_3$ | $C_{25}H_{52}$ | 202°–203°/1 | 1.4376 (15°) | 0.8592 (15°) | 3 | *153* |
| $C(OC_6H_{13})_4$ | | 184°–185°/2 | 1.4323 | 0.8767 | 3 | *6a* |

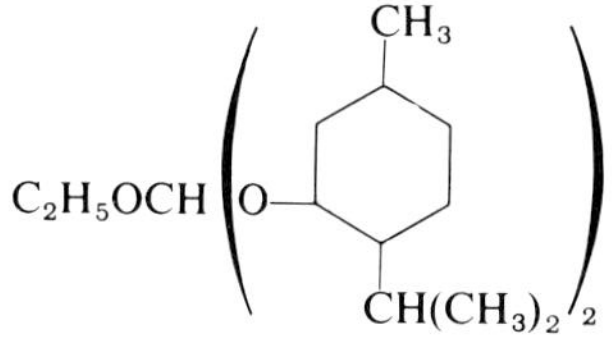

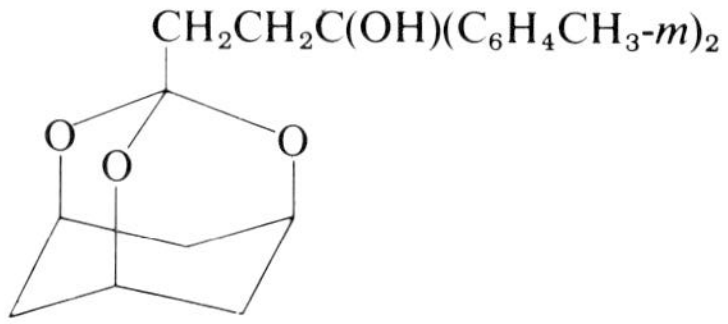

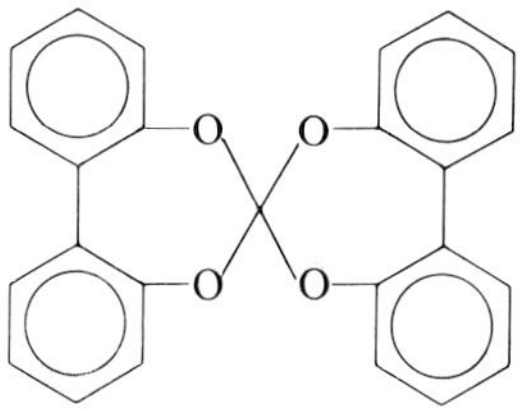

## TABLE III—continued

| Structure | B.p., °C/mm (M.p., °C) | Refractive index $n^a$ | Density $d^b$ | Method[c] | References |
|---|---|---|---|---|---|
| [imidazole]–C(OC$_6$H$_4$Cl-$p$)$_3$ | — | C$_{26}$H$_{17}$ — | — | 2 | *93b* |
| [tetracyclic structure with O, CN groups] | (238°) | C$_{26}$H$_{20}$ — | — | 2 | *93b* |
| (C$_4$H$_9$O)$_3$CC(OC$_4$H$_9$)$_3$ | 151°/0.01 | C$_{26}$H$_{54}$ — | — | 2 | *19, 20* |
| [benzimidazole]–C(OC$_6$H$_5$)$_3$ | — | C$_{27}$H$_{12}$ — | — | 11 | *152a* |

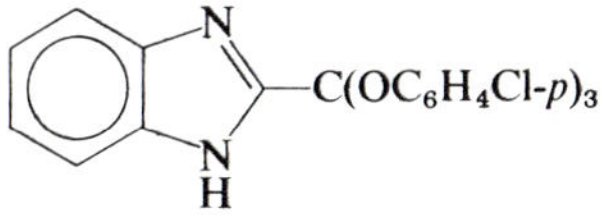

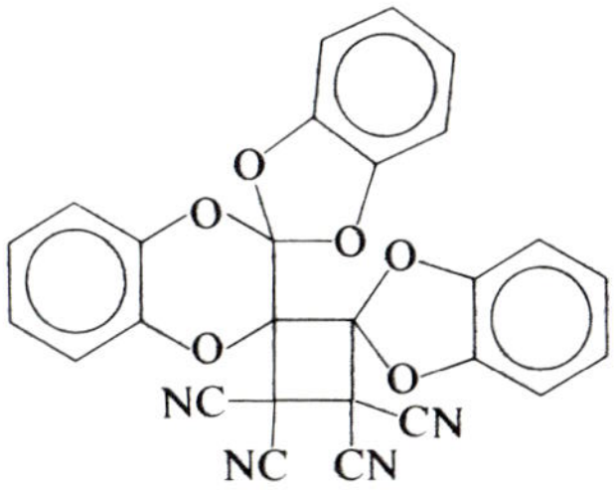

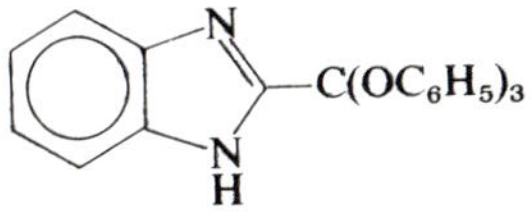

| Compound | bp/mp | Formula | $n_D$ | $d$ | | Ref. |
|---|---|---|---|---|---|---|
| C(OC$_6$H$_4$CH$_3$-$p$)$_3$ 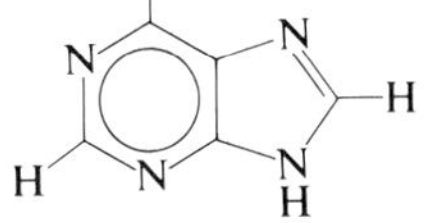 [with triazole] | (215°) | C$_{27}$H$_{24}$ | — | — | 2 | *63a* |
| C$_2$H$_5$C(OC$_8$H$_{17}$)$_3$ | 217°/3 | C$_{27}$H$_{56}$ | 1.4404 | 0.8635 | 3 | *7–9* |
| C$_6$H$_5$C[OCH$_2$(CF$_2$)$_5$CHF$_2$]$_3$ | 187°/0.1 | C$_{28}$H$_{14}$ | — | — | 2 | *133* |
| C[OCH$_2$(CF$_2$)$_5$CHF$_2$]$_4$ | 170°/0.008 | C$_{29}$H$_{12}$ | — | — | 2 | *133* |
| CH$_2$CH$_2$C(OH)($\beta$-C$_{10}$H$_7$)$_2$ 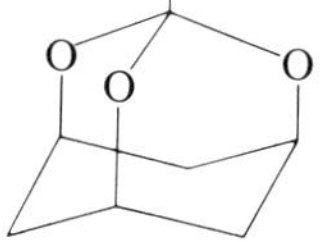 [bicyclic structure] | (185°) | C$_{30}$H$_{28}$ | — | — | 4 | *335* |
| ($\beta$-C$_{10}$H$_7$O)$_3$CH | (137°–150°) | C$_{31}$H$_{22}$ | — | — | 2 | *248* |
| C$_6$H$_5$CH$_2$C(OC$_2$H$_5$)$_2$OOC(C$_6$H$_5$)$_3$ | (72°–73°) | C$_{31}$H$_{32}$ | — | — | 5 | *319* |
| 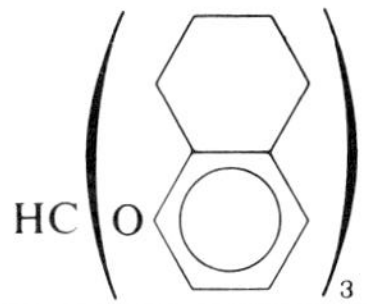 HC(naphthyloxy)$_3$ | (160°–162°) | C$_{31}$H$_{34}$ | — | — | 2 | *369* |

TABLE III—continued

| Structure | B.p., °C/mm (M.p., °C) | Refractive index $n^a$ | Density $d^b$ | Method$^c$ | References |
|---|---|---|---|---|---|
| [structure with H₃C, CH₃, CH₃ groups on bicyclic system, HC(O—)₃] | $C_{31}H_{52}$ (231°–234°) | — | — | 3 | 249 |
| [structure with H₃C, CH₃, CH₃ groups on cyclohexane, HC(O—)₃] | $C_{31}H_{58}$ (71.5°–72°) | — | — | 3 | 249 |
| C(OCH₂CH₂C₆H₅)₄ | $C_{33}H_{36}$ 109°–110°/20 | 1.5270 (20°) | 1.0379 (20°) | — | 111 |
| C(OC₈H₁₇)₄ | $C_{33}H_{68}$ 242°–245°/2 | 1.4425 | 0.8704 | 3 | 6a |

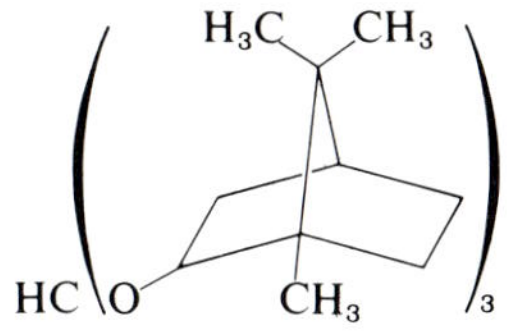

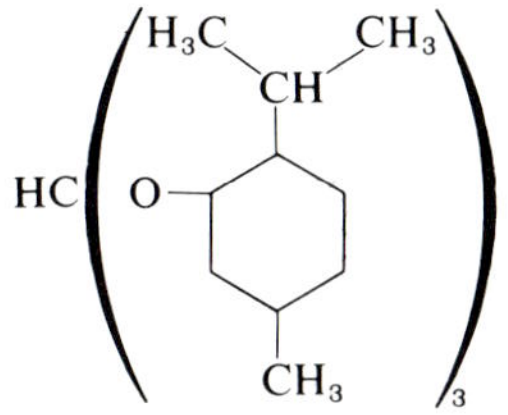

| Compound | | BP/mm (MP) | Formula | $n_D$ | $d_4$ | Method | Ref |
|---|---|---|---|---|---|---|---|
| [HC(O-[Br-substituted phenyl]-CH₂CH(CH₃)C₂H₅)]₃ | | (255°) | $C_{34}H_{40}$ | — | — | 2 | 181 |
| $C(OC_9H_{19})_4$ | | 237°–240°/1.5 | $C_{37}H_{76}$ | 1.4402 | 0.8742 | 3 | 6a |
| $HC(OC_6H_5CH_2C_6H_5\text{-}p)_3$ | | (126°–127°) | $C_{40}H_{34}$ | — | — | 2 | 181 |
| $C(OC_{10}H_{21})_4$ | | 288°–290°/2 | $C_{41}H_{84}$ | 1.4462 | 0.8692 | 3 | 6a |

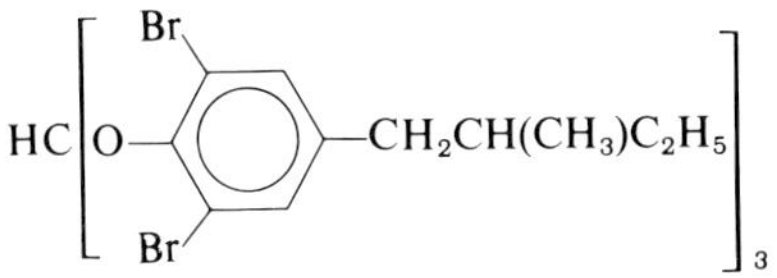

[a] For sodium D line at 25°C unless other temperature is specified.
[b] Measured at 25°C, unless otherwise specified.
[c] 1. Alcoholysis of nitriles and imidates.
  2. Alcoholysis of trihalomethyl compounds and α-halo ethers.
  3. Transesterification.
  4. Alteration of acyl or alkoxy substituents.
  5. Addition of alcohols to multiple bonds.
  6. Reaction of carbonium salts with alcohols.
  7. Addition of formates or lactones to epoxides.
  8. Cyclization of acyloxy compounds.
  9. Reaction of orthocarbonates with organometallic compounds.
  10. Diels-Alder additions of ketene acetals.
  11. Miscellaneous other reactions.

72. T. C. Daniels and R. E. Lyons, *J. Am. Chem. Soc.* **58**, 2646 (1936).
72a. R. Delange, *Compt. Rend.* **138**, 423 (1904).
72b. O. C. Dermer and F. B. Slezak, *J. Org. Chem.* **22**, 701 (1957).
72c. O. C. Dermer and F. B. Slezak, U.S. Patent 2,867,667 (1959); *Chem. Abstr.* **53**, 9066 (1959).
73. A. Deutsch, *Chem. Ber.* **12**, 115 (1879).
74. P. A. Diassi and J. Fried, U.S. Patent 3,073,817 (1963); *Chem. Abstr.* **58**, 12643 (1963).
74a. W. Dietsche, *Ann. Chem.* **712**, 21 (1968).
75. K. Dimroth and K. Schromm, *Angew. Chem. Intern. Ed. Engl.* **4**, 873 (1965).
76. O. Dobner and W. Stackmann, *Chem. Ber.* **9**, 1918 (1876).
77. W. von E. Doering and R. M. Haines, *J. Am. Chem. Soc.* **76**, 482 (1954).
78. W. von E. Doering and L. K. Levy, *J. Am. Chem. Soc.* **77**, 509 (1955).
79. L. C. Dolby, C. N. Lieske, D. R. Rosencrantz, and M. J. Schwartz, *J. Am. Chem. Soc.* **85**, 47 (1963).
80. I. B. Douglass and G. H. Warner, *J. Am. Chem. Soc.* **78**, 6070 (1956).
81. J. E. Driver, *J. Am. Chem. Soc.* **46**, 2090 (1924).
82. P. M. Duggleby and G. Holt, *J. Chem. Soc.* p. 3579 (1962).
83. J. P. Dusza and S. Bernstein, U.S. Patent 3,060,208 (1962); *Chem. Abstr.* **58**, 8014 (1963).
84. J. P. Dusza and S. Bernstein, U.S. Patent 3,069,419 (1962); *Chem. Abstr.* **58**, 10275 (1963).
85. J. P. Dusza and S. Bernstein, *J. Org. Chem.* **27**, 4677 (1962).
86. J. P. Dusza and S. Bernstein, *J. Org. Chem.* **28**, 760 (1963).
87. Dynamit-Nobel A.-G., Netherlands Patent Appl. 6,500,508 (1965); *Chem. Abstr.* **64**, 601 (1966).
87a. Dynamit Nobel A.-G., Netherlands Patent Appl. 6,609,612 (1967); *Chem. Abstr.* **67**, 21446 (1967).
88. G. M. Dyson, *in* "Organic Syntheses," 2nd ed., Coll. Vol. I, p. 506. Wiley, New York, 1951.
89. F. G. Eggart, P. Keller, C. Lehmann, and H. Wehrli, *Helv. Chim. Acta* **51**, 940 (1968).
90. G. Eglinton, E. R. H. Jones, B. L. Shaw, and M. C. Whiting, *J. Chem. Soc.* p. 1860 (1954).
91. H. Eilingsfeld, G. Neubauer, M. Seefelder, and H. Weidinger, *Chem. Ber.* **97**, 1232 (1964).
92. H. Eilingsfeld, M. Seefelder, and H. Weidinger, *Angew. Chem.* **72**, 836 (1960).
93. H. Eilingsfeld, M. Seefelder, and H. Weidinger, *Chem. Ber.* **96**, 2671 (1963).
93a. E. L. Eliel and C. A. Giza, *J. Org. Chem.* **33**, 3754 (1968).
93b. B. C. Ennis, G. Holan, and E. L. Samuel, *J. Chem. Soc., C* p. 30 (1967).
94. A. Ercoli and R. Gardi, Belgian Patent 618,831 (1962); *Chem. Abstr.* **59**, 11620 (1963).
95. A. Ercoli and R. Gardi, U.S. Patent 3,139,425 (1964); *Chem. Abstr.* **61**, 12065 (1964).
96. A. Ercoli and R. Gardi, British Patent 963,829 (1964); *Chem. Abstr.* **62**, 9209 (1965).
97. J. G. Erickson, *J. Org. Chem.* **20**, 1573 (1955).
98. U. Faas and H. Hilgert, *Chem. Ber.* **87**, 1343 (1954).
99. Feldmeuhle Papier and Zellstoffwerke A.-G., Belgian Patent 613,988 (1962); *Chem. Abstr.* **57**, 13623 (1962).

100. H. Feuer and W. H. Gardner, *J. Am. Chem. Soc.* **76**, 1375 (1954).
100a. L. F. Fieser and M. Fieser, "Reagents for Organic Synthesis," p. 1208. Wiley, New York, 1967.
101. V. E. B. Filmfabrik Agfa Wolfen, Belgian Patent 617,666 (1962); *Chem. Abstr.* **58**, 12425 (1963).
102. P. Fritsch, *Ann. Chem.* **297**, 315 (1897).
103. A. Frohling and J. F. Arens, *Rec. Trav. Chim.* **81**, 1009 (1962).
103a. Y. Gaoni, *J. Chem. Soc.* (C), p. 2925 (1968).
103b. Y. Gaoni, *J. Chem. Soc.* (C), p. 2934 (1968).
104. R. Gardi, *Hormonal Steroids, Proc. 1st Intern. Congr. Hormonal Steroids*, 1964, Vol. 2, p. 99. Academic Press, New York, 1965; *Chem. Abstr.* **64**, 14239 (1966).
105. R. Gardi, R. Vitali, and A. Ercoli, *Tetrahedron Letters* p. 448 (1961).
106. R. Gardi, R. Vitali, and A. Ercoli, *Gazz. Chim. Ital.* **93**, 413 (1963).
107. R. Gardi, R. Vitali, and A. Ercoli, *Gazz. Chim. Ital.* **93**, 431 (1963).
108. R. Gardi, R. Vitali, and A. Ercoli, *Tetrahedron* **21**, 179 (1965).
108a. H. Gilman and R. Wilder, *J. Am. Chem. Soc.* **77**, 6644 (1957).
109. Glaxo Group, Ltd., Netherlands Patent Appl. 6,405,165 (1964); *Chem. Abstr.* **62**, 11886 (1965).
110. Glaxo Group, Ltd., Belgian Patent 649,171 (1964); *Chem. Abstr.* **64**, 15958 (1966).
110a. Glaxo Group, Ltd., Netherlands Patent Appl. 6,608,762 (1966); *Chem. Abstr.* **67**, 100345 (1967).
111. V. P. Goguadze, *Soobshch. Gruzin. Filiala Akad. Nauk SSSR* **1**, 513 (1940); *Chem. Abstr.* **37**, 4710 (1943).
112. T. Goto, Y. Kishi, S. Takahashi, and Y. Hirata, *Tetrahedron* **21**, 2059 (1965).
112a. W. R. Grace and Co., British Patent 1,128,963 (1968); *Chem. Abstr.* **70**, 3272 (1969).
113. F. D. Greene, *J. Am. Chem. Soc.* **78**, 2250 (1956).
114. F. D. Greene and W. W. Rees, *J. Am. Chem. Soc.* **82**, 890 and 893 (1960).
116. H. Gross, *Angew. Chem.* **73**, 684 (1961).
116a. H. Gross, J. Freiberg, and B. Costisella, *Chem. Ber.* **101**, 1250 (1968).
117. H. Gross and A. Rieche, *Chem. Ber.* **94**, 538 (1961).
118. H. Gross, A. Rieche, and E. Hoft, *Chem. Ber.* **94**, 544 (1961).
118a. H. Gross and J. Rusche, *Chem. Ber.* **99**, 2625 (1966).
119. H. Gross, J. Rusche, and H. Bornowski, *Ann. Chem.* **675**, 142 (1964).
120. D. L. Hammick and H. E. Wilmut, *J. Chem. Soc.* p. 207 (1935).
121. W. E. Hanford and W. E. Mochel, U.S. Patent 2,229,651 (1941); *Chem. Abstr.* **35**, 2904 (1941).
121a. J. F. Harris, *J. Org. Chem.*, **32**, 2063 (1967).
122. R. H. Hartigan and J. B. Cloke, *J. Am. Chem. Soc.* **67**, 709 (1945).
123. E. G. E. Hawkins, *J. Chem. Soc.* p. 3463 (1955).
124. J. Hebky, *Collection Czech. Chem. Commun.* **13**, 442 (1948).
125. F. Heiber, *Chem. Ber.* **24**, 3677 (1891).
126. H. Henecka and P. Kurtz, *in* "Methoden der organischen Chemie" (E. Mueller, ed.), 4th ed., Vol. VIII, p. 504. Thieme, Stuttgart, 1952.
127. H. Henecka and P. Kurtz, *in* "Methoden der organischen Chemie" (E. Mueller, ed.), 4th ed., Vol. VIII, p. 697. Thieme, Stuttgart, 1952.
127a. K. R. Henery-Logan and T. L. Fridinger, *J. Am. Chem. Soc.* **89**, 5724 (1967).
128. J. H. Hennes, *Dissertation Abstr.* **17**, 36 (1957).
129. W. R. Hertler and R. E. Benson, *J. Am. Chem. Soc.* **84**, 3474 (1962).

130. L. Hesslinga, G. J. Katerberg, and J. F. Arens, *Rec. Trav. Chim.* **76**, 969 (1957).
131. H. Hibbert and M. E. Grieg, *Can. J. Res.* **4**, 254 (1931).
132. A. J. Hill and I. Rabinowitz, *J. Am. Chem. Soc.* **48**, 732 (1926).
133. M. E. Hill, D. T. Carty, D. Tegg, J. C. Butler, and A. F. Strong, *J. Org. Chem.* **30**, 411 (1965).
134. J. Hine, *J. Am. Chem. Soc.* **72**, 2438 (1950).
135. J. Hine, A. D. Ketley, and K. Tanabe, *J. Am. Chem. Soc.* **82**, 1398 (1960).
136. J. Hine, E. L. Pollitzer, and H. Wagner, *J. Am. Chem. Soc.* **75**, 5607 (1953).
137. J. Hine and J. J. Porter, *J. Am. Chem. Soc.* **79**, 5493 (1957).
138. J. Hine and K. Tanabe, *J. Am. Chem. Soc.* **80**, 3002 (1958).
139. D. Hodgson, E. Ritchie, and W. C. Taylor, *Australian J. Chem.* **13**, 385 (1960).
140. R. W. Hoffmann, *Angew. Chem. Intern. Ed. Engl.* **4**, 977 (1965).
140a. R. W. Hoffmann and U. Bressel, *Angew. Chem. Intern. Ed. Engl.* **6**, 808 (1967).
141. R. W. Hoffmann and H. Hauser, *Tetrahedron Letters* p. 197 (1964).
142. R. W. Hoffmann and H. Hauser, *Tetrahedron* **21**, 891 (1965).
142a. R. W. Hoffmann and J. Schneider, *Chem. Ber.* **100**, 3689 (1967).
142b. R. W. Hoffmann and J. Schneider, *Chem. Ber.* **100**, 3698 (1967).
142c. R. W. Hoffmann and J. Schneider, *Tetrahedron Letters* p. 4347 (1967).
143. R. W. Hoffmann, J. Schneider, and H. Hauser, *Chem. Ber.* **99**, 1892 (1966).
143a. G. Holan, E. L. Samuel, B. C. Ennis, and R. W. Hinde, *J. Chem. Soc., C* p. 20 (1967).
144. T. Holm, U. S. Patent 2,611,787 (1952); *Chem. Abstr.* **47**, 9997 (1953).
145. B. Holmberg, *Chem. Ber.* **40**, 1740 (1907).
146. J. Horak, *Collection Czech. Chem. Commun.* **28**, 2328 (1963).
147. J. Houben, *Chem. Ber.* **59**, 2878 (1926).
148. J. Houben and E. Pfankuch, *Chem. Ber.* **59**, 2392 (1926).
149. J. Houben and K. M. L. Schultze, *Chem. Ber.* **44**, 3235 (1911).
150. B. W. Howk and J. C. Sauer, U.S. Patent 2,840,613 (1958); *Chem. Abstr.* **52**, 17186 (1958).
151. B. W. Howk and J. C. Sauer, *J. Am. Chem. Soc.* **80**, 4607 (1958).
152. T. C. Huang and K. P. Sung, *J. Chinese Chem. Soc.* **2**, 1 (1934).
152a. R. Hull and R. Farrand, *Chem. Commun.* p. 164 (1965).
153. H. Hunter, *J. Chem. Soc.* **125**, 1389 (1924).
153a. C. K. Ikeda, R. A. Braun, and B. E. Sorenson, *J. Org. Chem.* **29**, 286 (1964).
154. C. K. Ingold and W. J. Powell, *J. Chem. Soc.* **119**, 1228 (1921).
155. J. I. Iotsitch and F. Kochelov, *J. Russ. Phys.-Chem. Soc.* **42**, 1491 (1910); *Bull. Soc. Chim.* [4] **10**, 1308 (1911).
156. Y. N. Ivaschenko and S. D. Moshchitskii, *Zh. Obshch. Khim.* **33**, 1412 (1963).
157. Y. N. Ivaschenko and S. D. Moshchitskii, *Zh. Obshch. Khim.* **33**, 3825 (1963).
158. J. Jadot, A. David, and K. Kasperczyk, *Bull. Soc. Roy. Sci. Liege* **29**, 196 (1960); *Chem. Abstr.* **55**, 11348 (1961).
160. R. Joly and C. Warnant, U.S. Patent 3,017,409 (1962); *Chem. Abstr.* **56**, 12988 (1962).
160a. J. Jonas, T. P. Forrest, M. Kratochvil, and H. Gross, *J. Org. Chem.* **33**, 2126 (1968).
161. R. G. Jones, *J. Am. Chem. Soc.* **73**, 5168 (1951).
162. A. Jonsson, *Acta Chem. Scand.* **7**, 596 (1953).
163. T. Kariyone and Y. Kimura, *J. Pharm. Soc. Japan* **500**, 746 (1923); *Chem. Abstr.* **18**, 386 (1923).
163a. J. Kashiro, M. Kanaoka, and A. Kosakada, Japanese Patent 3496 ('67); *Chem. Abstr.* **67**, 54117 (1967).

163b. J. R. Katz and J. Selman, *Z. Physik* **46**, 392 (1928).
164. W. Kaufman and E. Dreger, *in* "Organic Syntheses," Coll. Vol. I, p. 258. Wiley, New York, 1941.
165. J. D. Kendall and J. R. Majer, *J. Chem. Soc.* p. 687 (1948).
165a. G. Kesslin, A. C. Flisik, and R. W. Handy, U.S. Patent 3,258,496 (1966); *Chem. Abstr.* **65**, 7064 (1966).
165b. G. Kesslin and C. M. Orlando, *J. Org. Chem.* **31**, 2682 (1966).
165c. M. Kimura, M. Tohma, and I. Yoshizawa, *Chem. & Pharm. Bull (Tokyo)* **15**, 1204 (1967).
166. F. E. King, K. G. Latham, and M. W. Partridge, *J. Chem. Soc.* p. 4268 (1952).
167. H. King and E. V. Wright, *J. Chem. Soc.* p. 254 (1939).
168. A. Kiprianov, Z. P. Suitnik, and E. D. Suich, *Zh. Obshch. Khim.* **6**, 42 (1936).
169. A. Kiprianov, Z. P. Suitnik, and E. D. Suich, *Zh. Obshch. Khim.* **6**, 576 (1936).
170. W. Kirmse, "Carbene Chemistry," Chapter 9. Academic Press, New York, 1964.
171. A. Kirpal and H. Kunze, *Chem. Ber.* **62**, 2102 (1929).
172. A. V. Kirsanov and V. P. Molosnova, *Zh. Obshch. Khim.* **28**, 30 (1958).
173. A. V. Kirsanov and V. P. Molosnova, *Zh. Obshch. Khim.* **29**, 1684 (1959).
174. O. J. Kleinjot, *Inorg. Chem.* **2**, 825 (1963).
175. E. B. Knott, *J. Chem. Soc.* p. 686 (1945).
176. I. L. Knunyants, B. L. Dyatkin, L. S. German, and E. P. Mochalina, *Izv. Akad. Nauk SSSR, Otd. Khim. Nauk* p. 231 (1960).
177. W. G. Kofron, F. G. Kirby, and C. R. Hauser, *J. Org. Chem.* **28**, 873 (1963).
178. H. Kopp, *Jahresber. Chem.* p. 391 (1860).
179. F. Korte and K. Trautner, *Chem. Ber.* **95**, 281 (1962).
180. O. J. Kovacs, G. Schneider, and K. Lang, *Proc. Chem. Soc.* p. 374 (1963).
180a. O. J. Kovacs, G. Schneider, L. K. Lang, and A. Apjok, *Tetrahedron* **23**, 4181 (1967).
181. T. Koyama and T. Asou, *J. Pharm. Soc. Japan* **71**, 31 (1951).
182. R. Kuhn and D. Wieser, *Angew. Chem.* **69**, 371 (1957).
183. D. G. Kundiger and J. H. Hennes, U.S. Patent 2,814,647 (1957); *Chem. Abstr.* **52**, 5451 (1958).
184. D. G. Kundiger and J. H. Hennes, U.S. Patent 3,000,904 (1958); *Chem. Abstr.* **56**, 1459 (1962).
185. S. M. Kupchan, S. P. Erickson, and Y. T. S. Liang, *J. Am. Chem. Soc.* **88**, 347 (1966).
186. S. M. Kupchan, N. Gruenfeld, and N. Katsui, *J. Med. Pharm. Chem.* **5**, 690 (1962).
187. S. M. Kupchan and D. Lavie, *J. Am. Chem. Soc.* **77**, 683 (1955).
188. W. C. Kuryla and D. G. Leis, *J. Org. Chem.* **29**, 2773 (1964).
189. H. Kwart and M. B. Price, *J. Am. Chem. Soc.* **82**, 5123 (1960).
190. F. Lacasa, J. Pascual and L. V. del Arco, *Anales Real Soc. Espan. Fis. Quim. (Madrid)* **B52**, 549 (1956); *Chem. Abstr.* **51**, 5711 (1957).
191. G. D. Lander and F. T. Jewson, *J. Chem. Soc.* **83**, 766 (1903).
192. A. Lao, M.A. thesis, University of California, Santa Barbara, California (1965).
193. S. J. Lapporte, *J. Org. Chem.* **27**, 3098 (1962).
194. A. Y. Lazaris, E. N. Zilberman, and O. D. Strizhakov, *Zh. Obshch. Khim.* **32**, 900 (1962).
195. D. M. Lemal, E. P. Gosselink, and S. D. McGregor, *J. Am. Chem. Soc.* **88**, 582 (1966).
195a. A. Lenz, K. Hass, and H. Epler, German Patent 1,214,943 (1966); *Chem. Abstr.* **65**, 7064 (1966).

197. R. Levaillant, *Ann. Chim. (Paris)* [11] **6**, 552 (1936).
198. H. Limpricht, *Ann. Chem.* **135**, 87 (1865).
199. D. J. Loder and W. F. Gresham, U.S. Patent 2,409,699 (1946); *Chem. Abstr.* **41**, 1237 (1947).
199a. Lonza Elektrizitatswerke und Chemische Fabriken A.-G., British Patent 853,405 (1960); *Chem. Abstr.* **55**, 16426 (1961).
199b. C. J. Ludman and T. C. Waddington, *J. Chem. Soc.* (A), p. 1826 (1966).
200. McBane, M.S. thesis, University of Wisconsin (1941).
201. R. A. McDonald and R. A. Krueger, *J. Org. Chem.* **31**, 488 (1966).
202. S. M. McElvain and C. L. Aldridge, *J. Am. Chem. Soc.* **75**, 3987 (1953).
203. S. M. McElvain and C. L. Aldridge, *J. Am. Chem. Soc.* **75**, 3993 (1953).
204. S. M. McElvain, H. I. Anthes, and S. H. Shapiro, *J. Am. Chem. Soc.* **64**. 2525 (1942).
205. S. M. McElvain and A. N. Bolstad, *J. Am. Chem. Soc.* **73**, 1988 (1951).
206. S. M. McElvain and R. L. Clarke, *J. Am. Chem. Soc.* **69**, 2661 (1947).
207. S. M. McElvain, R. L. Clarke, and G. D. Jones, *J. Am. Chem. Soc.* **64**, 1966 (1942).
208. S. M. McElvain and D. H. Clemens, *J. Am. Chem. Soc.* **80**, 3915 (1958).
209. S. M. McElvain and M. J. Curry, *J. Am. Chem. Soc.* **70**, 3781 (1948).
210. S. M. McElvain and W. R. Davie, *J. Am. Chem. Soc.* **74**, 1816 (1952).
211. S. M. McElvain, E. R. Degginger, and J. D. Behun, *J. Am. Chem. Soc.* **76**, 5736 (1954).
212. S. M. McElvain and B. Fajardo-Pinzon, *J. Am. Chem. Soc.* **67**, 650 (1945).
213. S. M. McElvain and B. Fajardo-Pinzon, *J. Am. Chem. Soc.* **67**, 690 (1945).
214. S. M. McElvain, R. E. Kent, and C. L. Stevens, *J. Am. Chem. Soc.* **68**, 1922 (1946).
215. S. M. McElvain and D. Kundiger, *J. Am. Chem. Soc.* **64**, 254 (1942).
216. S. M. McElvain and G. R. McKay, *J. Am. Chem. Soc.* **77**, 5601 (1955).
217. S. M. McElvain and H. F. McShane, *J. Am. Chem. Soc.* **74**, 2662 (1952).
218. S. M. McElvain, S. B. Mirviss, and C. L. Stevens, *J. Am. Chem. Soc.* **73**, 3807 (1951).
219. S. M. McElvain and L. R. Morris, *J. Am. Chem. Soc.* **73**, 206 (1951).
220. S. M. McElvain and L. R. Morris, *J. Am. Chem. Soc.* **74**, 2657 (1952).
221. S. M. McElvain and R. D. Mullineaux, *J. Am. Chem. Soc.* **74**, 1811 (1952).
222. S. M. McElvain and J. W. Nelson, *J. Am. Chem. Soc.* **64**, 1825 (1942).
223. S. M. McElvain and J. P. Schroeder, *J. Am. Chem. Soc.* **71**, 40 (1949).
224. S. M. McElvain and J. P. Schroeder, *J. Am. Chem. Soc.* **71**, 47 (1949).
225. S. M. McElvain and R. E. Starn, *J. Am. Chem. Soc.* **77**, 4571 (1955).
226. S. M. McElvain and C. L. Stevens, *J. Am. Chem. Soc.* **68**, 1917 (1946).
227. S. M. McElvain and C. L. Stevens, *J. Am. Chem. Soc.* **69**, 2663 (1947).
228. S. M. McElvain and B. E. Tate, *J. Am. Chem. Soc.* **73**, 2233 (1951).
229. S. M. McElvain and J. T. Venerable, *J. Am. Chem. Soc.* **72**, 1661 (1950).
230. S. M. McElvain and P. M. Walters, *J. Am. Chem. Soc.* **64**, 1963 (1942).
231. S. M. McElvain and P. L. Weyna, *J. Am. Chem. Soc.* **81**, 2579 (1959).
232. J. F. McKenna and F. J. Sowa, *J. Am. Chem. Soc.* **60**, 124 (1938).
232a. D. Martin, *Chem. Ber.* **98**, 3286 (1965).
233. D. Martin, H. J. Herrman, S. Rackow, and K. Nadolski, *Angew. Chem. Intern. Ed. Engl.* **4**, 73 (1965).
234. S. G. Matsoyan, G. M. Pogosyan, and M. A. Eliazyan, *Vysokomolekul. Soedin.* **5** 777 (1963).
235. A. R. Mattocks, *J. Chem. Soc.* p. 1918 (1964).

236. H. Meerwein, in "Methoden der organischen Chemie" (E. Mueller, ed.), Vol. VI, Part 3, pp. 298–324. Thieme, Stuttgart, 1965.
237. H. Meerwein, in "Methoden der organischen Chemie" (E. Mueller, ed.), Vol. VI, Part 3, p. 306. Thieme, Stuttgart, 1965.
238. H. Meerwein, in "Methoden der organischen Chemie" (E. Mueller, ed.), Vol. VI, Part 3, p. 309. Thieme, Stuttgart, 1965.
239. H. Meerwein, in "Methoden der organischen Chemie" (E. Mueller, ed.), Vol. VI, Part 3, p. 361. Thieme, Stuttgart, 1965.
240. H. Meerwein, *Angew. Chem.* **67**, 374 (1955).
241. H. Meerwein, K. Bodenbenner, P. Borner, F. Kunert, and K. Wunderlich, *Ann. Chem.* **632**, 38 (1960).
242. H. Meerwein, P. Borner, O. Fuchs, H. J. Sasse, H. Schrodt, and J. Spille, *Chem. Ber.* **89**, 2060 (1956).
243. H. Meerwein and G. Hinz, *Ann. Chem.* **484**, 1 (1930).
244. H. Meerwein and H. Sohnke, *Chem. Ber.* **64**, 2375 (1931).
245. H. Meerwein and H. Sohnke, *J. Prakt. Chem.* [2] **137**, 295 (1933).
246. R. C. Mehrotra and R. P. Narain, *Indian J. Appl. Chem.* **28**, 53 (1965).
247. R. Meier and F. Bohler, *Chem. Ber.* **90**, 2342 (1957).
247a. N. A. Milas and R. J. Klein, *J. Org. Chem.* **33**, 848 (1968).
248. T. G. Miller and J. W. Thanassi, *J. Org. Chem.* **25**, 2009 (1960).
249. V. G. Mkhitaryan, *Zh. Obshch. Khim.* **8**, 1361 (1938).
250. V. G. Mkhitaryan, *Zh. Obshch. Khim.* **10**, 667 (1940).
251. W. E. Mochel, C. L. Agre, and W. E. Hanford, *J. Am. Chem. Soc.* **70**, 2268 (1948).
252. K. Moedritzer and J. R. Van Wazer, *J. Org. Chem.* **30**, 3925 (1965).
253. R. F. Moore and W. A. Waters, *J. Chem. Soc.* p. 238 (1953).
254. K. Morita, H. Nawa, and T. Miki, Japanese Patent 5827 ('65); *Chem. Abstr.* **62**, 16341 (1965).
255. F. Moulin, German Patent 1,126,854 (1959); *Chem. Abstr.* **55**, 16426 (1961).
255a. F. Moulin, Swiss Patent 420,100 (1967); *Chem. Abstr.* **67**, 537 and 707 (1967).
256. A. Mustafa, *J. Chem. Soc.* p. S83 (1949).
257. A. Mustafa, *Nature* **166**, 108 (1950).
258. A. Mustafa, *J. Chem. Soc.* p. 1034 (1951).
259. A. Mustafa, A. H. E. Harhash, A. K. E. Mansour, and S. M. A. E. Omran, *J. Am. Chem. Soc.* **78**, 4306 (1956).
260. A. Nagasaki, R. Oda, and S. Nukina, *J. Chem. Soc. Japan* **57**, 169 (1954).
261. R. P. Narain and R. C. Mehrotra, *Proc. Natl. Acad. Sci., India* **A33**, 45 (1963); *Chem. Abstr.* **59**, 5018 (1963).
261a. J. U. Nef, *Ann. Chem.* **298**, 350 (1897).
262. J. U. Nef, *Ann. Chem.* **308**, 329 (1899).
263. M. S. Newman and C. Courduvalis, *J. Am. Chem. Soc.* **88**, 781 (1966).
264. R. Nicoletti and L. Biocchi, *Ann. Chim.* (*Rome*) **50**, 1502 (1960).
264a. F. Nrdel, J. Buddrus, J. Windhoff, W. Bodrowski, D. Klamann, and K. Ulm, *Ann. Chem.* **710**, 77 (1967).
265. S. Oae, W. Tagaki, and A. Ohno, *Tetrahedron* **20**, 417 (1964).
265a. R. Ohme and E. Schmitz, *Ann. Chem.* **716**, 207 (1968).
266. J. M. Osbond, P. G. Philpott, and J. C. Wickens, *J. Chem. Soc.* p. 2779 (1961).
267. E. Pacsu, *Advan. Carbohydrate Chem.* **1**, 77 (1945).
268. J. J. Panouse, J. Schmitt, P. Cornu, A. Hallot, H. Pluchet, and P. Comoy, *Bull. Soc. Chim. France* p. 1753 (1963).
269. J. D. Park, W. M. Sweeney, and J. R. Lacher, *J. Org. Chem.* **21**, 1035 (1956).

270. F. Pauer, *Monatsh. Chem.* **58**, 1 (1931).
271. I. A. Pearl and D. L. Beyer, *J. Am. Chem. Soc.* **74**, 3189 (1952).
272. V. Petrow, O. Stephenson, and A. M. Wild, *J. Pharm. Pharmacol.* **12**, 37 (1960).
273. F. Piacenti, *Gazz. Chim. Ital.* **92**, 225 (1962).
273a. F. Piacenti, M. Bianchi, and P. Pino, *J. Org. Chem.* **33**, 3653 (1968).
274. R. M. Pike and A. M. Dewidar, *Rec. Trav. Chim.* **83**, 119 (1964).
275. A. Pinner, *Chem. Ber.* **16**, 352 (1883).
276. A. Pinner, *Chem. Ber.* **16**, 1643 (1883).
277. A. Pinner, *Chem. Ber.* **23**, 2917 (1890).
278. A. Pinner, "Die Iminoather und deren Derivative," Oppenheim, Berlin, 1892.
279. A. Pinner and F. Klein, *Chem. Ber.* **10**, 1889 (1877).
279a. J. Pirsch, *Monatsh. Chem.* **97**, 260 (1966).
280. H. W. Post, "Chemistry of the Aliphatic Ortho Esters," p. 26. Reinhold, New York, 1943.
281. H. W. Post and E. R. Erickson, *J. Am. Chem. Soc.* **55**, 3851 (1933).
282. H. W. Post and E. R. Erickson, *J. Org. Chem.* **2**, 260 (1937).
283. B. G. Ramsey and R. W. Taft, *J. Am. Chem. Soc.* **88**, 3058 (1966).
284. C. R. Redeman, S. W. Chaikin, and R. B. Fearing, *J. Am. Chem. Soc.* **70**, 3604 (1948).
286. H. Reitter and E. Hess, *Chem. Ber.* **40**, 3020 (1907).
287. A. Rieche, E. Schmitz, and E. Beyer, German (East) Patent 18,047 (1959); *Chem. Abstr.* **55**, 6444 (1961).
288. A. Rieche, E. Schmitz, and E. Beyer, *Chem. Ber.* **91**, 1935 (1958).
289. A. Rieche, E. Schmitz, and E. Beyer, *Chem. Ber.* **91**, 1942 (1958).
290. A. Rieche, E. Schmitz, and E. Beyer, German Patent 1,071,083 (1959); *Chem. Abstr.* **55**, 11368 (1961).
291. A. Rieche, E. Schmitz, W. Schade, and E. Beyer, *Chem. Ber.* **94**, 2926 (1961).
292. A. Rieche, H. E. Seyfarth, and A. Hesse, *Angew. Chem. Intern. Ed. Engl.* **5**, 253 (1965).
293. L. C. Rinzema, J. Stoffelsma, and J. F. Arens, *Rec. Trav. Chim.* **78**, 354 (1959).
294. T. R. Rix and J. F. Arens, *Koninkl. Ned. Akad. Wetenschap., Proc.* **B56**, 364 (1953); *Chem. Abstr.* **49**, 2299 (1955).
295. J. D. Roberts and R. E. McMahon, in "Organic Syntheses," Coll. Vol. IV, p. 457, Wiley, New York, 1963.
296. R. M. Roberts, *J. Org. Chem.* **14**, 277 (1949).
297. R. M. Roberts, J. Corse, R. Boschan, D. Seymour, and S. Winstein, *J. Am. Chem. Soc.* **80**, 1247 (1958).
298. R. M. Roberts, T. D. Higgins, and P. R. Noyes, *J. Am. Chem. Soc.* **77**, 3801 (1955).
299. A. Roedig and P. Bernemann, *Ann. Chem.* **600**, 1 (1956).
300. A. Roedig and E. Degener, *Chem. Ber.* **86**, 1469 (1953).
300a. A. Roedig, K. Grohe, and W. Mayer, *Chem. Ber.* **100**, 2946 (1967).
301. A. Roedig and G. Markl, *Ann. Chem.* **659**, 1 (1962).
302. R. Roger and D. G. Nielson, *Chem. Rev.* **61**, 179 (1961).
303. B. Rose, *Ann. Chem.* **205**, 249 (1880).
303a. R. Rossi, P. Pino, F. Piacenti, L. Lardicci, and G. Del Bino, *J. Org. Chem.* **32**, 842 (1967).
303b. R. Rossi, P. Pino, F. Piacenti, L. Lardicci, and G. Del Bino, *Gazz. Chim. Ital.* **97**, 1194 (1967).
304. Roussel-UCLAF, Belgian Patent 615,766 (1962); *Chem. Abstr.* **58**, 12641 (1963).
305. Roussel-UCLAF, French Patent M1407 (1962); *Chem. Abstr.* **59**, 1713 (1963).

# REFERENCES

306. Roussel-UCLAF, French Patent M1699 (1963); *Chem. Abstr.* **59**, 14067 (1963).
307. R. P. Ruh, U. S. Patent 2,737,530 (1956); *Chem. Abstr.* **50**, 10758 (1956).
308. W. Ruske and I. Hartmann, *J. Prakt. Chem.* [4] **18**, 146 (1962).
309. P. P. T. Sah, *J. Am. Chem. Soc.* **50**, 516 (1928).
310. P. P. T. Sah, *J. Am. Chem. Soc.* **53**, 1836 (1931).
311. P. P. T. Sah, *J. Chinese Chem. Soc.* **1**, 100 (1933); *Chem. Abstr.* **27**, 5729 (1933).
312. P. P. T. Sah and T. S. Mah, *J. Am. Chem. Soc.* **54**, 2964 (1932).
313. P. P. T. Sah, S. Y. Ma, and C. H. Kao, *J. Chem. Soc.* p. 305 (1938).
314. J. Santroch, Z. Valenta, and K. Wiesner, *Experientia* **21**, 730 (1965).
315. K. Sasaki, *Chem. & Pharm. Bull.* (*Tokyo*) **9**, 693 (1961).
315a. R. Scarpati, M. Cioffi, G. Scherillo, and R. A. Nicolaus, *Gazz. Chim. Ital.* **96**, 1164 (1966).
315b. R. Scarpati, D. Sica, and C. Santacroce, *Gazz. Chim. Ital.* **96**, 375 (1966).
315c. H. J. Schaeffer and R. Vince, *J. Med. Chem.* **11**, 15 (1965).
315d. J. W. Scheeren and W. Stevens, *Rec. Trav. Chim.* **85**, 793 (1966).
316. H. Scheibler, W. Beiser, and W. Krabbe, *J. Prakt. Chem.* [2] **133**, 131 (1932).
317. H. Scheibler and M. Depner, *Chem. Ber.* **68**, 2151 (1935).
318. H. Scheibler and M. Depner, *J. Prakt. Chem.* [4] **7**, 60 (1958).
319. E. Schmitz, A. Rieche, and E. Beyer, *Chem. Ber.* **94**, 2921 (1961).
319a. G. Schneider, *Tetrahedron Letters* p. 5921 (1966).
320. G. Schneider and K. J. Kovacs, *Chem. Commun.* p. 202 (1965).
320a. G. Schneider and L. K. Lang, *Chem. Commun.* p. 13 (1967).
321. A. Schonberg and R. Moubacher, *J. Chem. Soc.* p. 1430 (1939).
322. A. Schonberg and A. Mustafa, *J. Chem. Soc.* p. 997 (1947).
323. H. E. Seyfarth, *Chem. Ber.* **101**, 3499 (1968).
323a. H. E. Seyfarth, A. Hesse, and A. Rieche, *Chem. Ber.* **101**, 2069 (1968).
323b. H. E. Seyfarth and A. Hesse, *Chem. Ber.* **100**, 2491 (1967).
323c. H. E. Seyfarth, A. Hesse, and A. Rieche, *Chem. Ber.* **101**, 623 (1968).
323d. H. E. Seyfarth, A. Rieche, and A. Hesse, *Chem. Ber.* **100**, 624 (1967).
323e. E. Shapiro, L. Finkenor, H. Pluchet, L. Weber, C. H. Robinson, E. P. Oliveto, H. L. Herzog, F. A. Tabachnik, and E. Collins, *Steroids* **9**, 143 (1967).
324. E. C. Sherman and A. P. Dunlop, *J. Org. Chem.* **25**, 1309 (1960).
324a. V. I. Shevchenko and A. A. Koval, *Zh. Obshch. Khim.* **38**, 22 (1968).
325. D. A. Shirley, "Preparation of Organic Intermediates," p. 159. Wiley, New York, 1951.
325a. M. F. Shostakovskii, A. S. Atavin, A. N. Mirskova, G. A. Kalabin, and T. S. Proskurina, *Dokl. Akad. Nauk SSSR* **173**, 1360 (1967).
326. F. Sigmund and S. Herschdorfer, *Monatsh. Chem.* **58**, 280 (1931).
327. Y. A. Sinnema and J. F. Arens, *Rec. Trav. Chim.* **74**, 901 (1955).
328. B. Smith, *Acta Chem. Scand.* **10**, 1006 (1956).
329. B. Smith and S. Delin, *Svensk Kem. Tidskr.* **65**, 10 (1953).
330. L. L. Smith and M. Marx, U.S. Patent 2,966,486 (1960); *Chem. Abstr.* **55**, 9476 (1961).
331. L. L. Smith and M. Marx, *J. Am. Chem. Soc.* **82**, 4625 (1960).
332a. G. P. Sokolov and S. Hillers, *Khim. Geterosikl. Soedin.*, *Akad. Nauk Latv. SSR* p. 163 (1965).
333. H. Staudinger and G. Rathsam, *Helv. Chim. Acta* **5**, 645 (1922).
334. H. Stetter and K. H. Steinacker, *Chem. Ber.* **86**, 790 (1953).
335. H. Stetter and K. H. Steinacker, *Chem. Ber.* **87**, 205 (1954).
336. C. L. Stevens, D. Morrow, and J. Lawson, *J. Am. Chem. Soc.* **77**, 2341 (1955).

337. E. T. Stiller, U. S. Patent 2,422,598 (1947); *Chem. Abstr.* **41**, 5903 (1947).
338. A. Stoll and E. Seebeck, *Helv. Chim. Acta* **37**, 824 (1954).
339. P. R. Story and M. Saunders, *J. Am. Chem. Soc.* **84**, 4876 (1962).
340. F. Swarts, *Bull. Soc. Chim. Belges* **35**, 412 (1926).
340a. R. L. Talbot, *J. Org. Chem.* **32**, 834 (1967).
340b. C. Tamura, O. Amakasu, Y. Sasada, and K. Tsuda, *Acta Cryst.* **21**, 226 (1966).
341. D. S. Tarbell and W. E. Lovett, *J. Am. Chem. Soc.* **78**, 2259 (1956).
342. P. Tarrant and H. C. Brown, *J. Am. Chem. Soc.* **73**, 1781 (1951).
343. L. E. Tenenbaum and J. V. Scudi, U.S. Patent 2,636,884 (1953); *Chem. Abstr.* **48**, 5227 (1954).
344. H. Tieckelmann and H. W. Post, *J. Org. Chem.* **13**, 265 (1948).
345. F. Tiemann, *Chem. Ber.* **15**, 2685 (1882).
345a. H. D. A. Tigchelaar-Lutjeboer, H. Bootsma, and J. F. Arens, *Rec. Trav. Chim.* **79**, 888 (1960).
345b. J. Timmermans and M. Hennant-Roland, *J. Chim. Phys.* **56**, 984 (1959).
346. R. Tschesche and G. Snatzke, *Chem. Ber.* **88**, 1558 (1955).
347. K. Tsuda, C. Tamura, and S. Ikuna, *Chem. & Pharm. Bull. (Tokyo)* **12**, 634 (1964).
347a. N. J. Turro and J. R. Williams, *Tetrahedron Letters*, p. 321 (1969).
347b. H. Ulrich, "The Chemistry of Imidoyl Halides," Plenum Press, New York, 1968, p. 66.
348. Z. Valenta and K. Wiesner, *Experientia* **18**, 111 (1962).
349. Z. J. Vejdelek, K. Macek, and B. Budesinsky, *Chem. Listy* **50**, 603 (1956).
349a. J. G. Verkade, R. W. King, and C. W. Heitsch, *Inorg. Chem.* **3**, 884 (1964).
350. J. P. Vila and M. Ballester, *Anales Real. Soc. Espan. Fis. Quim. (Madrid)* **A42**, 1097 (1946); *Chem. Abstr.* **41**, 6549 (1947).
351. J. P. Vila and R. G. Jarque, *Anales Real. Soc. Espan. Fis. Quim. (Madrid)* **A40**, 248 (1944); *Chem. Abstr.* **39**, 3784 (1945).
351a. M. Vincent, J. Maillard, and M. Benard, *Bull. Soc. Chim. France* p. 1580 (1962).
351b. R. Vitali, R. Gardi and A. Ercoli, *Gazz. Chim. Ital.* **96**, 1115 (1966).
352. A. I. Vogel, *J. Chem. Soc.* p. 1850 (1948).
352a. A. I. Vogel, W. T. Cresswell, G. H. Jeffery, and J. Leicester, *J. Chem. Soc.* p. 514 (1952).
353. O. Vogl, B. C. Anderson and D. M. Simons, *J. Org. Chem.* **34**, 204 (1969).
353a. H. von Hartel, *Chem. Ber.* **60**, 1841 (1927).
354. J. Walter, *J. Prakt. Chem.* [2] **48**, 231 (1893).
355. P. M. Walters and S. M. McElvain, *J. Am. Chem. Soc.* **62**, 1482 (1940).
356. H. Wasserman and P. S. Wharton, *J. Am. Chem. Soc.* **82**, 1411 (1960).
357. O. W. Webster, M. Brown, and R. E. Benson, *J. Org. Chem.* **30**, 3223 (1965).
358. A. Weddige, *J. Prakt. Chem.* [2] **26**, 444 (1882).
359. H. Weidinger and H. Eilingsfeld, German Patent 1,122,936 (1960); *Chem. Abstr.* **57**, 4552 (1962).
360. H. A. Weidlich and W. Schulz, German Patent 919,465 (1954); *Chem. Abstr.* **52**, 14685 (1958).
360a. K. Wiesner, Z. Valenta, and J. A. Findlay, *Tetrahedron Letters* p. 221 (1967).
361. C. F. Wilcox and D. L. Nealy, *J. Org. Chem.* **28**, 3446 (1963).
362. A. W. Williamson and G. Kay, *Ann. Chem.* **92**, 346 (1854).
363. A. W. Williamson and G. Kay, *Proc. Roy. Soc.* **7**, 135 (1854).
364. S. Winstein and R. E. Buckles, *J. Am. Chem. Soc.* **65**, 613 (1943).
365. S. Winstein and L. Goodman, *J. Am. Chem. Soc.* **76**, 4368 (1954).

366. S. Winstein, C. R. Lindgren, H. Marshall, and L. L. Ingraham, *J. Am. Chem. Soc.* **75**, 147 (1953).
367. C. E. Wood and W. A. Comley, *J. Soc. Chem. Ind. (London)* **42**, 429T (1923).
368. R. B. Woodward and J. Z. Gougoutas, *J. Am. Chem. Soc.* **86**, 5030 (1964).
369. H. Wynberg and W. S. Johnson, *J. Org. Chem.* **24**, 1424 (1959).
370. B. G. Yasnitskii, S. A. Sarkisyants, and E. G. Ivanyuk, *Zh. Obshch. Khim.* **34**, 1940 (1964).
370a. C. M. S. Yoder and J. J. Zuckerman, *J. Heterocyclic Chem.* **4**, 166 (1967).
371. L. I. Zakharkin, *Izv. Akad. Nauk SSSR, Otd. Khim. Nauk* p. 1064 (1957).
372. V. I. Zaretskii, U.S.S.R. Patent 105,467 (1957); *Chem. Abstr.* **52**, 1207 (1958).
373. V. I. Zaretskii, *Zh. Prikl. Khim.* **37**, 218 (1964).
374. H. E. Zaugg, V. Papendick, and R. J. Michaels, *J. Am. Chem. Soc.* **86**, 1399 (1964).
376. J. Zemlicka and S. Chladek, *Tetrahedron Letters* p. 3057 (1965).
377. P. Zida and P. Viout, *Bull. Soc. Chim. France* p. 2309 (1965).

CHAPTER 2

# Reactions of Ortho Esters Which Result in Carbon–Oxygen and Carbon–Halogen Bond Formation

In this chapter are considered reactions of trialkyl orthocarboxylates and tetraalkyl orthocarbonates which result in the formation of new carbon–oxygen or carbon–halogen bonds. In most of these reactions the ortho ester (or an alkoxycarbonium ion derived from it) functions as an alkylating agent.

Many of these reactions are useful synthetic tools. These include formation and hydrolysis of cyclic ortho ester derivatives of diols and triols (which serve to protect the hydroxyl groups of the polyols, and provide means for their mild monoacylation), alkylation of phenols, enols and acids (to form ethers and esters), and conversion of carbonyl compounds to acetals and ketals. Other reactions, such as the alkylation of halogens by ortho esters, are of only limited synthetic utility, while still others, such as the acid-catalyzed hydrolyses of ortho esters, are of interest primarily because of the light they shed on the mechanisms of reactions of ortho esters with nucleophilic reagents.

## I. HYDROLYSIS OF CARBOXYLIC ORTHO ESTERS

Carboxylic ortho esters are inert, or nearly so, to aqueous alkali. Under acidic conditions, however, they hydrolyze rapidly to alcohols and ordinary carboxylate esters [Eq. (1)]. They are more reactive toward acid hydrolysis than almost any other class of compounds.

$$\text{RC(OR')}_3 + \text{H}_2\text{O} \xrightarrow{\text{H}^+} \text{RCO}_2\text{R'} + 2\,\text{R'OH} \tag{1}$$

The high hydrolytic reactivity of orthocarboxylates complicates their synthesis and storage. To prevent hydrolysis, ortho esters must be isolated under alkaline or anhydrous conditions. In spite of reasonable care to exclude atmospheric moisture, ortho ester samples which have been stored for any length of time are usually contaminated by hydrolysis products.

## A. Synthetic and Other Applications of Ortho Ester Hydrolyses

The stability of ortho esters under alkaline conditions, together with their facile hydrolysis under acidic conditions, makes the ortho ester function a useful protecting group. The ortho ester function of 3-substituted-2,4,10-trioxaadamantanes (**1**) has been used to protect the potential carboxyl group during reactions of the 3-substituent with organometallic compounds and metal hydrides (*39, 40, 212, 267, 268*). Conversion of carboxylic acids to 1-substituted 4-methyl-2,6,7-trioxabicyclo[2.2.2]octanes (**2**) (*24, 60*) by acid-catalyzed reactions with 2-hydroxymethyl-2-methyl-1,3-propanediol (a much more readily available triol than *cis*-phloroglucitol) should offer a more generally applicable method of protecting carboxylate groups. This possibility has not been explored.

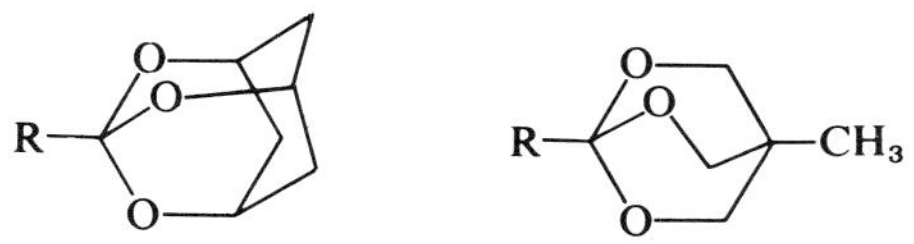

Compounds having two hydroxyl groups 1,2 or 1,3 to each other may often be converted to cyclic 2-alkoxy-1,3-dioxolanes and 2-alkoxy-1,3-dioxanes by transesterification with trialkyl orthocarboxylates. This serves to protect the two hydroxyl groups during reactions which alter other parts of the molecule. Formation of cyclic ortho esters (**3, 4,** and **5**) has been used extensively in syntheses involving 16,17-dihydroxy steroids (*35, 84, 89–91, 263*), 20-keto-17,21-dihydroxy steroids (*97, 98, 121–124, 124a, 254*), and 2',3'-dihydroxynucleoside and nucleotide derivatives and analogous compounds (*62, 95, 115, 115a, 128, 139, 140, 140a, 150, 174, 182a, 202a, 202b, 221a, 221b, 235, 245a, 245b, 286a, 306, 306a, 307*). Three of the hydroxyl groups of 3α,21-dihydroxy-11-oxo-20,20-bis(hydroxymethyl)-5β-pregnane were converted to a trioxabicyclo[2.2.2]octane function (**6**) by reaction with triethyl orthoformate to protect them during phosphorylation of the 3α-hydroxyl group (*238*). The final step in these syntheses is complete hydrolysis of the ortho ester to a di- or trihydroxy compound, or partial hydrolysis to hydroxyacyloxy compounds. The cyclic ethoxydioxolane grouping, introduced by reaction of *vic*-diols with triethyl orthoformate, is perhaps the most easily removed protecting group for diol functions so far reported. The dioxolane ring is cleaved in a few minutes by aqueous acetic acid at room temperature, and the remaining formate group hydrolyzes in a few minutes at room temperature at pH 9 (*150*).

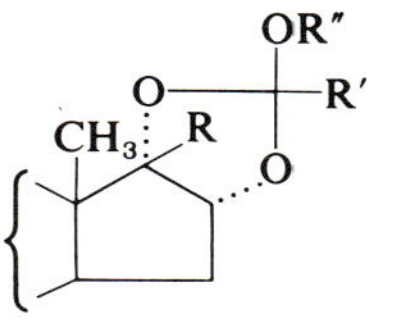

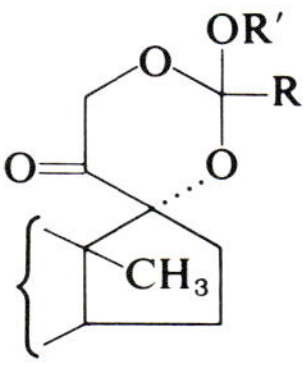

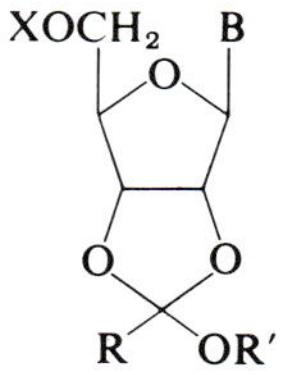

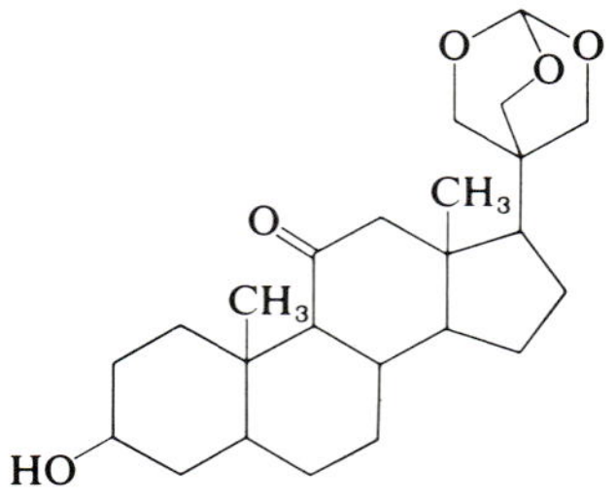

Conversion of diols to cyclic ortho esters (alkoxydioxolanes or alkoxydioxanes), followed by mild acid hydrolysis, provides a convenient means of monoacylating diols.

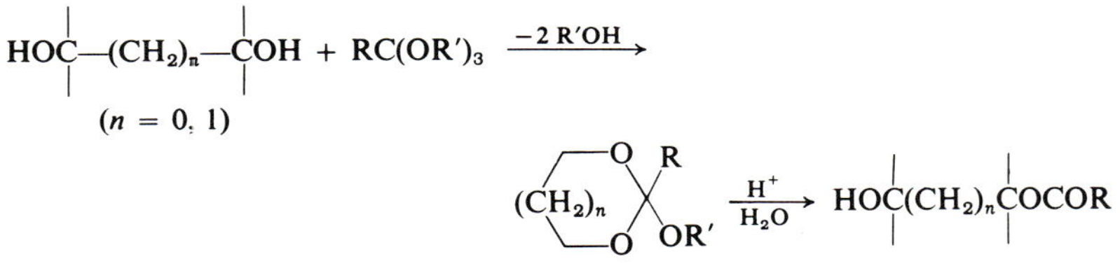

$$HOC-(CH_2)_n-COH + RC(OR')_3 \xrightarrow{-2 R'OH}$$
$$(n = 0, 1)$$

$$(CH_2)_n \underset{O}{\overset{O}{\bigtimes}} \underset{OR'}{\overset{R}{}} \xrightarrow[H_2O]{H^+} HOC(CH_2)_n COCOR \quad (2)$$

The usefulness of the procedure depends on the fact that under acidic conditions ortho esters hydrolyze much more readily than ordinary esters. Cyclic ortho esters derived from 17α,21-dihydroxy steroids hydrolyze under mild conditions to give mostly 17α-acyloxy-21-hydroxy derivatives from cyclic orthoacetates, orthopropionates, orthobutyrates, orthovalerates, and orthohexanoates (*120, 122, 124, 124a, 254*) and mixed 17α- and 21-formates from the cyclic orthoformates (*97, 98, 120*). Hydrolysis of 17α,21-cyclic ortho esters provides the only route to 17-acyloxy-21-hydroxy steroid derivatives (*105a*). Similarly, the orthoformate and the orthoacetate derivatives of a

16α,17α-dihydroxy steroid hydrolyze to 16-acyloxy-17-hydroxy derivatives *(90)*.

Cyclic orthoacetates, orthopropionates, and orthobenzoates derived from *cis*-2,3-dihydroxy-*trans*-decalin are converted by mild acid hydrolysis to mixtures of mono esters in which the *cis-axial*-acyloxy, *equatorial*-hydroxy isomer greatly predominates over the *cis-equatorial*-acyloxy, *axial*-hydroxy isomer *(158a)*. The stereospecificity of these hydrolysis reactions is probably due to greater steric crowding in the transition states for formation of the *equatorial*-acyloxy ester. This view is supported by the observation that hydrolysis of the cyclic orthoformate, in which steric crowding is less, leads to a mixture of the isomeric hydroxy formates. 2′,3′-Cyclic ortho ester derivatives of nucleosides yield mixtures of 2′- and 3′-formates *(95, 128, 221c)*, acetates *(115, 235)*, benzoates *(115, 127a, 235)*, and amino acid esters *(61a, 307, 308)* on acid hydrolysis. The cyclic orthoacetate derived from *cis*-1-hydroxy-2-hydroxymethylcyclohexane hydrolyzes rapidly to a mixture of monoacetates *(87)*, and mild acid hydrolysis of trioxaadamantane derivatives gives monoesters of *cis*-phloroglucitol.

Mixed divinyl nitroalkyl orthoacetates, prepared by the acid-catalyzed reaction of ketene divinyl acetal with nitroalkanols, are converted to nitroalkyl acetates by hydrolysis *(105, 236, 278)*.

2,2-Dialkoxy-2,3-dihydrofuran derivatives, prepared by reaction of diazomethyl ketones with ketene acetals, are hydrolyzed to substituted γ-keto esters *(245)*. 2-Methyl-2-alkoxy-1,3-dioxolan-4-ones hydrolyze to α-acetoxycarboxylic acids *(197)*. Triethyl α-chlorocyanoorthoacetate, prepared by reaction of trichloroacrylonitrile with ethanolic sodium ethoxide, hydrolyzes to ethyl α-chlorocyanoacetate *(237a)*.

Ortho ester hydrolysis provides a very effective means of removing water from organic materials. This fact has been used to advantage in synthesizing acetals (see below) and heterocyclic compounds *(22)*. Ortho ester hydrolysis has also been used to dry organic solvents *(157)* and jet fuels *(183)*, and to stabilize polyol diisocyanate varnish formulations *(104)*. Ortho ester hydrolysis is the basis of a method for quantitative estimation of ortho esters by Karl Fischer titrimetry *(70)*.

Hydrolysis of tetraethyl orthocarbonate in heavy water has been used as a convenient synthesis of $C_2H_5OD$ *(236, 278)*.

Transition metal aquo complexes react with triethyl orthoformate to form hexaethanol complexes, which can be prepared in no other way *(128aa, 172b)*.

## B. Kinetics and Mechanism of Ortho Ester Hydrolysis

The most extensively studied ortho ester hydrolysis reaction is that of triethyl orthoformate [Eq. (1), R = H, R′ = $C_2H_5$]. The only detectable catalysis of this reaction in aqueous buffer solutions is by hydronium ion (*49–51, 53, 134, 141, 258, 260*). Hydronium ion is more than twice as effective a catalyst in heavy water than in ordinary water (*49–51, 141*), and the reaction has small positive values for the entropy of activation (*81, 162*) and volume of activation (*162*).

The sensitivity of ortho ester hydrolysis to acid catalysis, plus the fact that the hydrolysis products are neutral and relatively stable, made this reaction a logical one to study in a search for catalysis by acids other than the solvated proton. The first examples of general acid catalysis to be discovered were hydrolysis reactions of ortho esters. Hydrolyses of triethyl orthoacetate, triethyl orthopropionate, and tetraethyl orthocarbonate are general acid-catalyzed reactions in water (*53*), and triethyl orthoformate hydrolysis exhibits general and acid catalysis in aqueous dioxane (*81, 82*). Trimethyl orthobenzoate hydrolysis is a general acid-catalyzed reaction in 70% methanol–30% water (*168*). Hydrogen ion-catalyzed hydrolyses of triethyl orthoacetate and trimethyl and triethyl orthobenzoates have small positive entropies of activation (*81, 118, 158*). In aqueous solutions the catalytic coefficient ratio of hydrogen ion in heavy and ordinary water ($k_{D^+}/k_{H^+}$) is 1.4 for tetraethyl orthocarbonate hydrolysis (*303*), almost 2.3 for trimethyl and triethyl orthobenzoate hydrolysis (*81, 118*) and about 2.7 for triethyl orthoformate hydrolysis (*51*). Rates of hydrolysis of trimethyl orthobenzoate (*118*) and tetraethyl orthocarbonate (*165*) are unaffected by added nucleophiles. Surface active agents accelerate the hydrolysis of several orthocarboxylates (*27, 117*).

Acid-catalyzed hydrolysis of most, if not all, orthocarboxylates involves rate-limiting formation of di- and trialkoxycarbonium ions. The fact that addition of hydroxylamine or semicarbazide to acidic aqueous solutions of trimethyl orthobenzoate does not affect the rate of disappearance of the ortho ester, but does change the nature of the reaction products, provides direct evidence for a carbonium ion intermediate in trimethyl orthobenzoate hydrolysis (*118*). Absence of acceleration of tetraethyl orthocarbonate hydrolysis by iodide ion, which is highly nucleophilic toward carbon, provides somewhat less direct evidence for rate-limiting carbonium ion formation in this reaction (*165*). The known stability of dialkoxycarbonium and trialkoxycarbonium ions makes it reasonable to suppose that they are intermediates in ortho ester hydrolyses. Meerwein prepared dialkoxycarbonium salts and their heterocyclic analogues by reaction of ortho esters with Lewis acids, by hydride abstraction from acetals, and by alkylation of lactones (*199, 200, 202*). Taft and Ramsey prepared several dialkoxy- and trialkoxycarbonium tetrafluoroborates by Meerwein's procedure, and provided indirect evidence for formation of trialkoxycarbonium bisulfates from tetraalkyl orthocarbonates in 96% sulfuric acid (*232*).

Conversion of di- and trialkoxycarbonium salts to hydrolysis products must be faster than their formation. There is no evidence for accumulation of these carbonium salts in aqueous solution, and in at least one instance rate of disappearance of ortho ester and rate of formation of carboxylate ester were shown to be the same (*292, 293*).

The minimum reaction scheme for acid-catalyzed ortho ester hydrolysis is given in Eq. (3).

$$RC(OR')_3 + HA \xrightarrow{Slow} R\overset{+}{C}(OR')_2 + R'OH + A^- \qquad (3)$$
$$R\overset{+}{C}(OR')_2 + H_2O \xrightarrow{Fast} RCO_2R' + R'OH + H^+$$

Conversion of ortho esters to dialkoxy and trialkoxy carbonium ions usually involves simultaneous processes catalyzed by weak acids and by solvated protons. Catalysis of carbonium ion formation by weak acids may involve either rate-limiting proton transfer from the acid to an ortho ester, or it may involve rate-limiting carbon–oxygen bond cleavage in a hydrogen-bonded complex of the acid and the ortho ester (*116*). Both of these processes are examples of bimolecular electrophilic displacements ($S_E2$) of dialkoxy-carbonium ions from alkoxy oxygen. Rate-limiting attack by the conjugate base of the catalyst on the conjugate acid of the ortho ester is excluded as a possible mechanism, since nucleophilic catalysis is incompatible with observed effects of structural changes in both the ortho ester and the catalyst on hydrolysis rate.

The rate of conversion of the carbonium ion to reaction products is usually faster than its rate of formation. Product formation could involve nucleophilic attack by water on an alkoxy carbon of the carbonium ion, which would produce the carboxylate ester and the conjugate acid of the alcohol, or it could involve formation of a dialkyl ortho ester, $RC(OR')_2OH$, as a reactive intermediate. There is kinetic evidence for the formation of an ortho acid ester intermediate of this type in the hydrolysis of the heterocyclic ortho ester 2,2-dimethoxy-3-phenyl-2$H$-chromene [Eq. (4)] (*166*), but there is no experimental evidence for formation of ortho ester acid intermediates in the hydrolysis reactions of acylic ortho esters.

$$\text{[chromene-OCH}_3\text{/OCH}_3\text{]} \xrightarrow[k_1]{\substack{H^+\\H_2O}} \text{[chromene-OH/OCH}_3\text{]} \xrightarrow[k_2]{\substack{H^+\\H_2O}} \text{[chromenone]} \qquad (4)$$

The mechanism of carbonium ion formation in the hydronium ion-catalyzed reaction is less obvious. A widely held dogma in the field of acid catalysis states that proton transfer from hydronium ion to substrate oxygen must be too fast to be rate-limiting (*28, 79*). This requires that an equilibrium must exist between the substrate and its conjugate acid. If preequilibria exist between ortho esters and their conjugate acids the most reasonable mechanism of formation of the carbonium ion intermediate in hydronium ion-catalyzed ortho ester hydrolysis is the A-1 mechanism of Winstein and Buckles [Eq. (5)] (298). This mechanism is strictly analogous to the accepted mechanism of acetal hydrolysis (72, 144, 188), and is consistent with observed solvent deuterium isotope effects, entropies of activation, and volumes of activation for ortho ester hydrolysis.

$$RC(OR')_3 + H^+ \underset{k_{-1}}{\overset{k_1}{\rightleftarrows}} RC(OR')_2\overset{+}{O}HR'$$

$$RC(OR')_2\overset{+}{O}HR' \xrightarrow[k_2]{\text{Slow}} R\overset{+}{C}(OR')_2 + R'OH \qquad (5)$$

$$R\overset{+}{C}(OR')_2 + H_2O \xrightarrow{\text{Fast}} RCO_2R' + R'OH + H^+$$

However, the effects of acyl substituents on ortho ester hydrolytic reactivity are incompatible with the A-1 mechanism. The rate-limiting step in this mechanism is dissociation of the conjugate acid of the ortho ester into an alkoxycarbonium ion and an alcohol. All other reactions which are known to involve rate-limiting carbonium ion formation, either from a substrate or its conjugate acid, are strongly accelerated by electron-releasing substituents at the carbonium center. Substituent effects on ortho ester hydrolysis are not only very much smaller in magnitude than would be predicted by analogy with acetal hydrolysis, but in some cases are in the opposite direction (Table I). Tetraethyl orthocarbonate and triethyl orthobenzoate, which are known from exchange experiments to form the most stable carbonium ions (80, 202), are hydrolyzed more slowly than triethyl orthoformate or triethyl orthoacetate. Further, hydrogen ion catalytic coefficients for para-substituted trimethyl orthobenzoates in aqueous methanol (168) correlate well with Hammet's $\sigma$-substituent constants (132), but not with Brown's $\sigma^+$-substituent constants (54), whereas the reverse is true for hydrolysis of acetals of substituted benzaldehydes (108).

In summary, although most of the evidence concerning hydronium ion-catalyzed ortho ester hydrolysis is consistent with an A-1 mechanism involving rate-limiting formation of di- and trialkoxycarbonium ions from ortho ester conjugate acids, effects of acyl substituents on reactivity are completely different from those observed for established A-1 hydrolyses of acetals and ketals.

There are two ways of rationalizing acyl substituent effects in terms of an A-1 mechanism. First, a saturation effect may be invoked: perhaps the transition state for dialkoxycarbonium ion formation is so stabilized by the two alkoxy groups that the acyl substituent does not significantly further

TABLE I

RELATIVE RATES OF HYDROLYSIS OF ACETALS AND ORTHO ESTERS, RR′C(OC$_2$H$_5$)$_2$, AT 25°C

| Acetals | | | Ortho esters | | |
|---|---|---|---|---|---|
| R | R′ | Relative rate$^a$ | R | R′ | Relative rate |
| H | H | 1.00$^b$ | H | C$_2$H$_5$O$^d$ | 1.00$^e$ |
| H | CH$_3$ | 6.00 × 10$^3$ | CH$_3$ | C$_2$H$_5$O$^d$ | 38.5 |
| H | C$_6$H$_5$ | 1.7 × 10$^5$ | C$_2$H$_5$ | C$_2$H$_5$O$^d$ | 24.3 |
| CH$_3$ | CH$_3$ | 1.83 × 10$^7$ | C$_6$H$_5$ | C$_2$H$_5$O$^f$ | 0.62 |
| C$_2$H$_5$O | CH$_3$ | 2 × 10$^{9c}$ | C$_2$H$_5$O | C$_2$H$_5$O$^d$ | 0.17 |

$^a$ Data for 50% dioxane solution (163a).
$^b$ $k_2 = 4.13 \times 10^{-5}$ liter mole$^{-1}$ sec$^{-1}$.
$^c$ Estimated.
$^d$ Data from Bronsted and Wynne-Jones (53).
$^e$ $k_2 = 5.38 \times 10^2$ liter mole$^{-1}$ sec$^{-1}$.
$^f$ Data from DeWolfe and Jensen (81).

stabilize it. Taft and co-workers found that a number of dialkoxy and trialkoxycarbonium ions have nearly identical appearance potentials, and presumably nearly identical stabilization energies, in the gas phase (195). Whether a similar saturation effect exists in aqueous solutions is not known. Such an effect has never been observed in unimolecular solvolyses, even in the case of arylmethyl halides, which form very stable carbonium ions (271).

A second possible explanation of substituent effects on hydronium ion-catalyzed hydrolytic reactivity assumes that, because dialkoxycarbonium ions are very much more stable than ordinary carbonium ions, the transition states for their formation could be such a short distance along the reaction coordinate that the acyl carbon of the transition state is nearly tetrahedral and hence not geometrically suited for resonance interaction with a substituent (81, 164). Again, there is no precedent among other carbonium ion reactions to support this argument. Formation of very stable triarylmethyl cations is aided by electron-releasing aryl substituents, and there are good linear free energy correlations between Brown's $\sigma^+$-substituent constants and rate or equilibrium constants in reactions of aryl alkyl systems (271, 276).

The most reasonable explanation for acyl substituent effects on ortho ester hydrolytic reactivity abandons the A-1 mechanism altogether. Bunton and DeWolfe (55) pointed out that ortho esters are very weak oxygen bases which hydrolyze rapidly in solutions of such low hydrogen ion concentration that the concentrations of their conjugate acids must be vanishingly small.

Therefore, the preequilibrium protonation required by the A-1 mechanism probably does not occur in ortho ester hydrolysis, and the hydronium ion-catalyzed reaction, like the weak acid-catalyzed reaction, involves rate-limiting proton transfer from catalyst to substrate. That is, the reaction occurs by the $S_E2$ mechanism.

The probability of rate-limiting proton transfers in the hydronium ion-catalyzed reaction is established by the following argument. Application of the steady state approximation to Eq. (5) leads to Eq. (6):

$$k_{H^+} = k_2/[RC(OR')_3][H_3O^+] = k_1k_2/(k_{-1} + k_2) \qquad (6)$$

where $k_{H^+}$ is the hydrogen ion catalytic coefficient. If the reaction occurs by the A-1 mechanism, i.e., if $k_{-1} \gg k_2$, Eq. (6) reduces to Eq. (7), where $K_a$ is the acidity constant of the ortho ester conjugate acid.

$$k_{H^+} = k_2k_1/k_{-1} = k_2/K_a$$
$$k_2 = k_{H^+}K_a \qquad (7)$$

Values of $k_2$ required by the A-1 mechanism can therefore be calculated from experimental values of $k_{H^+}$ and estimated values of $K_a$. The high reactivity of ortho esters precludes determination of their acidity constants by conventional means. These constants may be estimated, however, by assuming that Taft's $\rho^*$-constant for ortho ester protonation is similar to that for protonation of aliphatic amines (130), and then applying the $\rho^*\sigma^*$-linear free energy relationship (55, 275). The $pK_a$ values estimated in this way are $-5.4$ for dimethyl acetal, $-7$ for triethyl orthoformate, $-7.6$ for triethyl orthobenzoate, and $-8.5$ for tetraethyl orthocarbonate. These values are in excellent agreement with values calculated from infrared O–D stretching frequency measurements on deuteromethanol solutions of the ortho esters (222).

Using the measured $k_{H^+}$ and estimated $K_a$ values, Eq. (7) leads to $k_2 = 7$ sec$^{-1}$ for dimethyl acetal hydrolysis. This value is much less than the expected value for loss of a proton from a strong conjugate acid, and the A-1 mechanism is clearly applicable to acetal hydrolysis. For ortho ester hydrolysis, however, the situation is different. Calculated values of $k_2$ are $5 \times 10^9$ sec$^{-1}$ for triethyl orthoformate hydrolysis, $2 \times 10^{11}$ sec$^{-1}$ for triethyl orthoacetate hydrolysis, $3 \times 10^9$ sec$^{-1}$ for trimethyl orthobenzoate hydrolysis, and $1 \times 10^{10}$ sec$^{-1}$ for tetraethyl orthocarbonate hydrolysis. The A-1 mechanism therefore requires (since $k_{-1} \gg k_2$) that proton transfers from ortho ester conjugate acids to water occur at rates which are near to or faster than the limit for diffusion-controlled bimolecular processes in solution. It therefore seems reasonable to assume that these reactions do not occur by an A-1 mechanism, but rather by an $S_E2$ mechanism involving rate-limiting proton transfer from hydronium ion to the ortho ester.

Other evidence supports this mechanism. Observed substituent effects show that the rate-limiting transition states have little carbonium ion character. The relative reactivities recorded in Table I clearly suggest that the inductive effects of substituents on rate of hydronium ion attack at ortho ester oxygen outweigh their mesomeric effects on the stability of the transition state for dialkoxycarbonium ion formation. This is also suggested by Kwart's data on hydrolysis of para-substituted trimethyl orthobenzoates (*168*). Observed solvent isotope effects are consistent with the $S_E2$ mechanism if proton transfer from hydronium ion to the substrate is nearly complete in the transition state (*56*). The recently reported observation that the rate of hydrolysis of triphenyl orthoformate is more nearly proportional to hydronium ion concentration than to Hammett's acidity function provides further support for the $S_E2$ mechanism (*229a*).

It is probable that addition of the proton is concerted with breaking of the C—O bond. Ortho ester hydrolysis is mechanistically similar to mutarotation of glucose and dehydration of aldehyde hydrates, for which proton transfer to oxygen is concerted with carbon–oxygen bond heterolysis in either the forward or reverse step (*29–31, 144*, p. 689). An alternative possibility, that proton transfer to alkoxy oxygen is essentially complete before carbon–oxygen bond heterolysis begins, cannot be excluded on the basis of available evidence.

For hydrolysis of para-substituted trimethyl orthobenzoates in aqueous methanol, the catalytic effectiveness of chloroacetic acid relative to hydronium ion decreases sharply with decreasing mesomeric electron release from the para-substituent (*168*). This result is intelligible for an $S_E2$ mechanism. The driving force for this mechanism comes from both formation of the new O—H bond and breaking of the existing C—O bond. For a strong acid catalyst, formation of the new O—H bond can occur without much C—O bond breaking, and the transition state will have little carbonium ion character. For a weak acid catalyst, however, C—O bond breaking is more important, and reactivity should be influenced by the ability of the substituent to stabilize the incipient dialkoxycarbonium ion. Similar arguments account for the observation that in 60% dioxane at 30°C, acetic acid is 55 times more effective as a catalyst for triethyl orthoacetate hydrolysis than for triethyl orthoformate hydrolysis, while hydronium ion is only 42 times more effective (*81*), and for the fact that tetraethyl orthocarbonate hydrolysis in water at 25°C is less sensitive to hydronium ion catalysis than triethyl orthoformate hydrolysis, but more sensitive to catalysis by weak acids (*53*).

## C. Effect of Structure on Reactivity in Ortho Ester Hydrolysis

Carboxylic ortho esters are among the most reactive organic compounds known toward acid-catalyzed hydrolysis. Although a few kinetic studies of ortho ester hydrolysis have been reported, most of the published information on ortho ester hydrolytic reactivity is only qualitative.

Most ortho esters are inert or nearly so toward aqueous alkali. When hydrolysis does occur under alkaline conditions, it can usually be attributed to water functioning as a general acid catalyst (*53*).

The high hydrolytic reactivity of trialkyl ortho esters of aliphatic and benzoic acids in acidic solutions has been established by kinetic studies (*53, 81, 168*). Electron-withdrawing aryl substituents markedly decrease the rate of hydrolysis of trimethyl orthobenzoates (*168*). Trimethyl 2-chloro-2-(phenylmethanesulfonyl)orthoacetate [$C_6H_5SO_2CHClC(OCH_3)_3$] hydrolyzes slowly in dilute hydrochloric acid at room temperature (*37*), and the triethyl ortho ester of 2-carboxyimidazolium chloride is stable in water (*21*). Electron-withdrawing substituents on ortho ester oxygen appear not to affect reactivity as markedly as electron-withdrawing acyl substituents. For example, divinyl 2-nitroalkylorthoacetates hydrolyze readily in 1% $H_2SO_4$, and triphenyl and tris(*p*-tolyl) orthoformate hydrolyze readily in aqueous mineral acid (*18, 280*).

Most of the qualitative observations on hydrolytic reactivity of carboxylic ortho esters concern cyclic, bicyclic, and tricyclic ortho esters. 2-Alkoxy-1,3-dioxolanes hydrolyze rapidly in the presence of aqueous mineral acids (*19, 74, 299*). Various bicyclic and polycyclic dioxolanes also undergo facile acid hydrolysis. Thus, the cyclic orthobenzoate derivative of *cis*-3,4-dimethyl-3-cyclobuten-1,2-diol hydrolyzes rapidly in the presence of aqueous acid (*296*), as do the cyclic orthoacetate derivatives of *cis*- and *trans*-1,2-cyclohexanediol (*299*). 16,17-Alkoxyalkylenedioxy steroid derivatives (**3**) are readily hydrolyzed under acidic conditions (*35, 84, 90, 91, 95, 263*); however, the ortho ester function is stable under conditions which cause acid hydrolysis of a 3-enol ether function (*90*). 17,21-Cyclic ortho ester derivatives of steroids (**4**) hydrolyze rapidly in aqueous methanolic hydrochloric acid (*120–122*) or aqueous solutions of carboxylic acids (*97, 98, 124, 254*).

2′,3′-Cyclic ortho ester derivatives of nucleosides have half-lives of a few seconds to a few minutes at pH 2–3 (*115, 128, 150, 235, 306*). As expected, protonation of the amino group of a nucleoside 2′,3′-cyclic ortho ester

**7**           **8**           **9**

derivative of an α-amino acid greatly diminishes the hydrolytic reactivity of the ortho ester function (307).

Most 2,6,7-trioxabicyclo[2.2.2]octane derivatives (2) hydrolyze rapidly under acidic conditions (24, 86, 181). However, the perfluoro orthoacetate 7 is stable in 95% sulfuric acid at 125°C (277).

Trioxaadamantanes (1) and hexaoxadiamantane (8) (288) hydrolyze rapidly under acidic conditions, as does 2,10,11-trioxatricyclo[4.4.4.0$^{1,6}$]tetradecane (9) (185). The cyclic and spirocyclic ortho esters 10, 11, and 12 are also highly susceptible to acid hydrolysis (36, 88, 184, 200).

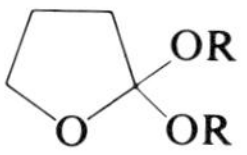

Quantitative data on reactivities of ortho esters toward acid-catalyzed hydrolysis are meager. Most of the published data concern the effects of acyl substituents on reactivity. These data (53, 55, 81, 168) show that hydrolytic reactivity is a function of the inductive, rather than the mesomeric effects of the acyl substituents (Table I). That is, in a series of related triethyl or trimethyl orthocarboxylates, hydrolytic reactivity closely parallels the estimated basicities of the ortho esters, rather than the relative stabilities of the dialkoxycarbonium ions which presumably are intermediates in their hydrolyses. This fact provides strong support for the $S_E2$ mechanism of hydrolysis proposed in the preceding section.

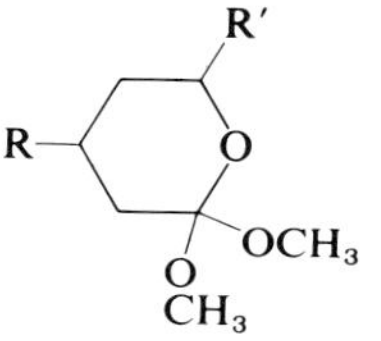

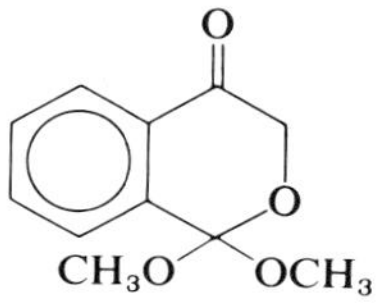

Effects of alkoxy substituents on hydrolytic reactivity of a series of trialkyl orthoformates were studied by Lao (171). Relative reactivities in acetic acid buffers at 35°C for the trimethyl, triethyl, and triisopropyl esters were 1:4:15 in 40% dioxane–60% water. Trineopentyl orthoformate is about as reactive as trimethyl orthoformate in 60% dioxane–40% water. The reactivities of the trimethyl, triethyl, and triisopropyl orthoformates correlate with σ* for the alkoxy substituents, with ρ* of log $(k/k_0) = \rho^*\sigma^*$ (275) being −5.7 [compared to $\rho^* = -8.0$ for hydrolysis of methylals in water at 25°C (259, 274)]. The effect of the neopentyloxy group on orthoformate reactivity is largely steric in origin. The magnitude and direction of alkoxy substituent effects are in accord with the proposed $S_E2$ mechanism.

A recently published kinetic study of triphenyl orthoformate hydrolysis (229a) revealed that this compound is less than a millionth as reactive as triethyl orthoformate ($k_{H^+} = 8.4 \times 10^{-4}$ l./mole second in aqueous 40% dioxane at 25°C.)

Cyclic acetals hydrolyze more slowly than their open-chain analogues, the difference in reactivity being due almost entirely to more negative entropies of activation for the cyclic compounds (*106, 108*). 2-Aryl-1,3-dioxolanes, for example, are only about one-thirtieth as reactive toward acid hydrolysis as the analogous benzaldehyde diethyl acetals (*108*). Lu studied the hydrolysis of a series of 1-substituted 4-methyl-2,6,7-trioxabicyclo[2.2.2]octanes (**2**) in aqueous dioxane buffer solutions, and found that while these bicyclic ortho esters are only about one-tenth as reactive as the analogous triethyl orthocarboxylates, acyl substituent effects in the two series are closely parallel (*181*). Relative reactivities of the bicyclic esters in water at 25°C, for 1-R = H, CH$_3$, C$_2$H$_5$, (CH$_3$)$_3$C, and C$_6$H$_5$ are 1:51:75:127:0.67 ($k_{H^+}$ = 67.9 liter/mole second at 25°C when R = H). The diminished reactivity of the bicyclic esters is due mainly to a more negative entropy of activation. As is true with open-chain esters, substituent effects follow an inductive rather than a mesomeric order.

Large gaps exist in the available data on ortho ester hydrolytic reactivity. A systematic study of the variation of weak acid catalytic coefficients for hydrolysis of several ortho esters should yield information concerning probable transition state structures for the hydrolysis reactions. No kinetic data have been published on hydrolytic reactivity of 2-alkoxy-1,3-dioxolanes, 2-alkoxy-1,3-dioxanes, or trioxaadamantanes. Such data would be of theoretical interest, and would also be pertinent to the use of cyclic ortho ester functions as protecting groups in organic syntheses. Few reactivity data for hydrolyses of triaryl orthocarboxylates have been published.

## II. REACTIONS OF CARBOXYLIC ORTHO ESTERS WHICH YIELD ETHERS

### A. Alkylation of Alcohols

Under mild conditions transesterification [Eq. (8)], rather than ether formation, is the principal reaction between alcohols and ortho esters. This transesterification reaction is an important method of synthesizing ortho esters, and is discussed in detail in Chapter 1.

$$RC(OR')_3 + R''OH \rightarrow RC(OR')_2OR'' + RC(OR'')_2OR' + RC(OR'')_3 + R'OH \quad (8)$$

At higher temperatures, or in the presence of strongly acidic catalysts, the dialkoxycarbonium ions derived from ortho esters may react with alcohols to form dialkyl ethers.

$$R\overset{+}{C}(OR')_2 + R''OH \rightarrow RCO_2R' + R''OR' \quad (9)$$

It is probable that heating a mixture of an ortho ester and an alcohol to sufficiently high temperatures in the presence of an acid catalyst would result in the formation of a mixture of ethers and carboxylate esters, but this reaction has not been developed into a practical synthesis of dialkyl ethers.

$3\beta$-Hydroxy-$\Delta^5$ steroids react rapidly with trialkyl orthoformates in 72% perchloric acid to form $3\beta$-alkoxy-$\Delta^5$ steroids in good yields (93). The actual alkylating agent under the strongly acidic reaction conditions is undoubtedly a dialkoxycarbonium perchlorate. The alkylation reaction appears to be facilitated by the double bond at position 5 of the steroid, since a saturated $3\beta$-hydroxy steroid gave a complex mixture of products containing only a small amount of the $3\beta$-alkoxy ether under the same reaction conditions. This indicates that homoallylic cations may be intermediates in alkylations of the $\Delta^5$ sterols, since these should react with alcohols under acidic conditions to give the observed products.

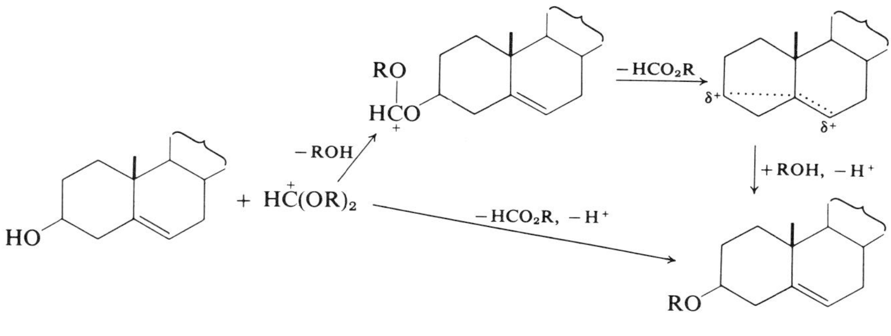

(10)

This reaction was used to prepare $3\beta$-methoxy, ethoxy, allyloxy, isobutyloxy, pentyloxy, and octadecyloxy derivatives of cholest-5-ene, stigmasta-5,22-diene, androst-5-ene, androst-5-ene-17$\beta$-ol, 17,17-dimethyl-18-norandrosta-5,13-diene, 17$\alpha$-ethynyl-17$\beta$-methyl-18-norandrosta-5,13-diene, methyl etiochol-5-enate, and methyl bis-norchol-5-enate. Cholestan-3$\beta$-ol reacted with trimethyl orthoformate under the same conditions to yield a mixture of products containing 9% of the 3$\alpha$-methoxy and 10% of the 3$\beta$-methoxy derivatives.

## B. Alkylation of Phenols

Walther reported in 1915 that 2,4-dinitrophenol and 2,4,6-trinitrophenol are converted to the corresponding ethyl aryl ethers in high yield by heating them with triethyl orthoformate (289).

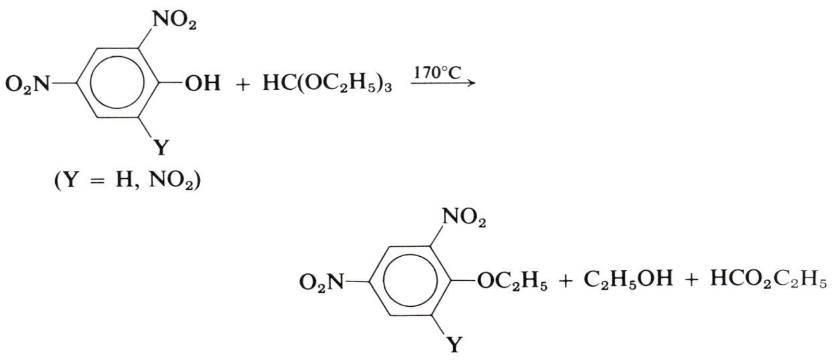

$$\text{O}_2\text{N}-\underset{\underset{(Y = H, NO_2)}{Y}}{\overset{NO_2}{\bigcirc}}-\text{OH} + \text{HC}(\text{OC}_2\text{H}_5)_3 \xrightarrow{170°C}$$

$$\text{O}_2\text{N}-\underset{Y}{\overset{NO_2}{\bigcirc}}-\text{OC}_2\text{H}_5 + \text{C}_2\text{H}_5\text{OH} + \text{HCO}_2\text{C}_2\text{H}_5 \quad (11)$$

Similarly, *p*-nitrophenol is converted to ethyl *p*-nitrophenyl ether in 67% yield by heating with triethyl orthoacetate *(261)*. Heating phenols with carboxylic ortho esters appears to be a general method for preparing alkyl aryl ethers *(138, 261)*. These reactions probably involve transesterification of the ortho esters to mixed dialkyl aryl orthocarboxylates, followed by decomposition of the mixed ortho esters to the alkyl aryl ethers, a carboxylate ester, and an alcohol. This hypothesis is supported by the fact that phenyldiethyl orthoacetate, prepared from phenol and triethyl orthoacetate, decomposes at 180°C to a mixture containing 62% phenetole plus ethanol, phenol, ethyl acetate, and phenyl acetate *(261)*.

Tetraalkyl orthocarbonates are superior reagents for alkylating phenols, a fact ascribed to the instability of aryl trialkyl orthocarbonates assumed to be intermediates in these reactions *(261)*.

$$\text{ArOH} + \text{C}(\text{OR})_4 \xrightarrow{-\text{ROH}} \text{ArOC}(\text{OR})_3 \xrightarrow{\Delta} \text{ArOR} + \text{ROCO}_2\text{R} \quad (12)$$

Smith *(261)* prepared anisole in 87% yield from phenol and tetramethyl orthocarbonate, and phenetole in 85% yield from phenol and tetraethyl orthocarbonate. Similarly, methyl *o*-cresyl ether was prepared in 54% yield, methyl *m*-cresyl ether in 57% yield, methyl 2,6-dimethylphenyl ether in 22% yield, and resorcinol monomethyl ether and dimethyl ethers in 16% and 66% yield. The low yield of methyl 2,6-dimethylphenyl ether, plus the fact that 2,6-di-*tert*-butylphenol does not react with tetraalkyl orthocarbonates, shows that the alkylation reaction is retarded by steric hindrance in the phenol and supports the hypothesis that mixed aryl trialkyl orthocarbonates are intermediates.

Hydroxyquinones are also alkylated by orthocarbonates. Thus, 6-hydroxy-3-formyl-1-ethyl-2,5-dimethyl-4,6[*H*]indole-4,6-dione was converted to the 6-ethoxy derivative by heating with tetraethyl orthocarbonate *(7)*.

## C. Formation of Ethers by Decomposition of Orthocarboxylates

When trialkyl orthocarboxylates or tetraalkyl orthocarbonates are heated to sufficiently high temperatures, particularly in the presence of acid catalysts, they decompose to dialkyl ethers and carboxylate (or carbonate) esters.

$$RC(OR')_3 \rightarrow RCO_2R' + R'OR' \qquad (13)$$

Analogous decomposition reactions are probably involved in alkylations of phenols by ortho esters (see above) (261). These reactions are catalyzed by Lewis acids such as boron trifluoride (186, 189), ferric chloride (281, 286), and zinc chloride (282). The Lewis acid-catalyzed reactions probably involve nucleophilic attack on an alkoxy group of an intermediate dialkoxycarbonium ion by an alcohol molecule or an alkoxy group of an alkoxyanion derived from the Lewis acid.

$$RC(OR')_3 + EX_n \rightarrow [R\overset{+}{C}(OR')_2R'O\overset{-}{E}X_n] \rightarrow RCO_2R' + R'OR' + EX_n \qquad (14)$$

Triethyl orthoformate has been decomposed to ethyl formate and diethyl ether by heating with boron trifluoride (189), ferric chloride (281, 286), and zinc chloride (282). Trimethyl orthobutyrate, when heated with boron trifluoride–ammonia complex, yields dimethyl ether and methyl butyrate (186). Tetraethyl orthocarbonate decomposed to diethyl carbonate and diethyl ether when heated (170).

## D. Conversion of Carbonyl Compounds to Enol Ethers

### 1. Formation of Monoenol Ethers

Ketones having hydrogen *alpha* to the carbonyl group can usually be converted to enol ethers by reaction with trialkyl orthoformates in the presence of acid catalysts.

$$R^1-CO-CHR^2R^3 + HC(OR^4)_3 \xrightarrow{H^+} R^1C(OR^4)=CR^2R^3 + R^4OH + HCO_2R^4 \qquad (15)$$

The enol ether could be formed by alkylation of the ketone or the isomeric enol by a dialkoxycarbonium ion derived from the ortho ester [Eqs. (16) and (17)]. Alternatively, it might be formed by dealcoholation of an initially formed ketal [Eq. (18)]. The fact that ketals were isolated prior to their conversion to enol ethers in a number of cases (8, 12, 48, 65, 75, 141a, 145, 159, 175, 203, 207–209, 214–216, 244, 248, 253, 260a) indicates that ketals are usually intermediates in enol ether syntheses by the ortho ester method, but does not preclude direct alkylation in some instances. It is interesting that the only aldehydes which have been converted to enol ethers by reaction with orthoformates are those which form unusually stable enols.

$$R^1-CO-CHR^2R^3 + H\overset{+}{C}(OR^4)_2 \xrightarrow{-HCO_2R^4}$$
$$R^1-\overset{+}{C}(OR^4)-CHR^2R^3 \xrightarrow{-H^+} R^1C(OR^4)=CR^2R^3 \quad (16)$$

$$R^1-C(OH)=CR^2R^3 + H\overset{+}{C}(OR^4)_2 \longrightarrow$$
$$R^1-C(OR^4)=CR^2R^3 + H^+ + HCO_2R^4 \quad (17)$$

$$R^1-CO-CHR^2R^3 + H\overset{+}{C}(OR^4)_2 \xrightarrow{-HCO_2R^4}$$
$$R^1-C(OR^4)_2-CHR^2R^3 \xrightarrow[-ROH]{+H^+} R^1\overset{+}{C}(OR^4)-CHR^2R^3$$
$$\downarrow -H^+$$
$$R^1-C(OR^4)=CR^2R^3 \quad (18)$$

Whether a ketal or an enol ether is obtained by reaction of a ketone with a trialkyl orthoformate is determined by the reaction conditions. Ketalization is carried out at relatively low temperatures (usually room temperature), and good yields of ketal require neutralization of the acid catalyst and/or distillation of the product at reduced pressures. Enol ether formation, on the other hand, usually occurs at elevated temperatures and is facilitated by distilling the product from the reaction mixture without neutralizing the acid catalyst. There are reports of enol ether syntheses in which no acid catalyst was added to the reaction mixtures. Traces of acids were probably present as impurities during the distillation of the reaction mixtures in these cases; in any event, the best results are obtained if a catalyst such as sulfuric acid or toluenesulfonic acid is used.

Esters of higher orthocarboxylic acids may be substituted for trialkyl orthoformates in the synthesis of enol ethers from ketones (241). There appears to be no advantage in using higher ortho esters, however, and the ready availability of orthoformates makes them the reagents of choice.

Enol ethers of higher alcohols may be prepared by heating the ketone with trimethyl or triethyl orthoformate, the higher alcohol, and an acid catalyst. Butyl cyclohexenyl ether was prepared in this manner from cyclohexanone, butanol, and triethyl orthoformate (190).

An indirect route to enol ethers whose general applicability has not been established involves enol etherification of β-keto esters, followed by saponification of the ester function and decarboxylation of the resulting unsaturated alkoxy acid (64).

$$R-CO-CH_2CO_2C_2H_5 \xrightarrow{HC(OC_2H_5)_3} RC(OC_2H_5)=CHCO_2C_2H_5 \xrightarrow[2.\ H^+]{1.\ OH^-}$$
$$RC(OC_2H_5)=CHCO_2H \xrightarrow[-CO_2]{\Delta} RC(OC_2H_5)=CH_2 \quad (19)$$

Carbonyl compounds which have been converted to enol ethers by reaction with trialkyl orthoformates, either in a one-step process or by dealcoholation of an initially formed ketal, include dialkyl ketones *(141a, 153a, 175)*, aryl alkyl ketones *(119)*, cycloalkanones *(12, 53a, 141a, 178a, 179a, 190, 207, 215, 216, 248)*, palmitaldehyde *(260a)*, cyclic β-keto aldehydes *(95a, 240)*, α-diketones *(286)*, β-diketones *(67, 198, 214, 286)*, β-ketonitriles *(21a)*, α-keto esters *(65, 67)*, β-keto esters *(14, 63, 64, 87a, 203)*, α-formyl esters *(22a)*, alkyl vinyl ketones *(8, 244)* and other α,β-unsaturated ketones *(65, 75, 159)*, aryl alkoxy acetaldehydes and aryl cyanoacetaldehydes *(241, 295)*, terpene ketones (pulegone, menthone, and camphor) *(9, 48)*, codeinone *(233)*, 3-keto steroids, *(145, 253)* and 20-keto steroids *(92, 208)*. In general, when it is possible for the carbonyl compound to yield mixtures of cis-trans or position isomeric enol ethers, mixtures of products are obtained *(8, 141a, 175, 244)*. Exocyclic formyl groups are etherified more readily than endocyclic keto groups: 2-formylcyclohexanones were converted to 2-ethoxymethylenecyclohexanones in good yields *(240)*.

2. Conversion of 3-Keto-$\Delta^4$-steroids to 3-Alkoxy-$\Delta^{3,5}$-Steroids

In 1938 Serini and Koster *(253)* and Schwenk, Fleischer, and Whitman *(251)* reported that steroidal 4-ene-3-ones react with triethyl orthoformate under mild conditions to form steroidal 3-ethoxy-3,5-dienes.

$$\text{(20)}$$

This reaction turned out to be very useful. Since mild acid hydrolysis of the dienol ether regenerates the original 4-ene-3-one system, dienol etherification provides a means of protecting the carbonyl group at position 3 during transformations of other parts of the steroid molecule. The double bond at position 5 of the dienol ether may be selectively reduced, yielding a monoenol ether hydrolyzable to a 3-ketone with ring A saturated. Perphthalate oxidation of dienol ethers, followed by hydrolysis, yields steroid 6-hydroxy-4-ene-3-ones *(262)*.

Steroid 4-ene-3-ones may be converted to alkyl dienol ethers by heating a solution of the unsaturated ketone in benzene-alcohol with a trialkyl orthoformate and hydrogen chloride *(99, 146, 242, 253)*, or by allowing a solution of the unsaturated ketone in dioxane–trialkyl orthoformate containing a little sulfuric or perchloric acid to stand at room temperature *(90, 154, 193, 262, 305)*. Sufficient tertiary amine to neutralize the acid catalyst is usually added to the reaction mixture prior to isolation of the product.

In all examples reported, the dienol ether was the first product isolated. There is insufficient evidence to justify speculation on the mechanism of the reaction. It is possible that unsaturated ketals are intermediates in these reactions, since these, if formed, should undergo facile acid-catalyzed dealcoholation.

Alcohol interchange reactions of steroid methyl or ethyl dienol ethers, prepared using trimethyl or triethyl orthoformate, yield dienol ethers of higher alcohols.

Specific examples of syntheses of 3-alkoxy-3,5-dienyl ethers from 3-keto-$\Delta^4$-steroids are described in the literature (*52, 57, 83, 89–91, 96, 99, 123b, 146, 152, 154, 179, 193, 238a, 242, 251, 253, 262, 273, 291, 294, 305*).

## III. REACTIONS OF CARBOXYLIC ORTHO ESTERS WHICH YIELD CARBOXYLATE ESTERS

In this section are discussed reactions in which an alkyl group or an alkoxy group is transferred from an orthocarboxylate to some other acyl derivative to yield a carboxylate ester. Ordinary carboxylate esters are among the products of hydrolysis, alcoholysis, phenolysis, and pyrolysis reactions of ortho esters. These reactions, which are discussed elsewhere, are of no practical utility as syntheses of carboxylate esters.

### A. Alkylation of Carboxylic Acids

In 1928 Petrenko-Kritchenko determined the relative reactivities of triethyl orthoformate and tetraethyl orthocarbonate toward acetic acid (*219*), but did not isolate the reaction products. It was not until 1965 that Cohen and Mier established that such reactions provide a general method of converting carboxylic acids to alkyl esters (*71*). They heated a number of carboxylic acids with excess triethyl orthoformate, and obtained fair to excellent yields of the corresponding ethyl esters [Eq. (21)]. The mechanism of the reaction was not established, but it probably involves alkylation of the carboxylate ion by the diethoxycarbonium ion of an intermediate diethoxycarbonium carboxylate ion pair.

$$HC(OC_2H_5)_3 + RCO_2H \rightarrow RCO_2C_2H_5 + HCO_2C_2H_5 + C_2H_5OH \quad (21)$$

The alkylation reactions were carried out by distilling ethanol and ethyl formate from the reaction mixtures as they formed, and then isolating the ethyl ester of the acid. Reaction times varied from a few minutes in the case of trinitrobenzoic acid to 1 or 2 hours with the aliphatic carboxylic acids. Although in most instances no catalyst was necessary, the ethylation of nicotinic and hippuric acids was facilitated by the addition of dimethylformamide or *p*-toluenesulfonic acid to the reaction mixtures. Esters prepared, and yields obtained, are as follows: ethyl benzoate, 90%; ethyl salicylate, 90%; ethyl 3,5-dinitrobenzoate, 94%; ethyl cinnamate, 94%; ethyl 2,4,6-trinitrobenzoate, 70%; ethyl caproate, 93%; ethyl adipate, 89%; ethyl nicotinate, 54%; ethyl hippurate, 68%; and ethyl levulinate, 82%.

## B. Reactions of Carboxylic Ortho Esters with Acyl Halides and Acetyl Cyanide

Acyl halides react with ortho esters to form mixtures of carboxylate ester and alkyl halides [Eq. (22)]. Acetyl cyanide undergoes a similar reaction with ethyl ortho esters to form ethyl acetate and $\alpha$-ketonitrile acetals [Eq. (23)] (*38*).

$$RCOX + R'C(OR'')_3 \rightarrow RCO_2R'' + R'CO_2R'' + R''X \quad (22)$$

$$RC(OC_2H_5)_3 + CH_3COCN \rightarrow CH_3CO_2C_2H_5 + RC(OC_2H_5)_2CN \quad (23)$$

A plausible mechanism for the reaction of acyl halides with orthocarboxylates involves reaction of the ortho ester with traces of hydrogen halide (present as an impurity in the acid halide, or produced by reaction of the halide with traces of alcohol present in the ortho ester) to form a dialkoxycarbonium halide plus alcohol. The alcohol reacts with acyl halide to form carboxylate ester plus hydrogen halide, and the hydrogen halide converts more of the ortho ester to carbonium salt plus alcohol. The dialkoxycarbonium halide could collapse to an unstable 1,1-dialkoxy-1-haloalkane, or decompose to an alkyl halide and a carboxylate ester by alkylation of halide ion by the carbonium ion.

$$RC(OR')_2{}^+X^- \rightarrow RCO_2R' + R'X \quad (24)$$

Triphenyl orthoformate, which yields a diphenoxycarbonium ion having little tendency to undergo nucleophilic attack by chloride ion at phenyl carbon, reacts with acetyl chloride or trichloroacetyl chloride to form phenyl acetate or phenyl trichloroacetate plus chlorodiphenoxymethane (*38*).

2-Ethoxy-1,3-dioxolanes react with acetyl chloride and acetyl bromide to yield, in addition to ethyl acetate, 2-haloethyl formates (*20*).

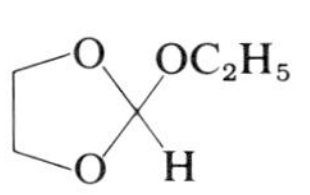

$$\underset{\substack{\diagdown O \diagup \diagdown H}}{\overset{\diagup O \diagdown \diagup OC_2H_5}{\phantom{X}}} + CH_3COX \longrightarrow CH_3CO_2C_2H_5 + HCO_2CH_2CH_2X \quad (25)$$

A number of reactions of acyl chlorides with trialkyl orthocarboxylates have been reported in the literature (*20, 38, 138, 169, 176, 177, 191, 224*).

## C. Reactions of Orthocarboxylates with Carboxylic Acid Anhydrides

Orthocarboxylates react with carboxylic acid anhydrides to form acyloxydialkoxyalkanes plus alkyl carboxylates [Eq. (26)]. These reactions are discussed in Chapter 1 (p. 52).

$$RC(OR')_3 + (R''CO)_2O \rightarrow R''CO_2R' + R''CO_2CR(OR')_2 \qquad (26)$$

## IV. CONVERSION OF CARBONYL COMPOUNDS TO HEMIACETALS, ACETALS, AND KETALS

### A. Synthesis of Hemiacetals

The only hemiacetal syntheses using ortho esters were reported by Post (226), who converted chloral hydrate to the ethyl, propyl, and butyl hemiacetals by reaction with the appropriate trialkyl orthoformates [Eq. (27)]. Chloral itself did not react with tripropyl orthoformate, even in the presence of an acid catalyst. The polycyclic ketone decachloropentacyclo-[5.3.0.0$^{2,6}$.0$^{4,10}$.0$^{5,9}$]decane-3-one yields a hemiketal when treated with ethanolic triethyl orthoformate (128a).

$$CCl_3CH(OH)_2 + HC(OR)_3 \rightarrow CCl_3CH(OH)OR + ROH + HCO_2R \qquad (27)$$
$$(R = C_2H_5, C_3H_7, C_4H_9)$$

### B. Synthesis of Acyclic Acetals and Ketals

Many aldehydes react with primary, secondary, and (sometimes) tertiary alcohols in the presence of acidic catalysts to form equilibrium mixtures containing significant amounts of acetals [Eq. (28), R = H].

$$RR'CO + 2 R''OH \underset{}{\overset{H^+}{\rightleftharpoons}} RR'C(OR'')_2 + H_2O \qquad (28)$$

The gain of 5 kcal/mole of bonding energy accompanying acetal formation is offset by a loss of entropy and by steric repulsions between the carbonyl and the alkoxy substituents of the acetal. The equilibrium constant for the reaction between acetaldehyde and ethanol is about 0.1; that for the reaction between isobutyraldehyde and isopropanol is about $4 \times 10^{-3}$ (237).

Direct reactions of ketones with alcohols [Eq. (28)] usually do not afford practical yields of ketals, presumably due to the greater steric crowding in ketals than in acetals.

Although most ketals are unobtainable by direct ketalization because of unfavorable equilibrium constants for reaction (28), many of them may be synthesized by means of another equilibrium reaction. This reaction, first

SYNTHESIS OF HEMIACETALS, ACETALS, AND KETALS 155

reported by Claisen in 1896 (*64*), involves treating the carbonyl compound with a trialkyl orthoformate in the presence of an acid catalyst and (usually) an alcohol [Eq. (29)]. It is applicable to most aldehydes and many ketones, and represents the only practical means of synthesizing most acyclic ketals.

$$RR'CO + HC(OR'')_3 \xrightarrow{H^+} RR'C(OR'')_2 + HCO_2R'' \qquad (29).$$

The greater success of reaction (29) than reaction (28) in preparing ketals is easily accounted for. Reaction (29) involves a more favorable entropy change and is accompanied by an increase in bonding energy of some 15 kcal/mole, due to the stabilization energy of the formate ester. These factors, taken together, account for the larger equilibrium constants for reaction (29) than for reaction (28), and therefore for the formation of isolable yields of many ketals which cannot be prepared by direct ketalization.

Discussions of the mechanism of reactions of carbonyl compounds with trialkyl orthoformates have been largely speculative (*2, 190, 225, 227, 239*). The reversibility of the reaction was established by Adkins and co-workers (*59, 220*), who found that in the case of diethyl ketals, equilibrium mixtures contained the same amounts of ethyl formate whether equilibrium was approached from the ketone–ethyl orthoformate side or from the ketal–ethyl formate side. No one has reported isolating an ortho ester from an equilibrium mixture prepared from a ketal and an alkyl formate, but in one case the ketone was isolated (*190*).

The orthoformate may merely serve to scavenge water produced by direct ketalization, or it may play a more direct role in conversion of the ketone to the ketal. The ortho ester may, of course, react by two or more pathways simultaneously; the equilibrium constant for the overall reaction would be the same in any case.

All of the plausible mechanisms for formation of ketals by reaction of ketones with orthoformates involve reaction of an alcohol with an alkoxydialkylcarbonium ion in the product-forming step [Eq. (34)]. This carbonium ion could be formed in several ways. It might be produced by loss of water from a protonated hemiketal, in which case the orthoformate is hydrolyzed by the water [Eq. (30)]. It could also result from loss of dialkyl orthoformate from the cationic adduct of the hemiketal and the dialkoxycarbonium ion derived from the orthoformate [Eq. (31)], or by reaction of alcohol with the adduct of the ketone and the dialkoxycarbonium ion [Eq. (32)]. Finally, alkylation of the ketone by the dialkoxycarbonium ion would yield the alkoxydialkylcarbonium ion and an alkyl formate [Eq. (33)]. The first three routes to the product-forming intermediate would be difficult to differentiate experimentally. They all differ from the fourth route in that they would lead to alkyl formate containing the carbonyl oxygen of the ketone, while direct alkylation of the carbonyl oxygen of the ketone would lead to ketal containing the carbonyl oxygen atom. Appropriate isotopic labeling experiments would distinguish between these possibilities.

$$R_2C=O + R'OH + H^+ \rightleftharpoons R_2C(OH_2^+)OR'$$
$$R_2C(OH_2^+)OR' \rightleftharpoons R_2\overset{+}{C}-OR' + H_2O \quad (30)$$
$$H_2O + HC(OR')_3 \rightleftharpoons HCO_2R' + 2\,R'OH$$

$$RC(OH)OR' + H\overset{+}{C}(OR')_2 \rightleftharpoons R_2C(OR')[\overset{+}{O}H-CH(OR')_2]$$
$$R_2C(OR')[\overset{+}{O}H-CH(OR')_2] \rightleftharpoons R_2\overset{+}{C}(OR') + HOCH(OR')_2 \quad (31)$$
$$HOCH(OR')_2 \rightleftharpoons HCO_2R' + R'OH$$

$$R_2CO + H\overset{+}{C}(OR')_2 \rightleftharpoons R_2\overset{+}{C}-OCH(OR')_2 \xrightarrow{R'OH}$$
$$R_2C(\overset{+}{O}HR')[OCH(OR')_2] \rightleftharpoons R_2\overset{+}{C}-OR' + HOCH(OR')_2 \quad (32)$$
$$HOCH(OR')_2 \rightleftharpoons HCO_2R' + R'OH$$

$$R_2CO + H\overset{+}{C}(OR')_2 \rightleftharpoons R_2\overset{+}{C}-OR' + HCO_2R' \quad (33)$$

$$R_2\overset{+}{C}-OR' + R'OH \rightleftharpoons R_2C(OR')_2 + H^+ \quad (34)$$

Although the position of the equilibrium for the reaction of carbonyl compounds with orthoformates [Eq. (29)] is further toward the right than that of the equilibrium for formation of the same acetal or ketal by direct reaction of the carbonyl compound with the alcohol [Eq. (29)], the reaction still may be far from completion at equilibrium. Although quantitative data are meager, it is apparent that reaction (29) is very sensitive to steric effects. Adkins and co-workers (59, 220) studied equilibrium conversions of ketones to ketals by reaction with triethyl orthoformate in acidic ethanol solution. Their results, for ethanol solutions originally 2 molar in both ketone and triethyl orthoformate, are summarized in Table II. There is a decrease in the value of the equilibrium constant of nearly four powers of ten on going from the least-hindered ketone, acetone, to the most hindered, di-*tert*-butyl ketone.

Most of the data concerning yields in conversions of ketones to ketals by reaction with trialkyl orthoformates were not obtained under uniform conditions, and hence are not comparable. The general trends, however, suggest that steric effects are very important. For example, yields of diethyl ketals from reactions of alkyl methyl ketones and triethyl orthoformate decrease in that order.

$$(CH_3)_2C(OC_2H_5)_2 > CH_3C(C_2H_5)(OC_2H_5)_2 > CH_3C(C_3H_7)(OC_2H_5)_2$$
$$> CH_3C[C(CH_3)_3](OC_2H_5)_2.$$

## TABLE II

Equilibrium Constants for Conversion of Ketones to Diethyl Ketals at 25°C[a]

$$RCOR' + HC(OC_2H_5)_3 \overset{K}{\rightleftharpoons} RC(OC_2H_5)_2R' + HCO_2C_2H_5$$

| R | R' | % Acetal at equilibrium[b] | K |
|---|---|---|---|
| $CH_3$ | $CH_3$ | 95 | 360 |
| $CH_3$ | $C_2H_5$ | 90 | 81 |
| $CH_3$ | $CH_2CH_2C_6H_5$ | 89 | 65 |
| $CH_3$ | $C_6H_5$ | 86 | 38 |
| $CH_3$ | $CH_2C(CH_3)_3$ | 84 | 28 |
| $CH(CH_3)_2$ | $CH(CH_3)_2$ | 65 | 3.4 |
| $CH_3$ | $C(CH_3)_3$ | 50 | 1 |
| $C_2H_5$ | $C(CH_3)_3$ | 36 | 0.33 |
| $C_6H_5$ | $C_6H_5$ | 34 | 0.26 |
| $CH(CH_3)_2$ | $C(CH_3)_3$ | 26 | 0.12 |
| $C(CH_3)_3$ | $C(CH_3)_3$ | 17 | 0.04 |

[a] Data from Carswell and Adkins (59) and Pfeiffer and Adkins (220).
[b] Initial ketone and orthoformate concentrations, 2.0 M. The catalyst was $7.5 \times 10^{-3}$ M HCl.

Similarly, $(C_2H_5)_2C(OC_2H_5)_2$ is formed in higher yield than

$$C_2H_5C(C_3H_7)(OC_2H_5)_2.$$

The yield of dialkyl acetals from acetaldehyde and trialkyl orthoformates decreases in the order

$$CH_3CH(OC_2H_5)_2 > CH_3CH(OC_3H_7)_2 > CH_3CH(OC_4H_9)_2,$$

and the yields of tolualdehyde diethyl acetals decrease in the order

$$p\text{-}CH_3C_6H_4CH(OC_2H_5)_2 > m\text{-}CH_3C_6H_4CH(OC_2H_5)_2$$
$$> o\text{-}CH_3C_6H_4CH(OC_2H_5)_2.$$

Cycloalkanones from cyclopentanone to cyclododecanone all form diethyl ketals in yields exceeding 80%, with the exception of cyclooctanone, which affords the ketal in 40% yield.

Chain branching in the alkoxy alkyl group of acetals and ketals has a profound effect on equilibrium conversions in carbonyl compound–orthoformate reactions. The diisopropyl acetal of acrolein and the di-*sec*-butyl acetal of acetaldehyde have been prepared in moderate yields by the orthoformate method, but the only ketal of a secondary alcohol reported in the literature is acetone dicyclohexyl ketal. Acetone and tri-*sec*-butyl orthoformate do not yield a ketal in isolable amounts (*190*).

The alcohol does not appear in the stoichiometric equation for reaction of a carbonyl compound with an orthoformate, and there are numerous reports of successful ketalization reactions carried out in the absence of added alcohol (*101, 136, 178, 192, 214, 284*). This does not rule out those mechanisms for the reaction which involve preliminary hemiketal formation, since some alcohol is always formed by reaction of the orthoformate with the acid catalyst, and some is usually present as an impurity in the orthoformate.

Most examples of ortho ester acetalization and ketalization reactions which are reported in the literature, however, were carried out in alcohol solutions. The reaction appears to proceed more rapidly and to give more reproducible results in the presence of the alcohol.

Ketalization by the orthoformate method is usually carried out by Claisen's original procedure (*57*), or a modification of it. He allowed mixtures of 1 mole of carbonyl compound, 1 mole of triethyl orthoformate, and 3 moles of alcohol to stand in the presence of an acid catalyst such as hydrogen chloride, ammonium chloride, or ferric chloride until equilibrium was attained, and then isolated the acetal or ketal. Acid catalysts which have been used include concentrated sulfuric acid, dry hydrogen chloride, concentrated hydrochloric acid, ammonium chloride, ammonium nitrate, arenesulfonic acids, and ferric chloride. Boron trifluoride would also be an effective catalyst, and acid ion exchange resins have been used (*17, 100*). Since many acetals and ketals are extremely susceptible to acid-catalyzed hydrolysis and dealcoholysis, the final work-up should be carried out under alkaline conditions, with the exclusion of moisture. A number of good general procedures for converting aldehydes and ketones to acetals and ketals by reaction with orthoformates and alcohols have been described in the literature (*6, 23, 67, 151, 160, 248, 285*).

Several variants of the ortho ester method of acetalization and ketalization have been described. Ketals derived from higher alcohols may be prepared by treating the ketone with a mixture of triethyl orthoformate, the higher alcohol, and an acid catalyst (*137, 190*). It was not established whether these reactions involve preliminary transesterification to give the higher homolog of the original ortho ester, whether transketalization of an initially formed diethyl ketal or mixed ketal occurs, or whether both of these processes occur simultaneously.

A large number of carbonyl compounds have been converted to dialkyl

# SYNTHESIS OF HEMIACETALS, ACETALS, AND KETALS 159

acetals and ketals by reaction with trialkyl orthoformates. Some of these are listed below (with references to the original literature) in order of increasing carbon content of the unsubstituted aldehyde or ketone.

$C_2$ **compounds**: acetaldehyde (*3, 225*) and glyoxal (diacetal) (*64*). $C_3$ **compounds**: propionaldehyde (*225*), 2,2-dibromopropanal (*67, 127, 151a, 187, 250, 255, 290*), 2-fluoro-3-hydroxypropanal (*156*), acetone (*9, 11, 59, 64, 67, 69, 107, 137, 143, 190, 220, 223*), chloroacetone (*9*), acetoxyacetone (*102*) 1,1-dichloroacetone (*301*), 1-acetoxy-3-hydroxyacetone (*8, 23a, 110*), acrolein (*111, 231, 247, 285*), 3,3-dibromo- and 3,3-diiodoacrolein (*126a*), 2-ketopropanal (diacetal) (*113*), 2-keto-3-bromopropanal (acetal) (*67*), ethyl 2-ketopropionate (*76, 269*), 2-chloromalonaldehyde (diacetal) (*73*), and 3-methoxy-2-phthalimidopropanal (*78*). $C_4$ **compounds**: 2-methylpropanal (*16, 131*), butanone (*11, 59, 101, 220*), 3-bromobutanal (*68*), 3-acetoxybutanal (*33*), 3-formyloxybutanal (*5*), 4-methylthiobutanal (*265*), 4-chloro-2-butanone (*220*), 4-ethoxy-2-butanone (*94, 114*), 4,4-dimethoxy-2-butanone (*272*), 2-butenal (*114, 160, 247, 285*), 2-bromo-2-butenal (*182*), 2,4,4,4-tetrachloro-2-butenal (*172a*), 2,3-epoxybutanal (*304*), 2,3-ethoxy-4,4-dimethoxybutanal (*304*), 2,3-butanedione (monoketal) (*136*), 2,3-butanedione (diketal) (*214*), 4-ethylenedithiobutanone (*211*), ethyl 2-ketobutyrate (*287*), ethyl 3-ketobutyrate (*203*), and 2,3-epoxy-4-oxobutyrate (*304*).

$C_5$ **compounds**: 2-pentanone (*11, 101*), 3-methyl-2-butanone (*11, 101, 220*), 3-pentanone (*59, 190*), 1-methoxy-3-pentanone (*8*), 1-ethoxy-3-pentanone (*244*), 5-chloro-2-pentanone (*264*), 3-hydroxypentanal (*5*), 3-formyloxypentanal (*5*), 2-bromo-2-bromomethylbutanal (*213*), cyclopentanone (*248*), methyl cyclopropyl ketone (*153a*), 4-pentenal (*163*), 2-methyl-2-butenal (*285*), 1-penten-3-one (*192*), 2,4-pentadienal (*230*), ethyl 2-ketopentanoate (*287*), ethyl 2-keto-3-methylbutyrate (*287*), diethyl 2-ketoglutarate (*153*), diethyl 3-ketoglutarate (297), furfural (*64, 67, 126, 279*), 5-ethoxyfurfural (*109*), 3-methylthio-2-thiophenecarboxaldehyde (*124b*), and 2-selenophenecarboxaldehyde (*217*). $C_6$ **compounds**: 2-hexanone (*11, 101*), 3,3-dimethyl-2-butanone (*16, 59, 220*), 6-formyloxyhexanal (*4, 5*), 2-methylcyclopentanone (*5*), cyclohexanone (*12, 151, 178, 190, 215, 248*), 2-methyl-1-penten-3-one (*8*), 3-methyl-3-penten-2-one (*207*), 5-ketohexanal (acetal) (*163*), 2-methyl-3-pentenedial (diacetal) (*192a*), 1,4-cyclohexanedione (diketal) (*151, 270*), ethyl 4,5-epoxy-6-oxo-2-hexenoate (*95a*), and 3-methylthio-5-methylthiophene-2-carboxaldehyde (*124b*), 3-*O*-benzyl-1,2,-*O*-isopropylidene-α-D-xylohexofuranos-5-ulose (*157a*). $C_7$ **compounds**: 2-heptanone (*284*), 4-heptanone (*11, 101*), 2,4-dimethyl-3-pentanone (*220*), 2-methylcyclohexanone (*141a, 248*), 4-methylcyclohexanone (*248*), cycloheptanone (*215, 248*), 6-ketoheptanal (acetal) (*163*), diethyl 4-ketopimelate (*173*), benzaldehyde (*64, 67, 108, 160, 225*), *m*-aminobenzaldehyde (*42*), *m*- and *p*-nitrobenzaldehyde (*108*), *p*-chlorobenzaldehyde (*108*), *o*-, *m*-, and *p*-hydroxybenzaldehyde (*34, 218*), *o*-, *m*-, and *p*-methoxybenzaldehyde (*67, 108, 151, 160*), *p*-dimethylaminobenzaldehyde (*160*), *p*-[*N*-methyl-*N*-(2-chloroethyl)]aminobenzaldehyde (*27a*), 3,4-methylenedioxybenzaldehyde (*67*), 2,4- and 3,4-dichlorobenzaldehyde (*160*), and 5-chloro-2-nitrobenzaldehyde (*1, 1a*), tetracyclo[3.2.0.0$^{2,7}$.0$^{4,6}$]heptan-3-one (*123a*). $C_8$ **compounds:** 2-octanone (*59*), 2,4,4-trimethyl-3-pentanone (*220*), cyclooctanone (*215, 248*), 7-octen-2-one (*175*), 1-acetylcyclohexene (*208*), 5-methyl-2-methylthiobenzaldehyde (*125*), *o*-phthalimidomethylbenzaldehyde (*41*), *o*-, *m*-, and *p*-tolualdehyde (*108, 160, 225*), acetophenone (*9, 59, 65, 67, 100, 107, 190, 220*), *p*-bromoacetophenone (*155*), α-bromoacetophenone (*159*), α-chloroacetophenone (*159*), and 3-hydroxy-2-methyl-5-phosphoryloxymethyl-4-pyridinecarboxaldehyde (*155a*).

$C_9$ **compounds:** 2,2,4,4-tetramethyl-3-pentanone (*220*), 4-methyl-3-vinylcyclohexanone (*206*), 4-acetyl-4-methyl-1-methoxycyclohexene (*209*), ethyl 3-oxo-2-phenylpropionate (*300*), cinnamaldehyde (*160*), *o*-nitrocinnamaldehyde (*160*), α-bromocinnamaldehyde (*6*), propiophenone (*107*), and ethyl *p*-acetylbenzoate (*246*). $C_{10}$ **compounds:** camphor (*10, 48*), 3,7-dimethyl-2,6-octadienal (*135*), 2,7-dimethyl-2,6-octadien-4-ynedial (diacetal) (*161*), 4-formyl-1,3,5-trimethylcyclohexene (*283*), dimethyl 4-acetyl-4-methylpimelate (*207*), and butyrophenone (*59*). $C_{11}$ **compounds:** 6-formyl-2,2,6-trimethylcycloheptanone (acetal) (*243*), and 2,2,6-trimethylcyclohexylideneacetaldehyde (*149*). $C_{12}$ **compounds:** cyclododecanone (*216, 248*), and 2,7,8-trimethyl-2,6-nonadien-4-ynal (*147*).

$C_{13}$ **compounds:** 4-(2,6,6-trimethylcyclohexenyl)-3-buten-2-one (*204, 205*) and benzophenone (*59, 103, 220*). $C_{14}$ **compounds:** benzyl phenyl ketone (*161*), 4,9-dimethyldodeca-2,4,8,10-tetraen-6-ynedial (diacetal) (*147*) and 4-(2,6,6-trimethylcyclohexenyl)-2-methyl-2-butenal (*148*). $C_{16}$ **compounds:** palmitaldehyde (*260a*), 6-(2,6,6-trimethylcyclohexenyl)-4-methyl-2,4-hexadienal (*148*), and nonyl phenyl ketone (*75*). The alkaloid codeinone (*233*), several 3-keto steroids (*99, 145, 253, 287*), and one 20-keto steroid (*208*) were also converted to dialkyl acetals by reaction with trialkyl orthoformates.

Acetals and ketals have also been prepared by treating the carbonyl compound with alcoholic solutions of alkyl formimidate hydrochlorides (*66*). Claisen attributed the results to reaction of the carbonyl compound with "nascent" ortho ester (formimidate salts react with alcohols to form orthoformates), but the detailed course of the reaction was not established.

Acyclic α,β-unsaturated ketones react with trialkyl orthoformates in acidic alcohol solutions to form 1,3,3-trialkoxyalkanes [Eq. (35)] (*8, 58, 94, 114, 207, 244*), as well as the expected unsaturated ketals (*116*).

$$R^1\text{—CO—}CR^2\text{=}CHR^3 + R^4OH + HC(OR^4)_3 \xrightarrow{H^+ \quad -HCO_2R^4} R^1\text{—}C(OR^4)_2CHR^2\text{—}CHR^3OR^4 \quad (35)$$

| R$^1$ | R$^2$ | R$^3$ | R$^4$ | References |
|---|---|---|---|---|
| C$_2$H$_5$ | H | H | C$_2$H$_5$ | *8, 244* |
| CH$_3$ | H | H | CH$_3$, C$_2$H$_5$, C$_4$H$_9$ | *58, 94, 114* |
| CH$_3$ | CH$_3$ | H, CH$_3$ | CH$_3$ | *207* |

The trialkoxy compound is probably formed by reaction of the conjugate acid of the ketone with the alcohol to form an alkoxyvinyl alcohol, followed by tautomerization to the alkoxy ketone, which then undergoes ketalization in the usual manner. The same product would be formed by acid-catalyzed dealcoholation of an initially formed $\alpha,\beta$-unsaturated ketal, followed by reaction of alcohol with the allylic cation and acid-catalyzed addition of alcohol to the resulting enol ether. Cyclic $\alpha,\beta$-unsaturated ketones, particularly 4-ene-3-one steroid derivatives, differ from the acyclic unsaturated ketones in reactions with trialkyl orthoformates, since they usually yield dienol ethers rather than ketals (see above).

## C. Synthesis of Cyclic Acetals and Ketals

Ortho esters have also been used in syntheses of cyclic ketals. In one group of reactions, ethylene ketal derivatives of the carbonyl compound were synthesized by treating the ketone with a mixture of triethyl orthoformate and ethylene glycol in the presence of an acid catalyst [Eq. (36)].

$$\text{RCOR}' + \text{HOCH}_2\text{CH}_2\text{OH} + \text{HC(OC}_2\text{H}_5)_3 \xrightarrow{\text{H}^+}$$

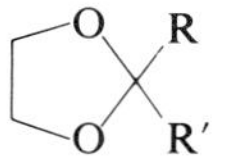

$$\underset{\text{O}}{\overset{\text{O}}{\diagdown}}\!\!\!\!\diagup\!\!\overset{\text{R}}{\underset{\text{R}'}{}} + 2\,\text{C}_2\text{H}_5\text{OH} + \text{HCO}_2\text{C}_2\text{H}_5 \qquad (36)$$

In this manner 2-ethyl-2-carbethoxydioxolane was prepared from ethyl 2-ketobutyrate (*287*), and 3,3-ethylenedioxy-17,20;20,21-bis(methylenedioxy)-$\Delta^5$-pregnene (**13**, Y = O) (*172*), 3-acetoxy-17,17-ethylenedioxyetiocholane (**14**) (*194*), its $\Delta^5$-derivative (*155b*), and 3-acetoxy-20,20-ethylenedioxycholane (**15**) (*294*) were prepared from the corresponding steroid ketones.

Nucleosides are converted to cyclic 2′,3′-O-ketals and acetals (**16**) by reaction with triethyl or trimethyl orthoformate and the appropriate aldehyde or ketone (*61, 133, 209a*). It was postulated that these reactions involve conversion of the ketone to a diethyl or dimethyl ketal, followed by reaction of this with the nucleoside.

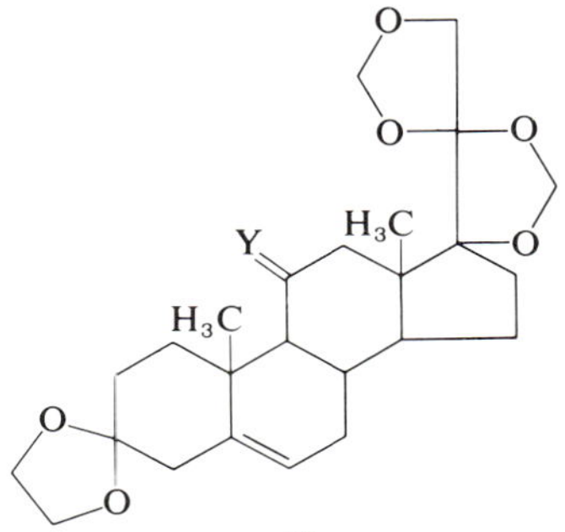

**13** 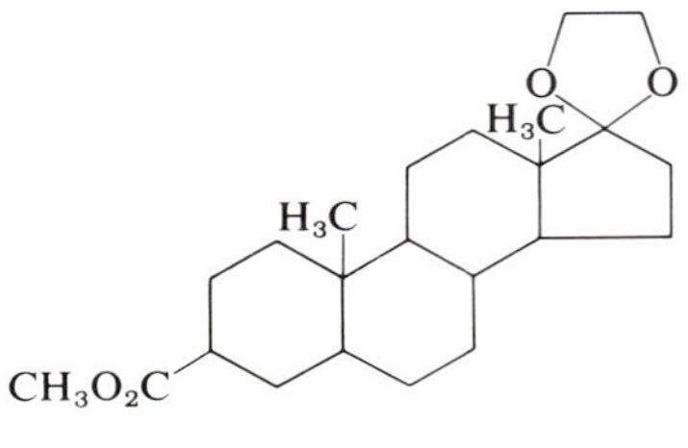   **14**

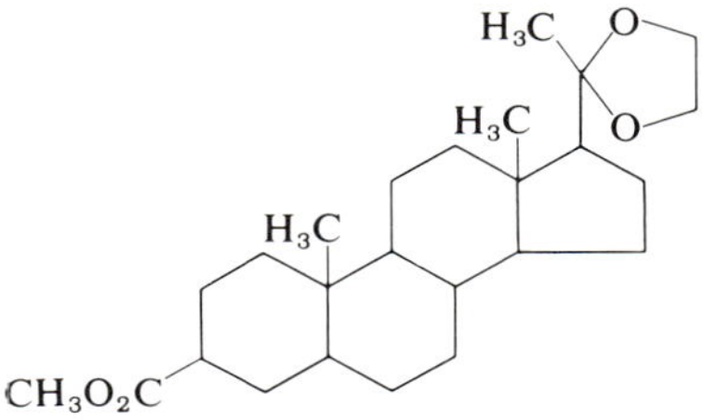

**15** 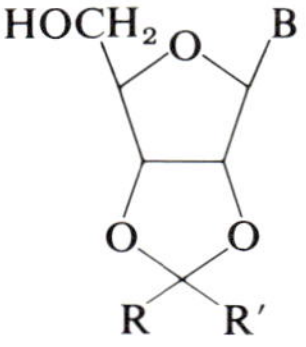   **16**

[B = uridyl, quanyl, adenyl.
R, R' = CH$_3$, CH$_3$; CH$_3$, C(CH$_3$)$_3$;
H, C$_2$H$_5$; H, CH$_3$CH=CH$_2$; H, C$_6$H$_5$; H,
p-CH$_3$OC$_6$H$_4$; –(CH$_2$)$_n$– (n = 4, 5, 6, 7)]

There are a few examples of reactions of ketones with cyclic ortho esters which yield ketals. Astle reported that methyl alkyl ketones are converted to 2-methyl-2-alkyldioxolanes in high yields by reactions with 2-ethoxy-1,3-dioxolane, but no specific examples were cited (*17*). Butane-2,3-dione is converted to the bis-ethyleneketal by 2-ethoxy-1,3-dioxolane (*19*), and benzaldehyde reacts with 2-ethoxy-4,5-benzo-1,3-dioxolane to form 2-phenyl-4,5-benzo-1,3-dioxolane (*129*).

α-Hydroxy ketones react with triethyl and trimethyl orthoformate to form dimeric ketals which are probably dialkoxy-1,4-dioxanes (*102, 112, 113*). The structures of the reaction products were not definitely established, but the reactions yielding them are probably described by Eq. (37).

$$2\,\text{RCOCH}_2\text{OH} + 2\,\text{HC(OR')}_3 \longrightarrow \text{[dialkoxy-1,4-dioxane]} + 2\,\text{HCO}_2\text{R'} + 2\,\text{R'OH} \quad (37)$$

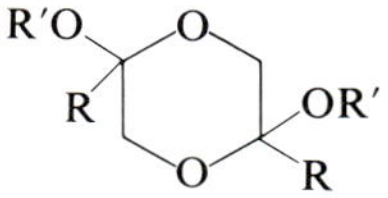

(R = CH$_3$, R' = C$_2$H$_5$; R = HOCH$_2$, R' = CH$_3$, C$_2$H$_5$)

Sugar glycosides, another class of cyclic acetals, can be synthesized using ortho esters. Bergmann first used this reaction in 1925 to prepare an ethyl 2,3-dideoxyglucopyranoside (*32*), and Freudenberg and Jacobs synthesized the methyl glucoside of 2,3,6-tri-*O*-methyl-D-glucopyranose using trimethyl orthoformate (*114a*). 3,4,6-Tri-*O*-acetyl-D-glucal (*257a*) and 4,6-di-*O*-acetyl-2,3-didehydro-2,3-dideoxy-D-glucopyranose (*32a, 170b*) react with trimethyl and triethyl orthoformates to form methyl (and ethyl) 4,6-di-*O*-acetyl-2,3-didehydro-2,3-dideoxy-α-D-glucopyranosides. Wolfram and co-workers used trimethyl orthoformate in the presence of boron trifluoride to prepare methyl β-D-glucopyranoside, methyl β-D-galactopyranoside, methyl α-D-glucopyranoside, methyl α-D-mannopyranoside, and methyl β-D-xylopyranoside from the aldoses (*302*).

Trihalomethyl aldehydes and ketones react with ortho esters to give products other than the expected acetals and ketals. Trichloroacetaldehyde reacts with trialkyl orthoacetates to form 4,4-dialkoxy-2,4-bis(trichloromethyl)-5-alkyl-1,3-dioxanes [Eq. (38)] (*167*). The products are mixed acetal-ortho esters. The isolation in one instance of an intermediate β-hydroxy-γ,γ,γ-trichloroortho ester suggests that the first step in this reaction is an aldol-type condensation of the ortho ester and the halogenated aldehyde. 2,2,3-Trichlorobutyraldehyde undergoes an analogous reaction with trimethyl orthopropionate.

$$RCH_2C(OR')_3 + Cl_3CCHO \longrightarrow [Cl_3CCH(OH)CHRC(OR')_3] \longrightarrow$$

(R = CH$_3$, R' = CH$_3$, C$_2$H$_5$; R = H, C$_3$H$_7$, R' = CH$_3$)   (38)

Trialkyl orthoformates and hexafluoroacetone undergo a reaction at 150°C which produces 2,2,4,4-tetrakis(trifluoromethyl)-5,5-dialkoxy-1,3-dioxolanes [Eq. (39)] (*44–46*). Cyclic orthoformates undergo analogous reactions which yield spirocyclic ortho ester-ketals [Eq. (40)] (*47*). The products of these reactions are remarkably inert. They are unaffected by prolonged heating at 150°C, or by heating with alkali, thionyl chloride, boron trifluoride, or phosphorus pentachloride.

Reactions (39) and (40) are not solvent-dependent, and are unaffected by sources and scavengers of free radicals. They may involve electrophilic attack by the carbonyl carbon of hexafluoroacetone on an orthoformate oxygen atom, followed by decomposition of the adduct to form the hemiketal of the ketone and a dialkoxycarbene, which then reacts with two additional

$$HC(OR)_3 + 3\,(CF_3)_2C{=}O \xrightarrow{150°C} \underset{\underset{CF_3}{|}}{\overset{RO}{\underset{F_3C}{\diagup}}}\!\!\!\!\!\!\!\!\!\!\!\begin{array}{c}RO\\ \diagup\!\!\!\!\diagdown\\O\quad CF_3\\ \diagdown\!\!\!\!\diagup\\O\quad CF_3\end{array} + (CF_3)_2(OH)OR \qquad (39)$$

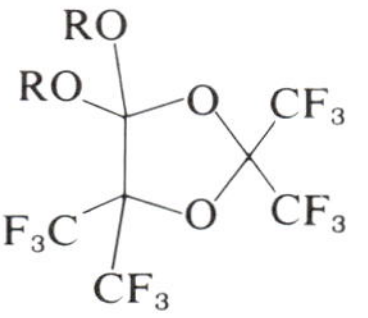

(R = $CH_3$, $C_2H_5$, $C_5H_{11}$, $C_4H_9CH(C_2H_5)CH_2$, $CH_2{=}CHCH_2$, $CH_2ClCH_2$)

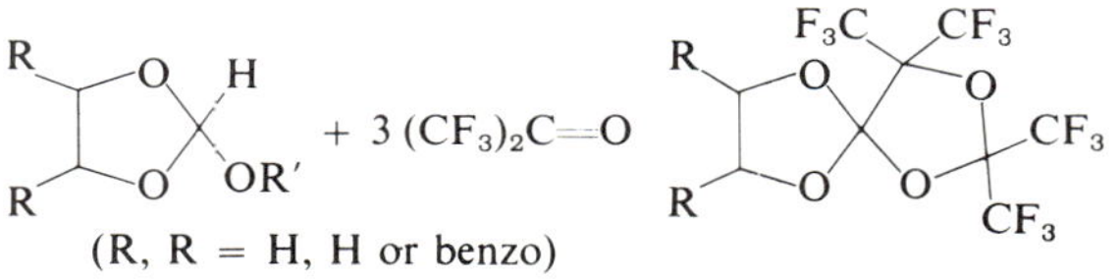

$$\underset{R}{\overset{R}{\diagup}}\!\!\!\!\begin{array}{c}O\quad H\\ \diagdown\!\!\!\!\diagup\\ \diagup\!\!\!\!\diagdown\\O\quad OR'\end{array} + 3\,(CF_3)_2C{=}O \longrightarrow \underset{R}{\overset{R}{\diagup}}\!\!\!\!\begin{array}{c}F_3C\quad CF_3\\ \diagdown\!\!\!\!\diagup\\O\quad\;\;O\\ \diagup\!\!\!\!\diagdown\!\!\!\!\diagup\!\!\!\!\diagdown CF_3\\O\quad\;\;O\\\;\;\;\;\;\;CF_3\end{array} + (CF_3)_2C(OH)OR' \qquad (40)$$

(R, R = H, H or benzo)

moles of the ketone to form the products (46). This hypothesis is supported by the observation that the bicyclic orthoformate 4-methyl-2,6,7-trioxabicyclo[2.2.2]octane, whose adduct with the ketone cannot eliminate a hemiketal, reacts with hexafluoroacetone to form 2,2-bis(trifluoromethyl)-5-methyl-5-formyloxymethyl-1,3-dioxane rather than an ortho ester-ketal.

## V. REACTIONS OF CARBOXYLIC ORTHO ESTERS WHICH YIELD ESTERS OF ACIDS OTHER THAN CARBOXYLIC AND CARBONIC ACIDS

### A. Borate Esters

Bassett reported in 1863 that triethyl orthoformate and tetraethyl orthocarbonate react with boron trioxide to form product mixtures containing triethyl borate (25, 26). The synthesis of dialkoxycarbonium and dioxolenium tetrafluoroborates, by reaction of ortho esters with boron trifluoride, yields trialkyl borates as by-products (85, 199–201, 249).

### B. Alkoxysilanes and Alkoxygermanes

Triethyl orthoformate reacts with silicon tetrafluoride in the presence of $TiCl_4$, $AlCl_3$, or $BF_3$ to form ethyl fluoride and unidentified silicate esters (195, 265). Organohalosilanes react with triethyl orthoformate to form ethoxysilanes and ethoxyhalosilanes [Eqs. (41)–(43)] (256, 257). This method of preparing alkoxysilanes is superior to the reaction of halosilanes with alcohols, since the by-products are ethyl halides and ethyl formate, rather than hydrogen halides.

$$RR'SiCl_2 \xrightarrow[AlCl_3]{HC(OC_2H_5)_3} RR'SiClOC_2H_5 + RR'Si(OC_2H_5)_2$$
$$+ RR'Si(OC_2H_5)OSi(OC_2H_5)RR' \qquad (41)$$

ESTERS OF ACIDS OTHER THAN CARBOXYLIC AND CARBONIC

$$\text{RSiHCl}_2 \xrightarrow{\text{HC(OC}_2\text{H}_5)_3} \text{RSiH(OC}_2\text{H}_5)_2 \tag{42}$$

$$\text{C}_6\text{H}_5\text{SiF}_3 \xrightarrow[\text{AlCl}_3]{\text{HC(OC}_2\text{H}_5)_3} \text{C}_6\text{H}_5\text{SiF(OC}_2\text{H}_5)_2 + \text{C}_6\text{H}_5\text{Si(OC}_2\text{H}_5)_3 \tag{43}$$

Germanium tetrachloride and trialkyl orthoformates react in the presence of aluminum chloride to form tetraalkoxy germanes in fair to good yields [Eq. (44)] (*221*). If the molar ratio of trialkyl orthoformate to germanium tetrachloride is less than 4 to 1, chloroalkoxygermanes are obtained.

$$\text{GeCl}_4 + 4\ \text{HC(OR)}_3 \xrightarrow{\text{AlCl}_3} \text{Ge(OR)}_4 + 4\ \text{RCl} + 4\ \text{HCO}_2\text{R} \tag{44}$$
$$(\text{R} = \text{C}_3\text{H}_7,\ \text{iso-C}_3\text{H}_7,\ \text{C}_4\text{H}_9,\ \text{iso-C}_4\text{H}_9)$$

### C. Nitrate Esters

Both nitric acid and dinitrogen trioxide react with triethyl orthoformate to form ethyl nitrate (*15*).

### D. Esters of Phosphorus-Containing Acids

Phosphorus pentoxide and triethyl orthoformate react to form product mixtures containing ethyl formate, triethyl phosphate, and tetraethyl pyrophosphate (*43*). If the same reactants are heated to higher temperatures, diethyl ether and ethyl formate (formed by pyrolysis of the ortho ester) are the principal products (*266*). Phosphorus pentasulfide reacts with triethyl orthoformate to form ethyl formate, ethyl thionformate, and a phosphorus-containing ester, $(\text{C}_2\text{H}_5)_3\text{PO}_2\text{S}_2$, of unknown structure (*43*).

Ethylphosphonic acid and phenylphosphonic acid are converted to their methyl and ethyl esters by reaction with trimethyl and triethyl orthoformates (*229*).

Early reactions of phosphorus trichloride (*15*), phosphorus pentachloride (*15*), and phosphorus triiodide (*25*) with triethyl orthoformate led to ethyl halides as the only definitely identified products, although esters of phosphorus and phosphoric acids must have been formed also. More recently, Arbusov reported that triethyl orthopropionate reacts with phosphorus trichloride to form mixtures of triethyl phosphite, diethoxychlorophosphine, and ethoxydichlorophosphine, along with ethyl chloride and ethyl propionate (*13*).

1,1,1-Trihalo-3-phospholines undergo stepwise reactions with triethyl orthoformate to form, first, 1-oxo-1-halophospholines and then 1-oxo-1-ethoxyphospholines (*142*).

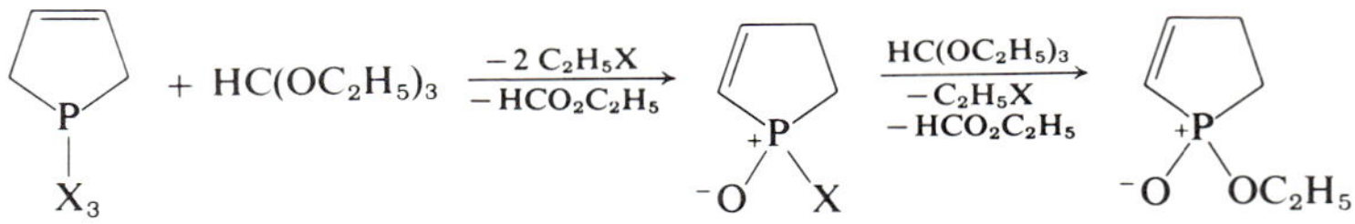

$$\underset{X_3}{\overset{|}{\underset{P}{\bigcirc}}} + HC(OC_2H_5)_3 \xrightarrow[-HCO_2C_2H_5]{-2\ C_2H_5X} \underset{-O}{\overset{+P}{\underset{}{\bigcirc}}}{}^X \xrightarrow[-HCO_2C_2H_5]{HC(OC_2H_5)_3 \atop -C_2H_5X} \underset{-O\quad OC_2H_5}{\overset{+P}{\underset{}{\bigcirc}}} \quad (45)$$

(X = Cl, Br)

Methyltriphenoxyphosphonium iodide reacts with triethyl orthocarboxylates to form ethyl carboxylates, ethyl iodide, phenetole, and diphenyl methylphosphonate [Eq. (46)] (210). In a similar reaction, diphenoxymethylphenylphosphonium iodide is converted to phenyl methylphenylphosphinate [Eq. (47)].

$$RC(OC_2H_5)_3 + CH_3\overset{+}{P}(OC_6H_5)_3I^- \rightarrow RCO_2C_2H_5 + C_2H_5I$$
$$+ C_6H_5OC_2H_5 + CH_3P(O)(OC_6H_5)_2 \quad (46)$$
$$(R = H, CH_3)$$

$$CH_3(C_6H_5)\overset{+}{P}(OC_6H_5)_2I^- + HC(OC_2H_5)_2 \rightarrow HCO_2C_2H_5 + C_6H_5OC_2H_5$$
$$+ CH_3P(C_6H_5)(OC_6H_5) \quad (47)$$

Phosphorylated formals are prepared by heating dialkyl phosphites with trialkyl orthoformates (234).

### E. Esters of Sulfur-Containing Acids

Ethylenesulfonic acid, toluenesulfonic acid, *p*-hydroxybenzenesulfonic acid, and methanesulfonic acid are converted to methyl and ethyl esters by reaction with trimethyl and triethyl orthoformates (229). Benzene- and toluenesulfonyl chlorides yield ethyl arenesulfonates when treated with triethyl orthoformate in the presence of zinc chloride at elevated temperatures (176, 177).

$$ArSO_2Cl + HC(OC_2H_5)_3 \xrightarrow{ZnCl_2} ArSO_3C_2H_5 + HCO_2C_2H_5 + C_2H_5Cl \quad (48)$$

Thionyl chloride reacts with trialkyl orthoformates to form mixtures of dialkyl sulfites and alkyl chlorosulfites, plus alkyl formates and alkyl chlorides (176, 177). Similarly, sulfuryl chloride and triethyl orthoformate yield mixtures of ethyl chlorosulfonate, diethyl sulfate, ethyl formate, and ethyl chloride (177).

A superior synthesis of ethyl thioncarboxylates involves treating triethyl orthocarboxylates with hydrogen sulfide in the presence of zinc chloride. $C_6H_5CSOC_2H_5$ and $CH_3CSOC_2H_5$ were prepared in yields of 80% and 52%, respectively, by this method (211a). Ethyl thionformate and ethylthionacetate are formed, along with triethyl borate, when the corresponding triethyl orthocarboxylates are refluxed with $B_2S_3$ (170a).

## VI. REACTIONS OF CARBOXYLIC ORTHO ESTERS WHICH YIELD ALKYL HALIDES

Trialkyl orthocarboxylates react with hydrogen halides, acyl halides, and acid halides of many inorganic acids, and with some binary inorganic halides to form product mixtures containing alkyl halides corresponding to the alkoxy groups of the ortho ester. In the majority of these reactions the products of interest are the esters or carbonium salts also formed, rather than the alkyl halides. For this reason, the various reactions which convert orthocarboxylates to alkyl halides are listed here without extensive discussion.

### A. Reactions of Orthocarboxylayes with Hydrogen Halides

Dry hydrogen chloride (*15*) and dry hydrogen iodide (*180*) convert triethyl orthoformate to ethyl chloride and ethyl iodide plus ethyl formate and ethanol. Similarly, propyl bromide was obtained from tripropyl orthoformate and hydrogen bromide (*180*).

### B. Reactions of Orthocarboxylates with Acyl Halides

Alkyl halides are formed, in addition to carboxylate esters, by the reaction of acyl halides with trialkyl orthocarboxylates [Eq. (22)] (*20, 38, 138, 169, 176, 177, 191, 224*).

Diethoxymethyl acetate has been reported to react with acetyl bromide to form ethoxybromomethyl acetate (*252*).

### C. Reactions of Orthocarboxylates with Other Acid Halides

Arene and alkanesulfonyl chlorides, thionyl chloride, and sulfuryl chloride react with trialkyl orthoformates to form alkyl chlorides (*176, 177*). 1,1,1-Trihalophospholines also react with triethyl orthoformate to form ethyl halides [Eq. (45)] (*142*).

### D. Reactions of Orthocarboxylates with Inorganic Binary Halides

Phosphorus trihalides and pentahalides convert trialkyl orthoformates to alkyl halides (*13, 15, 25, 128b*). Alkyl chlorides are among the products of reactions of trialkyl orthoformates with germanium tetrachloride (*221*).

A useful method of preparing ethyl fluoride involves reaction of triethyl orthoformate with silicon tetrafluoride (*196, 265*).

### E. Miscellaneous Reactions Which Yield Alkyl Halides from Trialkyl Orthocarboxylates

Ethyl halides are among the products of reactions of alkyl and aryl halosilanes with triethyl orthoformates (*256, 257*). Di- and triphenoxyphosphonium iodides react with trialkyl orthocarboxylates to form ethyl iodide and other products (*210*). There is an early report that bromine reacts with triethyl orthoformate to form ethyl bromide, ethyl formate, ethyl carbonate, and ethanol (*170*). More recently it was reported that triethyl orthoformate reacts with iodine in the presence of iron, magnesium, aluminum, or zinc to form ethyl iodide and ethyl formate (*77*). These reactions probably involve preliminary formation of metal iodides followed by reaction of these with the ortho ester. Dialkoxychlorophosphines and diarylchlorophosphines react with orthoformates to form alkyl halides and phosphonate esters (*84a*).

### REFERENCES

1. Aktieselskabet Grindstedvaerket, French Patent 1,482,641 (1967); *Chem. Abstr.* **68**, 105261 (1968).
1a. Aktieselskabet Grindstedvaerket, Netherlands Patent Appl. 6,608,039 (1966); *Chem. Abstr.* **66**, 105006 (1967).
2. E. R. Alexander, "Ionic Organic Reactions," pp. 216–217. Wiley, New York, 1950.
3. E. R. Alexander, H. M. Busch, and G. L. Webster, *J. Am. Chem. Soc.* **74**, 3173 (1952).
4. E. R. Alexander and E. N. Marvell, *J. Am. Chem. Soc.* **71**, 15 (1949).
5. E. R. Alexander and E. N. Marvell, *J. Am. Chem. Soc.* **72**, 3944 (1950).
6. C. H. F. Allen and C. O. Edens, *in* "Organic Syntheses," Coll. Vol. III, p. 731. Wiley, New York, 1955.
7. G. R. Allen and J. F. Poletto, U.S. Patent 3,226,397 (1965); *Chem. Abstr.* **65**, 3836 (1966).
8. M. F. Ansell, J. W. Lown, D. W. Turner, and D. A. Wilson, *J. Chem. Soc.* p. 3036 (1963).
9. A. Arbusov, *Chem. Ber.* **40**, 3303 (1907).
10. A. Arbusov, *Bull. Soc. Chim. France* [4] **8**, 102 (1910).
11. A. E. Arbusov, *Z. Physik. Chem.* **121**, 209 (1926).
12. A. E. Arbusov and B. M. Mikhialova, *Zh. Obshch. Khim.* **6**, 390 (1936).
13. B. A. Arbusov and N. P. Bogonostseva, *Izv. Akad. Nauk SSSR, Otd. Khim. Nauk* p. 484 (1953).
14. F. Arndt, L. Leowe, and M. Ozansoy, *Chem. Ber.* **73**, 779 (1940).
15. M. Arnhold, *Ann. Chem.* **240**, 192 (1887).
16. Z. Arnold and J. Zemlicka, *Collection Czech. Chem. Commun.* **24**, 786 (1959).
17. M. J. Astle, J. A. Zaslowsky, and P. G. Lafyatis, *Ind. Eng. Chem.* **46**, 787 (1954).
18. K. Auwers and M. Hessenland, *Ann. Chem.* **352**, 273 (1907) (footnote 2).
19. H. Baganz and L. Domaschke, *Chem. Ber.* **91**, 650 (1958).
20. H. Baganz and L. Domaschke, *Chem. Ber.* **91**, 653 (1958).
21. H. Baganz, L. Domaschke, J. Fock, and S. Rabe, *Chem. Ber.* **95**, 1832 (1962).

21a. B. R. Baker and J. H. Johnson, *J. Heterocyclic Chem.* **4**, 31 (1967).
22. B. R. Baker and M. A. Johnson, *J. Heterocyclic Chem.* **4**, 447 (1967).
22a. B. R. Baker and J. L. Kelley, *J. Med. Chem.* **11**, 688 (1968).
23. C. E. Ballou, *Biochem. Prep.* **7**, 45 (1960).
23a. C. E. Ballou and H. O. L. Fischer, *J. Am. Chem. Soc.* **78**, 1656 (1956).
24. R. A. Barnes, G. Doyle, and J. A. Hoffmann, *J. Org. Chem.* **27**, 90 (1962).
25. H. Bassett, *Chem. News* **7**, 158 (1863).
26. H. Bassett, *Ann. Chem.* **132**, 54 (1864).
27. M. T. A. Behme, J. G. Fullington, and E. H. Cordes, *J. Am. Chem. Soc.* **87**, 266 (1965).
27a. A. M. Belikova and N. I. Grineva, *Izv. Sibirsk. Otd. Akad. Nauk SSSR, Ser. Khim. Nauk* p. 79 (1966); *Chem. Abstr.* **68**, 13302 (1968).
28. R. P. Bell, *Trans. Faraday Soc.* **37**, 705 (1941).
29. R. P. Bell, "Acid-Base Catalysis," Chapter 4. Oxford Univ. Press, London and New York, 1941.
30. R. P. Bell and J. C. Clunie, *Proc. Roy. Soc.* **A212**, 33 (1952).
31. R. P. Bell and B. deB. Darwent, *Trans. Faraday Soc.* **46**, 34 (1950).
32. M. Bergmann, *Ann. Chem.* **443**, 223 (1925).
32a. M. Bergmann and W. Freudenberg, *Chem. Ber.* **64**, 158 (1931).
33. M. Bergmann and E. Kann, *Ann. Chem.* **438**, 278 (1924).
34. M. Bergmann and E. von Lippmann, *Ann. Chem.* **452**, 135 (1927).
35. S. Bernstein, R. P. Brownfield, R. H. Lenhard, S. Maines, and I. Bengler, *J. Org. Chem.* **27**, 690 (1962).
36. H. Bodenbenner, *Ann. Chem.* **623**, 183 (1959).
37. H. Boehme, H. Lohmeyer, and J. Wickop, *Ann. Chem.* **587**, 51 (1954).
38. H. Boehme and R. Neiderlein, *Chem. Ber.* **95**, 1859 (1962).
39. F. Bohlmann and W. Sucrow, *Chem. Ber.* **97**, 1839 (1964).
40. F. Bohlmann and W. Sucrow, *Chem. Ber.* **97**, 1846 (1964).
41. J. Bornstein, S. F. Bedell, P. E. Drummond, and C. L. Kosloski, *J. Am. Chem. Soc.* **78**, 83 (1956).
42. A. C. Bottomley, W. Cocker, and P. Nanney, *J. Chem. Soc.* p. 1891 (1937).
43. K. C. Brannock, *J. Am. Chem. Soc.* **73**, 4953 (1951).
44. R. A. Braun, *J. Am. Chem. Soc.* **87**, 5516 (1965).
45. R. A. Braun, Belgian Patent 652,696 (1965); *Chem. Abstr.* **64**, 9731 (1965).
46. R. A. Braun, *J. Org. Chem.* **31**, 1147 (1966).
47. R. A. Braun, *J. Org. Chem.* **31**, 2303 (1966).
48. M. Bredt-Savelsberg and C. Rumscheidt, *J. Prakt. Chem.* [2] **115**, 235 (1927).
49. F. Brescia and V. K. LaMer, *J. Am. Chem. Soc.* **60**, 1962 (1938).
50. F. Brescia and V. K. LaMer, *Ann. N. Y. Acad. Sci.* **39**, 395 (1940).
51. F. Brescia and V. K. LaMer, *J. Am. Chem. Soc.* **62**, 612 (1940).
52. British Drug House, Belgian Patent 610,054 (1962); *Chem. Abstr.* **57**, 13846 (1962).
53. J. N. Bronsted and W. F. K. Wynne-Jones, *Trans. Faraday Soc.* **25**, 59 (1929).
53a. C. Broquet, J. d'Angelo, and V. M. Thuy, *Bull. Soc. Chim. France* p. 341 (1968).
54. H. C. Brown and Y. Okamoto, *J. Am. Chem. Soc.* **80**, 4979 (1958).
55. C. A. Bunton and R. H. DeWolfe, *J. Org. Chem.* **30**, 1371 (1965).
56. C. A. Bunton and V. J. Shiner, *J. Am. Chem. Soc.* **83**, 3207, 3214 (1961).
57. D. Burn, G. Cooley, M. T. Davies, J. W. Ducker, B. Ellis, P. Feather, A. K. Hiscock, D. N. Kirk, A. P. Leftwick, V. Petrow, and D. M. Williamson, *Tetrahedron* **20**, 597 (1964).

58. W. H. Carothers and H. B. Dykstra, U.S. Patent 2,124,686 (1938); *Chem. Abstr.* **32**, 7055 (1938).
59. H. E. Carswell and H. Adkins, *J. Am. Chem. Soc.* **50**, 235 (1928).
60. Celanese Corp. of America, Netherlands Patent Appl., 6,412,635 (1965); *Chem. Abstr.* **63**, 16370 (1965).
61. S. Chladek and J. Smrt, *Collection Czech. Chem. Commun.* **28**, 1301 (1963).
61a. S. Chladek and J. Zemlicka, *Collection Czech. Chem. Commun.* **33**, 4298 (1968).
62. S. Chladek, J. Zemlicka, and F. Sorm, *Collection Czech. Chem. Commun.* **31**, 1785 (1966).
63. L. Claisen, *Chem. Ber.* **26**, 2729 (1893).
64. L. Claisen, *Chem. Ber.* **29**, 1005 (1896).
65. L. Claisen, *Chem. Ber.* **29**, 2931 (1896).
66. L. Claisen, *Chem. Ber.* **31**, 1010 (1898).
67. L. Claisen, *Chem. Ber.* **40**, 3903 (1907).
68. L. Claisen, *Chem. Ber.* **44**, 1161 (1911).
69. L. Claisen, *Chem. Ber.* **47**, 3171 (1914).
70. D. J. Clancey and D. E. Kramm, *Talanta* **13**, 531 (1966).
71. H. Cohen and J. D. Mier, *Chem. & Ind.* (London) p. 349 (1965).
72. E. H. Cordes, *Progr. Phys. Org. Chem.* **4**, 1 (1967).
73. J. W. Cornforth, E. Fawez, J. Goldsworthy, and R. Robinson, *J. Chem. Soc.* p. 1549 (1949).
74. G. Crank and F. W. Eastwood, *Australian J. Chem.* **17**, 1385 (1964).
75. R. J. Crawford and R. Raap, *Can. J. Chem.* **43**, 356 (1965).
76. T. Cuvigny, *Bull. Soc. Chim. France* p. 655 (1957).
77. M. T. Dangyan, *Bull. Armenian Branch Acad. Sci. USSR*, No. 2, p. 31 (1941); *Abstr.* **40**, 3398 (1946).
78. S. David and A. Veyrieres, *Compt. Rend.* **C264**, 1782 (1967).
79. J. N. E. Day and C. K. Ingold, *Trans. Faraday Soc.* **37**, 686 (1941).
80. R. H. DeWolfe, unpublished data.
81. R. H. DeWolfe and J. L. Jensen, *J. Am. Chem. Soc.* **85**, 3264 (1963).
82. R. H. DeWolfe and R. M. Roberts, *J. Am. Chem. Soc.* **76**, 4379 (1954).
83. P. A. Diassi, U.S. Patent 3,326,902 (1967); *Chem. Abstr.* **67**, 82332 (1967).
84. P. A. Diassi and J. Fried, U.S. Patent 3,073,817 (1963); *Chem. Abstr.* **58**, 12643 (1963).
84a. W. Dietsche, *Ann. Chem.* **712**, 21 (1968).
85. K. Dimroth and P. Heinrich, *Angew. Chem. Intern. Ed. Engl.* **5**, 676 (1966).
86. W. von E. Doering and L. K. Levy, *J. Am. Chem. Soc.* **77**, 509 (1955).
87. L. C. Dolby, C. N. Lieske, D. R. Rosencrantz, and M. J. Schwarcz, *J. Am. Chem. Soc.* **85**, 47 (1963).
87a. M. A. Dolliver, T. L. Gresham, G. B. Kistiakowsky, E. A. Smith, and W. E. Vaughan, *J. Am. Chem. Soc.* **60**, 440 (1938).
88. P. M. Duggleby and G. Holt, *J. Chem. Soc.* p. 3579 (1962).
89. J. P. Dusza and S. Bernstein, U.S. Patent 3,069,419 (1962); *Chem. Abstr.* **58**, 10275 (1963).
90. J. P. Dusza and S. Bernstein, *J. Org. Chem.* **27**, 4677 (1962).
91. J. P. Dusza and S. Bernstein, *J. Org. Chem.* **28**, 760 (1963).
92. J. P. Dusza, J. P. Joseph, and S. Bernstein, *J. Am. Chem. Soc.* **86**, 3908 (1964).
93. J. P. Dusza, J. P. Joseph, and S. Bernstein, *Steroids* **8**, 495 (1966).
94. H. B. Dykstra, *J. Am. Chem. Soc.* **57**, 2255 (1935).

95. C. Eckstein and F. Cramer, *Chem. Ber.* **98**, 995 (1965).
95a. R. C. Elderfield and H. H. Remberger, *J. Org. Chem.* **32**, 3809 (1967).
96. A. Ercoli and R. Gardi, *J. Am. Chem. Soc.* **82**, 748 (1960).
97. A. Ercoli and R. Gardi, U.S. Patent 3,139,425 (1964); *Chem. Abstr.* **61**, 12065 (1964).
98. A. Ercoli and R. Gardi, British Patent 963,829 (1964); *Chem. Abstr.* **62**, 9209 (1965).
99. A. Ercoli and P. Ruggieri, *J. Am. Chem. Soc.* **75**, 650 (1953).
100. M. E. Evans, F. W. Parrish, and L. Long, *Carbohydrate Res.* **3**, 453 (1967).
101. W. W. Ewlampieff, *J. Russ. Phys.-Chem. Soc.* **54**, 462 (1922).
102. W. W. Ewlampieff, *Chem. Ber.* **62**, 2386 (1929).
103. O. Exner, *Collection Czech. Chem. Commun.* **16**, 258 (1951).
104. Farbenfabriken Bayer A.-G., Netherlands Patent Appl. 6,404,866 (1964); *Chem. Abstr.* **62**, 12018 (1965).
105. H. Feuer and W. H. Gardner, *J. Am. Chem. Soc.* **76**, 1375 (1954).
105a. L. F. Fieser and M. Fieser, "Reagents for Organic Synthesis," p. 1208. Wiley, New York, 1967.
106. T. H. Fife, *J. Am. Chem. Soc.* **89**, 3228 (1967).
107. T. H. Fife and L. Hagopian, *J. Org. Chem.* **31**, 1772 (1966).
108. T. H. Fife and L. K. Jao, *J. Org. Chem.* **30**, 1492 (1965).
109. E. Fischer and H. von Neyman, *Chem. Ber.* **47**, 976 (1914).
110. H. O. L. Fischer and E. Baer, *Chem. Ber.* **65**, 345 (1932).
111. H. O. L. Fischer and E. Baer, *Helv. Chim. Acta* **18**, 514 (1935).
112. H. O. L. Fischer and H. Milbrand, *Chem. Ber.* **57**, 707 (1924).
113. H. O. L. Fischer and C. Taube, *Chem. Ber.* **57**, 1502 (1924).
114. R. L. Frank and R. P. Seven, *J. Am. Chem. Soc.* **71**, 2629 (1949).
114a. K. Freudenberg and W. Jakob, *Chem. Ber.* **74**, 162 (1941).
115. H. P. M. Fromageot, B. E. Griffin, C. B. Reese, and J. E. Sulston, *Tetrahedron* **23**, 2315 (1967).
115a. H. P. M. Fromageot, C. B. Reese, and J. E. Sulston, *Tetrahedron* **24**, 3533 (1968).
116. A. A. Frost and R. G. Pearson, "Kinetics and Mechanism," 2nd ed., p. 213. Wiley, New York, 1964.
117. J. G. Fullington and E. H. Cordes, *Proc. Chem. Soc.* p. 224 (1964).
118. J. G. Fullington and E. H. Cordes, *J. Org. Chem.* **29**, 970 (1964).
119. R. C. Fuson and D. M. Burness, *J. Am. Chem. Soc.* **68**, 1270 (1946).
120. R. Gardi, R. Vitali, and A. Ercoli, *Tetrahedron Letters* p. 448 (1961).
121. R. Gardi, R. Vitali, and A. Ercoli, *Gazz. Chim. Ital.* **93**, 413 (1963).
122. R. Gardi, R. Vitali, and A. Ercoli, *Gazz. Chim. Ital.* **93**, 431 (1963).
123. R. Gardi, R. Vitali, and A. Ercoli, *Tetrahedron* **21**, 179 (1965).
123a. P. G. Gassman and D. S. Patton, *J. Am. Chem. Soc.* **90**, 7276 (1968).
123b. V. Georgian and K. G. Holden, U.S. Patent 3,373,157 (1968); *Chem. Abstr.* **69**, 52422 (1968).
124. Glaxo Group Ltd., Belgian Patent 649,171 (1964); *Chem. Abstr.* **64**, 15958 (1965).
124a. Glaxo Group Ltd., Netherlands Patent Appl. 6,608,762 (1966); *Chem. Abstr.* **67**, 100345 (1967).
124b. Y. L. Goldfarb, M. A. Kalik, and M. L. Kirmalova, *Khim. Geterosikl. Soedin., Akad. Nauk Latv. SSR* p. 62 (1967); *Chem. Abstr.* **67**, 116772 (1968).
125. Y. L. Goldfarb, A. E. Skorova, and M. L. Kirmalova, *Izv. Akad. Nauk SSSR, Ser. Khim.* p. 1426 (1966).

126. L. M. Gomes, *Bull. Soc. Chim. France* p. 1753 (1967).
126a. A. Gorgues, *Compt. Rend.* **C265**, 1130 (1968).
127. M. Grard, *Ann. Chim. (Paris)* [10] **13**, 336 (1930).
127a. D. P. L. Green and C. B. Reese. *Chem. Commun.* p. 729 (1968).
128. B. E. Griffin, M. Jarman, C. B. Reese, and J. E. Sulston, *Tetrahedron* **23**, 2301 (1967).
128a. G. W. Griffin and A. K. Price, *J. Org. Chem.* **29**, 3192 (1964).
128aa. W. L. Groenenveld, *Inorg. Nucl. Chem. Letters* **3**, 145 (1967).
128b. H. Gross, J. Freiberg, and B. Costisella, *Chem. Ber.* **101**, 1250 (1968).
129. H. Gross and J. Rusche, *Chem. Ber.* **99**, 2625 (1966).
130. H. K. Hall, *J. Am. Chem. Soc.* **79**, 5441 (1957).
131. F. M. Hamer and R. J. Rathbone, *J. Chem. Soc.* p. 595 (1945).
132. L. P. Hammett, "Physical Organic Chemistry," p. 188. McGraw-Hill, New York 1940.
133. A. Hampton, J. C. Frantantoni, P. M. Carroll, and S. Wang, *J. Am. Chem. Soc.* **87**, 5481 (1965).
134. H. S. Harned and N. T. Samaras, *J. Am. Chem. Soc.* **54**, 1 (1932).
135. C. Harries and O. Schauwecker, *Chem. Ber.* **34**, 2987 (1901).
136. D. A. Harris, *J. Chem. Soc.* p. 2247 (1950).
137. E. G. E. Hawkins, *J. Chem. Soc.* p. 3463 (1955).
138. J. H. Hennes, *Dissertation Abstr.* **17**, 36 (1957).
139. A. Holy, *Collection Czech. Chem. Commun.* **32**, 3064 (1967).
140. A. Holy, J. Smrt, and F. Sorm, *Collection Czech. Chem. Commun.* **32**, 2980 (1967).
140a. A. Holy, J. Smrt, and F. Sorm, *Collection Czech. Chem. Commun.* **33**, 3809 (1968).
141. J. C. Hornel and J. A. V. Butler, *J. Chem. Soc.* p. 1360 (1936).
141a. H. O. House and V. Kramar, *J. Org. Chem.* **28**, 3362 (1963).
142. K. Hunger and F. Korte, *Tetrahedron Letters* p. 2855 (1964).
143. C. H. Hurd and M. A. Pollack, *J. Am. Chem. Soc.* **60**, 1905 (1938).
144. C. K. Ingold, "Structure and Mechanism in Organic Chemistry," p. 344. Cornell Univ. Press, Ithaca, New York, 1953.
145. H. H. Inhoffen, G. Kolling, G. Koch, and I. Nebel, *Chem. Ber.* **84**, 361 (1951).
146. H. H. Inhoffen, G. Stoeck, G. Kolling, and U. Stoeck, *Ann. Chem.* **568**, 52 (1960).
147. O. Isler, H. Guttman, H. Lindlar, M. Montavon, R. Ruegg, G. Ryser, and P. Zeller, *Helv. Chim. Acta* **39**, 463 (1956).
148. O. Isler, H. Lindlar, M. Montavon, R. Ruegg, and P. Zeller, *Helv. Chim. Acta* **39**, 249 (1956).
149. O. Isler, M. Montavon, R. Ruegg, and P. Zeller, *Helv. Chim. Acta* **39**, 259 (1956).
150. J. M. Jarman and C. B. Reese, *Chem. & Ind. (London)* p. 1493 (1964).
151. A. Johanassian and E. Akunian, *Bull. Univ. Etat RSS Armenie*, No. 5, p. 235 (1930); *Chem. Abstr.* **25**, 922 (1931).
151a. O. H. Johnson and J. R. Holum, *J. Org. Chem.* **23**, 738 (1958).
152. R. Joly and C. Warnant, U.S. Patent 3,017,409 (1962); *Chem. Abstr.* **56**, 12988 (1962).
153. R. G. Jones, *J. Am. Chem. Soc.* **77**, 4074 (1955).
153a. S. Julia, M. Julia, S. Y. Tchen, and P. Graffin, *Compt. Rend.* **253**, 678 (1961).
154. P. L. Julian, E. W. Meyer, W. J. Carpel, and W. Cole, *J. Am. Chem. Soc.* **73**, 1982 (1951).
155. C. E. Kaslow and W. R. Lawton, *J. Am. Chem. Soc.* **72**, 1723 (1950).

155a. M. Kawazu, M. Fujihara, and T. Danno, Japanese Patent 13,951 ('67); *Chem. Abstr.* **68**, 114443 (1968).
155b. R. W. Kelley and P. J. Sykes, *J. Chem. Soc.*, C p. 416 (1968).
156. P. W. Kent, D. R. Marshall, and N. F. Taylor, *J. Chem. Soc.*, C p. 1281 (1966).
157. G. Kesslin and R. Bradshaw, *Ind. Eng. Chem., Prod. Res. Develop.* **5**, 27 (1966).
157a. D. E. Kiely and H. G. Fletcher, *J. Org. Chem.* **33**, 3723 (1968).
158. M. Kilpatrick and M. L. Kilpatrick, *J. Am. Chem. Soc.* **53**, 3698 (1931).
158a. J. F. King and A. D. Allbut, *Tetrahedron Letters* p. 49 (1967).
159. S. K. Kitaeva, *Uch. Zap. Kazansk. Gos. Univ.* **101**, No. 3, *Sb. Stud. Rabot* No. 2, p. 59 (1941); *Chem. Abstr.* **40**, 552 (1946).
160. J. Klein and E. D. Bergmann, *J. Am. Chem. Soc.* **79**, 3452 (1957).
161. E. Knovenagel, *Chem. Ber.* **55**, 1929 (1922).
162. J. Koskikallio and E. Whalley, *Trans. Faraday Soc.* **55**, 798 and 809 (1959).
163. G. B. Kovalev, A. A. Shamshurin, and E. M. Altmark, *Zh. Organ. Khim.* **3**, 292 (1967).
163a. M. M. Kreevoy and R. W. Taft, *J. Am. Chem. Soc.* **77**, 5590 (1955).
164. M. M. Kreevoy and R. W. Taft, *J. Am. Chem. Soc.* **79**, 4020 (1965).
165. A. J. Kresge and R. J. Preto, *J. Am. Chem. Soc.* **87**, 4593 (1967).
166. R. Kuhn and D. Wieser, *Angew. Chem.* **69**, 371 (1957).
167. D. G. Kundiger and J. H. Hennes, U.S. Patent 3,000,904 (1958); *Chem. Abstr.* **56**, 1459 (1962).
168. H. Kwart and M. B. Price, *J. Am. Chem. Soc.* **82**, 5123 (1960).
169. A. Ladenberg, *Chem. Ber.* **4**, 728 (1871).
170. A. Ladenberg and H. Wichelhaus, *Ann. Chem.* **152**, 163 (1869).
170a. J. M. Lalancette and Y. Beauregard, *Tetrahedron Letters* p. 5169 (1967).
170b. S. Laland, W. G. Overend, and M. Stacey, *J. Chem. Soc.* p. 738 (1950).
171. A. Y. Lao, M.A. thesis, University of California, Santa Barbara, California (1965).
172. H. Laurent, G. Schulz, and R. Wiechert, *Chem. Ber.* **99**, 3051 (1966).
172a. A. LeCoq and E. Levas, *Compt. Rend.* **C266**, 723 (1968).
172b. P. W. N. M. van Leeuwen, *Rec. Trav. Chim.* **86**, 247 (1967).
173. K. Lempert, K. Simon-Ormai, and R. Markovits-Kornis, *Acta Chim. Acad. Sci. Hung.* **51**, 305 (1967).
174. N. J. Leonard and R. J. Laursen, *Biochemistry* **4**, 354 (1965).
175. P. LePerchec, F. Rouessac, and J.-M. Conia, *Bull. Soc. Chim. France* p. 822 (1967).
176. R. Levaillant, *Compt. Rend.* **195**, 882 (1932).
177. R. Levaillant, *Ann. Chim. (Paris)* [11] **6**, 552 (1936).
178. R. Y. Levina, S. G. Kulikov, and N. G. Parshikov, *Zh. Obshch. Khim.* **11**, 567 (1941).
178a. G. E. Lienhard and T. C. Wang, *J. Am. Chem. Soc.* **91**, 1146 (1968).
179. S. Liisberg, W. O. Godtfredsen, and S. Vangedal, *Tetrahedron* **9**, 149 (1960).
179a. D. G. Lindsay and E. B. Reese, *Tetrahedron* **21**, 1673 (1965).
180. W. Lippert, *Ann. Chem.* **276**, 177 (1893).
181. S.-C. Lu, M.A. thesis, University of California, Santa Barbara, California (1967).
182. J. C. Lunt and F. Sondheimer, *J. Chem. Soc.* p. 3361 (1950).
182a. J. R. McCarthy, R. K. Robins, and M. J. Robins, *J. Am. Chem. Soc.* **90**, 4992 (1968).

183. R. J. McCaully, W. F. Olds and P. A. Reiman, *U.S. Dept. Comm., Office Tech. Serv., A.D.* 256,758 (1961); *Chem. Abstr.* **58**, 13676 (1963).
184. S. M. McElvain, E. R. Degginger, and J. D. Behun, *J. Am. Chem. Soc.* **76**, 5736 (1954).
185. S. M. McElvain and G. R. McKay, *J. Am. Chem. Soc.* **77**, 5601 (1955).
186. S. M. McElvain and C. L. Stevens, *J. Am. Chem. Soc.* **69**, 2663 (1947).
187. B. Macierewicz, J. Oscapowicz, and J. Swiderski, *Roczniki Chem.* **36**, 673 (1962); *Chem. Abstr.* **58**, 8062 (1963).
188. D. McIntyre and F. A. Long, *J. Am. Chem. Soc.* **76**, 3204 (1954).
189. J. F. McKenna and F. J. Sowa, *J. Am. Chem. Soc.* **60**, 124 (1938).
190. C. A. MacKenzie and J. H. Stocker, *J. Org. Chem.* **20**, 1695 (1955).
191. S. Maira, Ph.D. dissertation, University of Buffalo, Buffalo, New York (1951); *Dissertation Abstr.* **11**, 532 (1951).
192. M. Maire, *Bull. Soc. Chim. France* [4] **3**, 280 (1908).
192a. S. M. Makin, V. M. Lihkosherstov, and M. I. Berezhnaya, *Zh. Organ. Khim.* **3**, 1419 (1967).
193. F. Mancini and R. Sciaki, *Gazz. Chim. Ital.* **97**, 431 (1967).
194. A. Marquet, M. Dvolaitzky, H. B. Kagan, L. Mamlok, C. Ouannes, and J. Jacques, *Bull. Soc. Chim. France* p. 1822 (1961).
195. R. H. Martin, F. W. Lampe, and R. W. Taft, *J. Am. Chem. Soc.* **88**, 1353 (1966).
196. G. Mason, J. A. Sperry, and E. S. Stern, *J. Chem. Soc.* p. 2558 (1963).
197. A. R. Mattocks, *J. Chem. Soc.* p. 1918 (1964).
198. E. G. Meek, J. H. Turnbùll, and W. Wilson, *J. Chem. Soc.* p. 811 (1953).
199. H. Meerwein, K. Bodenbenner, P. Borner, F. Kunert, and K. Wunderlich, *Ann. Chem.* **632**, 38 (1960).
200. H. Meerwein, P. Borner, O. Fuchs, H. J. Sasse, H. Schrodt, and J. Spille, *Chem. Ber.* **89**, 2060 (1956).
201. H. Meerwein, V. Hederich, H. Morschel, and K. Wunderlich, *Ann. Chem.* **635**, 1 (1960).
202. H. Meerwein, V. Hederich, H. Morschel, and K. Wunderlich, *Ann. Chem.* **635**, 6 (1960).
202a. M. P. Mertes, A. Holy and J. Smrt, *Collection Czech. Chem. Commun.* **33**, 3313 (1968).
202b. M. P. Mertes and J. Smrt, *Collection Czech. Chem. Commun.* **33**, 3304 (1968).
203. A. Michael, *J. Am. Chem. Soc.* **57**, 159 (1935).
204. B. M. Mikhailov and L. S. Povarov, U.S.S.R. Patent 175,945 (1965); *Chem. Abstr.* **64**, 8040 (1966).
205. B. M. Mikhailov and G. S. Ter-Sarkisyan, *Izv. Akad. Nauk SSSR, Ser. Khim.* p. 341 (1966).
206. K. I. Morita, M. Nishimura, and H. Hirose, *J. Org. Chem.* **30**, 3011 (1965).
207. K. Morita, M. Nishimura, and Z. Suzuki, *J. Org. Chem.* **30**, 533 (1965).
208. K. Morita, G. Slomp, and E. V. Jensen, *J. Am. Chem. Soc.* **84**, 3779 (1962).
209. K. I. Morita and Z. Suzuki, *Tetrahedron Letters* p. 263 (1964).
209a. C. G. D. Morley and H. P. C. Hogenkamp, *Arch. Biochem. Biophys.* **123**, 207 (1968).
210. L. V. Nesterov and A. Y. Kessel, *Zh. Obshch. Khim.* **36**, 1835 (1966).
211. E. E. Nifatev and V. N. Kulakov, *Zh. Organ. Khim.* **1**, 1955 (1965).
211a. A. Ohno, T. Koizumi, and G. Tsuchihashi, *Tetrahedron Letters* p. 2083 (1968).
212. J. M. Osbond, P. G. Philpott, and J. C. Wickens, *J. Chem. Soc.* p. 2779 (1961).

213. J. Oszczapowicz and J. Czerminski, *Roczniki Chem.* **37**, 969 (1963); *Chem. Abstr.* **60**, 3125 (1964).
214. L. N. Parfentev and A. M. Mirzaev, *Zh. Obshch. Khim.* **11**, 707 (1941).
215. W. E. Parham, R. W. Soeder, J. R. Throckmorton, K. Kuncl, and R. M. Dodson, *J. Am. Chem. Soc.* **87**, 321 (1965).
216. W. E. Parham and R. J. Sperley, *J. Org. Chem.* **32**, 926 (1967).
217. C. Paulmier and P. Pastour, *Bull. Soc. Chim. France* p. 4021 (1966).
218. H. Pauly and R. von Buttlar, *Ann. Chem.* **383**, 267 (1911).
219. P. Petrenko-Kritchenko, *Chem. Ber.* **61**, 845 (1928).
220. G. J. Pfeiffer and H. Adkins, *J. Am. Chem. Soc.* **53**, 1043 (1931).
221. R. M. Pike and A. M. Dewidar, *Rec. Trav. Chim.* **83**, 119 (1964).
221a. H. Pischel and A. Holy, *Collection Czech. Chem. Commun.* **34**, 89 (1969).
221b. H. Pischel and A. Holy, *Collection Czech. Chem. Commun.* **33**, 2066 (1968).
221c. H. Pischel and G. Wagner, *Z. Chem.* **8**, 178 (1968).
222. T. Pletcher and E. H. Cordes, *J. Org. Chem.* **32**, 2294 (1967).
223. H. W. Post, *J. Am. Chem. Soc.* **55**, 4167 (1933).
224. H. W. Post, *J. Org. Chem.* **1**, 231 (1936).
225. H. W. Post, *J. Org. Chem.* **5**, 244 (1940).
226. H. W. Post, *J. Org. Chem.* **6**, 830 (1941).
227. H. W. Post, "Chemistry of the Aliphatic Ortho Esters," pp. 44ff. Reinhold, New York, 1943.
228. H. W. Post and E. R. Erickson, *J. Org. Chem.* **2**, 260 (1937).
229. J. Preston and H. G. Clark, U.S. Patent 2,928,859 (1960); *Chem. Abstr.* **55**, 3522 (1961).
229a. M. Price, J. Adams, C. Lagenauer, and E. H. Corder, *J. Org. Chem.* **34**, 22 (1969).
230. R. Quelet and J. d'Angelo, *Bull. Soc. Chim. France* p. 3390 (1967).
231. R. Quelet, P. Bercot, and J. d'Angelo, *Bull. Soc. Chim. France* p. 3258 (1966).
232. B. G. Ramsey and R. W. Taft, *J. Am. Chem. Soc.* **88**, 3058 (1966).
233. H. Rapoport, C. H. Lovell, H. R. Reist, and M. E. Warren, *J. Am. Chem. Soc.* **89**, 1492 (1967).
234. A. I. Razumov and V. V. Moskva, U.S.S.R. Patent 175,961 (1965); *Chem. Abstr.* **64**, 9596 (1966).
235. C. B. Reese and J. B. Sulston, *Proc. Chem. Soc.* p. 214 (1964).
236. J. D. Roberts, C. M. Regan, and I. Allen, *J. Am. Chem. Soc.* **74**, 3679 (1952).
237. J. D. Roberts and M. C. Caserio, "Basic Principles of Organic Chemistry," p. 447. Benjamin, New York, 1964.
237a. A. Roedig, K. Grohe, and W. Mayer, *Chem. Ber.* **100**, 2946 (1967).
238. Roussel-UCLAF, Belgian Patent 615,766 (1962); *Chem. Abstr.* **58**, 12641 (1963).
238a. Roussel-UCLAF, British Patent 1,073,989 (1967); *Chem. Abstr.* **67**, 117092 (1967).
239. E. E. Royals, "Advanced Organic Chemistry," p. 360. Prentice-Hall, Englewood Cliffs, New Jersey, 1954.
240. E. E. Royals and K. C. Brannock, *J. Am. Chem. Soc.* **75**, 2050 (1953).
241. P. B. Russel and N. Whittaker, *J. Am. Chem. Soc.* **74**, 1310 (1952).
242. W. A. Ruyle, A. E. Erickson, A. Lovall, and E. M. Chamberlain, *J. Org. Chem.* **25**, 1260 (1960).
243. L. Ruzicka, C. F. Seidel, H. Schinz, and M. Pfeiffer, *Helv. Chim. Acta* **31**, 422 (1948).
244. L. H. Sarett, R. M. Lukes, and G. I. Poos, *J. Am. Chem. Soc.* **74**, 1393 (1952).

245. R. Scarpati, M. Cioffi, G. Scherillo, and R. A. Nicolaus, *Gazz. Chim. Ital.* **96**, 1164 (1966).
245a. H. J. Schaeffer and R. Vince, *J. Med. Chem.* **11**, 15 (1965).
245b. K.-H. Scheit, *Chem. Ber.* **101**, 1141 (1968).
246. L. Schmid, W. Swoboda, and M. Wichtl, *Monatsh. Chem.* **83**, 185 (1952).
247. R. Schmidt, German Patent 553,177 (1929); *Chem. Zentr.* **II**, 2108 (1932).
248. U. Schmidt and P. Grafen, *Ann. Chem.* **656**, 96 (1962).
249. G. Schneider and K. J. Kovacs, *Chem. Commun.* p. 202 (1965).
250. M. N. Schukina, *Zh. Obshch. Khim.* **18**, 1653 (1948).
251. W. Schwenk, G. Fleischer, and B. Whitman, *J. Am. Chem. Soc.* **60**, 1702 (1938).
252. L. Seed, British Patent 837,486 (1960); *Chem. Abstr.* **55**, 3526 (1961).
253. A. Serini and H. Koster, *Chem. Ber.* **71**, 1766 (1938).
254. E. Shapiro, L. Finkenor, H. Pluchet, L. Weber, C. H. Robinson, E. P. Oliveto, H. L. Herzog, F. A. Tabachnik, and E. Collins, *Steroids* **9**, 143 (1967).
255. J. C. Sheehan and C. A. Robinson, *J. Am. Chem. Soc.* **71**, 1436 (1949).
256. L. M. Shorr, *J. Am. Chem. Soc.* **76**, 1390 (1954).
257. L. M. Shorr, U.S. Patent 2,698,861 (1955); *Chem. Abstr.* **49**, 4328 (1955).
257a. M. F. Shostakovskii, V. M. Annenkova, E. A. Gaitseva, K. F. Lavrova, and A. I. Polyakov, *Izv. Sibirsk. Otd. Akad. Nauk SSSR, Ser. Khim. Nauk* p. 163 (1967); *Chem. Abstr.* **67**, 108930 (1967).
258. A. Skrabal, *Z. Elektrochem.* **33**, 322 (1927).
259. A. Skrabal and H. H. Eger, *Z. Physik. Chem.* **122**, 349 (1926).
260. A. Skrabal and O. Ringer, *Monatsh. Chem.* **42**, 9 (1921).
260a. A. J. Slotboom, G. H. DeHaas, and L. L. M. VanDeenen, *Chem. Phys. Lipids* **1**, 192 (1967).
261. B. Smith, *Acta Chem. Scand.* **10**, 1006 (1956).
262. L. L. Smith, J. J. Goodman, H. Mendelsohn, J. P. Dusza, and S. Bernstein, *J. Org. Chem.* **26**, 974 (1961).
263. L. L. Smith and M. Marx, *J. Am. Chem. Soc.* **82**, 4625 (1960).
264. J. C. Speck, D. J. Rynbrandt, and I. H. Kochevar, *J. Am. Chem. Soc.* **87**, 4979 (1965).
265. J. A. Sperry, British Patent 1,008,581 (1965); *Chem. Abstr.* **64**, 1955 (1966).
266. H. Staudinger and G. Rathsam, *Helv. Chim. Acta* **5**, 645 (1922).
267. H. Stetter and K. H. Steinacker, *Chem. Ber.* **86**, 790 (1953).
268. H. Stetter and K. H. Steinacker, *Chem. Ber.* **87**, 205 (1954).
269. C. L. Stevens and A. E. Sherr, *J. Org. Chem.* **17**, 1228 (1952).
270. R. Stolle, *Chem. Ber.* **34**, 1344 (1901).
271. A. Streitwieser, *Chem. Rev.* **56**, 571 (1956).
272. S. Sugasawa, Japanese Patent 1526 ('50); *Chem. Abstr.* **47**, 1730 (1953).
273. S. Swaminathan and M. S. Newman, *Tetrahedron* **2**, 88 (1958).
274. R. W. Taft, *J. Am. Chem. Soc.* **75**, 4231 (1953).
275. R. W. Taft *in* "Steric Effects in Organic Chemistry" (M. S. Newman, ed.), p. 556. Wiley, New York, 1956.
276. R. W. Taft and I. C. Lewis, *J. Am. Chem. Soc.* **81**, 5343 (1959).
277. R. L. Talbot, *J. Org. Chem.* **32**, 834 (1967).
278. D. S. Tarbell and W. E. Lovett, *J. Am. Chem. Soc.* **78**, 2259 (1956).
279. S. F. Thames and H. C. Odom, *J. Heterocyclic Chem.* **3**, 490 (1966).
280. F. Tiemann, *Chem. Ber.* **15**, 2685 (1882).

281. T. Tsukamoto and K. Suzuki, Japanese Patent 214 ('55); *Chem. Abstr.* **50**, 16835 (1956).
282. H. W. Underwood and O. L. Baril, *J. Am. Chem. Soc.* **57**, 2729 (1935).
283. Universal Oil Products, Netherlands Patent Appl. 300,576 (1965); *Chem. Abstr.* **64**, 9613 (1966).
284. Z. M. Utyusheva, *Uch. Zap. Kazansk. Gos. Univ.* **101**, No. 3, *Sb. Stud. Rabot* **2**, 64 (1941); *Chem. Abstr.* **40**, 545 (1946).
285. C. A. VanAllan, *in* "Organic Syntheses," Coll. Vol. IV, N. Rabjohn, Ed., p. 21. Wiley, New York, 1963.
286. J. Van Alphen, *Rec. Trav. Chim.* **49**, 492 (1930).
286a. R. Vince and J. Donovan, *J. Med. Chem.* **12**, 174 (1969).
287. E. Vogel and H. Schinz, *Helv. Chim. Acta* **33**, 116 (1950).
288. O. Vogl and B. C. Anderson, *Tetrahedron Letters*, p. 415 (1966).
289. R. F. von Walther, *J. Prakt. Chem.* [2] **91**, 258 (1915).
290. J. P. Ward and D. A. van Dorp, *Rec. Trav. Chim.* **85**, 117 (1966).
291. L. M. Weinstock, U.S. Patent 3,082,224 (1963); *Chem. Abstr.* **59**, 7615 (1963).
292. A. M. Wenthe and E. H. Cordes, *Tetrahedron Letters* p. 3163 (1964).
293. A. M. Wenthe and E. H. Cordes, *J. Am. Chem. Soc.* **87**, 3173 (1965).
294. P. Westerhof and J. Hartog, *Rec. Trav. Chim.* **86**, 235 (1967).
295. C. Weygand, *Chem. Ber.* **58**, 1473 (1925).
296. C. F. Wilcox and D. L. Nealy, *J. Org. Chem.* **28**, 3448 (1963).
297. R. Willstatter and R. Pummerer, *Chem. Ber.* **38**, 1461 (1905).
298. S. Winstein and R. E. Buckles, *J. Am. Chem. Soc.* **65**, 613 (1943).
299. S. Winstein and L. Goodman, *J. Am. Chem. Soc.* **76**, 4368 (1954).
300. W. Wislicenus and E. Bilhuber, *Chem. Ber.* **51**, 1366 (1918).
301. A. Wohl, *Chem. Ber.* **41**, 3599 (1908).
302. M. L. Wolfrom, J. W. Spoors, and R. A. Gibbons, *J. Org. Chem.* **22**, 1513 (1957).
303. W. F. K. Wynne-Jones, *Trans. Faraday Soc.* **34**, 245 (1938).
304. L. A. Yanovskaya, B. I. Kozyrkin, and V. F. Kucherov, *Izv. Akad. Nauk SSSR, Ser. Khim.* p. 1595 (1966).
305. K. Yasuda, *Chem. & Pharm. Bull (Tokyo)* **11**, 1167 (1963).
306. J. Zemlicka, *Chem. & Ind. (London)* p. 581 (1964).
306a. J. Zemlicka, *Collection Czech. Chem. Commun.* **33**, 3796 (1968).
307. J. Zemlicka and S. Chladek, *Tetrahedron Letters* p. 3057 (1965).
308. J. Zemlicka and S. Chladek, *Collection Czech. Chem. Commun.* **33**, 3293 (1968).

CHAPTER 3

# Reactions of Ortho Esters Which Result in Carbon–Nitrogen or Carbon–Phosphorus Bond Formation

## I. INTRODUCTION

Trialkyl orthocarboxylates react with many kinds of organic nitrogen compounds to yield acyclic and heterocyclic products. The great majority of these reactions involve substances containing primary or secondary amino groups: amines, amides, ureas, sulfonamides, hydrazines, hydrazides, sulfonhydrazides, and others. Most of the reactions discussed in this chapter involve acid-catalyzed replacements of ortho ester alkoxy groups by nitrogen-containing groups. A few involve alkylation of nitrogen by transfer of alkyl groups from ortho ester oxygen to nitrogen. Reactions of orthocarboxylates with phosphorus compounds are also considered.

The first step in nucleophilic attack by a nitrogen compound on an ortho ester probably involves attack by amino nitrogen on a dialkoxycarbonium ion derived from the ortho ester to yield an amide acetal or related compound [Eq. (1)]. The resulting amide acetals are apparently too reactive to be isolable unless the amino nitrogen bears an electronegative substituent such as arenesulfonyl or carbamido. Ordinarily they undergo either elimination of a molecule of alcohol to yield an imidic ester [Eq. (2)], or further substitution to form an ortho amide or related compound [Eq. (3)].

$$H^+ + RC(OR^1)_3 \longrightarrow R\overset{+}{C}(OR^1)_2 + R^1OH$$

$$R\overset{+}{C}(OR^1)_2 + R^2R^3NH \longrightarrow (R^1O)_2CRNR^2R^3 + H^+ \quad (1)$$

$$(R^1O)_2CRNHR^2 \overset{H^+}{\longrightarrow} RC(OR^1)=NR^2 + R^1OH \quad (2)$$

$$(R^1O)_2CRNR^2R^3 + 2\,R^2R^3NH \overset{H^+}{\longrightarrow} RC(NR^2R^3)_3 + 2\,R^1OH \quad (3)$$

The imidic esters and ortho amides may be converted to other products in subsequent reactions. Difunctional amino compounds often yield heterocyclic products.

Alkylations of nitrogen compounds by ortho esters probably involve nucleophilic attack by nitrogen on the alkoxy group of a dialkoxycarbonium ion derived from the ortho ester.

## II. REACTIONS WHICH YIELD ACYCLIC PRODUCTS

In this section are discussed reactions of nitrogen compounds with ortho esters which do not result in formation of heterocyclic rings.

### A. Amide Acetals and Ortho Amides

Amide acetals (**1**) and 1,1-diaminoalkyl ethers (**2**)

$$RC(OR^1)_2NR^2R^3 \qquad RC(NR^2R^3)_2OR^1$$
$$\mathbf{1} \qquad\qquad\qquad \mathbf{2}$$

are known classes of compounds. It should be possible to obtain them as products of reactions of ortho esters with secondary amines [Eqs. (2) and (4)], but no systematic attempt to do so has been reported. They are probably intermediates in the conversion of ortho esters to ketene aminals and ortho amides.

$$RC(OR^1)_3 + 2\,R^2R^3NH \rightarrow RC(NR^2R^3)_2OR^1 + 2\,R^1OH \quad (4)$$

Compounds closely related to amide acetals have been prepared using sulfonamides, ureas, and isocyanates rather than amines as starting materials. These reactions are discussed in Chapter 7.

Replacement of all three alkoxy groups of an ortho ester by amino groups yields an ortho amide [Eqs. (1) and (3)]. Preparation of orthoformamides and tris(acylamino)methanes by reaction of trialkyl orthoformates with secondary amines and with carboxamides is discussed in Chapter 7.

### B. Imidic Esters and Related Compounds

The first identifiable products from the complex series of reactions which occur between primary amines and carboxylic ortho esters are imidic esters. Most such reactions reported in the literature involve aromatic primary amines and trialkyl orthoformates. The preparation of alkyl *N*-arylformimidates from these starting materials [Eq. (5), R = H] is complicated by the fact that the initially formed imidic ester usually reacts rapidly with the primary amine to yield an $N,N'$-diarylformamidine [Eq. (6), R = H] and not a triaryl-orthoformamide, as reported by Giacolone (*61*) and by Lewis (*102*). Reaction (6) is usually much faster than reaction (5), so that $N,N'$-diarylformamidines rather than $N$-arylformimidic esters are obtained as products.

$$ArNH_2 + RC(OR')_3 \rightarrow ArN{=}CROR' + 2\,R'OH \qquad (5)$$

$$ArNH_2 + ArN{=}CROR' \rightarrow ArN{=}CR{-}NHAr + R'OH \qquad (6)$$

Claisen (*39*), who first prepared ethyl $N$-phenylformimidate by the reaction of aniline with triethyl orthoformate, observed that $N,N'$-diphenylformamidine is the first isolable product of the reaction and assumed that the formimidate is formed in a subsequent reaction of the amidine with ethyl orthoformate. In practice this is undoubtedly true, since the amidine is isolable in good yield from reaction mixtures which have been heated for a short time. Prolonged heating of diphenylformamidine with an excess of ethyl orthoformate gave a poor yield of the formimidate.

The low yield of $N$-arylformimidate afforded by Claisen's procedure was increased substantially by Hamer, who added aniline hydrochloride to the reaction mixture (*72*). Hamer claimed that the function of the aniline hydrochloride is to inhibit formation of phenyl isocyanide, a disagreeable minor by-product of the reaction. Later workers also added acidic substances to their reaction mixtures, without apparently being aware of the true function of the acids (*63, 95*).

Roberts showed that the reaction of $N,N'$-diarylformamidines with orthoformates [Eq. (7)] is dependent upon acid catalysis, and proposed that conversion of the amidine to the imidic ester involves alcoholysis of the amidine [the reverse of Eq. (6)] as its initial step (*144, 145*).

$$ArN{=}CH{-}NHAr + HC(OC_2H_5)_3 \xrightarrow{H^+} 2\,ArN{=}CHOC_2H_5 + C_2H_5OH \qquad (7)$$

The complex series of acid-catalyzed reactions which occur when an ortho ester is heated with a primary amine is summarized in Eqs. (8).

$$H^+ + RC(OR')_3 \rightleftharpoons R\overset{+}{C}(OR')_2 + R'OH \qquad (8a)$$

$$R\overset{+}{C}(OR')_2 + R''NH_2 \rightleftharpoons RC(OR')_2NHR'' + H^+ \qquad (8b)$$

$$RC(OR')_2NHR'' + H^+ \rightleftharpoons R\overset{+}{C}(OR')NHR'' + R'OH \qquad (8c)$$

$$R\overset{+}{C}(OR')NHR'' \rightleftharpoons RC(OR'){=}NR'' + H^+ \qquad (8d)$$

$$R\overset{+}{C}(OR')NHR'' + R''NH_2 \rightleftharpoons RC(OR')(NHR'')_2 + H^+ \qquad (8e)$$

$$RC(OR')(NHR'')_2 + H^+ \rightleftharpoons R\overset{+}{C}(NHR'')_2 + R'OH \qquad (8f)$$

$$R\overset{+}{C}(NHR'')_2 \rightleftharpoons R''N{=}CR{-}NHR'' + H^+ \qquad (8g)$$

Roberts and DeWolfe demonstrated that an *N*-arylformimidate [resulting from reactions (8a)–(8d)] rather than an *N,N'*-diarylformamidine is the initial product of the reaction of an orthoformate with an aromatic primary amine (*147*). Reaction (8e), the crucial step for amidine formation, should be kinetically first order in aromatic amine. It was therefore possible to suppress amidine formation by using a relatively high concentration of the orthoformate and a very low concentration of the aromatic amine. When a $6 \times 10^{-5}$ M solution of aniline in 4% triethyl orthoformate–96% heptane was allowed to stand at room temperature for 12 hours, the absorption spectrum of the resulting solution was identical with that of ethyl *N*-phenylformimidate. Formation of the imidic ester was accelerated by addition of 8 ppm of acetic acid to the reaction solution, and retarded by rinsing the reaction flask with ammonium hydroxide prior to drying and filling it with the reaction solution.

It was also found that while *N,N'*-diphenylformamidine ($4 \times 10^{-5}$ molar) does not react appreciably with 4% triethyl orthoformate in heptane in the presence of 16 ppm of acetic acid, it is slowly converted to ethyl *N*-phenylformimidate if 1.6% by volume of absolute ethanol is added to the solution. The alcohol essential for alcoholysis of the amidine [the reverse of (8e)] under the usual synthetic conditions is presumably present as an impurity in the ortho ester, or is formed by reactions (8a)–(8f). The mechanism outlined in Eq. (8) thus accounts for the known facts concerning reactions of trialkyl orthoformates with aromatic primary amines and with *N,N'*-diarylformamidines. Ainsworth (3) isolated triethyl orthoformate and *N,N'*bis(1,3,4-thiadiazol-2)-formamidine from a sample of ethyl *N*-(1,3,4-thiadiazol-2)formimidate which had been exposed to atmospheric moisture, and pointed out that this result provides direct chemical evidence for the mechanism outlined above.

Aromatic primary amines having sterically hindered amino groups yield alkyl *N*-arylformimidates which do not readily react with the hindered amines, with the result that the initially formed imidic ester accumulates in the reaction solution. For example, heating 2,6-diethylaniline with a slight excess of triethyl orthoformate afforded a 90% yield of ethyl *N*-(2,6-diethylphenyl)formimidate, even in the absence of added acid catalysts. No *N,N'*-bis(2,6-diethylphenyl)formamidine was isolated (*53*).

Optimum conditions for synthesizing alkyl *N*-arylformimidates from trialkyl orthoformates and aromatic primary amines were established by Roberts and co-workers (*148, 150*). Other syntheses of alkyl *N*-arylformimidates based on Roberts' procedure are described in the literature (*77, 171, 172*).

Aliphatic primary amines react with triethyl orthoformate much more slowly than do aromatic amines, and attempts to prepare *N*-alkylformimidic esters from orthoformates have thus far been unsuccessful (*186*). The low apparent reactivity of aliphatic primary amines is probably due to the fact that the effective catalyst in ortho ester–amine reactions is the conjugate acid of the amine (or an amine–Lewis acid complex); the conjugate acids of aliphatic amines are weaker than those of aromatic amines and should therefore be less effective in catalyzing the formation of the various carbonium ion intermediates of Eq. (8). Alcoholysis of $N,N'$-dialkylformamidines, which are the isolable products of these reactions, is probably slower than that of diarylformamidines for the same reason.

Alkyl $N$-arylacetimidates and alkyl $N$-arylpropionimidates are formed in high yields by simply heating equimolar amounts of aromatic primary amines and the trialkyl orthocarboxylates until the theoretical amount of alcohol distills off (53, 59, 103) [Eq. (5), R = $CH_3$, $C_2H_5$]. Trialkyl esters of higher ortho acids could probably be converted to $N$-arylimidic esters by the same procedure. Mukaiyama and Sato (123) synthesized a number of cyclic imidic esters (2-$N$-substituted iminotetrahydrofurans) by the reaction of primary amines with the cyclic ortho ester 2,2-diethoxytetrahydrofuran.

$$\underset{OC_2H_5}{\overset{OC_2H_5}{\bigcirc}} + RNH_2 \longrightarrow \underset{O}{\overset{NR}{\bigcirc}} + 2\ C_2H_5OH \qquad (9)$$

(R = $n$-$C_6H_{13}$, cyclo-$C_6H_{11}$, $C_6H_5CH_2$, $o$- and $p$-$ClC_6H_4$,
$o$- and $p$-$CH_3C_6H_4$, $C_6H_5$, $o$- and $p$-$CH_3OC_6H_4$, and 3,4-$Cl_2C_6H_3$)

Unlike $N$-arylformimidic esters, which undergo rapid acid-catalyzed reactions with aromatic amines to form $N,N'$-diarylformamidines, alkyl $N$-aryl esters of higher imidic acids react very sluggishly with aromatic amines. Amidine formation is so slow that the imidic esters accumulate in the reaction mixtures as they are formed. Apparently reactions (8a)–(8d) occur readily, but reaction (8e) is slow because of steric hindrance to formation of the alkyl diaminoalkyl ether intermediate.

In most cases it is not necessary to add an acid catalyst to mixtures of higher orthocarboxylates and aromatic primary amines in order to obtain high yields of imidic esters. This is true not because the reactions are uncatalyzed, but rather because they are so extremely sensitive to acid catalysis that traces of acidic impurities in the reactants suffice. The catalytic effect of acids is demonstrated by the fact that aniline and triethyl orthoacetate are converted to ethyl $N$-phenylacetimidate more rapidly if a small amount of acetic acid is added to the reaction mixture (186).

Compounds having primary amino groups which react with ortho esters to form $N$-substituted imidic esters include anilines, aryl hydrazines, carboxamides, ureas, carbamides, cyanamide, hydrazides, sulfonamides, and heterocyclic primary amines. Malondiamidine dihydrochloride yields $sym$-$N,N$-bis(ethoxymethylene)malondiamidine monohydrochloride when refluxed with a mixture of acetic anhydride and triethyl orthoformate *(139)*. Hydrogen cyanide tetramer and triethyl orthoformate yield

$$HN=C(CN)CH(CN)N=CHOC_2H_5$$

*(221)*.

The reaction of tetraalkyl orthocarbonates with primary amino compounds has not been thoroughly studied. There are only three published examples of formation of *N*-substituted diethyl imidocarbonates by reactions of primary amino compounds (aniline, *p*-toluenesulfonamide, and *p*-toluenesulfonhydrazide) with tetraalkyl orthocarbonates *(105, 106, 114, 197)*.

Ethyl *N*-substituted formimidates, $RN=CHOC_2H_5$, have been prepared from the following amino compounds: **Aromatic primary amines**, including aniline *(39, 53, 72, 144, 148–150)*, chloroanilines *(63, 76b, 77, 95)*, toluidines *(76b, 144, 172)*, *o*-anisidine *(76b)*, *o*-phenetidine *(77, 95)*, *o*-aminobenzhydrazide *(157)*, *o*-dimethyltriazolylaniline *(140)*, 2,4- and 2,5-dimethylanilines *(77)*, 2,6-diethylaniline, 2,5-dichloroaniline *(171)*, 2-methyl-4-methoxyaniline *(171)*, *o*-*N*-(tetra-*O*-acetyl-α-D-glucopyranosyl)aminoaniline *(112)*, α-naphthylamine *(129b)*, and tetrahydro-α-naphthylamine *(129b)*; **amides**, including urea *(212a)*, acetamide *(212a)*, benzamide *(212a)*, *o*-hydroxybenzamide *(78)*, ethyl carbamate *(66)*, and cyanamide *(66)*; **hydrazines**, including *o*- and *p*-nitrophenylhydrazine *(155)*, 2,4-dinitrophenylhydrazine *(155)*, a hydrazinoimidazopyridazine *(169c, 169d)*, a hydrazinotetrazolopyridazine *(95a)*, and a hydrazinoimidazopyridopyridazine *(169b)*; **hydrazones**, benzaldehyde hydrazone *(70, 71, 125a)*; **hydrazides**, including formhydrazide *(5)*, cyanacethydrazide *(157)*, thiosemicarbazide *(3, 82)*, and substituted benzhydrazides *(157)*; **sulfonamides**, including methanesulfonamide *(198)*, substituted benzenesulfonamides *(103, 105, 106, 158)*, benzenesulfonamide *(198)*, and substituted sulfonamidoquinazolines *(125)*; **heterocyclic primary amines**, including an aminopyrrolenine *(166)*, 2-aminopyrrole *(188a)*, 2-amino-3-carboxamido-4,5,6,7-tetrahydrobenzothiophene *(160aa)*, 2-aminopyridines *(1, 22, 104, 171)*, 3-aminopyridines *(22, 104, 171)*, 4-aminopyridines *(22, 104, 171, 190)*, 5-aminoisoxazole *(187a)*, 3-amino-pyrazoles *(1, 188–190, 192)*, 4-aminoimidazoles *(165, 189)*, a 3-aminoisothiazole *(73)*, a 4-aminopyrimidine *(192)*, 5-aminopyrimidines *(121, 195)*, aminopyrazolopyrimidines *(195a, 195b)*, an aminoimidazopyrimidine *(195b)*; and a 3-amino-$\Delta^4$-steroid *(52)*. A series of alkyl *N*-phenylformimidates was prepared by reaction of aniline with trimethyl, tripropyl, tributyl, triamyl, trihexyl, tricyclohexyl, and triheptyl orthoformates *(148)*.

Ethyl N-substituted acetimidates, $RN=C(CH_3)OC_2H_5$, were prepared by reaction of triethyl orthoacetate with aromatic primary amines *(53, 59)*, *N*-methylsemicarbazide *(97)*, *o*- and *p*-nitrophenylhydrazine *(155)*, *p*-toluenesulfonamide *(103)*, *p*-toluenesulfonhydrazide *(105, 106)*, 4-carboxamido-5-

aminoimidazoles (*138*), 4-chloro-5,6-diaminopyrimidine (*121*), 2-aminobenzoxazole and 2-aminobenzothiazole (*88*), a 2-amino-1,3,4-thiadiazole (*83*), and an aminopyrroloisoxazole (*60a*). Ethyl N-substituted propionimidates, $RN=C(C_2H_5)OC_2H_5$, were prepared from triethyl orthopropionate and aniline (*53*), o- and p-nitrophenylhydrazine (*155*), thiosemicarbazide (*3*), p-toluenesulfonhydrazide (*105, 106*), 4-chloro-5,6-diaminopyrimidine (*121*), and a 2-amino-1,3,4-thiadiazole (*83*). The same thiadiazole was converted to a butyrimidate and a valerimidate by reaction with triethyl orthobutyrate and triethyl orthovalerate (*83*).

Ethyl N-substituted benzimidates, $RN=C(C_6H_5)OC_2H_5$, were prepared from triethyl orthobenzoate and nitrophenylhydrazines (*155*), p-toluenesulfonhydrazide (*105, 106*), a 4-carboxamide-5-aminoimidazole (*138*), a 2-amino-1,3,4-thiadiazole (*83*), and diphenylphosphinic acid hydrazide (*170*).

### C. Acyclic Amidines

It was pointed out in the preceding section that while amidines are not the initial products of reaction of trialkyl orthocarboxylates with primary amines, they are often the only products which can be isolated. The amidine is formed by reaction of the amine with an initially formed imidic ester [Eq. (6)]. The net reaction is described by Eq. (10).

$$2 RNH_2 + R'C(OR'')_3 \rightarrow RN=CR'-NHR + 3 R''OH \qquad (10)$$

Although higher orthocarboxylates can be converted into amidines, the reaction occurs much more readily with trialkyl orthoformates, and most of the amidines which have been prepared in this manner are $N,N'$-disubstituted formamidines. A wide variety of compounds having primary amino groups have been converted to amidines by reaction with ortho esters. Aliphatic and aromatic primary amines, heterocyclic amines, amides, urea, hydrazines, hydrazides, and sulfonamides all yield amidines under appropriate reaction conditions.

Most aromatic primary amines yield $N,N'$-diarylformamidines when heated with triethyl orthoformate. The reaction [Eq. (10), R = Ar, R' = H, R'' = $C_2H_5$] may be driven to completion by distillation of ethanol from the reaction mixture as it is formed, and yields of the amidine are often nearly quantitative. A single recrystallization ordinarily yields a pure product. Aromatic amines having electron-withdrawing substituents form formamidines more readily than do those with electron-releasing substituents (*115*). Although the reactions which form the amidine are acid-catalyzed [Eq. (8)], it is usually unnecessary to add an acid to the reaction mixture: acidic impurities in the reactants and adsorbed on the glassware cause the reaction to occur at a feasible rate. All of the $N,N'$-diarylformamidines thus far reported are crystalline solids, most of which melt between 100° and 250°C. This fact, together with their ease of preparation, suggests that they should be useful derivatives for the characterization of aromatic primary amines.

Aromatic primary amines having bulky ortho substituents (such as 2,6-diethylaniline) may react with triethyl orthoformate to form $N$-arylformimidic esters rather than the expected formamidines, due to steric hindrance to reaction of the amine with the initially formed imidic ester [Eq. (6)] (*53*).

Aliphatic primary amines react much more slowly with triethyl orthoformate than do aromatic amines, probably because of the greater basicity of the aliphatic amine reaction mixtures. However, two procedures have been described which afford satisfactory yields of $N,N'$-dialkylformamidines. Several $N,N'$-dialkylformamidines were prepared by adding 2 moles of the aliphatic amine to a refluxing mixture of 1 mole each of triethyl orthoformate and acetic acid (*186*). This results in formation of the $N,N'$-dialkylformamidinium acetate, from which the free amidine is readily obtained. If ammonium acetate is substituted for the primary amine, formamidinium acetate is obtained (*185*).

A simpler procedure which gives $N,N'$-dialkylformamidines directly involves heating the amine and triethyl orthoformate with catalytic amounts of boron trifluoride etherate (*100*). Since the reaction is driven to completion by distillation of ethanol from the reaction mixture, this procedure is limited to amines that have boiling points higher than ethanol.

The acetic acid-catalyzed reaction also affords $N,N'$-dialkylacetamidines from triethyl orthoacetate and aliphatic primary amines (*186*). Bubbling dry ammonia through a refluxing mixture of triethyl orthoacetate and ammonium acetate gave a high yield of acetamidinium acetate (*185, 187*).

The unsymmetrical amidine $N,N$-dimethyl-$N'$-*tert*-butylformamidine was prepared in low yield by refluxing a mixture of dimethylammonium chloride, *tert*-butylamine and triethylorthoformate (*72a*).

Table I lists symmetrical $N,N'$-dialkyl- and $N,N'$-diarylformamidines which have been prepared from primary amines and triethyl orthoformate.

$N,N'$-Diarylacetamidines may be prepared by prolonged heating of 2 moles of the aromatic primary amine with 1 mole of triethyl orthoacetate [Eq. (17), R = Ar, R' = $CH_3$, R" = $C_2H_5$] (*53, 74, 153*). Under these conditions the initially formed ethyl $N$-arylacetimidates are converted to the amidines. If 1 mole of acetic acid is added to the reaction mixture, the reaction proceeds more rapidly, and yields the acetamidinium acetate (from which the free amidine may be obtained) (*186*).

The preparation of N,N'-symmetrically disubstituted amidines from primary amines and higher trialkyl orthocarboxylates is practically unknown. Anschutz and Stiepel obtained oxamidine derivatives in reactions of

TABLE I

Symmetrical N,N'-Disubstituted Formamidines, RNH—CH=NR, Prepared from RNH$_2$ and HC(OR')$_3$

| R | M.p. (°C) | References |
|---|---|---|
| H | a | 185 |
| CH$_3$ | a | 186 |
| n-C$_3$H$_7$ | b.p. 55°/0.65 mm | 186 |
| n-C$_4$H$_9$ | b.p. 85°/0.09 mm | 186 |
| cyclo-C$_6$H$_{11}$ | 104° | 100, 186 |
| n-C$_8$H$_{17}$ | 37° | 100 |
| C$_6$H$_5$CH$_2$ | 76° | 100, 186 |
| C$_6$H$_5$ | 139° | 14, 39, 61, 95, 115, 146, 210, 218a |
| o-CH$_3$C$_6$H$_4$ | 149° | 15, 49, 115, 137, 209, 210 |
| m-CH$_3$C$_6$H$_4$ | 125° | 15, 137, 143 |
| p-CH$_3$C$_6$H$_4$ | 143° | 15, 49, 115, 143, 209, 210 |
| o-ClC$_6$H$_4$ | 141° | 15, 95, 143, 146 |
| m-ClC$_6$H$_4$ | 118° | 15, 95, 136, 143 |
| p-ClC$_6$H$_4$ | 180° | 15, 95, 143, 146 |
| m-BrC$_6$H$_4$ | 135° | 115, 209, 210 |
| p-BrC$_6$H$_4$ | 170°, 184° | 48, 115 |
| o-O$_2$NC$_6$H$_4$ | 124° | 14, 209 |
| m-O$_2$NC$_6$H$_4$ | 200° | 115, 209, 210 |
| p-O$_2$NC$_6$H$_4$ | 237° | 115, 209, 210 |
| o-C$_2$H$_5$OC$_6$H$_4$ | 81° | 48 |
| m-C$_2$H$_5$OC$_6$H$_4$ | — | 54 |
| p-C$_2$H$_5$OC$_6$H$_4$ | 109° | 177 |
| o-CH$_3$OC$_6$H$_4$ | 105°, 115° | 48, 115 |
| p-CH$_3$OC$_6$H$_4$ | 123° | 177, 203 |
| p-IC$_6$H$_4$ | 175° | 50 |
| p-C$_2$H$_5$OCOC$_6$H$_4$ | 193°, 240° | 65, 156, 178 |
| p-(CH$_3$)$_2$NC$_6$H$_4$ | 157° | 50 |
| p-CH$_3$SC$_6$H$_4$ | 152° | 180 |
| m-HO$_2$CC$_6$H$_4$ | 250° | 65 |
| p-HO$_2$CC$_6$H$_4$ | 235° | 65 |
| p-C$_6$H$_5$C$_6$H$_4$ | 192° | 47 |
| p-H$_2$NSO$_2$C$_6$H$_4$ | — | 79 |
| p-C$_6$H$_5$CH$_2$OC$_6$H$_4$ | 153° | 50 |
| o-CH$_3$NHSO$_2$CH$_2$C$_6$H$_4$ | 185° | 196 |
| o-C$_6$H$_5$NHSO$_2$CH$_2$C$_6$H$_4$ | 178° | 196 |
| p-C$_6$H$_5$N=NC$_6$H$_4$ | 193° | 210 |
| 3,5-Br$_2$C$_5$H$_3$ | 195° | 177 |
| 2,4-Cl$_2$C$_6$H$_3$ | 95° | 178 |
| 2,5-(CH$_3$)$_2$C$_6$H$_3$ | 155° | 49, 179 |
| 2-C$_2$H$_5$O-4-O$_2$NC$_6$H$_3$ | 206° | 179 |
| 2-CH$_3$-4-IC$_6$H$_3$ | 169° | 50 |
| 2-CH$_3$-4-O$_2$NC$_6$H$_3$ | 166° | 16 |

## TABLE I—continued

| R | M.p. (°C) | References |
|---|---|---|
| $p\text{-}[(CH_3)_2NC_6H_4]_2CHC_6H_4$ | 200° | *137* |
| $2,4,6\text{-}Br_3C_6H_2$ | — | *209* |
| $2\text{-}CH_3\text{-}4\text{-}BrC_6H_3$ | 119° | *115* |
| $\alpha\text{-}C_{10}H_7$ | 202° | *48, 115, 178* |
| $\beta\text{-}C_{10}H_7$ | 186° | *115, 178* |

[a] Isolated as acetate salt.

methyl trimethoxyacetate with aniline and $p$-toluidine [Eq. (11)] (*10*). Sah treated a crude preparation of hexaethyl stilbene-4,4'-bis(orthocarboxylate) with ethanolic ammonia and isolated the corresponding bis amidine [Eq. (12)] (*159a*).

$$CH_3OCOC(OCH_3)_3 + 3 \, ArNH_2 \rightarrow ArNHCOC(NHAr)=NAr + 4 \, CH_3OH \quad (11)$$
$$(Ar = C_6H_5, p\text{-}CH_3C_6H_4)$$

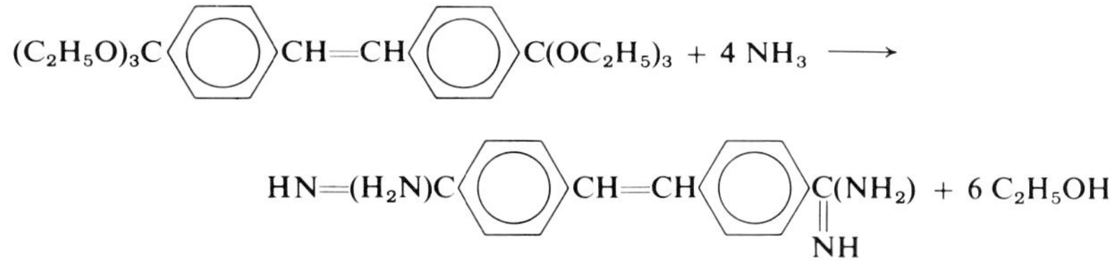

$(C_2H_5O)_3C\langle\bigcirc\rangle CH=CH\langle\bigcirc\rangle C(OC_2H_5)_3 + 4 \, NH_3 \longrightarrow$

$HN=(H_2N)C\langle\bigcirc\rangle CH=CH\langle\bigcirc\rangle \underset{NH}{\overset{\|}{C}}(NH_2) + 6 \, C_2H_5OH \quad (12)$

A number of amino-substituted derivatives of pyridines (*104, 176–179*), thiazoles (*178, 179*), benzothiazoles (*86, 175, 179*), isoxazoles (*84*), triazoles (*89*), and thiadiazoles (*3*) have been converted to N,N'-disubstituted formamidines by reaction with triethyl orthoformate.

There are several reports in the literature of syntheses of N,N'-bis(acyl)-formamidines [Eq. (13)] or N-acylformimidates from carboxamides and triethyl orthoformate. These reactions were carried out by heating the reactants with (*124*) or without (*127, 128, 154, 212a, 219*) an added acid catalyst. Analytical data are lacking for most of the reaction products, and conflicting values are reported for the melting point of N,N'-dibenzoylformamidine.

$$RCONH_2 + HC(OC_2H_5)_3 \rightarrow RCON=CH-NHCOR + 3 \, C_2H_5OH \quad (13)$$
$$(R = CH_3, C_2H_5, C_6H_5, NCCH_2, NCC_6H_4, C_6H_5N, 5\text{-nitrofuryl})$$

Doubt is cast on the amidine and imidic ester structures claimed for these products by reports of Bredereck and co-workers that acid-catalyzed reactions of a number of carboxamides with triethyl orthoformate, either without solvent or in toluene or dimethylformamide solutions, yield tris(acylamino)-methanes *(27–29)*. Bredereck's analytical data and cryoscopic molecular weight determinations support the tris(acylamino)methane structures. The products of reactions of carboxamides with triethyl orthoformate have very high melting or decomposition temperatures. The fact that the products described as diacylformamidines are reported to melt 8–17°C lower than Bredereck's products from the same reactants is therefore not too helpful in deciding whether these reactions yield two kinds of products or only the tris(acylamino)methanes claimed by Bredereck.

N-substituted ureas react with triethyl orthoformate under some conditions to yield *N,N'*-dicarbamylformamidines *(132, 215, 216)*.

$$2\ RNHCONH_2 + HC(OC_2H_5)_3 \rightarrow RNHCON{=}CHNHCONHR + 3\ C_2H_5OH \quad (14)$$
[R = H, $CH_3$, $C_2H_5$, $(CH_3)_3C$, $C_5H_{11}$, $C_7H_{15}$, $C_8H_{17}$, $C_6H_5CH_2$, $C_6H_5$, p-$C_2H_5C_6H_4$, 2,6-$(CH_3)_2C_6H_3$]

*N*-Alkylureas were converted to the dicarbamylformamidines simply by heating them with triethyl orthoformate *(215)*. However, when this procedure is applied to *N*-arylureas, *N*-arylcarbamates are the principal products. The *N,N'*-bis(arylcarbamyl)formamidines can be prepared in good yields by adding acetic anhydride to the reaction mixture. The fact that the bis(arylcarbamyl)formamidines react with alcohols to form *N*-arylcarbamates suggests that the principal function of the acetic anhydride is to react with the ethanol produced when the amidine is formed. Urea itself is converted by triethyl orthoformate either into *N,N'*-dicarbamylformamidine *(25)*, or to an equimolar complex of urea and 2,4-dihydroxy-1,3,5-triazine *(133)*, depending on reaction conditions. The dicarbamylformamidine is converted to the dihydroxy *s*-triazine by heating *(25)*.

Arylhydrazines and triethyl orthoformate react to form diaryl hydrazidines [Eq. (15)] *(39, 143, 155, 209, 210)*. Hydrazides yield analogous compounds [Eq. (16)] *(74, 157)*.

$$2\ ArNHNH_2 + HC(OC_2H_5)_3 \rightarrow ArNHNH{-}CH{=}NNHAr + 3\ C_2H_5OH \quad (15)$$
[Ar = $C_6H_5$, *o*- (and *p*-) $O_2NC_6H_4$, 2,4-$(O_2N)_2C_6H_3$]

$$2\ RCONHNH_2 + HC(OC_2H_5)_3 \rightarrow$$
$$RCONHN{=}CH{-}NHNHCOR + 3\ C_2H_5OH \quad (16)$$
[R = $NCCH_2$, *o*-$HOC_6H_4$, *o*- (*m*-, *p*-) $O_2NC_6H_4$, *p*-$CH_3OC_6H_4$, *p*-$ClC_6H_4$]

Runti reported that heating benzenesulfonamide with triethyl orthoformate yields *N,N'*-bis(benzenesulfonyl)formamidine,

$$C_6H_5SO_2N{=}CHNHSO_2C_6H_5$$

*(125)*. However, the proposed structure of the reaction product has been questioned by Hagedorn *(68)*, who prepared the amidine by a different method and fully characterized it. Hagedorn's product has a melting point 50°C higher than Runti's.

Tetraethyl orthocarbonate reacts with ammonia and with aniline to form guanidine and $N,N',N''$-triphenylguanidine, respectively *(76, 197)*.

$$3\ RNH_2 + C(OC_2H_5)_4 \rightarrow RN{=}C(NHR)_2 + 4\ C_2H_5OH \qquad (17)$$
$$(R = H, C_6H_5)$$

Two indirect routes to formamidine derivatives which utilize triethyl orthoformate as a starting material should be mentioned. The first is the preparation of unsymmetrical N,N'-disubstituted formamidines by reaction of imidic esters (prepared from primary amines and ethyl orthoformate) with ammonia or amines *(84, 104, 181, 216)*. Since the unsymmetrical formamidines produced by these reactions are highly susceptible to acid-catalyzed disproportionation reactions [which yield mixtures of the unsymmetrical and the two possible symmetrical N,N'-disubstituted formamidines *(145)*], care must be taken to exclude acidic impurities if this reaction is to be used for synthetic purposes.

N,N,N',N'-Tetrasubstituted formimidinium salts are obtained by the reaction of strong acids with the orthoformamides produced by heating secondary aromatic amines with triethyl orthoformate *(40, 41)*.

$$(RR'N)_3CH + H^+ \rightarrow RR'N{-}\overset{+}{C}H{-}NRR' + RR'NH \qquad (18)$$

## D. Ketene Aminals

Baganz and Domaschke *(18)* found that trialkyl orthocarboxylates having hydrogen alpha to the acyl carbon react with secondary amines to form ketene aminals [Eq. (19)]. The ketene aminals probably result from the elimination of alcohol or a secondary amine from an initially formed ethyl 1,1-diaminoalkyl ether or ortho amide.

$$RCH_2C(OC_2H_5)_3 + R'R''NH \rightarrow RCH{=}C(NR'R'')_2 + 3\ C_2H_5OH \qquad (19)$$
$$(R = H, CH_3;\ RR'' = {-}(CH_2)_5{-},\ {-}CH_2CH_2OCH_2CH_2{-})$$

## E. N-Acyl Compounds

A few reactions have been reported in which substances having primary amino groups are acylated by ortho esters. These reactions have not been studied in detail, and their mechanisms are unknown. In at least two instances, the observed products were probably formed by reactions of ordinary esters (from hydrolysis of the ortho esters) with the amino compound (*10, 107*).

$$\text{RC(OR')}_3 \xrightarrow[-2\,\text{R'OH}]{+\,\text{H}_2\text{O}} \text{RCO}_2\text{R'} \xrightarrow[-\,\text{R'OH}]{+\,\text{R''NH}_2} \text{RCONHR''} \quad (20)$$

Acyl derivatives of amino compounds have also been obtained in reactions where water was not intentionally present in the reaction mixtures. Thus, ethylenediamine was converted to the *N,N'*-diformyl derivative by heating with excess triethyl orthoformate in the presence of sulfuric acid, and arenesulfonylhydrazides were formylated by treatment with triethyl orthoformate and acetic anhydride (*125*). Similarly, *N,N*-dibutylacetamide was obtained by heating dibutylamine with triethyl orthoacetate (*108*). One possible route to the observed products in these reactions involves nucleophilic attack by the amino compound on the alkoxy alkyl group of the conjugate acid of an initially formed imidic ester [Eq. (21)]. Product and material balance studies which would support this hypothesis have not been reported.

$$\text{RC}(=\text{NR'})\overset{+}{\text{O}}\text{HR''} + \text{R'NH}_2 \rightarrow \text{R'R''NH} + \text{RCONHR'} \quad (21)$$

Tetraethyl orthocarbonate undergoes similar reactions. For example, *N,N'*-diphenylurea is one of the products formed when tetraethyl orthocarbonate is heated with aniline (*197*).

## F. Aminoethylene Compounds

Anilines, amides, and ureas react with mixtures of triethyl orthoformate and substances having activated methylene groups to form aminomethylene derivatives.

$$\text{RR'NH} + \text{HC}(\text{OC}_2\text{H}_5)_3 + \text{H}_2\text{CXY} \rightarrow \text{RR'NCH}=\text{CXY} + 3\,\text{C}_2\text{H}_5\text{OH} \quad (22)$$

The products of these reactions are useful synthetic intermediates, particularly in the preparation of nitrogen heterocycles. Since *N,N'*-disubstituted formamidines yield the same products when heated with the active methylene compounds, it is probable that either formamidines or formimidic esters are intermediates in most of these reactions (*136, 169*). This supposition is supported by the fact that an *N,N'*-diarylformamidine may be substituted for an aromatic primary amine in these reactions (*92, 101*). In the synthesis of *m*-chloroanilinomethylenemalonic ester, substitution of *N,N'*-bis(*m*-chlorophenyl)formamidine for *m*-chloroaniline results in a substantial improvement in yield (*101*). Ortho amides or amide acetals may be intermediates in condensations involving secondary amines. Aminomethylene compounds prepared according to Eq. (22) are listed in Table II.

## TABLE II
### RR′N—CH=CXY Prepared from Amino Compounds, Triethyl Orthoformate, and Active Methylene Compounds

| R | R′ | =CXY | References |
|---|---|---|---|
| $C_6H_5$, $CH_3NHCO$, $C_3H_7NHCO$ | H | $=C(CN)_2$ | 57, 215 |
| $p$-$CH_3OC_6H_4$, $m$-$CF_3C_6H_4$, $m$-$ClC_6H_4$, $p$-$C_2H_5OC_6H_4$, $H_2NCO$, $(CH_3)_2CHNHCO$ | H | $=C(CN)CO_2R''$ <br> ($R'' = CH_3, C_2H_5$) | 57, 169, 215 |
| $m$-$ClC_6H_4$, $H_2NCO$ | H | $=C(COCH_3)_2$ | 169, 215 |
| $m$-$ClC_6H_4$, $H_2NCO$, $CH_3NHCO$ | H | $=C(COCH_3)CO_2C_2H_5$ | 169, 215 |
| $m$-$ClC_6H_4$ | H | $=C(CO_2C_2H_5)_2$ | 101 |
| $(CH_3)_2CHNHCO$ | H | $=C(CN)CO_2H$ | 215 |
| $C_7H_{15}NHCO$ | H | $=C(CN)_2$ | 85 |
| $m$-$ClC_6H_4$ | H | $=C(NHR'')CO_2C_2H_5$ <br> ($R'' =$ alkyl, aryl, $-(CH_2)_nCN$) | 125 |
| RCO <br> ($R = CH_3, C_2H_5O, C_6H_5$) | H | $=C(CN)CO_2C_2H_5$ | 164 |
| $m$-$ClC_6H_4$ | H | $=C(NHR'')CN$ <br> ($R'' =$ alkyl, aryl) | 135 |
| $C_6H_5$ | $CH_3$ | $=C(CN)_2$ | 57 |
| $C_6H_5$ | H | 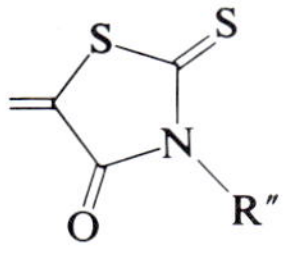 thiazolidine-2-thione-4-one ($R'' = CH_3, C_2H_5$) | 57 |
| $p$-$CH_3C_6H_4$ | H | 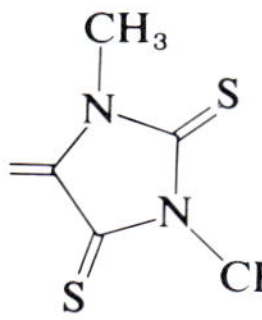 1,3-dimethyl-2-thione-5-thione imidazolidine | 57 |
| $C_6H_5$ | H | 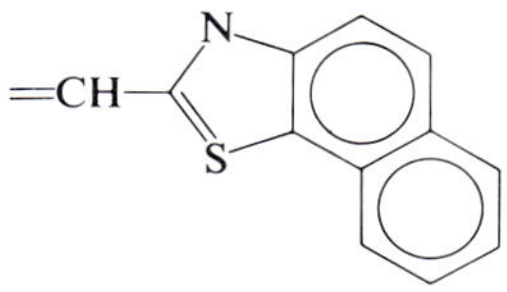 =CH-naphtho[2,1-d]thiazole | 91 |
| $o$-$H_2NCOC_6H_4$ | H | 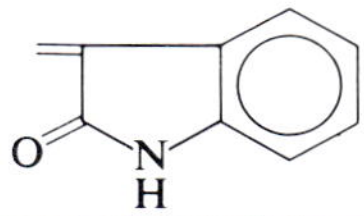 oxindole-3-ylidene | 52a |

TABLE II—continued

| R | R' | =CXY | References |
|---|---|---|---|
| C₆H₅, p-CH₃C₆H₄ | H, CH₃ | =CH—(benzoxazole) | 91 |
| C₆H₅ | H | =CH—(N-methylpyridinium) | 92, 129 |
| C₆H₅ | H | =CH—(N-methylquinolinium) | 129 |
| —C₆H₄—NHCH=CH—(3-ethylbenzoxazole) | H[a] | =CH—(3-ethylbenzoxazole) | 202 |

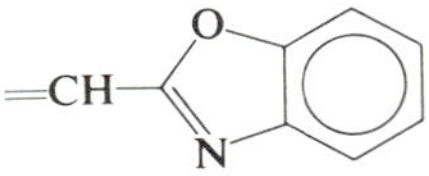

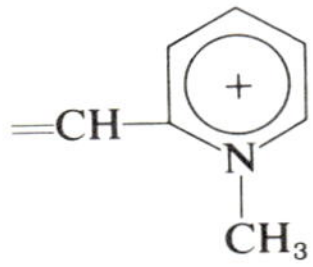

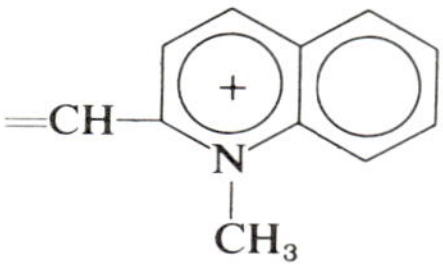

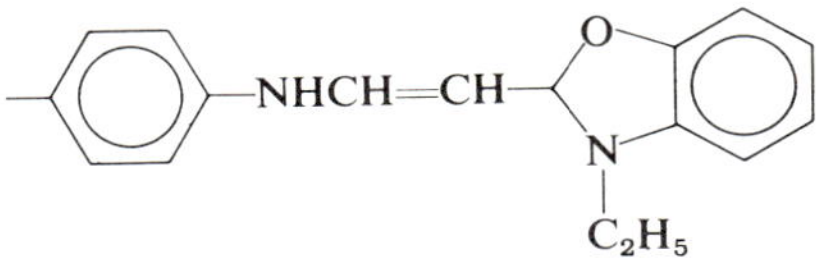

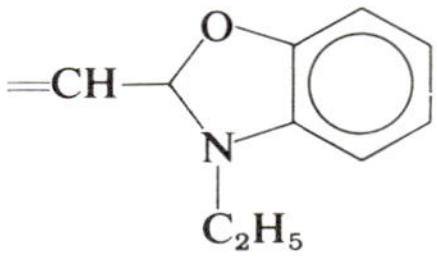

[a] Condensation of p-phenylenediamine and 2-methyl-3-ethylbenzoxazole with ethyl orthoformate.

### G. N-Alkylated Compounds

Several reactions have been reported in which compounds having primary or secondary amino groups are alkylated by ortho esters. Some of these reactions may involve nucleophilic attack by nitrogen on an alkoxy group of the dialkoxycarbonium ion derived from the ortho esters. Others involve isomerization of initially formed products such as imidic esters.

McElvain and Tate reported the first N-alkylation by an ortho ester. They obtained ethyldibutylamine as one product of the reaction of dibutylamine with triethyl orthoacetate at 220°C (*108*). Similarly, N-methylaniline reacts with triethyl orthoacetate in the presence of acids to form N-methyl-N-ethylaniline and ethyl acetate (*56*). Other examples of what may be direct N-alkylations by triethyl orthoformate are alkylations of a number of substituted 1,2,4-benzothiadiazine-1,1-dioxides [Eq. (23)] (*223*), and similar *N*-alkylations of *N,N'*-bis(alkanesulfonyl)formamidines [Eq. (24)] (*70*).

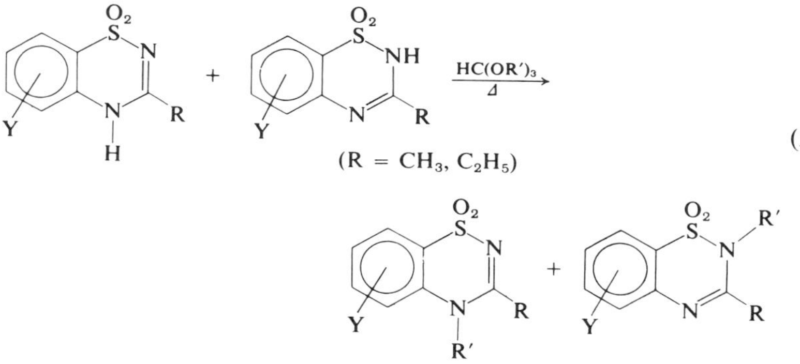

$$\text{(benzothiadiazine-NHR)} + \text{(benzothiadiazine=N-R)} \xrightarrow{HC(OR')_3, \Delta} \text{(benzothiadiazine-NRR')} + \text{(benzothiadiazine=N-R, N-R')} \quad (23)$$

(R = CH$_3$, C$_2$H$_5$)

$$RSO_2N{=}CHNHSO_2R + HC(OR')_3 \xrightarrow{H^+} RSO_2N{=}CHNR'SO_2R + HCO_2R' \quad (24)$$
$$(R = CH_3, C_2H_5; R' = CH_3, C_2H_5) \qquad\qquad\qquad + R'OH$$

Roberts observed that *N*-alkylformanilides are by-products of syntheses of *N*-arylformimidic esters from aromatic primary amines and trialkyl orthoformates, and found that the formanilides are produced by isomerization of the formimidates (*151*). The isomerization, which bears a formal resemblance to the Chapman rearrangement of alkyl *N*-arylbenzimidates (*37, 218*), differs from it in requiring acid catalysis and in not being strictly intramolecular.

It is not necessary to isolate the initially formed formimidic ester, and the reaction appears to be quite general. Heating 2 moles of the primary amine, 3 moles of a trialkyl orthoformate, and 0.04 mole of sulfuric acid to a final temperature of 170°–180°C affords 60%–80% yields of a number of *N*-alkylformanilides (*76b, 149, 151, 167*).

$$ArNH_2 + HC(OR)_3 \xrightarrow[\Delta]{H_2SO_4} ArNRCHO + ROH + HCO_2R \quad (25)$$
[Ar = C$_6$H$_5$, *m*-CH$_3$C$_6$H$_4$, *p*-ClC$_6$H$_4$; R = CH$_3$, C$_2$H$_5$, (CH$_3$)$_2$CHCH$_2$CH$_2$]

Since the resulting *N*-alkylformanilides are readily hydrolyzed to *N*-alkylanilines or reduced to *N*-methyl-*N*-alkylanilines (*77*), and since the trialkyl orthoformates are either commercially available or can be prepared by alcoholysis of trimethyl or triethyl orthoformate, this reaction affords a convenient route to pure *N*-alkylanilines and *N*-methyl-*N*-alkylanilines.

A closely related reaction was reported by Meyer, who found that diethyl *N*-(*p*-tolylsulfonyl)imidocarbonate, formed by reaction of *p*-toluenesulfonamide with tetraethyl orthocarbonate at 120°–160°C, rearranges smoothly to ethyl $N$-ethyl($p$-tolylsulfonyl)carbamate at 200°C (*114*). This reaction requires no added acid catalyst.

$$p\text{-CH}_3\text{C}_6\text{H}_4\text{SO}_2\text{NH}_2 + \text{C(OC}_2\text{H}_5)_4 \xrightarrow{160°\text{C}} p\text{-CH}_3\text{C}_6\text{H}_4\text{SO}_2\text{N}{=}\text{C(OC}_2\text{H}_5)_2 \xrightarrow{200°\text{C}}$$
$$p\text{-CH}_3\text{C}_6\text{H}_4\text{SO}_2\text{N(C}_2\text{H}_5)\text{CO}_2\text{C}_2\text{H}_5 \quad (26)$$

$N$-(Substituted sulfonyl)isocyanates react with ortho esters to form $N$-alkyl-$N$-(substituted sulfonyl)carbamates, probably by decomposition of initially formed $N$-(1,1-dialkoxyalkyl)carbamates (*21*).

$$\text{R}'\text{C(OR)}_3 + \text{R}''\text{SO}_2\text{NCO} \rightarrow \text{R}'\text{CO}_2\text{R} + \text{R}''\text{SO}_2\text{NRCO}_2\text{R} \quad (27)$$
[R = CH$_3$, C$_2$H$_5$, C$_4$H$_9$, C$_5$H$_{11}$; R' = H, CH$_3$, C$_6$H$_5$; R'' = Cl, C$_6$H$_5$, $p$-CH$_3$C$_6$H$_4$, N(C$_2$H$_5$)CO$_2$C$_2$H$_5$]

### H. Urethanes and Benzylidene Carbazates

Heating an $N$-alkyl or $N$-aryl urea with a large excess of triethyl orthoformate or triethyl orthoacetate affords moderate to high yields of $N$-alkyl or $N$-aryl urethanes (*214, 216*).

$$\text{RNHCONH}_2 + \text{R}'\text{C(OC}_2\text{H}_5)_3 \rightarrow \text{RNHCO}_2\text{C}_2\text{H}_5 \quad (28)$$
[R = CH$_2$=CHCH$_2$, C$_6$H$_5$, $p$-ClC$_6$H$_4$, $p$-CH$_3$OC$_6$H$_4$, $p$-(C$_2$H$_5$)$_2$NC$_6$H$_4$, 5-Cl-2-CH$_3$C$_6$H$_3$, $\alpha$-C$_{10}$H$_7$, 2-pyridyl; R' = H, CH$_3$]

Whitehead postulated that formamidine or acetamidine are by-products of these reactions, but did not establish this experimentally. Since phenylurethane was formed by heating $N,N'$-bis(phenylcarbamyl)formamidine with ethanol, and since $N,N'$-bis(arylcarbamyl)formamidines rather than aryl urethanes are obtained by heating $N$-alkyl ureas with a mixture of acetic anhydride and triethyl orthoformate, it seems likely that the urethanes are alcoholysis products of initially formed $N,N'$-bis(carbamyl)formamidines. In the presence of acetic anhydride, ethyl acetate rather than ethanol is the by-product of amidine formation, and the amidines can be isolated. A similar reaction of benzaldehyde semicarbazone yields ethyl benzylidenecarbazates (*216*).

$$\text{ArCH}{=}\text{NNHCONH}_2 + \text{HC(OC}_2\text{H}_5)_3 \rightarrow \text{ArCH}{=}\text{NNHCO}_2\text{C}_2\text{H}_5 \quad (29)$$
(Ar = C$_6$H$_5$, $p$-O$_2$NC$_6$H$_4$, $o$-O$_2$NC$_6$H$_4$, $o$-CH$_3$C$_6$H$_4$)

### I. Azacarbocyanine Dyes

1-Methyl-2-aminopyridinium methosulfate, 1-methyl-2-aminoquinolinium methosulfate, and 2-amino-3-alkylbenzothiazolium iodide react with triethyl orthoformate in the presence of tertiary amines to form diazacarbocyanine dyes (*30, 58, 90*).

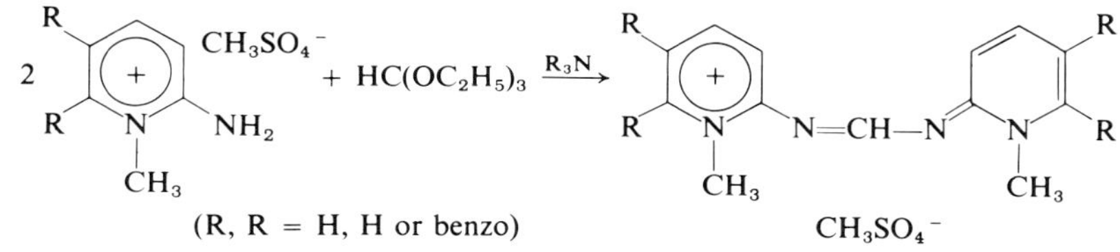

[Scheme (30): 2 [pyridinium with R, R substituents, N-CH₃, NH₂, CH₃SO₄⁻] + HC(OC₂H₅)₃ →^{R₃N} [pyridinium]–N=CH–N–[pyridine with R, R substituents, N-CH₃], CH₃SO₄⁻  (R, R = H, H or benzo)]

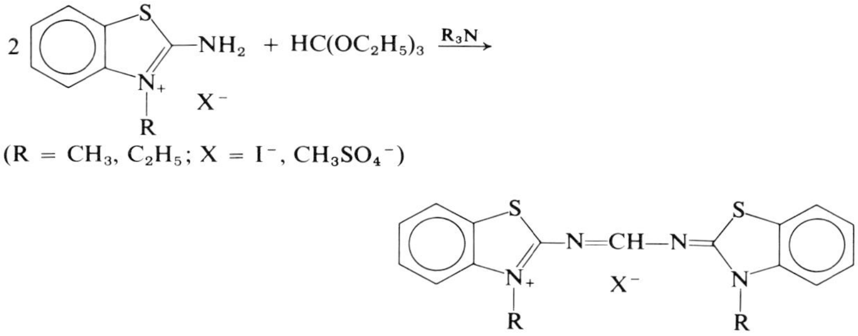

[Scheme: 2 [benzothiazolium-NH₂, N⁺-R, X⁻] + HC(OC₂H₅)₃ →^{R₃N}  (R = CH₃, C₂H₅; X = I⁻, CH₃SO₄⁻)]

[Scheme (31): benzothiazolium–N=CH–N=benzothiazole, X⁻, N–R on each]

## III. REACTIONS WHICH YIELD NITROGEN HETEROCYCLES

Ortho esters are useful reagents for synthesizing many kinds of heterocyclic compounds. The following discussion is concerned with the formation of nitrogen heterocycles by reactions of trialkyl orthocarboxylates with compounds having, in most cases, at least one amino group and some other nucleophilic functional group such as amino, hydroxy, or mercapto.

Cyclization reactions of ortho esters with bifunctional compounds are usually carried out by heating a mixture of the reactants, with or without an acid catalyst. In many instances superior results are obtained if acetic anhydride (which converts ortho esters into reactive 1,1-dialkoxyalkyl acetates) is also a component of the reaction mixture. Most of the syntheses discussed below are multistep reactions which involve amidines, amide acetals, or imidic esters as transient intermediates. In the majority of instances, however, there are no experimental data which warrant speculation on the detailed mechanism of heterocycle formation.

In addition to reactions which lead directly to heterocyclic products, syntheses of heterocycles from amidines, imidic esters, and aminomethylene compounds, which are often prepared from ortho esters, are useful reactions. These reactions are not discussed here.

## A. Products with One Hetero Ring

### 1. Quinolines

Dimethyloxasulfonium-(2-methylamino)benzoylmethylide is converted by refluxing with triethylorthoformate and a small amount of acetic acid to 1-methyl-3-methylsulfinyl-4(1H)quinoline *(204a)*.

### 2. Oxazoles and Thiazoles

N-Acylglycines are converted by heating with trialkyl orthocarboxylates and acetic anhydride into 2-substituted 4-alkoxyalkylene-1,3-oxazol-4-ones [Eq. (32)] *(13, 19, 43, 87)*. N-Thiocarbalkoxyglycines and N-dithiocarbalkoxyglycines yield thiazolinone derivatives when treated similarly [Eq. (32)] *(93, 94, 94a–94c)*. These reactions differ from most of the others considered in this section in that the acyl carbon of the ortho ester is not part of the hetero ring of the product, but appears in the exocyclic alkoxyalkylene group.

$$\text{RCYNHCH}_2\text{CO}_2\text{H} + \text{R}'\text{C}(\text{OR}'')_3 \xrightarrow{\text{Ac}_2\text{O}} \text{R}\!-\!\!\underset{\text{N}}{\overset{\text{Y}}{\diagup\!\!\diagdown}}\!\!\overset{\text{O}}{\diagdown\!\!\diagup}\text{CR}'\text{OR}'' \quad (32)$$

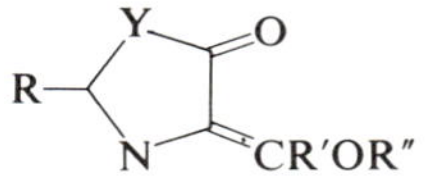

| Y | R | R' | R | References |
|---|---|---|---|---|
| O | $C_6H_5$ | H, $CH_3$, $C_2H_5$ | $CH_3$, $C_2H_5$, $C_3H_7$ | *13, 19, 43, 87* |
| O | Ar | H | $C_2H_5$ | *19, 43* |
| O | $C_5H_{11}$ | H | $C_2H_5$ | *19, 43* |
| O | RCH=CH | H | $C_2H_5$ | *43* |
| O | ArCH=CH | H | $C_2H_5$ | *43* |
| S | $C_6H_5CH_2S$ | $CH_3$ | $C_2H_5$ | *94c* |
| S | $C_2H_5S$ | H, $CH_3$ | $C_2H_5$ | *94a, 94b* |
| S | $C_8H_{17}S$ | H | $C_2H_5$ | *93* |
| S | $C_{14}H_{29}O$, $C_8H_{17}O$, $C_2H_5O$ | H, $CH_3$, $C_2H_5$ | $C_2H_5$ | *12a, 13, 93, 94* |

In similar reactions, thiohippuric acid and triethyl orthoformate, when heated in toluene containing acetic anhydride, yield 2-phenyl-4-ethoxymethylene-1,3-thiazol-4-one *(81)*, and carboxymethyl-N-(2-pyridyl)dithiocarbamate is converted to 5-(1-ethoxyethylidene)-3-(2-pyridyl)rhodanine by heating with triethyl orthoacetate and acetic anhydride *(94)*. If N-thioacylglycines are heated with triethyl orthocarboxylates and acetic anhydride in the absence of a solvent, oxonol dyes having two thiazolone groups linked at the 4-positions by a methine bridge are formed *(43, p. 847; 81, 93)*.

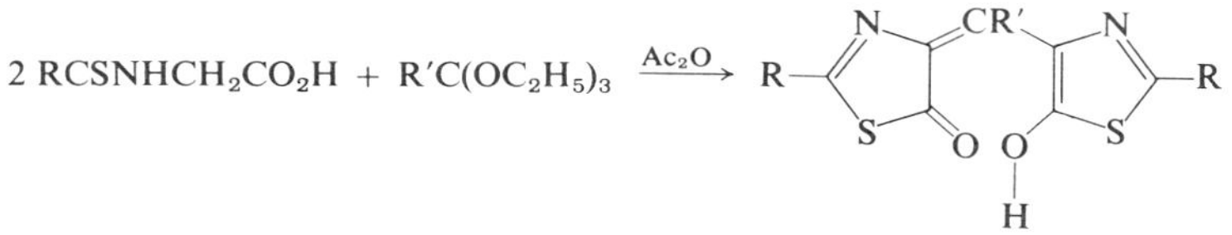

$$2\ RCSNHCH_2CO_2H + R'C(OC_2H_5)_3 \xrightarrow{Ac_2O} \text{[structure]} \quad (33)$$

*o*-Aminophenols and *o*-aminothiophenol yield 2-substituted 1,3-benzoxazoles and 2-substituted 1,3-benzothiazoles when heated with ortho esters in the presence of catalytic amounts of sulfuric acid (*80, 142*).

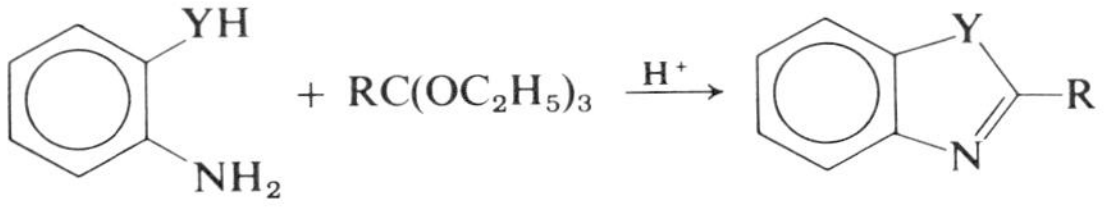

$$\text{[structure with YH, NH}_2\text{]} + RC(OC_2H_5)_3 \xrightarrow{H^+} \text{[benzoxazole/thiazole]-R} \quad (34)$$

(Y = O, S; R = H, $CH_3$, $C_2H_5$)

*N*-Methyl-*o*-aminothiophenol and triethyl orthoacetate yield 2-methylene-3-methylbenzothiazole in a similar reaction (*99*).

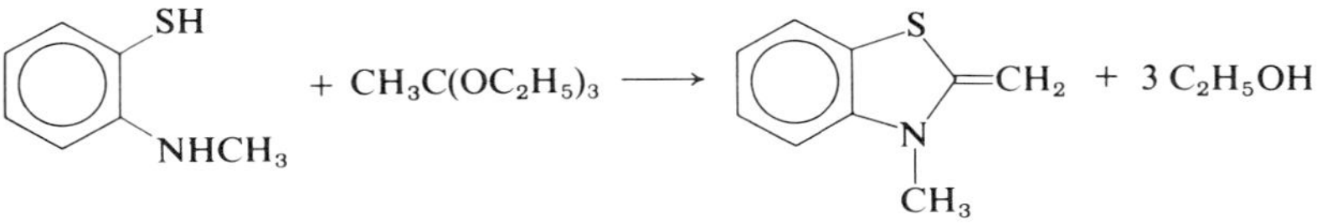

$$\text{[structure with SH, NHCH}_3\text{]} + CH_3C(OC_2H_5)_3 \longrightarrow \text{[benzothiazole]}=CH_2 + 3\ C_2H_5OH \quad (35)$$

2-Methyl-6-hydroxy-7-aminobenzothiazole is converted to 2-methyloxazolo[5,4-*a*]benzothiazole by heating with triethyl orthoformate (*60*), and 5-amino-4-hydroxybenzo-2,1,3-thiadiazole is similarly transformed to oxazolo[5,4-*e*]-2,1,3-benzothiadiazole (*131a*).

Triethyl orthoformate converts 2-*N*-*p*-toluenesulfonylaminoethanol to 2-ethoxy-*N*-(*p*-toluenesulfonyl)oxazolidine (*44*).

3. IMIDAZOLES AND IMIDAZOLINES

Ethylenediamine reacts with a variety of trialkyl orthocarboxylates to form 2-substituted imidazolines [Eq. (36)] (*17, 75, 153, 186*). Alkyl-*N*-(2-aminoethyl)imidic esters are intermediates in these reactions (*17*).

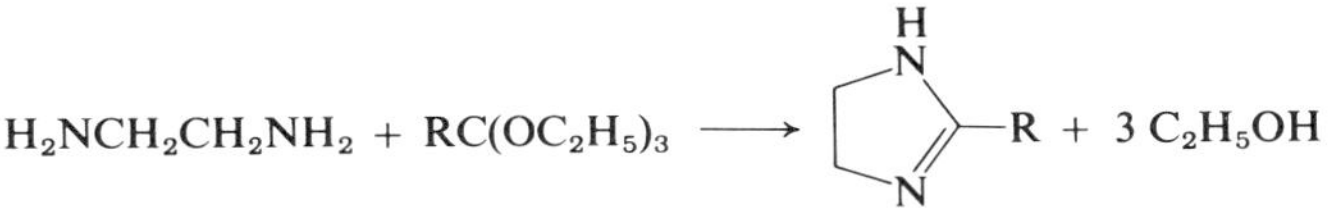

$$H_2NCH_2CH_2NH_2 + RC(OC_2H_5)_3 \longrightarrow \text{[imidazoline]-R} + 3\ C_2H_5OH \quad (36)$$

(R = $CH_3$, $C_2H_5$, $C_6H_5$, *m*-$CH_3C_6H_4$, *p*-$CH_3C_6H_4$, *p*-$C_6H_5C_6H_4$, *p*-$CH_3OC_6H_4$, *p*-$C_2H_5OC_6H_4$, α-$C_{10}H_7$)

*N*-Propyl-1,2-diaminoethane and a series of *N*-(3-silylpropyl)-1,2-diaminoethanes were converted to 1-propylimidazolines [and 1-(3-silylpropylimidazolines] by heating with triethyl orthoformate at 180°–200°C (*159*).

*o*-Diaminobenzene and N-substituted *o*-diaminobenzenes are converted to 1- and 2-substituted and 1,2-disubstituted benzimidazoles by heating with triethyl orthocarboxylates (*11, 51, 112, 142, 153*).

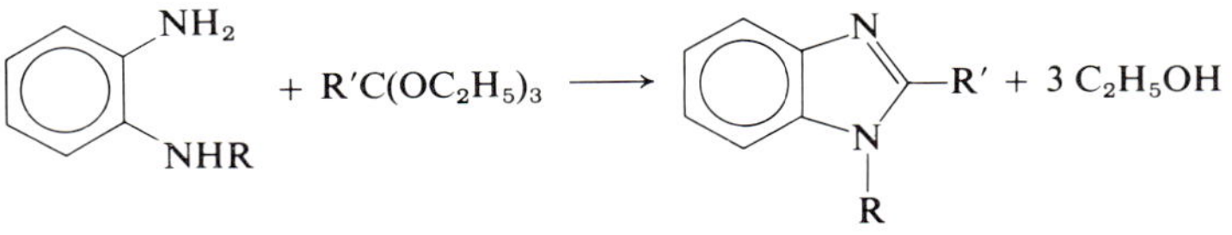

$$\text{[o-diaminobenzene with NH}_2\text{ and NHR]} + R'C(OC_2H_5)_3 \longrightarrow \text{[benzimidazole with N, NR]}-R' + 3\,C_2H_5OH \quad (37)$$

| R | R' | References |
|---|---|---|
| H | H | *51, 153* |
| H | $CH_3, C_2H_5$ | *142, 153* |
| Tetraacetyl-glucopyranosyl | H, $CH_3$ | *11, 112* |

α-Amino acid amides react with triethyl orthocarboxylates to form substituted 5(4*H*)-imidazolones (*31, 62*).

$$RNHCH_2CONH_2 + R'C(OC_2H_5)_3 \xrightarrow{\text{tr. HOAc}} \text{[imidazolone]}-R' \quad (38)$$

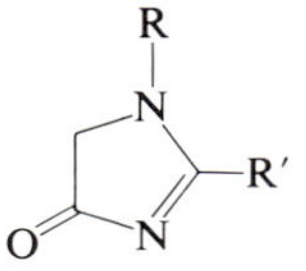

(R = $C_2H_5, C_6H_{11}, C_6H_5$; R' = $CH_3$)

$$RCH(NH_2)CONHR'' + R'C(OC_2H_5)_3 \xrightarrow{\text{tr. HOAc}} \text{[imidazolone]}-R' \quad (39)$$

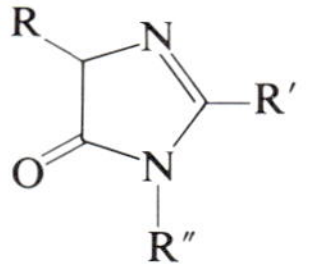

| R | R' | R'' |
|---|---|---|
| $CH_3$ | $CH_3, C_2H_5$ | $C_3H_7$ |
| $CH_2CH(CH_3)_2$, $CH_2C_6H_5$ | $CH_3$ | $C_3H_7$ |
| H | H, $CH_3, C_2H_5$ | $C_2H_5, C_4H_9, C_6H_5$, $CH_2C_6H_5$ |

α-Methylamino-β-amino-propionic acid reacts with triethyl orthoformate in the presence of hydrochloric acid to form 1-methyl-2-imidazoline-5-carboxylic acid (*112a*),

2-Substituted 4,5-dicyanoimidazoles are obtained by heating hydrogen cyanide tetramer with a trialkyl orthocarboxylate in anisole solution at 100°C, and then adding a sodium alkoxide and heating to 150°C (*206a*).

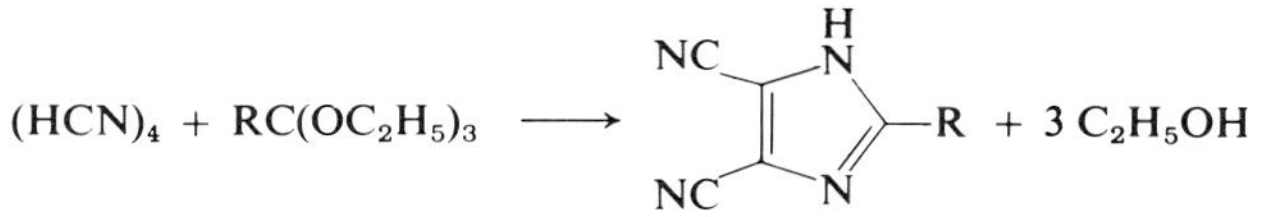

$$(HCN)_4 + RC(OC_2H_5)_3 \longrightarrow \underset{NC}{\overset{NC}{\Large{\diagup}}}\hspace{-2pt}\underset{N}{\overset{H}{\underset{\|}{N}}}\hspace{-6pt}\diagdown\hspace{-2pt}R + 3\,C_2H_5OH \qquad (40)$$

(R = H, CH$_3$, C$_2$H$_5$)

Aminomalonamidine dihydrochloride reacts with ortho esters in refluxing dimethyl sulfoxide to form 2-substituted 4(ethoxyalkylene)aminoimidazole-5-carboxamides (*138*).

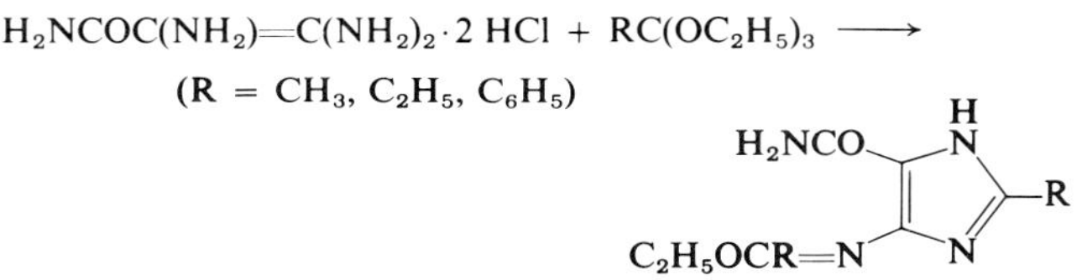

$$H_2NCOC(NH_2)\!\!=\!\!C(NH_2)_2 \cdot 2\,HCl + RC(OC_2H_5)_3 \longrightarrow$$

(R = CH$_3$, C$_2$H$_5$, C$_6$H$_5$)

$$\underset{C_2H_5OCR=N}{\overset{H_2NCO}{\Large{\diagup}}}\hspace{-2pt}\underset{N}{\overset{H}{\underset{\|}{N}}}\hspace{-6pt}\diagdown\hspace{-2pt}R \qquad (41)$$

Refluxing aryl isocyanates for 12–24 hours with an excess of triethyl orthoformate affords moderate to high yields of 1,3-diaryl-5,5-diethoxy-(3*H*,5*H*)imidazolediones (*217*).

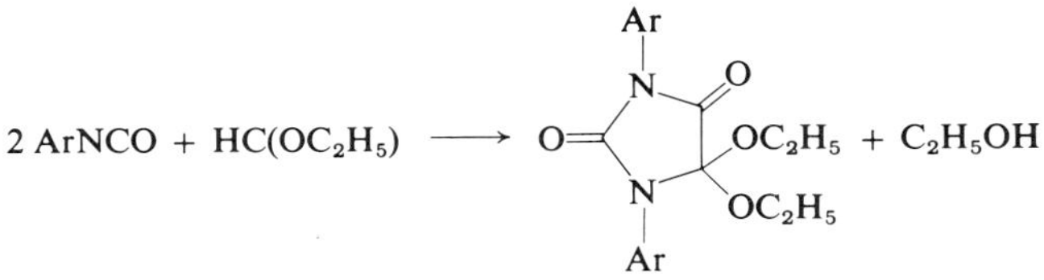

$$2\,ArNCO + HC(OC_2H_5) \longrightarrow \text{(imidazolidine-2,4-dione with N-Ar, N-Ar, and 5,5-(OC}_2\text{H}_5)_2\text{)} + C_2H_5OH \qquad (42)$$

(Ar = C$_6$H$_5$, *p*-ClC$_6$H$_4$, *p*-BrC$_6$H$_4$, *o*-O$_2$NC$_6$H$_4$, *p*-C$_6$H$_5$C$_6$H$_4$)

This reaction is unusual in that it involves replacement of the formyl hydrogen atom of the orthoformate as well as an ethoxy group. The proposed structure of the reaction product is consistent with elementary composition data and absorption spectra, and with the fact that hydrolysis of the product from phenyl isocyanate yields 1,3-diphenylparabanic acid. Whitehead and Traverso suggested that the initial step of the reaction is addition of the orthoformate C—H group across the carbon–nitrogen double bond of the isocyanate to form an anilidohemiorthooxalate which reacts with a second mole of the isocyanate to give a 1-(triethoxyacetyl)-1,3-diarylurea (*217*). Reaction of α-naphthyl isocyanate with ethyl orthoformate gave a low yield of a product whose composition and properties are in accord with a triethoxyacetyldinaphthylurea structure.

The fact that the formyl proton of orthoformates is not sufficiently acidic to undergo exchange in the presence of sodium ethoxide at 50°C (*126*) suggests that this reaction does not involve a direct C—H addition to the isocyanate. Von Brachel's observation that isocyanates are converted by triethyl orthoformate in the presence of acids to ethyl N-substituted *N*-diethoxymethylcarbamates (*206a, 206b*) indicates that these substances are intermediates in the formation of the diaryldiethoxyimidazolediones. A reaction sequence which would lead to the observed products is outlined in Eq. (43). If the reaction actually follows this course, the intermediate assumed by

$$HC(OC_2H_5)_3 + ArNCO \longrightarrow ArN(CO_2C_2H_5)CH(OC_2H_5)_2 \xrightarrow[-C_2H_5OH]{+H^+}$$

$$ArN(CO_2C_2H_5)\overset{+}{C}HOC_2H_5 \xrightarrow{-H^+} ArN(CO_2C_2H_5)\overset{..}{C}OC_2H_5 \xrightarrow{ArNCO}$$

$$ArN(CO_2C_2H_5)\overset{+}{C}(OC_2H_5)CO\overset{-}{N}Ar \xrightarrow{+C_2H_5OH}$$

$$ArN(CO_2C_2H_5)C(OC_2H_5)_2CONHAr \longrightarrow \underset{\underset{Ar}{|}}{\overset{\overset{Ar}{|}}{\underset{N}{\overset{N}{\diagdown}}}}\!\!\!\!\!\!\!\!\!\!\! \begin{array}{c}C_2H_5O\\C_2H_5O\end{array}\!\!\!\!\!\!\!\!\!\!\!\!\!\!\!\!\!\!\!\!\!\!\!\! =O + C_2H_5OH \quad (43)$$

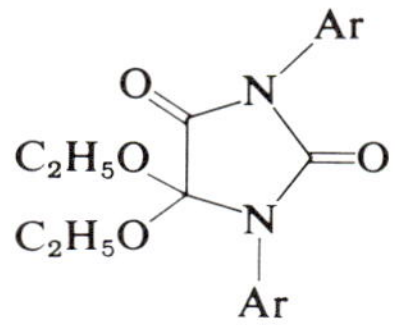

Whitehead and Traverso to be a triethoxyacetyldiarylurea was actually $C_2H_5OCON(C_{10}H_7)C(OC_2H_5)_2CONHC_{10}H_7$. These compounds are isomeric, should have similar spectral properties, and could both yield α-naphthylisatin, the product formed when the intermediate was treated with acid.

Another reaction which apparently involves a carbene intermediate is the synthesis of bis(1,3-diphenylimidazolidenylidene-2) from *N*,*N*'-diphenyl-1,2-diaminoethane and triethyl orthoformate (*211, 212*).

$$2\ C_6H_5NHCH_2CH_2NHC_6H_5 + 2\ HC(OC_2H_5)_3 \xrightarrow{200°C}$$

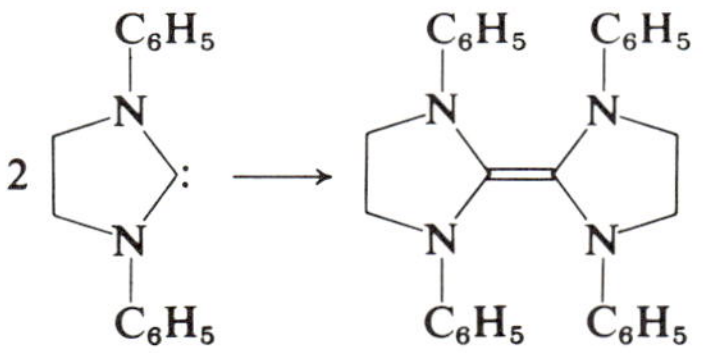

$$2 \underset{\underset{C_6H_5}{|}}{\overset{\overset{C_6H_5}{|}}{\underset{N}{\overset{N}{\diagup\diagdown}}}}\!\!: \longrightarrow \underset{\underset{C_6H_5}{|}\ \underset{C_6H_5}{|}}{\overset{\overset{C_6H_5}{|}\ \overset{C_6H_5}{|}}{\underset{N}{\overset{N}{\diagup\diagdown}}}\!\!=\!\!\underset{N}{\overset{N}{\diagup\diagdown}}} \quad (44)$$

The *N*,*N*'-bis(methanesulfonamide) and *N*,*N*'-bis(*p*-toluenesulfonamide) of ethylenediamine react with triethyl orthoformate to form 2-ethoxy-1,3-bis-(substituted sulfonyl)imidazolidines (*44*).

## 4. Oxadiazoles and Thiadiazoles

2-Substituted and 2,5-disubstituted 1,3,4-oxadiazoles may be synthesized by heating an acyl hydrazide with a trialkyl orthocarboxylate.

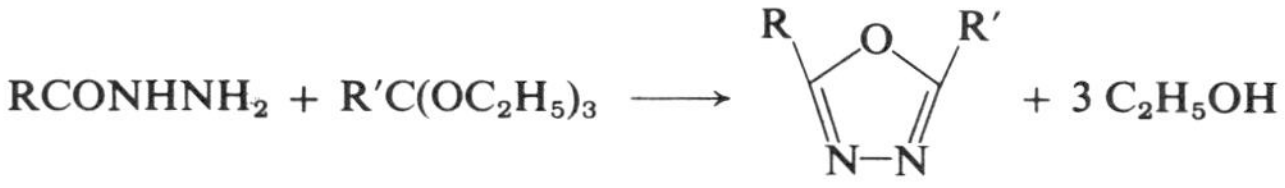

$$RCONHNH_2 + R'C(OC_2H_5)_3 \longrightarrow \underset{N-N}{\overset{R\;\;O\;\;R'}{\diagdown/\diagdown}} + 3\, C_2H_5OH \quad (45)$$

| R | R' | References |
|---|---|---|
| H | H | 5 |
| H | $NCCH_2$, $C_6H_5CH_2$ | 205 |
| H | Aryl | 1, 2, 4, 110, 111, 157, 205 |
| H | 2-(3,4)Pyridyl, 2-quinolinyl | 1, 4 |
| $CH_3$, $C_2H_6$, $C_6H_5$ | 4-Pyridyl | 1 |
| $C_6H_5$ | $C_6H_5$ | 55 |

$N$-Carboxamido imidic esters ($RCONHN\!=\!CR'OC_2H_5$) are probably intermediates in these reactions. In at least two instances they were actually isolated and converted to the oxadiazoles by further heating. The dihydrazide of succinic acid was converted to 1,2-bis(1,3,4-oxadiazol-2-yl)ethane by heating with an excess of triethyl orthoformate (157).

$p$-Chlorobenzamide oxime reacts with triethyl orthoformate in acetic anhydride to form 3-$p$-chlorophenyl-1,2,4-oxadiazole (33).

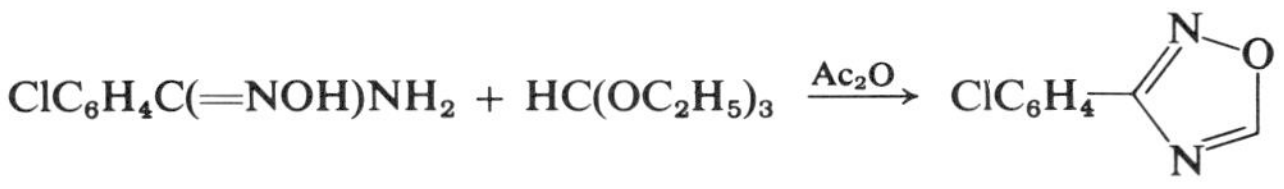

$$ClC_6H_4C(\!=\!NOH)NH_2 + HC(OC_2H_5)_3 \xrightarrow{Ac_2O} ClC_6H_4\overset{N\diagdown O}{\underset{N}{\diagup}} \quad (46)$$

Thiohydrazides react with trialkyl orthocarboxylates to form 2-substituted and 2,5-disubstituted 1,3,4-thiadiazoles [Eq. (47)] (1, 2, 4). Thiosemicarbazide and 4-substituted thiosemicarbazides undergo analogous reactions which yield 2-amino-5-substituted-1,3,4-thiadiazoles [Eq. (48)] (3, 82, 216). Thiosemicarbazide itself may, depending on reaction conditions, yield products derived from the aminothiadiazole such as 2-acetamido-1,3,4-thiadiazol-2-yl)formamidine (3). 4-Aminothiosemicarbazones,

$$RR'C\!=\!NNHCSNHNH_2,$$

react with ortho esters to form 1,3,4-thiadiazolyl hydrazones [Eq. (48), R = R''R'''C=N–] (160).

$$\text{RCSNHNH}_2 + \text{R'C(OC}_2\text{H}_5)_3 \longrightarrow \underset{\text{N-N}}{\overset{R\quad S\quad R'}{\diagdown\diagup}} + 3\,\text{C}_2\text{H}_5\text{OH} \qquad (47)$$

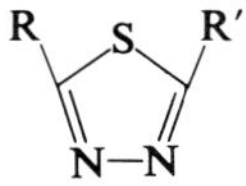

$(R = C_6H_5;\ R' = H, CH_3)$

$$\text{RNHCSNHNH}_2 + \text{R'C(OC}_2\text{H}_5)_3 \longrightarrow \underset{\text{N-N}}{\overset{RNH\quad S\quad R'}{\diagdown\diagup}} + 3\,\text{C}_2\text{H}_5\text{OH} \qquad (48)$$

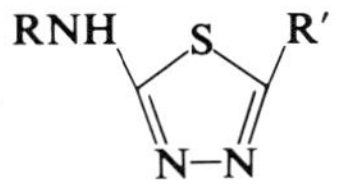

$(R = H, C_2H_5, C_3H_5, C_4H_9, C_5H_{11}, C_8H_{17}, C_6H_5;\ R' = H, CH_3, C_2H_5)$

## 5. Triazoles

Hydrazines, hydrazides, semicarbazides, carbohydrazides, thiosemicarbazides, and thiocarbohydrazide react with ortho esters to form triazole derivatives. Arylhydrazine hydrochlorides and triethyl orthoformate yield 4-arylamino-1-aryl-1$H$-1,2,4-triazolium chlorides along with smaller amounts of diaryltetrazines (*155*).

$$2\,\text{ArNHNH}_3{}^+\text{Cl}^- + 2\,\text{HC(OC}_2\text{H}_5)_3 \longrightarrow \underset{\underset{\text{Ar}}{\text{N}}}{\text{ArNH}\overset{+}{-}\text{N}\diagdown}\quad \text{Cl}^- + 6\,\text{C}_2\text{H}_5\text{OH} \qquad (49)$$

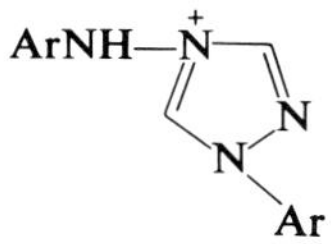

$(Ar = C_6H_5,\ o\text{-}CH_3C_6H_4,\ m\text{-}CH_3C_6H_4,\ p\text{-}CH_3C_6H_4)$

Formhydrazide, when refluxed with triethyl orthoformate, gave a mixture of $N,N'$-diformylhydrazine and 4-formylamino-4$H$-1,2,4-triazole. Presumably $N,N'$-bis(formamido)formamidine is formed as an intermediate; the water produced when this cyclizes to the triazole reacts with the amidine (or with ethyl formate formylhydrazone, its probable precursor) to give the diformylhydrazine. The overall result is described by Eq. (50) (*5*).

$$3\,\text{HCONHNH}_2 + 2\,\text{HC(OC}_2\text{H}_5)_3 \longrightarrow$$

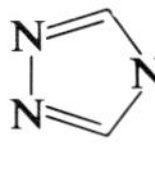

$$\underset{\text{N}}{\overset{\text{N}}{\diagdown}}\text{N—NHCHO} + \text{NCONHNHCHO} + 6\,\text{C}_2\text{H}_5\text{OH} \qquad (50)$$

Semicarbazide and substituted semicarbazides are converted to 1,2,4-triazolones by heating with ortho esters.

$$\text{H}_2\text{NCONRNHR'} + \text{R''C(OC}_2\text{H}_5)_3 \longrightarrow \underset{\underset{R\quad R'}{\text{N—N}}}{\overset{O\quad N\quad R''}{\diagdown\diagup}} + 3\,\text{C}_2\text{H}_5\text{OH} \qquad (51)$$

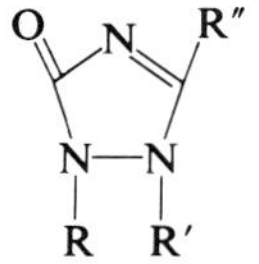

| R | R' | R | References |
|---|---|---|---|
| H | H, $C_2H_5$ | H | 45, 154 |
| H | $C_6H_5$ | H, $CH_3$ | 216 |
| H | p-$HO_2CC_6H_4$ | H | 216 |
| $CH_3$ | $CH_3$ | H | 110 |

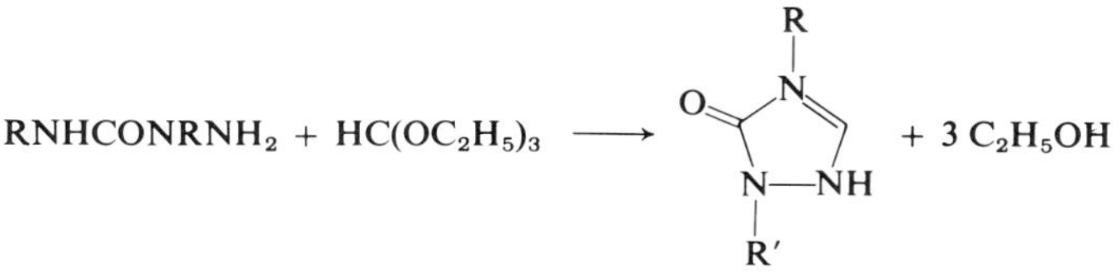

$$RNHCONRNH_2 + HC(OC_2H_5)_3 \longrightarrow \underset{R'}{\underset{|}{N}}\overset{R}{\overset{|}{N}}\text{-ring-}O + 3\,C_2H_5OH \quad (52)$$

| R | R' | References |
|---|---|---|
| H | $CH_3$, $C_2H_5$ | 97 |
| $CH_3$ | H, $CH_3$ | 97 |
| $C_2H_5$ | $C_2H_5$ | 97 |

Curtius and Heidenreich claimed in 1894 that carbohydrazide reacts with triethyl orthoformate to form tetrahydro-s-tetrazinone (45, 46). Kroger recently showed that the products from this and related reactions with higher trialkyl orthocarboxylates are actually 4-aminotriazolones rather than tetrazinones (96).

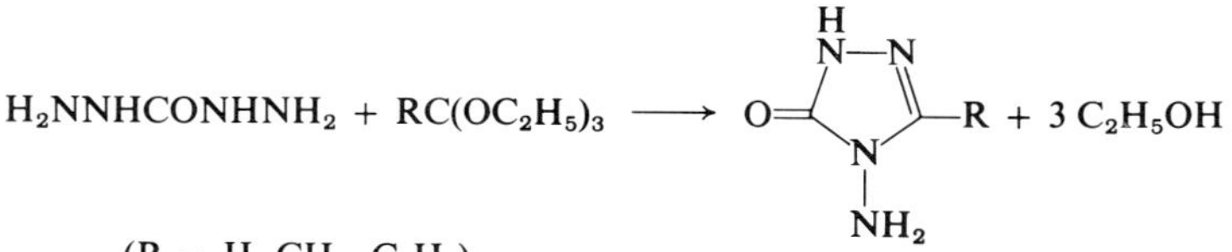

$$H_2NNHCONHNH_2 + RC(OC_2H_5)_3 \longrightarrow \text{triazolone-}R + 3\,C_2H_5OH \quad (53)$$

(R = H, $CH_3$, $C_2H_5$)

Thiosemicarbazide and thiocarbohydrazide react with ortho esters in the same manner as their oxygen analogues. Thiosemicarbazide reacts with triethyl orthoacetate to form a 1-ethoxyethylidene derivative which yields a mixture of approximately equal amounts of 2-amino-5-methyl-1,3,4-thiadiazole and 2-mercapto-5-methyl-1,3,4-triazole when heated (3). Thiocarbohydrazide is converted by heating with triethyl orthocarboxylates to 3-substituted 4-amino-5-mercapto-4H-triazoles (23, 174).

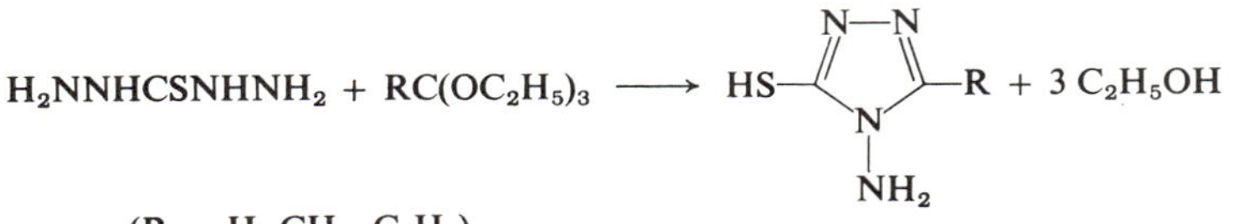

$$H_2NNHCSNHNH_2 + RC(OC_2H_5)_3 \longrightarrow HS\underset{NH_2}{\overset{N-N}{\underset{|}{\diagdown\diagup}}}R + 3\ C_2H_5OH \qquad (54)$$

(R = H, CH$_3$, C$_2$H$_5$)

## 6. Oxazines and Thiazines

5-Cyano-2-ethylthio-4$H$-1,3-thiazine-4-one was obtained by refluxing $N$-cyanoacetyldithiocarbamate with a mixture of triethyl orthoformate and acetic anhydride [Eq. (55)] (*12*). The thiazine was probably formed by cyclization of an unisolated ethoxymethylene derivative

$$[NCC(=CHOC_2H_5)CONHCS_2C_2H_5]$$

of the dithiocarbamate. A similar substance is probably the initial product of reaction of *o*-hydroxybenzamide with triethyl orthoacetate, but in this case the product isolated was 2-methyl-4$H$-1,3-benzoxazin-4-one, formed by loss of alcohol from the 2-ethoxyoxazine [Eq. (56)] (*78*).

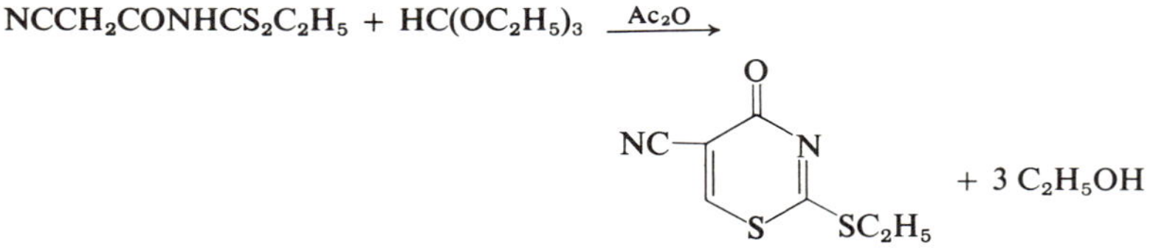

NCCH$_2$CONHCS$_2$C$_2$H$_5$ + HC(OC$_2$H$_5$)$_3$ $\xrightarrow{Ac_2O}$

[structure of 5-cyano-2-ethylthio-1,3-thiazin-4-one] + 3 C$_2$H$_5$OH (55)

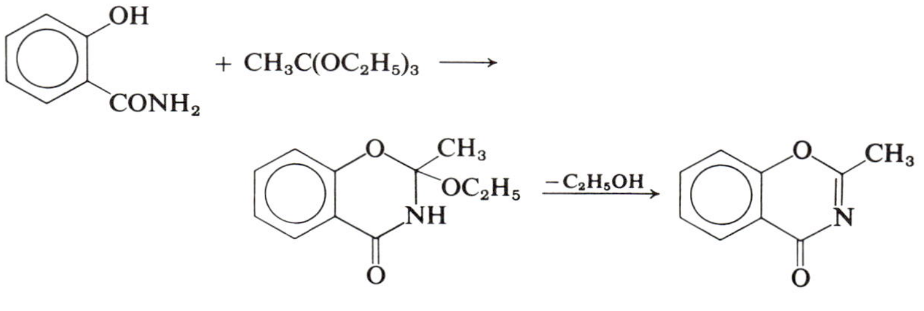

[o-hydroxybenzamide] + CH$_3$C(OC$_2$H$_5$)$_3$ ⟶

[2-ethoxy-2-methyl intermediate] $\xrightarrow{-C_2H_5OH}$ [2-methyl-4$H$-1,3-benzoxazin-4-one] (56)

## 7. Diazines

1,3-Diaminopropane is converted to 2-substituted tetrahydropyrimidines by heating with orthocarboxylates in the presence of an acid catalyst [Eq. (57)] (*17*). A similar reaction between malondiamide and triethyl orthoformate yields 4,6-dihydroxypyrimidine (*154*).

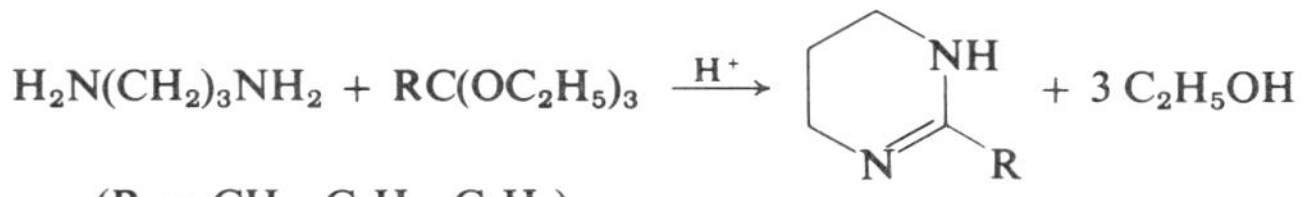

$$H_2N(CH_2)_3NH_2 + RC(OC_2H_5)_3 \xrightarrow{H^+} \underset{N}{\overset{NH}{\bigcirc}}\!\!R + 3\ C_2H_5OH \quad (57)$$

(R = $CH_3$, $C_2H_5$, $C_6H_5$)

3-*p*-Tolyl-6-methyl-3,4-dihydroquinazoline is formed in 60% yield when 2-amino-5-methyl-*N*-(*p*-tolyl)benzylamine is heated with triethyl orthoformate [Eq. (58)] (*207*). *o*-Aminobenzamides and *o*-aminobenzhydrazides yield 4-hydroxyquinazolones and 3-substituted-(3,4*H*)-4-quinazolones when heated with triethyl orthoformate [Eqs. (59)–(61)]. 3-substituted-4-quinazolones are also obtained by treatment of isatoic anhydride with primary amines and triethyl orthoformate (*39b*). *o*-Aminobenzamides are probably intermediates in these reactions (*18d*).

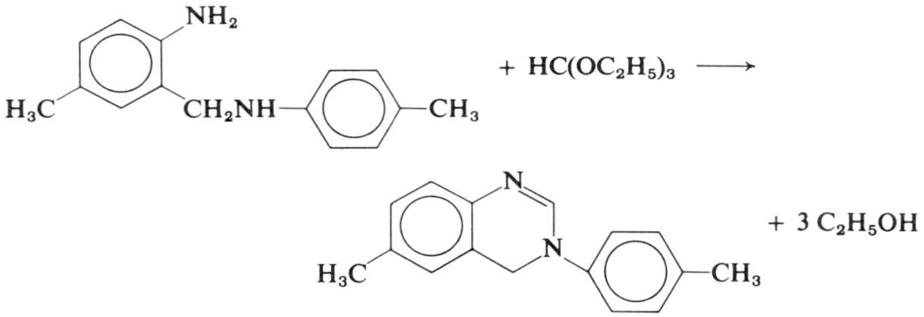

$$\text{(58)}$$

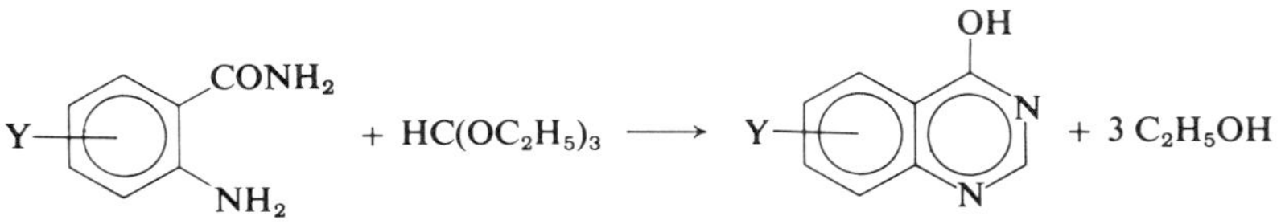

[Y = H, 4-Cl (*109, 154*)]   (59)

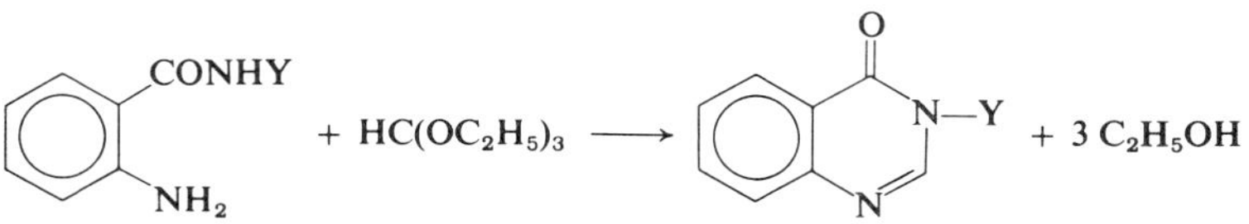

[Y = $C_6H_5$ (*39b, 208*), *p*-$CH_3C_6H_4$ (*39b*), *p*-$CH_3OC_6H_4$ (*39b*), $C_6H_{11}$ (*18d*), $NH_2$ (*205*)]   (60)

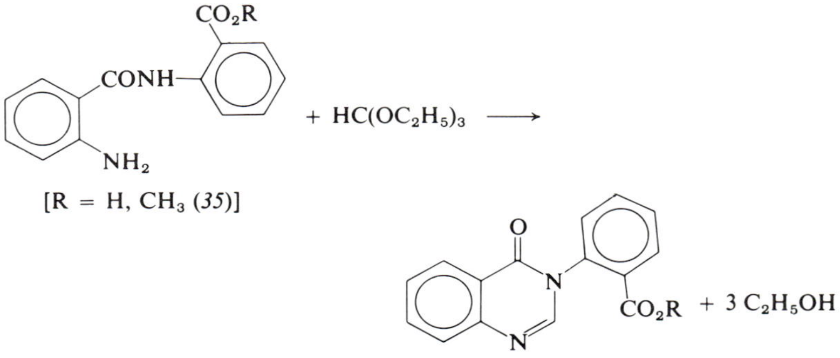

[Structure: o-aminobenzamide derivative]

$$\text{o-H}_2\text{N-C}_6\text{H}_4\text{-CONH-C}_6\text{H}_4\text{-CO}_2\text{R} + \text{HC(OC}_2\text{H}_5)_3 \longrightarrow$$

[R = H, CH₃ (35)]

[Structure: 3-(2-carboalkoxyphenyl)-4-quinazolone] $+ 3\,\text{C}_2\text{H}_5\text{OH}$ (61)

Heating a mixture of methyl *o*-aminobenzoate, triethyl orthoformate, and an aromatic amine results in formation of moderate to good yields of 3-aryl-4-quinazolones [Eq. (62)] (*156*). This reaction probably involves *N*-arylimidic esters and/or *N,N'*-diarylformamidines as intermediates, since the quinazolones may be prepared by heating methyl *o*-aminobenzoate with *N*-arylformimidates or *N,N'*-diarylformamidines.

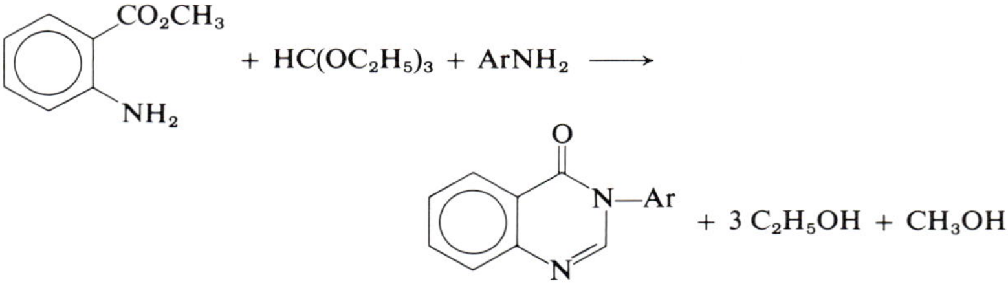

$$\text{o-H}_2\text{N-C}_6\text{H}_4\text{-CO}_2\text{CH}_3 + \text{HC(OC}_2\text{H}_5)_3 + \text{ArNH}_2 \longrightarrow$$

[Structure: 3-aryl-4-quinazolone] $+ 3\,\text{C}_2\text{H}_5\text{OH} + \text{CH}_3\text{OH}$ (62)

[Ar = C₆H₅, ClC₆H₄ (*m*- and *p*), *p*-BrC₆H₄, CH₃C₆H₄ (*m*- and *p*), *p*-CH₃OC₆H₄, 2-pyridyl]

1,4-Diphenyl-6-ethoxypyrimidine-2,4-dithione is the principal product obtained by refluxing a mixture of triethyl orthoacetate and phenyl isothiocyanate [Eq. (63)] (*217*). Ketene diethylacetal (formed by elimination of ethanol from the ortho ester) may be an intermediate in this reaction, since heating ketene acetal with phenyl isothiocyanate yields the same product.

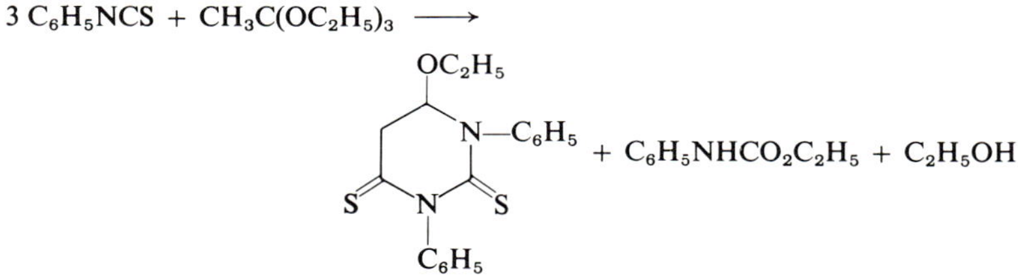

$$3\,\text{C}_6\text{H}_5\text{NCS} + \text{CH}_3\text{C(OC}_2\text{H}_5)_3 \longrightarrow$$

[Structure: 1,4-diphenyl-6-ethoxypyrimidine-2,4-dithione] $+ \text{C}_6\text{H}_5\text{NHCO}_2\text{C}_2\text{H}_5 + \text{C}_2\text{H}_5\text{OH}$ (63)

α-(Substituted aryl)-β-ketonitriles are converted by successive treatment with trialkyl orthoformates and guanidine to 2,4-diamino-5-aryl-6-substituted-pyrimidines (*18a, 158a*). Enol ethers formed from the ketonitrile are probably intermediates in these reactions.

## 8. Thiadiazines

*o*-Aminobenzenesulfonamide and some of its derivatives have been converted to substituted 2*H*-1,2,4-benzothiadiazine-1,1-dioxides by reaction with triethyl orthoformate.

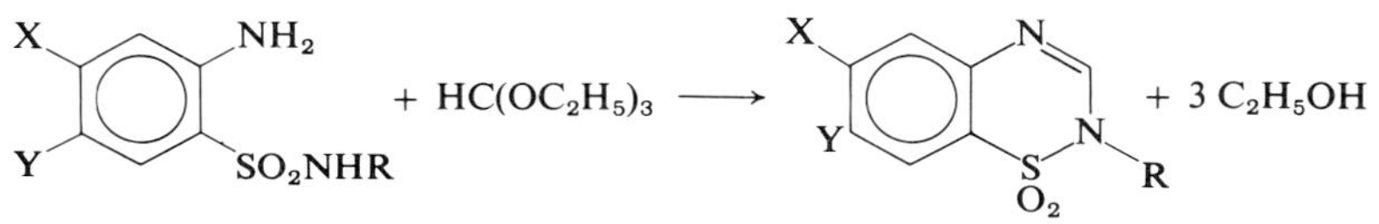

$$\underset{Y}{\overset{X}{\bigcirc}}\!\!\!\!\!\!\!\!\!\!\!\!\!\!\!\!\!\!\!\!\!\!\!\!\!\!\!\!\underset{SO_2NHR}{\overset{NH_2}{}} + HC(OC_2H_5)_3 \longrightarrow \underset{Y}{\overset{X}{\bigcirc}}\!\!\!\!\!\!\!\!\!\!\!\!\!\!\!\!\!\!\!\!\!\!\!\!\!\!\!\!\underset{S,O_2}{\overset{N}{}}\!\!\!\!\!\!\!\!\!\!\!\!\underset{R}{\overset{N}{}} + 3\,C_2H_5OH \quad (64)$$

| X | Y | R | References |
|---|---|---|---|
| Cl, CF$_3$ | C$_2$H$_5$OCH=NSO$_2$ | H | *158, 223* |
| H | H | Aryl | *59, 125* |
| Cl | CH$_3$ | CH$_3$O | *213* |
| Cl | CH$_3$ | C$_2$H$_5$, C$_3$H$_7$, C$_4$H$_9$, C$_5$H$_{11}$ | *60b* |

Triethyl orthoformate and 1-(1-methylhydrazino)-2-propanethiol undergo an acid-catalyzed reaction which forms 5,6-dihydro-4,6-dimethyl-4*H*-1,3,4-thiadiazine (*199*).

$$CH_3CH(SH)CH_2N(CH_3)NH_2 + HC(OC_2H_5)_3 \xrightarrow{H^+}$$

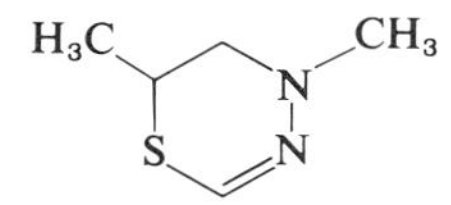

$$\underset{S}{\overset{H_3C}{\bigcirc}}\!\!\!\!\!\!\!\!\!\!\!\!\underset{N}{\overset{CH_3}{}} + 3\,C_2H_5OH \quad (65)$$

## 9. Triazines

The initial product of reaction of urea with triethyl orthoformate, *N,N'*-dicarbamylformamidine, undergoes cyclization to 2,4-dihydroxy-*s*-triazine when heated or treated with sodium hydroxide [Eq. (66)] (*24, 25, 132, 133*). The same product is obtained by heating *N*-carbamylurea with triethyl orthoformate in the presence of sulfuric acid (*158*).

$(H_2N)_2CO + HC(OC_2H_5)_3 \longrightarrow H_2NCONHCH=NCONH_2 \longrightarrow$ [triazine with OH groups] (66)

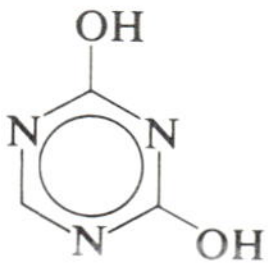

1-(1-Alkylhydrazino)-2-aminoalkanes and triethyl orthocarboxylates react to form 1,2,4,5-tetrahydro-*as*-triazines *(200)*.

$R^1CH(NH_2)CH_2NR^2NH_2 + R^3C(OC_2H_5)_3 \longrightarrow$ [tetrahydrotriazine ring with $R^1$, $R^2$, $R^3$] $+ 3 C_2H_5OH$ (67)

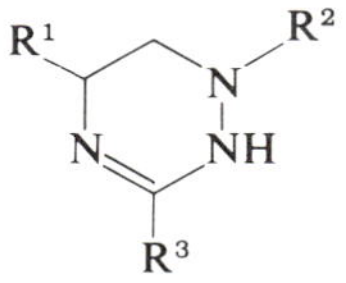

($R^1$ = H, $CH_3$; $R^2$ = H, $CH_3$; $R^3$ = H, $CH_3$, $C_2H_5$)

Guanylurea reacts with triethyl orthoformate in dimethylformamide to form 5-azacytosine (4-amino-1,2-dihydro-1,3,5-triazin-2-one) in 86% yield *(131b)*. *N*-Phenylbiguanidine hydrochloride is converted to 2-amino-4-(*N*-phenylamino)-1,3,5-triazine by reaction with triethyl orthoformate *(18b)*. Treating *N*-phenylbiguanidine hydrochloride with mixtures of triethyl orthoformate and ketones yields 2,4-diamino-5-phenyl-6,6-disubstituted-5,6-dihydro-1,3,5-triazines *(18b, 18c)*. Although the formyl carbon of the orthoformate does not appear in the products, the reaction does not occur in the absence of the ortho ester. The ortho ester may drive the reaction to completion by acting as a water scavenger. 2-Methyl-4-substituted-isobiurets and triethyl orthocarboxylates react to form 1,6-disubstituted 4-methoxy-1,2-dihydro-1,3,5-triazin-2-ones *(132a)*.

Treatment of *o*-haloanilines with triethyl orthoformates and then with substituted hydrazines yields 1-substituted 1,2-dihydro-1,2,4-benzotriazines *(129a)*. Alkyl *N*-arylformimidates are intermediates in these reactions.

## 10. Tetrazines

Hydrazine and triethyl orthoformate, when heated in a sealed tube at 120°C, form 1,4-dihydro-1,2,4,5-tetrazine. A similar reaction between phenylhydrazine hydrochloride and triethyl orthoformate gave a mixture of products which contained 1,4-dihydro-1,4-diphenyl-1,2,4,5-tetrazine (along with 4-anilino-1-phenyl-1,2,4-triazole) *(155, 173)*.

$RNHNH_2 + 2 HC(OC_2H_5)_3 \longrightarrow$ [tetrazine ring R—N, N—R] $+ 6 C_2H_5OH$ (68)

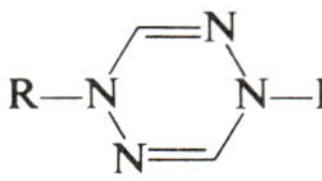

(R = H, $C_6H_5$)

## 11. Seven-Membered and Larger Heterocycles

2,2-Diaminobiphenyl reacts with ortho esters to form 2-substituted 5H-dibenzo[d,f][1,3]-diazepines (*101a, 141*).

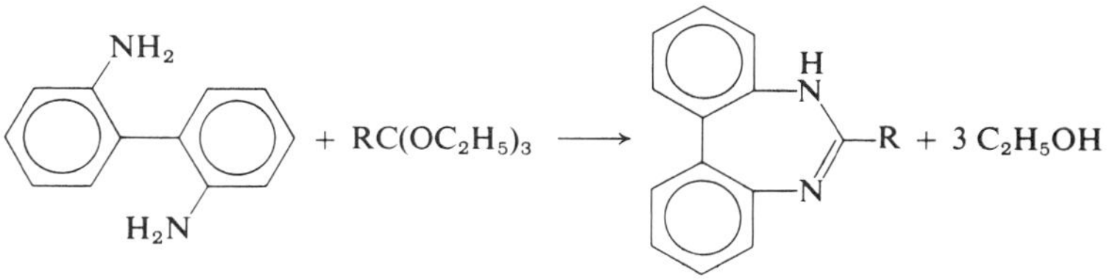

[Structure: 2,2-diaminobiphenyl + RC(OC$_2$H$_5$)$_3$ → 5H-dibenzo[d,f][1,3]-diazepine + 3 C$_2$H$_5$OH] (69)

(R = H, CH$_3$, C$_2$H$_5$, C$_2$H$_5$O, C$_6$H$_5$)

2,5-Dihydro-1,2,4-benzothiadiazepine-1,1-dioxide and 1,3-dihydro-2,3,5-benzothiadiazepine-2,2-dioxide are obtained by heating o-aminomethylbenzenesulfonamide and 2-aminobenzylsulfonamide with triethyl orthoformate (*38, 196*).

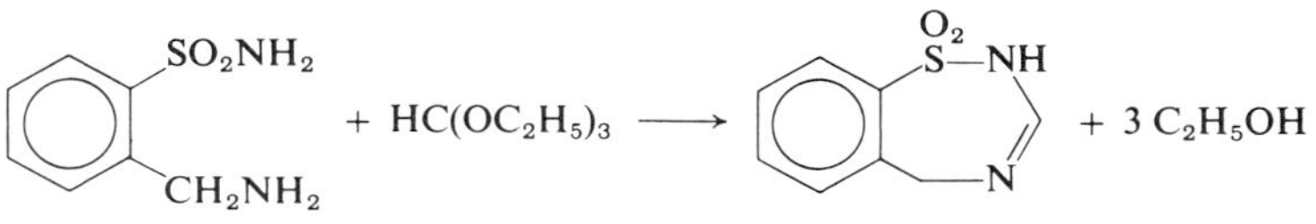

[Structure: o-aminomethylbenzenesulfonamide + HC(OC$_2$H$_5$)$_3$ → benzothiadiazepine dioxide + 3 C$_2$H$_5$OH] (70)

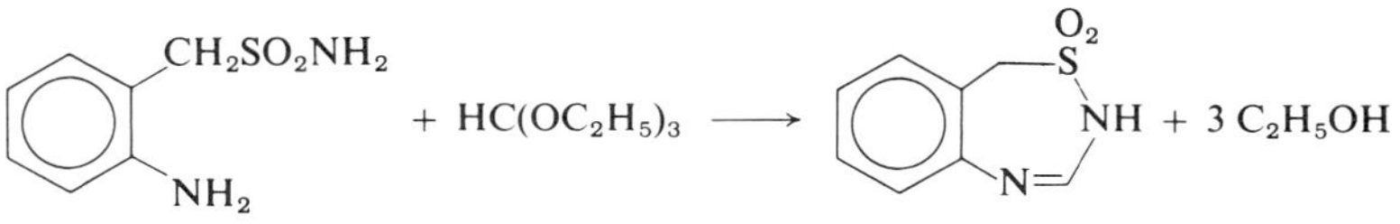

[Structure: 2-aminobenzylsulfonamide + HC(OC$_2$H$_5$)$_3$ → benzothiadiazepine dioxide + 3 C$_2$H$_5$OH] (71)

### B. Products with Two or More Fused Hetero Rings

#### 1. Purines

A variety of purine derivatives have been synthesized by heating mixtures of orthocarboxylates and 4,5-diaminopyrimidines. Imidic esters derived from the aminopyrimidines are probably intermediates in these reactions. In most cases they were cyclized to the purines without being isolated. Orthoformates have been most widely used in purine syntheses, although higher orthocarboxylates are also effective. If triethyl orthoformate is used, it is advantageous to add an acid catalyst to the reaction mixture (*121*). Superior results are obtained if the cyclization is brought about with a mixture of triethyl orthoformate and acetic anhydride, or with presynthesized diethoxymethyl acetate.

Purines prepared from orthocarboxylates and diaminopyrimidines [Eq. (72)] include purine and 2-chloropurine (*116, 121*), 6-chloropurine and 2,6-dichloropurine (*64, 116, 118, 121*), 9-ethylpurine (*119*), several 6-chloro-9-substituted purines (*71a, 76a, 120, 152, 160a, 161, 169a, 193, 195*), 6-carbethoxymethylpurine (*117*), 6-hydroxypurine (*121, 183a*), 2-chloro-6-methylpurine (*134*), 6-chloro-8-alkylpurines (*121*), 6-dimethylaminopurine (*64*), 6-dimethylamino-2-methylpurine (*64*), 6-dimethylamino-2-methylthiopurine (*64*), 6-hydroxy-2-aminopurine (*64*), 6-hydroxy-2-methylthiopurine (*64*), 2-chloro-6,9-dimethylpurine (*30a*), 6-[*N*-(2-hydroxyethyl)]aminopurine (*151a*), 6-(*N*-furfuryl)aminopurine (*76a*) and 6-chloro-9-(3,4,5-trihydroxytetrahydropyran-2-yl)-methylpurine (*204b*).

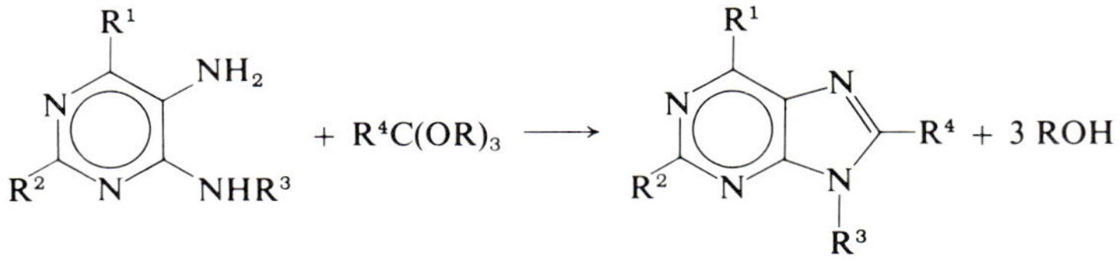

$$\text{pyrimidine} + R^4C(OR)_3 \longrightarrow \text{purine} + 3\,ROH \qquad (72)$$

Triethyl orthoformate converts 1,3-dimethyl-4,5-diaminopyrimidine-2,4-dione to 1,3-dimethylpurine-2,6-dione (theophyllin) (*142*). 9-Isopropyl- and 9-cyclohexyl-1,3-dimethylpurine-2,6-diones and 1-substituted 3-(*sec*-amino)-purine-2,5-diones were similarly prepared (*32, 84a*).

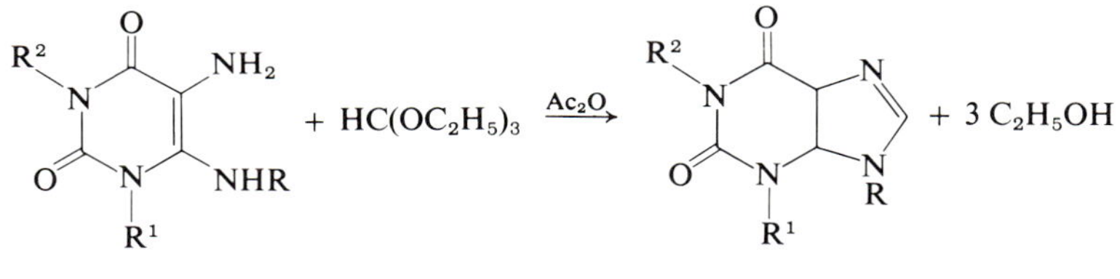

$$\text{pyrimidinedione} + HC(OC_2H_5)_3 \xrightarrow{Ac_2O} \text{purinedione} + 3\,C_2H_5OH \qquad (73)$$

[R = H, CH(CH$_3$)$_2$, C$_6$H$_{11}$, R$^1$ = R$^2$ = CH$_3$; R = H, R$^1$ = N(CH$_3$)$_2$, N(CH$_2$C$_6$H$_5$)$_2$, piperidino, morpholino, R$^2$ = H, CH$_3$]

Aminomalonamidine dihydrochloride is converted to 6-hydroxypurine (hypoxanthine) and 2,8-disubstituted 6-hydroxypurines by prolonged refluxing with trialkyl orthocarboxylates in dimethylformamide solution (*138*).

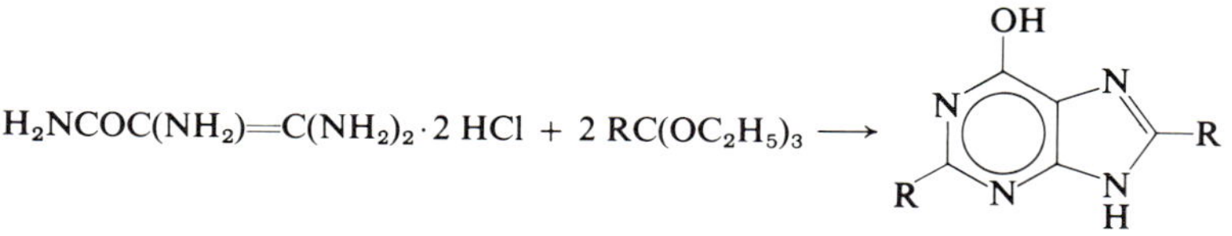

$$H_2NCOC(NH_2)\!\!=\!\!C(NH_2)_2 \cdot 2\,HCl + 2\,RC(OC_2H_5)_3 \longrightarrow \text{6-hydroxypurine} \qquad (74)$$

(R = H, CH$_3$, C$_2$H$_5$)

2-Substituted-4-$N$-ethoxyalkylene-5-carboxamidoimidazoles are formed as intermediates. 4-Amino-5-carboxamidoimidazoles, in fact, may be converted to substituted purines by heating them with triethyl orthoformate [Eq. (75)] (*138*, *184*, *201*). 6-Hydroxypurine is formed when a mixture of triethyl orthoformate, ammonium acetate and ethyl acetamidocyanoacetate are heated at 180°C for 8 hours (*183*).

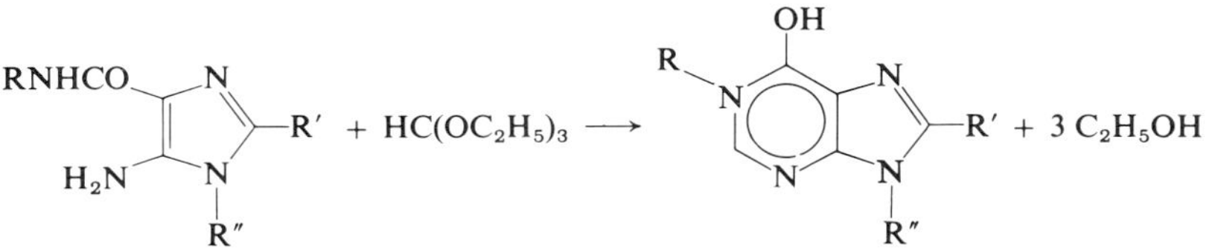

$$\text{RNHCO} \underset{\underset{R''}{|}}{\overset{N}{\underset{H_2N}{\bigcirc}}} -R' + HC(OC_2H_5)_3 \longrightarrow \underset{\underset{R''}{|}}{\overset{OH}{\underset{N}{\bigcirc}}}-R' + 3\,C_2H_5OH \quad (75)$$

[R = H, R' = CH$_3$, R" = H; R = OH, R' = H, R" = H;
R = H, R' = CH(CH$_3$)$_2$, R" = CH$_2$CH(CH$_3$)$_2$]

2. PTERIDINES

Condensation of 3-aminopyrazine-2-carboxamide and several of its derivatives with triethyl orthoformate–acetic anhydride yields pteridine and substituted pteridines.

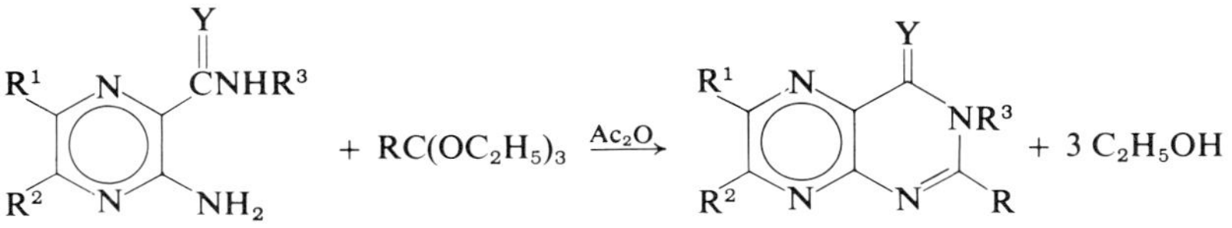

$$\text{(pyrazine-2-carboxamide)} + RC(OC_2H_5)_3 \xrightarrow{Ac_2O} \text{(pteridine product)} + 3\,C_2H_5OH \quad (76)$$

| R | R$^1$ | R$^2$ | R$^3$ | Y | References |
|---|---|---|---|---|---|
| H | H | H | H | O, S | 6, *113* |
| H | H | H | HO, –N=C(CH$_3$)$_2$, –N=CHC$_6$H$_5$ | O | *191*, *222* |
| H | C$_6$H$_5$ | C$_6$H$_5$ | H, CH$_2$C$_6$H$_5$, C$_4$H$_9$ | O, S | *182* |
| H, CH$_3$ | H | H, CH$_3$ | OH | O | *39a* |

3. OTHER HETEROCYCLIC COMPOUNDS WITH TWO OR MORE FUSED RINGS

A number of more or less complex heterocyclic compounds have been synthesized by reaction sequences whose last step involved condensation of a precursor having two nucleophilic functional groups with an orthocarboxylate. Triethyl orthoformate or triethyl orthoformate–acetic anhydride were used in most of these ring closures, but higher orthocarboxylates have also been used.

Among the fused-ring heterocycles synthesized using ortho esters are s-triazolo[3,4-b]-1,3,4-thiadiazoles (83), 1,2,4-triazolo[4,3-a]pyrimidines (131, 220), 7-oxo-7,8-dihydro-s-triazolo[4,3-a]pyrimidines (7–9), 5-oxo-5,8-dihydro-s-triazolo[4,3-a]pyrimidines (7, 8), 2-phenyl-7-hydroxy-1,2,3-triazolo[4,5-d]-pyrimidines (139), 2-phenyl-7-mercapto-1,2,3-triazolo[4,5-d]pyrimidine (139), 7-chloro-8-ethoxymethyleneamino-s-triazolo[4,3-c]pyrimidine (195a), 1-phenyl-4-hydroxy-1,2,3-triazolo[5,4-d]pyrimidine (54c), 7-amino-1,2,5-oxadiazolo[3,4-d]pyrimidine-1-oxide (194a), 3-substituted isoxazolo[4,5-d]pyrimidine-4(5H)-ones (136a, 187a), 7-hydroxy-2-methylthiazolo[5,4-d]pyrimidine (42), pyrazolo[1,5-d]as-triazin-4(5H)-one (1), 6-chloroimidazo[4,5-c]-pyridazines (98), imidazo[b]pyrazines (122, 163), 3-amino-4-hydroxy-7-carboxypyrazolo[4,3-c]pyridines (188), 6,7-dihydro-6,7,7-trimethyl-5H-pyrrolo[3,4-d]pyrimidines (188a), 5-chlorodihydropyrimido[4,5-e]-as-triazines (195), 7-benzyl-7H-s-triazolo[3,4-i]purine (194), 9-benzyl-9H-s-triazolo[3,4-i]purine (194), imidazo[1,2-b]s-triazolo[3,4-f]pyridazine (169c, 169d), substituted 6-cyano-8-methoxy-1H-pyrimido[4,5,6-i,j][2,7]naphthyridines (11a), s-triazolo[4,3-a]quinoline (155), s-triazolo[3,4-a]isoquinoline (168), 2-(2-benzothiazolyl)-4-carbethoxy-1-oxo-1,10-dihydro[2,1-b]benzothiazole (204), imidazo[b]quinoxalines (162), N,N',N''-tris(p-toluenesulfonyl)-2,6,7 triazabicyclo-[2.2.2]octane (44), 4,6-dichloro-2,8-bis(methylthio)pyrimido[4,5-b:5'4'-e]-pyridine (26), 2,4,6,8-tetraoxaoctahydropyrimido[4,5-b:5',4'-e]pyridine (26), imidazo[1,2-b]pyrido[3,2-d]s-triazolo[3,4-b]pyridazine (169b), imidazo-[1,2-b]pyrido[2,3-d]s-triazolo[3,4-b]pyridazine (169b), 6,12-dihydro-6,12-dioxoindolo[2,1-b]quinazoline (35), 8H-quinazolino[4,3-b]quinazolines (34, 130), 13H-s-triazolo[4',3':3,4]phthalazino[1,2-b]-13-quinazolone (206), s-triazolo[4,3-a]pyrimidin-5-one (20), pyrimido[5,4-c]pyridazine-8-one (123a), pyrimido[4,5-d]pyridazine-4-one (89a), pyrimido-[4,5-c]pyridizine-5-one (123a), 6-methyl-7-oxo-1,2,4-triazolo[4,3-b]-1,2,4-triazine (81a), 8-chloro-3,4-dihydropyrimido[5,4-e]-as-triazine (194b), 5-aminothiazolo[5,4-d]pyrimidine (174a), and 6,7,7-trimethyl-6,7-dihydro-5H-pyrrolo[3,4-d]pyrimidine-4-thiol (36).

## IV. REACTIONS RESULTING IN CARBON–PHOSPHORUS BOND FORMATION

Several reactions of acyclic and heterocyclic ortho esters with inorganic and organic phosphorus compounds result in formation of carbon–phosphorus bonds. Although the mechanism of none of these reactions has been investigated, it is possible in some instances to postulate reasonable reaction schemes.

Diethoxycarbonium tetrafluoroborate (formed from triethyl orthoformate and boron trifluoride) reacts with triethyl phosphite to form ethyltriethoxyphosphonium tetrafluoroborate [Eq. (77)] (54b). This reaction, which involves alkylation of phosphorus, differs from the others discussed below, which involve bond formation between phosphorus and ortho ester acyl carbon.

$$:P(OC_2H_5)_3 + [HC(OC_2H_5)_2]^+BF_4^- \rightarrow HCO_2C_2H_5 + C_2H_5P(OC_2H_5)_3^+BF_4^- \quad (77)$$

Dialkyl dialkoxymethylphosphonates, dialkyl diethoxymethylthiophosphonate, dialkyl 1,1-dialkoxyethylphosphonate, and dialkyl triethoxymethylphosphonate were prepared in low conversions but moderate yields by heating dialkyl phosphonates with trialkyl orthoformates, trialkyl orthoacetates, and tetraethyl orthocarbonate (121a, 121b, 136b, 136c).

$$(RO)_2PHY + HC(OR')_3 \rightarrow (RO)_2P(Y)CH(OR')_2 + R'OH \quad (78)$$
$$(R = CH_3, C_2H_5, R' = CH_3, C_2H_5, Y = O; R = R' = C_2H_5, Y = S)$$

$$(RO)_2PHO + CH_3C(OR')_3 \rightarrow (RO)_2P(O)C(CH_3)(OR')_2 + R'OH \quad (79)$$
$$[R = CH_3, R' = C_2H_5, C_3H_7; R = C_2H_5, R' = CH_3, C_2H_5, C_3H_7; R = C_3H_7, R' = CH_3, C_2H_5, C_3H_7; R = (CH_3)_2CH, C_4H_9, R' = C_2H_5]$$

$$(RO)_2PHO + C(OC_2H_5)_4 \rightarrow (RO)_2P(O)C(OC_2H_5)_3 + C_2H_5OH \quad (80)$$
$$[R = CH_3, C_2H_5, C_3H_7, (CH_3)_2CH, C_4H_9]$$

These reactions were carried out at elevated temperatures in sealed tubes. They are not facilitated by acid catalysts, although the reactions of orthoacetates with phosphonates are accelerated by tertiary amines and alkoxides (121a). This suggests that the orthoacetate reactions may involve nucleophilic additions of the phosphonates to ketene acetals, formed by loss of alcohol from the orthoacetates. The reactions of orthoformates and orthocarbonates may involve preliminary isomerization of the dialkyl phosphonates to dialkyl hydrogen phosphites, $(RO)_2P(OH)$, followed by reaction of these acid esters with the ortho esters.

Dialkyl phosphochloridites yield dialkyl-(1,1-dialkoxyethyl)phosphonates $[R_2P(O)C(CH_3)(OR')_2]$ when treated with trialkyl orthoacetates (121c).

Diethyl diethoxymethylphosphonate is also formed by reaction of triethyl orthoformate with diethyl phosphinyl chloride, with phosphorus trichloride or phosphorus trichloride-triethyl phosphite mixtures, with phosphorus tribromide, with dichlorodiethylaminophosphine, and with diethyl acetylphosphite (54a, 67, 97a).

$$(C_2H_5O)_2PCl + HC(OC_2H_5)_3 \longrightarrow$$
$$(C_2H_5O)_2P(O)CH(OC_2H_5)_2 + C_2H_5Cl \quad (81)$$

$$PCl_3 + 3\ HC(OC_2H_5)_3 \xrightarrow[ZnCl_2]{0°C}$$
$$(C_2H_5O)_2P(O)CH(OC_2H_5)_2 + C_2H_5Cl + HCO_2C_2H_5 \quad (82)$$

$$2\ P(OC_2H_5)_3 + PCl_3 + 3\ HC(OC_2H_5)_3 \longrightarrow$$
$$3\ (C_2H_5O)_2P(O)CH(OC_2H_5)_2 + 3\ C_2H_5Cl \quad (83)$$

2-Methoxy-2-substituted-1,3-dioxolanes yield 2-dimethylphosphonyl-2-substituted-1,3-dioxolanes when treated with phosphorus trichloride and trimethyl phosphite in the presence of zinc chloride at 0°C (*60b*).

$$PCl_3 + P(OCH_3)_3 + \underset{R\phantom{xx}O}{\underset{CH_3O\phantom{xx}O}{\diagdown\diagup}} \xrightarrow[ZnCl_2]{0°C} \underset{R\phantom{xx}O}{\underset{(CH_3O)_2P(O)\phantom{xx}O}{\diagdown\diagup}} \quad (84)$$

(R = CH$_3$, CHCl$_2$)

2-Chloro-5,5-dimethyl-1,3,2-dioxaphosphorinane and trimethyl orthoformate form 2-diethoxymethyl-2-oxo-5,5-dimethyl-1,3,2-dioxaphosphorinane in a similar reaction (*54a*).

$$\underset{H_3C}{\underset{H_3C}{\diagdown}}\!\!\!\diagup\!\!\!\overset{O}{\underset{O}{\diagdown}}\!\!P\!-\!Cl + HC(OC_2H_5)_3 \longrightarrow \underset{H_3C}{\underset{H_3C}{\diagdown}}\!\!\!\diagup\!\!\!\overset{O}{\underset{O}{\diagdown}}\!\!\overset{\overset{O}{\|}}{P}\!-\!CH(OC_2H_5)_2 + C_2H_5Cl \quad (85)$$

Alkyl ethylphosphinates and ethyl *p*-chlorophenylphosphinate react with trialkyl orthoformates to form alkyl ethyldialkoxymethylphosphinates and ethyl diethoxymethyl-*p*-chlorophenylphosphinate (*136c*).

$$R(R'O)PHO + HC(OR'')_2 \rightarrow R(R'O)P(O)CH(OR'')_2 + R''OH \quad (86)$$

(R = C$_2$H$_5$, R' = CH$_3$, C$_2$H$_5$, R'' = CH$_3$, C$_2$H$_5$; R = *p*-ClC$_6$H$_4$, R' = R'' = C$_2$H$_5$)

Diethoxymethyl disubstituted phosphine oxides are products of reactions of diphenylchlorophosphine (*54a*) or diethylphosphine oxide (*136c*) with triethyl orthoformate [Eqs. (87), (88)]. Diphenylchlorophosphine reacts with 2-alkoxy-1,3-dioxanes to form 2-diphenylphosphinyl-1,3-dioxanes [Eq. (89)] (*54a*).

$$(C_6H_5)_2PCl + HC(OC_2H_5)_3 \rightarrow (C_6H_5)_2P(O)CH(OC_2H_5)_2 + C_2H_5Cl \quad (87)$$

$$(C_2H_5)_2PHO + HC(OC_2H_5)_3 \rightarrow (C_2H_5)_2P(O)CH(OC_2H_5)_2 + C_2H_5OH \quad (88)$$

$$\underset{R}{\underset{R}{\diagdown}}\!\!\!\diagup\!\!\!\overset{O}{\underset{O}{\diagdown}}\!\!\overset{OC_2H_5}{\underset{H}{\diagdown}} + (C_6H_5)_2PCl \longrightarrow \underset{R}{\underset{R}{\diagdown}}\!\!\!\diagup\!\!\!\overset{O}{\underset{O}{\diagdown}}\!\!\overset{H}{\underset{P(O)(C_6H_5)_2}{\diagdown}} + C_2H_5OH \quad (89)$$

Reactions of compounds having chlorine bonded to trivalent phosphorus with orthoformates appear to involve initial formation of unstable phosphonium chlorides, which decompose either to derivatives of pentavalent phosphorus or to carboxylate esters and derivatives of trivalent phosphorus, depending on the site of nucleophilic attack by chloride ion in the product-forming step [Eq. (90)] (*54a*). The formation of carboxylate esters predominates at elevated temperatures, and is the only reaction observed between ortho esters and 2-chloro-1,3,2-benzodioxaphosphole or between orthoacetates or orthocarbonates and diphenylchlorophosphine (54a).

$$R_2PCl + R'C(OR'')_3 \rightarrow [R_2\overset{+}{P}(OR'')CR'(OR'')_2]Cl^- \begin{array}{c} \xrightarrow{-R''Cl} R_2P(O)CR'(OR'')_2 \\ \xrightarrow{-R''Cl} R_2POR'' + R'CO_2R'' \end{array} \qquad (90)$$

## REFERENCES

1. C. Ainsworth, *J. Am. Chem. Soc.* **77**, 1148 (1955).
2. C. Ainsworth, U.S. Patent 2,702,803 (1955); *Chem. Abstr.* **50**, 1088 (1956).
3. C. Ainsworth, *J. Am. Chem. Soc.* **78**, 1973 (1956).
4. C. Ainsworth, U.S. Patent 2,733,245 (1956); *Chem. Abstr.* **50**, 12115 (1956).
5. C. Ainsworth, *J. Am. Chem. Soc.* **87**, 5800 (1965).
6. A. Albert, D. J. Brown, and G. Cheeseman, *J. Chem. Soc.* p. 474 (1951).
7. C. H. F. Allen, H. R. Beilfuss, D. M. Burness, G. A. Reynolds, J. F. Tinker, and J. A. Van Allan, *J. Org. Chem.* **24**, 787 (1959).
8. C. H. F. Allen, H. R. Beilfuss, D. M. Burness, G. A. Reynolds, J. F. Tinker, and J. A. Van Allan, *J. Org. Chem.* **24**, 793 (1959).
9. C. H. F. Allen, G. A. Reynolds, J. F. Tinker, and L. A. Williams, *J. Org. Chem.* **25**, 361 (1960).
10. R. Anschutz and Stiepel, *Ann. Chem.* **306**, 8 (1899).
11. H. Antaki and V. Petrow, *J. Chem. Soc.* p. 2873 (1951).
11a. J. D. Atkinson and M. C. Johnson, *J. Chem. Soc., C* p. 1252 (1968).
12. M. R. Atkinson, G. Shaw, K. Schaffner, and R. N. Warner, *J. Chem. Soc.* p. 3847 (1956).
12a. P. Aubert and E. B. Knott, U.S. Patent 2,692,829 (1954); *Chem. Abstr.* **49**, 760 (1955).
13. P. Aubert, E. B. Knott, and L. A. Williams, *J. Chem. Soc.* p. 2185 (1951).
14. H. J. Backer and W. L. Wanmaker, *Rec. Trav. Chim.* **67**, 257 (1948).
15. H. J. Backer and W. L. Wanmaker, *Rec. Trav. Chim.* **68**, 247 (1949).
16. R. H. Backer and J. G. Van Oot, *J. Am. Chem. Soc.* **71**, 3060 (1949).
17. H. Baganz and L. Domaschke, *Chem. Ber.* **95**, 1840 (1962).
18. H. Baganz and L. Domaschke, *Chem. Ber.* **95**, 2095 (1962).
18a. B. R. Baker, P. C. Huang, and R. B. Meyer, *J. Med. Chem.* **11**, 475 (1968).
18b. B. R. Baker and M. A. Johnson, *J. Heterocyclic Chem.* **4**, 447 (1967).
18c. B. R. Baker and M. A. Johnson, *J. Heterocyclic Chem.* **4**, 507 (1967).
18d. B. R. Baker, M. U. Querry, A. F. Kadish and J. H. Williams, *J. Org. Chem.* **17**, 35 (1952).
19. H. J. Barber, R. Slack, C. E. Stickings, D. F. Elliott, and J. Attenburrow, British Patent 585,089 (1947); *Chem. Abstr.* **41**, 3822 (1947).
20. A. H. Beckett, R. G. W. Spickett, and S. H. B. Wright, *Tetrahedron* **24**, 2848 (1968).
21. H. Beiner, *Ann. Chem.* **686**, 102 (1965).
22. P. Benko and L. Pallos, Hungarian Patent 153,520 (1967); *Chem. Abstr.* **68**, 12864 (1968).
23. H. Beyer and C.-F. Kroger, *Ann. Chem.* **637**, 135 (1960).

24. H. Bredereck, F. Effenberger, and A. W. Hofmann, *Angew. Chem. Intern. Ed. Engl.* **1**, 331 (1962).
25. H. Bredereck, F. Effenberger, and A. W. Hofmann, *Chem. Ber.* **97**, 61 (1964).
26. H. Bredereck, F. Effenberger, and R. Sauter, *Chem. Ber.* **95**, 2049 (1962).
27. H. Bredereck, F. Effenberger, and H. J. Treiber, *Chem. Ber.* **96**, 1505 (1963).
28. H. Bredereck, R. Gompper, F. Effenberger, H. Keck, and H. Heise, *Chem. Ber.* **93**, 1398 (1960).
29. H. Bredereck, R. Gompper, H. Rempfer, K. Klemm, and H. Keck, *Chem. Ber.* **92**, 329 (1959).
30. L. G. S. Brooker, F. L. White, and R. H. Sprague, *J. Am. Chem. Soc.* **73**, 1087 (1951).
30a. D. J. Brown, B. T. England, and J. M. Lyall, *J. Chem. Soc.*, C p. 226 (1966).
31. J. Brunken and G. Bach, *Chem. Ber.* **89**, 1363 (1956).
32. E. Buehler and W. Pfleiderer, *Chem. Ber.* **100**, 492 (1967).
33. W. E. Buting and C. Ainsworth, U.S. Patent 3,279,988 (1966); *Chem. Abstr.* **66**, 28776 (1967).
34. H. W. Butler, S. A. Slorak, and H. J. Vipond, *J. Chem. Soc.* p. 3673 (1964).
35. K. Butler, M. W. Partridge, and J. A. Waite, *J. Chem. Soc.* p. 4970 (1960).
36. J. F. Cavallo, N. E. Webb, and J. A. D. Willis, *J. Chem. Soc.*, C p. 698 (1967).
37. A. W. Chapman, *J. Chem. Soc.* p. 1743 (1927).
38. G. Cignarella and U. Teotino, *J. Am. Chem. Soc.* **82**, 1594 (1960).
39. L. Claisen, *Ann. Chem.* **287**, 360 (1895).
39a. J. Clark and G. Neath, *J. Chem. Soc.*, C p. 919 (1968).
39b. R. H. Clark and E. C. Wagner, *J. Org. Chem.* **9**, 55 (1944).
40. D. H. Clemens and W. D. Emmons, *J. Am. Chem. Soc.* **83**, 2588 (1961).
41. D. H. Clemens, E. Y. Shropshire, and W. D. Emmons, *J. Org. Chem.* **27**, 3664 (1962).
42. A. H. Cook and E. Smith, *J. Chem. Soc.* p. 2329 (1949).
43. J. W. Cornforth, *in* "The Chemistry of Penicillin," H. T. Clarke, J. R. Johnson and R. Robinson, Eds., pp. 743 and 838. Princeton Univ. Press, Princeton, New Jersey, 1949.
44. G. Crank and F. W. Eastwood, *Australian J. Chem.* **18**, 1967 (1965).
45. T. Curtius and K. Heidenreich, *Chem. Ber.* **27**, 2684 (1894).
46. T. Curtius and K. Heidenreich, *J. Prakt. Chem.* [2] **52**, 454 (1895).
47. F. B. Dains, T. M. Andrews, and M. E. Roberts, *Univ. Kansas Sci. Bull.* **20**, 173 (1932); *Chem. Abstr.* **26**, 5935 (1932).
48. F. B. Dains and E. W. Brown, *J. Am. Chem. Soc.* **31**, 1148 (1909).
49. F. B. Dains and A. E. Daily, *Univ. Kansas Sci. Bull.* **19**, 215 (1930); *Chem. Abstr.* **26**, 427 (1932).
50. F. B. Dains, O. O. Malleis, and J. T. Meyer, *J. Am. Chem. Soc.* **35**, 970 (1913).
51. C. S. Davis, *J. Pharm. Sci.* **51**, 1111 (1962).
52. P. De Ruggieri, C. Gandolfi, and U. Guzzi, *Gazz. Chim. Ital.* **96**, 152 (1966).
52a. G. DeStevens and M. Dughi, *J. Am. Chem. Soc.* **83**, 3087 (1961).
53. R. H. DeWolfe, *J. Org. Chem.* **27**, 490 (1962).
54. R. H. DeWolfe, unpublished data.
54a. W. Dietsch, *Ann. Chem.* **712**, 21 (1968).
54b. K. Dimroth and A. Nurenbach, *Chem. Ber.* **93**, 1649 (1960).
54c. A. Dornow and J. Helberg, *Chem. Ber.* **93**, 2001 (1960).
55. H. Eilingsfeld, M. Seefelder, and H. Weidinger, *Chem. Ber.* **96**, 2671 (1963).
56. C. Feugeas, D. Olschwang, and M. Chatzopoulas; *Compt. Rend.* **C265**, 113 (1967).

57. H. Fischer, German Patent 834,104 (1952); *Chem. Abstr.* **50**, 402 (1956).
58. N. I. Fisher and F. M. Hamer, *J. Chem. Soc.* p. 907 (1937).
59. J. H. Freeman and E. C. Wagner, *J. Org. Chem.* **16**, 815 (1951).
60. S. G. Fridman and D. K. Golub, *Zh. Obshch. Khim.* **31**, 3394 (1961).
60a. S. M. Gadekar, S. N. Nibi, B. D. Johnson, and J. R. Cummings, *J. Med. Chem.* **11**, 453 (1968).
60b. M. Ghelardoni, F. Russo, and M. G. Salmon, *Ann. Chim. (Rome)* **56**, 1083 (1966).
61. A. Giacolone, *Gazz. Chim. Ital.* **62**, 577 (1932).
62. S. Ginsberg, *J. Org. Chem.* **27**, 4062 (1962).
63. S. A. Glickman, U.S. Patent 2,684,976 (1954); *Chem. Abstr.* **49**, 11005 (1955).
64. L. Goldman, J. W. Marisco, and A. L. Gazzola, *J. Org. Chem.* **21**, 599 (1956).
65. C. Goldschmidt, *Chemiker Ztg.* **26**, 743 (1902).
66. R. Gompper, H. E. Noppel, and H. Schafer, *Angew. Chem. Intern. Ed. Engl.* **2**, 686 (1963).
67. H. Gross, J. Freiberg, and B. Costisella, *Chem. Ber.* **101**, 1250 (1968).
68. I. Hagedorn, H. Etting, and K. E. Lichtel, *Chem. Ber.* **99**, 520 (1966).
70. I. Hagedorn, K. E. Lichtel, and H. D. Winkelmann, *Angew. Chem. Intern. Ed. Engl.* **4**, 702 (1965).
71. I. Hagedorn and H. Winkelmann, *Chem. Ber.* **99**, 850 (1966).
71a. H. C. Hamann, V. T. Spaziano, T. C. Chou, C. C. Price, and H. H. Lin, *Can. J. Chem.* **46**, 419 (1968).
72. F. M. Hamer, R. J. Rathbone, and B. S. Winton, *J. Chem. Soc.* p. 954 (1947).
72a. D. L. Harris and K. M. Wellman, *Tetrahedron Letters*, p. 5225 (1968).
73. K. Hartke and L. Peshkar, *Angew. Chem. Intern. Ed. Engl.* **6**, 83 (1967).
74. L. Heslinga, G. J. Katerberg, and J. F. Arens, *Rec. Trav. Chim.* **76**, 969 (1957).
75. A. J. Hill and J. V. Johnson, *J. Am. Chem. Soc.* **76**, 922 (1954).
76. A. W. Hofmann, *Ann. Chem.* **139**, 114 (1866).
76a. R. Hull, *J. Chem. Soc.* p. 2746 (1958).
76b. F. A. Hussein and K. S. Al-Dulaimi, *J, Chem. U.A.R.* **9**, 287 (1966); *Chem. Abstr.* **66**, 73311 (1967).
77. F. A. Hussein and S. Y. Kazandji, *J. Indian Chem. Soc.* **43**, 663 (1966).
78. T. Irie, E. Kurasawa, and T. Hanada, *Nippon Kagaku Zasshi* **79**, 1401 (1958); *Chem. Abstr.* **54**, 5657 (1960).
79. R. C. Iris, R. D. Leyva, and C. Ramirez, *Rev. Inst. Salubridad Enfermedades Trop. (Mex.)* **6**, 85 (1945); *Chem. Abstr.* **40**, 1471 (1946).
80. G. S. Jenkins, A. M. Knevel, and C. S. Davis, *J. Org. Chem.* **26**, 274 (1961).
81. J. B. Jepson, A. Lawson, and V. D. Lawton, *J. Chem. Soc.* p. 1791 (1955).
81a. K. Kalfus, *Collect. Czech. Chem. Commun.* **33**, 2513 (1968).
82. M. Kanaoka, *J. Pharm. Soc. Japan* **75**, 1149 (1955); *Chem. Abstr.* **50**, 5647 (1956).
83. M. Kanaoka, *Pharm. Bull. (Tokyo)* **5**, 385 (1957); *Chem. Abstr.* **52**, 5390 (1958).
84. H. Kano and E. Yamazaki, *Tetrahedron* **20**, 159 (1964).
84a. H. G. Kazmirowski, G. Dietz, and E. Carstens, *J. Prakt. Chem.* [4] **19**, 162 (1963).
85. B. B. Kehm and C. W. Whitehead, in "Organic Syntheses" (N. Rabjohn, ed.), Coll. Vol. IV, p. 515. Wiley, New York, 1963.
86. J. D. Kendall, British Patent 447,038 (1936); *Chem. Abstr.* **30**, 6958 (1936).
87. J. D. Kendall and G. F. Duffin, British Patent 633,736 (1949); *Chem. Abstr.* **44**, 7175 (1950).
88. J. D. Kendall and D. J. Frye, U.S. Patent 2,394,067 (1946); *Chem. Abstr.* **40**, 2401 (1946).

89. J. D. Kendall and H. G. Suggate, U.S. Patent 2,534,914 (1950); *Chem. Abstr.* **45**, 2350 (1951).
89a. T. Kinoshita and R. N. Castle, *J. Heterocyclic Chem.* **5**, 845 (1968).
90. A. I. Kiprianov and F. A. Mikhailenko, *Ukr. Khim. Zh.* **30**, 1309 (1964).
91. E. B. Knott, British Patent 609,812 (1948); *Chem. Abstr.* **43**, 2642 (1949).
92. E. B. Knott, British Patent 609,814 (1948); *Chem. Abstr.* **43**, 4164 (1949).
93. E. B. Knott, *J. Chem. Soc.* p. 2399 (1952).
94. E. B. Knott, U.S. Patent 2,691,581 (1954); *Chem. Abstr.* **49**, 83 (1955).
94a. E. B. Knott, British Patent 938,179 (1955); *Chem. Abstr.* **52**, 939 (1958).
94b. E. B. Knott, U.S. Patent 2,728,766 (1955); *Chem. Abstr.* **50**, 10580 (1956).
94c. E. B. Knott, U.S. Patent 2,743,273 (1956); *Chem. Abstr.* **51**, 907 (1957).
95. E. B. Knott and R. A. Jeffreys, *J. Org. Chem.* **14**, 879 (1949).
95a. A. Kovacic, B. Stanovnik, and M. Tisler, *J. Heterocyclic Chem.* **5**, 351 (1968).
96. C.-F. Kroger, L. Hummel, M. Mutscher, and H. Beyer, *Chem. Ber.* **98**, 3025 (1965).
97. C.-F. Kroger, P. Selditz, and M. Mutscher *Chem. Ber.* **98**, 3034 (1965).
97a. S. S. Krokhina, R. Y. Pyrkin, Y. A. Levin, and B. E. Ivanov, *Izv. Akad. Nauk. SSSR, Otd. Khim. Nauk.* p. 484 (1953).
98. T. Kuraishi and R. N. Castle, *J. Heterocyclic Chem.* **3**, 218 (1966).
99. H. Larive and R. Dennilauler, *Chimia (Aarau)* **15**, 115 (1961); *Chem. Abstr.* **61**, 12116 (1964).
100. G. Lehmann, H. Seefluth, and G. Hilgetag, *Chem. Ber.* **97**, 299 (1964).
101. L. Levai, C. Ritvay-Emandity, and G. Csepreghy, *J. Org. Chem.* **31**, 4003 (1966).
101a. Y. A. Levin, A. P. Mokova, and V. I. Kukhtin, *Zh. Obshch. Khim.* **31**, 1573 (1961).
102. C. D. Lewis, R. G. Krupp, H. Tieckelmann, and H. W. Post, *J. Org. Chem.* **12**, 303 (1947).
103. B. Loev and M. F. Kormandy, *Can. J. Chem.* **42**, 176 (1964).
104. R. R. Lorenz, B. F. Tullar, C. F. Koelsch, and S. Archer, *J. Org. Chem.* **30**, 2531 (1965).
105. R. M. McDonald and R. A. Kruger, *Tetrahedron Letters* p. 857 (1965).
106. R. M. McDonald and R. A. Kruger, *J. Org. Chem.* **31**, 488 (1966).
107. S. M. McElvain and R. E. Starn, *J. Am. Chem. Soc.* **77**, 4571 (1955).
108. S. M. McElvain and B. E. Tate, *J. Am. Chem. Soc.* **67**, 202 (1945).
109. M. M. McKee, R. L. McKee, and R. W. Bost, *J. Am. Chem. Soc.* **69**, 184 (1947).
110. J. Maillard, M. Vincent, and M. Benard, *Bull. Soc. Chim. France* p. 529 (1961).
111. J. Maillard, M. Vincent, and V. V. Tri, *Bull. Soc. Chim. France* p. 376 (1966).
112. P. Mamalas, V. Petrow, and B. Sturgeon, British Patent 690,119 (1953); *Chem. Abstr.* **48**, 6470 (1954).
112a. R. Martin, H. R. Mathews, H. Rapoport and G. Hyagarajan, *J. Org. Chem.* **33**, 3758 (1968).
113. H. G. Mautner, *J. Org. Chem.* **23**, 1450 (1959).
114. R. F. Meyer, *J. Org. Chem.* **28**, 2902 (1963).
115. Y. Mizuno and M. Nishimura, *J. Pharm. Soc. Japan* **68**, 58 (1948).
116. J. A. Montgomery, *J. Am. Chem. Soc.* **78**, 1928 (1956).
117. J. A. Montgomery and K. Hewson, *J. Org. Chem.* **30**, 1528 (1965).
118. J. A. Montgomery and L. B. Holum, *J. Am. Chem. Soc.* **80**, 404 (1958).
119. J. A. Montgomery and C. Temple, *J. Am. Chem. Soc.* **79**, 5238 (1957).
120. J. A. Montgomery and C. Temple, *J. Am. Chem. Soc.* **80**, 409 (1958).
121. J. A. Montgomery and C. Temple, *J. Org. Chem.* **25**, 395 (1960).

# REFERENCES

121a. V. V. Moskva, A. I. Maikova, and A. I. Razumov, *Zh. Obshch. Khim.* **37**, 1623 (1967).
121b. V. V. Moskva, A. I. Maikova, and A. I. Razumov, *Zh. Obshch. Khim.* **37**, 2137 (1967).
121c. V. V. Moskva, A. I. Razumova and A. I. Maikova, *Zh. Obshch. Khim.* **38**, 202 (1968).
122. F. L. Muehlmann and A. R. Day, *J. Am. Chem. Soc.* **78**, 242 (1956).
123. T. Mukaiyama and K. Sato, *Bull. Chem. Soc. Japan* **36**, 99 (1963).
123a. T. Nakagome, R. N. Castle and H. Murakami, *J. Heterocyclic Chem.* **5**, 523 (1968).
124. H. Nakano, A. Sugihara, and M. Ito, Japanese Patent 3897 ('66); *Chem. Abstr.* **65**, 686 (1966).
125. F. C. Novello, S. C. Bell, E. L. A. Abrams, C. Ziegler, and J. M. Sprague, *J. Org. Chem.* **25**, 970 (1960).
125a. H. Neunhoeffer and H. Hennig, *Chem. Ber.* **101**, 3947 (1968).
126. S. Oae, W. Tagaki, and A. Ohno, *Tetrahedron* **20**, 417 (1961).
127. R. Oda, M. Nomura, and S. Tanimoto, *Bull. Inst. Chem. Res., Kyoto Univ.* **32**, 231 (1954); *Chem. Abstr.* **50**, 7112 (1956).
128. R. Oda, K. Teremura, S. Tanimoto, M. Nomura, H. Suda, and K. Matsuko, *Bull. Inst. Chem. Res., Kyoto Univ.* **33**, 117 (1955); *Chem. Abstr.* **51**, 11355 (1957).
129. T. Ogata, *J. Chem. Soc. Japan* **55**, 394 (1934); *Chem. Abstr.* **28**, 5816 (1934).
129a. L. Pallos and P. Benko, *Ind. Chim. Belge* **32**, 1334 (1967).
129b. L. Pallos, P. Banks, F. Ordogh and B. Rotdy, Hung. Patent 154661; *Chem. Abstr.* **69**, 43662 (1968).
130. M. W. Partridge, S. A. Slorak, and H. J. Vipond, *J. Chem. Soc.* p. 3673 (1964).
131. W. W. Paudler and L. S. Helmic, *J. Heterocyclic Chem.* **3**, 269 (1966).
131a. V. G. Pesin and I. A. Belenkaya, *Khim. Geterotsikl. Soedin, Akad. Nauk Latv. SSR* p. 289 (1967).
131b. A. Piskala, *Collection Czech. Chem. Commun.* **32**, 3966 (1967).
132. A. Piskala and J. Gut, *Collection Czech. Chem. Commun.* **26**, 2519 (1961).
132a. A. Piskala and J. Gut, *Collection Czech. Chem. Commun.* **28**, 1681 (1963).
133. A. Piskala and J. Gut, *Collection Czech. Chem. Commun.* **28**, 2376 (1963).
134. R. N. Prasad, C. W. Noell, and R. K. Robins, *J. Am. Chem. Soc.* **81**, 193 (1959).
135. C. C. Price and V. Boekelheide, *J. Am. Chem. Soc.* **68**, 1246 (1946).
136. C. C. Price, N. J. Leonard, and H. F. Herbrandsen, *J. Am. Chem. Soc.* **68**, 1252 (1946).
136a. P. Rajagopalan and C. N. Talaty, *Tetrahedron* **23**, 3541 (1967).
136b. A. I. Razumov and V. V. Moskva, *Zh. Obshch. Khim.* **34**, 3125 (1960).
136c. A. I. Razumov and V. V. Moskva, *Zh. Obshch. Khim.* **35**, 1595 (1965).
137. F. Reitzenstein and G. Bonitsch, *J. Prakt. Chem.* [2] **86**, 58 (1912).
138. E. Richter, J. E. Leffler, and E. C. Taylor, *J. Am. Chem. Soc.* **82**, 3144 (1960).
139. E. Richter and E. C. Taylor, *J. Am. Chem. Soc.* **78**, 5848 (1956).
140. W. Ried and H. Lohwasser, *Ann. Chem.* **699**, 88 (1966).
141. W. Ried and W. Storbeck, *Chem. Ber.* **95**, 459 (1962).
142. W. Ried, W. Storbeck, and E. Schmidt, *Arch. Pharm.* **295**, 143 (1962).
143. R. M. Roberts, *J. Org. Chem.* **14**, 277 (1949).
144. R. M. Roberts, *J. Am. Chem. Soc.* **71**, 3848 (1949).
145. R. M. Roberts, *J. Am. Chem. Soc.* **72**, 3603 (1950).
146. R. M. Roberts, *J. Am. Chem. Soc.* **72**, 3608 (1950).
147. R. M. Roberts and R. H. DeWolfe, *J. Am. Chem. Soc.* **76**, 2411 (1954).
148. R. M. Roberts, T. D. Higgins, and P. R. Noyes, *J. Am. Chem. Soc.* **77**, 3801 (1955).

149. R. M. Roberts and P. J. Vogt, *in* "Organic Syntheses" (N. Rabjohn, ed.), Coll. Vol. IV, p. 420. Wiley, New York, 1963.
150. R. M. Roberts and P. J. Vogt, *in* "Organic Syntheses" (N. Rabjohn, ed.), Coll. Vol. IV, p. 464. Wiley, New York, 1963.
151. R. M. Roberts and P. J. Vogt, *J. Am. Chem. Soc.* **78**, 4778 (1956).
151a. R. K. Robins, *in* "Heterocyclic Compounds" (R. C. Elderfield, ed.), Vol. 8, pp. 162ff. Wiley, New York, 1967.
152. R. K. Robins and H. H. Lin, *J. Am. Chem. Soc.* **79**, 490 (1957).
153. D. M. Rockwell, Ph.D. dissertation, Yale University (1931).
154. C. Runti, V. D'Osualdo, and F. Ulian, *Ann. Chim. (Rome)* **49**, 1668 (1959).
155. C. Runti and C. Nisi, *J. Med. Chem.* **7**, 814 (1964).
156. C. Runti, C. Nisi, and L. Sindellari, *Ann. Chim. (Rome)* **51**, 719 (1961).
157. C. Runti, L. Sindellari, and C. Nisi, *Ann. Chim. (Rome)* **49**, 1649 (1957).
158. C. Runti, L. Sindellari, and F. Ulian, *Ann. Chim. (Rome)* **50**, 847 (1960).
158a. P. B. Russell and N. Whittaker, *J. Am. Chem. Soc.* **74**, 1310 (1952).
159. J. C. Saam and H. M. Bank, *J. Org. Chem.* **30**, 3350 (1965).
159a. P P. T. Sah, *J. Am. Chem. Soc.* **64**, 1487 (1942).
160. J. Sandstrom, *Acta Chem. Scand.* **14**, 1037 (1960).
160aa. F. Sauter, *Monatsh. Chem.* **99**, 1507 (1968).
160a. H. J. Schaeffer and R. N. Johnson, *J. Med. Chem.* **11**, 21 (1968).
161. H. J. Schaeffer and R. Vince, *J. Med. Chem.* **10**, 689 (1967).
162. E. Schipper and A. R. Day, *J. Am. Chem. Soc.* **73**, 5672 (1951).
163. E. Schipper and A. R. Day, *J. Am. Chem. Soc.* **74**, 350 (1952).
164. G. Shaw, *J. Chem. Soc.* p. 1834 (1955).
165. G. Shaw and D. N. Butler, *J. Chem. Soc.* p. 4040 (1959).
166. T. Sheradsky and P. L. Southwick, *J. Heterocyclic Chem.* **4**, 1 (1965).
167. T. H. Siddall, *Tetrahedron Letters* p. 2027 (1966).
168. G. S. Sidhu, S. Naqui, and D. S. Iyengar, *J. Heterocyclic Chem.* **3**, 158 (1966).
169. H. R. Snyder and R. E. Jones, *J. Am. Chem. Soc.* **68**, 1253 (1946).
169a. V. T. Spaziano, H. C. Shah, T. C. Chou, C. C. Price, and H. H. Lin, *J. Chem. Soc., C* p. 915 (1968).
169b. B. Stanovnik, A. Krbavcic, and M. Tisler, *J. Org. Chem.* **32**, 1139 (1967).
169c. B. Stanovnik and M. Tisler, *Tetrahedron Letters* p. 2403 (1966).
169d. B. Stanovnik and M. Tisler, *Tetrahedron* **23**, 387 (1968).
170. E. Steininger, *Monatsh. Chem.* **97**, 1195 (1966).
171. J. A. Stephens, U.S. Patent 2,909,553 (1959); *Chem. Abstr.* **54**, 5577 (1960).
172. Sterling Drug., Inc., Netherlands Patent Appl. 6,414,916 (1965); *Chem. Abstr.* **64**, 8152 (1966).
173. R. Stolle, *J. Prakt. Chem.* [2] **68**, 464 (1904).
174. R. Stolle and P. E. Bowles, *Chem. Ber.* **41**, 1099 (1908).
174a. S. Sugihara, E. Suzuki, and S. Inoue, *Chem. Pharm. Bull. (Tokyo)* **16**, 745 (1968).
175. T. Takahashi, T. Goto, H. Taniyama, and K. Sasaki, *J. Pharm. Soc. Japan* **66**, 29 (1946).
176. T. Takahashi, Z. Ichikawa, and T. Yatsuko, *J. Pharm. Soc. Japan* **64**, No. 2A, 14 (1945); *Chem. Abstr.* **45**, 8531 (1951).
177. T. Takahashi and H. Saikachi, *J. Pharm. Soc. Japan* **64**, No. 11A, 55 (1944); *Chem. Abstr.* **45**, 8530 (1951).
178. T. Takahashi and K. Sataki, *J. Pharm. Soc. Japan* **74**, 135 (1954); *Chem. Abstr.* **49**, 1718 (1955).

# REFERENCES

179. T. Takahashi, S. Senda, and H. Zenno, *J. Pharm. Soc. Japan* **69**, 104 (1949); *Chem. Abstr.* **44**, 3450 (1950).
180. T. Takahashi, Y. Yoshikawa, and Z. Ichikawa, *J. Pharm. Soc. Japan* **66**, 1 (1942); *Chem. Abstr.* **45**, 8531 (1951).
181. E. C. Taylor and J. G. Berger, *Angew. Chem. Intern. Ed. Engl.* **5**, 131 (1966).
182. E. C. Taylor, J. A. Carbon, and D. R. Hoff, *J. Am. Chem. Soc.* **75**, 1904 (1953).
183. E. C. Taylor and C. C. Cheng, *Tetrahedron Letters* No. 12, 9 (1959).
183a. E. C. Taylor and C. C. Cheng, *J. Org. Chem.* **25**, 148 (1960).
184. E. C. Taylor, C. C. Cheng, and O. Vogl, *J. Org. Chem.* **24**, 2019 (1959).
185. E. C. Taylor and W. A. Erhart, *J. Am. Chem. Soc.* **82**, 3138 (1960).
186. E. C. Taylor and W. A. Erhart, *J. Org. Chem.* **28**, 1108 (1963).
187. E. C. Taylor, W. A. Erhart, and M. Kawanisi, *Org. Syn.* **46**, 39 (1966).
187a. E. C. Taylor and E. E. Garcia, *J. Org. Chem.* **29**, 2116 (1964).
188. E. C. Taylor and K. S. Hartke, *J. Am. Chem. Soc.* **81**, 2456 (1959).
188a. E. C. Taylor and R. W. Hendess, *J. Am. Chem. Soc.* **87**, 1995 (1965).
189. E. C. Taylor and P. K. Loefler, *J. Am. Chem. Soc.* **82**, 3147 (1960).
190. E. C. Taylor, A. McKillop, and S. Vromen, *Tetrahedron* **23**, 885 (1967).
191. E. C. Taylor, O. Vogl, and P. K. Loefler, *J. Am. Chem. Soc.* **81**, 2479 (1959).
192. E. C. Taylor, S. Vromen, R. V. Ravindranathan, and A. McKillop, *Angew. Chem. Intern. Ed. Engl.* **5**, 308 (1966).
193. C. Temple, C. L. Kussner, and J. A. Montgomery, *J. Med. Pharm. Chem.* **5**, 866 (1962).
194. C. Temple, C. L. Kussner, and J. A. Montgomery, *J. Org. Chem.* **30**, 3601 (1965).
194a. C. Temple, C. L. Kussner, and J. A. Montgomery, *J. Org. Chem.* **33**, 2086 (1968).
194b. C. Temple, C. L. Kussner, and J. A. Montgomery, *J. Heterocyclic Chem.* **5**, 581 (1968).
195. C. Temple, R. L. McKee, and J. A. Montgomery, *J. Org. Chem.* **28**, 923 (1963).
195a. C. Temple, R. L. McKee, and J. A. Montgomery, *J. Org. Chem.* **28**, 2257 (1963).
195b. C. Temple, B. H. Smith, and J. A. Montgomery, *J. Org. Chem.* **33**, 530 (1968).
196. U. M. Teotino and G. Cignarella, *J. Am. Chem. Soc.* **81**, 4935 (1959).
197. H. Tieckelman and H. W. Post, *J. Org. Chem.* **13**, 268 (1948).
198. G. Tosolini, *Chem. Ber.* **94**, 2731 (1961).
199. D. L. Trepanier and P. E. Krieger, *J. Heterocyclic Chem.* **4**, 254 (1967).
200. D. L. Trepanier, J. Richman, and A. D. Rudzik, *J. Med. Chem.* **10**, 228 (1967).
201. G. E. Trout and P. R. Levy, *Rec. Trav. Chim.* **85**, 1254 (1966).
202. A. Van Dormael and A. DeCat, *Bull. Soc. Chim. Belges* **58**, 487 (1949).
203. A. Van Dormael, J. Jaeken, and J. Nys, *Bull. Soc. Chim. Belges* **58**, 477 (1949).
204. A. Van Dormael and J. Nys, *Chemie Ind.* **63**, 483 (1950); *Chem. Abstr.* **47**, 57 (1953).
204a. A. M. Van Leusen and E. C. Taylor, *J. Org. Chem.* **33**, 66 (1968).
204b. R. Vince and J. Donovan, *J. Med. Chem.* **12**, 174 (1969).
205. M. Vincent, J. Maillard, and M. Benard, *Bull. Soc. Chim. France* p. 1580 (1962).
206. C. E. Volker, J. Marth, and H. Beyer, *Chem. Ber.* **100**, 875 (1967).
206a. H. von Brachel, German Patent 1,156,780 (1963); *Chem. Abstr.* **60**, 5344 (1964).
206b. H. von Brachel and R. Merten, *Angew. Chem. Intern. Ed. Engl.* **1**, 592 (1962).
207. R. von Walther and R. Ramberg, *J. Prakt. Chem.* [2] **73**, 209 (1906).
208. E. C. Wagner, *J. Org. Chem.* **5**, 133 (1940).
209. R. Walther, *J. Prakt. Chem.* [2] **52**, 429 (1895).
210. R. Walther, *J. Prakt. Chem.* [2] **53**, 472 (1896).
211. H. Wanzlick, F. Esser, and H. J. Kleiner, *Chem. Ber.* **96**, 1208 (1963).

212. H. Wanzlick and H. J. Kleiner, *Angew. Chem.* **73**, 493 (1961).
212a. A. S. Wasfi, *J. Indian Chem. Soc.* **45**, 750 (1968).
213. P. H. L. Wei, S. C. Bell, and S. J. Childress, *J. Heterocyclic Chem.* **3**, 1 (1966).
214. C. W. Whitehead, *J. Am. Chem. Soc.* **75**, 671 (1953).
215. C. W. Whitehead and J. J. Traverso, *J. Am. Chem. Soc.* **75**, 671 (1953).
216. C. W. Whitehead and J. J. Traverso, *J. Am. Chem. Soc.* **77**, 5872 (1955).
217. C. W. Whitehead and J. J. Traverso, *J. Am. Chem. Soc.* **80**, 962 (1958).
218. K. B. Wiberg and B. I. Rowland, *J. Am. Chem. Soc.* **77**, 2205 (1955).
218a. H. Wichelhaus, *Chem. Ber.* **2**, 115 (1869).
219. H. Wichelhaus, *Chem. Ber.* **3**, 2 (1870).
220. L. A. Williams, *J. Chem. Soc.* p. 1829 (1960).
221. D. W. Woodward, U.S. Patent 2,534,331 (1951); *Chem. Abstr.* **45**, 5191 (1951).
222. W. B. Wright and J. M. Smith, *J. Am. Chem. Soc.* **77**, 3927 (1955).
223. H. L. Yale and J. T. Sheehan, *J. Org. Chem.* **26**, 4315 (1961).

CHAPTER 4

# Reactions of Ortho Esters Which Involve Carbon–Carbon or Carbon–Hydrogen Bond Formation

## I. INTRODUCTION

This chapter is primarily concerned with reactions in which the acyl carbon atoms of ortho esters form new bonds to carbon or hydrogen. A number of useful reactions of carboxylic ortho esters involve replacement of an ortho ester alkoxy group by a hydrocarbon group or a substituted hydrocarbon group. In some instances the alkoxy group is displaced by a carbanionoid reagent such as an organometallic compound. In others, the entering group is derived from a carbon acid—a substance having an active methylene group or a methyl group activated by a quaternary nitrogen atom, a terminal alkyne, or hydrogen cyanide. Carbon–carbon bonds are also formed in reactions of ortho esters with diazo ketones and diazo esters, and by electrophilic addition and substitution reactions of dialkoxycarbonium ions derived from ortho esters.

Reactions of metal hydrides with ortho esters, which are mechanistically similar to reactions of ortho esters with organometallic compounds, are also considered.

## II. REACTIONS OF ORTHO ESTERS WITH ORGANOMETALLIC COMPOUNDS

### A. Reactions of Ortho Esters with Grignard Reagents

1. REACTIONS OF ORTHOFORMATES WITH GRIGNARD REAGENTS

a. *Formation of Acetals and Aldehydes.* In the period immediately following Victor Grignard's discovery of ether-solvated organomagnesium halides (*246*), new synthetic applications of these versatile reagents were sought in many laboratories. In *1904* Boudroux in France and Chichibabin in Russia independently and almost simultaneously reported that Grignard reagents reacted with triethyl orthoformate to form acetals, which are easily hydrolyzed to the corresponding aldehydes (*53–55, 114, 115, 246*).

$$HC(OC_2H_5)_3 + RMgX \longrightarrow RCH(OC_2H_5)_2 + MgXOC_2H_5$$

$$RCH(OC_2H_5)_2 + H_2O \xrightarrow{H^+} RCHO + 2\ C_2H_5OH \tag{1}$$

This procedure affords a method of converting organic halides and terminal acetylenes to aldehydes having one more carbon atom than the starting material [Eqs. (1) and (2)].

$$RC{\equiv}CH \xrightarrow[-R'H]{R'MgX} RC{\equiv}CMgX \xrightarrow{HC(OC_2H_5)_3} RC{\equiv}CCH(OC_2H_5)_2 \xrightarrow{H_2O} RC{\equiv}CCHO \tag{2}$$

The synthesis of aldehydes from Grignard reagents and orthoformates is now known as the Boudroux-Chichibabin reaction. It is interesting that Chichibabin discovered the reaction while seeking a better aldehyde synthesis than his previously discovered reaction between formate esters and Grignard reagents (*113*), whereas Boudroux obtained benzaldehyde while attempting to prepare triphenylmethane from phenylmagnesium bromide and triethyl orthoformate (*53*).

The Boudroux-Chichibabin reaction has been carried out using alkylmagnesium halides, arylmagnesium halides, vinylmagnesium halides, 1-alkynylmagnesium halides, and heterocyclic organomagnesium halides. From the standpoints of availability of starting materials and general applicability, it is the most satisfactory of the several reactions used for synthesizing aldehydes from Grignard reagents (*530*).

Yields of acetals (or aldehydes) from the Boudroux-Chichibabin reaction depend to some extent on the structure of the Grignard reagent used, but the most important factor influencing yields seems to be the detailed procedure used in carrying out the reaction and working up the products. Yields exceeding 50%, based on the Grignard reagent, are probably quite generally attainable in properly conducted reactions, and yields exceeding 85% have been reported in several instances.

The detailed course of the reaction of Grignard reagents with orthoformates has not been studied, but qualitative observations indicate that it must be complex. Simply adding the orthoformate to the ethereal organomagnesium halide solution and then working up the mixture affords only low yields of acetals. Yields are improved considerably by refluxing the reaction mixture for several hours before work-up. The highest yields are obtained by distilling the ether from the reaction solution and then cautiously heating the residue until a highly exothermic reaction begins. The exothermicity of this reaction may create problems in large-scale syntheses; however, maximum yields are obtained only if this exothermic stage is reached but not allowed to get out of control. In several syntheses described in the literature, benzene or toluene was added to the reaction mixture prior to the final heating period (*37, 295, 394, 413, 495*). There appears to be no particular advantage to doing this, except possibly to moderate the exothermic reaction.

The work-up procedure depends on whether the acetal or the aldehyde is the desired reaction product. To obtain the aldehyde, the reaction mixture is treated with aqueous mineral acid. To obtain the acetal, the work-up must be carried out under milder conditions, such as treatment with aqueous ammonium chloride or acetic acid solution. Several superior detailed procedures for the Boudroux-Chichibabin reaction are described in the literature (*23, 396, 530*).

In the vast majority of published examples of the Boudroux-Chichibabin reaction, triethyl orthoformate was the starting material. Trimethyl orthoformate is equally satisfactory (*45, 342, 514, 616*). Orthoformates having alkoxy groups other than methoxy and ethoxy apparently have not been used in this reaction. 2-Methoxy-1,3-dioxolane (*560aa*) and 2-methoxy-1,3-dioxanes (*173aa*) have also been used in this reaction.

The mechanism of the Boudroux-Chichibabin reaction has not been fully studied, but the general course of the reaction can be postulated from knowledge of the properties of Grignard reagents and orthoformates. The first step of the reaction is probably complex formation due to interaction of the ortho ester oxygen atoms with electrophilic magnesium atoms of the various species present in the Grignard solution ($R_2Mg$, $RMgX$, and $MgX_2$, as an approximation). Products are probably formed either by reaction of the dialkoxycarbonium ion produced by dissociation of this complex with an organomagnesium species, or by a concerted displacement of the complexed alkoxy group by a carbanion group from the Grignard reagent. The known ability of orthoformates to form dialkoxycarbonium salts by reaction with boron trifluoride etherate and with antimony pentachloride (*402*) suggests that dialkoxycarbonium haloalkoxyalkylmagnesiate, alkoxydialkylmagnesiate, and dihaloalkoxymagnesiate ion pairs may be discrete intermediates in these reactions. The existence of more than one kind of intermediate could account for the fact that products appear to be formed by two reactions, one of which requires the higher temperatures and/or more polar conditions attained by distilling most of the ether from the reaction mixture. The acetal should be formed more readily by electrophilic attack by the dialkoxycarbonium ion on the organomagnesium portion of a dialkoxycarbonium alkoxydialkylmagnesiate or haloalkoxyalkylmagnesiate ion pair than by reaction of organomagnesium species with a solvated dialkoxycarbonium dihaloalkoxymagnesiate ion pair [Eqs. (3) and (4)].

$$HC(OC_2H_5)_3 + \text{``RMgX''} \longrightarrow$$
$$H\overset{+}{C}(OC_2H_5)_2 \cdot C_2H_5OMgXR^- + H\overset{+}{C}(OC_2H_5)_2 \cdot C_2H_5OMgR_2^-$$
$$\downarrow \qquad\qquad\qquad\qquad \downarrow$$
$$RCH(OC_2H_5)_2 \qquad\qquad RCH(OC_2H_5)_2 \quad (3)$$
$$+ \qquad\qquad\qquad\qquad +$$
$$MgXOC_2H_5 \qquad\qquad RMgOC_2H_5$$

$$HC(OC_2H_5)_3 + MgX_2 \longrightarrow H\overset{+}{C}(OC_2H_5)_2 \cdot C_2H_5OMgX_2^- \xrightarrow{RMgX}$$
$$RCH(OC_2H_5)_2 + MgX_2 + MgX(OC_2H_5) \quad (4)$$

The recent observation that 2-*axial*-alkoxy-1,3-dioxanes react readily with Grignard reagents, whereas their 2-*equatorial*-alkoxy isomers are inert, suggests that participation by unshared electron pairs on oxygen in expulsion of the alkoxy group is an important driving force for the formation of the carbonium ion intermediate (*173aa*).

Aldehydes whose acetals have been synthesized by the Boudroux-Chichibabin reaction are listed below in order of increasing carbon content of the aldehyde. In most instances the acetal was hydrolyzed to the aldehyde, frequently without being isolated.

$C_3$ **aldehydes:** propionaldehyde (*342, 373, 396, 618*) and propynal (*171*). $C_4$ **aldehydes:** butyraldehyde (*115, 342, 396, 618*), isobutyraldehyde (*396*), 2-butynal (*171, 371, 399a, 587, 588*), and 2-butyndial (*262, 283, 491, 521, 612, 616, 617*). $C_5$ **aldehydes:** valeraldehyde (*174a, 342, 382a, 524a*), isovaleraldehyde (*342, 396*), 4-methoxybutanal (*456*), 2-methyl-3-butenal, 3-pentenal (*281, 634*), 3-methyl-3-butenal (*373a, 531*), 2-pentynal (*171, 399a*), 4-penten-2-ynal (*108*), furfural (*235*), and 2-thiophenecarboxaldehyde (*244, 245, 539*). $C_6$ **aldehydes:** hexanal (*23, 342*), 4-methylpentanal (*101*), 2-ethylbutanal (*101*), 3-methylpentanal (*157*), 2-methylene-3-methyl-3-butenal (*461*), 4-methyl-2,3-pentadienal (*461*), 2-hexynal (*171, 399a*), 4-methyl-2-pentynal (*283*), 2,4-hexadiyndial (*496*), 5-methoxy-4-penten-2-ynal (*160*), 4-methyl-2-thiophenecarboxaldehyde (*244, 245*), and pyridinecarboxaldehydes (*279, 611*). $C_7$ **aldehydes:** heptanal (*342*), cyclohexancarboxaldehyde (*142, 618*), 3-cyclohexenecarboxaldehyde *(531)*, 2-methylene-3-methyl-3-pentenal *(461)*, 4-methyl-2,3-hexadienal *(461)*, 2-heptynal *(171, 373, 399a)*, 4,4-dimethyl-2-pentynal *(50)*, 4-methyl-4-penten-2-ynal *(416a)*, benzaldehyde *(115, 246, 373, 444, 529, 618)*, p-chlorobenzaldehyde *(55)*, and p-bromobenzaldehyde *(55, 115)*. $C_8$ **aldehydes:** cyclohexylacetaldehyde *(216)*, 3-methylcyclohexanecarboxaldehyde *(114, 116)*, 2-amylpropenal *(436)*, 2-octynal *(37, 171, 417, 418)*, 4-*tert*-butoxy-2-butynal *(399a)*, o- (*m*- and *p*-)tolualdehyde *(46, 529, 530)*, phenylacetaldehyde *(55, 115, 166, 618)*, p-fluorophenylacetaldehyde *(376)* and 4,4-diethoxy-2-butynal *(242)*.

$C_9$ **aldehydes:** 2-nonynal *(37, 417, 418)*, 4-*tert*-butoxy-2-pentynal *(399a)*, cyclohexylpropynal *(399a)*, 6,6-dimethyl-2,4-heptadiynal *(51)*, 2,8-nonadiynal *(52)*, 2,6-nonadiynal *(533)*, p-ethoxybenzaldehyde *(514)*, 2-phenylpropenal *(436)*, phenylpropynal *(107, 171, 371, 399a, 417, 418)*, and 4-(2-tetrahydropyranyloxy)-2-butynal *(295)*. $C_{10}$ **aldehydes:** 2-decynal *(37)*, 4-phenoxy-2-butynal *(399a)*, p-isopropylbenzaldehyde *(43, 44)*, trimethylbenzaldehydes *(530)*, 4-phenylbutyraldehyde *(595a)*, p-ethylphenylacetaldehyde *(46)*, dimethylphenylacetaldehydes *(46)*, p-allylphenylacetaldehyde *(495)*, 2-benzylpropenal *(436)*, and 3,4-methylenedioxyphenylpropynal *(394)*. $C_{11}$ **aldehydes:** 2-*trans*-decalincarboxaldehyde *(419)*, 2-undecynal *(37)*, tetramethylbenzaldehydes *(530)*, 2-methyl-2-p-toluenesulfonylpropynal *(204)*, α-naphthaldehyde *(55)*, and β-naphthaldehyde *(119)*. $C_{12}$ **aldehydes:** 2,8-dodecadiynal *(499)*, pentamethylbenzaldehyde *(128, 516, 530)*, 3-(p-isopropylphenyl)-propionaldehyde *(46)*, p-butylphenylacetaldehyde *(45)*, and p-cyclopentylbenzaldehyde *(103)*. $C_{13}$ **aldehyde:** 2-methyl-5-isopropylphenylacetaldehyde *(46)*. $C_{14}$ **aldehyde:** o-benzylbenzaldehyde *(41)*. $C_{15}$ **aldehydes:** 3,3-diphenylpropenal *(362, 397)*, 9-phenanthraldehyde *(159, 267, 413, 600)*. $C_{16}$ **aldehyde:** 4-isopropyl-1,6-dimethyl-2-naphthaldehyde *(247)*.

b. *Formation of Ethers and Hydrocarbons.* Ethers are formed by reactions of Grignard reagents with acetals at elevated temperatures *(120, 315, 373, 535)*.

$$RCH(OR')_2 + R''MgX \rightarrow RR''CHOR' + R'OMgX \qquad (5)$$

Under even more drastic conditions, the resulting ethers may be converted to hydrocarbons *(315)*. Ether and hydrocarbon formation are therefore potential competing reactions in the synthesis of acetals from orthoformates. For example, Chichibabin and Jelgasin obtained ethyl dinaphthylmethyl ether as the principal product of reaction of α-naphthylmagnesium bromide with triethyl orthoformate *(120)*. Since the conversion of acetals to ethers and hydrocarbons requires more drastic conditions than the conversion of orthoformates to acetals, these side reactions can be minimized by avoiding large excesses of the Grignard reagent and unnecessarily high reaction temperatures.

2. Reactions of Other Ortho Esters with Grignard Reagents

a. *Higher Trialkyl Orthocarboxylates.* Higher trialkyl orthocarboxylates react with Grignard reagents to form ketals, from which ketones are obtained by acid hydrolysis.

$$RC(OC_2H_5)_3 \xrightarrow{R'MgX} RC(OC_2H_5)_2R' \xrightarrow{H^+, H_2O} RCOR' \qquad (6)$$

The first reaction of this kind was reported by Blaise and Maire, who converted ethyl triethoxyacetate to 4-hydroxy-4-ethyl-3-hexanone by reaction with ethylmagnesium iodide and hydrolysis of the initial product (*49*). The reaction was studied in more detail by Barré and Ladouceur, who converted triethyl orthoacetate, triethyl orthopropionate, and triethyl orthovalerate into a series of ketones [Eq. (6): $R = CH_3$, $R' = C_2H_5$, $C_3H_7$, $C_4H_9$, $C_5H_{11}$; $R = C_2H_5$, $R' = C_2H_5$, $C_3H_7$, $(CH_3)_2CHCH_2$, $C_5H_{11}$, $(CH_3)_2CHCH_2CH_2$, $C_6H_{13}$, $C_{10}H_{21}$; $R = C_4H_9$, $R' = C_2H_5$, $C_5H_{11}$] (*33*). Yields of ketones ranged from 6% to 50%. Ortho esters are undoubtedly transient intermediates in the conversion of tetraethyl orthocarbonate to ketones by reaction with Grignard reagents (see below).

b. *Tetraalkyl Orthocarbonates.* The initial product of reaction of a Grignard reagent with tetraethyl orthocarbonate is a trialkyl orthocarboxylate.

$$C(OC_2H_5)_4 + RMgX \rightarrow RC(OC_2H_5)_3 + C_2H_5OMgX \qquad (7)$$

The synthesis of trialkyl orthocarboxylates by this reaction is complicated by subsequent reactions of the initially formed ortho esters with the Grignard reagents to form ketals and ethers, but a few successful examples have been reported [Eq. (7), $R = C_2H_5$, $C_6H_5$, $C_6H_5C \equiv C$, $(=CHC_6H_4-)_2$] (*117, 285, 513*). For a more detailed discussion of this reaction, see p. 44.

If an excess of the Grignard reagent is used, symmetrical diethyl ketals rather than ortho esters are the principal products of reactions of tetraethyl orthocarbonate with Grignard reagents (*33, 284, 398*).

$$C(OC_2H_5)_4 + 2\,RMgX \rightarrow R_2C(OC_2H_5)_2 + 2\,C_2H_5OMgX \qquad (8)$$

Triethyl orthocarboxylates, formed by reaction (7), are undoubtedly intermediates in these reactions. Barré and Ladouceur prepared several ketones in yields ranging from 15% to 57% by means of reaction (8), followed by hydrolysis of the ketals [$R_2CO$, $R = C_2H_5$, $C_3H_7$, $(CH_3)_2CH$, $C_4H_9$]) (*33*). The highest yields were obtained by using 2 moles of Grignard reagent per mole of orthocarbonate, and replacing the ether with benzene before the final heating period.

## B. Reactions of Orthoformates with Other Organometallic Compounds

1. REACTIONS WITH ORGANOZINC COMPOUNDS

Nearly 30 years before the discovery of the Boudroux-Chichibabin reaction, Ladenberg synthesized propionaldehyde diethylacetal by the reaction of diethylzinc with triethyl orthoformate (377). Due to the success and convenience of the Boudroux-Chichibabin reaction, this related acetal synthesis was not developed further.

A more useful reaction is the condensation of α-bromo esters with triethyl orthoformate in the presence of zinc. This variant of the Reformatsky reaction affords moderate yields of acetals of α-formyl esters (118, 142, 532, 640).

$$RR'CBrCO_2C_2H_5 + HC(OC_2H_5)_3 \xrightarrow{Zn} RR'C[CH(OC_2H_5)_2]CO_2C_2H_5 + ZnBrOC_2H_5 \quad (9)$$

| R | R' | % Yield | References |
|---|---|---|---|
| H | H | — | 118, 142, 532 |
| $CH_3$ | H | 44 | 142 |
| $C_2H_5$ | H | 51 | 142 |
| $C_3H_7$ | H | 56 | 142 |
| $C_4H_9$ | H | 46 | 142 |
| $CH_3$ | $CH_3$ | 58 | 142, 640 |
| $-(CH_2)_5-$ | | 52 | 142 |
| NC | H | — | 142 |

The condensation is conveniently carried out by adding the bromo ester and triethyl orthoformate to the zinc in refluxing benzene. Hydrolysis and decarboxylation of the formyl ester acetals yields aldehydes of the structure RR'CHCHO, and in at least two instances (R = H, R' = H, CN), heating the ester resulted in loss of alcohol to form ethoxymethylene derivatives, $C_2H_5OCH{=}CR'CO_2C_2H_5$ (142).

Propargyl bromide reacts with zinc and triethyl orthoformate to form 3-butynal diethylacetal in 27% yield. Substituted propargyl bromides yield acetals of formylallenes under the same conditions [Eqs. (10) and (11)] (223).

$$HC{\equiv}CCH_2Br + HC(OC_2H_5)_3 + Zn \rightarrow HC{\equiv}CCH_2CH(OC_2H_5)_2 + ZnBrOC_2H_5 \quad (10)$$

$$RCH_2C{\equiv}CCH_2Br + HC(OC_2H_5)_3 + Zn \rightarrow RCH_2C[CH(OC_2H_5)_2]{=}C{=}CH_2 + ZnBrOC_2H_5 \quad (11)$$

$$[R = C_6H_5, C_4H_9, (CH_3)_2CH]$$

## 2. Reactions with Organoaluminum Compounds

Triethyl orthoformate reacts with propargylaluminum bromide and with allylaluminum bromide in ether solution to form diethylketals of 3-butynal and 3-butenal, respectively (412).

## 3. Reaction of Alkyllithium Reagents with Orthoformates

Berlin and Rathmore reported an interesting reaction of alkyllithium reagents with orthoformates which yields *trans*-alkenes rather than acetals (42).

$$2\ RCH_2Li + HC(OR')_3 \rightarrow RCH_2\underset{H}{\overset{H}{C}}=C-R + 2\ R'OLi + R'OH \qquad (12)$$

$$(R = C_3H_7,\ C_5H_{11},\ C_6H_{13})$$

Since one of the carbons of the olefinic linkage is derived from the orthoformate, this reaction provides a method of synthesizing *trans*-alkenes having odd numbers of carbon atoms. Alkyllithium compounds are much more basic than Grignard reagents, organozinc compounds, and organoaluminum compounds. Presumably the first step in their reactions with orthoformates is an α-elimination to form a dialkoxycarbene, which then undergoes a series of electrophilic substitution–α-elimination reactions which finally yields the observed product. The last step of the reaction may be an intramolecular insertion of a carbene into a neighboring C—H bond, since such insertion reactions are known to yield predominantly *trans*-alkenes (341). A mechanism which rationalizes the observed results is outlined in Eq. (13). Yields of alkenes from trimethyl orthoformate were 33%–43%. However, the bicyclic orthoformate, 4-ethyl-2,6,7-trioxabicyclo[2.2.2]octane, afforded alkenes in yields of about 75%.

$$\begin{aligned}
RCH_2Li + HC(OR')_3 &\rightarrow :C(OR')_2 + RCH_3 + LiOR' \\
:C(OR')_2 + RCH_2Li &\rightarrow RCH_2\ddot{C}(OR')_2{}^-Li^+ \rightarrow RCH_2\ddot{C}OR' + LiOR' \\
RCH_2\ddot{C}OR' + RCH_2Li &\rightarrow (RCH_2)_2\ddot{C}OR'{}^-Li^+ \rightarrow (RCH_2)_2C: + LiOR' \qquad (13)\\
RCH_2\underset{H}{\overset{H}{\ddot{C}}}-C-R &\rightarrow RCH_2\underset{H}{\overset{H}{C}}=C-R
\end{aligned}$$

2-Alkoxy-1,3-dioxolanes yield carbonyl compounds derived from the parent glycol and dibutyl carbinol, rather than the expected alkenes, when treated with butyl lithium (295a).

## III. REACTIONS OF ORTHO ESTERS WITH METAL HYDRIDES

Although trialkyl orthocarboxylates are rather inert to catalytic hydrogenation (*301*), they are reduced to acetals by lithium aluminum hydride and by diisobutylaluminum hydride. Trimethyl and triethyl orthocarboxylates are reduced in good yields to the corresponding acetals by refluxing with 0.25 mole of $LiAlH_4$ in benzene-ether or ether solution for a few hours (*125–127, 505*).

$$RC(OR')_3 \xrightarrow{LiAlH_4} RCH(OR')_2 \tag{14}$$

($R = CH_3SCH_2CH_2$, $R' = CH_3, C_2H_5$; $R = H, CH_3, C_4H_9, CCl_2{=}CCl$, $R' = C_2H_5$)

In a similar reaction, 2-methoxy-2-methyl-1,3-dioxolane was converted to 2-methyl-1,3-dioxolane-2-D by refluxing with lithium aluminum deuteride in the absence of a solvent (*545*). Triethyl orthobenzoate is converted to benzaldehyde diethylacetal in 95% yield by reaction with diisobutylaluminum hydride in benzene solution, and triethyl orthobutyrate forms butyraldehyde diethylacetal in 90% yield when treated with the same reagent (*635*).

Dioxolenium salts derived from steroids and carbohydrates react with lithium borohydride to yield the ethylidene acetals formed by attack of borohydride on the less-hindered side of the dioxolane ring (*102, 139*). A specific example is the synthesis of methyl 3,4-*O*-ethylidene-β-L-arabinopyranoside (*102*).

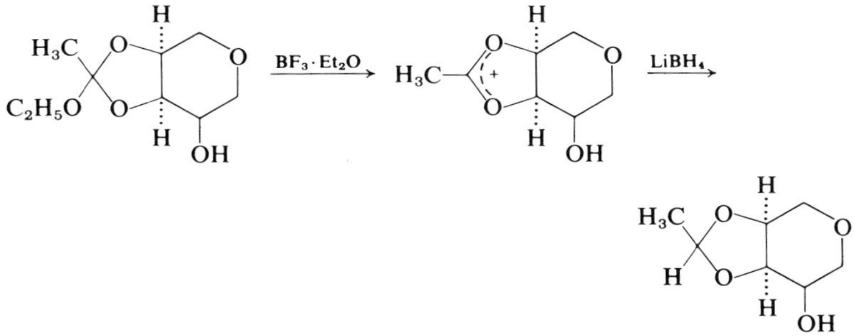

(15)

## IV. REACTIONS OF ORTHO ESTERS WITH COMPOUNDS WHICH HAVE ACTIVATED CARBON–HYDROGEN BONDS

### A. Condensation Reactions of Orthocarboxylates with Compounds Which Have Activated Methylene Groups

1. Introduction

Claisen reported in 1893 that acetyl acetone, ethyl acetoacetate, and diethyl malonate react with mixtures of triethyl orthoformate and acetic anhydride to yield ethoxymethylene derivatives (*122*). This reaction is general for trialkyl orthocarboxylates and compounds having activated methylene groups [Eq. (16); X and Y are $-I$, $-M$ groups such as cyano, acyl, or carbalkoxy; R = H, alkyl, aryl].

$$XYCH_2 + RC(OC_2H_5)_3 + 2\,(CH_3CO)_2O \rightarrow$$
$$XYC{=}CROC_2H_5 + 2\,CH_3CO_2H + 2\,CH_3CO_2C_2H_5 \quad (16)$$

The alkoxyalkylene derivatives produced by this condensation are useful intermediates in syntheses of heterocyclic compounds, cyanine dyes, and other substances.

Depending on the reaction conditions and the structure of the active methylene compound, 1,1-dialkoxyalkyl derivatives, or products in which two molecules of the active methylene compound are linked by a methine bridge derived from the ortho ester, may also be formed [Eqs. (17) and (18)].

$$XYCHR + R'C(OC_2H_5)_3 \rightarrow XYCR{-}CR'(OC_2H_5)_2 \quad (17)$$

$$2\,XYCH_2 + RC(OC_2H_5)_3 \rightarrow XYC{=}CR{-}CHXY \quad (18)$$

These reactions are discussed in this section, as are the reactions in which initially formed alkoxyalkylene derivatives undergo cyclization to other products. Coupling reactions analogous to (18) which yield cyanine dyes are discussed in a separate section.

## 2. Course and Mechanism of the Reaction

Condensation reactions of active methylene compounds with trialkyl orthocarboxylates are facilitated by the presence in the reaction mixture of acetic anhydride, and are usually carried out using at least 2 moles of acetic anhydride per mole of ortho ester. In the case of relatively unreactive methylene compounds such as diethyl malonate, it is also advantageous to add a catalyst such as zinc chloride to the reaction mixture.

Claisen assumed that the function of the acetic anhydride is to drive the reaction to completion by reacting with the ethanol produced in the condensation step [Eq. (19)]. Actually, the principal function of the acetic anhydride is to convert the ortho ester into the more reactive 1,1-diethoxyalkyl acetate [Eq. (20)] (*490*).

$$XYCH_2 + HC(OC_2H_5)_3 \rightleftharpoons XYC{=}CHOC_2H_5 + 2\,C_2H_5OH$$
$$2\,C_2H_5OH + 2\,(CH_3CO)_2O \rightarrow 2\,CH_3CO_2C_2H_5 + 2\,CH_3CO_2H \quad (19)$$

$$RC(OC_2H_5)_3 + (CH_3CO)_2O \rightarrow RC(OC_2H_5)_2OCOCH_3 + C_2H_5OH \quad (20)$$

1,1-Diethoxymethyl acetate is formed in good yield from triethyl orthoformate and acetic anhydride (*490*), and reacts directly with ethyl acetoacetate

(*490*) and diethyl malonate (*221*) to form the ethoxymethylene and diethoxymethyl derivatives, respectively. (Diethyl diethoxymethylmalonate was shown to lose alcohol when heated with zinc chloride to form diethyl ethoxymethylenemalonate.) Most attempts to demonstrate the reversibility of the condensation step have been unsuccessful. For example, no reaction occurs when ethyl ethoxymethyleneacetoacetate is refluxed with ethanol for 8 hours (*294*). Recently Tikhanova and co-workers obtained triethyl orthoformate in 98% yield by heating methyl ethoxymethylenenitroacetate with ethanol, thus demonstrating the reversibility of the reaction by means of which this substance was produced from methyl nitroacetate and triethyl orthoformate (*560b*).

Acetic anhydride is not essential to the condensation of active methylene compounds with trialkyl orthocarboxylates. Low to moderate yields of ethoxymethylene derivatives of ethyl acetoacetate, ethyl cyanoacetate, ethyl oxalacetate, and malononitrile were obtained by simply refluxing mixtures of the active methylene compounds and triethyl orthoformate for several hours at 140°C. Diethyl malonate did not react under these conditions, however (*294*). Benzoylacetonitrile and triethyl orthoformate also condense in the absence of acetic anhydride (*512*), as do malononitrile and triethyl orthoacetate (*589*). Indanedione (*346*) and several pyrazolones (*313*) react with triethyl orthocarboxylates in the absence of acetic anhydride to form ethoxymethylene derivatives.

Condensations of orthocarboxylates with active methylene compounds are facilitated by acidic catalysts, but in some cases they occur in the absence of added catalysts. When acetic anhydride is a component of the reaction mixture, the acetic acid formed by reaction of the anhydride with ethanol probably acts as a catalyst. Acetic acid has also been reported to catalyze these condensations in the absence of acetic anhydride (*177, 433*). The function of zinc chloride in catalyzing the formation of diethyl ethoxymethylenemalonate from diethyl malonate, triethyl orthoformate, and acetic anhydride is probably to catalyze elimination of ethanol from the initially formed diethoxymethyl derivative, rather than to catalyze reaction of the ortho ester with the active methylene compound (*221*). The fact that ethyl ethoxymethylenecyanoacetate is formed in about the same yield in the presence of sodium methoxide as in its absence indicates that these condensation reactions do not always require acid catalysis. No systematic study of the catalytic effects of acids on ortho ester-active methylene condensations has been reported.

The known facts about reactions of orthocarboxylates with active methylene compounds lead to the conclusion that these reactions are mechanistically complex, but shed little light on the detailed mechanism or mechanisms by which they occur. Under acidic conditions, one plausible mechanism involves reaction of a dialkoxycarbonium ion with the enolic form of the active methylene compound, followed by acid-catalyzed elimination of alcohol from the resulting 1,1-dialkoxyalkyl intermediate [Eq. (21), R' = alkyl or $CH_3CO$].

$$(C_2H_5O)_2CROR' + HA \longrightarrow (C_2H_5O)_2\overset{+}{C}R + R'OH + A^-$$

$$(C_2H_5O)_2\overset{+}{C}R + R''C(OH)=CHY \longrightarrow R''\overset{+}{C}(=OH)CHY \underset{-HA}{\overset{+A^-}{\longrightarrow}}$$
$$\qquad\qquad\qquad\qquad\qquad\qquad\qquad\qquad |$$
$$\qquad\qquad\qquad\qquad\qquad\qquad\qquad CR(OC_2H_5)_2 \qquad (21)$$

$$R''COCH[CR(OC_2H_5)_2]Y \underset{\substack{-C_2H_5OH \\ -A^-}}{\overset{+HA}{\longrightarrow}} R''COCY=CROC_2H_5$$

Dialkoxycarbonium ions are unlikely to be formed under neutral or basic conditions. One possible uncatalyzed mechanism would be a concerted reaction of the ortho ester with the enolic form of the active methylene compound, followed by elimination of alcohol from the dialkoxyalkyl intermediate [Eq. (22)]. Both of these mechanisms account for the observation that the ease of condensation increases as the acidity of the active methylene compound increases.

$$R'-CO-CHY-CR(OC_2H_5)_2 \underset{-C_2H_5OH}{\overset{B:}{\longrightarrow}} R'CO-CY=CROC_2H_5 \qquad (22)$$

### 3. Synthesis of Diethoxymethyl Derivatives

Very few reactions corresponding to Eq. (17) have been reported, and the general applicability of this acetal synthesis remains to be established. Diethyl malonate reacts with diethoxymethyl acetate to form diethyl diethoxymethylmalonate, but this substance loses alcohol readily, yielding the ethoxymethylene derivative (*221*). If there is only one proton on the activated carbon, the diethoxymethyl derivative is more stable. For example, phenylacetyl- and benzoylalanine are converted to acetals by reaction with triethyl orthoformate in the presence of acetic anhydride (*538*).

$$\begin{array}{c} \qquad\qquad\qquad\qquad\qquad\qquad\qquad\qquad CH(OC_2H_5)_2 \\ \qquad\qquad\qquad\qquad\qquad\qquad\qquad\qquad | \\ RCONHCHCO_2H + HC(OC_2H_5)_3 \overset{Ac_2O}{\longrightarrow} RCONH-C-CO_2H \qquad (23) \\ | \qquad\qquad\qquad\qquad\qquad\qquad\qquad\qquad\qquad | \\ CH_3 \qquad\qquad\qquad\qquad\qquad\qquad\qquad\qquad\qquad CH_3 \end{array}$$

$$(R = C_6H_5, C_6H_5CH_2)$$

Methyl nitroacetate reacts with triethyl orthoformate–acetic anhydride at 60°C to form methyl 2-nitro-3,3-diethoxypropionate in high yield [Eq. (17),

X = CO$_2$CH$_3$, Y = NO$_2$, R = H] (560b). The acetal loses ethanol to form methyl ethoxymethylenenitroacetate when heated with acetic anhydride at 120°C. 1,1-Dinitroethane and ethyl orthoformate–acetic anhydride react similarly at 80°C to form 3,3-dinitropropionaldehyde diethyl acetal (560b).

### 4. Synthesis of Acyclic Alkoxyalkylene Derivatives

Table I summarizes the published examples of condensation reactions of trialkyl orthocarboxylates with acyclic active methylene compounds. Yields of ethoxymethylene compounds as high as 90% have been reported, but yields in the 40%–60% range are more typical. In one instance, a single activating substituent sufficed to permit the preparation of an ethoxymethylene derivative: biacetyl was converted to ethoxymethylenebiacetyl by refluxing with triethyl orthoformate and acetic anhydride (154).

### 5. Synthesis of Cyclic Alkoxyalkylene Derivatives

A number of cyclic and heterocyclic compounds having activated methylene groups have been converted to alkoxyalkylene derivatives in reactions with trialkyl orthocarboxylates. These reactions are listed in Table II.

### 6. Ortho Ester–Active Methylene Condensations Which Are Accompanied by Heterocycle Formation

A number of glycine amides, glycine thioamides, o-hydroxyaryl ketones, and other polyfunctional compounds having activated methylene groups undergo complex condensation reactions with trialkyl orthocarboxylates to form oxazolones, thiazolones, isoflavones, and other heterocyclic products. The mechanisms of these reactions have not been studied. In some cases (oxazolone and thiazolone formation, for example), cyclization is probably independent of reaction of the ortho ester with the active methylene group. In other reactions, such as those involving isoflavone formation from o-hydroxyaryl ketones, an alkoxymethylene derivative is probably a precursor of the heterocyclic product.

Reactions of trialkyl orthocarboxylates with active methylene compounds which result in heterocycle formation are listed in Table III.

## B. Methine Bridge Formation by Ortho Esters

### 1. Introduction

In addition to organometallic compounds and active methylene compounds, whose reactions with ortho esters are discussed above, there are several other classes of substances which react with ortho esters to form new

TABLE I

Condensation Reactions of Trialkyl Orthocarboxylates with Active Methylene Compounds

$$XYCH_2 + RC(OR')_3 \xrightarrow{Z} XYC=CROR'$$

| X | Y | R | R' | Z | References |
|---|---|---|---|---|---|
| CN | CN | H, CH$_3$, C$_2$H$_5$, C$_6$H$_5$ | C$_2$H$_5$ | Ac$_2$O | 153, 173, 276, 277, 460, 560a, 566 |
| CN | CN | H, CH$_3$, C$_2$H$_5$ | C$_2$H$_5$ | AcOH | 433 |
| CN | CN | H, CH$_3$, C$_2$H$_5$ | C$_2$H$_5$ | None | 294, 589, 609 |
| CN | CH$_3$CO | C$_6$H$_5$ | CH$_3$ | Ac$_2$O | 592 |
| CN | C$_6$H$_5$CO | H | C$_2$H$_5$ | None | 512 |
| CN | C$_6$H$_5$CO | CH$_3$ | CH$_3$ | Ac$_2$O | 591 |
| CN | CO$_2$R (R = CH$_3$, C$_2$H$_5$, C$_3$H$_7$, C$_5$H$_{11}$) | H, CH$_3$, C$_2$H$_5$ | CH$_3$, C$_2$H$_5$ | Ac$_2$O | 141a, 141b, 492, 590 |
| CN | CO$_2$C$_2$H$_5$ | H, C$_2$H$_5$ | C$_2$H$_5$ | None | 294, 609 |
| CN | CO$_2$C$_2$H$_5$ | H, CH$_3$, C$_2$H$_5$ | C$_2$H$_5$ | AcOH | 433 |
| CN | CONHCOR | H | C$_2$H$_5$ | Ac$_2$O | 525 |
| CN | CON(CH$_3$)CO$_2$C$_2$H$_5$ | H | C$_2$H$_5$ | Ac$_2$O | 497 |
| CN | p-ClC$_6$H$_4$ | C$_2$H$_5$ | C$_2$H$_5$ | None | 610 |
| CN | 2-NH$_2$-3,5-(CN)$_2$-6-Cl-C$_5$N | H | C$_2$H$_5$ | Ac$_2$O | 13a |
| CH$_3$CO | CH$_3$CO | H, C$_2$H$_5$OCO | C$_2$H$_5$ | Ac$_2$O | 122, 293 |

| | | | | |
|---|---|---|---|---|
| $CH_3CO$ | $CH_3CO$ | H | $C_2H_5$ | AcOH | 433 |
| $CH_3CO$ | $CO_2CH_3$ | H | $CH_3, C_2H_5$ | $Ac_2O$ | 123 |
| $CH_3CO$ | $CO_2C_2H_5$ | H | $C_2H_5$ | $Ac_2O$ | 122, 123, 293 |
| $CH_3CO$ | $CO_2C_2H_5$ | H | $C_2H_5$, DEMA[a] | None | 294, 490 |
| $CH_3CO$ | $CO_2C_2H_5$ | H, $CH_3$, $C_2H_5$ | $C_2H_5$ | AcOH | 433 |
| $CH_3CO$ | $CO_2C_2H_5$ | $CH_3OCH_2$ | $C_2H_5$ | $Ac_2O$ | 544 |
| $CH_3CO$ | $COCH_2OC_2H_5$ | H | $C_2H_5$ | $Ac_2O$ | 544 |
| $CO_2CH_3$ | $CO_2CH_3$ | H | $CH_3$ | $Ac_2O + ZnCl_2$ | 123 |
| $CO_2C_2H_5$ | $CO_2C_2H_5$ | H | $C_2H_5$ | $Ac_2O + ZnCl_2$ | 122, 123, 221, 459 |
| $CO_2C_2H_5$ | $CO_2C_2H_5$ | H | $C_2H_5$ | AcOH | 433 |
| $CO_2C_2H_5$ | $CO_2C_2H_5$ | H, $CH_3$, $C_2H_5$ | $C_2H_5$ | $Ac_2O$ | 590, 601, 608 |
| $CO_2C_2H_5$ | $COCO_2C_2H_5$ | H | $C_2H_5$ | None | 294 |
| $CO_2C_2H_5$ | $COCO_2C_2H_5$ | H, $CO_2C_2H_5$ | $C_2H_5$ | $Ac_2O$ | 292, 293 |
| $CO_2C_2H_5$ | $XCH_2CO$ (X = Cl, Br) | H | $C_2H_5$ | $Ac_2O$ | 40 |
| $CO_2C_2H_5$ | $F_3CCO$ | H, $CO_2C_2H_5$ | $C_2H_5$ | $Ac_2O$ | 292, 293 |
| $CO_2C_2H_5$ | $C_6H_5CO$ | H | $C_2H_5$ | $Ac_2O$ | 194, 457, 458, 599 |
| $CO_2C_2H_5$ | $C_2H_5OCH_2CO$ | H | $C_2H_5$ | $Ac_2O$ | 175, 544 |
| $COCOCH_3$ | H | H | $C_2H_5$ | $Ac_2O$ | 154 |
| $COCO_2C_2H_5$ | $C_6H_5$ | H | $C_2H_5$ | $Ac_2O$ | 457 |
| $NO_2$ | $CO_2R$ (R = $CH_3, C_2H_5$) | H | $C_2H_5$ | $Ac_2O$ | 133, 299, 494, 560b |
| $NO_2$ | $CO_2C_2H_5$ | $C_6H_5$ | $C_2H_5$ | None | 286 |

[a] DEMA = diethoxymethyl acetate.

TABLE II
CONDENSATION REACTIONS OF TRIALKYL ORTHOCARBOXYLATES WITH CYCLIC AND HETEROCYCLIC ACTIVE METHYLENE COMPOUNDS

| Active methylene compound | Other reactants | Product | References |
|---|---|---|---|
| 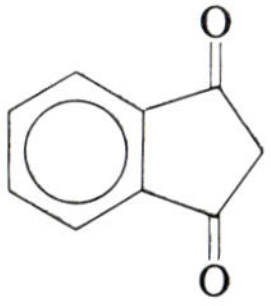 indane-1,3-dione | $HC(OC_2H_5)_3$ + AcOH, then $H_2O$ | 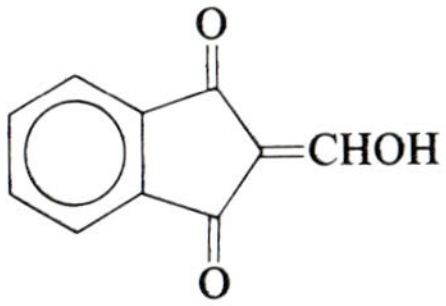 2-(hydroxymethylene)indane-1,3-dione (=CHOH) | 176, 177 |
| 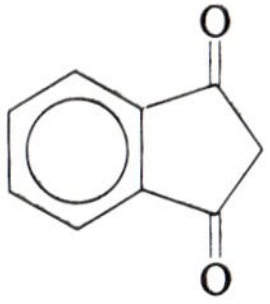 indane-1,3-dione | $CH_3C(OC_2H_5)_3$ | 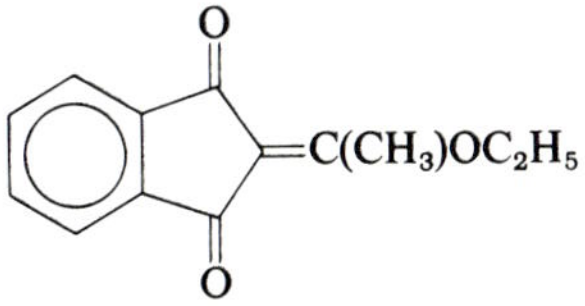 2-(1-ethoxyethylidene)indane-1,3-dione (=C(CH$_3$)OC$_2$H$_5$) | 346 |
| 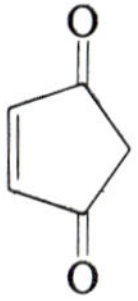 cyclopent-4-ene-1,3-dione | $RC(OC_2H_5)_3$ + $ZnCl_2$ + $Ac_2O$ (R = H, $C_2H_5$, $C_6H_5CH_2$) | 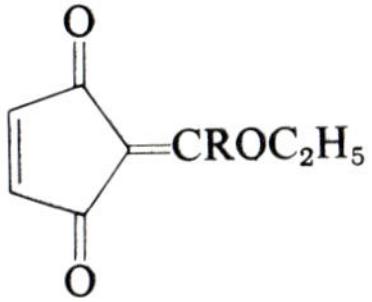 =CROC$_2$H$_5$ derivative | 412a |
| 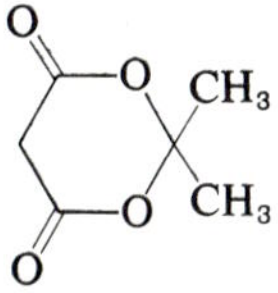 2,2-dimethyl-1,3-dioxane-4,6-dione | $HC(OR)_3$ (R = $CH_3$, $C_2H_5$) | 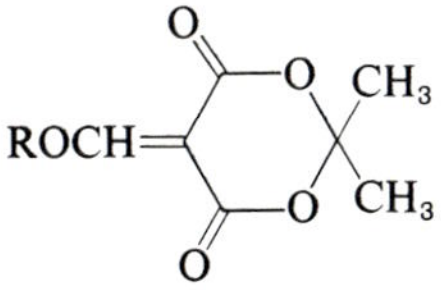 ROCH= derivative | 48 |

## REACTIONS WITH COMPOUNDS HAVING ACTIVATED C–H BONDS

| Substrate | Reagent | Product | Ref. |
|---|---|---|---|
| 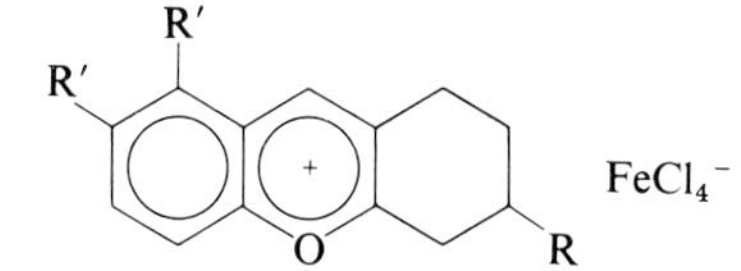 chromenylium FeCl$_4^-$ (R = H, CH$_3$; R', R' = H, H or benzo) | HC(OC$_2$H$_5$)$_3$ + Ac$_2$O | 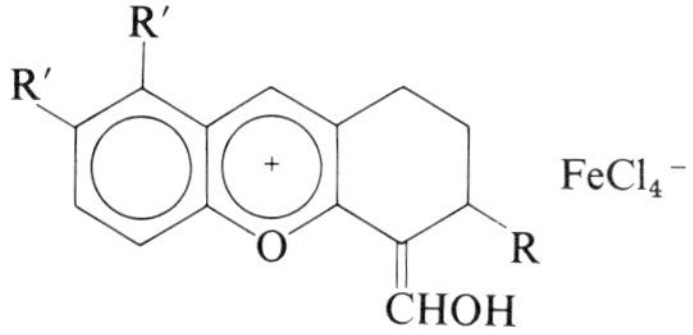 chromenylium with CHOH substituent, FeCl$_4^-$ | 543 |
| 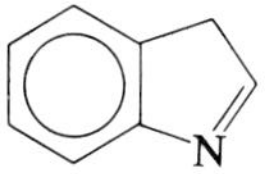 indoline | HC(OC$_2$H$_5$)$_3$ + NaOC$_2$H$_5$, then H$_2$O + HOAc | 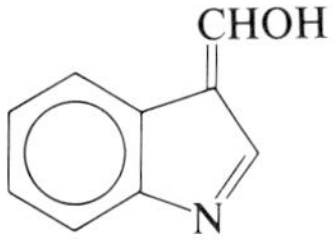 3-(hydroxymethylene)indoline (CHOH) | 489 |
| 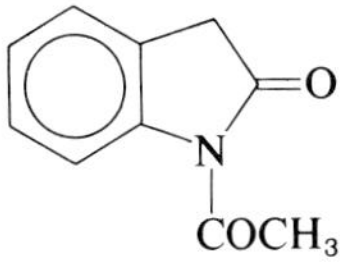 N-acetylisatin (COCH$_3$) | RC(OC$_2$H$_5$)$_3$ + Ac$_2$O (R = H, CH$_3$) | 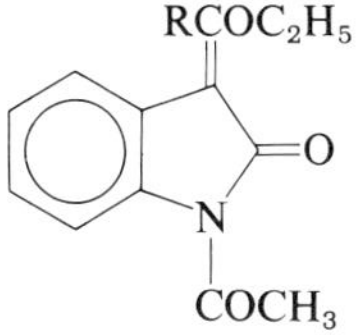 RCOC$_2$H$_5$ / N-acetyl product | 38 |
| 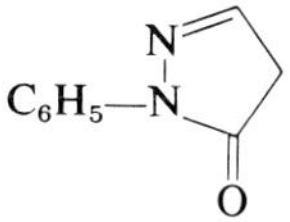 1-phenyl-3-pyrazolin-5-one (C$_6$H$_5$–N) | RC(OC$_2$H$_5$)$_3$ (R = H, CH$_3$, C$_2$H$_5$, C$_6$H$_5$) | 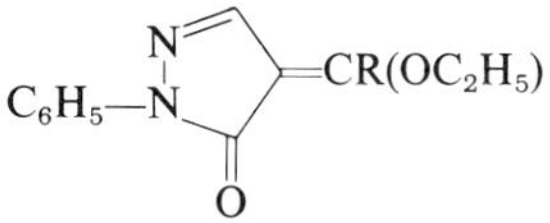 4-[CR(OC$_2$H$_5$)] pyrazolinone (C$_6$H$_5$–N) | 313 |
| 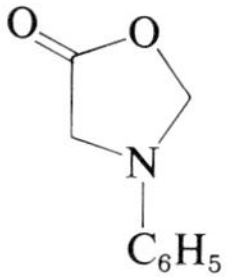 3-phenyl-oxazolidin-2-one (C$_6$H$_5$) | CH$_3$C(OC$_2$H$_5$)$_3$ + Ac$_2$O | 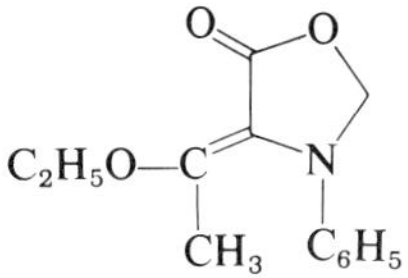 oxazolidinone with =C(CH$_3$)OC$_2$H$_5$, C$_6$H$_5$ | 346 |

TABLE II—continued

| Active methylene compound | Other reactants | Product | References |
|---|---|---|---|
| (oxazolone with C₆H₅) | RC(OC₂H₅)₃ + Ac₂O (R = H, CH₃) | (oxazolone with C₆H₅ and =C(R)OC₂H₅) | 138, 350 |
| (oxazolone with SCH₂C₆H₅) | HC(OC₂H₅)₃ + Ac₂O | (oxazolone with SCH₂C₆H₅ and =CHOC₂H₅) | 131 |
| (oxazolone with SR) | CH₃C(OC₂H₅)₃ + Ac₂O | (oxazolone with SR and =C(CH₃)OC₂H₅) | 346 |
| (rhodanine with N-R) | R'C(OC₂H₅)₃ + Ac₂O (R = C₂H₅, C₃H₅, C₆H₅, C₃H₄NS, CH₃CO₂C₂H₅; R' = H, CH₃, C₂H₅) | (rhodanine with N-R and =C(R')OC₂H₅) | 165, 346, 304, 348, 349, 353 |

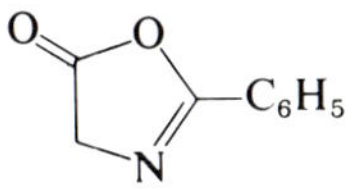

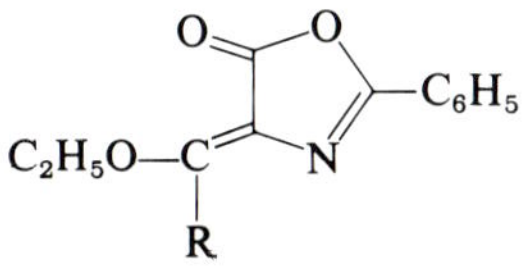

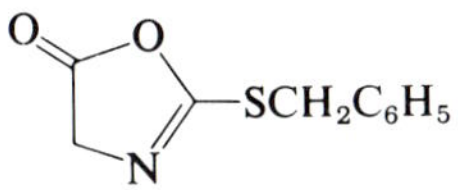

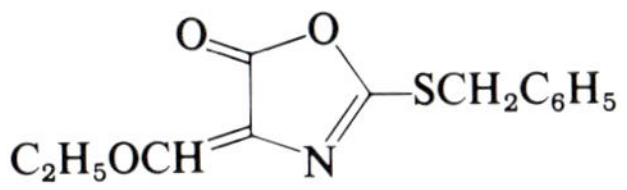

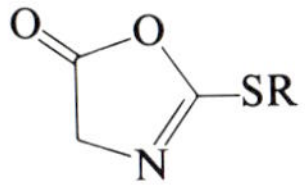

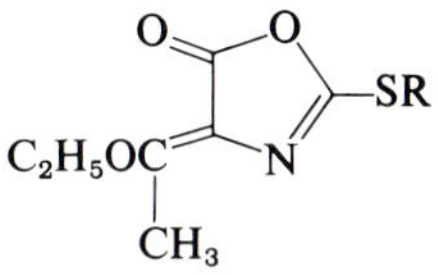

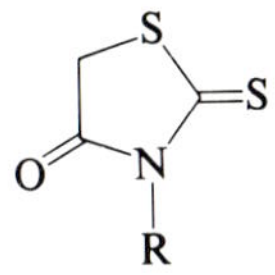

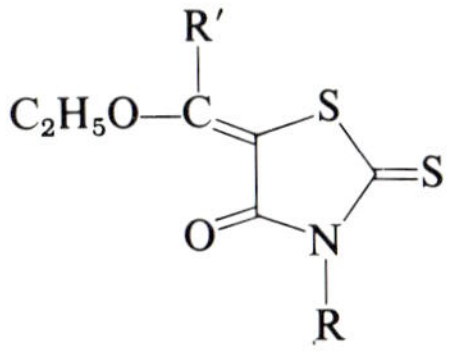

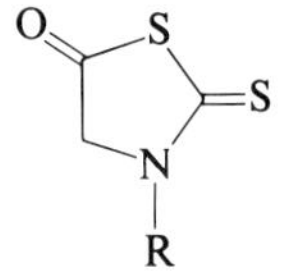

| | | |
|---|---|---|
| ![rhodanine] | HC(OR′)₃ + Ac₂O (R = H, alkyl, C₆H₅; R′ = CH₃, C₂H₅) | ![rhodanine with R′OCH=] 225, 393 |
| ![thiazolobenzimidazolone] | HC(OC₂H₅)₃ + Ac₂O | ![thiazolobenzimidazolone with =CHOC₂H₅] 163 |
| ![methoxybenzofuranone] | HC(OC₂H₅)₃ + Ac₂O | ![methoxybenzofuranone with =CHOC₂H₅] 520 |

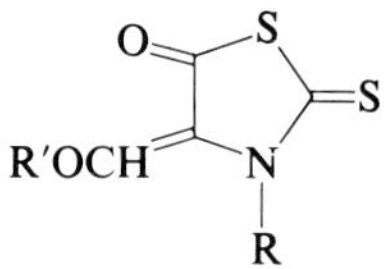

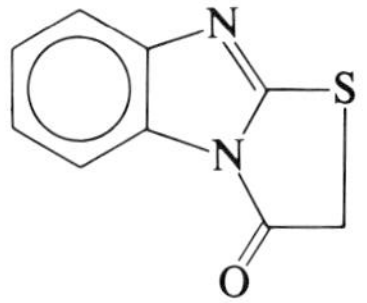

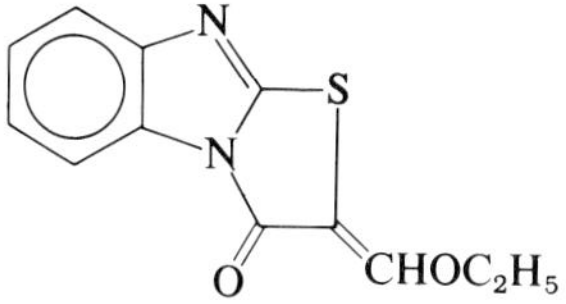

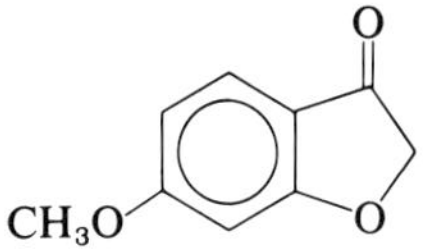

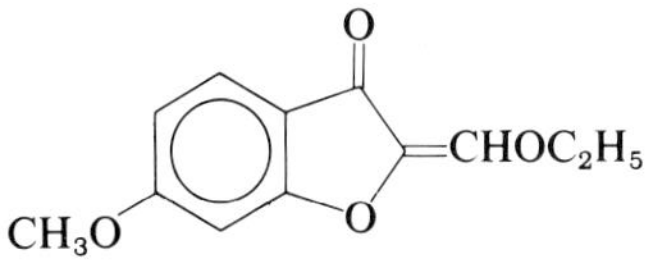

## TABLE III
### Condensation Reactions of Orthocarboxylates and Active Methylene Compounds Which Yield Heterocyclic Products

| Methylene compound | Ortho ester | Condensing agent | Product | References |
|---|---|---|---|---|
| $C_6H_5COCH_2CH_2CO_2H$ | $HC(OC_2H_5)_3$ | $DMF \cdot SO_3$ | 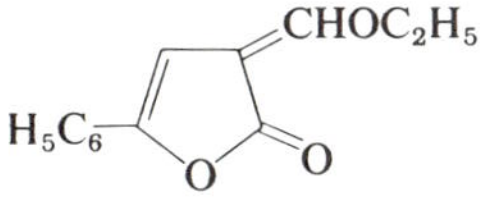 (furanone with =CHOC$_2$H$_5$, H$_5$C$_6$) | 498 |
| $RCONHCH_2CO_2H$ (R = alkyl, aryl) | $HC(OC_2H_5)_3$ | $Ac_2O$ | 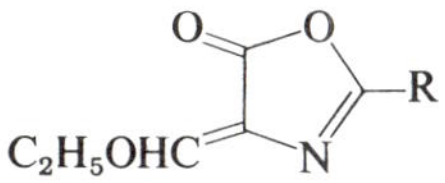 (oxazolone with R, =CHOC$_2$H$_5$) | 138 |
| $C_6H_5CONHCH_2CO_2H$ | $RC(OC_2H_5)_3$ (R = CH$_3$, C$_2$H$_5$) | $Ac_2O$ | 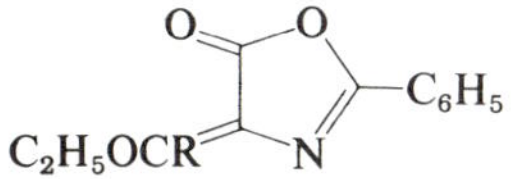 (oxazolone with C$_6$H$_5$, =CROC$_2$H$_5$) | 15, 308 |
| 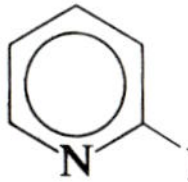 (pyridyl)NHCS$_2$CH$_2$CO$_2$H | $CH_3C(OC_2H_5)_3$ | $Ac_2O$ | 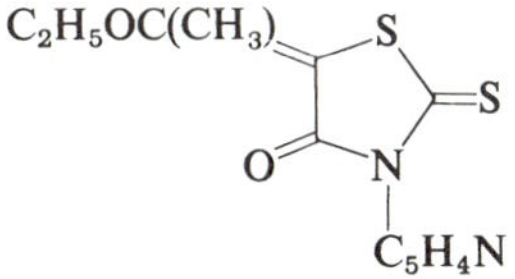 (thiazolidinone with =C(OC$_2$H$_5$)CH$_3$, N-C$_5$H$_4$N, S) | 353 |
| $HO_2CCH_2NHCS_2CO_2C_2H_5$ | $HC(OC_2H_5)_3$ | $Ac_2O$ | 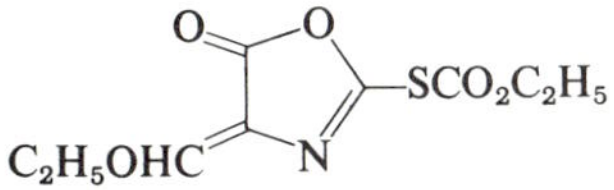 (ring with SCO$_2$C$_2$H$_5$, =CHOC$_2$H$_5$) | 131 |

| | | | |
|---|---|---|---|
| $C_6H_5CSNHCH_2CO_2H$ | $HC(OC_2H_5)_3$ | $Ac_2O$ | 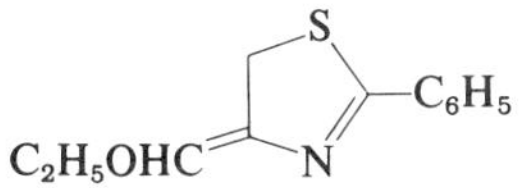 (2-phenyl-4-(ethoxymethylene)thiazolin-5-one structure), $C_6H_5$, $C_2H_5OHC$ — 290 |
| $RSCSNHCH_2CO_2H$ | $R'C(OC_2H_5)_3$ (R' = H, $CH_3$, $C_2H_5$) | $Ac_2O$ | 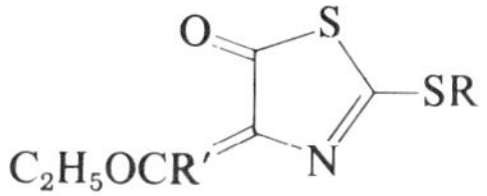 (2-SR-4-(ethoxymethylene)thiazolin-5-one structure), SR, $C_2H_5OCR'$ — 14, 132, 349–351 |
| 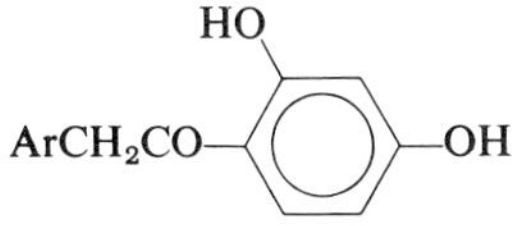 (2,4-dihydroxyphenyl-ArCH_2CO-) | $HC(OC_2H_5)_3$ | $R_3N$ | 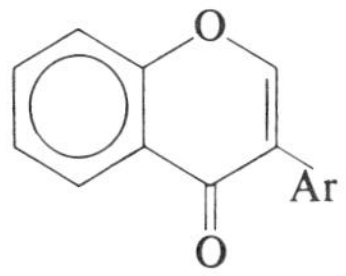 (3-Ar-chromone structure) — 302, 515 |
| 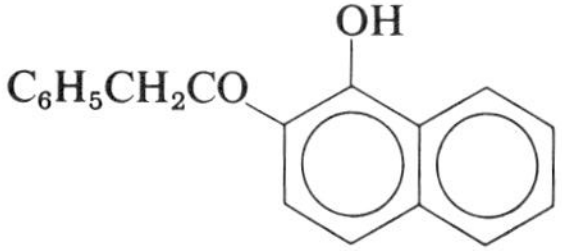 (1-hydroxy-2-(C_6H_5CH_2CO)-naphthalene) | $HC(OC_2H_5)_3$ | $R_3N$ | 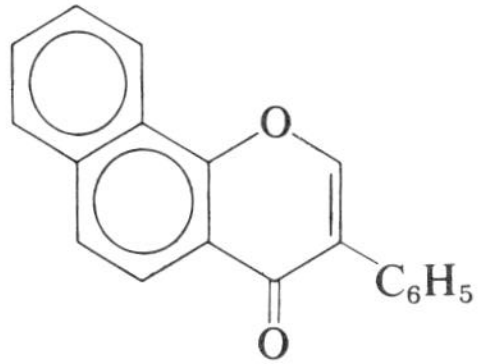 (3-phenyl-benzo[f]chromone structure) — 515 |
| 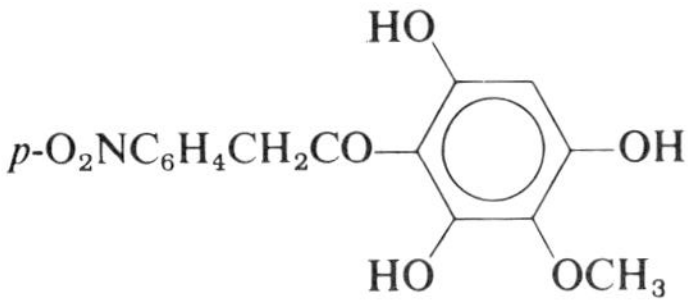 $p\text{-}O_2NC_6H_4CH_2CO$-(trihydroxy-methoxyphenyl) | $HC(OC_2H_5)$ | $R_3N$ | 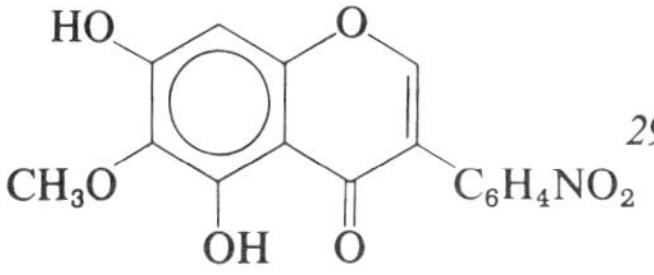 (3-(p-nitrophenyl)-5-OH-6-OCH_3-7-OH-chromone structure), $C_6H_4NO_2$ — 297 |

carbon–carbon bonds. These include compounds having methyl or methylene groups activated by positive formal charges (such as those of quaternary nitrogen atoms), and certain aromatic substances which are highly susceptible to electrophilic substitution.

The initial products of reactions of these compounds with ortho esters are alkoxyalkylene compounds, whose alkoxy groups are susceptible to displacement by nucleophilic reagents. If the nucleophilic reagent is a second molecule of the substance which reacted with the ortho ester to form the alkoxyalkylene compound, a product is formed in which two molecules of the substance have become linked by a methine or substituted methine bridge. Equation (24) represents methine bridge formation between two molecules of a substance having an activated methyl or methylene group, while Eq. (25) schematically represents formation of a methine bridge between two aromatic molecules.

$$\begin{array}{l} \diagup CH_2 + RC(OR')_3 \rightarrow \diagup C{=}CROR' + 2\ R'OH \\ \diagup C{=}CROR' + \diagup CH_2 \rightarrow \diagup C{=}CR{-}CH\diagdown + R'OH \end{array} \quad (24)$$

$$\begin{array}{l} \text{Ar–}CH + RC(OR')_3 + HA \rightarrow \text{Ar–}^+C{=}CROR'\ A^- + 2\ R'OH \\ \text{Ar–}CH + \text{Ar–}^+C{=}CROR'\ A^- \rightarrow \text{Ar–}^+C{=}CR{-}C\text{–Ar}\ A^- + R'OH \end{array} \quad (25)$$

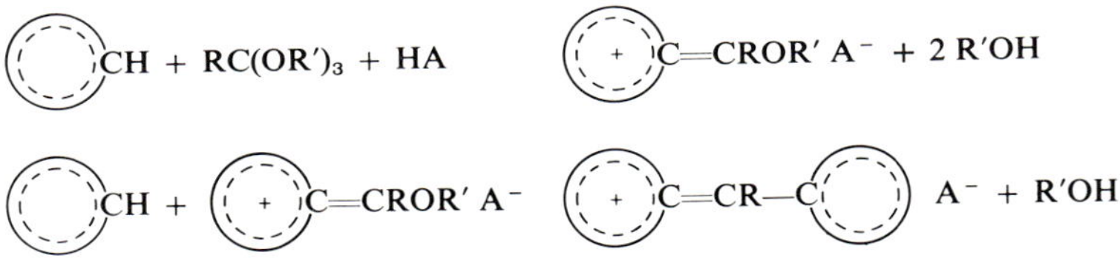

The first example of methine bridge formation by an ortho ester was described by Claisen in 1897 (*123*). He found that prolonged heating of ethyl acetoacetate with triethyl orthoformate and acetic anhydride gave a low yield of 2,4-diacetyl-2-pentenedioate [Eq. (26)]. This reaction and analogous reactions of benzoylacetonitrile (*143*) and *p*-toluylacetonitrile (*453*) are the only examples of formation of symmetrical, acyclic methine-bridged products from ortho esters and active methylene compounds.

$$2\ CH_3COCH_2CO_2C_2H_5 + HC(OC_2H_5)_3 \xrightarrow{Ac_2O} \begin{array}{c} CH_3CO{-}C{-}CO_2C_2H_5 \\ \parallel \\ CH \\ \mid \\ CH_3CO{-}CH{-}CO_2C_2H_5 \end{array} \quad (26)$$

2. CYANINE DYES

Most of the known methine bridging reactions involve coupling of cyclic or heterocyclic groups. The products of these reactions usually are colored substances having oxygen-, nitrogen-, sulfur-, or selenium-containing chromophores linked by conjugated systems of carbon–carbon double bonds. They are known collectively as cyanine dyes.

The cyanines are too unstable to be of much use as textile dyes. However, a number of them are uniquely effective at sensitizing silver halide photographic emulsions to light of longer wavelengths than the blue-violet portion of the spectrum. Modern photography—both black-and-white and color—largely owes its existence to the ability of cyanine dyes to extend the range of sensitivity of photographic emulsions. It is not surprising that many thousands of these dyes have been synthesized.

Cyanine dye chemistry, a broad and complex field, is the subject of several reviews and monographs (7, 87, 98, 158, 254, 255, 298, 403, 404). The following discussion is limited to those cyanine dye syntheses involving ortho ester condensations. Not discussed are the large number of cyanine dye syntheses which use reagents such as N,N'-diarylformamidines, N-arylimidic esters, and ethoxymethylene compounds, all of which are conveniently prepared using ortho esters as starting materials.

3. Monomethine Cyanines and Related Compounds

Many heterocyclic substances having active methylene groups, and a few cyclic and heterocyclic aromatic compounds, react with carboxylic ortho esters to form products in which two cyclic or heterocyclic nuclei are joined by a methine or substituted methine bridge. These products are classed as monomethine cyanines.

a. *Monomethineoxonols.* A variety of cyclic ketomethylene compounds react with triethyl orthoformate or diethoxymethyl acetate to form monomethineoxonol dyes.

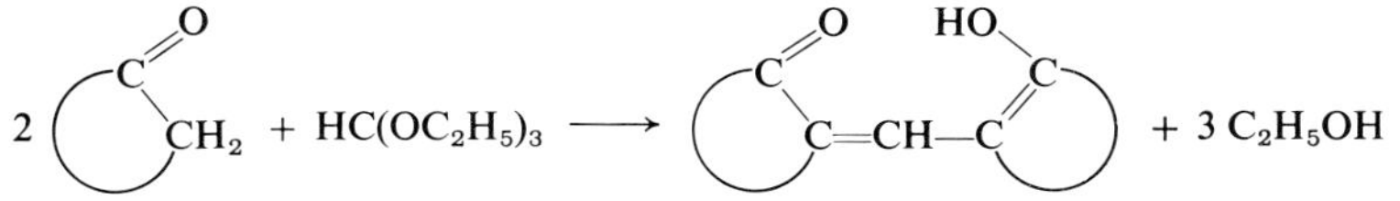

$$2 \text{(ring)}\text{CH}_2 + \text{HC}(\text{OC}_2\text{H}_5)_3 \longrightarrow \text{(ring)}\text{C}=\text{CH}-\text{C(ring)} + 3\,\text{C}_2\text{H}_5\text{OH} \quad (27)$$

These substances, which are vinylogs of carboxylic acids, are usually isolated as mesomeric salts:

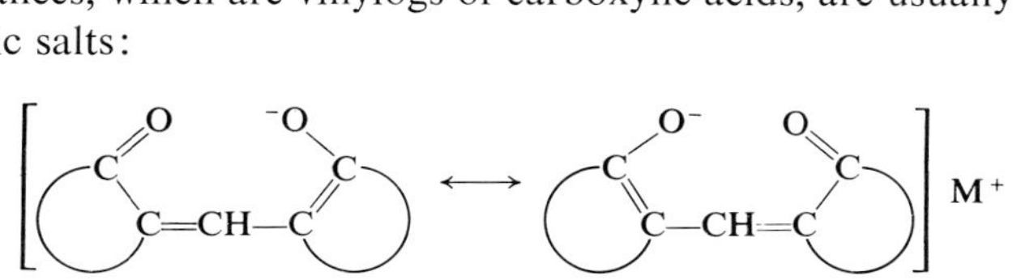

$$\left[\text{(ring)}\text{C}=\text{CH}-\text{C(ring)} \longleftrightarrow \text{(ring)}\text{C}-\text{CH}=\text{C(ring)}\right] \text{M}^+$$

The condensation reactions are carried out by heating the ketomethylene compound and the ortho ester, usually in the presence of acetic anhydride or pyridine. If diethoxymethyl acetate is used, a tertiary amine such as triethylamine is usually added to the reaction mixture, and the product is the triethylammonium oxonol salt. Dyes having ethylidene and propylidene bridges are prepared by using triethyl orthoacetate or triethyl orthopropionate in place of triethyl orthoformate (*290, 312*).

These reactions presumably involve conversion of 1 mole of the ketomethylene compound to an ethoxymethylene derivative, followed by reaction of this with a second mole of the ketomethylene compound to form the monomethine oxonol. This view receives support from the fact that unsymmetrical monomethine oxonols may be synthesized by reactions of ketomethylene compounds with ketoethoxymethylene compounds (*132, 307, 347*). In some instances the heterocyclic ketomethylene compound was synthesized *in situ* and converted to the oxonol dye without being isolated.

Heterocyclic ketomethylene compounds which have been converted to monomethineoxonol dyes [Eq. (27)] include 5,6-benzochroman-2,5-dione (*500*), 1-ethyloxindole (*143*), 1-phenyl-5-pyrazolones (*86, 143, 453*), 1-methyl-3-phenyl-5-pyrazolone (*501*), 1,2-diphenyl-3-imino-5-pyrazolidinone (*56*), 1,2-diphenylpyrazolidin-3,5-dione (*288*), 3:7a-diaza-1-indenone (*352, 585*), 3-ethyl-1-phenyl-2-thiohydantoin (*143*), 3-phenyl-5(4*H*)-isoxazolone (*86*), 2-substituted 5-thiazolones (*132, 138, 143, 290, 345*), 2,3-dihydrobenzofuran-3-one (*520*), 3-substituted rhodanines (*143, 312, 453*), and imidazopyridinone (*582*).

b. *Other Monomethine Dyes Derived from Heterocyclic Methylene Compounds.* A few cyanine dyes are derived from heterocyclic methylene compounds in which the methylene group is activated by a quaternary nitrogen atom or a carbon–nitrogen double bond. Examples are the dyes formed by reactions of orthoformates with 1-alkylpyridinium, indolium, indolizinium, benzothiazolothiazinium, and dibenzopyrrolothiazolium salts (*17, 266, 334, 489, 564a*).

c. *Monomethinecyanines Derived from Aromatic Compounds.* Triethyl orthoformate reacts with acids to yield diethoxycarbonium ion, which undergoes electrophilic substitution reactions with reactive aromatic compounds such as pyrroles, indoles, azulenes, and cyclopentadienides to form diethoxymethyl derivatives which may react with a second mole of the aromatic compound to form monomethine dyes.

Pyrroles having unsubstituted 2- or 5-positions react with triethyl orthoformate to form substituted bis(2-pyrrole)- or bis(5-pyrrole)methinecyanines. Compound **1** is one of several bis(2-pyrrole)methinecyanines prepared from pyrroles and triethyl orthoformate in the presence of acids (*26, 36, 488*).

If both the 2- and 5-positions of the pyrrole nucleus are already substituted, reaction of pyrroles with triethyl orthoformate may yield bis(3-pyrrole)-

methinecyanines. For example, 1-carboxyalkyl-2,5-dimethylpyrroles react with triethyl orthoformate in ethanolic sulfuric acid to yield **2** (*27, 334, 358*).

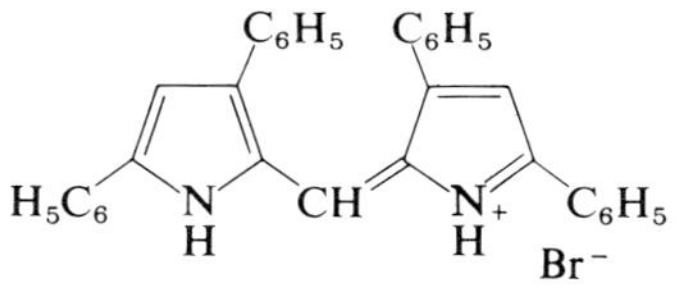

**1**

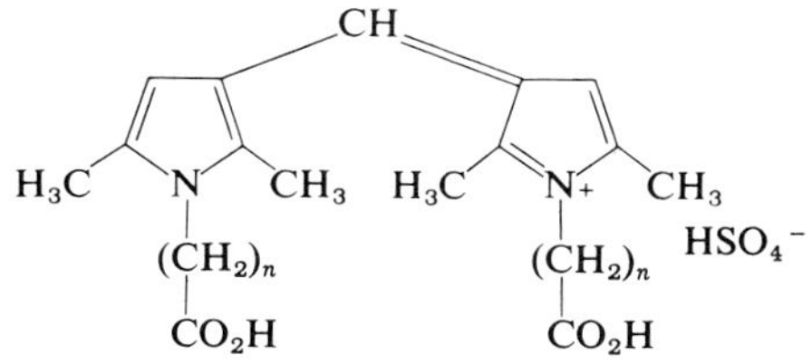

**2** (n = 1, 2)

Indoles (*363*) and pyrrocolines (*541, 564*) may be converted to monomethine dyes by treatment with triethyl orthoformate under acidic conditions. Examples are the indole dye **3** and the pyrrocoline dye **4**.

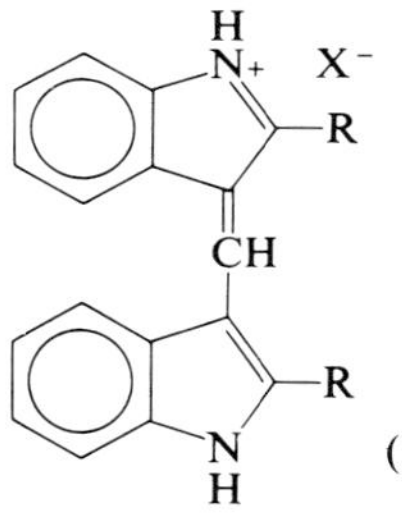

**3**  (R = H, CH$_3$)

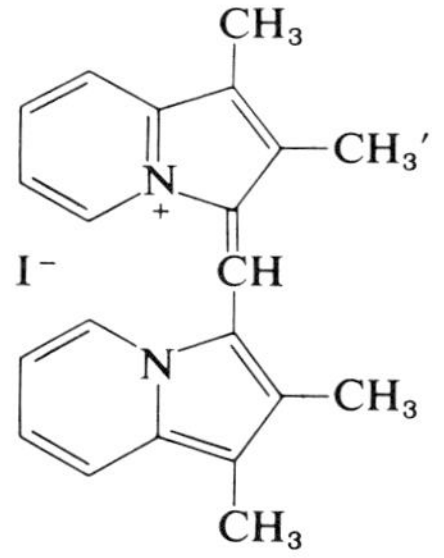

**4**

A new class of monomethine dyes, named phosphinines by their discoverers, are prepared by reaction of trialkyl orthocarboxylates with cyclopentadienylenetriphenylphosphorane (*145, 146, 438*).

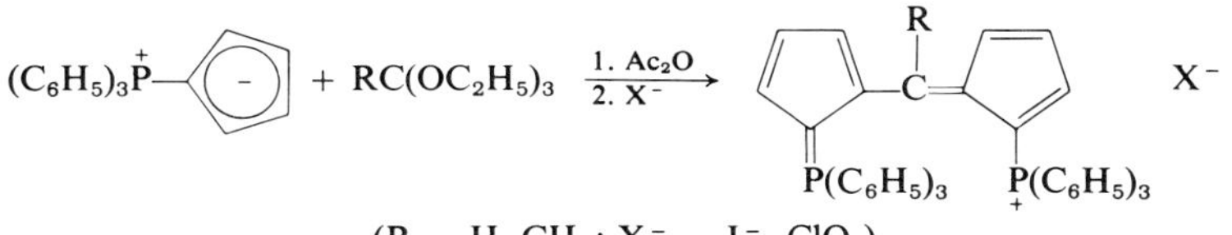

$$(C_6H_5)_3\overset{+}{P}-\underset{(-)}{\bigcirc} + RC(OC_2H_5)_3 \xrightarrow[2.\ X^-]{1.\ Ac_2O} \underset{\overset{|}{P(C_6H_5)_3}\ \ \overset{|}{\overset{+}{P}(C_6H_5)_3}}{\bigcirc\!\!-\!\!\underset{|}{\overset{R}{C}}\!\!=\!\!\bigcirc} \quad X^- \quad (28)$$

(R = H, CH$_3$; X$^-$ = I$^-$, ClO$_4$)

Azulene and 1-methylazulene condense with triethyl orthoformate in the presence of strong acids to form the only known examples of methine dyes in which the dye cation consists only of carbon and hydrogen [Eq. (29)] (*340, 540*). More highly alkylated azulenes yield ethoxymethylene derivatives rather than monomethinecyanines (*340*).

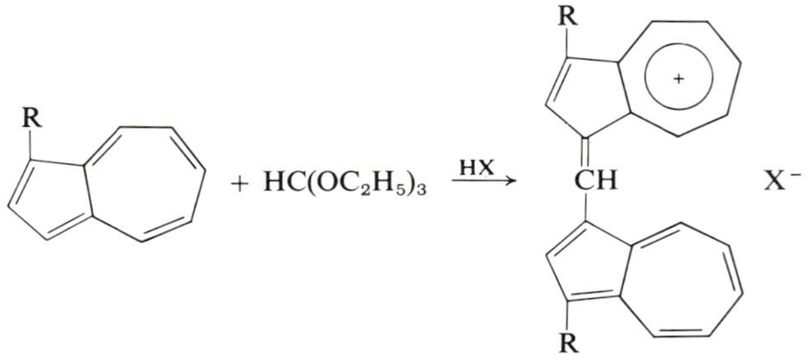

248    4. ORTHO ESTER REACTIONS FORMING C–C OR C–H BONDS

$$\text{[azulene with R]} + HC(OC_2H_5)_3 \xrightarrow{HX} \text{[bis-azulenyl methine cation]} \quad X^- \qquad (29)$$

(R = H, CH$_3$; X = I$^-$, ClO$_4^-$)

Bis(2,6-diphenyl-4-thiapyrolo)monomethinecyanine perchlorate was synthesized, along with a symmetrical trimethine dye, by treating 4-methyl-2,6-diphenylthiapyrylium perchlorate and 2,6-diphenylthiapyrone with triethyl orthoformate *(615)*.

### 4. MEROCYANINE AND OTHER DIMETHINE DYES

a. *Merocyanines.* Merocyanine dyes have a heterocyclic system containing a quaternary nitrogen atom linked to a carbonyl group by a conjugated system of carbon–carbon double bonds. The carbonyl portion of the molecule is usually derived from a cyclic ketomethylene compound. Structurally, they are hybrids of true cyanines and oxonols. They differ from the cyanines and oxonols in that the mesomeric chromophore bears no net formal charge. The simplest merocyanines are linked by a bridge consisting of two methine groups, and may be represented as resonance hybrids of the following generalized structures:

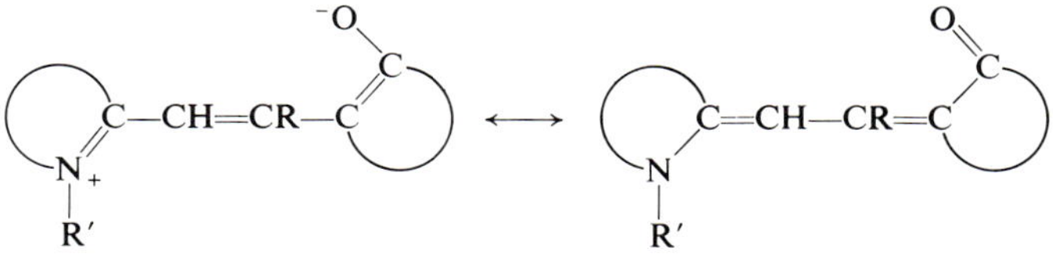

$$\text{[resonance structures of merocyanine dye]}$$

One important synthesis of merocyanines involves reaction of methyl-substituted heterocyclic quaternary salts with ethoxymethylene compounds. In many instances it is not necessary to isolate the ethoxymethylene compound: the merocyanine is prepared by simply heating a mixture of the quaternary salt, the ketomethylene compound, the ortho ester, and an amine or acetic anhydride.

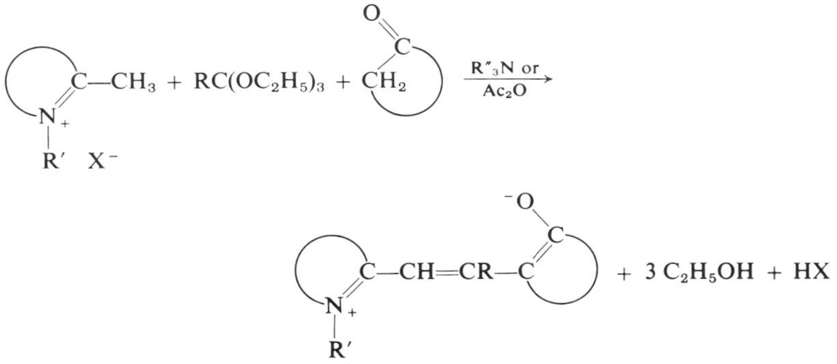

# REACTIONS WITH COMPOUNDS HAVING ACTIVATED C–H BONDS 249

$$\text{Het-C(CH}_3\text{)=N}^+\text{R'} \ X^- + RC(OC_2H_5)_3 + CH_2(CO-) \xrightarrow{R''_3N \text{ or } Ac_2O}$$

$$\text{Het-C(N}^+\text{R')-CH=CR-C(-O}^-\text{)=} + 3\,C_2H_5OH + HX \quad (30)$$

Merocyanine dyes synthesized according to Eq. (30) were prepared by reactions of ortho esters with mixtures of 2-methylquinolinium, 2-methylindolenium, 2-methylbenzoxazolium, or 2-methylbenzothiazolium salts and 1-indanone, 1,3-indandione, substituted pyrazolones, substituted thiazolones or substituted oxazolones (*81, 82, 91, 92, 161, 201–203, 305, 381, 382, 559, 586, 639*).

1-Ethyl-5-[6-(3-ethyl-2-benzothiazolinylidene)-3,5-neopentylene-2,4-hexadienylidene]-2-thiobarbituric acid, an oxonol dye having a six-carbon unsaturated chain linking the quaternary salt and ketomethylene portions of the chromophore, was synthesized as shown in Eq. (31) (*265*).

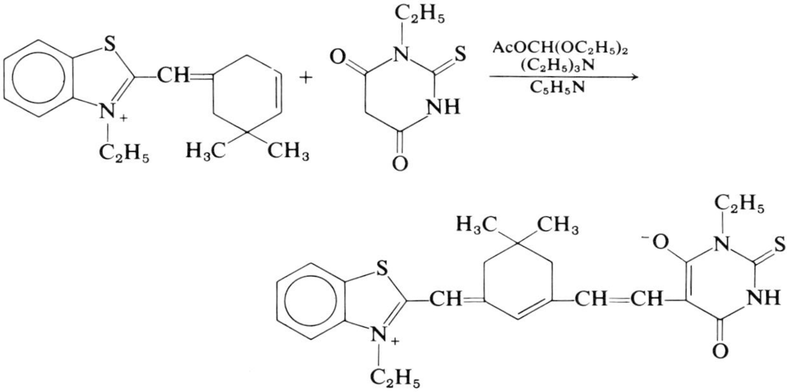

$$\text{[benzothiazolium-CH=cyclohexenyl(gem-diMe)]} + \text{[1-ethyl-2-thiobarbituric acid]} \xrightarrow[C_5H_5N]{AcOCH(OC_2H_5)_2 \ (C_2H_5)_3N}$$

$$\text{[benzothiazolium-CH=cyclohexenyl(gem-diMe)-CH=CH-thiobarbiturate]} \quad (31)$$

Thia analogues of merocyanine dyes have been prepared by heating 2-methylbenzoxazolium and 2-methylbenzothiazolium salts with pyrazolethiones and indolenthiones plus an ortho ester in pyridine (*198, 199, 311*).

b. *Other Dimethine Dyes.* A number of dyes having two heterocyclic nuclei linked by methyl-substituted dimethine bridges have been prepared from

2-methylbenzoxazolium or 2-methylbenzothiazolium salts, triethyl orthoacetate, and substituted pyrroles, indoles, or pyrrocolines (*105, 263, 264*). Dyes 5–7 are examples. Although these substances are classified here as dimethine dyes, they have chromophores closely related to those of the trimethinecyanines.

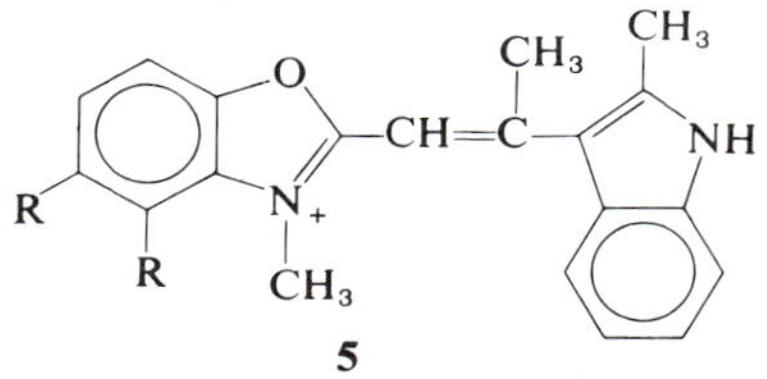

Substituted 2-methylquinolinium salts, *p*-amidobenzenesulfonamides, and triethyl orthoformate react to form quinohemicyanine dyes 8 (*485*).

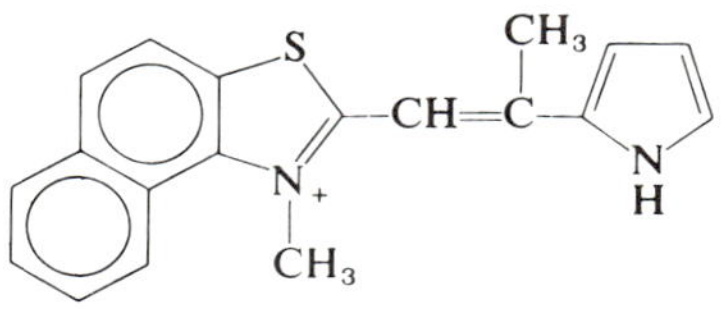

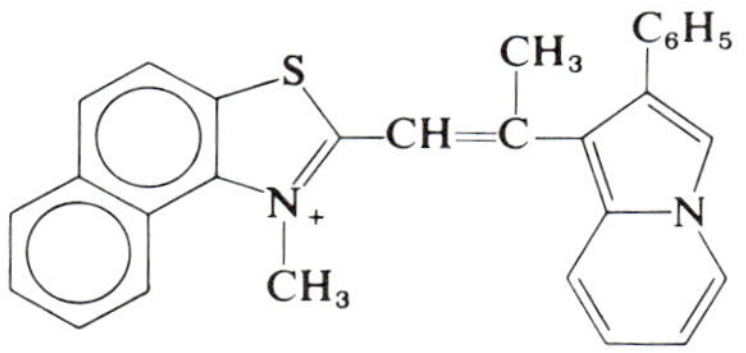

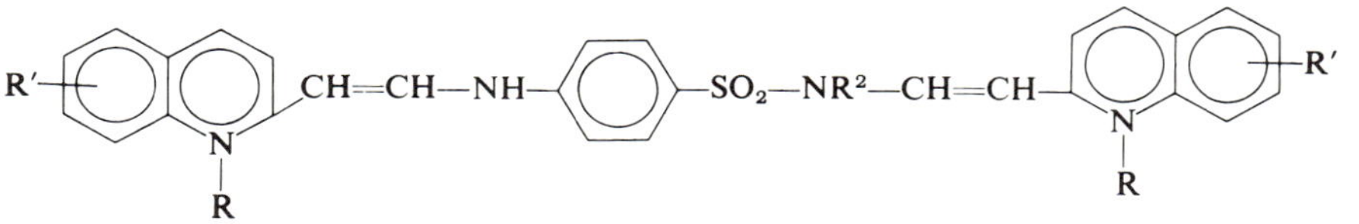

## 5. Symmetrical Trimethinecyanine Dyes

a. *Introduction.* Symmetrical trimethinecyanines (carbocyanines) are the largest and most important group of cyanine dyes. They possess chromophores in which two aromatic systems are linked by a chain of three methine or substituted methine groups. The aromatic residues are usually derived from alkyl-substituted heterocyclic quaternary salts.

The most generally applicable and widely used synthesis of symmetrical trimethinecyanines involves condensation of 2 moles of an alkyl-substituted heterocyclic or aromatic "onium" salt with 1 mole of a trialkyl orthocarboxylate. This reaction was first used by Konig in *1922* for the synthesis of bis(1-alkylquinoline-2)trimethinecyanine iodides (2,2'-carbocyanine iodides, 9) from 1-alkylquinaldinium iodides, triethyl orthoformate, and acetic anhydride (*364*).

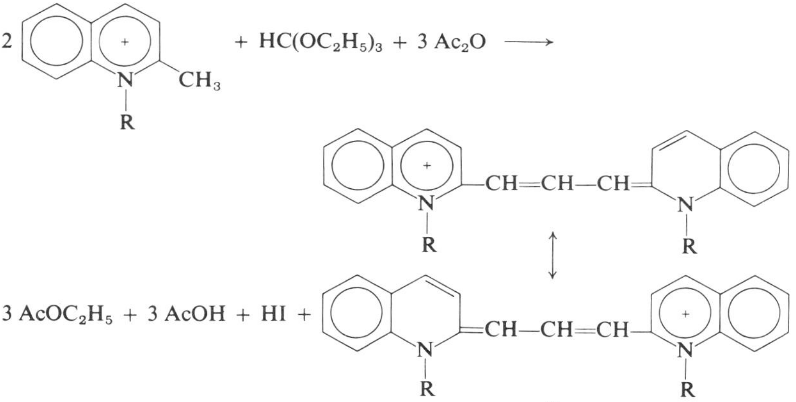

## REACTIONS WITH COMPOUNDS HAVING ACTIVATED C–H BONDS 251

$$2 \text{ [quinolinium-CH}_3\text{]} + \text{HC(OC}_2\text{H}_5)_3 + 3 \text{ Ac}_2\text{O} \longrightarrow$$

$$3 \text{ AcOC}_2\text{H}_5 + 3 \text{ AcOH} + \text{HI} + \text{[bis-quinolinium trimethine dye]} \quad (32)$$

**9**

Reactions of alkyl-substituted quaternary salts with ortho esters sometimes produce trinuclear pentamethine dyes as well as symmetrical trimethine dyes. These reactions are discussed in Section IV, B, 8.

Konig's ortho ester condensation proved to be generally applicable to methyl-substituted quaternary salts. In the period between 1922 and 1928, he used triethyl orthoformate to synthesize 4,4'-carbocyanines (**10**) (*366*), indocarbocyanines (**11**) (*365, 367*), oxacarbocyanines (**12**) (*369*), and thiacarbocyanines (**13**) (*368, 369*) from 1-alkyllepidinium, 1,2,3,3-tetramethylindoleninium, 2,3-dimethylbenzoxazolium, and 2,3-dimethylbenzothiazolium salts.

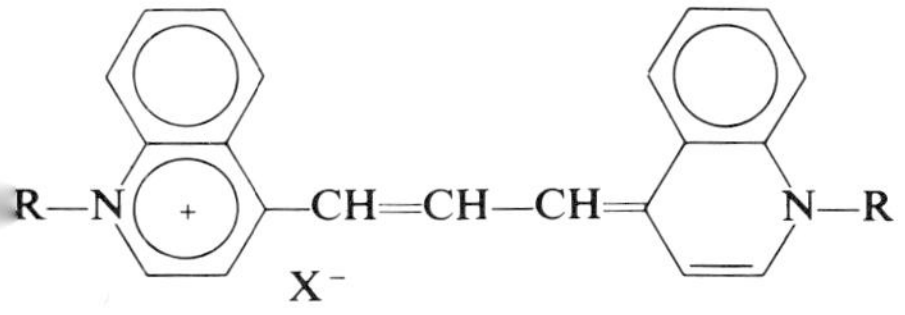

**10**

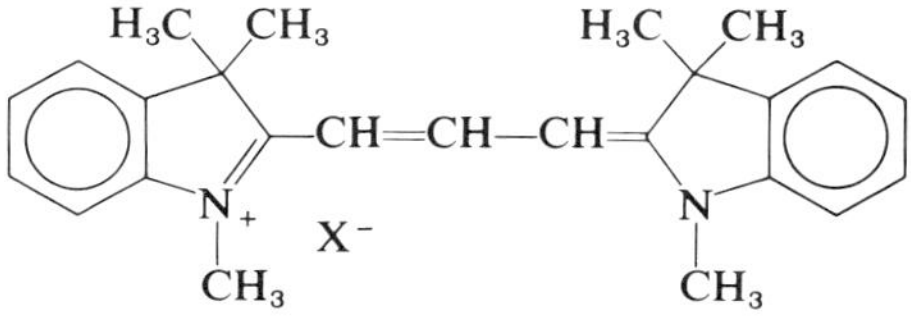

**11**

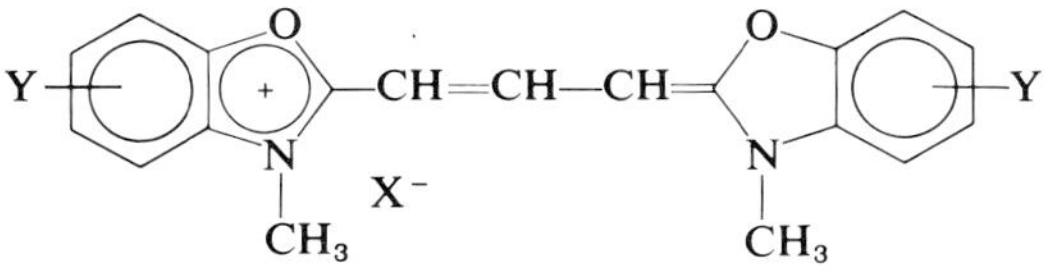

**12**

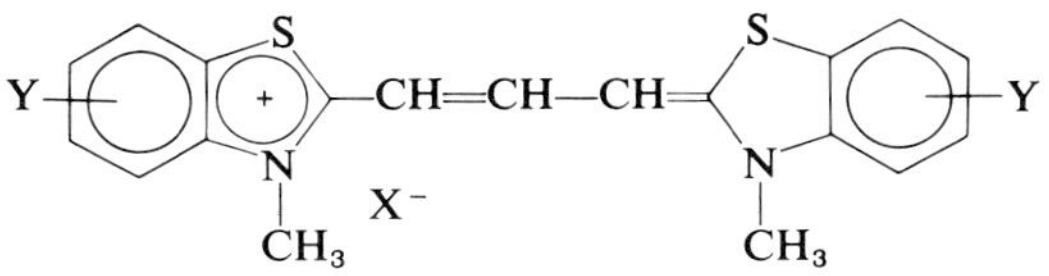

**13**

b. *Scope of the Reaction.* Most of the trimethine dyes synthesized by the ortho ester method were prepared from 1,2- and 1,4-disubstituted pyridinium salts, 1,2-disubstituted oxazolium, thiazolium, selenazolium, and imidazolium salts, and their benzo and other derivatives. These reactions are generalized in Eqs. (33) and (34).

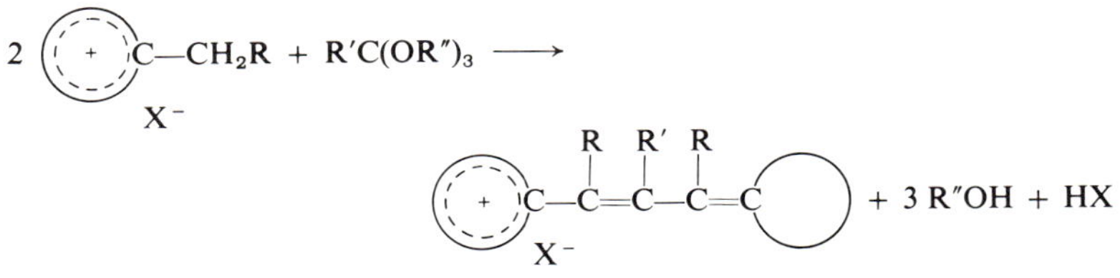

$$2 \; \text{[Het]}\!\!-\!\!\text{C}\!-\!\text{CH}_2\text{R} + \text{R}'\text{C}(\text{OR}'')_3 \longrightarrow$$
$$\text{X}^-$$

$$\text{[Het]}\!\!-\!\!\overset{\text{R}}{\text{C}}\!=\!\overset{\text{R}'}{\text{C}}\!-\!\overset{\text{R}}{\text{C}}\!=\!\text{C}\!-\![\text{Het}] + 3\,\text{R}''\text{OH} + \text{HX} \quad (33)$$
$$\text{X}^-$$

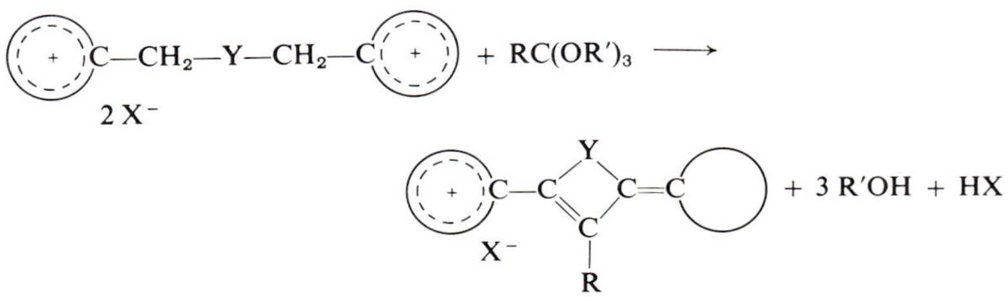

$$\text{[Het]}\!\!-\!\!\text{C}\!-\!\text{CH}_2\!-\!\text{Y}\!-\!\text{CH}_2\!-\!\text{C}\!-\![\text{Het}] + \text{RC}(\text{OR}')_3 \longrightarrow$$
$$2\,\text{X}^-$$

$$\text{[Het]}\!\!-\!\!\text{C}\!-\!\text{C}\overset{\text{Y}}{\underset{\underset{\text{R}}{\text{C}}}{\diagup\!\!\!\diagdown}}\text{C}\!=\!\text{C}\!-\![\text{Het}] + 3\,\text{R}'\text{OH} + \text{HX} \quad (34)$$
$$\text{X}^-$$

The structure of the trimethine chain of trimethinecyanines is determined by the alkyl substituent of the onium salt and by the acyl substituent of the ortho ester.

Dyes synthesized from methyl-substituted onium salts have trimethine chains without terminal substituents [Eq. (33), R = H]. Other primary alkyl-substituted salts yield dyes with substituents on the terminal carbons of the trimethine chain [Eq. (33), R = alkyl, aryl, alkoxy]. If two cationic residues are linked by a bridge of the type $-\text{CH}_2-\text{Y}-\text{CH}_2-$, the dye will have a trimethine system which is part of a cyclic structure [Eq. (34)]. There are also trimethine dyes in which the terminal carbons of the trimethine bridge are connected by bridges to the heterocyclic portions of the molecule.

Trimethine dyes synthesized from triethyl orthoformate or diethoxymethyl acetate have unsubstituted central methine carbons [Eq. (33), R' = H]. Higher ortho esters yield meso-substituted trimethinecyanines with substituents on the central methine carbon atom [Eq. (33), R' = alkyl, aryl, alkoxy].

c. *Reaction Conditions.* Trimethinecyanine dyes are usually synthesized by heating a mixture of the alkyl-substituted onium salt, the ortho ester, a solvent, and sometimes a catalyst.

Konig used acetic anhydride or nitrobenzene as solvents in most of his trimethinecyanine syntheses. The fact that a strong acid is a by-product of the reaction [Eq. (34)] led Hamer to use the basic solvent pyridine. In 1927 she described improved procedures for the preparation of 2,2'-, 4,4'-, thia-, oxa-, and indocarbocyanines from the appropriate quaternary salts and a threefold to fourfold excess of triethyl orthoformate in pyridine solution (*250*). The use of pyridine resulted in markedly improved yields, and pyridine is still usually the solvent of choice in syntheses of trimethinecyanines from quaternary salts and ortho esters.

1,3-Disubstituted 2-methylbenzimidazolium salts give poor yields of bis(1,3-disubstituted benzimidazole-2)trimethinecyanines in reactions with triethyl orthoformate in pyridine solution, but afford satisfactory yields when nitrobenzene is used as the solvent (*19, 20, 215, 230–233, 430, 511, 542, 583, 622, 623, 627–629, 648*). 2-Methyl-3-ethyl-4,5-(2',3'-thionaphtheno)-thiazole also yields a dye when heated with triethyl orthoformate in nitrobenzene solution, but not in pyridine solution (*414*). Many other solvent systems have been used in syntheses of trimethinecyanine dyes from ortho esters, but in most instances there is no published evidence that these solvents are superior to pyridine (see, however, ref. *619*).

Some heterocyclic quaternary salts which fail to react with triethyl orthoformate in pyridine (e.g., 1-ethyl-2-methylpyridinium iodide and 1-methyl-2-alkylisoquinolinium iodides) are converted to trimethinecyanines by diethoxymethyl acetate in the presence of pyridine or triethylamine (*144*). The reagent, which is prepared by reaction of triethyl orthoformate with acetic anhydride (*490*), has also been used to convert other quaternary salts to symmetrical trimethine dyes (*77–80, 83, 84, 90, 197, 266, 287, 359, 360, 361, 584*). Symmetrical trimethinecyanines have also been prepared by reaction of pyridinium (*452, 454, 466, 467, 475, 643*), indoleninium (*234*), oxazolium (*3, 100, 366, 442, 524, 555*), thiazolium (*6, 193, 211, 291, 300, 324, 325, 328, 330, 333, 335, 336, 359, 369, 440, 548, 551–553, 555, 567, 568, 571–573, 577, 585, 620, 621, 624, 625, 641, 649*), diazinium (*16, 222, 256, 378, 379, 510*), pyrazolopyrimidinium (*503*), and thiadiazolium (*226, 227, 237*) salts with 1,1-dialkoxyalkyl acetates formed *in situ* from trialkyl orthocarboxylates and acetic anhydride. A mixture of acetyl chloride and triethyl orthoformate was used to prepare a bis(1-substituted imidazoline-2)trimethinecyanine dye (*440*).

d. *Mechanism of Formation of Trimethinecyanines from Ortho Esters and Alkyl-Substituted Quaternary Salts.* The media in which trimethinecyanines are formed always include substances which are acids and bases in the Brönsted sense. The acid may be the conjugate acid of an amine, a carboxylic acid, or a phenol; the base is usually a tertiary amine, although acetic anhydride and nitrobenzene may function as Brönsted bases.

The probable first step in trimethinecyanine formation is abstraction of a proton from the alkyl-substituted quaternary salt to form a methylene base [Eq. (35)] (*76a*).

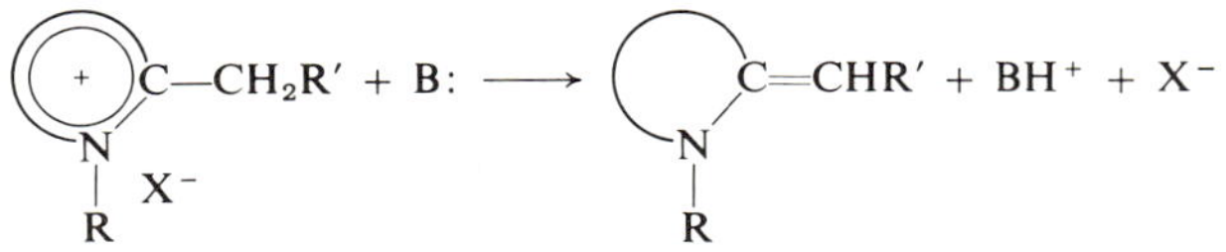

$$\text{(ring)}\overset{+}{\text{N}}\text{C}-\text{CH}_2\text{R}' + \text{B}: \longrightarrow \text{(ring)}\text{NC}=\text{CHR}' + \text{BH}^+ + \text{X}^- \quad (35)$$

There is indirect evidence for this proton abstraction in the case of quinaldinium (*406, 407*) and benzothiazolium (*380, 405*) salts, and such methylene bases have been synthesized independently (*380, 420*). Physical and chemical evidence suggests that methylene bases derived from 2-methyl-3-alkylbenzothiazolium salts exist as unsymmetrical dimers (*380, 598*) [Eq. (36)]. Larive and Dennilauler suggest that in thiacarbocyanine syntheses, the dimeric form of the methylene base is an intermediate in trimethinecyanine formation (*380*). Formation of the observed products can be accounted for more simply in terms of the monomeric methylene base, as is shown in Eq. (39).

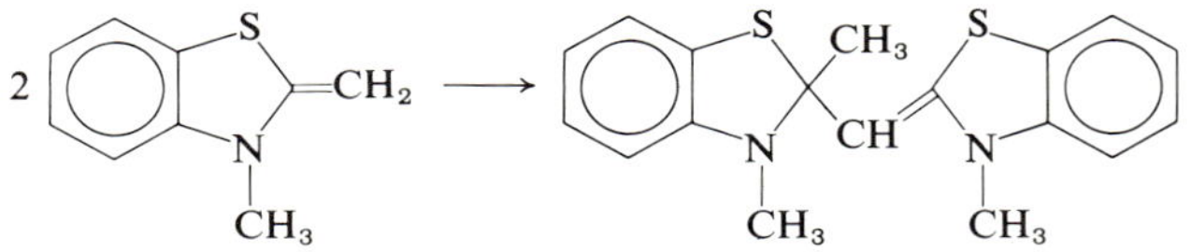

$$2 \;\text{(benzothiazoline)}=CH_2 \longrightarrow \text{(dimer structure with CH}_3\text{ and CH)} \quad (36)$$

Subsequent steps involve formation of a dialkoxycarbonium salt from the ortho ester, and reaction of this with the methylene base to form a 2-(2-alkoxy)vinyl derivative of the quaternary salt [Eq. (37)].

$$\text{R}''\text{C}(\text{OC}_2\text{H}_5)_3 + \text{BH}^+ \longrightarrow \text{R}''\overset{+}{\text{C}}(\text{OC}_2\text{H}_5)_2 + \text{B} + \text{C}_2\text{H}_5\text{OH}$$

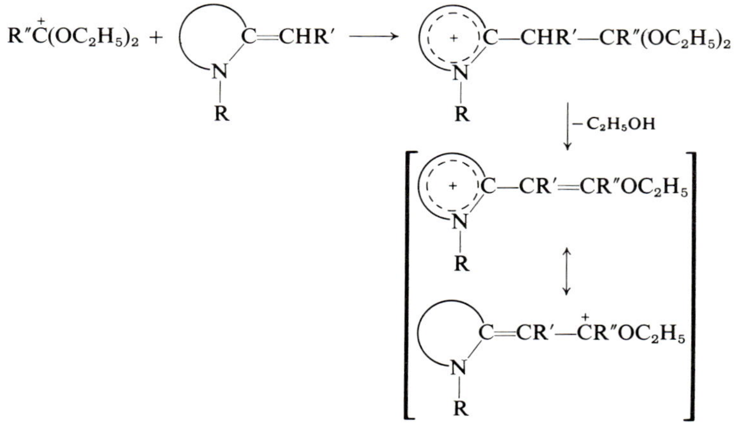

$$\text{R}''\overset{+}{\text{C}}(\text{OC}_2\text{H}_5)_2 + \text{(ring)NC}=\text{CHR}' \longrightarrow \overset{+}{\text{(ring)}}\text{NC}-\text{CHR}'-\text{CR}''(\text{OC}_2\text{H}_5)_2$$

$$\Big\downarrow -\text{C}_2\text{H}_5\text{OH}$$

$$\left[ \overset{+}{\text{(ring)}}\text{NC}-\text{CR}'=\text{CR}''\text{OC}_2\text{H}_5 \;\updownarrow\; \text{(ring)NC}=\text{CR}'-\overset{+}{\text{CR}''}\text{OC}_2\text{H}_5 \right] \quad (37)$$

REACTIONS WITH COMPOUNDS HAVING ACTIVATED C–H BONDS 255

The fact that 2-(2-alkoxy)vinyl derivatives of benzoxazoles, benzothiazoles, and benzoselenazoles can be synthesized and isolated (*21, 22, 75a, 383*) supports the proposed mechanism.

The final steps in trimethinecyanine formation probably involve electrophilic addition of the 2-(2-alkoxyvinyl) quaternary salt to the exocyclic methylene group of the methylene base or its dimer, followed by loss of alcohol [Eq. (38)]. Synthesis of an unsymmetrical trimethinecyanine dye in good yield from a methylene base and an ethoxyvinyl quaternary salt provides support for this sequence of events [Eq. (39)] (*380*).

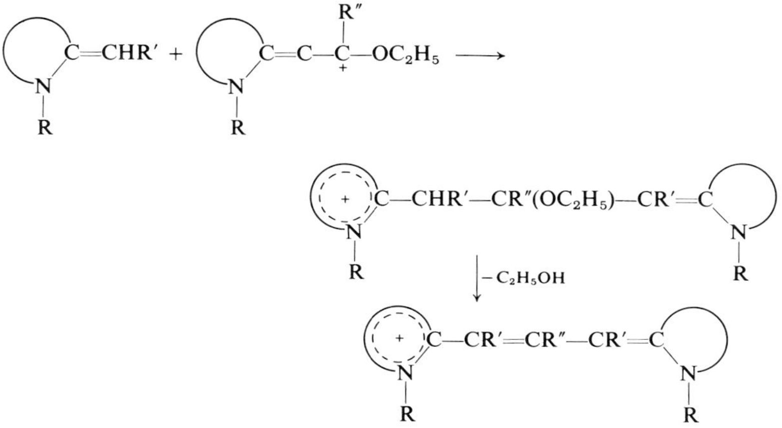

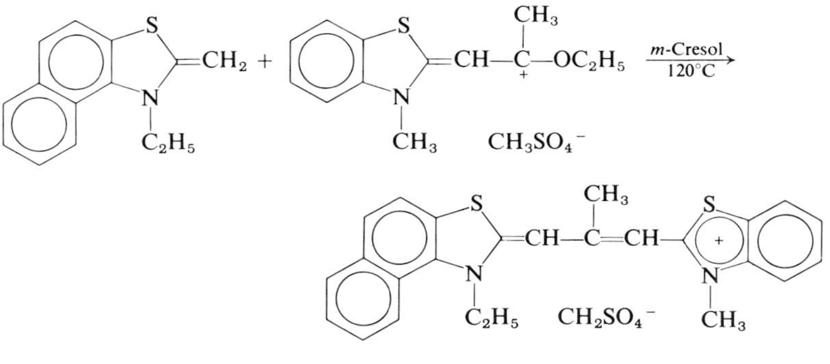

*e. Examples of Symmetrical Trimethinecyanines Synthesized from Ortho Ester and Onium Salts*

i. *Symmetrical bis(pyridine-2)trimethinecyanines, bis(quinoline-2)trimethinecyanines, and related compounds.* Trimethinecyanines derived from 2-alkylpyridinium, 2-alkylquinolinium, 2-alkylisoquinolinium, 2-alkylbenzoquinolinium, and related heterocyclic quaternary salts have the general formula **14**, where the aryl substituents may represent fused ring systems. These dyes include derivatives of pyridine (*144*), thiapyranopyridine (*642a*), benzothiapyranopyridine (*1*), quinoline (*4, 85, 144, 250, 256, 322, 364, 374, 416, 445, 454, 462, 463, 463a, 465, 466, 472–474, 476–478, 480, 483, 484, 506, 507, 572, 637, 638*), isoquinoline (*89, 97, 144, 606, 607*), benzoquinoline (*109–111, 372, 462, 463, 463a, 464, 475, 476, 478, 479, 481, 482, 507*), phenanthridine (*81, 384*), phenanthroline (*452*), 1,8-naphthyridine (*454*), and miscellaneous quinoline derivatives (*1, 32, 78, 140a*).

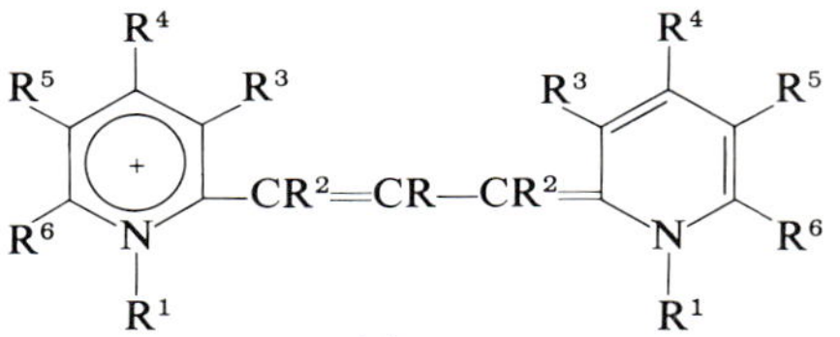

**14**

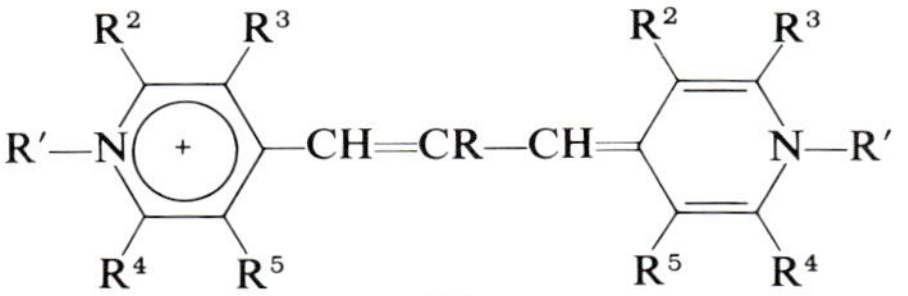

**15**

ii. *Symmetrical bis(pyridine-4)trimethinecyanines and related compounds.* Trimethinecyanine dyes derived from 4-alkylpyridinium, 4-alkylquinolinium, thienopyridinium, and related quaternary salts have the general formula **15**, where the aryl substituents may represent fused ring systems. These dyes include derivatives of pyridine (*144*), quinoline (*65, 77, 80, 144, 250, 251, 269, 399, 410, 450, 468–471*), benzoquinoline (*1a, 112, 410, 467*), acridine (*546*), thienopyridine (*642, 643*), thianaphthenopyridine (*1a, 644*), pyridophenothiazine (*77, 79*), and miscellaneous polycyclic quinoline derivatives (*303*).

iii. *Symmetrical trimethinecyanines derived from 2-methyloxazolinium and 2-methyloxazolium salts.* Apparently the only bis(oxazoline-2)trimethine dye reported thus far is compound **16**, prepared from 2,4-dimethyl-3-ethyl-

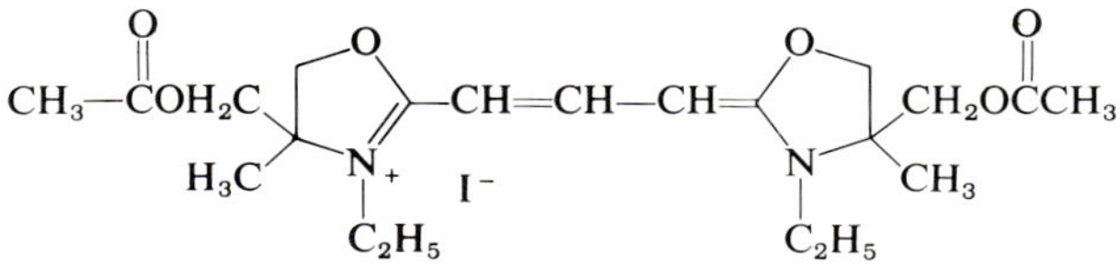

**16**

4-acetoxymethyloxazolinium iodide and triethyl orthoformate plus acetic anhydride (*440*, *442*). Bis(oxazole-2)trimethinecyanines have been prepared from 4-phenyl- and 4,5-diphenyl-2-methyl-3-alkyloxazolinium iodides and triethyl orthoformate plus pyridine and acetic anhydride (*555*).

iv. *Symmetrical trimethine dyes derived from benzoxazolium salts and related compounds.* Symmetrical trimethinecyanines derived from 2-alkylbenzoxazolium salts have the general formula **17**, where the aryl substituents may

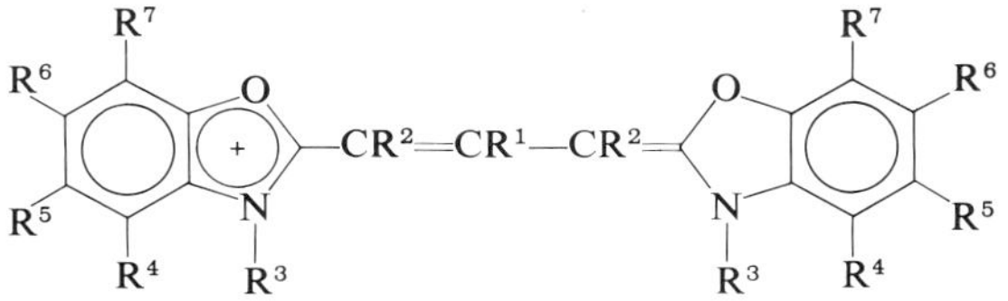

**17**

represent fused ring systems. These dyes include derivatives of benzoxazole (*3, 9, 12, 20, 31, 66, 72, 83, 93, 94, 96, 99, 100, 144, 169, 170, 185–187, 188, 190, 192, 197b, 205, 206, 236, 239, 240, 250, 270, 271, 291, 359, 369, 439, 443, 477, 524, 562, 585, 593–595*), phenanthroxazole (*168*), dioxolobenzoxazole (*35*), benzoxazolothiadiazole (*8, 140a*), and bis(benzoxazole) (*229*).

v. *Symmetrical bis(thiazoline-2)trimethinecyanines and related compounds.* Dyes having the general formula **18** were prepared from orthocarboxylates

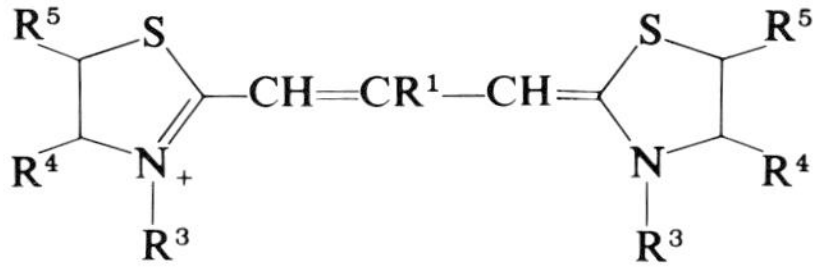

**18**

and 2-alkylthiazolinium salts. These include derivatives of thiazoline (*36, 63, 73–75, 129, 162, 370, 441, 558, 594*), hexahydrobenzothiazole (*86*), and 4,9-diketotetrahydronaphtho[2,3-*d*]thiazoline (*324*).

vi. *Symmetrical bis(thiazole-2)trimethinecyanines and related compounds.* Dyes having the general formula **19** were synthesized from 2-alkylthiazolium

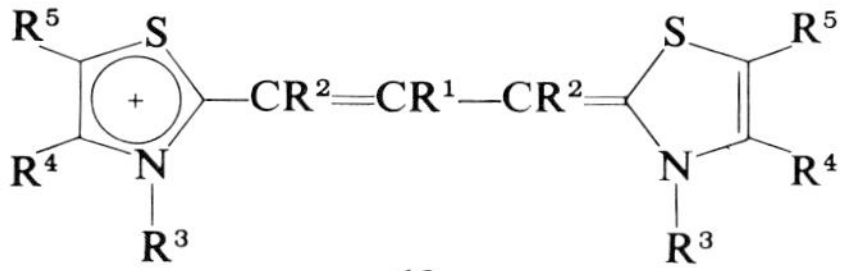

**19**

and related quaternary salts. These include derivatives of thiazole (*31, 36, 58, 68, 70, 88, 95, 141, 144, 151, 167, 211a, 224, 289, 317, 370, 409, 443, 549, 550, 552, 553, 555, 556, 569, 594, 595, 646*), 4,5-polymethylenethiazoles (*31, 88, 95, 147–149*), dihydronaphthothiazoles (*150, 554*), indenothiazole (*557*), thienothiazoles (*641, 645*), pyranothiazole (*536*), thiapyranothiazoles (*536*), thiazoloquinolines (*155*), thiazoloazulene (*211*), pyridothiazole (*559a*), and pyrazolothiazole (*416e*).

vii. *Symmetrical trimethinecyanine dyes derived from benzothiazolium and related salts.* The largest group of symmetrical trimethine dyes is that derived from 2-alkylbenzothiazolium salts and more complex quaternary salts related to them. Dyes of this structural class have the general formula **20**,

[Structure **20**: bis(benzothiazolium) trimethine dye with substituents $R^1$–$R^7$ and bridge $-CR^2=CR^1-CR^2=$]

where the aryl substituents may represent fused ring systems. They include derivatives of benzothiazole (*2, 5, 6, 11, 18, 31, 39, 41, 57, 64, 65, 67, 76, 80, 84, 85, 88, 91, 92, 94, 96, 99, 144, 180, 187, 188, 190, 193, 196, 207, 208, 210, 219, 219a, 220, 220a, 228, 238, 249, 250, 252, 270, 291, 300, 316, 318, 319, 321, 322, 326–333, 335–339, 359, 368–370, 385, 386, 388–392, 401, 437, 439, 504, 509, 522, 523, 547, 551, 560, 565, 567, 568, 570–576, 578–580, 585, 586, 594, 595, 598, 613, 620, 621, 623–625, 638, 649*), naphthothiazoles (*18, 60–62, 69, 88, 93, 94, 96, 144, 180, 200, 253, 258, 282, 356, 357, 381, 383, 508, 523, 591*), tetrahydronaphthothiazole (*387*), thiazoloquinolines (*32, 156, 577*), thienobenzothiazoles (*191, 257, 258*), dioxolobenzothiazoles (*10, 13, 626*), dioxanobenzothiazoles(*213*), oxazolobenzothiazoles (*320*), imidazolobenzothiazoles (*213*), oxazolobenzothiazole (*214*), thiazolobenzothiazoles (*121, 195, 241, 502*), thia- and selenadiazolobenzothiazoles (*140a, 212*), phenanthrothiazole (*314*), naphthothiazolopyridine (*522*), acenaphthenothiazole (*183*), benzofurobenzothiazole (*183*), pyrenothiazole (*431*), and bis(benzothiazole) (*455*).

viii. *Symmetrical trimethinecyanines prepared from selenazolines.* Two patents describe the synthesis of a number of bis(selenazoline-2)trimethinecyanines, **21**, from 2-methyl-3-alkylselenazolinium salts and trialkyl orthocarboxylates (*602, 604*).

[Structure **21**: bis(selenazoline) trimethinecyanine with bridge $-CH=CR-CH=$ and N-substituent $R'$]

ix. *Symmetrical trimethine dyes derived from benzoselenazolium salts and related compounds.* Trimethine dyes prepared from ortho esters and benzoselenazolium salts, **22**, include derivatives of benzoselenazole (*94, 96, 124, 179, 181, 182, 184, 188, 189, 197a, 370, 382, 508a, 594, 603, 605*) and naphthoselenazole (*60, 61*).

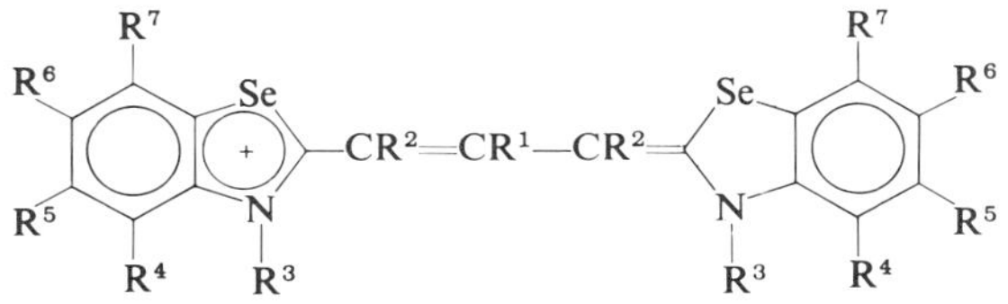

**22**

x. *Symmetrical trimethinecyanines derived from benzimidazolium and other imidazolium salts.* Dyes prepared from triethyl orthoformate and benzimidazolium and related salts have the general formula **23**. They include deriva-

**23**

tives of benzimidazole (*19, 20, 197, 219, 220a, 230, 231, 233, 236, 430, 542, 571, 583, 584, 622, 623, 625a, 625b, 627–629, 647, 648*), pyrrolobenzimidazole (*232*), oxazinobenzimidazole (*232*), furanobenzimidazole (*622*), dioxolobenzimidazole (*622*), imidazoquinoxaline (*90, 215, 287, 288a, 360, 361*), imidazobenzotriazole (*215*), imidazothiadiazoles and selenadiazoles (*140a, 215*) and imidazophenazine (*511*).

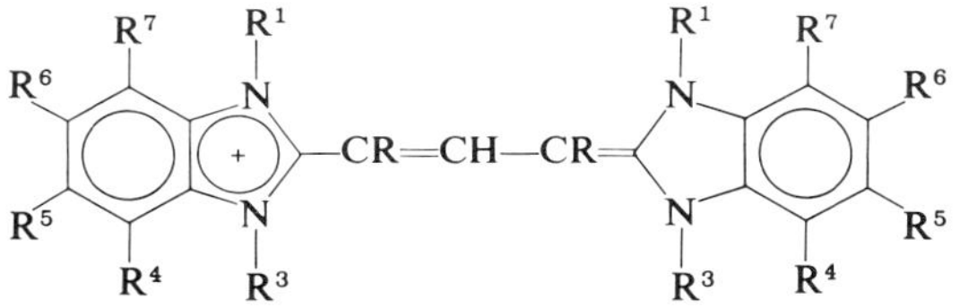

xi. *Symmetrical trimethinecyanines having pyrrolenine, indolenine, and pyrrolopyridine ring systems.* Dihydropyrroleninium iodides were converted to trimethine dyes **24** by reaction with triethyl orthoformate (*218, 527, 528*).

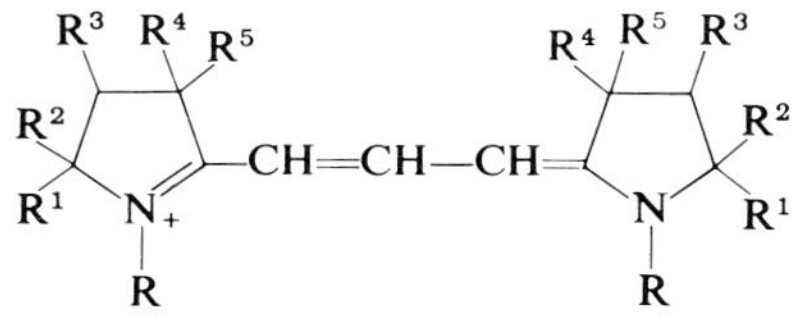

**24**

The bis(pyrrolenine-2) dye 25 was prepared similarly from 1,2,3,5,5-pentamethylpyrroleninium iodide (*217*). A number of 2-methyl-3*H*-indolium salts have been converted to trimethine dyes having the general formula 26 by reaction with orthocarboxylates (*178, 187, 200, 234, 250, 278, 365, 367, 594*). Structurally related dyes were synthesized from triethyl orthoformate and pyrrolopyridinium salts (*200, 203*).

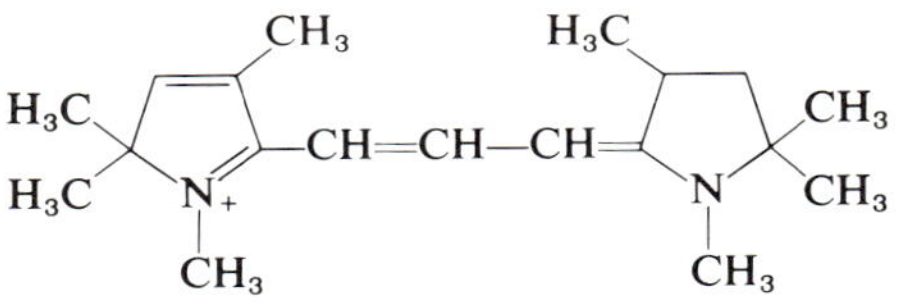

xii. *Miscellaneous symmetrical trimethine dyes.* Symmetrical trimethine dyes possessing unusual structures have been synthesized from trialkyl orthocarboxylates and tropylium (*249*), pyrylium (*59, 408, 614*), thiapyrylium (*615*), pyrazolium (*309*), dibenzonaphthyridinium (*152*), pyridazinium (*164*), cinnolinium (*378, 379*), quinazolinium (*256, 314, 366*), pyrazinium (*222*), quinoxalinium (*135, 399*), benzoquinoxalinium (*16, 399*), thiazinium (*162, 259, 597*), naphthothiazinium (*260, 296*), pyrenothiazinium (*432*), benzodiazepinium (*421*), thiazepinium (*423*), benzo- and naphthothiazepinium *422, 423*), pyrazolopyrimidinium (*503*), pyrimidobenzimidazolium (*34*), thiadiazolium (*88, 226, 227, 237, 561*), triazinium (*130*), imidazotriazinium (*400*), pyrazolotriazinium (*310*), tetrazolium (*411, 526*), imidium (*323, 354*), and diarylcarbonium (*395*) salts.

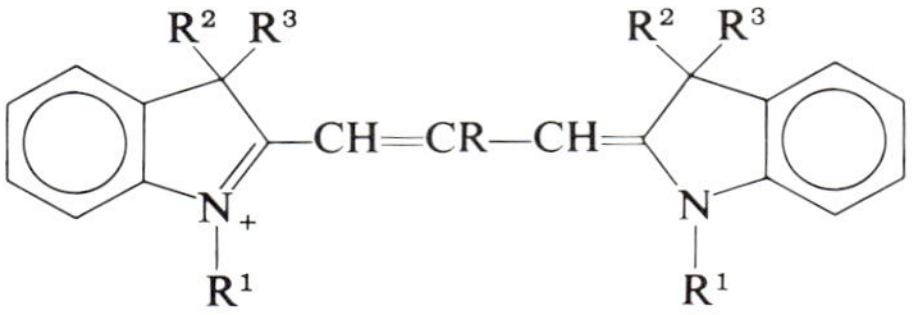

6. UNSYMMETRICAL TRIMETHINE DYES

Triethyl orthoformate has been used in syntheses of a number of unsymmetrical trimethine dyes. Substituted pyrrolium salts are converted to trimethine salts 27 by reaction with methyl ketones and triethyl orthoformate in acetic acid containing hydrogen bromide (*30, 134*). Since a symmetrical

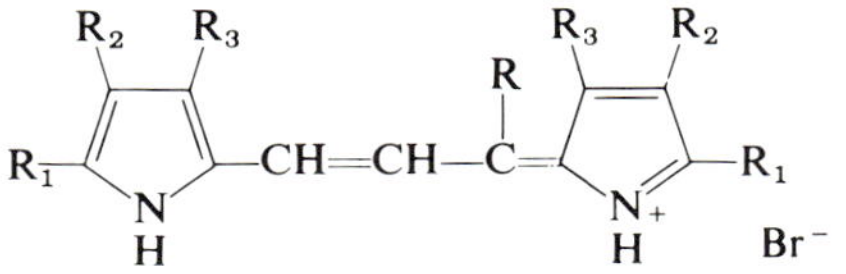

monomethine salt can be isolated as an initial reaction product and subsequently converted to the trimethine salt by heating with the ketone in acetic acid–HBr solution, it is probable that formation of the trimethinecyanine involves condensation of the ketone and the monomethinecyanine, followed by rearrangement and dehydration of the adduct (30).

The trimethine dye **28** was prepared by reaction of 4,5-dimethyl-1-*p*-tolyl-1*H*-tetrazolium iodide and 3-methyl-1-phenyl-5-pyrazolone with triethyl orthoformate (596). 3-Methyl-5-isoxazolone and 3-methyl-4-isopropylidene-5-oxazolone react with triethyl orthoformate in pyridine to form the trimethineoxonol dye **29** (581).

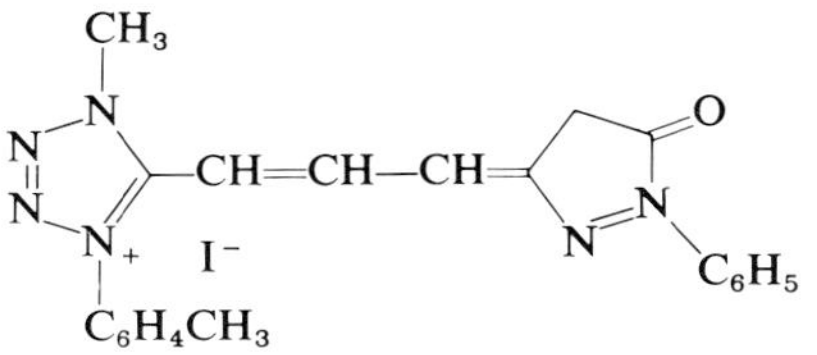

**28**

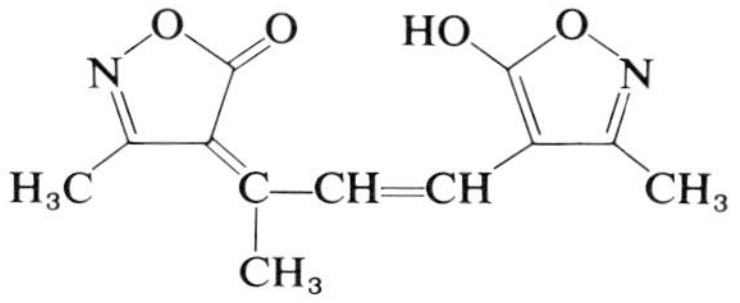

**29**

### 7. Pentamethine and Higher Methine Dyes

Perchloric acid in acetic acid or acetic anhydride catalyzes the reaction of 1-*p*-dimethylaminophenyl-1-arylethylenes with triethyl orthoformate to form the pentamethine dyes **30** (563).

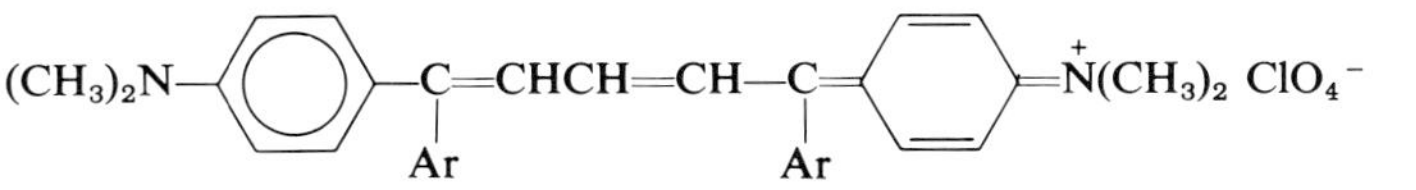

[Ar = *p*-(CH$_3$)$_2$NC$_6$H$_4$, C$_6$H$_5$, *p*-ClC$_6$H$_4$, 2,4-Cl$_2$C$_6$H$_3$]

**30**

2(2-Anilinovinyl)-1,3,3-trimethylindolenium iodide is converted to the pentamethine dye **31** by reaction with triethyl orthoformate in acetic anhydride (448).

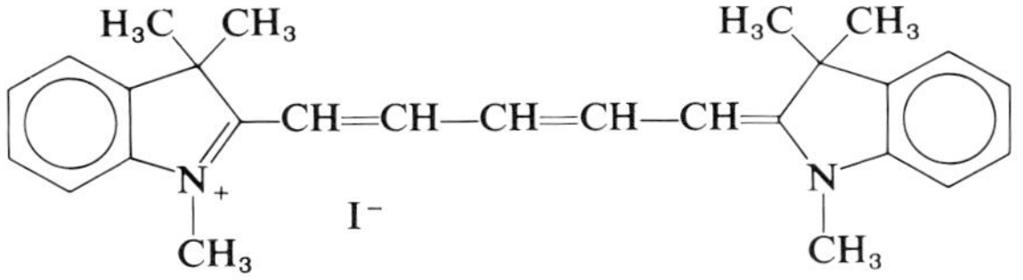

**31**

2,3-Dimethyl-5-anilino-1,3,4-thiadiazole reacts with trimethyl orthoacrylate in pyridine to form the bis(thiadiazolyl)pentamethinecyanine **32** (561).

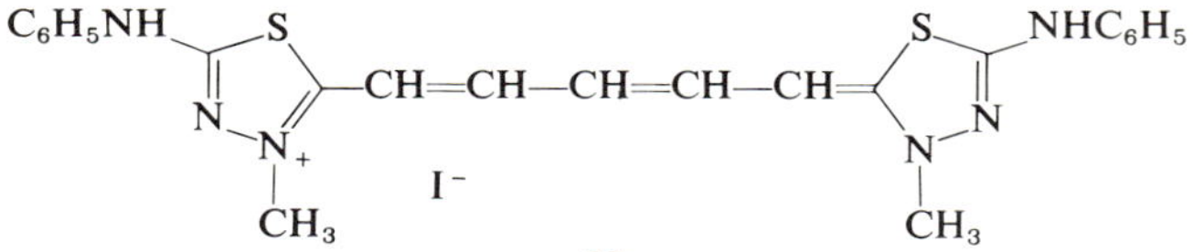

$$\underset{\underset{\mathrm{CH_3}}{|}}{\overset{C_6H_5NH}{\underset{N}{\bigvee}}\underset{N_+}{\overset{S}{\bigvee}}}-CH=CH-CH=CH-CH=\underset{\underset{\mathrm{CH_3}}{|}}{\overset{S}{\underset{N}{\bigvee}}\overset{NHC_6H_5}{\underset{N}{\bigvee}}} \quad I^-$$

**32**

An unsymmetrical hexamethinemerocyanine was synthesized by reaction of a 2-(cyclohexenylidene)methinebenzothiazolium salt with 1-ethyl-2-thiobarbituric acid and diethoxymethyl acetate in pyridine containing triethylamine *(265)*. Diethoxymethyl acetate in pyridine–triethylamine was also used in the preparation of bis(2-benzothiazolyl)heptamethinecyanine and bis(2-benzothiazolyl)nonamethinecyanine dyes *(265)*.

1-Substituted 2-methylbenzothiazolium salts react with γ-pyrone and triethyl orthoformate in the presence of bases to form ethoxy-substituted bis(2-benzothiazolyl)heptamethinecyanine dyes *(305a)*. The hexamethyl tris(acetal) of 1,3,5-pentanetrione, formed by reaction of γ-pyrone with triethyl orthoformate *(612a)* is probably an intermediate in these reactions.

## 8. TRINUCLEAR CYANINE DYES

The reaction of lepidine alkiodides with triethyl orthoformate in pyridine *(250)* produces, in addition to the expected bis(4-quinoline)trimethinecyanines, complex dyes which are powerful photographic sensitizers in the near infrared *(166a)*. These dyes, named neocyanines, were first incorrectly assigned the structure **33** *(251)*. Formulas analogous to **33** were subsequently

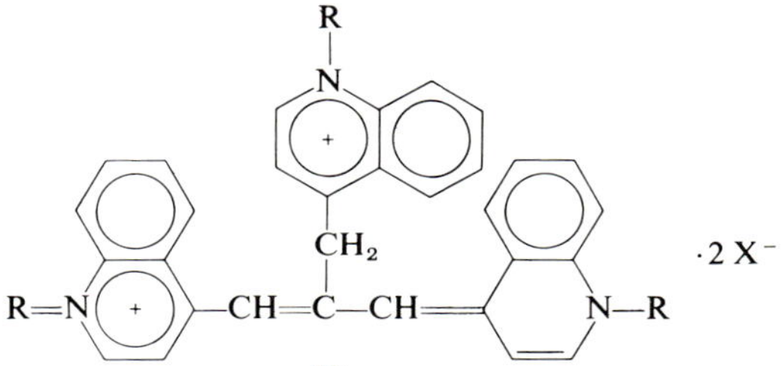

**33**

assigned to trinuclear dyes prepared from other heterocyclic quaternary salts and triethyl orthoformate *(446, 447, 449, 451)*.

The correct structure of neocyanines (**34**) was proposed by Konig in 1935 *(368a)*, and confirmed by Hamer, Rathbone, and Winton in 1947 *(261)*.

$$\text{R—N}^+ \underset{34}{\bigcirc}\text{—CH=CH—}\overset{\underset{|}{\overset{R}{\underset{|}{N^+}}}}{\text{C}}\text{—CH—CH=}\bigcirc\text{N—R} \quad \cdot 2\text{X}^-$$

Neocyanine is a bis(quinoline-4)pentamethinecyanine with a 4-quinolyl substituent on the central carbon of the pentamethine chain. Trinuclear dyes derived from other heterocyclic quaternary salts are assigned analogous structures.

Both symmetrical trimethinecyanines and trinuclear pentamethinecyanines are formed in reactions of heterocyclic quaternary salts with triethyl orthoformate. The fact that the trimethinecyanines can be converted to trinuclear dyes [by heating with triethyl orthoformate and a methyl-substituted quaternary salt in the presence of acetic anhydride, for example (272)] suggests that they are intermediates in the formation of the trinuclear dyes.

The first step in the formation of a trinuclear cyanine from a dinuclear cyanine probably involves reaction of the trimethinecyanine with diethoxycarbonium ion (from triethyl orthoformate) to form an ethoxymethylene derivative. The corresponding ethylthiomethylene compounds, prepared from triethyl orthothioformate, have been isolated (306). The ethoxymethylenetrimethinecyanine cation then adds to the methylene base derived from the quaternary salt to form an intermediate which loses ethanol to form the trinuclear dye cation [Eq. (40)].

Trinuclear pentamethinecyanines have been prepared from triethyl orthoformate and heterocyclic quaternary salts by heating the reactants in pyridine solution (71, 147, 251, 261), in acetic anhydride solution (169), or with acetic acid (414), chloroacetic acid (270), succinic acid (447, 449), or chloromalonic acid (270). They have also been synthesized from quaternary salts and diethoxymethyl acetate in the presence of acetic anhydride, pyridine, acetic acid, or dioxane (144). Heating symmetrical trimethinecyanines with acetic anhydride and triethyl orthoformate results in their partial conversion to trinuclear dyes (450, 451). 2-Methylbenzothiazolium salts react with bis(benzothiazole-2)trimethinecyanines and triethyl orthoformate in acetic anhydride to form trinuclear pentamethinecyanines (272). This reaction should be useful for preparing trinuclear dyes having more than one kind of heterocyclic nucleus. The relative amount of trinuclear dye obtained from triethyl orthoformate and a quaternary heterocyclic salt is increased by increasing the ratio of quaternary salt to orthoformate (144), and by increasing the reaction temperature (415).

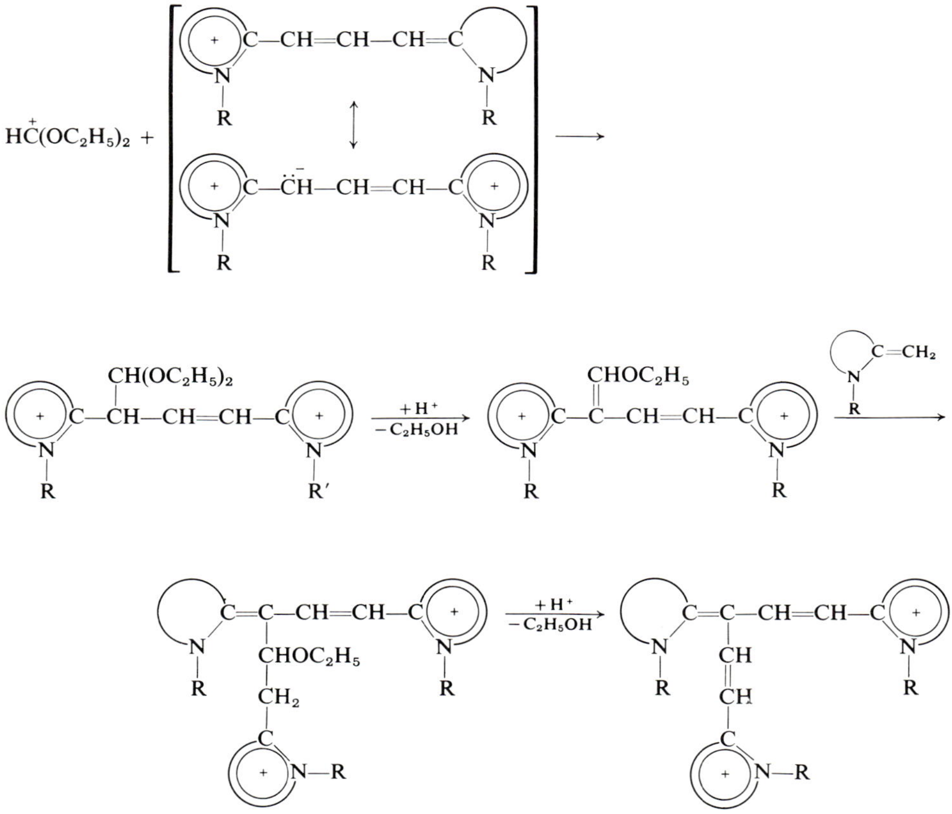

(40)

Trinuclear pentamethinecyanine dyes (35) have been synthesized from triethyl orthoformate and salts of the following heterocyclic quaternary cations: 1-alkyl-2-methylpyridinium (*144*, *446*), 1,4-dimethylpyridinium (*144*), quinaldinium (*446*), lepidinium (*144*, *251*, *269*, *415*, *450*), 6-alkoxy-lepidinium (*268*), 1,2,3-trimethylbenzimidazolium (*447*), 2,3-dimethyl-4-phenyloxazolium (*144*), 2-methyl-3-alkylthiazolium (*17*, *447*), 3-alkyl-2,4-dimethylthiazolium (*167*), 2-methyl-3-ethyl-5,6-dihydro-4*H*-cyclopentathiazolium (*147*), 2-methyl-3-alkyltetrahydrobenzothiazolium (*189a*), 2-methyl-3-alkyltetrahydrobenzoselenazolium (*189a*), 2-methyl-3-alkylbenzoxazolium (*144*, *270*), 2-methyl-3-alkylbenzothiazolium (*144*, *270*, *272*, *447*, *449*), 2,3-dimethylnaphthothiazolium (*144*), 2-methyl-3-alkylbenzoselenazolium (*144*), and 2-methyl-3-alkyl-1,3,4-thiadiazolium (*226*).

# REACTIONS WITH COMPOUNDS HAVING AC[...]

## 9. OTHER REACTIONS INVOLVING METH[...]

1,2-Dimethyltropylium tetrafluorob[...] tetrafluoroborate by triethyl orthofor[...] (249).

$$\underset{H_3C}{\overset{H_3C}{\diagdown}}\!\!\!\bigcirc^{+}\quad BF_4^- \ +\ HC(OC_2H_5)[\ ]$$

1,3-Diethylbarbituric acid and 2-thi[...] complex reactions with triethyl orthoac[...] to products formulated as **36** (28). The o[...] condensation–cyclization reaction of 2 ethylidene derivative of the barbituric a[...]

[structure of diethylbarbituric acid derivative with $C_2H_5$, $H_5C_2$, N, N, O, Y groups]

Ethyl 2-benzothiazolylacetate reacts [...] anhydride to form 3,5-di-2-benzothiazo[...]

## C. Reactions of Ortho Esters with Acetyle[...]

Howk and Sauer found that termin[...] carboxylates and tetraethyl orthocarb[...] such as zinc chloride, zinc iodide, cadm[...] mercuric bromide to form acetylenic [...] (42)] (273–275). This reaction has also b[...] by Epsztein, Marszak, and Holand (17[...]

$$RC\equiv CH\ +\ R'C(OC_2H_5)_3\ \xrightarrow{ZnX_2}\ [\ ]$$

(R = H, alkyl, alkenyl, alkynyl, cycloalkeny[...]

The carbon–carbon bond-forming st[...] attack by a dialkoxycarbonium ion on t[...] thus an electrophilic substitution. Yiel[...]

---

BLE BONDS   267

ng several catalysts and a variety % (375, 632). Boron trifluoride thoformate to 2-methylpropene ethylacetal (136a). 1-Diethoxy- methyl-4-ethoxycyclopentene are d cyclopentadiene.

(Y[...] )iketene

'43, 534) and diphenylketene (537) ions are catalyzed by zinc chloride hyl orthocarbonate does not add to 34).

$$_2H_5O)_2CH\text{—}CR_2\text{—}CO_2C_2H_5 \qquad (45)$$
$$_6H_5)$$

n the presence of boron trifluoride acts of triethyl 3,3-diethoxyortho- thobutyrate is probably formed by he diethylacetal, produced by loss of .e in the presence of boron trifluoride hyl 3-ethoxycrotonate.

$$\rightarrow CH_3C(OC_2H_5)=CHCO_2C_2H_5 \qquad (46)$$

'rs

ers to enol ethers yield bis acetals of valuable synthetic intermediates.

$$\rightarrow RC(OR')_2\text{—}\overset{|}{\underset{|}{C}}\text{—}\overset{|}{C}(OR')OR'' \qquad (47)$$

Additions of ortho esters to enol ethers are catalyzed by Lewis acids (e.g., $BF_3$, $ZnCl_2$, $AlCl_3$, $TiCl_4$), and probably involve addition of dialkoxy-carbonium salts to the double bonds of the enol ethers.

Most additions of ortho esters to enol ethers which have been described in the literature involve orthoformates. The ease with which addition occurs, and the yield of bis acetal obtained, depend on the structure of the enol ether. Vinyl ethers, $CH_2=CHOR$, generally give malonaldehyde bis acetals

## 4. ORTHO ESTER REACTIONS FORMING C–C OR C–H BONDS

in yields of from 50% to 80% when boron trifluoride etherate is used as the catalyst (*493*). β-Substituted vinyl ethers, RCH=CHOR′, are converted to bis acetals in 70%–80% yields when R is primary alkyl or halogen, but in only 20%–30% yields when R is isopropyl. $(CH_3)_2C=CHOC_2H_5$ and triethyl orthoformate afford only 2%–30% yields of 2,2-dimethyl-1,1,3,3-tetraethoxypropane (*632*). Conversion of α-substituted vinyl ethers, $CH_2=CR′OR″$, to bis acetals in satisfactory yield requires use of zinc chloride as catalyst and use of a large excess of the orthoformate (*632*).

Formation of 1,1,3,5,5-pentaalkoxypentanes (by addition of malonaldehyde bis acetals to vinyl ethers) is not ordinarily a serious competing reaction in malonaldehyde bis acetal synthesis, since acetals yield carbonium ions much less readily than ortho esters in the presence of Lewis acids (*493*). However, 1,1,3,5,5-pentaethoxypentane was a major product of the reaction between ethyl vinyl ether and triethyl orthoformate when titanium tetrachloride was the catalyst (*632*).

1,1,3,3-Tetraalkoxyalkanes have been prepared by Lewis acid-catalyzed additions of trialkyl orthoformates to alkyl vinyl ethers (*136, 428, 429, 493, 631–633*), divinyl ether (*136*), alkyl propenyl ethers (*343, 429, 632, 636*), ethyl butenyl ether (*343*), alkyl alkoxyvinyl ethers (*24*), alkyl halovinyl ethers (*355, 632*), ethyl isopropenyl ether (*429, 632*), ethyl chloroethoxyvinyl ether (*25*), and 2-ethoxy-1-alkenes (*630, 632*).

Cyclic enol ethers also react with orthoformates. 1-Ethoxycyclohexene reacted with an excess of triethyl orthoformate in the presence of zinc chloride to form 1,1-diethoxy-2(diethoxymethyl)cyclohexane in 30% yield (*632*). Triethyl orthoformate adds to dihydrofuran (*137, 633*) and dihydro-4H-pyran (*633*) to form 2-ethoxy-3-diethoxymethyltetrahydrofuran and 2-ethoxy-3-diethoxymethyltetrahydropyran, respectively.

The cyclic orthoformate, 2-ethoxy-1,3-dioxolane, reacts with methyl vinyl ether in the presence of boron trifluoride etherate to form 2-(2-methoxy-2-ethoxyethyl)-1,3-dioxolane (*136*).

A 17-acetyl steroid reacts with trialkyl orthoformates in the presence of perchloric acid to yield products which hydrolyze to 17-(1-ethoxy-2-methyl-3-oxo-1-propene) derivatives (*172*). Addition of the orthoformates to intermediate enol ethers is probably involved in these reactions.

Probably because of the ease with which higher orthocarboxylates undergo decomposition and self-condensation reactions in the presence of Lewis acids, orthoformates are the only ortho esters whose additions to enol ethers have been thoroughly studied. However, triethyl orthoacetate reacts with ethyl vinyl ether in the presence of a mixed boron trifluoride etherate–zinc chloride catalyst to produce a 30% yield of 1,1,3,3-tetraethoxybutane (*632*), and tetramethyl orthocarbonate has been reported to undergo a stannic chloride-catalyzed reaction with methyl vinyl ether to form 1,1,3,3-tetramethoxybutane (*136*). (The expected product of this reaction is trimethyl 3,3-dimethoxyorthopropionate.)

Methyl 2,2,3,3-tetramethoxypropionate is among the products obtained by treating trimethyl orthoformate with boron trifluoride and diisopropylamine. The last step of this reaction presumably involves addition of dimethoxycarbonium ion or dimethoxycarbene to tetramethoxyethylene (*453a*).

## D. Addition of Orthoformates to Enol Acetates

Trimethyl and triethyl orthoformates react with vinyl acetate in the presence of acidic catalysts to form mixtures of malonaldehyde derivatives (*280, 296a, 562a–562c, 632*).

$$HC(OR)_3 + CH_2=CHOCOCH_3 \xrightarrow{E} \begin{array}{c} (RO)_2CHCH_2CH(OR)OCOCH_3 \\ + \\ CH_2[CH(OR)_2]_2 \\ + \\ CH_2[CH(OR)OCOCH_3]_2 \end{array} \quad (48)$$

$$(R = CH_3, C_2H_5)$$

The malonaldehyde bis acetals and 1,3-dialkoxy-1,3-propanediol diacetates presumably are formed by disproportionation and alcoholysis reactions of initially formed 1,3,3-trialkoxypropyl acetates.

Equimolar amounts of vinyl acetate and triethyl or trimethyl orthoformates react at room temperature in the presence of mixtures of boron trifluoride and mercuric oxide to form mixtures of malonaldehyde derivatives in total yields of 80%–90% (*562a–562c*). Higher reaction temperatures apparently lower the yields (*632*). The reaction has also been carried out with ferric chloride as the catalyst and a 2:1 molar ratio of orthoformate to vinyl acetate (*280, 296*).

Attempts to add triethyl orthoformate to 1-acetoxycyclohexene were unsuccessful (*632*).

## VI. AROMATIC SUBSTITUTION REACTIONS OF ORTHO ESTERS

Phenols and aromatic tertiary amines react with triethyl orthoformate in the presence of Lewis acids. Products of these reactions include formyl derivatives of the aromatic compound and triarylmethanes. In most instances the initial step of the reaction probably involves attack by diethoxycarbonium ion on an activated position of the aromatic substrate to yield a substituted benzaldehyde diethylacetal. This may, depending upon reaction conditions, be converted to the free aldehyde or to a triarylmethane.

*N,N*-Dimethylaniline *(209)*, *N,N*-diethyl-α-naphthylamine *(435)*, *N,N*-dimethyl-*m*-toluidine *(434)*, and *N,N*,3,5-tetramethylaniline *(434)* are converted to triarylmethanes by heating with triethyl orthoformate and zinc chloride. Substitution occurs para to the dialkylamino group.

$$\text{C}_6\text{H}_5\text{-NR}_2 + \text{HC(OC}_2\text{H}_5)_3 \xrightarrow{\text{ZnCl}_2} \text{HC}[\text{C}_6\text{H}_4\text{-NR}_2]_3 \quad (49)$$

2,6-Dimethylphenol reacts with triethyl orthoformate at room temperature in the presence of magnesium bromide to form 4,4′,4″-trihydroxy-3,3′,3″,5,5′,5″-hexamethyltriphenylmethane in high yield *(104)*.

Eight phenols were converted to substituted *o*- and *p*-hydroxybenzaldehydes in 40%–96% yields by treatment with triethyl orthoformate and aluminum chloride in methylene chloride solution *(248)*. Formylation occurs para to the hydroxyl group with 2,5-dimethylphenol and pyrocatechol, ortho to the hydroxyl group with 3,4-dimethylphenol and β-naphthol, and at positions both ortho and para to the phenolic hydroxyl with *m*-cresol, 3,5-dimethylphenol, thymol, and α-naphthol.

Aryloxymagnesium halides having an unsubstituted ortho position react with triethyl orthoformate to yield, after hydrolysis, *o*-hydroxyaromatic aldehydes, with no detectable amount of the *p*-hydroxy isomers *(106)*. The reaction is sensitive to the electronic and steric properties of substituents on the aromatic ring of the phenol. Activating alkyl or alkoxy substituents facilitate the substitution reaction. Although phenol is converted to *o*-hydroxybenzaldehyde in only 7% yield, *o*-, *m*-, and *p*-cresol; 2,5-, 3,4-, and 3,5-dimethylphenol; *p*-*tert*-butylphenol; thymol; and *m*-methoxyphenol are converted to hydroxybenzaldehydes in yields of 15%–55%. The hindered phenol, carvacrol (2-methyl-5-isopropylphenol), gives less than 5% of 2-hydroxy-3-methyl-6-isopropylbenzaldehyde. Phenols with electron-withdrawing substituents are relatively unreactive. *o*- and *p*-Chlorophenols are converted to the chlorobenzaldehydes in yields of less than 5%, and carbomethoxy and nitrophenols do not react at all.

The specificity of this reaction suggests that it involves electrophilic attack on the ortho position of the phenol by the acyl carbon of an orthoformate molecule in an aryloxymagnesium halide–orthoformate complex, or by a diethoxycarbonium ion of an ion pair derived from such a complex.

In similar reactions, a number of phenols were treated with triethyl orthoformate and magnesium bromide to form mixtures of *o*- and *p*-hydroxybenzaldehydes *(104)*.

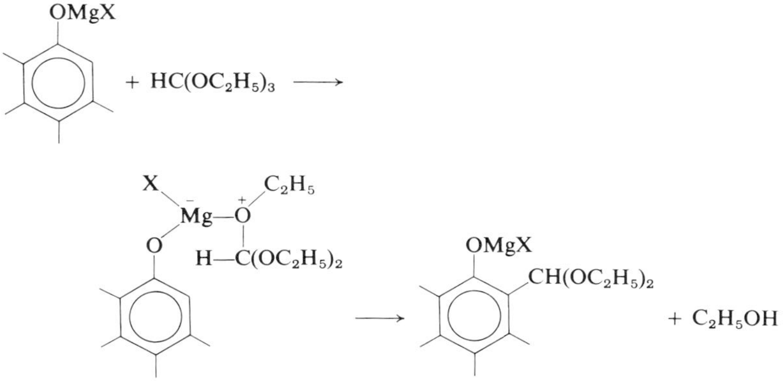

$$\text{OMgX-C}_6\text{H}_3(\text{CH}_3)_2 + \text{HC}(\text{OC}_2\text{H}_5)_3 \longrightarrow$$

$$\begin{array}{c}\text{X}^-\text{Mg}^+\text{—O—C}_2\text{H}_5 \\ \text{O—C}_6\text{H}_3(\text{CH}_3)_2 \\ \text{H—C}(\text{OC}_2\text{H}_5)_2\end{array} \longrightarrow \begin{array}{c}\text{OMgX} \\ \text{C}_6\text{H}_2(\text{CH}_3)_2\text{CH}(\text{OC}_2\text{H}_5)_2\end{array} + \text{C}_2\text{H}_5\text{OH} \quad (50)$$

$\beta$-Naphthol is converted to $\alpha$-anilinomethylene-$\beta$-naphthol by heating with $N,N'$-diphenylformamidine and triethyl orthoformate (344). The fact that the same product is obtained from ethyl $N$-phenylformimidate suggests that the function of the ortho ester may be to convert the diphenylformamidine to the formimidic ester.

Azulene is converted to 1-formylazulene by treatment with trimethyl or triethyl orthoformate in the presence of boron trifluoride etherate, followed by hydrolysis of the acetal. Similar treatment of guaiazulene yields 3-formyl-1-isopropyl-5-methylazulene (560c). Substituted 8-methylazulenes react with triethyl orthoformate in the presence of hydrogen halides to form 1-ethoxymethylene-8-methylazulenium salts (340).

## VII. REACTIONS OF ORTHO ESTERS WITH ETHYL DIAZOACETATE AND WITH DIAZO KETONES

When a diazo ketone or ethyl diazoacetate is added to an ether–boron trifluoride solution of a trialkyl orthocarboxylate, a spontaneous reaction occurs in which nitrogen is evolved and an alkoxyacetal is formed (517–520).

$$\text{RCOCHN}_2 + \text{R}'\text{C}(\text{OR}'')_3 \xrightarrow[-\text{N}_2]{\text{BF}_3} \text{RCO—CH}(\text{OR}'')\text{—CR}'(\text{OR}'')_2 \quad (51)$$

| R | R' | R" | % Yield |
|---|----|----|---------|
| $C_2H_5O$ | H | $CH_3, C_2H_5$ | 78–95 |
| $C_2H_5O$ | $CH_3$ | $C_2H_5$ | 77 |
| $C_2H_5O$ | $C_2H_5$ | $C_2H_5$ | 63 |
| $CH_3$ | H | $C_2H_5$ | 61 |
| $p\text{-}O_2NC_6H_4$ | H | $CH_3$ | 100 |
| $o\text{-}CH_3C_6H_4$ | H | $CH_3$ | — |

Dialkoxycarbonium salts formed by reaction of the ortho esters with boron trifluoride are probably intermediates in these reactions. This view is supported by the fact that the reactions are spontaneous and exothermic even in the dark, and that catalytic amounts of boron trifluoride result in formation of high yields of products, although boron trifluoride is destroyed by the diazo compounds in the absence of ortho esters. A plausible mechanism for the reaction is outlined in Eq. (52).

$$R'C(OR'')_3 + BF_3 \rightarrow R'\overset{+}{C}(OR'')_2\ \overset{-}{B}F_3OR''$$

$$RCOCHN_2 + R'\overset{+}{C}(OR'')_2 \rightarrow RCO-\overset{+}{C}H-CR'(OR'')_2 + N_2 \quad (52)$$

$$RCO-\overset{+}{C}H-CR'(OR'')_2 + \overset{-}{B}F_3OR'' \rightarrow RCO-CH(OR'')-CR'(OR'')_2 + BF_3$$

## VIII. MISCELLANEOUS ORTHO ESTER ADDITION REACTIONS WHICH FORM CARBON–CARBON BONDS

### A. Benzoyl Peroxide-Catalyzed Addition of Triethyl Orthoformate to Diethyl Maleate

Free radical reactions of ortho esters have received little attention. One exception is the benzoyl peroxide-catalyzed reaction of triethyl orthoformate with diethyl maleate, which produces a complex mixture of products containing derivatives of levulinic and succinic acids (*424–427*). The presence in the product mixture of diethyl 2-triethoxymethylsuccinate indicates that triethoxymethyl radical is an intermediate in these reactions.

### B. Cobalt Tetracarbonyl-Catalyzed Hydroformylation of Ortho Esters

Piacenti, Pino, and co-workers have developed an aldehyde and acetal synthesis involving treatment of trialkyl orthocarboxylates with hydrogen and carbon monoxide at high pressures in the presence of cobalt tetracarbonyl. The reaction involves attachment of a dialkoxymethyl or formyl group to the alkyl residue of the alkoxy group of the ortho ester. Orthoformates have been most extensively used in this reaction; higher orthocarboxylates may also be used, and are in fact more reactive than orthoformates (*461c*). Aldehydes are obtained when the reaction is carried out in benzene or dioxane solution. Acetals are the products if no solvent is used, or if the reaction takes place in alcohol solution (487). The reaction, for trialkyl orthoformates without solvent, is described by Eq. (53).

$$2 \text{ HC(OR)}_3 + \text{CO} + \text{H}_2 \xrightarrow{\text{Co}_2(\text{CO})_8} \text{RCH(OR)}_2 + 2 \text{ HCO}_2\text{R} + \text{ROH} \quad (53)$$

| R | References |
|---|---|
| $CH_3$ | 461a |
| $C_2H_5$ | 461a, 461b |
| $C_3H_7$ | 461a, 461b |
| $C_4H_9$ | 461a |
| (S)-$C_2H_5CH(CH_3)CH_2$ | 461a, 461b |
| (S)-$C_2H_5CH(CH_3)CH(CH_3)$ | 508c, 516a |
| $(C_2H_5)_2CHCH_2$ | 461a |

The fact that optically active orthoformates having alkoxy groups with asymmetric centers at carbons beta to alkoxy oxygen yield acetals of high optical purity rules out product formation exclusively by hydroformylation of olefins produced from the orthoformates (461b, 508a, 508b). Such olefins do account for the formation of acetals isomeric to the main products in some cases, however. A possible mechanism for the formation of acetals without olefin intermediates is outlined in Eq. (54).

$$\text{HC(OR)}_3 + \text{HCo(CO)}_4 \longrightarrow \text{HCO}_2\text{R} + \text{ROH} + \text{RCo(CO)}_4 \xrightarrow{\text{CO}}$$

$$\text{RCOCo(CO)}_4 \xrightarrow{\text{H}_2} \text{RCHO} + \text{HCo(CO)}_4 \xrightarrow[-\text{HCO}_2\text{R}]{\text{HC(OR)}_3} \text{RCH(OR)}_2 \quad (54)$$

$\alpha,\beta$-Unsaturated carbonyl compounds react with triethyl orthoformate under cobalt carbonyl hydroformylation conditions to form $\beta$-diethoxymethyl derivatives (486, 486a). Ethyl crotonate was converted to ethyl 3-methyl-4,4-diethoxybutyrate and crotonaldehyde to 2-methylsuccinaldehyde tetraethyl diacetal by this reaction.

## C. Conversion of α,β-Unsaturated Ketones to Bicyclooctane Derivatives

Morita and co-workers discovered a useful reaction which differs from others discussed in this chapter in that the carbon–carbon bonds formed involve neither the acyl nor the alkoxy carbon of the ortho ester. When $\alpha,\beta$-unsaturated ketones are treated with trimethyl orthoformate in the presence of phosphoric acid, bicyclooctane derivatives are often the principal reaction products [Eq. (55)]. Several examples of this reaction, which involves

Diels-Alder dimerization and cyclization of an intermediate dienol ether, were described (*416b–416d*).

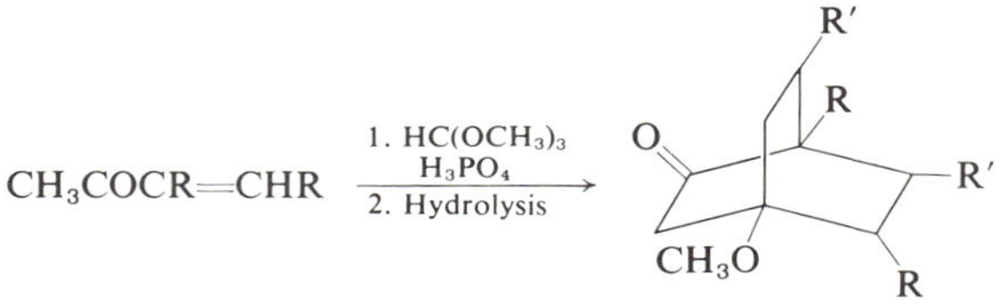

$$\text{CH}_3\text{COCR}=\text{CHR} \xrightarrow[\text{2. Hydrolysis}]{\text{1. HC(OCH}_3)_3 \\ \text{H}_3\text{PO}_4} \quad \text{[bicyclic product with CH}_3\text{O, O, R, R']} \quad (55)$$

### D. Other Addition Reactions of Ortho Esters

Chloroacetone reacts with triethyl orthoacetate in phenol solution to form ethyl levulinate and other products. The ethyl levulinate is probably produced by reaction of ketene diethyl acetal (formed from the orthoacetate) with chloroacetone (*375a*).

Treatment of β-diketones with triethyl orthoformate in the presence of Lewis acids and sulfuric acid produces 3,5-dialkylphenetoles. Enol ethers of the β-dicarbonyl compounds are intermediates in these reactions (*223a*).

## IX. ELIMINATION REACTIONS OF ORTHO ESTERS—SYNTHESIS OF KETENE ACETALS

Elimination reactions of trialkyl orthocarboxylates provide the most useful and generally applicable route to ketene acetals (*395a*). Two types of elimination reactions have been used for this purpose: dealcoholation [Eq. (56)] and dealkoxybromination [Eq. (57)].

$$\text{RR'CH—C(OR'')}_3 \rightarrow \text{RR'C}=\text{C(OR'')}_2 + \text{R''OH} \quad (56)$$

$$\text{RCHBr—C(OR')}_3 + 2\,\text{Na} \rightarrow \text{RCH}=\text{C(OR')}_2 + \text{NaBr} + \text{NaOR'} \quad (57)$$

### A. Dealcoholation of Ortho Esters

The first reaction to produce a ketene acetal by loss of alcohol from an ortho ester was carried out unintentionally by Reitter and Weindel in 1907 (*500a*). They attempted to prepare ethyl 3,3,3-triethoxypropionate (tetraethyl hemiorthomalonate) from ethyl cyanoacetate by the Pinner synthesis, but obtained ethyl 3,3-diethoxyacrylate instead [Eq. (56), R = H, R' = $CO_2C_2H_5$, R'' = $C_2H_5$]. Staudinger and Rathsam later prepared phenylketene diethyl acetal by pyrolysis of triethyl phenylorthoacetate [Eq. (56), R = H, R' = $C_6H_5$, R'' = $C_2H_5$] (*537*), but reported that attempts to prepare ketene acetals from triethyl orthoacetate and triethyl orthopropionate yielded ethyl carboxylates and diethyl ether instead. The synthesis of ketene acetals by dealcoholation of ortho esters was studied extensively by McElvain and co-workers (*395a, 395b, 395d, 396a–396d, 397a, 398a–398d*).

Although some ortho esters [trialkyl phenylorthoacetates, for example (*398c, 537*)] decompose to ketene acetals and alcohol on being heated without added catalysts, ortho ester dealcoholation reactions are accelerated by proton acids, Lewis acids, and by strong bases. In the presence of acids, reaction (56) is reversible; it may be displaced toward the right by distilling off the alcohol as it is formed (*398a*). Acid-catalyzed dealcoholation is a practical means of preparing ketene acetals only from ortho esters having activating α-substituents such as CN, $CO_2R$, or $C_6H_5$. Such ortho esters can be converted to ketene acetals in fair yields (*398a, 398c*). Ordinary carboxylate esters, ethers, and alkenes are by-products of these reactions. Ortho esters lacking activating substituents yield alkyl carboxylates, ethers, and alkenes, but no ketene acetals, when heated in the presence of acids. A simpler explanation for these observations than that originally proposed by McElvain (*395a*) is to assume that all of the pyrolysis products are formed from the same dialkoxycarbonium ion. Isolable amounts of ketene acetals are formed in the case of ortho esters having activating α-substituents because such substituents stabilize the transition states for ketene acetal formation.

$$RR'CHC(OC_2H_5)_3 \xrightarrow[-C_2H_5OH]{+H^+}$$

$$RR'CH-C\begin{array}{c}OC_2H_5\\+\\OC_2H_5\end{array} \begin{array}{c}\xrightarrow{C_2H_5OH} RR'CHCO_2C_2H_5 + (C_2H_5)_2O + H^+\\ \longrightarrow RR'CHCO_2C_2H_5 + CH_2=CH_2 + H^+ \\ \longrightarrow RR'C=C(OC_2H_5)_2 + H^+\end{array} \quad (58)$$

The acid-catalyzed dealcoholation of ortho esters is further limited by the fact that the necessary ortho esters are often difficult to prepare (*29*).

Ortho esters having electron-withdrawing α-substituents are also converted to ketene acetals by reaction with carbanion bases.

$$RR^1CH-C(OR^2)_3 + M^+R^{3-} \rightarrow R^3H + M^+OR^{2-} + RR^1C=C(OR^2)_2 \quad (59)$$

Tetraethoxyethylene was prepared in 39% yield by reaction of triethyl diethoxyorthoacetate with ethylsodium in ligroin [Eq. (59), $R = R^1 = C_2H_5O$, $R^2 = R^3 = C_2H_5$, M = Na] (*395d*). Similarly, trimethyl phenylorthoacetate was converted to phenylketene dimethyl acetal in 60% yield by treatment with ethylsodium, butyllithium, or cinnamylsodium (*398d*). Trialkyl orthoisobutyrates and trialkyl orthoacetates did not yield ketene acetals when treated with ethylsodium, however. Apparently the ethylcarbanion preferentially abstracted protons from the alkoxy groups of these ortho esters (*398d*).

The most effective reagents for converting ortho esters to ketene acetals by removal of alcohols are substances such as mesitylmagnesium bromide, aluminum methoxide, and aluminum *tert*-butoxide, which can function both as bases and as Lewis acids (*398d*). Of these, the reagent of choice is aluminum *tert*-butoxide (*396b*), which affords higher yields of ketene acetals at lower temperatures than does aluminum methoxide. Table IV summarizes syntheses of ketene acetals from ortho esters and aluminum alkoxides.

TABLE IV

ALUMINUM ALKOXIDE-CATALYZED DEALCOHOLATION REACTIONS OF ORTHO ESTERS

$$RR^1CHC(OR^2)_3 \xrightarrow{Al(OR^3)_3} RR^1C=C(OR^2)_2 + R^2OH$$

| R | $R^1$ | $R^2$ | $R^3$ | Temp. (°C) | Yield of ketene acetal (%) |
|---|---|---|---|---|---|
| H | H | $C_2H_5$ | $C(CH_3)_3$ | 190° | 65 |
| $CH_3$ | H | $C_2H_5$ | $C(CH_3)_3$ | 190° | 83 |
| $C_2H_5$ | H | $CH_3$ | $C(CH_3)_3$ | 190° | 53 |
| $C_6H_5$ | H | $C_2H_5$ | $C(CH_3)_3$ | 170° | 85 |
| $C_6H_5CH_2$ | H | $C_2H_5$ | $C(CH_3)_3$ | 140° | 97 |
| NC | H | $CH_3$ | $C(CH_3)_3$ | 185° | 69 |
| NC | H | $CH_3$ | $CH_3$ | 260° | 86 |
| $NCCH_2$ | H | $CH_3$ | $CH_3$ | 260° | 79 |
| $NCCH_2CH_2$ | H | $CH_3$ | $CH_3$ | 260° | 75 |
| $C_2H_5O$ | H | $C_2H_5$ | $C(CH_3)_3$ | 180° | 61 |
| $CH_3$ | $CH_3$ | $C_2H_5$ | $C(CH_3)_3$ | 180° | 62 |
| $-CH_2CH_2CH_2CH_2-$ | | $CH_3$ | $C(CH_3)_3$ | 180° | 52 |
| $-CH_2CH_2CH_2CH_2CH_2-$ | | $CH_3$ | $C(CH_3)_3$ | 180° | 40 |

Heterocyclic ketene acetals **37–40** were prepared by heating the corresponding heterocyclic ortho esters with aluminum *tert*-butoxide (*395b, 396c, 516a*).

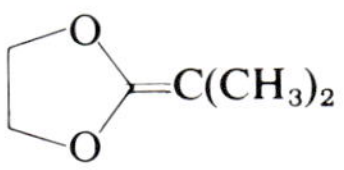

**37** 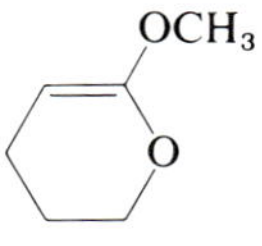     **38** 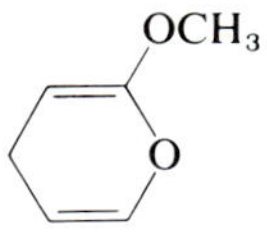     **39** 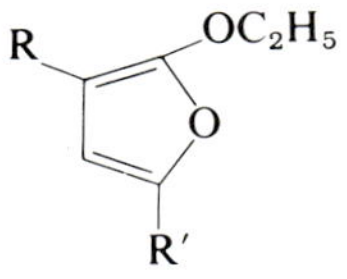     **40**

## B. Dealkoxybromination of α-Bromo Ortho Esters

α-Bromo ortho esters, prepared by bromination of ortho esters having α-hydrogens, may in some instances be converted to ketene acetals by treatment with sodium metal in refluxing benzene [Eq. (57)] (*395c, 396, 598a*). Diethylacetals of ketene, methylketene, *n*-propylketene, and isopropylketene were prepared in yields of 65%–80% by this reaction. However, an attempt to convert trimethyl phenylorthoacetate to phenylketene dimethyl acetal by this reaction was unsuccessful (*398c*).

This reaction is preferable to the synthesis of ketene acetals by dehydrohalogenation of α-bromoacetals, since it yields a single elimination product. The ketene acetals are quite stable under the reaction conditions. Ketene acetals of the type $RR^1C{=}C(OR^2)_2$ have not been prepared by reaction (57) due to lack of success in preparing the α-bromo ortho esters (*395a*).

Attempts to dealkoxybrominate α-bromo ortho esters under other reaction conditions were unsuccessful. Treatment of triethyl bromoorthoacetate and triethyl dibromoorthoacetate with ethanolic sodium ethoxide yielded triethyl orthoacetate rather than ketene acetals. Apparently the initially formed ketene acetals added ethanol under the reaction conditions (*398e*). Attempted conversion of triethyl haloorthoacetates to ketene diethylacetal by reaction with zinc or magnesium gave only polymeric products (*47*).

There is one example of a possible 1,4-dealkoxybromination reaction of an ortho ester. Treatment of triethyl γ-bromocrotonate with magnesium in tetrahydrofuran yielded products which may have been derived from vinylketene diethylacetal (*521a*).

## X. OTHER ELIMINATION REACTIONS OF ORTHO ESTERS

Crank and Eastwood discovered that heating 2-alkoxy-1,3-dioxolanes with catalytic amounts of a carboxylic acid results in decomposition of the cyclic orthoformates to ethanol, carbon dioxide and alkenes. The reaction is highly stereospecific: *cis*-4,5-disubstituted 2-alkoxy-1,3-dioxolanes yield only *cis*-alkenes, while *trans*-4,5-disubstituted dioxolanes yield *trans*-alkenes (*139a, 295a*). The reaction thus provides a potentially useful stereospecific synthesis of alkenes from 1,2-diols, from which the dialkoxydioxolanes are easily prepared (see Ch. 1).

Tricrotyl orthoformate is converted to a mixture of 1,3-butadiene, crotyl alcohol and potassium formate by heating with potassium *tert*-butoxide. A mechanism involving concerted 1,4-elimination was proposed for this reaction (*313a*).

## REFERENCES

1. P. I. Abramenko and N. I. Semnikova, U.S.S.R. Patent 200,426 (1967); *Chem. Abstr.* **68**, 88212 (1968).
1a. P. I. Abramenko and V. G. Zhiryakov, *Zh. Organ. Khim.* **1**, 1132 (1965).
2. M. A. Alperovich, Y. A. Naumov, and I. K. Ushenko, *Zh. Obshch. Khim.* **31**, 1344 (1961).
3. M. A. Alperovich and L. T. Bogolyubskaya, *Zh. Obshch. Khim.* **34**, 645 (1964).
4. M. A. Alperovich and I. K. Ushenko, *Zh. Obshch. Khim.* **29**, 3384 (1959).
5. M. A. Alperovich, I. K. Ushenko, and L. N. Tiurina, *Zh. Obshch. Khim.* **28**, 2538 (1958).
6. M. A. Alperovich, I. K. Ushenko, and L. N. Tiurina, *Zh. Obshch. Khim.* **29**, 3376 (1959).
7. G. deW. Anderson, in "Chemistry of Carbon Compounds," E. H. Rodd, Ed., Vol. IV, Part B, pp. 1053ff. Elsevier, Amsterdam, 1959.
7a. K. J. M. Andrews and B. P. Tong, *J. Chem. Soc.*, (C) p. 1753 (1968).
8. E. Animali, D. Dal Monte, and E. Sandri, *Boll. Sci. Fac. Chim. Ind. Bologna* **22**, 48 (1964); *Chem. Abstr.* **62**, 1766 (1965).
9. A. W. Anish, U.S. Patent 2,441,342 (1948); *Chem. Abstr.* **42**, 5362 (1948).
10. A. W. Anish, British Patent 627,521 (1949); *Chem. Abstr.* **44**, 7170 (1950).
11. A. W. Anish, U.S. Patent 2,508,325 (1950); *Chem. Abstr.* **45**, 486 (1951).
12. A. W. Anish, U.S. Patent 2,521,959 (1950); *Chem. Abstr.* **44**, 10558 (1950).
13. A. W. Anish and L. C. Hensley, U.S. Patent 2,423,217 (1947); *Chem. Abstr.* **41**, 6828 (1947).
13a. J. D. Atkinson and M. C. Johnson, *J. Chem. Soc.*, C p. 1252 (1968).
14. P. Aubert and E. B. Knott, U.S. Patent 2,692,829 (1954); *Chem. Abstr.* **49**, 760 (1955).
15. P. Aubert, E. B. Knott, and L. A. Williams, *J. Chem. Soc.* p. 2185 (1951).
16. F. S. Babichev, *Ukr. Khim. Zh.* **16**, 188 (1950).
17. F. S. Babichev and V. K. Kibirev, *Zh. Obshch. Khim.* **33**, 3646 (1963).
18. F. S. Babichev and E. Shchetinskaya, *Zh. Obshch. Khim.* **34**, 2441 (1964).
19. G. Bach, German Patent 1,121,925 (1960); *Chem. Abstr.* **56**, 15078 (1962).
20. G. Bach, German (East) Patent 19,683 (1960); *Chem. Abstr.* **56**, 11113 (1962).
21. G. Bach, *Mitt. Forschungslab. Agfa Leverkusen-Muenchen* **9**, 94 (1961); *Chem. Abstr.* **57**, 6787 (1962).
22. G. Bach and W. Felgner, German (East) Patent 20,781 (1961); *Chem. Abstr.* **56**, 15080 (1962).
23. G. B. Bachman, in "Organic Syntheses," Coll. Vol. II, A. H. Blatt, Ed., p. 323. Wiley, New York, 1943.
24. H. Baganz and K. Praefke, *Chem. Ber.* **96**, 2661 (1963).
25. H. Baganz and K. Praefke, *Chem. Ber.* **96**, 2666 (1963).
26. J. Bailey, Belgian Patent 649,903 (1963); *Chem. Abstr.* **64**, 11375 (1966).
27. J. Bailey, British Patent 781,000 (1957); *Chem. Abstr.* **52**, 938 (1958).
28. J. Bailey, Belgian Patent 649,100 (1964); *Chem. Abstr.* **64**, 8361 (1966).
29. J. E. Baldwin and L. E. Walker, *J. Org. Chem.* **31**, 3985 (1966).
30. P. Bamfield, A. W. Johnson, and J. Leng, *J. Chem. Soc.* p. 7001 (1965).
31. H. C. Barany and M. Pianka, *J. Chem. Soc.* p. 2217 (1953).
32. E. Barni, G. DiModica, and A. Gasco, *Boll. Sci. Fac. Chim. Ind. Bologna* **25**, 87 (1967); *Chem. Abstr.* **68**, 60516 (1968).
33. R. Barré and B. Ladouceur, *Can. J. Res.* **B27**, 61 (1949).

## REFERENCES

34. L. Basaglia and B. Mariani, *Ann. Chim.* (*Rome*) **53**, 755 (1963).
35. L. Basaglia and B. Mariani, *Chim. Ind.* (*Milan*) **46**, 633 (1964).
36. P. Basignana and M. Gandino, *Compt. Rend. 31st Congr. Intern. Chim. Ind., Liege, 1958*; *Chem. Abstr.* **54**, 14232 (1960).
37. P. Z. Bedoukian, U.S. Patent 3,268,594 (1966); *Chem. Abstr.* **65**, 20011 (1966).
38. H. Behringer and H. Weissauer, *Chem. Ber.* **85**, 774 (1952).
39. B. Beilenson and F. M. Hamer, *J. Chem. Soc.* p. 1225 (1936).
40. E. Benary and F. Ebert, *Chem. Ber.* **56**, 1897 (1923).
41. E. Bergmann, *J. Org. Chem.* **4**, 1 (1939).
42. K. D. Berlin and B. S. Rathore, *Tetrahedron Letters* p. 2547 (1964).
43. L. Bert, *Bull. Soc. Chim. France* [4] **37**, 1408 (1925).
44. L. Bert, *Bull. Soc. Chim. France* [4] **37**, 1588 (1925).
45. L. Bert and Dauphin, *Bull. Soc. Chim. France* [4] **49**, 663 (1931).
46. L. Bert and C. Moureu, *Compt. Rend.* **186**, 699 (1928).
47. F. Beyerstedt and S. M. McElvain, *J. Am. Chem. Soc.* **59**, 1273 (1937).
48. G. A. Bihlmayer, G. Derflinger, J. Derkosch, and O. E. Polansky, *Monatsh. Chem.* **98**, 564 (1967).
49. E. E. Blaise and M. Maire, *Ann. Chim.* (*Paris*) [8] **15**, 564 (1908).
50. F. Bohlmann, *Chem. Ber.* **86**, 63 (1953).
51. F. Bohlmann, *Chem. Ber.* **86**, 657 (1953).
52. F. Bohlmann, H. H. Inhoffen, and P. Herbst, *Chem. Ber.* **90**, 1661 (1957).
53. F. Boudroux, *Compt. Rend.* **138**, 92 (1904).
54. F. Boudroux, *Compt. Rend.* **138**, 700 (1904).
55. F. Boudroux, *Bull. Soc. Chim. France* [3] **31**, 585 (1904).
56. H. Bredereck, F. Effenberger, and W. Resemann, *Chem. Ber.* **95**, 2796 (1962).
57. N. Bregant, *Arkiv Kemi* **23**, 188 (1952).
58. N. Bregant, *Arkiv Kemi* **23**, 192 (1952).
59. H. Brockmann, H. Junge, and R. Muhlmann, *Chem. Ber.* **77**, 529 (1944).
60. L. G. S. Brooker, British Patent 354,264 (1929); *Chem. Abstr.* **26**, 5269 (1932).
61. L. G. S. Brooker, British Patent 359,463 (1929); *Chem. Abstr.* **27**, 239 (1933).
62. L. G. S. Brooker, British Patent 378,870 (1932); *Chem. Abstr.* **27**, 3889 (1933).
63. L. G. S. Brooker, British Patent 385,320 (1932); *Chem. Abstr.* **27**, 4412 (1933).
64. L. G. S. Brooker, U.S. Patent 1,934,657 (1933); *Chem. Abstr.* **28**, 652 (1934).
65. L. G. S. Brooker, U.S. Patent 1,934,659 (1933); *Chem. Abstr.* **28**, 653 (1934).
66. L. G. S. Brooker, U.S. Patent 1,939,201 (1933); *Chem. Abstr.* **28**, 1545 (1934).
67. L. G. S. Brooker, British Patent 394,691 (1934); *Chem. Abstr.* **28**, 56 (1934).
68. L. G. S. Brooker, British Patent 408,273 (1934); *Chem. Abstr.* **28**, 5356 (1934).
69. L. G. S. Brooker, U.S. Patent 1,969,447 (1934); *Chem. Abstr.* **28**, 6075 (1934).
70. L. G. S. Brooker, U.S. Patent 1,973,462 (1934); 1,994,563 (1935).
71. L. G. S. Brooker, U.S. Patent 1,994,652-3 (1935); *Chem. Abstr.* **29**, 2871 (1935).
72. L. G. S. Brooker, British Patent 439,798 (1935); *Chem. Abstr.* **30**, 3345 (1936).
73. L. G. S. Brooker, *J. Am. Chem. Soc.* **58**, 662 (1936).
74. L. G. S. Brooker, U.S. Patent 2,441,558 (1948); *Chem. Abstr.* **42**, 6684 (1948).
75. L. G. S. Brooker, U.S. Patent 2,552,252 (1951); *Chem. Abstr.* **46**, 2434 (1952).
75a. L. G. S. Brooker, French Patent 1,470,163 (1967); *Chem. Abstr.* **68**, 3916 (1968).
76. L. G. S. Brooker and H. W. J. Cressman, U.S. Patent 2,398,999 (1946); *Chem. Abstr.* **40**, 3691 (1946).
76a. L. G. S. Brooker, S. G. Dent, D. W. Heseltine, and E. Van Lare, *J. Am. Chem. Soc.* **75**, 4335 (1953).

77. L. G. S. Brooker and D. W. Heseltine, U.S. Patent 2,646,430 (1953); *Chem. Abstr.* **48**, 1184 (1954).
78. L. G. S. Brooker and D. W. Heseltine, British Patent 713,255 (1954); *Chem. Abstr.* **49**, 2913 (1955).
79. L. G. S. Brooker and D. W. Heseltine, British Patent 713,923 (1954); *Chem. Abstr.* **49**, 2916 (1955).
80. L. G. S. Brooker and D. W. Heseltine, U.S. Patent 2,776,280 (1957); *Chem. Abstr.* **51**, 5605 (1957).
81. L. G. S. Brooker and G. H. Keyes, *J. Am. Chem. Soc.* **58**, 659 (1936).
82. L. G. S. Brooker and G. H. Keyes, *J. Am. Chem. Soc.* **73**, 5356 (1951).
83. L. G. S. Brooker and G. H. Keyes, U.S. Patent 2,917,516 (1959); *Chem. Abstr.* **54**, 9575 (1960).
84. L. G. S. Brooker, G. H. Keyes, and D. W. Heseltine, *J. Am. Chem. Soc.* **73**, 5350 (1951).
85. L. G. S. Brooker, G. H. Keyes, and W. W. Williams, *J. Am. Chem. Soc.* **64**, 199 (1942).
86. L. G. S. Brooker, G. H. Keyes, R. H. Sprague, R. H. Van Dyke, E. Van Lare, G. Van Zandt, F. L. White, H. W. Cressman, and S. G. Dent, *J. Am. Chem. Soc.* **73**, 5332 (1951).
87. L. G. S. Brooker, *in* "Recent Progress in the Chemistry of Natural and Synthetic Colouring Matters and Related Fields" (T. S. Gore *et al.*, eds.), pp. 573ff. Academic Press, New York, 1962.
88. L. G. S. Brooker, A. L. Sklar, and H. W. J. Cressman, *J. Am. Chem. Soc.* **67**, 1875 (1945).
89. L. G. S. Brooker and R. H. Sprague, U.S. Patent 2,525,520 (1950); *Chem. Abstr.* **45**, 2345 (1951).
90. L. G. S. Brooker and E. J. Van Lare, French Patent 1,401,594 (1965); *Chem. Abstr.* **64**, 19852 (1966).
91. L. G. S. Brooker and G. Van Zandt, British Patent 647,542 (1950); *Chem. Abstr.* **45**, 4159 (1951).
92. L. G. S. Brooker and G. Van Zandt, British Patent 649,725 (1951); *Chem. Abstr.* **45**, 6103 (1951).
93. L. G. S. Brooker and F. L. White, *J. Am. Chem. Soc.* **57**, 547 (1935).
94. L. G. S. Brooker and F. L. White, *J. Am. Chem. Soc.* **57**, 2480 (1935).
95. L. G. S. Brooker and F. L. White, U.S. Patent 2,336,463 (1943); *Chem. Abstr.* **38**, 2891 (1944).
96. L. G. S. Brooker and F. L. White, U.S. Patent 2,478,366 (1949); *Chem. Abstr.* **44**, 6311 (1950).
97. L. G. S. Brooker and F. L. White, *J. Am. Chem. Soc.* **73**, 1094 (1951).
98. L. G. S. Brooker and F. L. White, *J. Phot. Sci.* **5**, 71 (1957).
99. J. Brunken and G. Bach, German (East) Patent 11,108 (1956); *Chem. Abstr.* **53**, 4983 (1959).
100. J. Brunken and J. Muller, German (East) Patent 15,119 (1958); *Chem. Abstr.* **54**, 9577 (1960).
101. H. Brunner and E. H. Farmer, *J. Chem. Soc.* p. 1039 (1946).
102. J. G. Buchanan and A. R. Edgar, *Chem. Commun.* p. 29 (1967).
103. P. Cagniant and A. Deluzarche, *Compt. Rend.* **224**, 473 (1947).
104. B. Cardillo, G. Casnati, A. Ricca, and L. Velo, *Chim. Ind. (Milan)* **47**, 1336 (1965).
105. B. H. Carroll and J. E. Jones, U.S. Patent 2,688,545 (1954); *Chem. Abstr.* **49**, 88 (1955).

## REFERENCES

106. G. Casnati, M. Grisafulli, and A. Ricca, *Tetrahedron Letters* p. 243 (1965)
107. J. Chauvelier, *Bull. Soc. Chim. France* [5] **17**, 272 (1950).
108. L. F. Chelpanova, A. A. Petrov, G. P. Bondarev, and V. D. Nemirovskii, *Zh. Obshch. Khim.* **32**, 2487 (1962).
109. I. N. Chernyuk, G. T. Pilyugin, A. I. Gorelikov, and M. I. Rogovik, *Zh. Obshch. Khim.* **34**, 3330 (1964).
110. I. N. Chernyuk, G. T. Pilyugin, and A. I. Gorelikov, *Zh. Organ. Khim.* **1**, 923 (1965).
111. I. N. Chernyuk, G. T. Pilyugin, and A. V. Zlochevskaya, *Zh. Organ. Khim.* **1**, 1129 (1965).
112. I. N. Chernyuk, G. T. Pilyugin, and A. V. Zlochevskaya, *Khim. Geterosikl. Soedin., Akad. Nauk Latv. SSR* p. 590 (1966).
113. A. E. Chichibabin, *J. Russ. Phys.-Chem. Soc.* **35**, 222 (1903).
114. A. E. Chichibabin, *J. Russ. Phys.-Chem. Soc.* **36**, 418 (1904).
115. A. E. Chichibabin, *Chem. Ber.* **37**, 186 (1904).
116. A. E. Chichibabin, *Chem. Ber.* **37**, 850 (1904).
117. A. E. Chichibabin, *Chem. Ber.* **38**, 561 (1905).
118. A. E. Chichibabin, *J. Prakt. Chem.* [2] **73**, 326 (1906).
119. A. E. Chichibabin, *Chem. Ber.* **44**, 447 (1911).
120. A. E. Chichibabin and S. A. Jelgasin, *Chem. Ber.* **47**, 48 (1914).
121. U. Chiodoni and B. Mariani, *Ann. Chim.* (*Rome*) **53**, 741 (1963).
122. L. Claisen, *Chem. Ber.* **26**, 2729 (1893).
123. L. Claisen, *Ann. Chem.* **297**, 1 (1897).
124. L. M. Clark, *J. Chem. Soc.* p. 2313 (1928).
125. C. J. Claus and J. L. Morgenthau, *J. Am. Chem. Soc.* **73**, 5005 (1951).
126. C. J. Claus and J. L. Morgenthau, U.S. Patent 2,786,872 (1957); *Chem. Abstr.* **51**, 12130 (1957).
127. C. J. Claus and J. L. Morgenthau, U.S. Patent 2,830,092 (1958); *Chem. Abstr.* **52**, 13777 (1958).
128. H. Clement, *Ann. Chim.* (*Paris*) [11] **13**, 304 (1940).
129. M. Coenen and O. Riester, *Ann. Chem.* **633**, 110 (1960).
130. C. Cogrossi, B. Mariani, and R. Sgarbi, *Chim. Ind.* (*Milan*) **46**, 530 (1964).
131. A. H. Cook, G. Harris, I. M. Heilbron, and G. Shaw, *J. Chem. Soc.* p. 1056 (1948).
132. A. H. Cook, G. Harris, and G. Shaw, *J. Chem. Soc.* p. 1435 (1949).
133. A. H. Cook and I. M. Heilbron, *in* "The Chemistry of Penicillin," H. T. Clarke, J. R. Johnson, and R. Robinson, Eds., p. 942, Princeton Univ. Press, Princeton, New Jersey, 1949.
134. A. H. Cook and J. R. Majer, *J. Chem. Soc.* p. 482 (1944).
135. A. H. Cook and C. A. Perry, *J. Chem. Soc.* p. 394 (1943).
136. J. W. Copenhaver, U.S. Patent 2,527,533 (1950); *Chem. Abstr.* **45**, 1622 (1951).
136a. J. W. Copenhaver, U.S. Patent 2,677,708 (1954); *Chem. Abstr.* **49**, 1812 (1955).
137. J. W. Copenhaver, U.S. Patent 2,517,543 (1950); *Chem. Abstr.* **45**, 1166 (1951).
138. J. W. Cornforth, *in* "The Chemistry of Penicillin," H. T. Clarke, J. R. Johnson, and R. Robinson, Eds., pp. 743, 803, 838, and 847. Princeton Univ. Press, Princeton, New Jersey, 1949.
139. J. M. Coxon, M. P. Hartshorn, and D. N. Kirk, *Tetrahedron* **20**, 2547 (1964).
139a. G. Crank and F. W. Eastwood, *Australian J. Chem.* **17**, 1392 (1964).
140. T. Cuvigny, *Bull. Soc. Chim. France* p. 655 (1957).
140a. D. DalMonte, E. Sandri, and P. Mozzarachio, *Boll. Sci. Fac. Chim. Ind. Bologna* **25**, 3 (1967); *Chem. Abstr.* **68**, 96789 (1968).

141. B. Das and M. K. Rout, *J. Indian Chem. Soc.* **34**, 505 (1957).
141a. E. G. deBollemont, *Compt. Rend.* **128**, 1338 (1899).
141b. E. G. deBollemont, *Bull. Soc. Chim. France* [3] **25**, 18 (1901).
142. N. C. Deno, *J. Am. Chem. Soc.* **69**, 2233 (1947).
143. S. G. Dent and L. G. S. Brooker, U.S. Patent 2,533,206 (1950); *Chem. Abstr.* **45**, 3272 (1951).
144. S. G. Dent and L. G. S. Brooker, U.S. Patent 2,537,880 (1951); *Chem. Abstr.* **45**, 3741 (1951).
145. H. Depoorter, J. Nys, and A. Van Dormael, *Tetrahedron Letters* p. 199 (1961).
146. H. Depoorter, J. Nys, and A. Van Dormael, *Bull. Soc. Chim. Belges* **73**, 921 (1964).
147. G. DeStevens, U.S. Patent 2,882,160 (1959); *Chem. Abstr.* **53**, 14795 (1959).
148. G. DeStevens, U.S. Patent 2,892,835 (1959); *Chem. Abstr.* **54**, 1138 (1960).
149. G. DeStevens, U.S. Patent 2,892,836 (1959); *Chem. Abstr.* **54**, 1138 (1960).
150. G. DeStevens, U.S. Patent 2,905,666 (1959); *Chem. Abstr.* **54**, 10612 (1960).
151. G. DeStevens and R. H. Sprague, U.S. Patent 2,892,834 (1959); *Chem. Abstr.* **54**, 1139 (1960).
152. M. J. S. Dewar, *J. Chem. Soc.* p. 615 (1944).
153. O. Diels, H. Gartner, and R. Kaak, *Chem. Ber.* **55**, 3441 (1922).
154. O. Diels and J. Petersen, *Chem. Ber.* **55**, 3449 (1922).
155. G. DiModica and E. Barni, *Gazz. Chim. Ital.* **93**, 679 (1963).
156. G. DiModica, E. Barni, and G. Nasimi, *Gazz. Chim. Ital.* **94**, 314 (1964).
157. C. Djerassi and L. E. Geller, *J. Am. Chem. Soc.* **81**, 2789 (1959).
158. M. A. Doja, *Chem. Rev.* **11**, 273 (1932).
159. C. A. Dornfield and G. H. Coleman, in "Organic Syntheses," Coll. Vol. III, E. C. Horning, Ed., p. 701. Wiley, New York, 1955.
160. A. Dornow and F. Ische, *Angew. Chem.* **67**, 653 (1955).
161. F. P. Doyle, T. F. W. Lawrence, and J. D. Kendall, British Patent 662,775 (1951); *Chem. Abstr.* **46**, 6020 (1952).
162. P. D. Dreyfuss and B. Gaspar, U.S. Patent 2,677,683 (1954); *Chem. Abstr.* **49**, 1458 (1955).
163. G. F. Duffin and J. D. Kendall, *J. Chem. Soc.* p. 361 (1956).
164. G. F. Duffin and J. D. Kendall, *J. Chem. Soc.* p. 3789 (1959).
165. G. F. Duffin, J. D. Kendall, and H. R. J. Waddington, *J. Chem. Soc.* p. 3799 (1959).
166. C. Dufraise and R. Chaux, *Bull. Soc. Chim. France* [4] **39**, 915 (1926).
166a. M. L. Dundon, A. L. Schoen, and R. M. Briggs, *J. Opt. Soc. Am.* **12**, 397 (1926).
167. E. I. DuPont de Nemours and Co., British Patent 571,077 (1945); *Chem. Abstr.* **44**, 5739 (1950).
168. E. I. DuPont de Nemours and Co., British Patent 633,158 (1949); *Chem. Abstr.* **44**, 5743 (1950).
169. E. I. DuPont de Nemours and Co., British Patent 631,848 (1949); *Chem. Abstr.* **44**, 7172 (1950).
170. E. I. DuPont de Nemours and Co., British Patent 675,360 (1952); *Chem. Abstr.* **47**, 431 (1953).
171. M. H. Durand, *Bull. Soc. Chim. France* p. 2387 (1961).
172. J. P. Dusza, J. P. Joseph, and S. Bernstein, *J. Am. Chem. Soc.* **86**, 3908 (1964).
173. H. D. Edwards and F. P. Doyle, U.S. Patent 2,721,799 (1955); *Chem. Abstr.* **50**, 1505 (1956).
173aa. E. L. Eliel and F. Nader, *J. Am. Chem. Soc.* **91**, 536 (1969).

173a. R. Epsztein and S. Holand, *Compt. Rend.* **255**, 2778 (1962).
173b. R. Epsztein, S. Holand, and I. Marszak, *Compt. Rend.* **252**, 1803 (1961).
173c. R. Epsztein and I. Marszak, *Bull. Soc. Chim. France* p. 313 (1968).
174. J. G. Erickson, *J. Am. Chem. Soc.* **73**, 1338 (1951).
174a. H. Erlenmeyer and J. P. Jung, *Helv. Chim. Acta* **32**, 35 (1949).
175. G. Errera, *Chem. Ber.* **31**, 1682 (1898).
176. G. Errera, *Gazz. Chim. Ital.* **32**, Part II, 330 (1902).
177. G. Errera, *Gazz. Chim. Ital.* **33**, Part I, 417 (1903).
178. I. G. Farbenindustrie A.-G., British Patent 402,521 (1931).
179. I. G. Farbenindustrie A.-G., British Patent 410,481 (1931).
180. I. G. Farbenindustrie A.-G., British Patent 403,845 (1932).
181. I. G. Farbenindustrie A.-G., British Patent 421,015 (1932).
182. I. G. Farbenindustrie A.-G., French Patent 730,966 (1932); *Chem. Zentr.* **103**, Part II, 2916 (1932).
183. I. G. Farbenindustrie A.-G., British Patent 396,217 (1933); *Chem. Abstr.* **28**, 56 (1934).
184. I. G. Farbenindustrie A.-G., British Patent 403,974 (1933); *Chem. Abstr.* **28**, 3323 (1934).
185. I. G. Farbenindustrie A.-G., British Patent 432,969 (1933).
186. I. G. Farbenindustrie A.-G., British Patent 415,949 (1933).
187. I. G. Farbenindustrie A.-G., French Patent 742,639 (1933); *Chem. Abstr.* **27**, 3618 (1933).
188. I. G. Farbenindustrie A.-G., French Patent 752,563 (1933); *Chem. Abstr.* **28**, 1198 (1934).
189. I. G. Farbenindustrie A.-G., British Patent 405,309 (1934); *Chem. Abstr.* **28**, 3995 (1934).
189a. I. G. Farbenindustrie A.-G., British Patent 501,803 (1936).
190. I. G. Farbenindustrie A.-G., French Patent 811,947 (1937); *Chem. Abstr.* **32**, 1200 (1938).
191. I. G. Farbenindustrie A.-G., British Patent 506,536 (1937).
192. I. G. Farbenindustrie A.-G., British Patent 512,494 (1939); *Chem. Abstr.* **35**, 1238 (1941).
193. Farbwerke Hoechst A.-G., Netherlands Patent Appl. 6,505,474 (1964); *Chem. Abstr.* **64**, 12856 (1966).
194. F. Feist, D. Delfs, and B. Langenkamp, *Chem. Ber.* **59**, 2958 (1926).
195. Ferrania Societa per Azione, French Patent 1,394,744 (1965); *Chem. Abstr.* **64**, 16036 (1966).
196. Ferrania Societa per Azione, Belgian Patent 663,390 (1965); *Chem. Abstr.* **65**, 5575 (1966).
197. Ferrania Societa per Azione, Italian Patent 663,182 (1964); *Chem. Abstr.* **64**, 11356 (1966).
197a. Ferrania Societa per Azioni, French Patent 1,480,882 (1967); *Chem. Abstr.* **68**, 14101 (1968).
197b. Ferrania Societa per Azioni, French Patent 1,483,213 (1967); *Chem. Abstr.* **68**, 31052 (1968).
198. G. E. Ficken, U.S. Patent 3,164,587 (1965); *Chem. Abstr.* **62**, 14868 (1965).
199. G. E. Ficken, British Patent 994,301 (1963); *Chem. Abstr.* **60**, 9399 (1964).
200. G. E. Ficken and J. D. Kendall, *J. Chem. Soc.* p. 3202 (1959).
201. G. E. Ficken and J. D. Kendall, British Patent 841,588 (1960); *Chem. Abstr.* **55**, 2319 (1961).

202. G. E. Ficken and J. D. Kendall, British Patent 870,753 (1961); *Chem. Abstr.* **55**, 25560 (1961).
203. G. E. Ficken and J. D. Kendall, *J. Chem. Soc.* p. 584 (1961).
204. L. Field, J. R. Holsten, and R. D. Clark, *J. Am. Chem. Soc.* **81**, 2572 (1959).
205. J. C. Firestine, U.S. Patent 2,521,705 (1950); *Chem. Abstr.* **45**, 3273 (1951).
206. J. C. Firestine, U.S. Patent 2,609,371 (1952); *Chem. Abstr.* **47**, 431 (1953).
207. J. C. Firestine, U.S. Patent 2,647,050 (1953); *Chem. Abstr.* **47**, 11056 (1953).
208. J. C. Firestine, German Patent 1,112,403 (1960); *Chem. Abstr.* **56**, 4295 (1962).
209. O. Fischer and G. Korner, *Chem. Ber.* **17**, 99 (1884).
210. N. I. Fisher and F. M. Hamer, *J. Chem. Soc.* p. 2502 (1930).
211. G. Fisscher and E. J. Poppe, *Mitt. Forschungslab. Agfa Leverkusen-Muenchen* **10**, 123 (1965); *Chem. Abstr.* **66**, 66723 (1967).
211a. W. Franke and W. Bauriedel, German Patent 1,237,895 (1967); *Chem. Abstr.* **67**, 101026 (1967).
212. S. G. Fridman, *Zh. Obshch. Khim.* **31**, 1096 (1961).
213. S. G. Fridman, *Zh. Obshch. Khim.* **33**, 207 (1963).
214. S. G. Fridman and D. K. Golub, *Zh. Obshch. Khim.* **31**, 3394 (1961).
215. S. G. Fridman and L. I. Kotova, *Zh. Obshch. Khim.* **32**, 2871 (1962).
216. J. Fried and R. C. Elderfield, *J. Org. Chem.* **6**, 574 (1941).
217. D. J. Fry, B. A. Lea, and J. D. Kendall, U.S. Patent 2,979,501 (1961); *Chem. Abstr.* **55**, 16237 (1961).
218. D. J. Fry, B. A. Lea, and J. D. Kendall, U.S. Patent 2,984,663 (1961); *Chem. Abstr.* **55**, 21931 (1961).
219. Fuji Photo Film Co., Ltd., Belgian Patent 668,894 (1965); *Chem. Abstr.* **65**, 17098 (1966).
219a. Fuji Photo Film Co., Ltd., Belgian Patent 669,133 (1965); *Chem. Abstr.* **67**, 74484 (1967).
220. Fuji Photo Film Co., Ltd., Belgian Patent 669,934 (1966); *Chem. Abstr.* **66**, 105906 (1967).
220a. Fuji Photo Film Co., Ltd., British Patent 1,091,739 (1967); *Chem. Abstr.* **68**, 50982 (1968).
221. R. C. Fuson, W. E. Parham, and L. J. Reed, *J. Org. Chem.* **11**, 194 (1946).
222. C. Gastaldi and E. Princivalle, *Ann. Chim. Appl.* **26**, 450 (1936).
223. R. Gelin, S. Gelin, and A. Arcis, *Compt. Rend.* **C263**, 499 (1966).
223a. S. Gelin, J. Rouet, and R. Gelin, *Compt. Rend.* **C265**, 1483 (1967).
224. N. Gerencevic and M. Prostenik, *Kem. Ind.* (*Zagreb*) **13**, 98 (1964); *Chem. Abstr.* **62**, 2852 (1965).
225. R. A. Gershtein, Z. P. Sytnik, and E. B. Lifshits, *Zh. Organ. Khim.* **2**, 543 (1966).
226. Gevaert Photo-Producten N.V., British Patent 457,450 (1934).
227. Gevaert Photo-Producten N.V., British Patent 490,729 (1938); *Chem. Abstr.* **33**, 1511 (1939).
228. Gevaert Photo-Producten N.V., British Patent 511,940 (1939); *Chem. Abstr.* **35**, 42 (1941).
229. Gevaert Photo-Producten N.V., British Patent 632,641 (1949); *Chem. Abstr.* **45**, 6945 (1951).
230. Gevaert Photo-Producten N.V., Belgian Patent 615,547 (1962); *Chem. Abstr.* **57**, 16049 (1962).
231. Gevaert Photo-Producten N.V., Belgian Patent 615,549 (1962); *Chem. Abstr.* **57**, 14623 (1962).

232. Gevaert Photo-Producten N.V., Belgian Patent 618,235 (1962); *Chem. Abstr.* **58**, 14164 (1963).
233. Gevaert Photo-Producten N.V., Belgian Patent 640,453 (1964); *Chem. Abstr.* **63**, 1924 (1965).
234. E. Ghigi, *Gazz. Chim. Ital.* **63**, 698 (1933).
235. H. Gilman and R. Franz, *Rec. Trav. Chim.* **51**, 991 (1932).
236. H. Glockner and K. Kueffner, German Patent 1,159,577 (1963); *Chem. Abstr.* **60**, 10851 (1964).
237. J. Goerdler and H. W. Hammen, *Chem. Ber.* **97**, 1134 (1964).
238. J. Goetze and M. Hase, German Patent 1,053,309 (1957); *Chem. Abstr.* **56**, 5570 (1962).
239. J. Goetze, German Patent 1,051,115 (1959); *Chem. Abstr.* **55**, 21928 (1961).
240. J. Goetze, U.S. Patent 3,044,875 (1962); *Chem. Abstr.* **57**, 13339 (1962).
241. G. Grandolini and A. Fravolini, *Gazz. Chim. Ital.* **92**, 1344 (1962).
242. J. Grard, *Compt. Rend.* **189**, 541 (1929).
243. W. F. Gresham, U.S. Patent 2,449,471 (1948); *Chem. Abstr.* **43**, 1055 (1949).
244. E. Grichkevich-Trokhimovsky, *Bull. Soc. Chim. France* [4] **10**, 1602 (1911).
245. E. Grichkevich-Trokhimovsky, *Bull. Soc. Chim. France* [4] **12**, 364 (1912).
246. V. Grignard, *Compt. Rend.* **130**, 1322 (1900).
247. J. Gripenberg and R. G. Lindahl, *Acta Chem. Scand.* **3**, 256 (1948).
248. H. Gross, A. Rieche, and G. Matthey, *Chem. Ber.* **96**, 308 (1963).
249. K. Hafner, H. W. Riedel, and M. Danielisz, *Angew. Chem. Intern. Ed. Engl.* **2**, 215 (1963).
250. F. M. Hamer, *J. Chem. Soc.* p. 2796 (1927).
251. F. M. Hamer, *J. Chem. Soc.* p. 1472 (1928).
252. F. M. Hamer, *J. Chem. Soc.* p. 3160 (1928).
253. F. M. Hamer, *J. Chem. Soc.* p. 2598 (1929).
254. F. M. Hamer, *Quart. Rev. (London)* **4**, 327 (1950).
255. F. M. Hamer, "Chemistry of Heterocyclic Compounds. Vol. XVIII. The Cyanine Dyes and Related Compounds," Wiley (Interscience), New York, 1964.
256. F. M. Hamer, I. M. Heilbron, J. H. Reade, and H. N. Walls, *J. Chem. Soc.* p. 251 (1932).
257. F. M. Hamer and E. B. Knott, British Patent 593,023 (1947); *Chem. Abstr.* **43**, 513 (1949).
258. F. M. Hamer and E. B. Knott, U.S. Patent 2,447,332 (1948); *Chem. Abstr.* **42**, 8685 (1948).
259. F. M. Hamer and R. J. Rathbone, *J. Chem. Soc.* p. 243 (1943).
260. F. M. Hamer and R. J. Rathbone, *J. Chem. Soc.* p. 487 (1943).
261. F. M. Hamer, R. J. Rathbone, and B. S. Winton, *J. Chem. Soc.* p. 1437 (1947).
262. K. Henkel and F. Weygand, *Chem. Ber.* **76**, 812 (1943).
263. D. W. Heseltine and L. G. S. Brooker, U.S. Patent 2,666,761 (1954); *Chem. Abstr.* **49**, 757 (1955).
264. D. W. Heseltine and L. G. S. Brooker, U.S. Patent 2,719,151 (1955); *Chem. Abstr.* **50**, 5435 (1956).
265. D. W. Heseltine, U.S. Patent 2,734,900 (1956); *Chem. Abstr.* **51**, 913 (1957).
266. D. W. Heseltine and L. L. Lincoln, U.S. Patent 3,094,418 (1963); *Chem. Abstr.* **60**, 3136 (1964).
267. C. L. Hewett, *J. Chem. Soc.* p. 193 (1938).
268. Y. Hiho and Y. Chifu, *Rept. Sci. Res. Inst. (Tokyo)* **25**, 227 (1949); *Chem. Abstr.* **45**, 2342 (1951).

269. Y. Hishiki, *Rept. Sci. Res. Inst. (Tokyo)* **28**, 405 (1952); *Chem. Abstr.* **47**, 11053 (1953).
270. Y. Hishiki, *Rept. Sci. Res. Inst. (Tokyo)* **30**, 39 (1954); *Chem. Abstr.* **50**, 93 (1956).
271. Y. Hishiki, *J. Sci. Res. Inst. (Tokyo)* **48**, 130 (1954); *Chem. Abstr.* **49**, 4295 (1955).
272. Y. Hishiki, *Kagaku Kenkyusho Hokoku* **33**, 65 (1957); *Chem. Abstr.* **52**, 6029 (1958).
273. B. W. Howk and J. C. Sauer, *J. Am. Chem. Soc.* **80**, 4607 (1958).
274. B. W. Howk and J. C. Sauer, U.S. Patent 2,840,613 (1958); *Chem. Abstr.* **52**, 17186 (1958).
275. B. W. Howk and J. C. Sauer, in "Organic Syntheses," Coll. Vol. IV, N. Rabjohn, Ed., p. 801. Wiley, New York, 1963.
276. W. Huber, *J. Am. Chem. Soc.* **65**, 2222 (1943).
277. W. Huber and H. A. Holscher, *Chem. Ber.* **71**, 87 (1938).
278. G. K. Hughes and F. Lions, *J. Proc. Roy. Soc. N. S. Wales* **71**, 494 (1938); *Chem. Abstr.* **33**, 588 (1939).
279. J. P. Huls and H. G. P. Van der Voort, *Rec. Trav. Chim.* **71**, 798 (1952).
280. M. E. Hultquist, U.S. Patent 2,459,076 (1949); *Chem. Abstr.* **43**, 4291 (1949).
281. H. H. Inhoffen, F. Bohlmann, and E. Reinfeld, *Chem. Ber.* **82**, 313 (1949).
282. International Polaroid Corp., Netherlands Patent Appl. 6,502,038 (1966); *Chem. Abstr.* **66**, 11873 (1967).
283. J. I. Iotsitch, V. L. Breitfous, R. J. Roudolf, N. Stassevitch, N. A. Elmanovitch, N. V. Kondyref, and D. A. Fomine, *J. Soc. Phys.-Chim. Russe* **39**, 652 (1907).
284. J. I. Iotsitch and F. Kochelov, *J. Russ. Phys.-Chem. Soc.* **42**, 1018 (1910).
285. J. I. Iotsitch and F. Kochelov, *J. Russ. Phys.-Chem. Soc.* **42**, 1491 (1910).
286. R. G. Jarque and A. G. Santos, *Anales Real Soc. Espan. Fis. Quim. (Madrid)* **B46**, 309 (1950); *Chem. Abstr.* **46**, 7043 (1952).
287. P. W. Jenkins and L. G. S. Brooker, French Patent 1,400,756 (1965); *Chem. Abstr.* **64**, 12858 (1966).
288. P. W. Jenkins and L. G. S. Brooker, French Patent 1,401,588 (1965); *Chem. Abstr.* **66**, 11874 (1967).
288a. P. W. Jenkins and L. G. S. Brooker, U.S. Patent 3,326,688 (1967); *Chem. Abstr.* **67**, 65435 (1967).
289. J. J. Jennen and H. Michaelis, U.S. Patent 2,495,260 (1950); *Chem. Abstr.* **44**, 7171 (1950).
290. J. B. Jepson, A. Lawson, and V. D. Lawton, *J. Chem. Soc.* p. 1791 (1955).
291. J. E. Jones, U. S. Patent 2,972,539 (1961); *Chem. Abstr.* **55**, 11152 (1961).
292. R. G. Jones, *J. Am. Chem. Soc.* **73**, 3684 (1951).
293. R. G. Jones, *J. Am. Chem. Soc.* **73**, 5168 (1951).
294. R. G. Jones, *J. Am. Chem. Soc.* **74**, 4889 (1952).
295. R. G. Jones and M. J. Mann, *J. Am. Chem. Soc.* **75**, 4048 (1953).
295a. J. S. Josan and F. W. Eastwood, *Australian J. Chem.* **21**, 2013 (1968).
296. H. V. Joy and M. T. Bogert, *J. Org. Chem.* **1**, 236 (1936).
296a. R. Justoni and R. Pessina, *Gazz. Chim. Ital.* **85**, 34 (1955).
297. S. A. Kagal and S. S. Karmarkar, *Proc. Indian Acad. Sci.* **A44**, 36 (1956).
298. P. Kainrath, *Angew. Chem.* **60**, 36 (1948).
299. M. J. Kamlet, *J. Org. Chem.* **24**, 714 (1959).
300. H. Kampfer, O. Riester, and J. Goetze, Belgian Patent 662,796 (1965); *Chem. Abstr.* **65**, 4008 (1966).
301. T. Kariyone and Y. Kimura, *J. Pharm. Soc. Japan* **500**, 746 (1923); *Chem. Abstr.* **18**, 386 (1924).

302. S. S. Karmarkar, *J. Sci. Ind. Res. (India)* **20B**, 334 (1961).
303. G. Kempter, H. Dost, and W. Schmidt, *Chem. Ber.* **98**, 945 (1965).
304. G. Kempter, W. Schmidt, and H. Dost, *Chem. Ber.* **98**, 955 (1965).
305. J. D. Kendall, British Patent 519,895 (1940); *Chem. Abstr.* **36**, 715 (1942).
305a. J. D. Kendall, British Patent 526,684 (1939).
306. J. D. Kendall and F. P. Doyle, U.S. Patent 2,533,816 (1950); *Chem. Abstr.* **45**, 2803 (1951).
307. J. D. Kendall and G. F. Duffin, British Patent 633,735 (1947); *Chem. Abstr.* **44**, 7175 (1950).
308. J. D. Kendall and G. F. Duffin, British Patent 633,736 (1949); *Chem. Abstr.* **44**, 7175 (1950).
309. J. D. Kendall and G. F. Duffin, British Patent 730,489 (1955); *Chem. Abstr.* **49**, 15580 (1955).
310. J. D. Kendall, G. F. Duffin, and H. R. J. Waddington, British Patent 862,825 (1961); *Chem. Abstr.* **56**, 15076 (1962).
311. J. D. Kendall and G. E. Ficken, British Patent 829,584 (1960); *Chem. Abstr.* **54**, 12847 (1960).
312. J. D. Kendall and D. J. Fry, British Patent 540,577 (1941); *Chem. Abstr.* **36**, 4041 (1942).
313. J. D. Kendall and D. J. Fry, British Patent 544,647 (1942); *Chem. Abstr.* **37**, 1046 (1943).
313a. G. Kesslin and C. M. Orlando, *J. Org. Chem.* **31**, 2682 (1966).
314. G. H. Keyes and L. G. S. Brooker, *J. Am. Chem. Soc.* **59**, 74 (1937).
315. M. S. Kharasch and O. Reinmuth, "Grignard Reactions of Non-metallic Substances," Chapter XV. Prentice-Hall, Englewood Cliffs, New Jersey, 1954.
316. S. A. Kheifets and N. N. Sveshnikov, *Tr. Vses. Nauchn.-Issled. Kinofotoinst.* (No. 40), 12 (1960); *Chem. Abstr.* **58**, 14154 (1963).
317. A. I. Kiprianov and F. I. Asnina, *Zh. Obshch. Khim.* **18**, 165 (1948).
318. A. I. Kiprianov and B. I. Dashevskaya, *Zh. Obshch. Khim.* **19**, 1158 (1949).
319. A. I. Kiprianov and I. P. Fedorova, *Zh. Obshch. Khim.* **28**, 1023 (1958).
320. A. I. Kiprianov and G. M. Golubushina, *Ukr. Khim. Zh.* **29**, 1173 (1963).
321. A. I. Kiprianov, Z. M. Ivanova, and S. G. Fridman, *Ukr. Khim. Zh.* **20**, 641 (1954).
322. A. I. Kiprianov and F. A. Mikhailenko, *Zh. Obshch. Khim.* **31**, 786 (1961).
323. A. I. Kiprianov and Y. L. Slominskii, *Zh. Organ. Khim.* **3**, 168 (1967).
324. A. I. Kiprianov and A. V. Stetsenko, *Ukr. Khim. Zh.* **19**, 508 (1953).
325. A. I. Kiprianov and A. V. Stetsenko, *Ukr. Khim. Zh.* **19**, 517 (1953).
326. A. I. Kiprianov, Z. P. Suitnik, and E. D. Sych, *Zh. Obshch. Khim.* **6**, 42 (1936).
327. A. I. Kiprianov, Z. P. Suitnik, and E. D. Sych, *Zh. Obshch. Khim.* **6**, 576 (1936).
328. A. I. Kiprianov and M. G. Suleimanova, *Ukr. Khim. Zh.* **31**, 1281 (1965).
329. A. I. Kiprianov and E. D. Sych, *Trans. Inst. Chim. Charkov* **2**, 15 (1936); *Chem. Abstr.* **32**, 4167 (1938).
330. A. I. Kiprianov and E. D. Sych, *Trans. Inst. Chim. Charkov* **2**, 25 (1936); *Chem. Abstr.* **32**, 4167 (1938).
331. A. I. Kiprianov, I. K. Ushenko, and E. D. Sych, *Zh. Obshch. Khim.* **15**, 200 (1945).
332. A. I. Kiprianov and I. K. Ushenko, *Zh. Obshch. Khim.* **15**, 207 (1945).
333. A. I. Kiprianov and I. K. Ushenko, *Zh. Obshch. Khim.* **20**, 134 (1950).
334. A. I. Kiprianov and T. M. Verbovskaya, *Zh. Obshch. Khim.* **32**, 3946 (1962).
335. A. I. Kiprianov and L. M. Yagupolskii, *Zh. Obshch. Khim.* **20**, 2111 (1950).
336. A. I. Kiprianov and L. M. Yagupolskii, *Zh. Obshch. Khim.* **22**, 2209 (1952).

337. A. I. Kiprianov and I. N. Zhmurova, *Dokl. Akad. Nauk. SSSR* **85**, 789 (1952).
338. A. I. Kiprianov and I. N. Zhmurova, *Zh. Obshch. Khim.* **23**, 493 (1953).
339. A. I. Kiprianov and I. N. Zhmurova, *Zh. Obshch. Khim.* **23**, 874 (1953).
340. E. C. Kirby and D. H. Reid, *J. Chem. Soc.* p. 1724 (1961).
341. W. Kirmse, "Carbene Chemistry," p. 55. Academic Press, New York, 1964.
342. A. Kirrman, *Ann. Chim. (Paris)* [10] **11**, 262 (1929).
343. V. T. Klimko and A. P. Skoldinov, *Zh. Obshch. Khim.* **29**, 4027 (1959).
344. E. B. Knott, U.S. Patent 2,515,878 (1950); *Chem. Abstr.* **44**, 8953 (1950).
345. E. B. Knott, *J. Chem. Soc.* p. 2399 (1952).
346. E. B. Knott, *J. Chem. Soc.* p. 1482 (1954).
347. E. B. Knott, U.S. Patent 2,691,581 (1954); *Chem. Abstr.* **49**, 83 (1955).
348. E. B. Knott, U.S. Patent 2,713,579 (1955); *Chem. Abstr.* **50**, 710 (1956).
349. E. B. Knott, U.S. Patent 2,728,766 (1955); *Chem. Abstr.* **50**, 10580 (1956).
350. E. B. Knott, U.S. Patent 2,743,273 (1956); *Chem. Abstr.* **51**, 907 (1957).
351. E. B. Knott, British Patent 738,197 (1955); *Chem. Abstr.* **52**, 939 (1958).
352. E. B. Knott, *J. Chem. Soc.* p. 1360 (1956).
353. E. B. Knott, U.S. Patent 2,839,404 (1958); *Chem. Abstr.* **53**, 946 (1959).
354. I. L. Knunyants and L. V. Razvadovskaya, *Zh. Obshch. Khim.* **9**, 557 (1939).
355. E. Kobayashi, Japanese Patent 2169 ('56); *Chem. Abstr.* **51**, 9677 (1957).
356. Kodak-Pathé, French Patent 712,995 (1931); *Chem. Abstr.* **26**, 1531 (1932).
357. Kodak-Pathé, French Patent 713,047 (1931); *Chem. Abstr.* **26**, 1531 (1932).
358. Kodak-Pathé, French Patent 1,150,715 (1958); *Chem. Abstr.* **54**, 12847 (1960).
359. Kodak Soc. Anon., Belgian Patent 598, 522 (1961); *Chem. Abstr.* **55**, 18403 (1961).
360. Kodak Soc. Anon., Belgian Patent 660,253 (1965); *Chem. Abstr.* **64**, 3744 (1966).
361. Kodak Soc. Anon., Belgian Patent 660,254 (1965); *Chem. Abstr.* **64**, 3745 (1966).
362. E. P. Kohler and R. G. Larsen, *J. Am. Chem. Soc.* **57**, 1448 (1935).
363. W. Konig, *J. Prakt. Chem.* [2] **84**, 216 (1911).
364. W. Konig, *Chem. Ber.* **55**, 3293 (1922).
365. W. Konig, *Chem. Ber.* **57**, 685 (1924).
366. W. Konig, German Patent 410,487 (1922).
367. W. Konig, French Patent 578,435 (1924); *Chem. Zentr.* **I**, 1454 (1925).
368. W. Konig, *Chem. Ber.* **61**, 2065 (1928).
368a. W. Konig, *Z. Wiss. Phot., Photophysik Photochem.* **34**, 1535 (1935).
369. W. Konig and W. Meier, *J. Prakt. Chem.* [2] **109**, 324 (1925).
370. M. Koral, A. Leifer, M. Collins, P. Dougherty, A. J. Fusco, and J. E. Lavalle, *J. Chem. Eng. Data* **10**, 67 (1965).
371. L. M. Koutcheroff, *J. Russ. Phys.-Chem. Soc.* **38**, 1176 (1906).
372. Z. Y. Krainer and P. F. Gudz, *Nauk. Zap. Kremenes'k. Derzh. Inst.* **7**, 129 (1962); *Chem. Abstr.* **61**, 5819 (1964).
373. A. L. Kranzfelder and R. R. Vogt, *J. Am. Chem. Soc.* **60**, 1714 (1938).
373a. D. Kritchevsky, *J. Am. Chem. Soc.* **65**, 487 (1943).
374. A. Kroutzberger, *Arch. Pharm.* **299**, 984 (1966); *Chem. Abstr.* **66**, 96205 (1967).
375. V. F. Kucherov and S. S. Yufit, *Vopr. Khim. Terpenov i Terpenoidov, Akad. Nauk Lit. SSR, Tr. Vses. Sovesbchanya, Vilnyus,* p. 197 (1959); *Chem. Abstr.* **55**, 15395 (1961).
375a. D. G. Kundiger and E. E. Richardson, *J. Am. Chem. Soc.* **77**, 2897 (1955).
376. Laboratoires Dausse S. A., French Patent 1,327,160 (1963); *Chem. Abstr.* **60**, 460 (1964).
377. A. Ladenberg, *Chem. Ber.* **5**, 752 (1872).
378. A. B. Lal, *J. Indian Chem. Soc.* **36**, 64 (1959).

379. A. B. Lal, *Chem. Ber.* **94**, 1723 (1961).
380. H. Larive and R. Dennilauler, *Chimia (Aarau)* **15**, 115 (1961); *Chem. Abstr.* **61**, 12116 (1964).
381. H. Larive and E. M. Geiger, U.S. Patent 2,909,067 (1960); *Chem. Abstr.* **54**, 9576 (1960).
382. H. Larive and E. M. Geiger, U.S. Patent 2,921,067 (1960); *Chem. Abstr.* **54**, 9576 (1960).
382a. R. A. Letch and R. P. Linstead, *J. Chem. Soc.* p. 443 (1932).
383. I. I. Levkoev, M. V. Deichmeister, and N. N. Sveshnikov, *Dokl. Akad. Nauk SSSR* **129**, 331 (1959).
384. I. I. Levkoev and V. V. Durmashkina, *Tr. Vses. Nauchn.-Issled. Kinofotoinst.* (No. 40), p. 21 (1960); *Chem. Abstr.* **58**, 14154 (1963).
385. I. I. Levkoev, S. A. Kheifets, and N. S. Barvyn, *Zh. Obshch. Khim.* **21**, 1340 (1951).
386. I. I. Levkoev and B. S. Portnaya, *Zh. Obshch. Khim.* **21**, 2050 (1951).
387. I. I. Levkoev and N. N. Sveshnikov, *Zh. Obshch. Khim.* **16**, 1655 (1946).
388. I. I. Levkoev, N. N. Sveshnikov, and S. A. Kheifets, *Zh. Obshch. Khim.* **16**, 1489 (1946).
389. I. I. Levkoev, N. N. Sveshnikov, and E. Z. Kulik, *Zh. Obshch. Khim.* **27**, 3097 (1957).
390. I. I. Levkoev, Z. P. Sytnik, and R. S. Shusher, *Zh. Obshch. Khim.* **24**, 2034 (1954).
391. I. I. Levkoev, A. F. Vompe, and N. N. Sveshnikov, *Zh. Obshch. Khim.* **22**, 879 (1952).
392. I. I. Levkoev, V. G. Zhiryakov, N. N. Sveshnikov, and N. S. Barvyn, *Sb. Statei Obshch. Khim., Akad. Nauk SSSR* **2**, 1263 (1953); *Chem. Abstr.* **49**, 5443 (1955).
393. C. P. Lo and W. J. Croxall, *J. Am. Chem. Soc.* **76**, 4166 (1954).
394. H. Lohaus, *J. Prakt. Chem.* [2] **119**, 245 (1928).
395. H. Lorenz and R. Wizinger, *Helv. Chim. Acta* **28**, 600 (1945).
395a. S. M. McElvain, *Chem. Rev.* **45**, 453 (1949).
395b. S. M. McElvain and C. L. Aldridge, *J. Am. Chem. Soc.* **75**, 3993 (1953).
395c. S. M. McElvain, H. I. Anthes, and S. H. Shapiro, *J. Am. Chem. Soc.* **64**, 2525 (1942).
395d. S. M. McElvain and R. L. Clarke, *J. Am. Chem. Soc.* **69**, 2661 (1947).
396. S. M. McElvain, R. L. Clarke, and C. D. Jones, *J. Am. Chem. Soc.* **64**, 1966 (1942).
396a. S. M. McElvain and D. H. Clemens, *J. Am. Chem. Soc.* **80**, 3915 (1958).
396b. S. M. McElvain and W. R. Davie, *J. Am. Chem. Soc.* **73**, 1400 (1951).
396c. S. M. McElvain and G. R. McKay, *J. Am. Chem. Soc.* **77**, 5601 (1955).
396d. S. M. McElvain and H. F. McShane, *J. Am. Chem. Soc.* **74**, 2662 (1952).
397. S. M. McElvain and L. R. Morris, *J. Am. Chem. Soc.* **74**, 2657 (1952).
397a. S. M. McElvain and R. D. Mullineaux, *J. Am. Chem. Soc.* **74**, 1811 (1952).
398. S. M. McElvain and J. W. Nelson, *J. Am. Chem. Soc.* **64**, 1825 (1942).
398a. S. M. McElvain and J. P. Schroeder, *J. Am. Chem. Soc.* **71**, 47 (1949).
398b. S. M. McElvain and R. E. Starn, *J. Am. Chem. Soc.* **77**, 4571 (1955).
398c. S. M. McElvain and C. L. Stevens, *J. Am. Chem. Soc.* **68**, 1917 (1946).
398d. S. M. McElvain and J. T. Venerable, *J. Am. Chem. Soc.* **72**, 1661 (1950).
398e. S. M. McElvain and P. M. Walters, *J. Am. Chem. Soc.* **64**, 1963 (1942).
399. L. Mandacescu, L. Stoicescu-Crivat, I. Gobe, S. Lica, and M. Stefanescu, *Acad. Rep. Populare Romine, Filiala Iasi, Studii Cercetari Stiint., Chim.* **13**, 115 (1962); *Chem. Abstr.* **59**, 4069 (1963).
399a. R. Mantione, M. L. Martin, G. J. Martin, and H. Normant, *Bull. Soc. Chim. France* p. 2912 (1967).

400. B. Mariani and R. Sgarbi, *Chim. Ind. (Milan)* **46**, 630 (1964); *Chem. Abstr.* **61**, 12117 (1964).
400a. I. Marszak and R. Epsztein, French Patent 1,455,862 (1966); *Chem. Abstr.* **67**, 53687 (1967).
401. E. L. Martin, U.S. Patent 2,647,052 (1953); *Chem. Abstr.* **47**, 11057 (1953).
402. H. Meerwein, K. Bodenbenner, P. Borner, F. Kunert, and K. Wunderlich, *Ann. Chem.* **632**, 38 (1960).
403. C. E. K. Mees, "The Theory of the Photographic Process," Chapter XXIV. Macmillan, New York, 1942.
404. C. E. K. Mees and T. H. James, "The Theory of the Photographic Process," 3rd Ed. Chapter 11. Macmillan, New York, 1966.
405. J. Metzger, H. Larive, R. Dennilauler, R. Baralle, and C. Gaurat, *Bull. Soc. Chim. France* p. 2888 (1964).
406. J. Metzger, H. Larive, R. Dennilauler, R. Baralle, and C. Gaurat, *Bull. Soc. Chim. France* p. 40 (1967).
407. J. Metzger, H. Larive, E. J. Vincent, and R. Dennilauler, *Bull. Soc. Chim. France* p. 46 (1967).
408. C. Michaelidis and R. Wizinger, *Helv. Chim. Acta* **34**, 1776 (1951).
409. E. B. Middleton, U.S. Patent 2,524,674 (1950); *Chem. Abstr.* **45**, 2348 (1951).
410. E. B. Middleton and G. A. Dawson, U.S. Patent 2,201,816 (1940).
411. E. B. Middleton and G. A. Dawson, U.S. Patent 2,424,483 (1947); *Chem. Abstr.* **42**, 1141 (1948).
412. L. Miginiac-Groizeleau, *Ann. Chim. (Paris)* [13] **6**, 1071 (1961).
412a. T. Miki, K. Hiraga, T. Asako, and H. Masuya, *Chem. & Pharm. Bull. (Tokyo)* **15**, 670 (1967).
413. H. F. Miller and G. B. Bachman, *J. Am. Chem. Soc.* **57**, 766 (1935).
414. Z. I. Miroshnichenko and M. A. Alperovich, *Zh. Obshch. Khim.* **34**, 241 (1964).
415. Y. Mizuno and M. Nishimura, *J. Pharm. Soc. Japan* **68**, 54 (1948); *Chem. Abstr.* **44**, 331 (1950).
416. Y. Mizuno and Y. Tanabe, *J. Pharm. Soc. Japan* **73**, 227 (1951); *Chem. Abstr.* **48**, 475 (1954).
416a. V. B. Mochalin and N. G. Ivanova, *Zh. Obshch. Khim.* **31**, 3896 (1961).
416b. K. Morita, M. Nishimura, and Z. Suzuki, *J. Org. Chem.* **30**, 533 (1965).
416c. K. Morita, G. Slomp, and E. V. Jensen, *J. Am. Chem. Soc.* **84**, 3779 (1962).
416d. K. Morita and Z. Suzuki, *Tetrahedron Letters* p. 263 (1964).
416e. Z. I. Moskalenko and M. A. Alperovich, *Khim. Geterosikl. Soedin., Akad. Nauk Latv. SSR* p. 492 (1967); *Chem. Abstr.* **68**, 14063 (1968).
417. C. Moureu and R. Delange, *Bull. Soc. Chim. France* [3] **31**, 1331 (1904).
418. C. Moureu and R. Delange, *Compt. Rend.* **138**, 1339 (1904).
419. M. Mousseron and R. Granger, *Compt. Rend.* **217**, 483 (1943).
420. O. Mumm, H. Hinz, and J. Diedericksen, *Chem. Ber.* **72**, 2107 (1939).
421. L. K. Mushkalo, *Nauk Zap., Kiivs'k. Derzh. Univ.* **16**, No. 15, 133 (1957); *Chem. Abstr.* **53**, 18057 (1959).
422. L. K. Mushkalo and N. K. Mikhailyuchenko, *Ukr. Zh. Khim.* **30**, 202 (1964).
423. L. K. Mushkalo and F. I. Skokol, *Zh. Obshch. Khim.* **31**, 3069 (1961).
424. A. Nagasaka, S. Nakamura, and R. Oda, *J. Chem. Soc. Japan, Ind. Chem. Sect.* **58**, 460 (1955); *Chem. Abstr.* **50**, 4789 (1956).
425. A. Nagasaka, S. Nakamura, and R. Oda, *Bull. Inst. Chem. Res., Kyoto Univ.* **33**, 85 (1955); *Chem. Abstr.* **50**, 7063 (1956).

426. A. Nagasaka and R. Oda, *Bull. Inst. Chem. Res., Kyoto Univ.* **32**, 238 (1954); *Chem. Abstr.* **50**, 7062 (1956).
427. A. Nagasaka, R. Oda, and S. Nukina, *J. Chem. Soc. Japan, Ind. Chem. Sect.* **57**, 169 (1954); *Chem. Abstr.* **49**, 11626 (1955).
428. F. Nagasawa and E. Kobayashi, Japanese Patent 4321 ('52); *Chem. Abstr.* **48**, 5211 (1954).
429. I. N. Nazarov, S. M. Makin, and B. P. Kruptsov, *Zh. Obshch. Khim.* **29**, 3683 (1959).
430. O. Neunhoeffer and A. Keiler, *Chem. Ber.* **91**, 122 (1958).
431. O. Neunhoeffer and W. Weigel, *Ann. Chem.* **647**, 101 (1961).
432. O. Neunhoeffer and W. Weigel, *Ann. Chem.* **647**, 108 (1961).
433. L. Nicholl, P. J. Tarsio, and H. Blohm, U.S. Patent 2,824,121 (1948); *Chem. Abstr.* **52**, 11909 (1958).
434. E. Noelting, *Chem. Ber.* **24**, 553 (1898).
435. E. Noelting, *Chem. Ber.* **37**, 1916 (1904).
436. H. Normant, *Compt. Rend.* **240**, 1435 (1955).
437. J. Nys and H. Depoorter, German Patent 1,081,311 (1960); *Chem. Abstr.* **57**, 328 (1962).
438. J. Nys and H. Depoorter, Belgian Patent 583,922 (1960); *Chem. Abstr.* **57**, 331 (1962).
439. J. Nys and H. Depoorter, British Patent 886,271 (1962); *Chem. Abstr.* **57**, 12008 (1962).
440. J. Nys and J. Libeer, *Bull. Soc. Chim. Belges* **65**, 377 (1956).
441. J. Nys and M. J. Libeer, Belgian Patent 568,840 (1958); *Chem. Abstr.* **54**, 24051 (1960).
442. J. Nys and J. Libeer, U.S. Patent 2,954,376 (1960); *Chem. Abstr.* **55**, 7114 (1961).
443. J. Nys and J. Libeer, British Patent 886,270 (1962); *Chem. Abstr.* **57**, 12010 (1962).
444. S. Oae, W. Tagaki, and A. Ohno, *Tetrahedron* **20**, 417 (1964).
445. T. Ogata, *Proc. Imp. Acad. (Tokyo)* **3**, 336 (1927); *Chem. Abstr.* **21**, 3201 (1927).
446. T. Ogata, *Proc. Imp. Acad. (Tokyo)* **8**, 503 (1932); *Chem. Abstr.* **27**, 1631 (1933).
447. T. Ogata, *Proc. Imp. Acad. (Tokyo)* **9**, 602 (1933); *Chem. Abstr.* **28**, 2007 (1934).
448. T. Ogata, *Proc. Imp. Acad. (Tokyo)* **10**, 572 (1934); *Chem. Abstr.* **29**, 1815 (1935).
449. T. Ogata, *J. Chem. Soc. Japan, Pure Chem. Sect.* **55**, 394 (1934); *Chem. Abstr.* **28**, 5816 (1934).
450. T. Ogata, *J. Soc. Sci. Phot. Japan* **13**, No. 2, 24 (1950); *Chem. Abstr.* **46**, 6531 (1952).
451. T. Ogata, *Proc. Japan Acad.* **25**, No. 11, 17 (1950); *Chem. Abstr.* **46**, 4531 (1952).
452. T. Ogata, K. Kawasaki, and M. Masuda, *Bull. Inst. Phys. Chem. Res. (Tokyo)* **13**, 486 (1934).
453. T. Ogata, R. Tanno, and K. Nishida, *Rept. Sci. Res. Inst. (Tokyo)* **28**, 259 (1952); *Chem. Abstr.* **47**, 5284 (1953).
453a. R. A. Olofson, S. W. Walinsky, J. P. Mariano, and J. L. Jernow, *J. Am. Chem. Soc.* **90**, 6554 (1968).
454. M. Pailer and E. Kuhn, *Monatsh. Chem.* **84**, 85 (1953).
455. M. Pailer and E. Renner-Kuhn, *Monatsh. Chem.* **85**, 601 (1954).
456. M. H. Palomaa and T. K. Kaski, *Chem. Ber.* **72**, 317 (1939).
457. L. Panizzi, *Gazz. Chim. Ital.* **73**, 13 (1943).
458. L. Panizzi, *Gazz. Chim. Ital.* **77**, 283 (1947).
459. W. E. Parham and L. J. Reed, *in* "Organic Syntheses," Coll. Vol. III, E. C. Horning, Ed., p. 395. Wiley, New York, 1955.

460. T. Passalacqua, *Gazz. Chim. Ital.* **43**, Part II, 566 (1913).
461. Y. Pasternak and J. C. Traynard, *Bull. Soc. Chim. France* p. 356 (1966).
461a. F. Piacenti, *Gazz. Chim. Ital.* **92**, 225 (1962).
461b. F. Piacenti, C. Coni, and P. Pino, *Chem. & Ind. (London)* p. 1240 (1960).
461c. F. Piacenti and P. P. Neggiani, *Chim. Ind. (Milan)* **44**, 1396 (1962); *Chem. Abstr.* **60**, 11888 (1964).
462. G. T. Pilyugin, *Izv. Akad. Nauk SSSR, Otd. Khim. Nauk* p. 519 (1952).
463. G. T. Pilyugin, *Izv. Akad. Nauk SSSR, Otd. Khim. Nauk* p. 1068 (1953).
463a. G. T. Pilyugin, *Zh. Obshch. Khim.* **25**, 793 (1955).
464. G. T. Pilyugin and I. N. Chernyuk, *Zh. Obshch. Khim.* **32**, 1404 (1962).
465. G. T. Pilyugin and I. N. Chernyuk, *Zh. Obshch. Khim.* **34**, 201 (1964).
466. G. T. Pilyugin, A. V. Dombrovskii, B. M. Gutsulyak, and N. I. Ganuschak, *Zh. Obshch. Khim.* **32**, 1411 (1963).
467. G. T. Pilyugin, Y. O. Gorichok, B. M. Gutsulyak, and S. I. Gorichok, *Khim. Geterosikl. Soedin., Akad. Nauk Latv. SSR* p. 896 (1965); *Chem. Abstr.* **64**, 19838 (1966).
468. G. T. Pilyugin and B. M. Gutsulyak, *Zh. Obshch. Khim.* **29**, 3076 (1959).
469. G. T. Pilyugin and B. M. Gutsulyak, *Zh. Obshch. Khim.* **31**, 623 (1961).
470. G. T. Pilyugin and B. M. Gutsulyak, *Zh. Obshch. Khim.* **32**, 1050 (1962).
471. G. T. Pilyugin, B. M. Gutsulyak, and Y. O. Gorichok, *Zh. Obshch. Khim.* **34**, 1992 (1964).
472. G. T. Pilyugin and Z. Y. Krainer, *Dokl. Akad. Nauk. SSSR* **81**, 609 (1951).
473. G. T. Pilyugin and Z. Y. Krainer, *Zh. Obshch. Khim.* **23**, 634 (1953).
474. G. T. Pilyugin and Z. Y. Krainer, *Zh. Obshch. Khim.* **25**, 2271 (1955).
475. G. T. Pilyugin and E. P. Opanasenko, *Zh. Obshch. Khim.* **27**, 1015 (1957).
476. G. T. Pilyugin and E. P. Opanasenko, *Zh. Obshch. Khim.* **30**, 1303 (1960).
477. G. T. Pilyugin and E. P. Opanasenko, *Zh. Obshch. Khim.* **31**, 1233 (1961).
478. G. T. Pilyugin, E. P. Opanasenko, and A. M. Isak, *Zh. Obshch. Khim.* **32**, 1398 (1962).
479. G. T. Pilyugin, E. P. Opanasenko, and O. E. Petrenko, *Zh. Obshch. Khim.* **33**, 3228 (1963).
480. G. T. Pilyugin, O. E. Petrenko, and E. P. Opanasenko, *Zh. Obshch. Khim.* **34**, 3333 (1964).
481. G. T. Pilyugin, O. E. Petrenko, and E. P. Opanasenko, *Zh. Obshch. Khim.* **34**, 3337 (1964).
482. G. T. Pilyugin and O. E. Petrenko, *Zh. Organ. Khim.* **1**, 1143 (1965).
483. G. T. Pilyugin, O. E. Petrenko, and E. P. Opanasenko, *Probl. Organ. Sinteza, Akad. Nauk SSSR, Otd. Obshch. i Tekhn. Khim.* p. 173 (1965); *Chem. Abstr.* **64**, 19837 (1966).
484. G. T. Pilyugin and N. A. Tsvetkova, *Zh. Obshch. Khim.* **34**, 3341 (1964).
485. G. T. Pilyugin and L. E. Zhviglazova, *Zh. Vses. Khim. Obshchestva im. D.I. Mendeleeva* **12**, 593 (1967); *Chem. Abstr.* **68**, 70160 (1968).
486. P. Pino, *Gazz. Chim. Ital.* **81**, 625 (1951).
486a. P. Pino, F. Piacenti, and E. Mantica, *Chim. Ind. (Milan)* **38**, 34 (1956); *Chem. Abstr.* **50**, 12828 (1956).
487. P. Pino, F. Piacenti, and P. P. Neggiani, *Chim. Ind. (Milan)* **44**, 1367 (1962); *Chem. Abstr.* **59**, 429 (1963).
488. H. Pleininger, H. Bauer, and A. R. Katrizky, *Ann. Chem.* **654**, 165 (1962).
489. H. Pleininger and D. Wild, *Chem. Ber.* **99**, 3063 (1966).
490. H. W. Post and E. R. Erickson, *J. Org. Chem.* **2**, 260 (1937).

491. N. A. Preobrazhenskii, *Zh. Obshch. Khim.* **15**, 952 (1945).
492. C. C. Price, N. J. Leonard, and H. F. Herbrandson, *J. Am. Chem. Soc.* **68**, 1252 (1946).
493. T. V. Protopopova and A. P. Skoldinov, *Zh. Obshch. Khim.* **27**, 57 (1957).
494. M. Prystas and J. Gut, *Collection. Czech. Chem. Commun.* **28**, 2501 (1963).
495. R. Quelet, *Bull. Soc. Chim. France* [4] **45**, 255 (1929).
496. A. Quilico, G. Gaudiano, and L. Merlini, *Tetrahedron* **2**, 359 (1958).
497. R. K. Ralph and G. Shaw, *J. Chem. Soc.* p. 1877 (1956).
498. Y. S. Rao and R. Filler, *Chem. & Ind. (London)* p. 280 (1964).
499. R. A. Raphael and F. Sondheimer, *J. Chem. Soc.* p. 2693 (1951).
500. P. C. Rath, B. K. Sabata, and M. K. Rout, *J. Indian Chem. Soc.* **41**, 803 (1964).
500a. H. Reitter and A. Weindel, *Chem. Ber.* **40**, 3358 (1907).
501. M. Ridi and S. Checchi, *Gazz. Chim. Ital.* **83**, 36 (1953).
502. W. Ried and R. Bender, *Chem. Ber.* **91**, 2798 (1958).
503. W. Ried and E. Köcher, *Ann. Chem.* **647**, 144 (1961).
504. O. Riester, Belgian Patent 650,828 (1965); *Chem. Abstr.* **64**, 17760 (1966).
505. A. Roedig and E. Degener, *Chem. Ber.* **86**, 1469 (1953).
506. M. I. Rogovik, *Nauk. Zap., Chernivets'k. Derzh. Univ.* **53**, 74 (1961); *Chem. Abstr.* **61**, 4518 (1964).
507. M. I. Rogovic and G. T. Pilyugin, *Zh. Obshch. Khim.* **34**, 3326 (1964).
508. A. E. Rosenoff, U.S. Patent 3,177,210 (1965); *Chem. Abstr.* **62**, 16423 (1965).
508a. A. E. Rosenoff, U.S. Patent 3,354,170 (1967); *Chem. Abstr.* **68**, 105207 (1968).
508b. R. Rossi, P. Pino, F. Piacenti, L. Lardicci, and G. DelBino, *J. Org. Chem.* **32**, 842 (1967).
508c R. Rossi, P. Pino, F. Piacenti, L. Lardicci, and G. DelBino, *Gazz. Chim. Ital.* **97**, 1194 (1967).
509. C. B. Roth and L. Horsitz, German Patent 1,081,756 (1960); *Chem. Abstr.* **59**, 8914 (1963).
510. F. M. Rowe and H. J. Twitchett, *J. Chem. Soc.* p. 1704 (1936).
511. Y. S. Rozum and V. A. Gladkaya, *Ukr. Khim. Zh.* **32**, 1200 (1966).
512. P. B. Russell and N. Whittaker, *J. Am. Chem. Soc.* **74**, 1310 (1952).
513. P. P. T. Sah, *J. Am. Chem. Soc.* **64**, 1487 (1942).
514. P. P. T. Sah, *J. Chinese Chem. Soc.* **13**, 89 (1946); *Chem. Abstr.* **41**, 5869 (1947).
515. V. R. Sathe and K. Vankataraman, *Current Sci. (India)* **18**, 373 (1949).
516. J. Savard and R. Hosgut, *Rev. Fac. Sci. Univ. Istanbul* **3**, 164 (1938); *Chem. Abstr.* **32**, 5795 (1938).
516a. R. Scarpati, M. L. Graziano, and R. A. Nicolaus, *Gazz. Chim. Ital.* **97**, 1317 (1967)
517. A. Schoenberg and K. Praefke, *Tetrahedron Letters* p. 2043 (1964).
518. A. Schoenberg and K. Praefke, *Chem. Ber.* **99**, 196 (1966).
519. A. Schoenberg, K. Praefke, and J. Kohtz, *Chem. Ber.* **99**, 2433 (1966).
520. A. Schoenberg, K. Praefke, and J. Kohtz, *Chem. Ber.* **99**, 3076 (1966).
521. C. Schopf and W. Arnold, *Ann. Chem.* **558**, 117 (1947).
521a. W. M. Schubert, W. A. Lanka, and T. H. Liddicoet, *Science* **116**, 124 (1962).
522. G. Schwarz and P. DeSmet, British Patent 615,205 (1949); *Chem. Abstr.* **43**, 8293 (1949).
523. G. Schwarz and P. F. DeSmet, U.S. Patent 2,592,196 (1952); *Chem. Abstr.* **46**, 6022 (1952).
524. G. Schwarz and M. Shcowenaars, British Patent 625,245 (1949); *Chem. Abstr.* **45**, 3269 (1951).

524a. H. Seifert, *Monatsh. Chem.* **79**, 198 (1948).
525. G. Shaw, *J. Chem. Soc.* p. 1834 (1955).
526. A. Sieglitz, L. Berlin, and H. Hamal, U.S. Patent 2,770,620 (1956); *Chem. Abstr.* **51**, 5606 (1957).
527. A. Sieglitz, L. Berlin, and P. Heimke, German Patent 883,025 (1953); *Chem. Abstr.* **53**, 7840 (1959).
528. A. Sieglitz, L. Berlin, P. Heimke, and B. Schacke, German Patent 902,290 (1954); *Chem. Abstr.* **52**, 13495 (1958).
529. L. I. Smith and M. Bayliss, *J. Org. Chem.* **6**, 437 (1941).
530. L. I. Smith and J. Nichols, *J. Org. Chem.* **6**, 489 (1941).
531. W. Sobeki, *Chem. Ber.* **43**, 1038 (1910).
532. R. Sokolovsky, *J. Russ. Phys.-Chem. Soc.* **37**, 886 (1905).
533. F. Sondheimer, *J. Am. Chem. Soc.* **74**, 4040 (1952).
534. F. Sorm and J. Smrt, *Chem. Listy* **47**, 413 (1953).
535. H. Spath, *Chem. Ber.* **47**, 766 (1914).
536. R. H. Sprague, *J. Am. Chem. Soc.* **79**, 2275 (1957).
537. H. Staudinger and G. Rathsam, *Helv. Chim. Acta* **5**, 645 (1922).
538. H. E. Stavely and M. Berestecki, *J. Am. Chem. Soc.* **73**, 3448 (1951).
539. W. Steinkopf and H. Schmitt, *Ann. Chem.* **533**, 264 (1938).
540. F. N. Stepanov and N. A. Aldanova, *Zh. Obshch. Khim.* **29**, 339 (1954).
541. F. N. Stepanov and L. I. Lukashina, *Zh. Obshch. Khim.* **29**, 2792 (1959).
542. A. V. Stetsenko and L. I. Filileeva, *Ukr. Khim. Zh.* **32**, 853 (1966).
543. W. Stevens and R. Wizinger, *Helv. Chim. Acta* **44**, 1708 (1961).
544. E. T. Stiller, U.S. Patent 2,422,598 (1947); *Chem. Abstr.* **41**, 5903 (1947).
545. P. R. Story and M. Saunders, *J. Am. Chem. Soc.* **84**, 4876 (1962).
546. K. Sugimoto, *Rept. Sci. Res. Inst. (Tokyo)* **25**, 265 (1949); *Chem. Abstr.* **45**, 7124 (1951).
547. J. Swiderski and J. Oszczapowics, *Roczniki Chem.* **37**, 1059 (1963); *Chem. Abstr.* **60**, 8165 (1964).
548. E. D. Sych, *Ukr. Khim. Zh.* **18**, 148 (1952).
549. E. D. Sych, *Ukr. Khim. Zh.* **22**, 217 (1956).
550. E. D. Sych, *Ukr. Khim. Zh.* **22**, 80 (1956).
551. E. D. Sych, *Ukr. Khim. Zh.* **24**, 79 (1958).
552. E. D. Sych, *Ukr. Khim. Zh.* **24**, 372 (1958).
553. E. D. Sych and E. D. Smaznaya-Ilina, *Zh. Obshch. Khim.* **33**, 74 (1963).
554. E. D. Sych and L. P. Umanskaya, *Ukr. Khim. Zh.* **28**, 714 (1962).
555. E. D. Sych and L. P. Umanskaya, *Ukr. Khim. Zh.* **30**, 201 (1964).
556. E. D. Sych and L. P. Umanskaya, *Zh. Obshch. Khim.* **34**, 2068 (1964).
557. E. D. Sych, L. P. Umanskaya, and E. K. Perkovskaya, *Ukr. Khim. Zh.* **33**, 68 (1967).
558. Z. P. Sytnik, I. L. Lezkoev, T. G. Gnevysheva, and K. I. Pokrovskaya, U.S.S.R. Patent 104,295 (1958); *Chem. Abstr.* **60**, 8174 (1964).
559. R. C. Taber and L. G. S. Brooker, Belgian Patent 663,379 (1965); *Chem. Abstr.* **66**, 19847 (1967).
559a. T. Takahashi and T. Goto, *J. Pharm. Soc. Japan* **66**, 2 (1946); *Chem. Abstr.* **45**, 8531 (1951).
560. I. K. Taki, *J. Sci. Res. Inst. (Tokyo)* **45**, 95 (1951); *Chem. Abstr.* **46**, 4530 (1952).
560aa. S. Tanimoto, S. Shimojo, and R. Oda, *Yuki Gosei Kagaku Kyokai Shi* **26**, 435 (1968); *Chem. Abstr.* **69**, 77137 (1968).
560a. E. C. Taylor and E. E. Garcia, *J. Org. Chem.* **29**, 2116 (1964).

560b. N. A. Tikhonova, K. K. Babievskii, and V. M. Belikov, *Izv. Akad. Nauk SSSR, Ser. Khim.* p. 877 (1967).
560c. W. Treibe, *Tetrahedron Letters* p. 4707 (1967).
561. P. B. Tripathy and M. K. Rout, *J. Indian Chem. Soc.* **39**, 103 (1962).
562. A. T. Troschenko, *Zh. Obshch. Khim.* **9**, 1661 (1930).
562a. T. Tsukamoto, Japanese Patent 4510 ('56); *Chem. Abstr.* **51**, 17985 (1957).
562b. T. Tsukamoto and T. Suzuki, Japanese Patent 3071 ('55); *Chem. Abstr.* **51**, 16519 (1957).
562c. T. Tsukamoto, T. Suzuki, K. Heijo, S. Takebi, G. Sudo, and Y. Tanaka, U.S. Patent 2,823,226 (1958); *Chem. Abstr.* **52**, 10161 (1958).
563. W. B. Tuemmler and B. S. Wildi, *J. Am. Chem. Soc.* **80**, 3772 (1958).
564. G. Y. Turchinovich, *Tr. Kievsk. Politekhn. Inst.* **43**, 81 (1963); *Chem. Abstr.* **62**, 10559 (1965).
564a. G. Y. Turchinovich, *Vestn. Kievsk. Politekhn. Inst., Ser. Khim. Mashinostr. Teckhnol.* **2**, 26 (1966); *Chem. Abstr.* **67**, 118074 (1967).
565. N. F. Turitsyna and I. I. Levkoev, *Zh. Obshch. Khim.* **22**, 309 (1952).
566. Y. Urishibara and M. Takebayashi, *Bull. Chem. Soc. Japan* **11**, 557 (1936).
567. I. K. Ushenko, *Ukr. Khim. Zh.* **16**, 450 (1950).
568. I. K. Ushenko, *Zh. Obshch. Khim.* **22**, 711 (1952).
569. I. K. Ushenko, *Zh. Obshch. Khim.* **22**, 870 (1952).
570. I. K. Ushenko, *Ukr. Khim. Zh.* **21**, 738 (1955).
571. I. K. Ushenko, *Ukr. Khim. Zh.* **22**, 76 (1956).
572. I. K. Ushenko, *Zh. Obshch. Khim.* **31**, 2854 (1961).
573. I. K. Ushenko, *Zh. Obshch. Khim.* **31**, 2861 (1961).
574. I. K. Ushenko, *Nauchn. Zap., Donetsku Inst. Sov. Torgovli, Ser. Tavarovedno-Tekhnol. Nauk* p. 153 (1962); *Chem. Abstr.* **62**, 7902 (1965).
575. I. K. Ushenko and I. P. Dmitrenko, *Sb. Statei Obshch. Khim., Akad. Nauk SSSR* **1**, 650 (1953); *Chem. Abstr.* **49**, 1023 (1955).
576. I. K. Ushenko and S. E. Gornostaeva, *Zh. Obshch. Khim.* **28**, 1668 (1958).
577. I. K. Ushenko and V. A. Portnyagina, *Ukr. Khim. Zh.* **21**, 744 (1955).
578. I. K. Ushenko, F. Z. Rodova, and V. I. Korystov, *Zh. Obshch. Khim.* **32**, 3650 (1962).
579. I. K. Ushenko and K. D. Zhikerava, *Zh. Obshch. Khim.* **32**, 3656 (1962).
580. I. K. Ushenko, K. D. Zhikareva, and F. Z. Rodova, *Zh. Obshch. Khim.* **33**, 798 (1963).
581. A. L. Van der Auwera, Netherlands Patent Appl. 6,605,280 (1966); *Chem. Abstr.* **65**. 20260 (1966).
582. A. Van Dormael, *Bull. Soc. Chim. Belges* **58**, 167 (1949).
583. A. Van Dormael and J. Libeer, *Sci. Ind. Phot.* [2] **20**, 451 (1949); *Chem. Abstr.* **44**, 1839 (1950).
584. E. J. Van Lare, U.S. Patent 2,739,149 (1956).
585. E. J. Van Lare and L. G. S. Brooker, U.S. Patent 2,691,652 (1954); *Chem. Abstr.* **49**, 2916 (1955).
586. G. Van Zandt and L. G. S. Brooker, U.S. Patent 2,485,679 (1949); *Chem. Abstr.* **44**, 7694 (1950).
587. P. Viguer, *Compt. Rend.* **152**, 1490 (1911).
588. P. Viguer, *Ann. Chim. (Paris)* [8] **28**, 450 (1913).
589. J. P. Vila and M. Ballester, *Anales Real Soc. Espan. Fis. Quim. (Madrid)* **B45**, 87 (1949).
590. J. P. Vila and R. G. Jarque, *Anales Fis. Quim. (Madrid)* **40**, 946 (1944).

591. J. P. Vila and F. Serratosa, *Chem. Ber.* **85**, 686 (1952).
592. J. P. Vila and F. Serratosa, *Anales Real Soc. Espan. Fis. Quim. (Madrid)* **B50**, 471 (1954).
593. W. H. Vinton, U.S. Patent 2,647,053 (1953); *Chem. Abstr.* **47**, 11057 (1953).
594. W. H. Vinton, U.S. Patent 2,647,115 (1953); *Chem. Abstr.* **48**, 65 (1954).
595. W. H. Vinton, U.S. Patent 2,647,116 (1953); *Chem. Abstr.* **48**, 66 (1954).
595a. J. von Braun and O. Kruber, *Chem. Ber.* **45**, 393 (1912).
596. H. R. Waddington, G. F. Duffin, and J. D. Kendall, British Patent 785,334 (1957); *Chem. Abstr.* **52**, 6030 (1958).
597. H. R. Waddington, G. F. Duffin, and J. D. Kendall, British Patent 848,016 (1960); *Chem. Abstr.* **55**, 5205 (1961).
598. H. Wahl and J. J. Vorsanger, *Bull. Soc. Chim. France* p. 3359 (1965).
598a. P. M. Walters and S. M. McElvain, *J. Am. Chem. Soc.* **62**, 1482 (1940).
599. R. Weiss and K. Woidich, *Monatsh. Chem.* **47**, 427 (1926).
600. C. Weizmann, E. Bergmann, and T. Berlin, *J. Am. Chem. Soc.* **60**, 1331 (1938).
601. H. L. Wheeler and C. O. Johns, *Am. Chem. J.* **40**, 233 (1908).
602. F. L. White, British Patent 392,410 (1933); *Chem. Abstr.* **27**, 5550 (1933).
603. F. L. White, British Patent 408,277 (1934); *Chem. Abstr.* **28**, 5677 (1934).
604. F. L. White, U.S. Patent 1,957,869 (1934); *Chem. Abstr.* **28**, 3995 (1934).
605. F. L. White, U.S. Patent 1,990,681 (1935); *Chem. Abstr.* **29**, 2103 (1935).
606. F. L. White and L. G. S. Brooker, British Patent 625,347 (1949); *Chem. Abstr.* **44**, 7168 (1950).
607. F. L. White and L. G. S. Brooker, U.S. Patent 2,466,523 (1949); *Chem. Abstr.* **44**, 7172 (1950).
608. C. W. Whitehead, *J. Am. Chem. Soc.* **74**, 4267 (1952).
609. C. W. Whitehead and J. J. Traverso, *J. Am. Chem. Soc.* **78**, 5294 (1956).
610. N. Whittaker, British Patent 743,221 (1956); *Chem. Abstr.* **51**, 2038 (1957).
611. J. P. Wibaut and R. Huls, *Rec. Trav. Chim.* **71**, 1021 (1952).
612. J. P. Wibaut, A. L. Van Hulsenbeek, and C. M. Siegmann, *Koninkl. Ned. Akad. Wetenschap., Proc.* **53**, 989 (1950).
612a. R. Willstatter and R. Pummerer, *Chem. Ber.* **38**, 1461 (1905).
613. C. D. Wilson, U.S. Patent 2,425,772 (1947); *Chem. Abstr.* **42**, 49 (1948).
614. R. Wizinger and U. Arni, *Chem. Ber.* **92**, 2309 (1959).
615. R. Wizinger and P. Ulrich, *Helv. Chim. Acta* **39**, 217 (1956).
616. A. Wohl and E. Bernreuther, *Ann. Chem.* **481**, 1 (1930).
617. A. Wohl and B. Mylo, *Chem. Ber.* **45**, 322 (1912).
618. C. A. Wood and W. A. Comley, *J. Soc. Chem. Ind. (London)* **42**, 429T (1923).
619. H. Wuhrmann, *Sci. Ind. Phot.* [2] **21**, 441 (1950); *Chem. Abstr.* **45**, 1444 (1951).
620. L. M. Yagupolskii and B. E. Gruz, *Zh. Obshch. Khim.* **31**, 3955 (1961).
621. L. M. Yagupolskii and A. I. Kiprianov, *Zh. Obshch. Khim.* **22**, 2216 (1952).
622. L. M. Yagupolskii, G. I. Klyushnik, and V. I. Troitskaya, *Zh. Obshch. Khim.* **34**, 307 (1964).
623. L. M. Yagupolskii, N. V. Kondratenko, and Y. A. Fialkov, *Zh. Obshch. Khim.* **36**, 828 (1966).
624. L. M. Yagupolskii and M. S. Marenets, *Zh. Obshch. Khim.* **25**, 1771 (1955).
625. L. M. Yagupolskii and V. I. Troitskaya, *Zh. Obshch. Khim.* **27**, 518 (1957).
625a. L. M. Yagupolskii and V. I. Troitskaya, *Zh. Obshch. Khim.* **29**, 2730 (1959).
625b. L. M. Yagupolskii and V. I. Troitskaya, *Zh. Obshch. Khim.* **29**, 2409 (1959).
626. L. M. Yagupolskii and V. I. Troitskaya, *Zh. Obshch. Khim.* **31**, 628 (1961).

627. L. M. Yagupolskii, V. I. Troitskaya, and B. E. Gruz, *Zh. Obshch. Khim.* **35**, 1644 (1965).
628. L. M. Yagupolskii, V. I. Troitskaya, B. E. Gruz, and N. V. Kondratenko, *Zh. Obshch. Khim.* **35**, 1644 (1965).
629. L. M. Yagupolskii, V. I. Troitskaya, I. I. Levkoev, E. B. Lifshits, P. A. Yufa, and N. S. Barvyn, *Zh. Obshch. Khim.* **37**, 191 (1967).
630. L. A. Yanovskaya and V. F. Kucherov, *Izv. Akad. Nauk SSSR, Otd. Khim. Nauk* p. 2184 (1960).
631. L. A .Yanovskaya, V. F. Kucherov, and B. A. Rudenko, *Izv. Akad. Nauk SSSR, Otd. Khim. Nauk* p. 2182 (1962).
632. L. A. Yanovskaya, S. S. Yufit, and V. F. Kucherov, *Izv. Akad. Nauk SSSR, Otd. Khim. Nauk* p. 1246 (1960).
633. F. G. Young, U.S. Patent 2,556,312 (1951); *Chem. Abstr.* **46**, 1031 (1952).
634. W. G. Young and J. D. Roberts, *J. Am. Chem. Soc.* **68**, 649 (1946).
635. L. I. Zakharkin and I. M. Khorlina, *Izv. Akad. Nauk SSSR, Otd. Khim. Nauk* p. 2255 (1959).
636. P. Zeller and O. Isler, *Helv. Chim. Acta* **42**, 844 (1959).
637. H. Zenno, *J. Soc. Sci. Phot. Japan* **14**, 44 (1951); *Chem. Abstr.* **47**, 3733 (1953).
638. H. Zenno, *J. Pharm. Soc. Japan* **72**, 1628 (1952); *Chem. Abstr.* **47**, 4610 (1953).
639. H. Zenno, *J. Pharm. Soc. Japan* **72**, 1630 (1952); *Chem. Abstr.* **47**, 4611 (1953).
640. M. Zhdanowitz, *J. Russ. Phys.-Chem. Soc.* **42**, 1279 (1910).
641. V. G. Zhiryakov, *Zh. Obshch. Khim.* **34**, 2034 (1964).
642. V. G. Zhiryakov and P. I. Abramenko, *Zh. Obshch. Khim.* **35**, 150 (1965).
642a. V. G. Zhiryakov and P. I. Abramenko, *Khim. Geterosikl. Soedin., Akad. Nauk Latv. SSR* p. 621 (1967); *Chem. Abstr.* **68**, 96791 (1968).
643. V. G. Zhiryakov, P. I. Abramenko, and G. F. Kurepina, U.S.S.R. Patent 159,726 (1963); *Chem. Abstr.* **61**, 8449 (1964).
644. V. G. Zhiryakov, P. I. Abramenko, and G. F. Kurepina, U.S.S.R. Patent 159,909 (1964); *Chem. Abstr.* **60**, 14651 (1964).
645. V. G. Zhiryakov, P. I. Abramenko, and G. F. Kurepina, U.S.S.R. Patent 168,991 (1965); *Chem. Abstr.* **63**, 1923 (1965).
646. V. M. Zubarovskii and R. N. Moskaleva, *Zh. Obshch. Khim.* **32**, 570 (1962).
647. V. M. Zubarovskii, R. N. Moskaleva, and M. P. Bachurina, *Ukr. Khim. Zh.* **30**, 80 (1964).
648. V. M. Zubarovskii, R. N. Moskaleva, and M. P. Bachurina, *Khim. Geterosikl. Soedin., Akad. Nauk Latv. SSR* p. 571 (1965); *Chem. Abstr.* **64**, 3732 (1966).
649. V. M. Zubarovskii, T. M. Verbovskaya, and A. I. Kiprianov, *Zh. Obshch. Khim.* **31**, 3056 (1961).

CHAPTER 5

# Carbohydrate Ortho Esters

## I. INTRODUCTION

This chapter deals with the synthesis, properties, and applications of ortho esters derived from carbohydrates. Most of these substances are bicyclic compounds in which an ortho ester 2-alkoxydioxolane ring is fused to the furanose or pyranose ring of a carbohydrate. Usually the fusion involves the anomeric carbon and a neighboring carbon of the carbohydrate (**1**). Carbohydrate ortho esters in which the anomeric carbon is not part of the alkoxy-

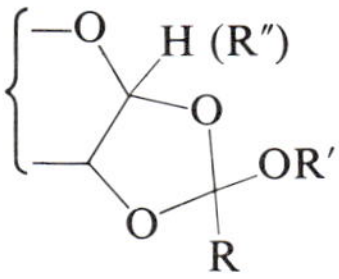

**1**

dioxolane ring are known, and there are a number of tricyclic ortho esters in which all three of the ortho ester oxygens are bonded to the carbohydrate skeleton.

Carbohydrate ortho esters are interesting substances for a number of reasons. Historically, their mode of formation and unusual chemical properties contributed significantly to the development of modern theories of substitution reactions of carbohydrate derivatives. The inertness of the ortho ester function under alkaline conditions coupled with the ease of its hydrolytic removal under acidic conditions makes the ortho ester function a useful protecting group for the two esterified hydroxyl groups of the carbohydrate. Mild acid hydrolysis of carbohydrate ortho esters yields monoacyloxy carbohydrates and their derivatives. Dioxolenium salts derived from carbohydrate ortho esters promise to be intermediates in a number of useful syntheses.

The only comprehensive review of carbohydrate ortho esters was published in 1945 (*106*). References (*57*), (*76*), (*116*), (*130*), and (*131*) contain brief reviews of this class of compounds.

## II. SYNTHESIS AND PROPERTIES OF CARBOHYDRATE ORTHO ESTERS

### A. Reactions of Carbohydrate Derivatives Which Form Ortho Ester Functions

1. Conversion of *O*-Acylglycosyl Halides to Ortho Esters

*O*-Acylglycosyl halides are acylated carbohydrate derivatives having a halogen atom bonded to the anomeric carbon. They are prepared by reaction of saccharides with acyl halides, by reaction of per-*O*-acylated saccharides with hydrogen halides, certain metal halides, and methyl dichloromethyl ether (*36*), and by other methods. [For a review of the preparation and properties of acylglycosyl halides, see refs. (*130*), Chap. 6; (*75*), (*41*).] The first carbohydrate ortho ester to be prepared was an unexpected by-product of the conversion of an acylglycosyl halide to an acyl glycoside by the procedure of Koenigs and Knorr (*74*). Most of the known carbohydrate ortho esters were synthesized using acylglycosyl halides as starting materials.

a. *The Koenigs-Knorr Reaction.* A general method of synthesizing glycosides involves reaction of an acylglycosyl chloride or bromide with an alcohol or other hydroxyl compound in the presence of silver carbonate or silver oxide, and sometimes an inert solvent [Eq. (1)]. This synthesis, known as the Koenigs-Knorr reaction (*74*), occurs with predominant inversion of configuration at the anomeric carbon.

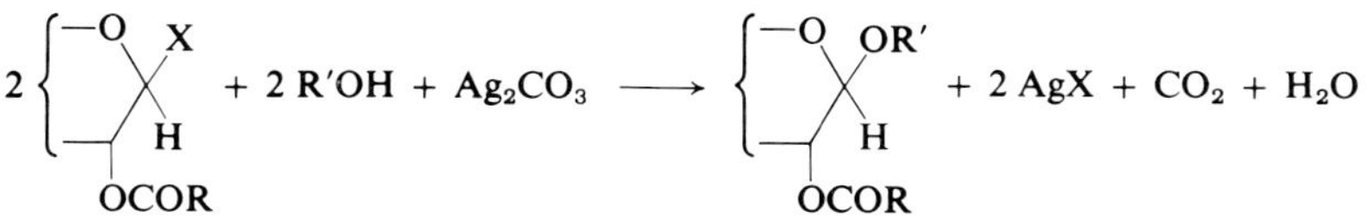

$$2\left\{\begin{matrix}-O & X \\ & \diagdown \\ & H \\ OCOR \end{matrix}\right\} + 2\,R'OH + Ag_2CO_3 \longrightarrow \left\{\begin{matrix}-O & OR' \\ & \diagdown \\ & H \\ OCOR \end{matrix}\right\} + 2\,AgX + CO_2 + H_2O \quad (1)$$

Fischer, Bergmann, and Rabe in 1920 (*17*) applied the Koenigs-Knorr reaction to 2,3,4-tri-*O*-acetyl-L-rhamnopyranosyl bromide. They obtained three, rather than the expected two, isomeric products whose compositions corresponded to methyl tri-*O*-acetylrhamnopyranoside. One of these, the "γ" isomer, differed from the other two in possessing only two saponifiable acetyl groups. The alkali-resistant "acetyl" group, together with the "glycosidic" methyl group, was easily removed by acid hydrolysis. In 1924 Dale reported a similar reaction of 2,3,4,6-tetra-*O*-acetyl-D-mannopyranosyl bromide which yielded a methyl glycoside having three, rather than four, saponifiable acetyl groups (*12*), and Freudenberg and co-workers in 1925 reported a similarly anomolous methyl glycoside of heptaacetyl maltose (*25*).

The true structures of these abnormal products of Koenigs-Knorr reactions were not recognized until 1930, when Freudenberg and Braun (*4*, *21*) and Bott, Haworth, and Hirst (*3*) independently and almost simultaneously suggested that they were cyclic methyl orthoacetates rather than methyl glycosides. The reactions producing them are formulated in Eqs. (2)–(4).

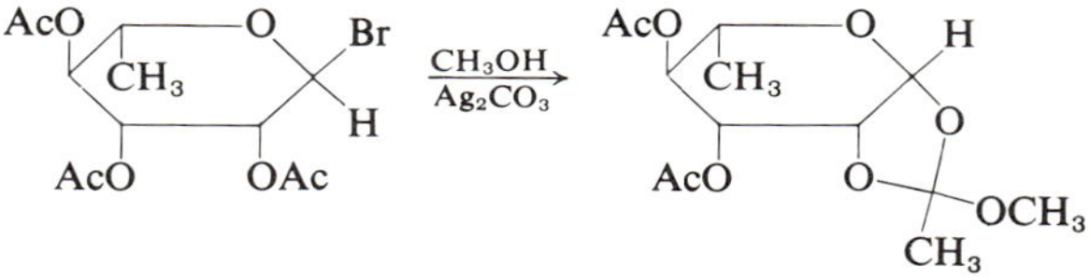

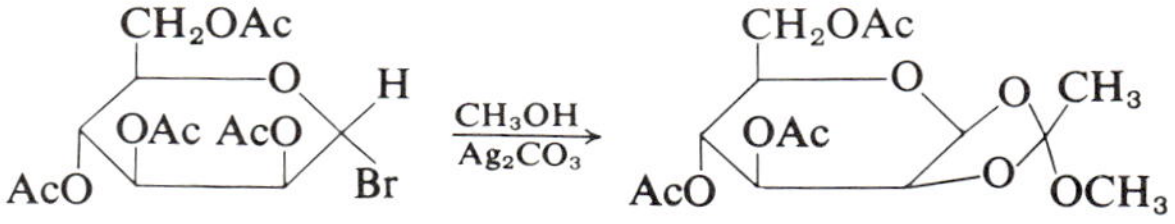

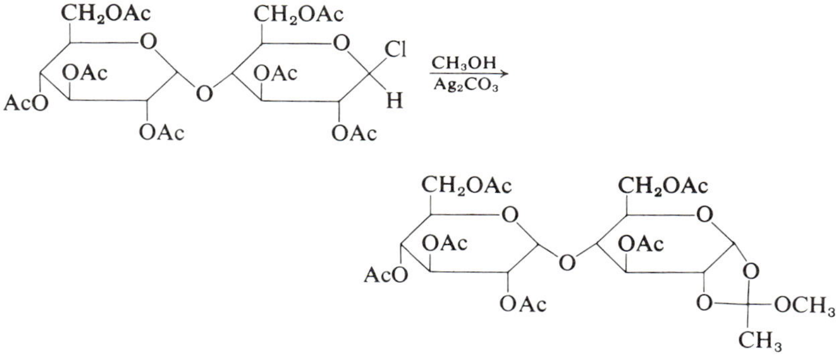

The ortho ester structures account for the fact that alkaline hydrolysis of these compounds gives products which retain the constituent atoms of one acetyl group, and for their great sensitivity to acid-catalyzed hydrolysis.

The two-step hydrolysis of the methyl orthoacetate derivative of L-rhamnose [3,4-di-*O*-acetyl-6-deoxy-1,2-*O*-(1-methoxyethylidene)-L-mannopyranose] is described by Eq. (5).

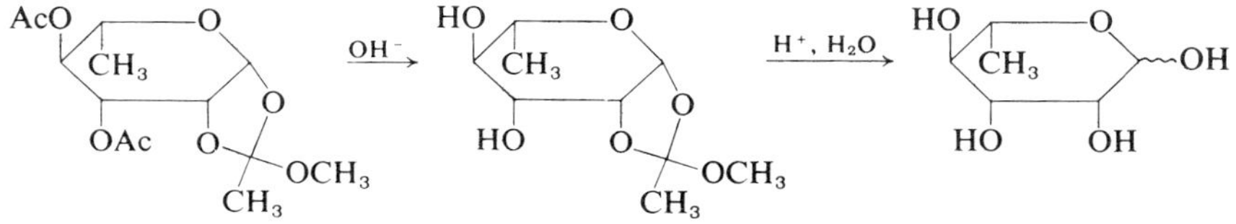

$$\text{(structures shown)} \tag{5}$$

The structure of the triacetyl D-mannose methyl orthoacetate [3,4,6-tri-*O*-acetyl-1,2-*O*-(1-methoxyethylidene)-D-mannopyranose] was established by degradation and synthesis experiments (*3*), and that of 3,4-di-*O*-acetyl-1,2-*O*-(1-methoxyethylidene)-L-rhamnopyranose by the ultraviolet absorption spectrum of its saponification product (*4*) and by its chemical properties (*37, 90*).

b. *Mechanism and Stereochemistry of Ortho Ester Formation from Acylglycosyl Halides.* One member of each anomeric pair of acylglycosyl halides is more stable than the other and is usually the only isomer obtained in syntheses of these substances. In the case of acylated D-aldohexopyranosyl halides the stable anomer is that having the α-configuration, with the substituent at C-5 trans to the halogen at C-1. The stable anomer of acylated aldopentopyranosyl halides is that in which the anomeric halogen atom is trans to the acyloxy group at C-3. Similar generalizations apply to acylated aldofuranosyl halides (*130*, p. 200).

Under the original conditions of Koenigs and Knorr, only 1,2-*trans*-*O*-acylglycosyl halides yield ortho esters as major or minor alcoholysis products. The stable anomers of acetylglycosyl halides derived from the pyranose forms of D-mannose, D-talose, D-lyxose, and D-ribose satisfy this condition, and all of them have been converted to cyclic orthoacetates by the Koenigs-Knorr reaction.

The stereochemistry of ortho ester formation from acylglycosyl halides is readily accounted for if it is assumed that the acyloxy group at C-2 anchimerically assists departure of halogen from C-1, with formation of a dioxolenium ion intermediate. Reaction of the alcohol at the acyl carbon of this ion yields the ortho ester. Dissociation of the halide ion is undoubtedly also assisted by electrophilic interaction with silver ion from the silver carbonate or silver oxide present.

Analogous mechanisms are firmly established for solvolysis reactions of *trans*-1,2-disubstituted cyclohexanes (*119, 140, 141*). The reasonableness of dioxolenium ions as intermediates in ortho ester formation from acylglycosyl halides is also supported by recent syntheses of dioxolenium tetrafluoroborate and hexachloroantimonate derivatives of carbohydrates (6, 15a, 109a–112).

A process involving acetoxy participation and dioxolenium ion formation also receives strong support from a study by Isabell and Frush (58) of the methanolysis of tetra-O-acetyl-α-D-glucosyl bromide and tetra-O-acetyl-α-D-mannosyl bromide in the presence of silver carbonate. The acetylglucosyl bromide, which has the cis-1,2-configuration, yielded the tetra-O-acetyl-β-methyl glucoside (the product of inverted configuration at C-1) almost exclusively. On the other hand, the acetylmannosyl bromide, which has the trans-1,2-configuration, formed mostly the tri-O-acetyl methyl 1,2-O-orthoacetate, plus some of the methyl tetra-O-acetyl-β-D-mannopyranoside. Since the methyl mannoside is that of inverted configuration at C-1, it must result from a process competing with dioxolenium ion formation. When ether was added to the reaction mixture, the yield of ortho ester diminished, and a corresponding amount of tetra-O-acetyl-α-methyl mannoside was formed, in addition to approximately the same amount of β-methyl mannoside as was formed in the absence of ether. This result is rationalized by assuming that ether reacts with the dioxolenium ion intermediate to form an unstable alkylated ortho ester cation. This cation reacts with methanol at acyl carbon to form the cyclic methyl orthoacetate, and with methanol at the anomeric carbon to form the α-methyl mannoside of retained configuration. Equation (6) outlines the probable course of methanolysis of 2,3,4,6-tetra-O-acetyl-α-D-mannopyranosyl bromide under the conditions described above.

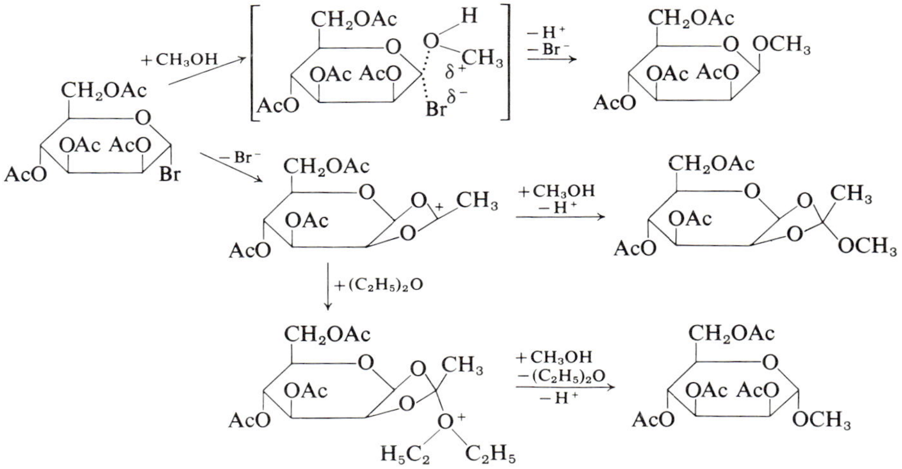

(6)

The ortho ester acyl carbon atom in carbohydrate ortho esters is an asymmetric center, and reactions producing these compounds yield mixtures of diastereomers. In a few instances both members of the diastereomeric pair of ortho esters prepared from acylglycosyl halides were actually isolated. Examples are the ortho esters formed by reaction of 1,2,3,4-tetra-O-acetyl-β-D-glucopyranose with 2,3,4,6-tetra-O-acetyl-α-D-mannopyranosyl bromide (*134*), one diastereomer of which is shown (**2**), and the diastereomeric 3,4,6-tri-O-acetyl-1,2-O-(1-methoxyethylidene)-β-D-mannopyranoses and 3,4,6-tri-O-acetyl-1,2-O-(1-benzyloxyethylidene)-β-D-mannopyranoses (*113*). In other cases the relative amounts of the two diastereomers were estimated by nuclear magnetic resonance spectrometry (*82, 92*).

**2**

The diastereomer formed by attack by the alcohol on the least-hindered side of the dioxolenium ion is the major product. Typically this is the isomer in which the ortho ester alkoxy group has the exo configuration. Many of the crystalline bicyclic carbohydrate ortho esters reported in the literature are probably reasonably pure samples of the predominant exo diastereomers. In many instances where amorphous or syrupy ortho esters were obtained, it seems likely that mixtures of both diastereomers were isolated.

c. *Modifications of the Koenigs-Knorr Ortho Ester Synthesis.* In the Koenigs-Knorr reaction, silver carbonate or silver oxide is used to neutralize the hydrogen halide produced by alcoholysis of the acylglycosyl halide. Silver ion serves the additional function of assisting dissociation of the halogen from C-1. Electrophilic catalysis by silver ion is more important to glycoside formation than for dioxolenium ion formation, since the principal driving force for the latter reaction is intramolecular attack by the acyloxy group at C-2 on the backside of C-1. It is not surprising, therefore, that yields of ortho esters from *trans*-1,2-acyloxyglycosyl halides are usually substantially improved by buffering the reaction mixture with bases which have no electrophilic counterion. Pyridine and its derivatives are the most frequently used bases in syntheses of ortho esters from acylglycosyl halides. Pyridine (*44, 50, 78, 80, 81, 83, 104, 108, 138*) and quinoline (*17, 37–39, 75, 93, 98*) often afford satisfactory results, but in some cases react with the glycosyl halide to form quaternary pyridinium glycosides (*81, 83*). The best results are obtained using hindered bases such as 2,6-lutidine (*20, 66c, 69, 71, 71a, 73, 92*) or 2,4,6-collidine (*44, 49, 75, 82, 123–125*), which do not attack the anomeric carbon of the acylglycosyl halide. Hindered pyridine bases are especially superior to silver carbonate as acid binding agents in reactions of acylglycosyl halides with less-reactive alcohols such as substituted benzyl alcohols (92).

There is another advantage to using hindered nitrogen bases rather than silver salts. In many cases the *cis*-1,2-acylglycosyl halides are the only anomers which are readily available, and these yield only glycosides under Koenigs and Knorr's reaction conditions. However, they frequently afford reasonable yields of ortho esters when pyridine bases are substituted for the silver salts (*49, 81, 83, 124, 125*). Thus 2,3,4,6-tetra-*O*-acetyl-α-D-glucopyranosyl bromide and 2,3,4,6-tetra-*O*-benzoyl-α-D-glucopyranosyl bromide are converted to cyclic orthoacetates (*81, 83, 124, 125*) and orthobenzoates (*49*) by treatment with alcohol–collidine–nitromethane or alcohol–pyridine mixtures. These reactions probably involve preliminary bromide ion-catalyzed epimerization of the α-acylglucosyl bromide to the β-anomer, which is converted to the ortho ester. This hypothesis is supported by the observation that yields are improved by addition of tetrabutylammonium bromide to the reaction mixtures (*78, 82*). By using tetrabutylammonium bromide as the epimerization catalyst and collidine as the base, almost quantitative yields of a number of 1,2-*O* ortho esters of acylated glucopyranoses were obtained (*82*). The reaction sequence of Eq. (7), developed by Lemieux and Morgan (*82*), should prove to be a versatile synthesis of carbohydrate ortho esters.

[R = CH$_3$, C$_6$H$_5$, R' = CH(CH$_3$)$_2$; R = C(CH$_3$)$_3$, CH$_3$O, R' = CH$_3$]

(7)

Zinc oxide in benzene–benzyl alcohol was used in the conversion of 1,3,4,6-tetra-*O*-benzoyl-D-fructofuranosyl bromide to a tribenzoyl benzyloxybenzylidenefructofuranose of unknown structure (*43*).

2,3,4,6-Tetra-*O*-acetyl-α-D-glucopyranosyl bromide and 2,3,4,6-tetra-*O*-acetyl-α-D-galactopyranosyl bromide are converted to alkyl 1,2-*O* orthoacetates in moderate yield by refluxing with a suspension of lead carbonate and anhydrous calcium sulfate in alcohol-ethyl acetate or alcohol-benzene mixtures (*64, 68, 71, 137*). This reaction appears to offer no advantage over the tetrabutylammonium bromide–lutidine procedure described above.

d. *Formation of Tricyclic Ortho Esters from Acylglycosyl Halides.* An acylglycosyl halide having a suitably oriented hydroxyl group on the carbohydrate skeleton should yield a dioxolenium ion capable of cyclizing to a tricyclic ortho ester. There are two examples of this type of reaction, both of which involve di-*O*-benzoylribofuranosyl chlorides. In one, 3,5-di-*O*-benzoyl-D-ribofuranosyl chloride was converted by stirring with mercuric acetate in benzene to 5-*O*-benzoyl-1,2,3-*O*-benzylidyne-α-D-ribofuranose [Eq. (8)] (*96*). The same benzoylbenzylidyneribose was obtained by stirring 2,5-di-*O*-benzoyl-D-ribofuranosyl chloride with silver benzoate in benzene [Eq. (8)] (*97*). The formation of this highly strained tricyclic orthobenzoate is attributable to the propinquity of the 2- or 3-hydroxyl group to the acyl carbon in the bicyclic dioxolenium ions derived from the glycosyl chlorides.

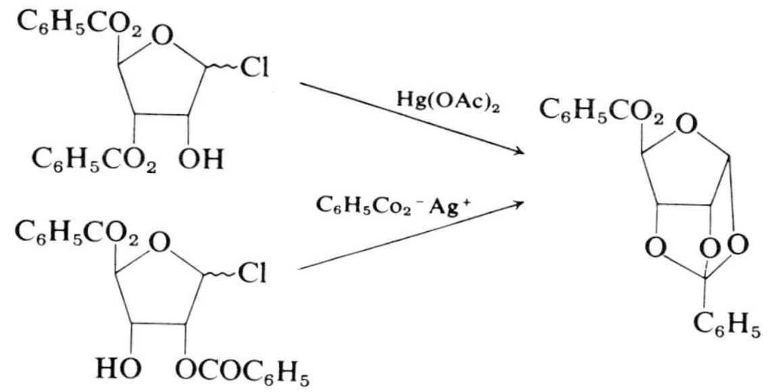

$$\text{(8)}$$

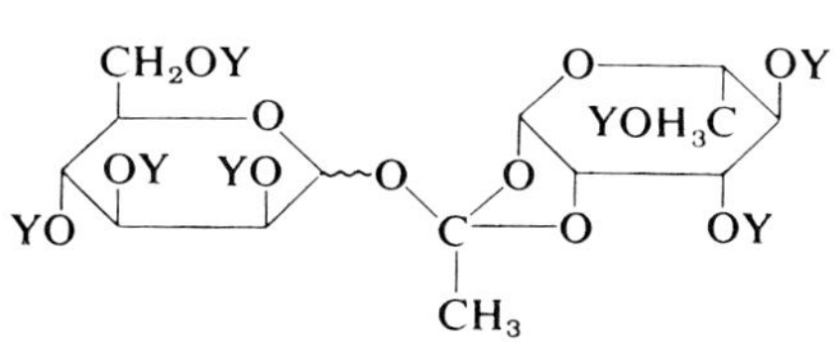

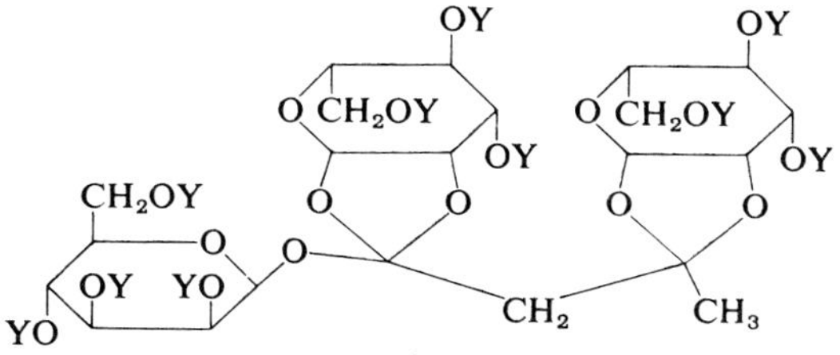

e. *Ortho Esters Formed by Silver Oxide-Catalyzed Oligomerization of Acylglycosyl Halides.* Goldschmid, Perlin, and Giam (*29, 31*) found that heating 2,3,4,6-tetra-*O*-acetyl-α-D-mannopyranosyl bromide with silver oxide in benzene results in formation of ortho esters **3** and **4** (Y = Ac). The dimeric orthoacetate may be formed by reaction of an anion derived from the halide and silver oxide with a second acylglycosyl halide molecule or a dioxolenium cation derived from it. The trimeric ortho ester is probably formed by addition of a dioxolenium cation to a ketene acetal formed by loss of a proton from a dioxolenium cation.

2. REACTIONS OF CARBOHYDRATES WITH ACYCLIC ORTHO ESTERS

a. *Formation of Bicyclic Carbohydrate Ortho Esters.* Carbohydrate derivatives having two cis-vicinal hydroxyl groups on a pyranose or furanose ring react smoothly with trialkyl orthocarboxylates to form bicyclic ortho esters.

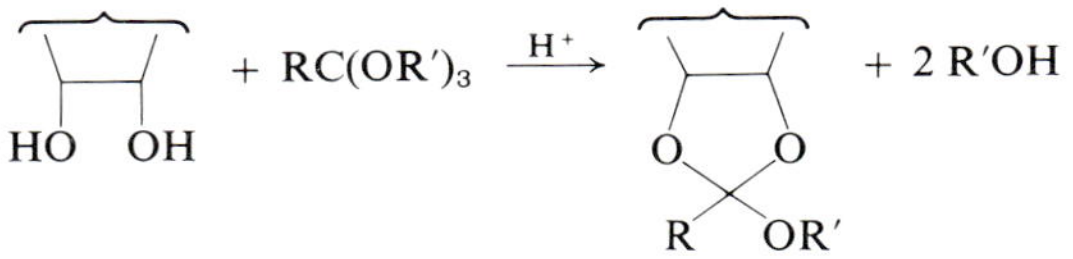

$$\text{\raisebox{0pt}{\shortstack{HO OH}}} + RC(OR')_3 \xrightarrow{H^+} \text{bicyclic ortho ester} + 2\,R'OH \qquad (9)$$

Since *trans*-1,2-cyclohexanediol reacts with triethyl orthoformate (*11*) and triethyl orthoacetate (*139*) to form bicyclic ortho esters, it is probable that pyranoses having trans-vicinal hydroxyl groups would also yield bicyclic ortho esters by transesterification. No examples have been reported, however.

Ortho ester formation by transesterification is a useful reaction in syntheses of nucleoside, nucleotide, and oligonucleotide derivatives. Introduction of an ortho ester function at the 2′,3′-position of the ribose moiety blocks the 2′- and 3′-hydroxyls with a group which is inert under basic conditions, yet is easily removed by acid hydrolysis. Ortho esterification followed by partial hydrolysis provides a very mild means of monoacylating the 2′- and 3′-hydroxyls of nucleoside derivatives. Reactions yielding nucleoside ortho esters are summarized by Eq. (10).

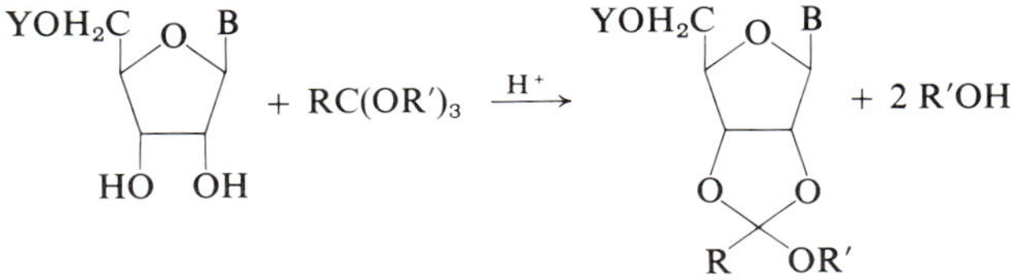

$$\text{nucleoside diol} + RC(OR')_3 \xrightarrow{H^+} \text{nucleoside ortho ester} + 2\,R'OH \qquad (10)$$

[R = H, $CH_3$, $C_6H_5$, $C_6H_5CH_2OCONHCH_2$; R′ = $CH_3$, $C_2H_5$; Y = H, $HOPO_2$, $ROPO_2$; B = 1-cytosyl, 1-uracilyl, 9-adenyl, 9-quanyl, 9-hypoxanthyl, and substitution products thereof]

Many ribonucleoside and ribonucleotide 2′,3′-*O*-orthoformates and their substitution products are described in the literature (*7a, 9, 15, 26, 27, 33, 34, 51–54, 84, 89, 92a–92c, 115a, 115b, 117, 121b, 129, 129a, 129b, 144, 144a,*

*147*) as well as a few 2′,3′-*O*-orthoacetates (*28, 29*), two 2′,3′-*O*-orthobenzoates (*26, 32a*), a 2′,3′-*O*-orthocarbonate (*121*), a 2′,3′-*O*-orthoglycinate (*145, 146*), and several 2′,3′-*O*-(*N*-carbobenzyloxy)orthoglycinates (*7, 7a, 7b, 145, 146, 146a*).

The transesterification procedure, in spite of its simplicity and convenience, has found little application in the synthesis of ortho esters of carbohydrates other than nucleosides and their derivatives. Methyl β-L-arabinopyranoside reacts with triethyl orthoacetate in the presence of toluenesulfonic acid to form a mixture of diastereomeric methyl 3,4-*O*-(1-ethoxyethylidene)-β-L-arabinopyranosides containing approximately 80% of the exo isomer and 20% of the endo isomer (*6*). Methyl β-D-ribopyranoside undergoes a similar reaction with triethyl orthoformate in the presence of trichloroacetic acid (*115c*).

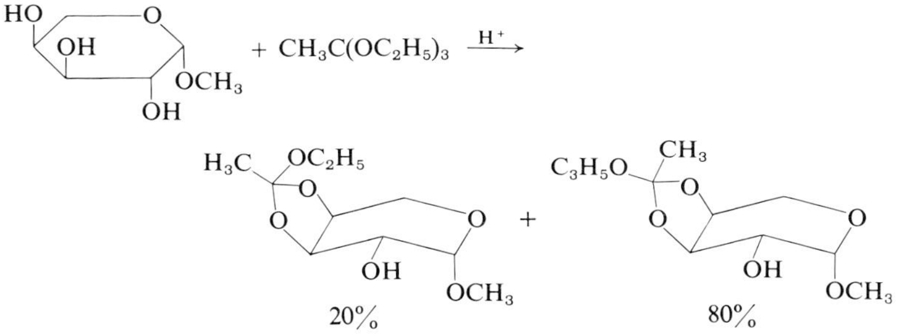

$$\text{(11)}$$

In a similar reaction the antibiotic lincomycin was converted to the cyclic ortho ester **5**.

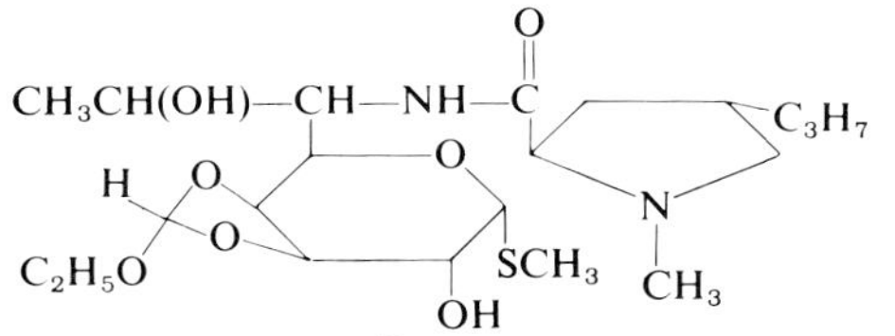

Methyl 2,6-dideoxy-α-D-galactopyranoside was converted to a mixture of the 3-*O*-acetyl and 4-*O*-acetyl derivatives by treating it first with trimethyl orthoacetate and then with water in the presence of acid (*5*). Methyl 2,6-dideoxy-3,4-*O*-(1-methoxyethylidene)-α-D-galactopyranoside was undoubtedly an intermediate in this acetylation process, although it was not isolated.

A recent paper describes the synthesis of bicyclic carbohydrate ortho esters in which the two rings are not fused. Treatment of 1,2-*O*-isopropylidene-3-alkoxy-α-D-glucofuranose with triethyl orthoformate in the presence of acetic acid yielded 1,2-O-isopropylidene-3-alkoxy-5,6-O-ethoxymethylene-α-D-glucofuranoses (63).

b. *Formation of Tricyclic and Tetracyclic Ortho Esters.* The 3,5- and 6-hydroxyls of D-glucofuranose derivatives are so situated that these substances can be converted into tetracyclic orthoformates of type **6** (R = H) by acid-catalyzed reaction with triethyl orthoformate. Orthoformates derived from 1,2-O-isopropylidene-α-D-glucofuranose (23, 42), 1,2-O-benzylidene-α-D-glucofuranose (47), and μ-thiolglucoxazoline (125a) have been reported. 1,2,5-O-Methylidene-myo-inositol (**7**) was prepared by heating the inositol with triethyl orthoformate and toluenesulfonic acid in toluene (88).

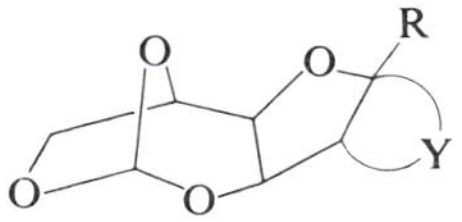

[Y = —O—C(CH$_3$)$_2$—O—, —O—CHC$_6$H$_5$—O—, C$_1$—N=C(SH)—O—C$_2$]

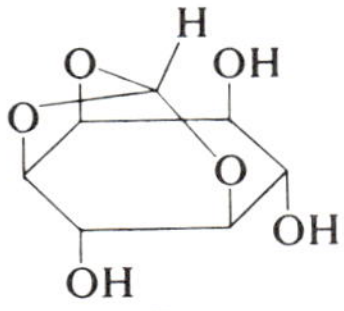

3. FORMATION OF CARBOHYDRATE ORTHO ESTERS BY TRANSESTERIFICATION REACTIONS OF OTHER CARBOHYDRATE ORTHO ESTERS

Cyclic ortho esters having 2-alkoxydioxolane structures undergo transesterification reactions with alcohols having alkyl groups different from that of the alkoxy group of the dioxolane. 1,4,6-Tri-O-benzoyl-2,3-O-(1-benzyloxybenzylidene)-β-D-fructofuranose reacts at room temperature with ethanol in dioxane containing a trace of hydrogen chloride to form 1,4,6-tri-O-benzoyl-2,3-O-(1-ethoxybenzylidene)-β-D-fructofuranose (**32**, Y = C$_6$H$_5$CO, R = C$_6$H$_5$, R' = C$_2$H$_5$) (94). When 3,4,6-tri-O-acetyl-1,2-O-(1-alkoxyethylidene)-α-D-glucopyranoses are refluxed with a solution of cholesterol in 1,2-dichloroethane or ethyl acetate containing toluenesulfonic acid or mercuric bromide, 3,4,6-tri-O-acetyl-1,2-O-(1-cholesteryloxyethylidene)-α-D-glucopyranose is formed (65, 68, 71).

4,6-O-Benzylidene-1,2-O-(1-methoxyethylidene)-α-D-glucopyranose undergoes a unique transesterification reaction with itself in which 3 moles of the monosaccharide ortho ester cyclize, with elimination of 3 moles of methanol, to form the macrocyclic triorthoester **8**. The structure of this substance was assigned on the basis of its composition, molecular weight, nuclear magnetic resonance spectrum, and chemical properties (66b).

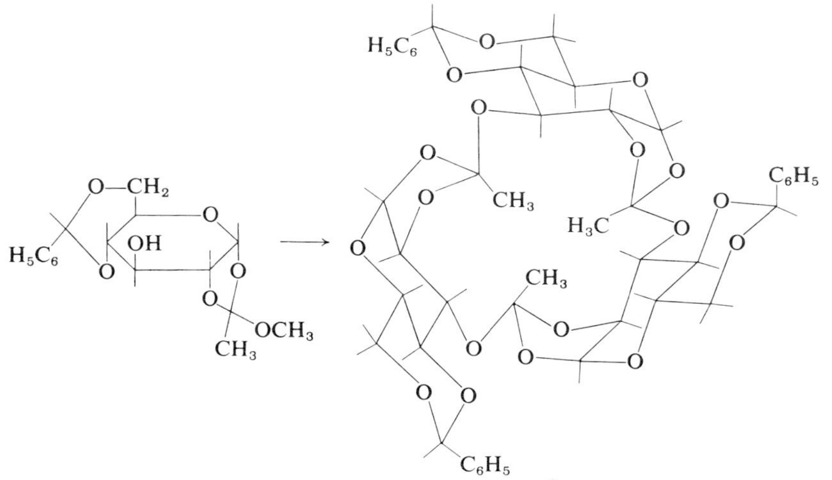

Bicyclic carbohydrate ortho esters having a free hydroxyl group suitably situated on the furanose or pyranose ring may undergo intramolecular transesterification reactions to form tricyclic ortho esters. For example, 1,2,4-*O*-benzylidyne-α-D-ribopyranose is formed when 1,2-*O*-benzyloxybenzylidene-α-D-ribopyranose is exposed to moisture or very dilute aqueous acid [Eq. (12)] (*18*). In similar reactions, 1,2-*O*-methoxybenzylidene-β-L-arabinofuranose was converted to 1,2,5-*O*-benzylidyne-β-L-arabinofuranose (*73*), and 2,3-*O*-benzyloxybenzylidene-β-D-fructofuranose formed 2,3,6-*O*-benzylidyne-β-D-fructofuranose (*43, 48*).

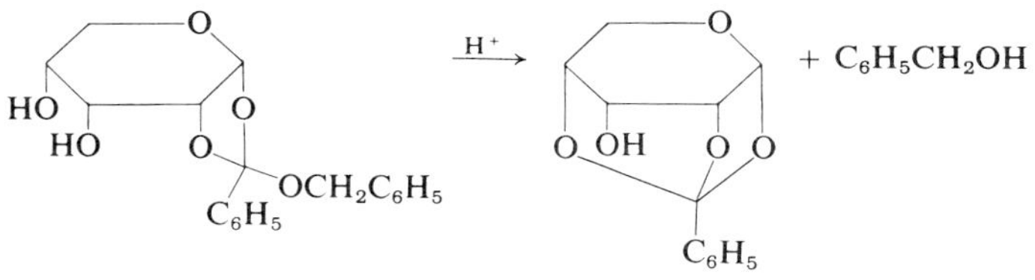

### 4. Formation of Carbohydrate Ortho Esters by Reaction of Dioxenium Salts with Alcohols

1,2,3-Tri-*O*-acetyl-α-D-idopyranose-4,6-dioxenium hexachloroantimonate, which can be prepared from α- or β-2,3,4,6-tetra-*O*-acetyl-D-glucopyranosyl chlorides or penta-*O*-acetyl-α-D-glucopyranose and antimony pentachloride in methylene chloride solution, reacts with ethanol in pyridine solution to form 1,2,3-tri-O-acetyl-4,6-O-(1-ethoxyethylidene)-α-D-idopyranose [Eq. (13)] (*111, 112*). Formation of the dioxenium hexachloroantimonate presumably involves a series of acetyl migrations proceeding via dioxolenium ion intermediates. Other carbohydrate dioxenium and dioxolenium salts, which can be prepared by similar reactions (*15a, 109a, 109b*), could probably also be converted to ortho esters.

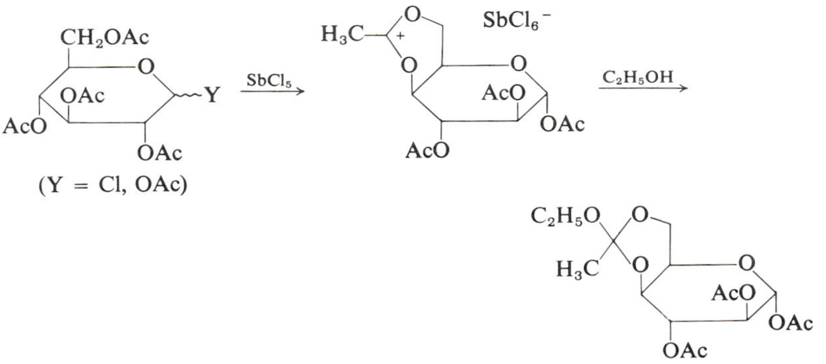

## 5. Conversion of Carbohydrate Thiono Esters and Xanthate Derivatives to Ortho Esters

1,2-O-Isopropylidene-α-D-glucofuranose 3,5,6-O-(S-alkylmonothio)orthocarbonates (**10**) were prepared by alkylation of 1,2-O-isopropylidene-α-D-glucofuranose 5,6-O-thionocarbonate (**9**) with methyl iodide and benzyl bromide in a suspension of silver oxide in dimethylformamide or dioxane (*127, 128*).

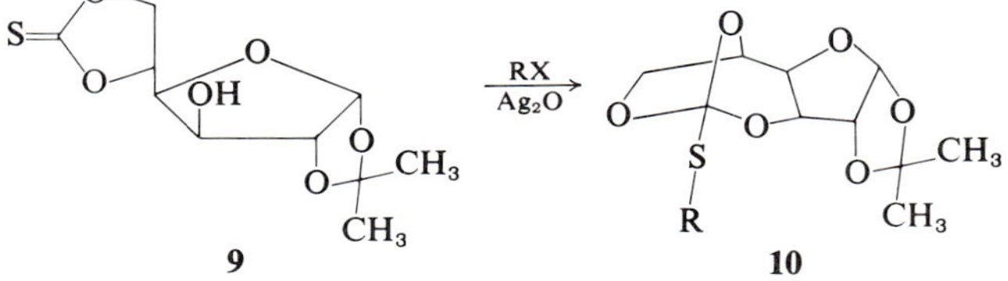

(R = CH₃, CH₂C₆H₅)

Methylation of 1,2-O-isopropylidene-5,6-dithio-β-L-idofuranose 5,6-trithiocarbonate (**11**) with methyl iodide in a dimethylformamide suspension of barium oxide yielded a mixture which probably contained 1,2-O-isopropylidene-5,6-dithio-β-L-idofuranose 3-O-5,6-S-(S-methyltrithio)orthocarbonate (**12**), although the thioortho ester was not isolated (*128*).

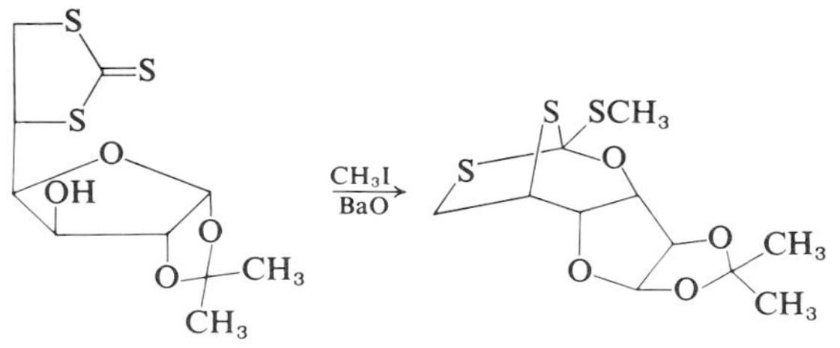

(15)

**11** → **12**

Tetracyclic orthocarbonate **10** and thioorthocarbonate **12** could arise by cyclization of the 2-alkylthiodioxolenium ion or 2-alkylthiodithiolenium ion formed by alkylation of the thione sulfur atom of the starting material, or they might be formed by alkylation of the sulfhydryl group of a cyclic thioorthocarbonic acid (such as **13**) present in equilibrium with the thionocarbonate. An equilibrium between **9** and **13** would account for the formation of bis(1,2-*O*-isopropylidene-α-D-glucofuranose 3,5,6-*O*-orthocarbonyl) disulfide (**14**) when **9** is oxidized by lead tetraacetate (*13*).

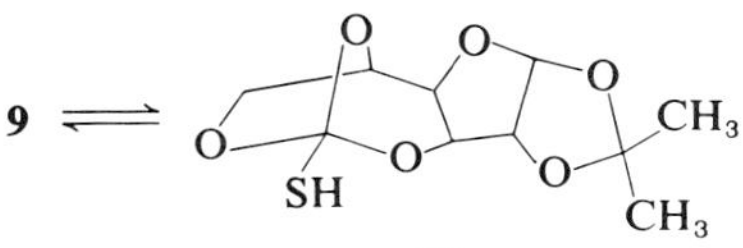

**9** ⇌ **13** 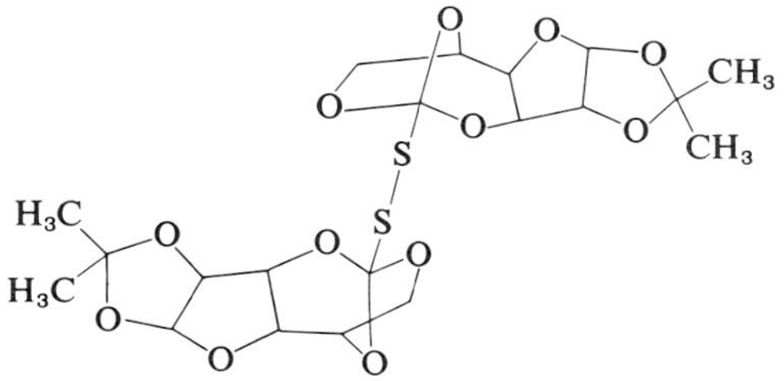     **14**

Reaction of 5'-*O*-trityluridine-2',3'-*O*-thionocarbonate with methanol in the presence of silver carbonate yielded a product whose physical properties, composition, and acid hydrolysis to a 2',3'-*O*-carbonate suggest that it is 5'-*O*-trityluridine-2',3'-*O*-(dimethyl orthocarbonate) (*121*).

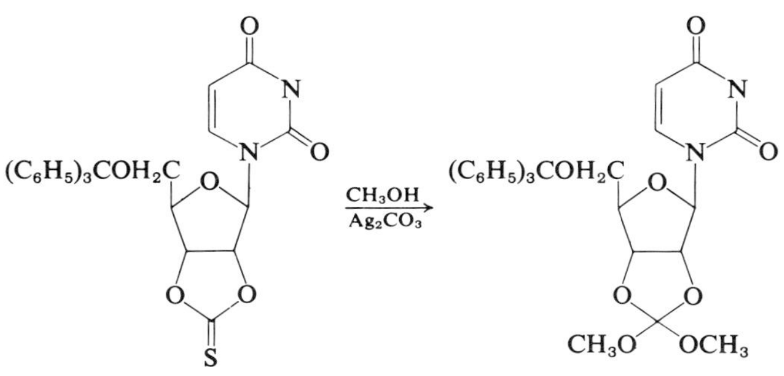

(16)

Methyl 3,5-di-*O*-*p*-toluenesulfonyl-2-*O*-thionobenzoyl-α-D-xylopyranoside yielded a dimeric product assigned bicyclic thioorthobenzoate structure **15** (Y = $C_6H_5CO_2$) when treated with sodium benzoate in dimethylformamide. Methyl 5-deoxy-2-*O*-thionobenzoyl-3-*O*-*p*-toluenesulfonyl-α-D-xylofuranoside underwent a similar reaction to form **15**, Y = H (*121a*).

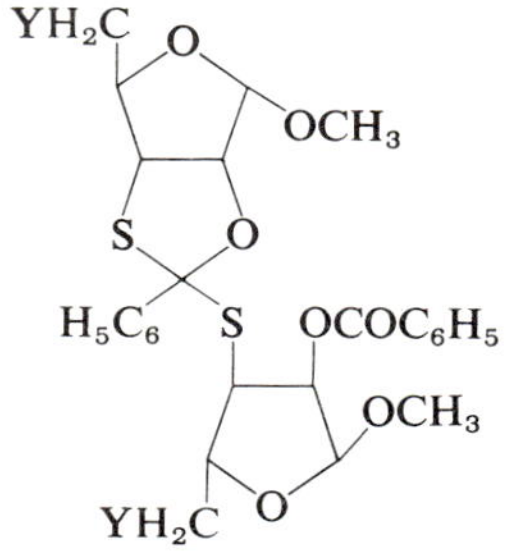

**15**

When bis(2,3:5,6-di-*O*-isopropylidene-1-thiocarbonyl-D-mannitol) disulfide (**16**) (prepared by iodine oxidation of the xanthate derivative of 2,3:5,6-di-*O*-isopropylidene-D-mannitol) was allowed to stand in pyridine, it decomposed to give sulfur, carbon disulfide, and bis(2,3:5,6-di-*O*-isopropylidene-D-mannitol) orthocarbonate, **17** (*14, 126*). Compound **17** was also formed

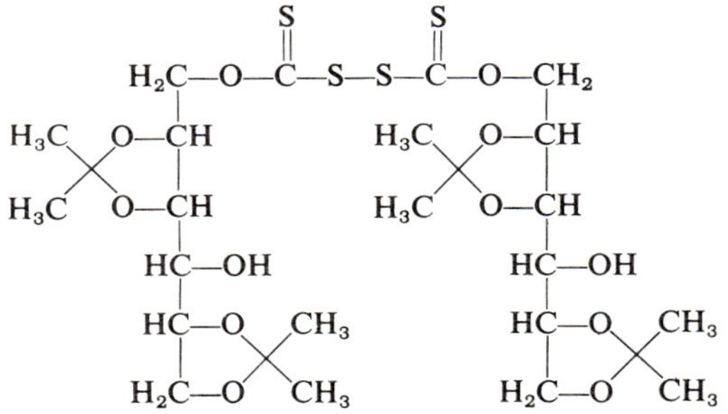

**16**

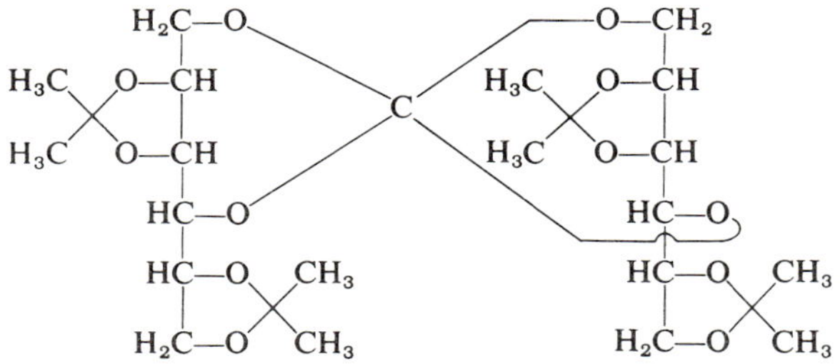

**17**

when 2,3:5,6-di-*O*-isopropylidene-D-mannitol was treated with thiophosgene in pyridine, or when the sodium salt of diisopropylidenemannitol was treated with thiophosgene.

## 6. Peroxy Ortho Esters from Glycosyl Halides and Acylated Aldonyl Chlorides

Penta-*O*-acetyl-D-gluconoyl chloride and *tert*-butylhydroperoxide react on warming to form 2-*tert*-butylperoxy-2-methyl-5-(D-arabino-1,2,3,4-tetra-acetoxybutyl)-1,3-dioxolan-4-one, **18** (*122*).

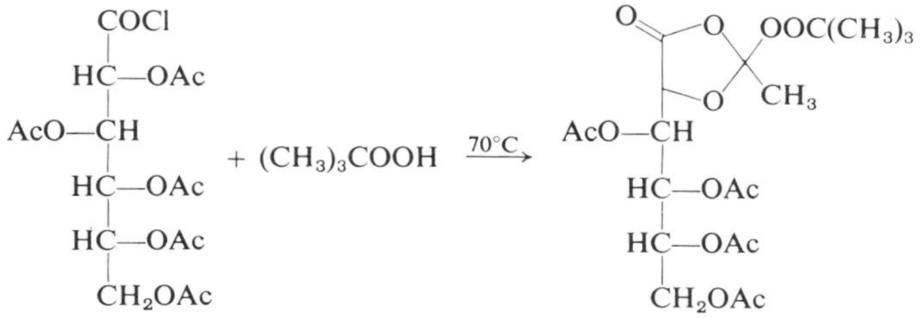

$$\begin{array}{c} \text{COCl} \\ | \\ \text{HC-OAc} \\ | \\ \text{AcO-CH} \\ | \\ \text{HC-OAc} \\ | \\ \text{HC-OAc} \\ | \\ \text{CH}_2\text{OAc} \end{array} + (\text{CH}_3)_3\text{COOH} \xrightarrow{70°C} \begin{array}{c} \text{(structure 18)} \\ \text{AcO-CH} \\ | \\ \text{HC-OAc} \\ | \\ \text{HC-OAc} \\ | \\ \text{CH}_2\text{OAc} \end{array} \qquad (17)$$

**18**

Tetra-*O*-acetylmucoyl dichloride similarly yields the bis(*tert*-butylperoxy) ortho ester **19** (*122*).

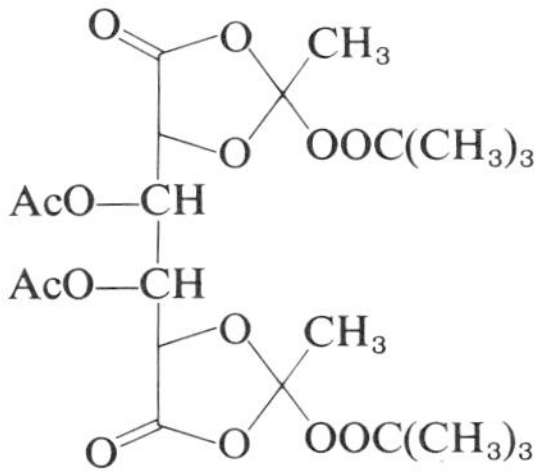

**19**

These are the only two reported examples of this type of ortho ester synthesis in the carbohydrate series. The reaction is probably a general one, at least for the synthesis of 2-alkylperoxy-1,3-dioxolan-4-ones, since aliphatic α-acetoxy acid chlorides undergo analogous cyclizations to peroxy ortho esters (*120*).

A number of acetyl glycosyl halides were converted to 1,2-*O*-(1-*tert*-butylperoxyethylidene) derivatives in similar reactions (*123, 123a,*)

7. Purported Syntheses of Carbohydrate Ortho Ester Acids, Anhydrides, and Halides

There are a number of claims in the literature of syntheses of 2-hydroxy-, 2-acyloxy-, and 2-halo-1,3-dioxolane derivatives of carbohydrates. All of these substances should be unstable and highly reactive. While some of them may be transient intermediates in certain transformations of acylated sugars, it now appears that no authentic examples of these classes of compounds have been isolated.

a. *2-Hydroxy-1,3-dioxolanes.* Several claims have been published of syntheses of ortho ester acid derivatives of carbohydrates. These substances contain 2-hydroxy-2-alkyl or 2-hydroxy-2-aryl dioxolane systems such as **20**.

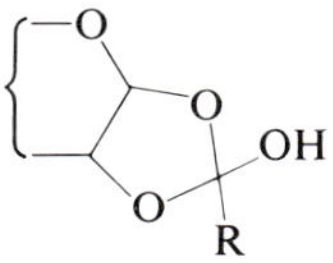

**20**

The product of base-catalyzed isomerization of 1,2,3,4-tetra-*O*-acetyl-β-D-glucopyranose was assigned a 1,6-*O*-(1-hydroxyethylidene) structure by Haworth, Hirst, and Teece (*40*), but was later shown to be 1,2,3,6-tetra-*O*-acetyl-β-D-glucopyranose (*2, 46*). Richtmeyer and Hudson claimed an acid ortho ester structure for one of the products obtained by treating hepta-*O*-acetylceltrobiosyl chloride with silver carbonate in aqueous acetone (*118*), but this has not been confirmed.

Another substance postulated (*106*, p. 110) to be an acidic ortho ester is the 1,2-*O*-isopropylidene-D-glucofuranose monobenzoate prepared by Ohle (*99, 100*). Pigman and Isbell assigned an acidic ortho ester structure to the D-talopyranose monobenzoate obtained by perbenzoic acid oxidation of D-galactal (*115*), but the infrared spectrum of this compound showed it to be a normal benzoate (*61*). Ness and Fletcher proposed a 3,5-di-*O*-benzoyl-1,2-*O*-(1-hydroxybenzylidene)-α-D-ribofuranose structure for their alkali-labile ribose tribenzoate (*93*), but later showed that this compound is actually 1,3,5-tri-*O*-benzoyl-α-D-ribofuranose (*95*). These authors point out that the chemical properties of all acylated sugar derivatives previously assumed to be acidic 1,2-*O* ortho esters can be accounted for equally well by assigning them *cis*-1-acyloxy-2-hydroxy structures, and conclude that there is no good evidence for the existence of cyclic ortho acid structures in the sugar series.

More recently an acid ortho ester structure was claimed for a D-glucopyranose tetraacetate obtained by heating pentaacetyl-β-D-mannopyranose with dry acetic acid (*133*).

A monoacetyl derivative of nojirimycin (5-amino-5-deoxy-D-glucopyranose; 1,2,3,4-tetrahydroxy-5-hydroxymethylpiperidine) was assigned a 1,5-*O*-(1-hydroxyethylidene) structure on the basis of its chemical and spectroscopic properties (*54a*).

Although sugar 2-hydroxydioxolane derivatives are probably too unstable to be isolable, they are almost certainly intermediates in acid-catalyzed ortho ester hydrolyses (*38*), and in acyl migrations of partially acylated sugars. Hydrolysis reactions are discussed below. For examples of acyl migrations in the sugar series, and leading references, see references (*2*), (*16*), (*40*), (*46*), (*59*), (*77*), [(*106*), p. 111], [(*130*), p. 180], [(*131*), p. 304], and (*132*).

b. *2-Acyloxy-1,3-dioxolane Derivatives of Carbohydrates.* The only putative ortho ester acid anhydride is a maltose octaacetate prepared by Freudenberg, Hochstetter, and Engels from hepta-*O*-acetyl-maltosyl chloride and silver acetate in benzene (*25*). An ortho ester anhydride structure, **21**, has been proposed for this substance (*106*, p. 80). Since the octaacetate was amorphous, and since its chemical and physical properties have not been studied, the ortho ester anhydride structure has not been confirmed experimentally. More recently, **21** was proposed as an intermediate in the mutarotation of hepta-*O*-acetyl-β-maltose in the presence of acetic acid (*142*).

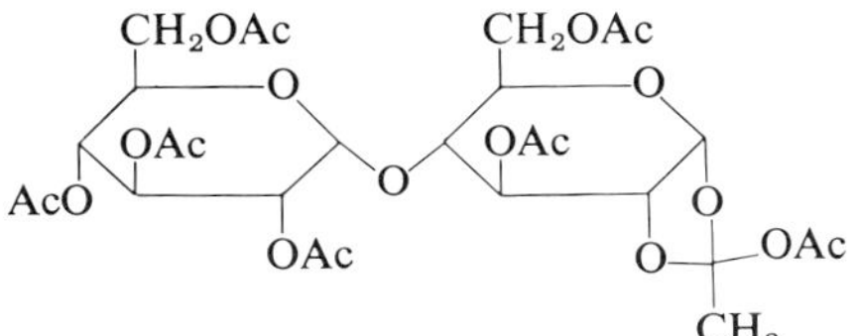

**21**

An octaacetate of turanose (3-α-D-glucopyranosyl-D-fructose) prepared from heptaacetylturanosyl bromide was originally assigned an ortho ester anhydride structure by Pacsu (*101*), who later concluded that the substance is not an ortho ester derivative (*109*).

c. *Orthoacyl Halide Derivatives of Carbohydrates.* Ortho acid halide structures have been claimed for three haloacyl derivatives of sugars. A hepta-acetylchloromaltose prepared by Freudenberg and co-workers (*22, 25*) was assigned the orthoacyl chloride structure **22**, and a 1,2-acetobromo derivative of 5,6-diacetyl-3-mesyl-glucofuranose was assigned structure **23** by Helferich and Jochinke (*45*). The product of reaction of penta-*O*-acetyl-α-D-glucopyranose with aluminum chloride was also assigned an orthoacyl halide structure (*148*).

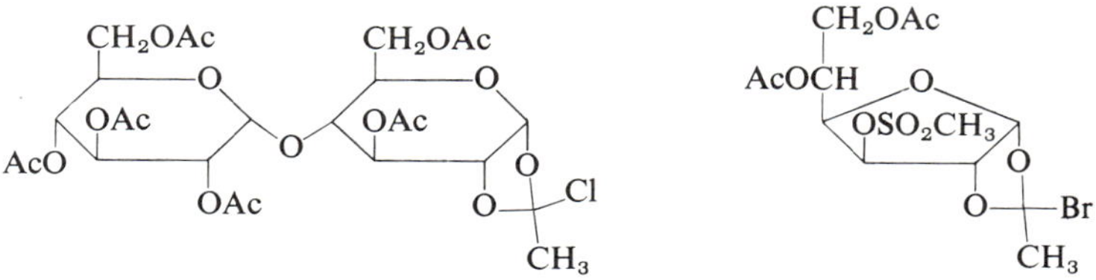

The nuclear magnetic resonance spectra of the maltose and glucofuranose glycosyl halides indicate that they are open-chain isomers of **22** and **23** having no dioxolane ring systems (*50*). Korytnik and Mills concluded on the basis of other evidence that the acetylmaltosyl chloride is probably a normal acylglycosyl chloride (*75*).

### B. Syntheses Involving Alteration of the Carbohydrate or Acyl Groups of Other Ortho Esters

In this section are discussed reactions by means of which one carbohydrate ortho ester may be converted into another without altering the skeletal framework of the ortho ester function. These syntheses involve substitution and elimination reactions of the carbohydrate moiety of the ortho ester, or transformations of the ortho ester acyl substituent.

1. DEACYLATION OF ACYLATED ORTHO ESTERS

The ortho ester function is inert or nearly so toward nucleophilic reagents under basic conditions. All of the usual methods of deacylation of sugar esters which employ basic media can be used to remove acyl groups from acylated carbohydrate ortho esters. Deacylation yields hydroxy ortho esters which are often sufficiently stable to be isolable. A specific example is the deacetylation of 3,4,6-tri-*O*-acetyl-1,2-*O*-(1-methoxyethylidene)-β-D-mannopyranose, which yields 1,2-*O*-(1-methoxyethylidene)-β-D-mannopyranose (*3, 85, 136, 139*).

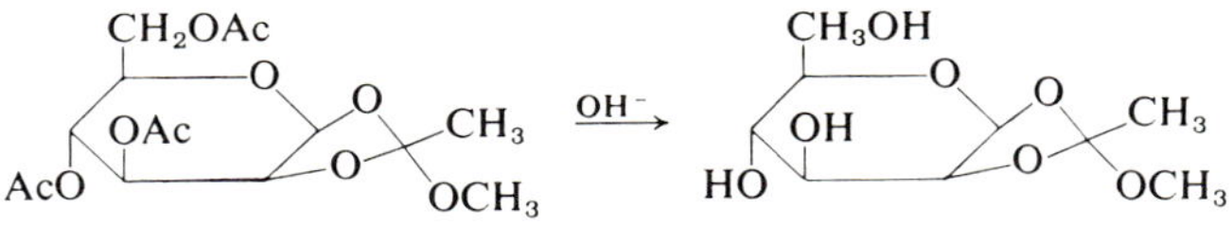

Carbohydrate ortho esters which have been prepared by deacylation of partially or completely acylated derivatives include **3**, **4**, and **24–34** (Y = H).

For names of these compounds and references to the original literature, see Table I.

**24**

**25**

**26**

**27**

**28**

**29**

**30**

**31**

**32**

**33**

**34**

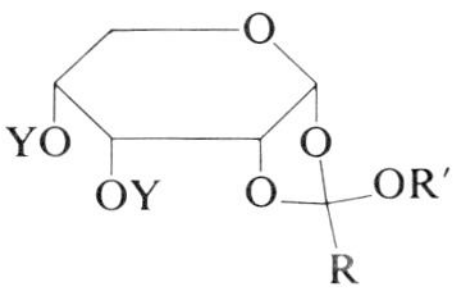

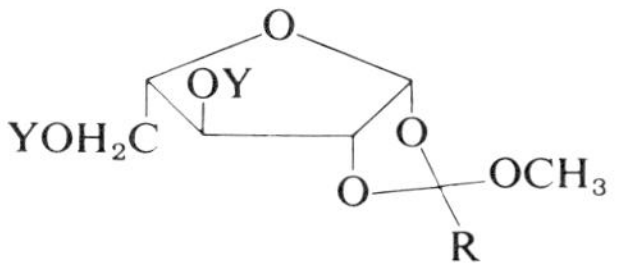

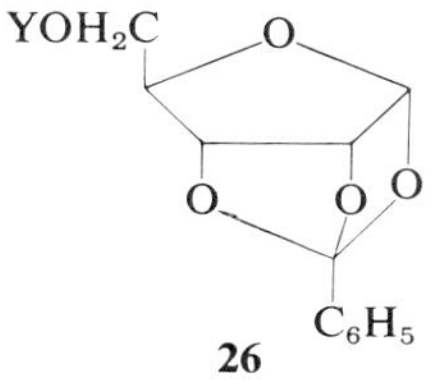

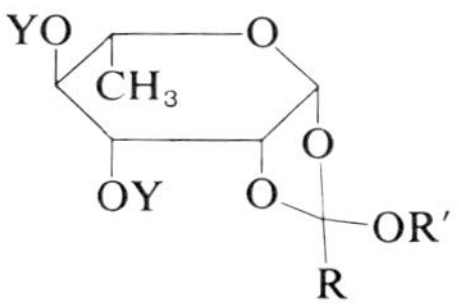

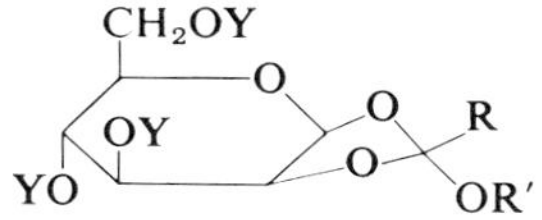

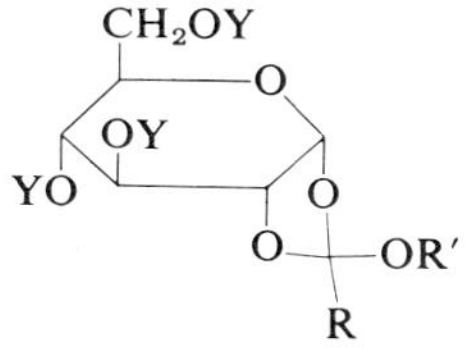

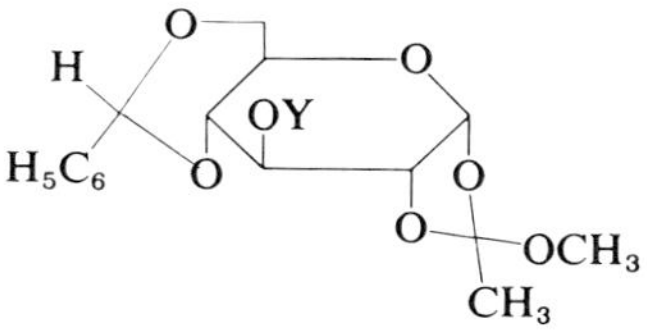

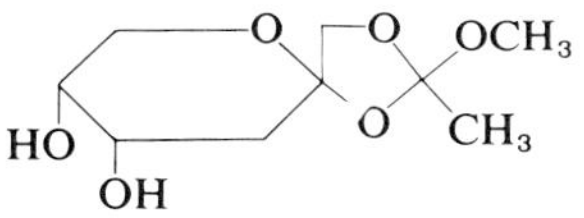

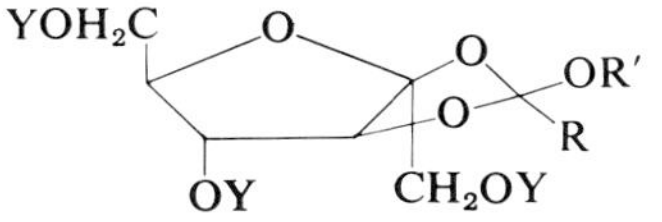

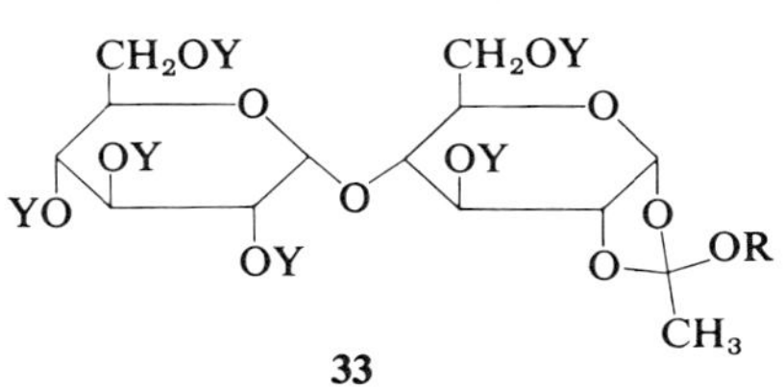

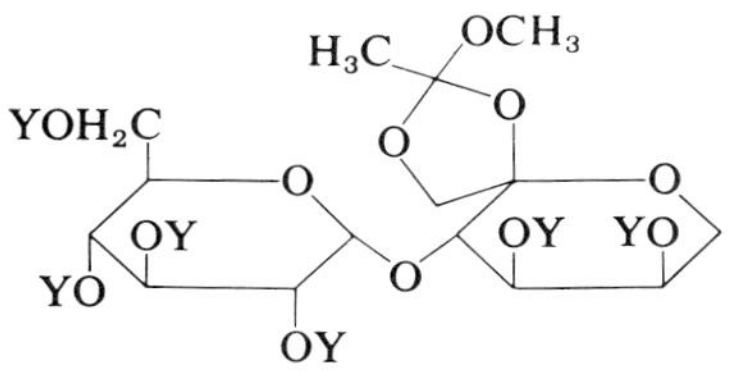

## 2. Acylation of Hydroxy Ortho Esters

The ortho ester function of carbohydrate ortho esters remains intact during acylation of free hydroxyl groups with acid anhydrides or acyl chlorides in the presence of pyridine, and a number of acylated carbohydrate ortho esters have been prepared from hydroxy precursors. Acylation of hydroxy ortho esters is particularly useful for preparing acylated tricyclic ortho esters and acylated ortho esters in which the acyloxy acyl substituent differs from the ortho ester acyl substituent, since these compounds cannot be obtained directly from acylglycosyl halides. Acylated ortho esters which have been prepared in this way include compounds **26, 27, 28, 32, 35,** and **36** (Y = $CH_3CO$) and **35–37** (Y = $C_6H_5CO$). Table I gives the names of these compounds and references to the original literature.

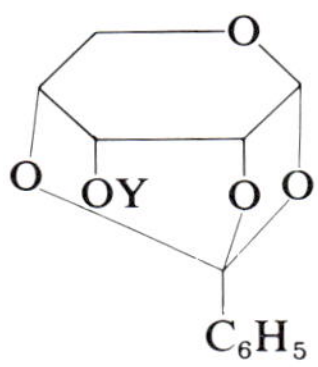

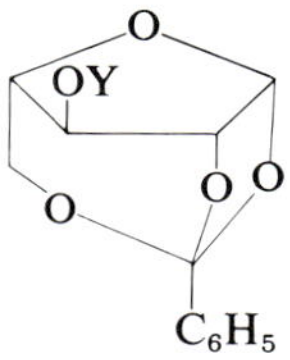

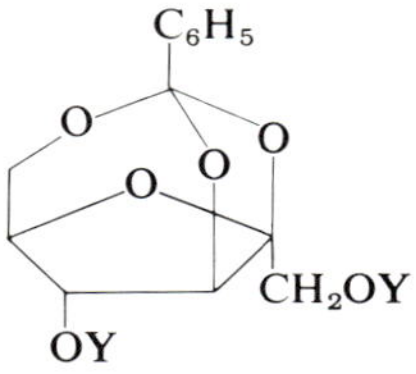

## 3. Tosylation and Mesylation of Hydroxy Ortho Esters

Carbohydrate ortho esters having free hydroxyl groups react with methanesulfonyl chloride and *p*-toluenesulfonyl chloride in pyridine solution to form mesylate and tosylate esters. Sulfonate esters which have been prepared by tosylation and mesylation of carbohydrate hydroxy ortho esters include **26, 32,** and **37** (Y = $CH_3SO_2$); **32** and **37** (Y = *p*-$CH_3C_6H_4SO_2$); and 5'-*O*-*p*-toluenesulfonyl-2',3'-*O*-ethoxymethyleneadenosine, **38** (*89*). See Table I for literature citations and names of **26, 32,** and **37**.

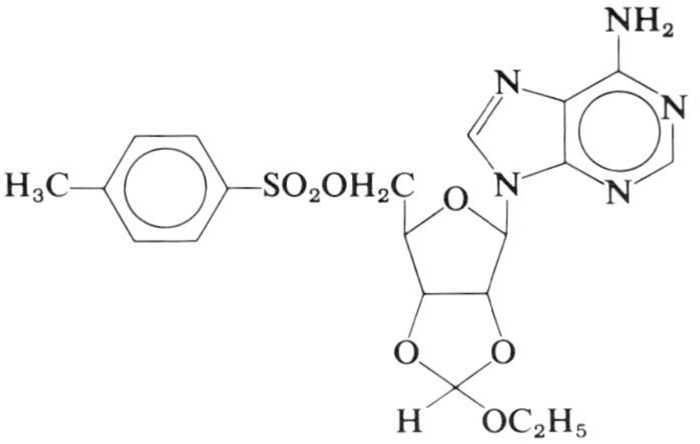

## 4. ALKYLATION OF HYDROXY ORTHO ESTERS

Carbohydrate ortho esters having free hydroxyl groups can be converted to methyl ethers by treatment with methyl iodide and silver oxide (*3*, *18*, *37*, *75*) or dimethyl sulfate (*3*). Three 3,4,6-tri-*O*-benzyl-1,2-*O*-(1-alkoxyethylidene)-β-D-mannopyranoses were prepared by refluxing the 3,4,6-tri-*O*-acetyl ortho esters with benzyl chloride and potassium hydroxide in tetrahydrofuran (*19*, *20*). Saponification of the acetyl groups presumably preceded benzylation. Trityl ethers of 2′,3′-*O*-ethoxymethylene nucleosides (**39**) have been prepared from the ortho esters and mono-*p*-methoxytrityl chloride in pyridine (*15*).

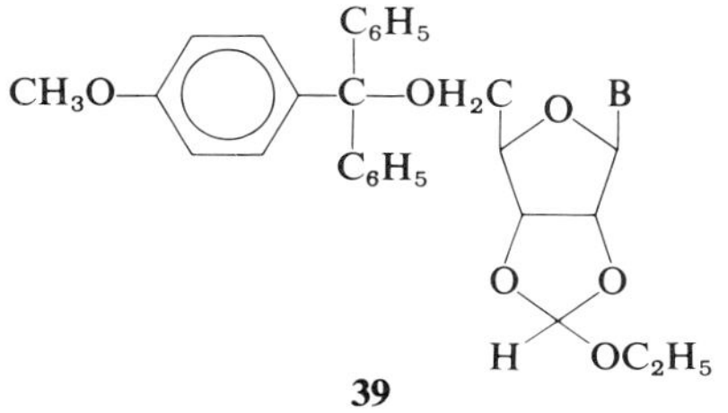

**39**

Other ethers which have been prepared by alkylation of carbohydrate hydroxy ortho esters include **27**, **28**, **30**, and **35** (Y = CH$_3$) and **28** (Y = C$_6$H$_5$CH$_2$). See Table I for names of these compounds and literature citations.

## 5. OTHER REACTIONS INVOLVING ALTERATION OF THE CARBOHYDRATE PORTION OF THE ORTHO ESTER

a. *Miscellaneous Substitution Reactions.* 6-Iodo-6-deoxy-1,4-di-*O*-methanesulfonyl-2,3-*O*-(1-benzyloxybenzylidene)-β-D-fructofuranose (**40**) was prepared from the 1,4,6-trimesyl ortho ester and sodium iodide in acetone (*48*). Hydrogen peroxide oxidation converts μ-thiolglucoxazoline [**6**, Y = –N=C(SH)—O–] to μ-hydroxyglucoxazoline [**6**, Y = –N=C(OH)—O–], in effect substituting a hydroxyl for a thiol group at position 2 of the oxazoline ring (*125a*).

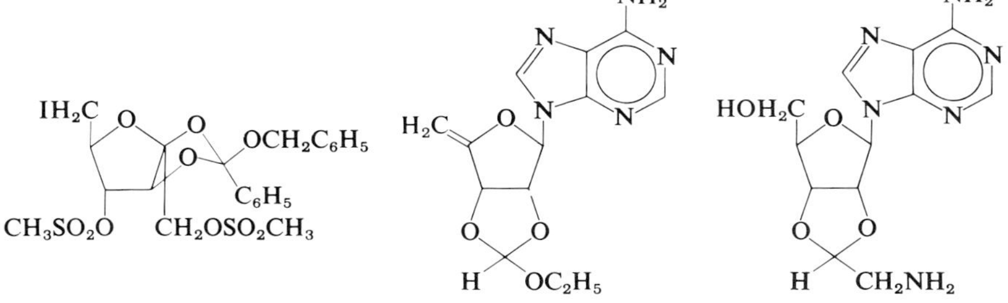

**40** **41** **42**

TABLE I
CARBOHYDRATE ORTHO ESTERS

| Compound | Method of synthesis[a] | Type formula | M.p. | $[\alpha]_D^b$ | References |
|---|---|---|---|---|---|
| 3,5-Di-O-benzoyl-1,2,-O-(1-benzyloxybenzylidene)-α-D-ribofuranose | 1 | 43 | — | — | 93 |
| 3,4-Di-O-acetyl-1,2-O-(1-methoxyethylidene)-α-D-ribopyranose | 1 | 24 | 77°–78° | +2.7° | 86 |
| 3,4-Di-O-acetyl-1,2-O-[1-(3-acetoxy-2-oxopropoxy)-ethylidene]-α-D-ribopyranose | 1 | 24 | 97°–98° | −11.8° | 66a |
| 3,4-Di-O-acetyl-1,2-O-[1-(3-acetoxy-2-oxopropoxy)-ethylidene]-α-L-ribopyranose | 1 | 44 | 97°–98° | +11.8° | 66a |
| 1,2-O-(1-Benzyloxybenzylidene)-α-D-ribopyranose | 7 | 24 | 101°–103° | +6.6° | 18 |
| 1,2,3-O-Benzylidyne-α-D-ribofuranose | 7 | 26 | 106°–108° | +41.0° | 96 |
| 5-O-Acetyl-1,2,3-O-benzylidyne-α-D-ribofuranose | 8 | 26 | 117°–118° | +42° | 96 |
| 5-O-Benzoyl-1,2,3-O-benzylidyne-α-D-ribofuranose | 1 | 26 | 182°–183° | +43.4° | 96, 97 |
| 5-O-Methanesulfonyl-1,2,3-O-benzylidyne-α-D-ribofuranose | 8 | 26 | 119°–124° | +42° | 96 |
| 1,2,4-O-Benzylidyne-α-D-ribopyranose | 4 | 35 | 151°–152° | +68.4° | 18 |
| 3-O-Acetyl-1,2,4-O-benzylidyne-α-D-ribopyranose | 8 | 35 | 96°–97° | +83.5° | 18 |
| 3-O-Benzoyl-1,2,4-O-benzylidyne-α-D-ribopyranose | 8 | 35 | 170°–171° | +78.1° | 18 |
| 3-O-Methyl-1-,2,4-O-benzylidyne-α-D-ribopyranose | 9 | 35 | 103°–105° | +84.3° | 18 |
| 1,2-O-(1-Methoxybenzylidene)-β-L-arabinofuranose | 7 | 25 | — | −5° | 31a, 73 |
| 3,5-Di-O-benzoyl-1,2-O-(1-methoxybenzylidene)-β-L-arabinofuranose | 1 | 25 | — | +18° | 71, 73, 73a |
| 3,5-Di-O-benzoyl-1,2-O-(1-methoxyethylidene)-β-L-arabinofuranose | 1 | 25 | 120°–121° | 0° | 31a |
| 3,5-Di-O-benzoyl-1,2-O-(1-methoxy-p-nitrobenzylidene)-β-L-arabinofuranose | 1 | 25 | 155°–159° | +36° | 31a |
| 3,4-Di-O-acetyl-1,2-O-(1-methoxyethylidene)-β-L-arabinopyranose | 1 | 45 | — | +52° | 71, 73a |

| Compound | | | m.p. | [α] | Ref. |
|---|---|---|---|---|---|
| 3,4-Di-O-benzoyl-1,2-O-(1-methoxybenzylidene)-β-L-arabinopyranose | 1 | 45 | — | — | 71, 73a |
| Methyl 3,4-O-(1-ethoxyethylidene)-β-L-arabinopyranoside (80% exo, 20% endo) | 2 | 46 | b.p. 120° at 0.01 mm | +100° (py) | 6 |
| 1,2,5-O-Benzylidyne-β-L-arabinofuranose | 4 | 36 | 148°–149° | +28° | 66d, 73 |
| 3-O-Acetyl-1,2,5-O-benzylidyne-β-L-arabinofuranose | 8 | 36 | 128°–130° | +62° | 66d, 73 |
| 3-O-Benzoyl-1,2,5-O-benzylidyne-β-L-arabinofuranose | 8 | 36 | 143°–145° | +75° | 73 |
| 3,4-Di-O-acetyl-1,2-O-(1-methoxyethylidene)-α-D-lyxopyranose | 1 | 47 | 90° | −103.5° | 87 |
| 3,5,6-Tri-O-acetyl-1,2,-O-(1-ethoxyethylidene)-α-D-glucofuranose | 1 | 48 | 90° | +27° | 138 |
| 5,6-Di-O-acetyl-3-O-methanesulfonyl-1,2-O-(1-methoxyethylidene)-α-D-glucofuranose | 1 | 48 | 160°–161° | +13.1° | 45 |
| 5,6-Di-O-acetyl-3-O-methanesulfonyl-1,2-O-(1-ethoxyethylidene)-α-D-glucofuranose | 1 | 48 | 65°–66° | +14.5° | 50 |
| 1,2-O-(1-tert-Butylperoxyethylidene)-α-D-glucopyranose | 7 | 29 | 121° | +22.8° (H$_2$O) | 125 |
| 1,2-O-[1-(2-N-Carbobenzyloxyamino-2-carbomethoxy)-ethylidene]-α-D-glucopyranose | 7 | 29 | — | — | 137 |
| 3,4,6-Tri-O-acetyl-1,2-O-(1-methoxyethylidene)-α-D-glucopyranose | 1 | 29 | — | +34.4° | 64, 68, 71, 78, 79, 81, 83 |
| 3,4,6-Tri-O-acetyl-1,2-O-(1-ethoxyethylidene)-α-D-glucopyranose | 1 | 29 | 97°–97.5° | +32° | 64, 68, 71, 80, 82, 125, 138 |
| 3,4,6-Tri-O-acetyl-1,2-O-(1-propoxyethylidene)-α-D-glucopyranose | 1 | 29 | 92°–94.5° 69° | +39.5° +30° | 80 125 |
| 3,4,6-Tri-O-acetyl-1,2-O-{1-[2-(2,4-dinitrophenylamino-2-carbomethoxy-1-methylethoxy)]ethylidene}-α-D-glucopyranose | 1 | 29 | — | +28.2° | 12a |
| 3,4,6-Tri-O-acetyl-1,2-O-(1-(2'-benzyloxycarbonylamino-2'-carbomethoxy)-ethylidene]-α-D-glucopyranose | 1 | 29 | — | — | 12b |
| 3,4,6-Tri-O-acetyl-1,2-O-(1-isopropoxyethylidene)-α-D-glucopyranose (exo isomer) | 1 | 29 | 116°–117° | +30° | 1c, 44, 82 |

TABLE I—continued

| Compound | Method of synthesis[a] | Type formula | M.p. | $[\alpha]_D$[b] | References |
|---|---|---|---|---|---|
| 3,4,6-Tri-O-acetyl-1,2-O-(1-tert-butoxyethylidene)-α-D-glucopyranose | 1 | 29 | 153°–154.5° | +34° | 1d, 82, 125 |
| 3,4,6-Tri-O-acetyl-1,2-O-(1-phenoxyethylidene)-α-D-glucopyranose (68% exo, 32% endo) | 1 | 29 | — | — | 1d, 82 |
| 3,4,6-Tri-O-acetyl-1,2-O-[1-(2-N-carbobenzylamino-2-carbomethoxy)ethylidene]-α-D-glucopyranose | 1 | 29 | — | — | 137 |
| 3,4,6-Tri-O-acetyl-1,2-O-(1-cholesteryloxyethylidene)-α-D-glucopyranose | 1, 3 | 29 | 98°–100° | +2° | 64, 68, 71 |
| 3,4,6-Tri-O-acetyl-1,2-O-(1-isopropoxybenzylidene)-α-D-glucopyranose | 1 | 29 | — | — | 82 |
| 3,4,6-Tri-O-acetyl-1,2-O-(1-methoxy-2,2-dimethylpropylidene)-α-D-glucopyranose | 1 | 29 | — | — | 82 |
| 3,4,6-Tri-O-acetyl-1,2-O-(dimethoxymethylidene)-α-D-glucopyranose | 1 | 29 | — | — | 82 |
| 3,4,6-Tri-O-acetyl-1,2-O-[(2-tetrahydrofuryl)methylidene]-α-D-glucopyranose | 3 | 65 | exo: —<br>endo: 117° | +42°<br>+85° | 80a |
| 4,6-Di-O-acetyl-3-O-p-toluenesulfonyl-1,2-O-(1-methoxyethylidene)-α-D-glucopyranose (?) | 1 | 29 | — | — | 22 |
| 3-O-Acetyl-4,6-O-benzylidene-1,2-O-(1-methoxyethylidene)-α-D-glucopyranose | 1 | 29 | 148°–149° | +36° | 75 |
| 3,4,6-Tri-O-benzoyl-1,2-O-(1-methoxybenzylidene)-α-D-glucopyranose | 1 | 29 | — | +11.5° | 49 |
| 3,4,6-Tri-O-benzoyl-1,2-O-(1-benzyloxybenzylidene)-α-D-glucopyranose | 1 | 29 | — | +3.2° | 49 |
| 4,6-O-Benzylidene-3-O-methyl-1,2-O-(1-methoxyethylidene)-α-D-glucopyranose | 9 | 29 | 130°–131° | −63.8° | 75 |

| Compound | | | m.p. | [α] | Refs. |
|---|---|---|---|---|---|
| 4,6-O-Benzylidene-1,2-O-(1-methoxyethylidene)-α-D-glucopyranose | 1 | 29 | 140°–149° | — | 66e |
| 1,2-O-Isopropylidene-3,5,6-O-methylidene-α-D-glucofuranose | 2, 10 | 6 | 200°–201° | −40.9° | 23, 42 |
| 1,2-O-Benzylidene-3,5,6-O-methylidene-α-D-glucofuranose | 2 | 6 | 166°–167° | −31.2° | 47 |
| 3,5,6-O-Methylidene-μ-thiolglucoxazoline | 2 | 6 | 255° | +20° (0.1N OH⁻) | 125a |
| 3,5,6-O-Methylidene-μ-hydroxyglucoxazine | 10 | 6 | 220°–225° | +32° | 125a |
| 1,2-O-Isopropylidene-3,5,6-O-(S-methylmonothioorthocarbonyl)-α-D-glucofuranose | 6 | 10 | 178°–180° | −3.3° | 127, 128 |
| 1,2-O-Isopropylidene-3,5,6-O-(S-benzylmonothioorthocarbonyl)-α-D-glucofuranose | 6 | 10 | 169°–170° | +25.4° | 128 |
| Bis(1,2-O-isopropylidene-α-D-glucofuranose-3,5,6-orthocarbonyl) disulfide | 6 | 14 | 268°–270° | +68° | 13 |
| 6-Deoxy-1,2-O-(1-methoxyethylidene)-β-L-mannopyranose | 7 | 27 | 143°–144° | +16.3° | 4, 17, 37 |
| 3,4-Di-O-acetyl-6-deoxy-1,2-O-(1-methoxyethylidene)-β-L-mannopyranose | 1, 8 | 27 | 84°–86° | +34.7° | 10, 17, 21, 37, 38, 60, 90, 92 |
| 3,4-Di-O-acetyl-6-deoxy-1,2-O-(1-tert-butylperoxyethylidene)-β-L-mannopyranose | 1 | 27 | 122° | +35.6° | 123 |
| 3,4-Di-O-benzoyl-6-deoxy-1,2-O-(1-methoxybenzylidene)-β-L-mannopyranose | 1 | 27 | 174°–175° | +37.8° | 98 |
| 6-Deoxy-1,2-O-(1-methoxyethylidene)-3,4-di-O-methyl-β-L-mannopyranose | 9 | 27 | 67° | +36° | 37 |
| 1,2-O-(1-Methoxyethylidene)-β-D-mannopyranose | 7 | 28 | — | −6° ($H_2O$) | 3, 85, 136 |
| 3,4,6-Tri-O-acetyl-1,2-O-(1-methoxyethylidene)-β-D-mannopyranose | 1 | 28 | 106° (exo) | −22° (exo) +12° (endo) | 3, 12, 60, 85, 92, 113, 138 |
| 3,4,6-Tri-O-acetyl-1,2-O-(1-ethoxyethylidene)-β-D-mannopyranose | 1 | 28 | 81°–82° | −28° | 85 |
| 3,4,6-Tri-O-acetyl-1,2-O-(1-tert-butoxyethylidene)-β-D-mannopyranose | 1 | 28 | 112°–113° | −13.7° | 123 |
| 3,4,6-Tri-O-acetyl-1,2-O-(1-tert-butylperoxyethylidene)-β-D-mannopyranose | 1 | 28 | 129° | −28.9° | 123, 124 |

TABLE I—continued

| Compound | Method of synthesis[a] | Type formula | M.p. | $[\alpha]_D$[b] | References |
|---|---|---|---|---|---|
| 3,4,6-Tri-O-acetyl-1,2-O-(1-benzyloxyethylidene)-β-D-mannopyranose | 1 | 28 | 146°–147° | −1.0° | 92, 113 |
| 3,4,6-Tri-O-acetyl-1,2-O-(1-p-bromobenzyloxyethylidene)-β-D-mannopyranose | 1 | 28 | 182°–182.5° | +1.5° | 92 |
| 3,4,6-Tri-O-acetyl-1,2-O-(1-o-bromobenzyloxyethylidene)-β-D-mannopyranose | 1 | 28 | 119°–121° | +3.9° | 91 |
| 3,4,6-Tri-O-acetyl-1,2-O-(1-p-iodobenzyloxyethylidene)-β-D-mannopyranose | 1 | 28 | 178°–179° | +0.9° | 92 |
| 3,4,6-Tri-O-acetyl-1,2-O-(1-diphenylmethoxyethylidene)-β-D-mannopyranose | 1 | 28 | 150°–151° | +10.4° | 92 |
| 3,4,6-Tri-O-acetyl-1,2-O-{1-[5-(2-phenyl-1,3-dioxolanyloxy)]-ethylidene}-β-D-mannopyranose | 1 | 28 | 176°–176.5° | −8.4° | 92 |
| 3,4,6-Tri-O-acetyl-1,2-O-(1-isopropoxyethylidene)-β-D-mannopyranose | 1 | 28 | 106° | −13° ($CH_2Cl_2$) | 20 |
| 3,4,6-Tri-O-acetyl-1,2-O-(1-cyclohexyloxyethylidene)-β-D-mannopyranose | 1 | 28 | 141° | −11.4° ($CH_2Cl_2$) | 20 |
| 3,4,6-Tri-O-benzyl-1,2-O-(1-isopropoxyethylidene)-β-D-mannopyranose | 9 | 28 | 100° | +13° ($CH_2Cl_2$) | 20 |
| 3,4,6-Tri-O-benzyl-1,2-O-(1-cyclohexyloxyethylidene)-β-D-mannopyranose | 9 | 28 | 90° | +13.6° | 20 |
| 3,4,6-Tri-O-acetyl-1,2-O-(1-methoxybenzylidene)-β-D-mannopyranose | 1 | 28 | — | −61° | 114 |
| 1,2-O-(1-Benzyloxyethylidene)-β-D-mannopyranose | 7 | 28 | 130° (exo) 117° (endo) | +26° ($H_2O$) −15° ($H_2O$) | 113 |

| Compound | | | | | | |
|---|---|---|---|---|---|---|
| 3,4,6-Tri-O-benzyl-1,2-O-(1-methoxyethylidene)-β-D-mannopyranose | 9 | 28 | 78°–81° | +12° (CH$_2$Cl$_2$) | | 19, 20 |
| 1,2-O-(1-Methoxyethylidene)-3,4,6-tri-O-methyl-β-D-mannopyranose | 9 | 28 | b.p. 120° at 0.01 mm | −20° (H$_2$O) | | 3 |
| 1,2-O-(1-Methoxybenzylidene)-β-D-mannopyranose | 7 | 28 | — | — | | 32 |
| 3,4,6-Tri-O-benzoyl-1,2-O-(1-methoxybenzylidene)-β-D-mannopyranose | 1 | 28 | — | −124° (CH$_3$OH) | | 32 |
| 1,2,3-Tri-O-acetyl-4,6-O-(1-ethoxyethylidene)-β-D-idopyranose | 5 | 49 | — | +47.2° | | 111, 112 |
| 3,4,6-Tri-O-acetyl-1,2-O-(1-methoxyethylidene)-β-D-idopyranose (exo) | 1 | 62 | 99°–101° | +35.0° | | 15a |
| 3,4,6-Tri-O-acetyl-1,2-O-(1-ethoxyethylidene)-β-D-idopyranose | 1 | 62 | 113°–115° | +34.0° | | 15a |
| 3,5,6-Tri-O-acetyl-1,2-O-(1-methoxyethylidene)-α-D-galactofuranose | 1 | 51 | — | +24° | | 1b, 69, 71 |
| 3,4,6-Tri-O-acetyl-1,2-O-(1-ethoxyethylidene)-α-D-galactopyranose | 1 | 52 | — | +78° | | 64, 68, 71 |
| 3,4,5,6-Tetra-O-acetyl-1,2-O-(1-ethoxyethylidene)-1-ethylthio-β-D-galactose | 1 | 50 | 125°–126° | +54° | | 143 |
| 3,4,6-Tri-O-acetyl-1,2-O-(1-methoxyethylidene)-β-D-talopyranose | 1 | 53 | 91°–92.5° | +3.7° | | 115 |
| 2,3-O-(1-Benzyloxybenzylidene)-β-D-fructofuranose | 7 | 32 | — | +17.3° | | 43 |
| 2,3-O-(1-Cyclohexyloxybenzylidene)-β-D-fructofuranose | 7 | 32 | — | +22.3° | | 43 |
| 1,4,6-Tri-O-acetyl-2,3-O-(1-benzyloxybenzylidene)-α-D-fructofuranose | 8 | 32 | 67°–68° | +7.3° | | 43 |
| 1,4,6-Tri-O-benzoyl-2,3-O-(1-methoxybenzylidene)-α-D-fructofuranose | 1 | 32 | 91°–92° | −1° | | 43 |
| 1,4,6-Tri-O-benzoyl-2,3-O-(1-ethoxybenzylidene)-α-D-fructofuranose | 1 | 32 | 126°–127° | −2.3° | | 94 |
| 1,4,6-Tri-O-benzoyl-2,3-O-(1-propoxybenzylidene)-α-D-fructofuranose | 1 | 32 | 140°–141° | −4.2° | | 43 |

TABLE I—continued

| Compound | Method of synthesis[a] | Type formula | M.p. | $[\alpha]_D$[b] | References |
|---|---|---|---|---|---|
| 1,4,6-Tri-O-benzoyl-2,3-O-(1-cyclohexyloxybenzylidene)-α-D-fructofuranose | 1 | 32 | 111°–112° | −9.4° | 43 |
| 1,4,6-Tri-O-benzoyl-2,3-O-(1-benzyloxybenzylidene)-α-D-fructofuranose | 1 | 32 | 145°–147° | +5.8° | 43 |
| 2,3-O-(1-Benzyloxybenzylidene)-3,4,6-tri-O-methanesulfonyl-α-D-fructofuranose | 8 | 32 | 96°–97° | +17.4° | 43 |
| 2,3-O-(1-Benzyloxybenzylidene)-3,4,6-tri-O-p-toluenesulfonyl-α-D-fructofuranose | 8 | 32 | 95°–96° | +21° | 43 |
| 2,3-O-(1-Benzyloxybenzylidene)-6-deoxy-6-iodo-1,4-di-O-methanesulfonyl-α-D-fructofuranose | 10 | 32 | 100°–102° | −7.4° | 48 |
| 2,3,6-O-Benzylidyne-α-D-fructofuranose | 4 | 37 | — | — | 43, 48 |
| 1,4-Di-O-benzoyl-2,3,6-O-benzylidyne-α-D-fructofuranose | 8 | 37 | 136°–137° | −52.6° | 43 |
| 2,3,6-O-Benzylidyne-1,4-di-O-p-toluenesulfonyl-α-D-fructofuranose | 8 | 37 | 101°–103° | −43.9° | 48 |
| 2,3,6-O-Benzylidyne-1,4-di-O-methanesulfonyl-α-D-fructofuranose | 8 | 37 | 96°–97° | −12.9° | 48 |
| 1,2-O-(1-methoxyethylidene)-α-D-fructofuranose | 7 | 54 | — | −12.7° | 104 |
| 3,4,5-Tri-O-acetyl-1,2-O-(1-methoxyethylidene)-α-D-fructofuranose | 1 | 54 | — | −13.6° | 104 |
| 1,4,5-Tri-O-acetyl-2,3-O-(1-methoxyethylidene)-β-L-sorbopyranose | 1 | 55 | — | — | 105 |
| 1,4,5-Tri-O-acetyl-2,3-O-(1-ethoxyethylidene)-β-L-sorbopyranose | 1 | 55 | — | — | 105 |
| Methyl 3,4-di-O-acetyl-1,2-O-(1-methoxyethylidene)-α-D-glucopyranuronate | 1 | 57 | 118° | +54° | 30 |
| 5-O-Acetyl-1,2-O-(1-methoxyethylidene)-α-D-glucopyranurono-3,6-lactone | 1 | 56 | 110°–111° | +112.5° | 30, 75 |

| Compound | | | mp | [α] | References |
|---|---|---|---|---|---|
| 2-*tert*-Butylperoxy-2-methyl-5-(D-arabino-1,2,3,4-tetraacetoxybutyl)-1,3-dioxolane-4-one | 10 | 18 | 162°–163° | −8.0° | *122* |
| Erythro-1,2-diacetoxy-1,2-bis[5-(2-*tert*-butylperoxy-2-methyl-4-oxo-1,3-dioxolanyl)]ethane | 10 | 19 | 211°–213° | — | *122* |
| 1,2,5-*O*-Methylidene-*myo*-inositol | 2 | 7 | — | — | 88 |
| 3,4-*O*-Ethoxymethylenelincomycin | 2 | 5 | 174°–177° | — | 91 |
| Bis(2,3:5,6-di-*O*-isopropylidene-D-mannitol) orthocarbonate | 6 | 17 | 246°–247° | — | 14, *126* |
| 3,4,6,7-Tetra-*O*-acetyl-1,2-*O*-(1-methoxyethylidene)-D-glycero-β-L-taloheptopyranose | 1 | 58 | 106° | +3.2° | 28 |
| 3,4,6,7-Tetra-*O*-acetyl-1,2-*O*-(1-methoxyethylidene)-D-glycero-α-D-guloheptopyranose | 1 | 59 | 112° | +43° | 39 |
| 3,4,6-Tri-*O*-acetyl-1,2-*O*-[1-(2,3,4,6-tetra-*O*-acetyl-β-D-mannopyranosyl)ethylidene]-β-D-mannopyranose | 1 | 3 | 152°–154° | −23.6° | 31 |
| 1,2-*O*-[1-(β-D-Mannopyranosyl)ethylidene]-β-D-mannopyranose | 7 | 3 | 120° | — | 31 |
| 3,4,6-Tri-*O*-acetyl-1,2-*O*-[1-(1,2,3,4-tetra-*O*-acetyl-6-β-D-glucopyranosyl)ethylidene]-β-D-mannopyranose (two diastereomers) | 1 | 2 | 169° 174° | +17° −27.6° | 134 |
| 3-*O*-α-D-Glucopyranosyl-1,2-*O*-(1-methoxyethylidene)-α-D-fructofuranose | 7 | 34 | 137° | +115° | 102, *109* |
| 4,5-Di-*O*-acetyl-1,2-*O*-(1-methoxyethylidene)-3-*O*-(2,3,4,6-tetra-*O*-acetyl-α-D-glucopyranosyl)-α-D-fructopyranose | 1 | 34 | 162°–167° | +80° | 102, *109* |
| 3,6-Di-*O*-acetyl-1,2-*O*-(1-methoxyethylidene)-4-*O*-(2,3,4,6-tetra-*O*-acetyl-β-D-glucopyranosyl)-β-D-mannopyranose | 1 | 60 | 167° | −12.7° | 55, 56 |
| 4-*O*-α-D-Glucopyranosyl-1,2-*O*-(1-methoxyethylidene)-α-D-glucopyranose | 7 | 33 | — | — | 24 |
| 3,6-Di-*O*-acetyl-1,2-*O*-(1-methoxyethylidene)-4-*O*-(2,3,4,6-tetra-*O*-acetyl-α-D-glucopyranosyl)-α-D-glucopyranose | 1 | 33 | 163°–164° | +101.6° | 21, 24, 25 |
| 3,6-Di-*O*-acetyl-1,2-*O*-((1-ethoxyethylidene)-4-*O*-(2,3,4,6-tetra-*O*-acetyl-α-D-glucopyranosyl)-α-D-glucopyranose | 1 | 33 | 164°, 133°–134° | +98.8° | 25, *50*, *108* |
| 3,6-Di-*O*-benzoyl-1,2-*O*-(1-methoxybenzylidene)-4-*O*-(2,3,4,6-tetra-*O*-benzoyl-β-D-glucopyranosyl)-α-D-glucopyranose | 1 | 64 | 173° | +32.4° | *71a* |

TABLE I—continued

| Compound | Method of synthesis[a] | Type formula | M.p. | $[\alpha]_D$[b] | References |
|---|---|---|---|---|---|
| 3,4-Di-O-acetyl-1,2-O-(1-methoxyethylidene)-6-O-(2,3,4-tri-O-acetyl-α-D-galactopyranosyl)-α-D-glucopyranose | 1 | 63 | — | +118° | 66c |
| 3,6-Di-O-acetyl-1,2-O-(1-methoxyethylidene)-4-O-(2,3,4,6-tetra-O-acetyl-β-D-galactopyranosyl)-β-D-altropyranose | 1 | 61 | 121°–122° | +25.3° | 28 |
| Trimannose ortho ester | 7 | 4 (Y = H) | 156°–157° | −5.4° | 29 |
| Deca-O-acetyl trimannose ortho ester | 10 | 4 (Y = Ac) | 213°–215° | −29° | 29 |

[a] Methods of synthesis:
  1. From acylglycosyl halides
  2. Reaction of carbohydrates with acyclic ortho esters
  3. Intermolecular transesterification
  4. Intramolecular transesterification
  5. From dioxocarbonium salts
  6. From thionocarbonates
  7. Deacylation of acylated ortho esters
  8. Esterification of hydroxy ortho esters
  9. Alkylation of hydroxy ortho esters
  10. Miscellaneous other reactions

[b] Rotations measured in chloroform solutions, unless otherwise specified.

## STRUCTURES REFERRED TO IN TABLE I

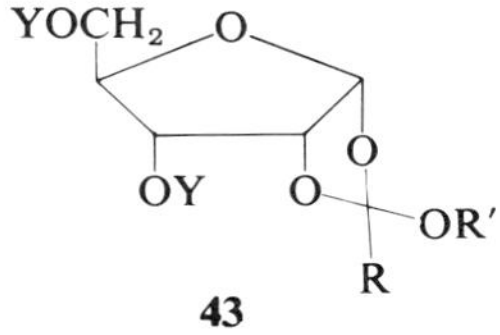

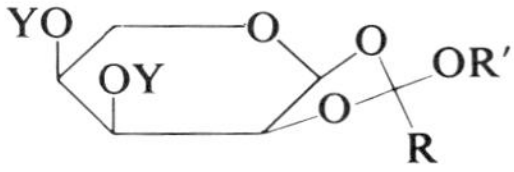

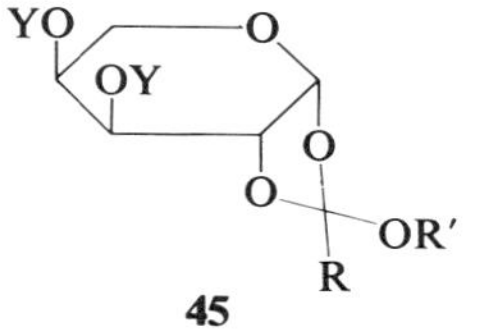

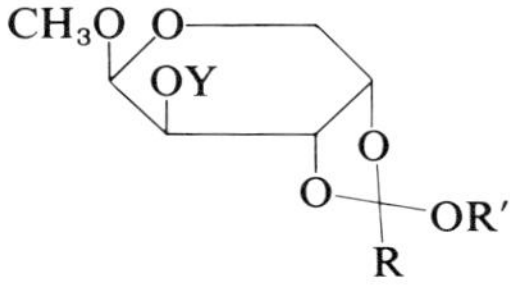

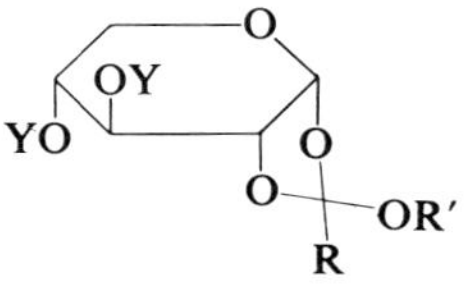

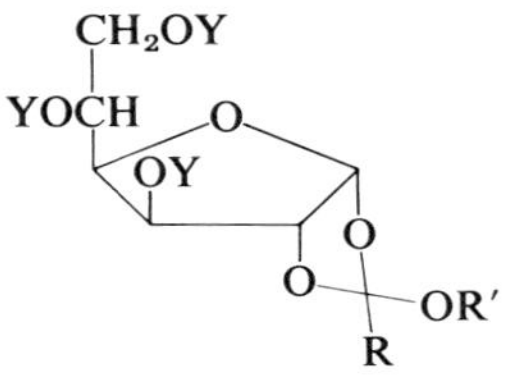

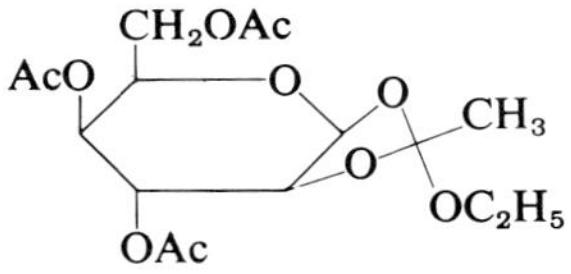

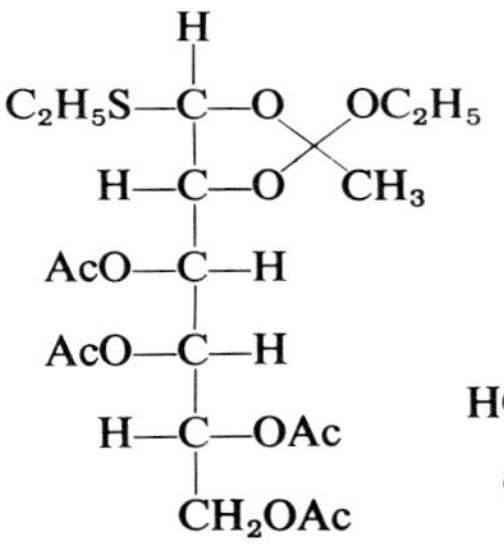

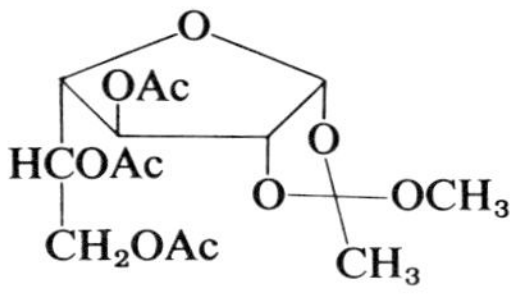

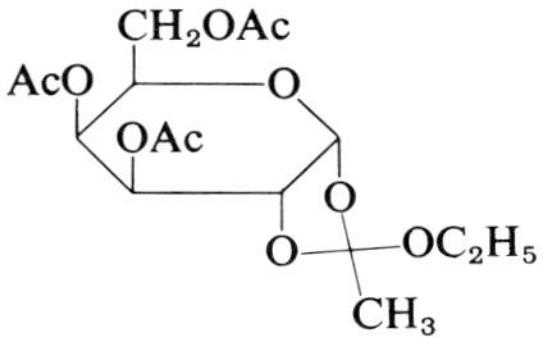

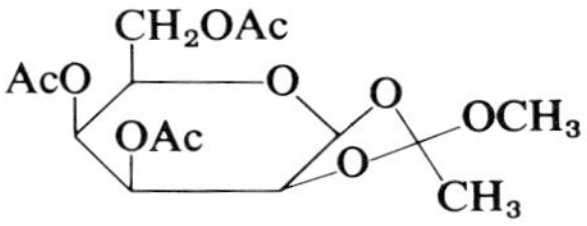

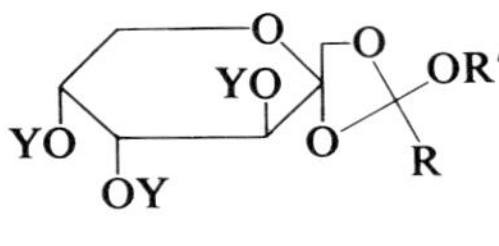

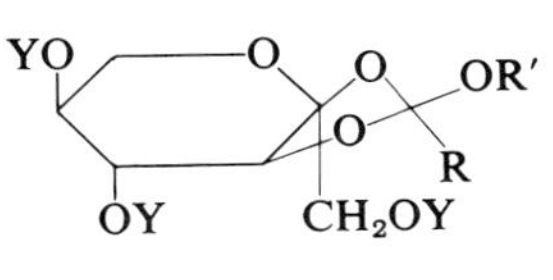

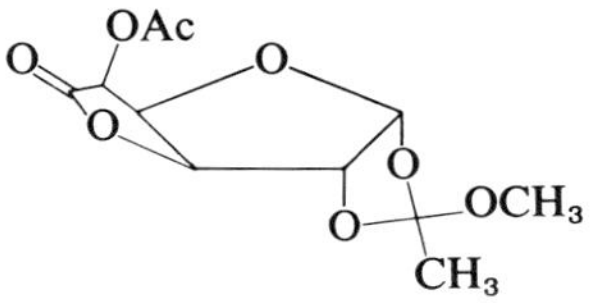

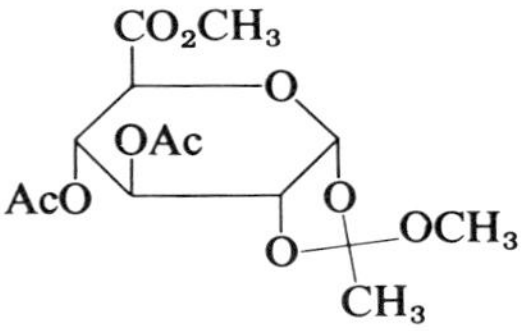

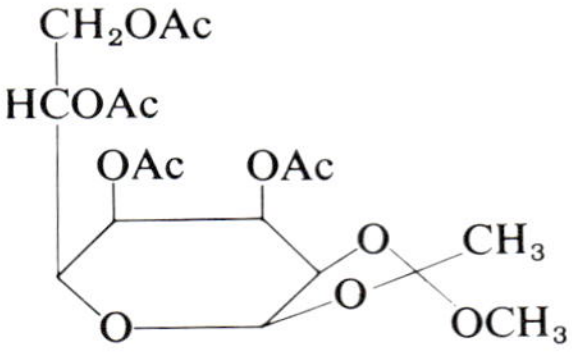

**58**

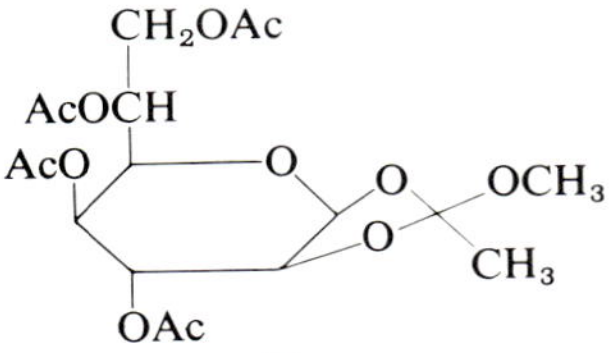

**59**

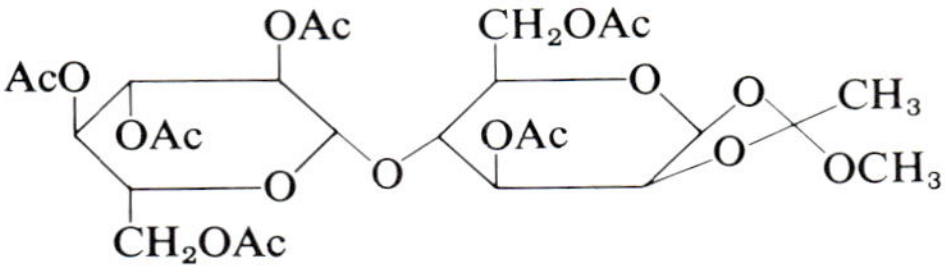

**60**

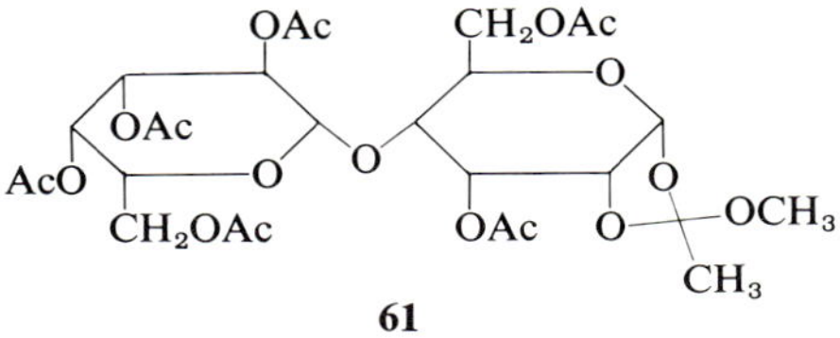

**61**

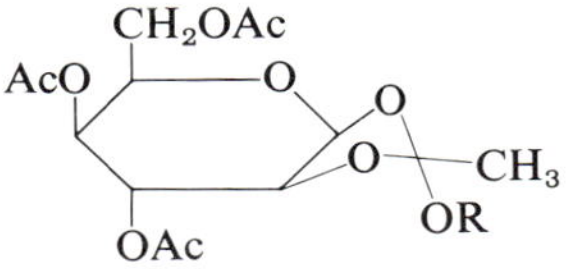

**62**

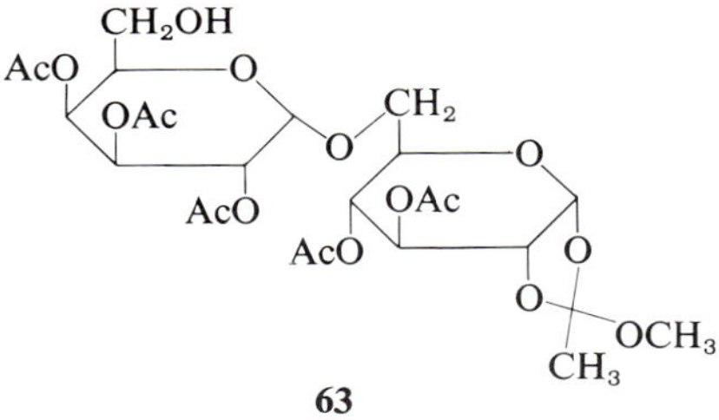

**63**

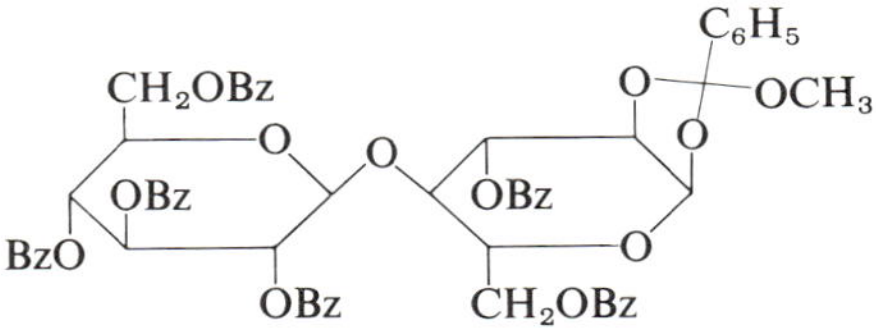

**64**

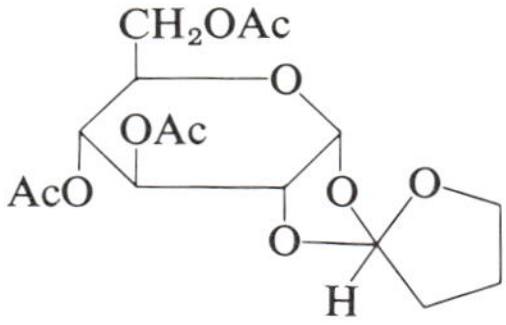

**65**

b. *Elimination Reactions.* Treatment of 5′-*O*-*p*-toluenesulfonyl-2′,3′-*O*-ethoxymethyleneadenosine with potassium *tert*-butoxide in *tert*-butyl alcohol results in elimination of a proton and tosylate ion to form the unsaturated ortho ester **41** (*89*). A similar elimination reaction was used in the conversion of a tricyclic orthobenzoate of psicofuranine to angustomycin A (*89a*).

c. *Miscellaneous Reactions of 2′,3′-O-Alkoxymethylene Nucleoside Derivatives.* Nucleosides and nucleoside derivatives having 2′,3′-*O*-alkoxymethylene protecting groups undergo a variety of reactions which leave the ortho ester protecting group intact. These include substitution (*9, 52–54, 147*) and partial hydrolysis (*52*) of the purine or pyrimidine moiety, coupling of 2′,3′-*O*-alkoxymethylene-5′-phosphates with 2′,5′-blocked nucleosides (*35*), and coupling of 2′,5′-blocked nucleotides with 2′,3′-*O*-ethoxymethylene nucleosides (*7, 9, 52–54, 129, 147*).

6. Reactions Involving Modification of the Ortho Ester Acyl Substituent

2′,3′-*O*-(1-Ethoxy-2-aminoethylidene)adenosine (**42**) was prepared by hydrogenolysis of the *N*-benzyloxycarbonyl derivative (*145, 146*). This is the only orthoglycinate ester which has been reported.

### C. Sources and Properties of Carbohydrate Ortho Esters

Table I (begins on p. 320) lists the sources and properties of carbohydrate ortho esters. The table also includes references to type formulas for the ortho esters, which appear either in the preceding text or immediately following the table. The esters of Table I are named according to rules of carbohydrate nomenclature adopted by the American Chemical Society (*1*).

The only physical properties recorded in Table I are melting points and optical rotations. Spectrometric data and chromatographic properties are frequently described in the original literature. Infrared spectra of carbohydrate ortho esters are discussed by Tipson and Isbell (*135*).

### III. REACTIONS OF CARBOHYDRATE ORTHO ESTERS WHICH YIELD NON-ORTHO ESTER PRODUCTS

Reactions which transform one carbohydrate ortho ester into another are discussed in Section II, B. The reactions considered below are those in which carbohydrate ortho esters are converted into non-ortho-ester products. These reactions include hydrolysis, alcoholysis, acidolysis, and isomerization reactions, as well as dioxolenium salt formation.

## A. Hydrolysis of Carbohydrate Ortho Esters

The ortho ester groups in carbohydrate ortho esters, like those of other orthocarboxylates, are practically inert to aqueous alkali. The inertness of these compounds toward aqueous hydroxide solutions is illustrated by the fact that 6-deoxy-1,2-O-(1-methoxyethylidene)-3,4-di-O-methyl-β-L-mannopyranose (**27**, Y = R = R' = CH$_3$) does not hydrolyze appreciably in 0.1 N sodium hydroxide at 70°C during 90 minutes, and hydrolyzes only slowly at higher temperatures (*38*). The very slow hydrolysis which occurs in alkaline solution is probably due to catalysis by water functioning as a general acid.

Carbohydrate ortho esters also resemble other ortho esters in being highly susceptible to acid-catalyzed hydrolysis. When the ortho ester grouping includes the anomeric carbon of the carbohydrate moiety, the initial hydrolysis product is a carboxylate ester having the acyloxy group on a carbon other than the anomeric carbon. Thus, carbohydrate 1,2-O ortho esters yield 2-acyloxy carbohydrates as initial hydrolysis products (*31, 75, 80, 103*). These may lose the 2-acyl group in a slower subsequent reaction.

$$\left\{\begin{array}{c}-O\\-O\end{array}\right\}\!\!\begin{array}{c}H\\O\\OR'\\R\end{array} \xrightarrow[H_2O]{H^+} \left\{\begin{array}{c}-O\\-\end{array}\right\}\!\!\begin{array}{c}H\\OH\\OCOR\end{array} \xrightarrow[H_2O]{H^+} \left\{\begin{array}{c}-O\\-\end{array}\right\}\!\!\begin{array}{c}\\(H_3OH)\\OH\end{array} \quad (19)$$

Polarimetric studies of acid-catalyzed carbohydrate ortho ester hydrolyses typically reveal a rapid reaction followed by a slower one (*38, 103*). This led Haworth, Hirst, and Samuels to propose that the initial hydrolysis product is an ortho ester acid (**20**), which is hydrolyzed to the acyloxy product more slowly than it is formed (*38*). Pacsu later found that the fast reaction of 1,2-O-(1-methoxyethylidene)-α-D-maltose (**33**, Y = H, R = R' = CH$_3$) in aqueous acid produces 2-O-acetyl-α-D-maltose, which undergoes a slower mutarotation to a mixture of α- and β-2-acetylmaltose (*103*). He pointed out that earlier kinetic studies could be rationalized by consecutive hydrolysis and mutarotation reactions, and concluded that ortho ester acids (**20**) hydrolyze more rapidly than the original ortho esters (*106*, pp. 100–103).

Any detailed mechanism of carbohydrate 1,2-O ortho ester hydrolysis must account for the retention of configuration at the anomeric carbon. If it is assumed that the rate-limiting step involves formation of a resonance-stabilized carbonium ion intermediate (see Chap. 2), the stereochemistry excludes a 1,2-O dioxolenium ion intermediate. Such an ion would yield a 2-acyloxy carbohydrate of inverted configuration at C-1 if it underwent attack by water at the anomeric carbon, and should yield a mixture of epimeric 2-acyloxy sugars if it underwent unimolecular opening to a 2-acyloxytetrahydropyrylium or 2-acyloxytetrahydrofurylium ion. Formation of 2-acyloxy carbohydrates of retained configuration at the anomeric carbon is accounted for if the acid catalyst transfers a proton to the ortho ester oxygen bonded to the anomeric carbon, yielding an open alkoxy-$O$-glycosyl carbonium ion (**66**). A series of rapid reactions with water would convert this open carbonium ion to an alcohol and the 2-acyloxy sugar.

$$\left\{\begin{array}{c}-\text{O}\\ \text{O}\\ \text{R}\end{array}\right\}\!\!\!\overset{\text{H}}{\underset{\text{OR'}}{\diagup}} \xrightarrow{\text{H}^+} \left\{\begin{array}{c}-\text{O}\\ \text{O}\\ \text{C}^+\\ \text{R }\ \text{OR'}\end{array}\right\}\!\!\!\overset{\text{H}}{\underset{\text{OH}}{\diagup}} \xrightarrow{\text{H}_2\text{O}} \left\{\begin{array}{c}-\text{O}\\ \\ \text{OCOR'}\end{array}\right\}\!\!\!\overset{\text{H}}{\underset{\text{OH}}{\diagup}} + \text{R'OH} + \text{H}^+ \quad (20)$$

**66**

This mechanism is postulated on the basis of very limited experimental evidence: one instance of definite retention of configuration at the anomeric carbon (*103*), and one instance of probable retention of configuration (*38, 106*, p. 102). Open carbonium ions such as **66** are also probable intermediates in the acid-catalyzed methanolysis of carbohydrate ortho esters (*114*). Possibly **66** is formed in preference to a dioxolenium ion because formation of the latter would be accompanied by an increase in steric strain, while formation of an open ion is accompanied by a relief of strain and a decrease in steric crowding. These factors apparently combine to render the activation energy for formation of the open ion less than that for formation of the dioxolenium ion.

The kinetics of carbohydrate ortho ester hydrolysis have received little attention. In water at 20°C the hydrogen ion catalytic coefficients for hydrolysis of 1,2-$O$-(1-methoxyethylidene)-α-D-maltose (**33**, Y = H, R = R' = CH$_3$) and 6-deoxy-1,2-$O$-(1-methoxyethylidene)-β-L-mannopyranose (**27**, Y = H, R = R' = CH$_3$) are approximately 50 and 10 liter/mole second, respectively (*38, 103*). In 95% dioxane at 23.5°C, the hydrogen ion catalytic coefficients for hydrolysis of 3,4,6-tri-$O$-acetyl-1,2-$O$-(1-methoxyethylidene)-α-D-glucopyranose (**29**, Y = Ac, R = R' = CH$_3$) and 3,4,6-tri-$O$-acetyl-1,2-$O$-(1-methoxyethylidene)-β-D-mannopyranose (**28**, Y = Ac, R = R' = CH$_3$) are 1 and 5 × 10$^{-2}$ liter/mole second, respectively (*79*). Carbohydrate orthoacetates are considerably less reactive towards acid hydrolysis than are acyclic orthoacetates. This is probably due in part to the presence of the alkoxydioxolane ring system, and in part to their smaller basicity. A thorough study of catalysis, structural effects on reactivity, and products for acid hydrolysis of a series of carbohydrate ortho esters would yield useful information about the mechanism and stereochemistry of these reactions.

There is little published information on acid hydrolyses of carbohydrate ortho esters whose ortho ester groups do not include the anomeric carbon atom of the carbohydrate. Hydrolysis of 1,2-$O$-isopropylidene-3,5,6-$O$-methylidene-α-D-glucofuranose (**6**, Y = –O—C(CH$_3$)$_2$—O–) yields 6-$O$-formyl-1,2-$O$-isopropylidene-α-D-glucofuranose (*42*). 1,2-$O$-Isopropylidene-3,4,6-$O$-($S$-methylmonothioorthocarbonyl)-α-D-glucofuranose (**10**, R = CH$_3$) is converted to α-D-glucofuranose-5,6-$O$-carbonate by acid hydrolysis (*127*). Mild acid hydrolysis of 2′,3′-$O$ ortho ester derivatives of ribonucleosides and ribonucleotides yields mixtures of 2′-$O$- and 3′-$O$-acyl derivatives (*8, 15, 62, 117, 144–146*).

### B. Alcoholysis and Other Glycoside-Forming Reactions of Carbohydrate Ortho Esters

In acidic solutions containing alcohols carbohydrate ortho esters undergo a number of competing reactions which yield ortho esters, glycosides, acylated glycosides, and deacylated carbohydrates. In the case of a cyclic 1,2-$O$ ortho ester derivative of an aldose, reactions which occur are summarized by Eq. (21). If R′ and R″ are the same, transesterification is not observed, and the principal reactions are glycoside formation and deorthoacylation.

$$\left\{\begin{array}{c}-O\\-O\end{array}\!\!\!\!\bigg\rangle\!\!\!\!\begin{array}{c}H\\OR'\\R\end{array}\right\} \xrightarrow[H^+]{R''OH} \left\{\begin{array}{c}-O\\-O\end{array}\!\!\!\!\bigg\rangle\!\!\!\!\begin{array}{c}H\\OR''\\R\end{array}\right\} + \left\{\begin{array}{c}-O\\-O\end{array}\!\!\!\!\bigg\rangle\!\!\!\!\begin{array}{c}OR''\\H\\OCOR\end{array}\right\} +$$

$$\left\{\begin{array}{c}-O\\-O\end{array}\!\!\!\!\bigg\rangle\!\!\!\!\begin{array}{c}OR''\\H\\OH\end{array}\right\} + \left\{\begin{array}{c}-O\\-O\end{array}\!\!\!\!\bigg\rangle(H,OH)\\OH\end{array}\right\} + \left\{\begin{array}{c}-O\\-O\end{array}\!\!\!\!\bigg\rangle\!\!\!\!\begin{array}{c}H\\OR''\\OH\end{array}\right\} \quad (21)$$

Early reports on carbohydrate ortho ester alcoholysis reactions are misleading because only one or two of the reaction products were isolated. Thus, Dale obtained a low yield of methyl 2,3,4,6-tetra-$O$-acetyl-β-D-mannopyranoside by reaction of methanol with 3,4,6-tri-$O$-acetyl-1,2-$O$-(1-methoxyethylidene)-β-D-mannopyranose (**28**, Y = Ac, R = R′ = CH$_3$) (*12*), while Isbell isolated a hexaacetate rather than a glucoside from methanolysis of hexa-$O$-acetyl-β-D-glucopyranosyl-β-D-mannopyranose 1,2-$O$ methyl orthoacetate (**60**) (*56*), and McPhillamy and Elderfield obtained both methyl

2,3,4-tri-*O*-acetyl-6-deoxy-β-L-mannopyranoside and 3,4-di-*O*-acetyl-6-deoxy-β-L-mannopyranose by heating 3,4-di-*O*-acetyl-6-deoxy-1,2-*O*-(1-methoxyethylidene)-β-L-mannopyranose (**27**, Y = Ac, R = R' = CH$_3$) with methanol containing hydrogen chloride (*90*).

In recent studies of acid-catalyzed methanolysis reactions of carbohydrate 1,2-*O* methyl orthoacetates more care was taken to identify minor reaction

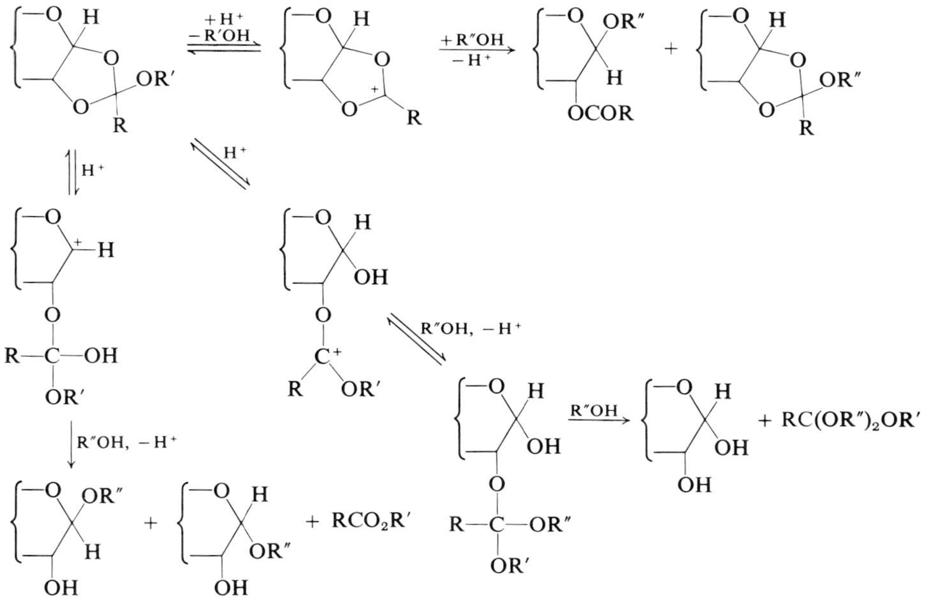

(22)

products. Methanolysis of 3,4,6-tri-*O*-acetyl-1,2-*O*-(1-methoxyethylidene)-α-D-glucopyranose (**29**, Y = Ac, R = R' = CH$_3$), yields approximately equal amounts of 3,4,6-tri-*O*-acetyl-D-glucose and methyl 3,4,6-tri-*O*-acetyl-β-D-glucopyranoside, plus a small amount of methyl 3,4,6-tri-*O*-acetyl-α-D-glucopyranoside (*78*, *114*). Methanolysis of 3,4,6-tri-*O*-acetyl-1,2-*O*-(1-methoxyethylidene)-β-D-mannopyranose (**28**, Y = Ac, R = R' = CH$_3$) yields mainly 3,4,6-tri-*O*-acetyl-D-mannose and a mixture of α- and β-mannosides (*114*). Methanolysis of 3,4,6-tri-*O*-benzyl-1,2-*O*-(1-methoxyethylidene)-β-D-mannopyranose (**28**, Y = C$_6$H$_5$CH$_2$, R = R' = CH$_3$) in methylene chloride solution containing *p*-toluenesulfonic acid formed 74% methyl 2-*O*-acetyl-3,4,6-tri-*O*-benzyl-α-D-mannopyranoside, 20% methyl 3,4,6-tri-*O*-benzyl-α-D-mannopyranoside, and 6% methyl 3,4,6-tri-*O*-benzyl-β-D-mannopyranoside (*19*).

Perlin proposed a mechanism which accounts for the products obtained in carbohydrate ortho ester alcoholysis reactions (*114*). Protonation of the ortho ester alkoxy oxygen atom leads to formation of a dioxolenium ion which reacts with alcohol to form the 2-acyloxy glycoside of inverted configuration at the anomeric carbon. Protonation of the ortho ester anomeric oxygen atom results in formation of two different carbonium ions. Cleavage of the anomeric carbon–oxygen bond yields a 2-*O*-(1-hydroxy-1-alkoxyethylidene)tetrahydropyrylium ion which reacts with alcohol and loses methyl acetate to form a mixture of 2-hydroxy α- and β-glycosides. Cleavage of the bond between the ortho ester acyl carbon and the anomeric oxygen atom, followed by reaction of the open carbonium ion with methanol, yields a 2-glycosyl-dimethyl orthoacetate, which reacts with more methanol to form trimethyl orthoacetate and the 1,2-deacylated carbohydrate. This reaction scheme is summarized in Eq. (22). Isolation of methyl acetate and trimethyl orthoacetate from a reaction mixture would provide support for the proposed mechanism.

Three alcoholysis reactions of tricyclic carbohydrate ortho esters have been reported. Benzyl 3,5-di-*O*-benzoyl-β-D-ribofuranoside is formed when 5-*O*-benzoyl-1,2,3-*O*-benzylidyne-α-D-ribofuranose (**26**, Y = $C_6H_5CO$) is heated with benzyl alcohol at 140°C (*97*). Benzyl alcoholysis of 1,2,4-*O*-benzylidyne-α-D-ribopyranose (**35**, Y = H), followed by benzoylation of the product, yields benzyl 2,3,4-tri-*O*-benzoyl-β-D-ribopyranoside, and methanolysis of the same ortho ester followed by barium methoxide debenzoylation of the product yields methyl β-D-ribopyranoside (*18*).

Kochetkov, Khorlin, and Bochkov found that acylated carbohydrate ortho esters react with alcohols or partially blocked carbohydrates, in refluxing nitromethane solutions containing mercuric bromide, to produce glycosides in moderate to high yields (*65, 66, 66c, 67–72, 73a*). The reaction is stereospecific, giving the *trans*-1,2-glycoside almost exclusively. In addition to being more stereospecific, this reaction gives higher yields of glycosides than the Koenigs-Knorr glycosylation reaction. This fact, coupled with recent improvements in syntheses of carbohydrate ortho esters from acylglycosyl halides, makes the ortho ester glycosylation procedure particularly attractive for the synthesis of oligosaccharides.

The mechanism of this glycosylation reaction is probably quite similar to that of hydrogen ion-catalyzed alcoholysis of sugar ortho esters, except that 1,2-*O* dioxolenium ions appear to play a more important role. The fact that glycosylation predominates over transesterification only when the mole ratio of mercuric bromide to hydroxy compound exceeds about 0.01 suggests that the glycosides are formed by nucleophilic attack of the hydroxy compound on the anomeric carbon of a 1,2-*O* dioxolenium alkoxydibromomercurate ion pair. Possibly because of the greater stability of 2-phenyl than 2-methyl dioxolenium ions, carbohydrate orthobenzoates are more reactive glycosylating agents than the corresponding orthoacetates, and give higher yields of glycosides (*71, 73a*).

Oligosaccharides and other glycosides synthesized by the mercuric bromide-catalyzed reaction of hydroxy compounds with carbohydrate ortho esters include cholesteryl 2,3,4,6-tetra-*O*-acetyl-β-D-glucopyranoside (45% yield), 1,2:3,4-di-*O*-isopropylidene-6-*O*-(2,3,4,6-tetra-*O*-acetyl-β-D-glucopyranosyl)-α-D-galactopyranose (51.5%), 1,2:3,4-di-*O*-isopropylidene-6-*O*-(2,3,4,6-tetra-*O*-acetyl-β-D-galactopyranosyl)-α-D-galactopyranose (64%), 1,2,3,4-tetra-*O*-acetyl-6-*O*-(2,3,4,6-tetra-*O*-acetyl-β-D-glucopyranosyl)-β-D-glucopyranose (35%), 1,2,3,4-tetra-*O*-acetyl-6-*O*-(2,3,4-tri-*O*-acetyl-6-deoxy-α-L-mannopyranosyl)-β-D-glucopyranose (45%), 1,2,3,4-tetra-*O*-acetyl-6-*O*-(2,3,4-tri-*O*-acetyl-α-L-arabinopyranosyl)-β-D-glucopyranose (54.5%), 1,2,3,4-tetra-*O*-acetyl-6-*O*-(2,3,6,2',3',4',6'-hepta-*O*-acetyl-β-maltosyl)-β-D-glucopyranose (55%), 1,2,3,4-tetra-*O*-acetyl-6-*O*-(2,3,4-tri-*O*-benzoyl-α-L-arabinopyranosyl)-β-D-glucopyranose (93%), 1,2-*O*-isopropylidene-3,6-di-*O*-acetyl-5-*O*-(2,3,4,6-tetra-*O*-acetyl-β-D-glucopyranosyl)-α-D-glucofuranose (10%), methyl 2-*O*-(2,3,4,6-tetra-*O*-acetyl-β-D-glucopyranosyl)-4,6-*O*-benzylidene-α-D-glucopyranoside (21%), methyl 2-*O*-acetyl-3-*O*-(2,3,4,6-tetra-*O*-acetyl-β-D-glucopyranosyl)-4,6-*O*-benzylidene-α-D-glucopyranoside (7%), 1,2,5,6-tetra-*O*-benzoyl-3-*O*-(2,3,5,6-tetra-*O*-acetyl-β-D-galactofuranosyl)-D-mannitol (28.3%), 1,2,3,4-tetra-*O*-acetyl-6-*O*-(2,3,5-tri-*O*-benzoyl-α-L-arabinofuranosyl)-β-D-glucopyranose (90%), *O*-(2,3,4,6-tetra-*O*-acetyl-β-D-galactofuranosyl)-*N*-benzyloxycarbonyl-L-serine methyl ester (17.3%) (*71*), 2,3-di-*O*-benzoyl-1-*O*-(2,3,5,6-tetra-*O*-acetyl-β-D-galactofuranosyl)-D-glyceritol (*1b*), branched arabinose polymers (*66d*), and trisaccharide (*71a*) and heteropolysaccharide (*66c*) derivatives.

A reaction which is related mechanistically to alcoholysis of carbohydrate alkyl orthocarboxylates is the isomerization of these ortho esters to alkyl acyloxyglycosides.

$$\left\{\begin{array}{l}-\text{O} \quad \text{H} \\ \phantom{-}\text{O} \\ \phantom{-}\text{O} \quad \text{OR'} \\ \phantom{--}\text{R}\end{array}\right. \xrightarrow{\text{H}^+ \text{ or HgBr}_2} \left\{\begin{array}{l}-\text{O} \quad \text{OR'} \\ \phantom{-}\text{H} \\ \text{OCOR}\end{array}\right. + \left\{\begin{array}{l}-\text{O} \quad \text{OR'} \\ \phantom{-}\text{H} \\ \text{OH}\end{array}\right. \quad (23)$$
(mostly) (some)

Helferich and Weiss reported that treatment of 3,4,6-tri-*O*-benzoyl-1,2-*O*-(1-methoxybenzylidene)-α-D-glucopyranose with mercuric bromide and dry hydrogen chloride in nitromethane solution isomerizes the ortho ester to methyl 2,3,4,6-tetra-*O*-benzoyl-β-D-glucopyranoside (*49*) and Kochetkov and co-workers obtained similar results with 3,4,6-tri-*O*-acetyl-1,2-*O*-(1-ethoxyethylidene)-α-D-glucopyranose and other glucose 1,2-O orthoacetates (*1c, 12a, 71*). This isomerization reaction is the main reaction competing with the ortho ester glycosylation reaction described above. When 3,4,6-tri-O-acetyl-1,2-O-(1-methoxyethylidene)-β-D-mannopyranose was treated with toluenesulfonic acid in methylene chloride solution, 82% of the isomeric methyl tetraacetyl α-mannoside and 7% of methyl 3,4,6-tri-O-acetyl-α-D-mannopyranoside was formed (*19*). 1,2-O-(1-*tert*-Butylperoxyethylidene) ortho ester derivatives of 3,4,6-tri-O-acetyl-β-D-mannopyranose and 3,4-di-O-acetyl-6-deoxy-β-L-mannopyranose are isomerized to the *tert*-butylperoxy 2-O-acetyl α-mannopyranosides in high yield by treatment with silver perchlorate in benzene-ether solution containing dry hydrogen chloride, or by treatment with boron trifluoride etherate in methylene chloride (*123*).

Isomerization of an intermediate 1,2-O-(1-ethoxyethylidene) ortho ester has been proposed to account for formation of ethyl hepta-O-acetyl-β-D-cellobioside when 1,3,6,2',3',4',6'-hepta-O-acetyl-α-D-cellobiose is alkylated with ethyl iodide in a refluxing dioxane suspension of silver oxide (*9a*).

### C. Conversion of Carbohydrate Ortho Esters to Acylglycosyl Halides

Carbohydrate 1,2-O ortho esters react with hydrogen chloride in chloroform or methylene chloride solution to form acylglycosyl chlorides [Eq. (24)]. These reactions probably involve 1,2-O dioxolenium ion intermediates, and probably yield mostly *trans*-1,2-acylglycosyl chlorides, although in most reported examples the configuration at the anomeric carbon was not established.

$$\left\{ \begin{array}{c} -O \\ \\ O \end{array} \right\}_{R}^{H} \hspace{-0.5em} \underset{OR'}{\bigvee} \hspace{1em} \longrightarrow \hspace{1em} \left\{ \begin{array}{c} -O \\ \\ \end{array} \right\}_{OCOR}^{Cl} \hspace{3em} (24)$$

Treatment of 3,4,6-tri-O-acetyl-1,2-O-(1-methoxyethylidene)-β-D-talopyranose with hydrogen chloride in chloroform yields 2,3,4,6-tetra-O-acetyltalopyranosyl chloride (*115*). Similarly, 3,4-di-O-acetyl-1,2-O-[1-(3-acetoxyacetonyl)ethylidene]-α-D-ribopyranose yields 2,3,4-tri-O-acetyl-D-ribopyranosyl chloride (*66a*), and hexa-O-acetyl-1,2-O-(1-methoxyethylidene)-α-D-maltose yields the heptaacetylmaltosyl chloride (*55, 56*). 3,4,6-Tri-O-acetyl-1,2-O-[1-(1,2,3,4-tetra-O-acetyl-6-β-D-glucopyranosyl)ethylidene]-β-D-mannopyranose undergoes rapid mutarotation in chloroform solutions of either hydrogen chloride or hydrogen bromide (*127*). These reactions probably involved acetylglycosyl halide formation, although the reaction products were not isolated. 5-O-Benzoyl-1,2,3-O-benzylidyne-α-D-ribofuranose reacts with hydrogen chloride in chloroform to form 2,6-di-*O*-benzoyl-D-ribofuranosyl chloride (*97*). 3,4,6,7-Tetra-*O*-acetyl-1,2-*O*-(1-methoxyethyldien)-D-glycero-β-L-taloheptopyranose (**58**) is converted to the 2,3,4,6,7-penta-*O*-acetylheptopyranosyl chloride by hydrogen chloride in chloroform (*28*).

Carbohydrate 1,2-*O* ortho esters are also converted to acylglycosyl chlorides by titanium tetrachloride in anhydrous chloroform solution. Titanium tetrachloride probably reacts with the ortho esters to form complex 1,2-*O* dioxolenium alkoxytetrachlorotitanate ion pairs, which isomerize to titanium alkoxytrichloride complexes of the acylglycosyl chloride. The acylglycosyl chloride is obtained by hydrolyzing the complex with cold water. Hexa-*O*-acetyl-1,2-*O*-(1-methoxyethylidene)-α-D-maltose (*107*) and 3,6-di-*O*-acetyl-4-*O*-(2,3,4,6-tetra-*O*-acetyl-α-D-glucopyranosyl)-1,2-*O*-(1-methoxyethylidene)-β-D-mannopyranose (*55*, *56*) were converted to the hepta-*O*-acetylglycosyl chlorides by this procedure, and 2,3,4-tri-*O*-acetyl-1,2-*O*-(1-methoxyethylidene)-D-fructofuranose formed tetraacetylfructosyl chloride (*107*).

Acylated carbohydrate 1,2-*O* ortho esters are converted to *trans*-1,2-acylglycosyl bromides by reaction with hydrogen bromide in acetic acid. 3,4-Di-*O*-acetyl-6-deoxy-1,2-*O*-(1-methoxyethylidene)-β-L-mannopyranose was converted to 2,3,4-tri-*O*-acetyl-6-deoxy-α-L-mannopyranosyl bromide in this manner (*1a*), and 2,3,4-tri-*O*-acetyl-1,2-*O*-(1-methoxyethylidene)-D-fructopyranose similarly yielded tetraacetylfructosyl bromide (*107*).

### D. Reactions of Carbohydrate Ortho Esters with Carboxylic Acids and Acid Anhydrides

1,2-*O* and 1,2,3-*O* Carbohydrate ortho esters react with carboxylic acids to form 1-acyloxy carbohydrates in which the configuration at the anomeric carbon has been inverted. The first step of the reaction is probably formation of a 1,2-*O* dioxolenium carboxylate ion pair, which undergoes nucleophilic attack at C-1 by the carboxylic acid or its conjugate base to form the products.

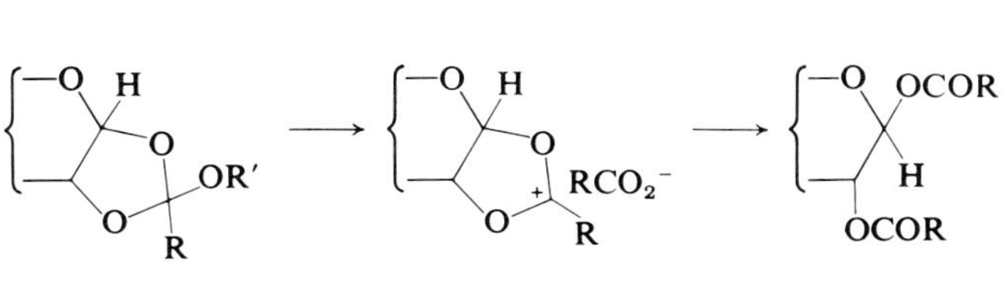

$$\begin{matrix} \text{(structure with OR')} & \longrightarrow & \text{(dioxolenium + RCO}_2^-\text{)} & \longrightarrow & \text{(product with OCOR, H, OCOR)} \end{matrix} \qquad (25)$$

Examples of this reaction are the conversion of 3-*O*-acetyl-4,6-*O*-benzylidene-1,2-*O*-(1-methoxyethylidene)-α-D-glucopyranose to 1,2,3-tri-*O*-acetyl-

4,6-O-benzylidene-β-D-glucopyranose (75), formation of 1,2,3,4,6-penta-O-acetyl-β-D-glucopyranose from 3,4,6-tri-O-acetyl-1,2-O-(1-ethoxyethylidene)-α-D-glucopyranose (80), and synthesis of 1-O-acetyl-2,5-O-dibenzoyl-β-D-ribofuranose and 1,2,5-tri-O-benzoyl-β-D-ribofuranose by reaction of acetic and benzoic acids with 5-O-benzoyl-1,2,3-O-benzylidyne-α-D-ribofuranose (97).

5-O-Benzoyl-1,2,3-O-benzylidyne-α-D-ribofuranose reacts with warm acetic anhydride to form 1,3-di-O-acetyl-2,5-di-O-benzoyl-β-D-ribofuranose (97).

### E. Formation of Dioxolenium Salts from Carbohydrate Ortho Esters

Carbohydrate ortho esters should react, as do other ortho esters, with Lewis acids such as boron trifluoride or antimony pentachloride, to form dioxolenium salts. Apparently the only published example of such a reaction is the conversion of methyl 3,4-O-(1-ethoxyethylidene)-β-L-arabinopyranoside to the 3,4-O-dioxolenium tetrafluoroborate by reaction with boron trifluoride etherate (6).

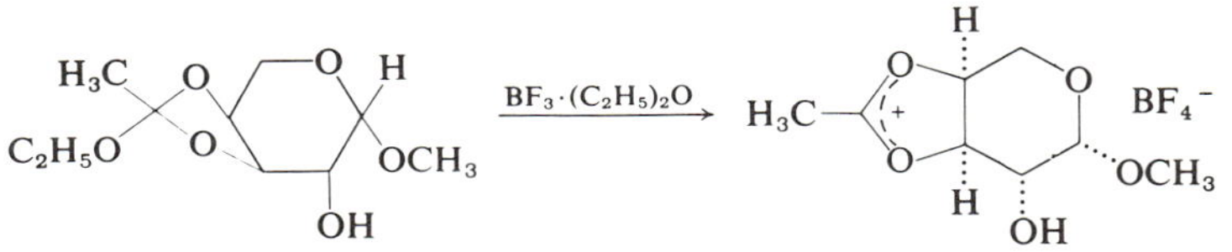

$$\text{structure} \xrightarrow{BF_3 \cdot (C_2H_5)_2O} \text{structure} \quad BF_4^- \qquad (26)$$

Dioxolenium hexachloroantimonate derivatives of carbohydrates, prepared from acylglycosyl chlorides and antimony pentachloride, undergo more or less extensive isomerization to mixtures of dioxolenium salts (15a, 109a–112). Salts prepared from carbohydrate ortho esters should undergo similar isomerizations when the carbohydrate moiety has acyloxy groups trans to the points of attachment of the dioxolenium ring to the rest of the ion.

### F. Conversion of Carbohydrate Ortho Esters to Cyclic Acetals

Lemieux and Detert recently synthesized 3,4,6-tri-O-acetyl-1,2-O-isopropylidene-α-D-glucopyranose, 3,4,6-tri-O-acetyl-1,2-O-cyclopentylidene (and cyclohexylidene)-α-D-glucopyranose, 3,4,6-tri-O-acetyl-1,2-O-benzylidene-α-D-glucopyranose, and 3,4,6-tri-O-acetyl-1,2-O-[(2-tetrahydrofuryl)methylidene]-α-D-glucopyranose by reaction of 3,4,6-tri-O-acetyl-1,2-O-(1-ethoxyethylidene)-α-D-glucopyranose with the appropriate carbonyl compounds or their dialkyl acetals in trimethyl orthoformate–dimethyl formamide solutions (*80a*). This reaction could probably be applied to other 1,2-ortho esters also.

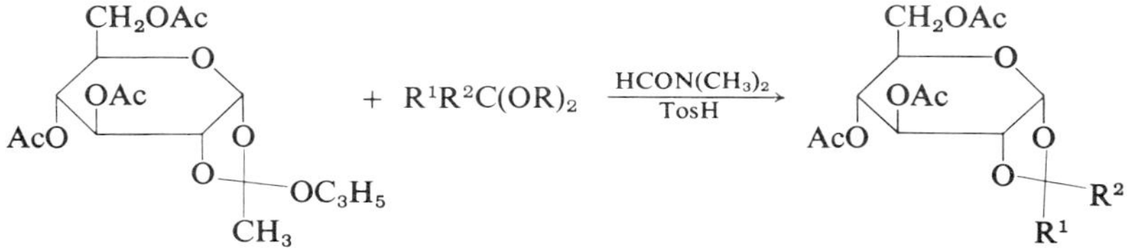

$$\text{[carbohydrate ortho ester]} + R^1R^2C(OR)_2 \xrightarrow[\text{TosH}]{\text{HCON(CH}_3)_2} \text{[product]} \quad (27)$$

| $R^1$ | $R^2$ |
|---|---|
| $CH_3$ | $CH_3$ |
| H | $C_6H_5$ |
| [———$(CH_2)_n$———] | ($n = 4, 5$) |
| [—$CH_2CH_2CH_2O$—] | |

### G. The Ortho Ester Function in Synthetic Carbohydrate Chemistry

Formation and subsequent reactions of carbohydrate ortho esters offer many useful synthetic sequences to the carbohydrate chemist.

Carbohydrate derivatives having two suitably situated hydroxyl groups (usually cis-vicinal groups on a pyranose or furanose ring) undergo trans-esterification reactions with acyclic ortho esters to form cyclic carbohydrate ortho esters. The ortho ester function thus introduced serves as a base-stable, acid-labile protecting group for the two esterified hydroxyls. Mild acid hydrolysis produces a mixture of monoacyloxy derivatives of the carbohydrate, and complete hydrolysis regenerates the dihydroxy system. This type of reaction has been extensively used to protect the 2′,3′-hydroxyls of nucleosides and nucleotides during condensation reactions.

Acylated carbohydrate bicyclic ortho esters in which the ortho ester grouping includes the anomeric carbon atom are readily prepared from acylglycosyl halides, and undergo numerous useful reactions. They react with hydroxy compounds under appropriate conditions to form acyloxy-*trans*-1,2-glycosides, which can be saponified to the deacylated glycoside. Saponification of the acyloxy ortho ester, followed by alkylation, acylation, or alkane- or arenesulfonation of the freed hydroxyl groups provides routes to carbohydrate derivatives which would be difficult to prepare by other means. Reaction of the acyloxy ortho esters with carboxylic acids yields peracylated carbohydrates with a trans configuration at the anomeric carbon atom and its neighbor. Conversion of the acylated ortho esters to dioxolenium salts offers interesting possibilities for conversion of a carbohydrate into one or more of its diastereomers. This brief list by no means exhausts the possibilities for transformations of carbohydrate ortho esters into other carbohydrate derivatives.

A recently discovered procedure for degrading aldoses to their next lower homologues involves formation and isomerization of *tert*-butylperoxy ortho esters as the initial step *(123)*. Equation (28) illustrates this degradation for the conversion of 6-deoxy-L-mannose to 5-deoxy-L-arabinose.

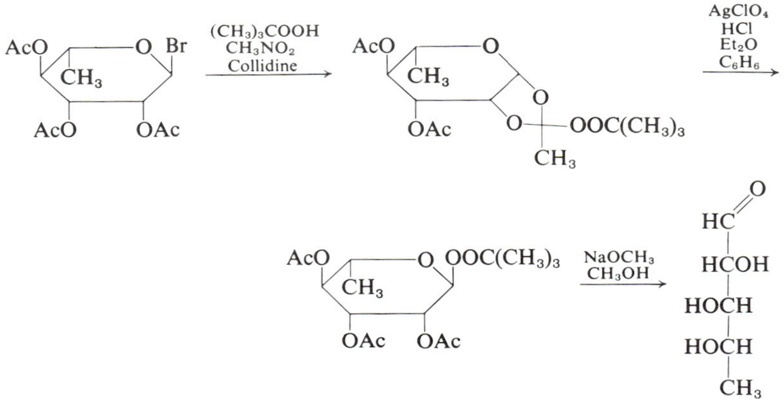

$$\text{(28)}$$

## REFERENCES

1. Anonymous, *J. Org. Chem.* **28**, 281 (1963).
1a. M. Bergmann and F. Beck, *Chem. Ber.* **54**, 1574 (1921).
1b. H. F. G. Beving, H. B. Boren, and P. J. Garegg, *Acta Chim. Scand.* **21**, 2083 (1967).
1c. A. F. Bochkov, V. I. Snyatkova, and N. K. Kochetkov, *Izv. Akad. Nauk SSSR, Ser. Khim.* p. 2684 (1967).
1d. A. F. Bochkov, T. A. Sokolskaya, and N. K. Kochetkov, *Izvest. Akad. Nauk SSSR, Ser. Khim.* p. 1570 (1968).
2. W. A. Bonner, *J. Am. Chem. Soc.* **80**, 3697 (1958).
3. H. G. Bott, W. N. Haworth, and E. L. Hirst, *J. Chem. Soc.* p. 1395 (1930).
4. E. Braun, *Chem. Ber.* **63**, 1972 (1930).
5. J. S. Brimacombe and D. Portsmouth, *Carbohydrate Res.* **1**, 128 (1965).
6. J. G. Buchanan and A. R. Edgar, *Chem. Commun.* p. 29 (1967).
7. S. Chladek and J. Zemlicka, *Collection Czech. Chem. Commun.* **32**, 1776 (1967).
7a. S. Chladek and J. Zemlicka, *Collection Czech. Chem. Commun.* **33**, 232 (1968).
7b. S. Chladek and J. Zemlicka, *Collection Czech. Chem. Commun.* **33**, 4298 (1968).
8. S. Chladek, J. Zemlicka, and F. Sorm, *Biochem. Biophys. Res. Commun.* **22**, 554 (1966).
9. S. Chladek, Z. Zemlicka, and F. Sorm, *Collection Czech. Chem. Commun.* **31**, 1785 (1966).
9a. W. M. Corbett, J. Kidd, and A. M. Liddle, *J. Chem. Soc.* p. 616 (1960).
10. B. Coxon and L. D. Hall, *Tetrahedron* **20**, 1685 (1964).
11. G. Crank and F. W. Eastwood, *Australian J. Chem.* **17**, 1392 (1964).

## REFERENCES

12. J. K. Dale, *J. Am. Chem. Soc.* **46**, 1046 (1924).
12a. V. A. Derevitskaya, E. M. Klimov, and N. K. Kochetkov, *Carbohydrate Res.* **7**, 7 (1968).
12b. V. A. Derevitskaya, M. G. Vafina, and N. K. Kochetkov, *Carbohydrate Res.* **3**, 377 (1967).
13. W. M. Doane, B. S. Shasha, C. R. Russell, and C. E. Rist, *J. Org. Chem.* **30**, 3071 (1965).
14. W. M. Doane, B. S. Shasha, C. R. Russell, and C. E. Rist, *J. Org. Chem.* **32**, 1080 (1967).
15. F. Eckstein and F. Cramer, *Chem. Ber.* **98**, 995 (1965).
15a. F. G. Espinosa, W. P. Trautwein, and H. Paulsen, *Chem. Ber.* **101**, 191 (1968).
16. E. Fischer, *Chem. Ber.* **53**, 1624 (1920).
17. E. Fischer, M. Bergmann, and A. Rabe, *Chem. Ber.* **53**, 2362 (1920).
18. H. G. Fletcher and R. K. Ness, *J. Am. Chem. Soc.* **77**, 5337 (1955).
19. N. E. Franks and R. Montgomery, *Carbohydrate Res.* **3**, 511 (1967).
20. N. E. Franks and R. Montgomery, *Carbohydrate Res.* **6**, 286 (1968).
21. K. Freudenberg and E. Braun, *Naturwissenschaften* **18**, 393 (1930).
22. K. Freudenberg and O. Ivers, *Chem. Ber.* **55**, 929 (1922).
23. K. Freudenberg and W. Jacob, *Chem. Ber.* **80**, 325 (1947).
24. K. Freudenberg and H. Scholze, *Chem. Ber.* **63**, 1969 (1930).
25. K. Freudenberg, H. von Hochstetter, and H. Engels, *Chem. Ber.* **58**, 666 (1925).
26. H. P. M. Fromageot, B. E. Griffin, C. B. Reese, and J. E. Sulston, *Tetrahedron* **23**, 2315 (1967).
27. H. P. M. Fromageot, C. B. Reese, and J. E. Sulston, *Tetrahedron* **24**, 3533 (1968).
28. H. L. Frush and H. S. Isbell, *J. Res. Natl. Bur. Std.* **27**, 413 (1941).
29. C. S. Giam, H. R. Goldschmid, and A. S. Perlin, *Can. J. Chem.* **41**, 3074 (1963).
30. W. F. Goebel and F. H. Babers, *J. Biol. Chem.* **110**, 707 (1935).
31. H. R. Goldschmid and A. S. Perlin, *Can. J. Chem.* **39**, 2025 (1961).
31a. P. A. J. Gorin, *Can. J. Chem.* **40**, 275 (1962).
32. P. A. J. Gorin and A. S. Perlin, *Can. J. Chem.* **39**, 2474 (1961).
32a. D. P. L. Green and C. B. Reese, *Chem. Commun.* p. 729 (1968).
33. B. E. Griffin, M. Jarman, and C. B. Reese, *Tetrahedron* **24**, 639 (1968).
34. B. E. Griffin, M. Jarman, C. B. Reese, and J. E. Sulston, *Tetrahedron* **23**, 2301 (1967).
35. B. E. Griffin and C. B. Reese, *Tetrahedron Letters* p. 2925 (1964).
36. H. Gross and I. Farkas, *Chem. Ber.* **93**, 95 (1960).
37. W. N. Haworth, E. L. Hirst, and E. J. Miller, *J. Chem. Soc.* p. 2469 (1929).
38. W. N. Haworth, E. L. Hirst, and H. Samuels, *J. Chem. Soc.* p. 2861 (1931).
39. W. N. Haworth, E. L. Hirst, and M. Stacey, *J. Chem. Soc.* p. 2864 (1931).
40. W. N. Haworth, E. L. Hirst, and E. G. Teece, *J. Chem. Soc.* p. 1405 (1930).
41. L. J. Haynes and F. H. Newth, *Advan. Carbohydrate Chem.* **10**, 207 (1955).
42. E. J. Hedgley and O. Meresz, *Proc. Chem. Soc.* p. 399 (1964).
43. B. Helferich and L. Bottenbruch, *Chem. Ber.* **86**, 651 (1953).
44. B. Helferich, A. Doppstadt, and A. Gottschlich, *Naturwissenschaften* **40**, 441 (1953).
45. B. Helferich and H. Jochinke, *Chem. Ber.* **74**, 719 (1941).
46. B. Helferich and A. Mueller, *Chem. Ber.* **63**, 2142 (1930).
47. B. Helferich and A. Porck, *Ann. Chem.* **582**, 233 (1953).
48. B. Helferich and W. Schulte-Hurmann, *Chem. Ber.* **87**, 977 (1954).
49. B. Helferich and K. Weiss, *Chem. Ber.* **89**, 314 (1956).

50. K. Heyns, W. P. Trautwein, F. G. Espinosa, and H. Paulsen, *Chem. Ber.* **99**, 1183 (1966).
51. A. Holy, *Collection Czech. Chem. Commun.* **32**, 3064 (1967).
52. A. Holy and K. H. Scheit, *Chem. Ber.* **99**, 3778 (1966).
53. A. Holy and J. Smrt, *Collection Czech. Chem. Commun.* **31**, 3800 (1966).
53a. A. Holy, J. Smrt, and F. Sorm, *Collection Czech. Chem. Commun.* **33**, 3809 (1968).
54. A. Holy, J. Smrt, and F. Sorm, *Collection Czech. Chem. Commun.* **32**, 2980 (1967).
54a. S. Inouye, T. Tsuruoka, T. Ito, and T. Nida, *Tetrahedron* **23**, 2125 (1968).
55. H. S. Isbell, *J. Am. Chem. Soc.* **52**, 5298 (1930).
56. H. S. Isbell, *J. Res. Natl. Bur. Std.* **7**, 1115 (1931).
57. H. S. Isbell, *Ann. Rev. Biochem.* **9**, 65 (1940).
58. H. S. Isbell and H. L. Frush, *J. Res. Natl. Bur. Std.* **43**, 161 (1949).
59. H. S. Isbell and H. L. Frush, *J. Am. Chem. Soc.* **71**, 1579 (1949).
60. H. S. Isbell, F. A. Smith, E. C. Creitz, H. L. Frush, J. D. Moyer, and J. E. Stewart, *J. Res. Natl. Bur. Std.* **59**, 41 (1957).
61. H. S. Isbell, J. E. Stewart, H. L. Frush, J. D. Moyer, and F. A. Smith, *J. Res. Natl. Bur. Std.* **57**, 179 (1956).
62. M. Jarman and C. B. Reese, *Chem. & Ind. (London)* p. 1493 (1964).
63. J. S. Josan and F. W. Eastwood, *Carbohydrate Res.* **7**, 161 (1968).
64. A. Y. Khorlin, A. F. Bochkov, and N. K. Kochetkov, *Izv. Akad. Nauk SSSR, Ser. Khim.* p. 2214 (1964).
65. A. Y. Khorlin, A. F. Bochkov, and N. K. Kochetkov, *Khim. Prirodn. Soedin. Akad. Nauk SSSR, Inst. Khim. Prirodn. Soedin.* p. 6 (1966).
66. A. Y. Khorlin, A. F. Bochkov, and N. K. Kochetkov, *Izv. Akad. Nauk SSSR, Ser. Khim.* p. 168 (1966).
66a. C. W. Klingensmith and W. L. Evans, *J. Am. Chem. Soc.* **61**, 3012 (1939).
66b. N. K. Kochetkov and A. F. Bochkov, *Tetrahedron Letters* p. 4669 (1967).
66c. N. K. Kochetkov, A. F. Bochkov, and I. G. Yazlovetsky, *Carbohydrate Res.* **5**, 243 (1967).
66d. N. K. Kochetkov, A. F. Bochkov, and I. G. Yazlovetsky, *Carbohydrate Res.* **9**, 49 (1969).
66e. N. K. Kochetkov and A. F. Bochkov, *Carbohydrate Res.* **9**, 61 (1969).
67. N. K. Kochetkov, A. Y. Khorlin, and A. F. Bochkov, *Izv. Akad. Nauk SSSR, Ser. Khim.* p. 2234 (1963).
68. N. K. Kochetkov, A. Y. Khorlin, and A. F. Bochkov, *Tetrahedron Letters* p. 289 (1964).
69. N. K. Kochetkov, A. Y. Khorlin, and A. F. Bochkov, *Dokl. Akad. Nauk SSSR* **161**, 1342 (1965).
70. N. K. Kochetkov, A. Y. Khorlin, and A. F. Bochkov, *Dokl. Akad. Nauk SSSR* **162**, 104 (1965).
71. N. K. Kochetkov, A. Y. Khorlin, and A. F. Bochkov, *Tetrahedron* **23**, 693 (1967).
71a. N. K. Kochetkov, A. Y. Khorlin, A. F. Bochkov, L. B. Demushkina, and I. O. Zolotukhin, *Zh. Obshch. Khim.* **37**, 1272 (1967).
72. N. K. Kochetkov, A. Y. Khorlin, A. F. Bochkov, and I. G. Yazlovetskii, *Izv. Akad. Nauk SSSR, Ser. Khim.* p. 385 (1966).
73. N. K. Kochetkov, A. Y. Khorlin, A. F. Bochkov, and I. G. Yazlovetskii, *Izv. Akad. Nauk SSSR, Ser. Khim.* p. 2030 (1966).
73a. N. K. Kochetkov, A. Y. Khorlin, and A. F. Bochkov, *Zh. Obshch. Khim.* **37**, 338 (1967).
74. W. Koenigs and E. Knorr, *Chem. Ber.* **34**, 957 (1901).

75. W. Korytnik and J. A. Mills, *J. Chem. Soc.* p. 636 (1959).
76. R. U. Lemieux, *Advan. Carbohydrate Chem.* **9**, 1–57 (1954).
77. R. U. Lemieux, *in* "Molecular Rearrangements" (P. DeMayo, ed.), Part II, p. 765. Wiley (Interscience), New York, 1964.
78. R. U. Lemieux, *Chem. Can.* **16**, 14 (1964); *Chem. Abstr.* **62**, 4104 (1965).
79. R. U. Lemieux and C. Brice, *Can. J. Chem.* **33**, 109 (1955).
80. R. U. Lemieux and J. D. T. Cipera, *Can. J. Chem.* **34**, 906 (1956).
80a. R. U. Lemieux and D. H. Detert, *Can. J. Chem.* **46**, 1039 (1968).
81. R. U. Lemieux and A. R. Morgan, *J. Am. Chem. Soc.* **85**, 1889 (1963).
82. R. U. Lemieux and A. R. Morgan, *Can. J. Chem.* **43**, 2199 (1965).
83. R. U. Lemieux and A. R. Morgan, *Can. J. Chem.* **43**, 2214 (1965).
84. N. J. Leonard and R. J. Larsen, *Biochemistry* **4**, 354 (1965).
85. P. A. Levene and H. Sobotka, *J. Biol. Chem.* **67**, 771 (1926).
86. P. A. Levene and R. S. Tipson, *J. Biol. Chem.* **92**, 109 (1931).
87. P. A. Levene and M. L. Wolfrom, *J. Biol. Chem.* **78**, 525 (1928).
88. A. V. Luk'yanov and O. N. Tolkachev, U.S.S.R. Patent 184,841 (1966); *Chem. Abstr.* **66**, 95365 (1967).
89. J. R. McCarthy, M. J. Robins, and R. K. Robins, *Chem. Commun.* p. 536 (1967).
89a. J. R. McCarthy, R. K. Robins, and M. J. Robins, *J. Am. Chem. Soc.* **90**, 4992 (1968).
90. H. B. McPhillamy and R. C. Elderfield, *J. Org. Chem.* **4**, 150 (1939).
91. B. J. Magerlein, U.S. Patent 3,262,926 (1966); *Chem. Abstr.* **65**, 13814 (1966).
92. M. Mazurek and A. S. Perlin, *Can. J. Chem.* **43**, 1918 (1965).
92a. M. P. Mertes, A. Holy, and J. Smrt, *Collection Czech. Chem. Commun.* **33**, 3313 (1968).
92b. M. P. Mertes and J. Smrt, *Collection Czech. Chem. Commun.* **33**, 3304 (1968).
92c. J. Moravec and J. Smrt, *Collection Czech. Chem. Commun.* **33**, 1768 (1968).
93. R. K. Ness and H. G. Fletcher, *J. Am. Chem. Soc.* **76**, 1663 (1954).
94. R. K. Ness and H. G. Fletcher, *J. Am. Chem. Soc.* **78**, 1001 (1956).
95. R. K. Ness and H. G. Fletcher, *J. Am. Chem. Soc.* **78**, 4710 (1956).
96. R. K. Ness and H. G. Fletcher, *J. Org. Chem.* **22**, 1465 (1957).
97. R. K. Ness and H. G. Fletcher, *J. Org. Chem.* **22**, 1470 (1957).
98. R. K. Ness, H. G. Fletcher, and C. S. Hudson, *J. Am. Chem. Soc.* **73**, 296 (1951).
99. H. Ohle, *Chem. Ber.* **57**, 403 (1924).
100. H. Ohle and E. Dickhauser, *Chem. Ber.* **58**, 2593 (1925).
101. E. Pacsu, *J. Am. Chem. Soc.* **54**, 3649 (1932).
102. E. Pacsu, *J. Am. Chem. Soc.* **55**, 2451 (1933).
103. E. Pacsu, *J. Am. Chem. Soc.* **57**, 537 (1935).
104. E. Pacsu, *J. Am. Chem. Soc.* **57**, 745 (1935).
105. E. Pacsu, *J. Am. Chem. Soc.* **61**, 2669 (1939).
106. E. Pacsu, *Advan. Carbohydrate Chem.* **1**, 77 (1945).
107. E. Pacsu and F. B. Cramer, *J. Am. Chem. Soc.* **59**, 1059 (1937).
108. E. Pacsu and F. V. Rich, *J. Am. Chem. Soc.* **57**, 587 (1935).
109. E. Pacsu, E. J. Wilson, and L. Graf, *J. Am. Chem. Soc.* **61**, 2675 (1939).
109a. H. Paulsen, F. G. Espinosa, and W. P. Trautwein, *Chem. Ber.* **101**, 186 (1968).
109b. H. Paulsen, F. G. Espinosa, W. P. Trautwein, and K. Heyns, *Chem. Ber.* **101**, 179 (1968).
110. H. Paulsen, W. P. Trautwein, F. G. Espinosa, and K. Heyns, *Tetrahedron Letters* p. 4131 (1966).

111. H. Paulsen, W. P. Trautwein, F. G. Espinosa, and K. Heyns, *Chem. Ber.* **100**, 2822 (1967).
112. H. Paulsen, W. P. Trautwein, F. G. Espinosa, and K. Heyns, *Tetrahedron Letters* p. 4137 (1966).
113. A. S. Perlin, *Can. J. Chem.* **41**, 399 (1963).
114. A. S. Perlin, *Can. J. Chem.* **41**, 555 (1963).
115. W. W. Pigman and H. S. Isbell, *J. Res. Natl. Bur. Std.* **19**, 189 (1937).
115a. H. Pischel and A. Holy, *Collection Czech. Chem. Commun.* **34**, 89 (1969).
115b. H. Pischel and A. Holy, *Collection Czech. Chem. Commun.* **33**, 2066 (1968).
115c. H. Pischel and G. Wagner, *Z. Chem.* **8**, 178 (1968).
116. A. L. Raymond, *in* "Organic Chemistry—an Advanced Treatise" (H. Gilman, ed.), p. 1610. Wiley, New York, 1943.
117. C. B. Reese and J. B. Sulston, *Proc. Chem. Soc.* p. 214 (1964).
118. N. K. Richtmyer and C. S. Hudson, *J. Am. Chem. Soc.* **58**, 2534 (1936).
119. R. M. Roberts, J. Corse, R. Boschan, D. Seymour, and S. Winstein, *J. Am. Chem. Soc.* **80**, 1247 (1958).
120. C. Ruechardt and G. Hamprecht, *Angew. Chem. Intern. Ed. Engl.* **6**, 949 (1967).
121. W. V. Ruyle, T. Y. Shen, and A. A. Patchett, *J. Org. Chem.* **30**, 4353 (1965).
121a. K. J. Ryan, E. M. Acton, and L. M. Goodman, *J. Org. Chem.* **33**, 1783 (1968).
121b. K. H. Scheit, *Chem. Ber.* **101**, 1141 (1968).
122. M. Schulz and P. Berlin, *Angew. Chem. Intern. Ed. Engl.* **6**, 950 (1967).
123. M. Schulz and H. Boeden, *Tetrahedron Letters* p. 2843 (1966).
123a. M. Schulz, H. Boeden, and P. Berlin, *Ann. Chem.* **715**, 172 (1968).
124. M. Schulz and H. Steinmaus, *Monatsh. Deut. Akad. Wiss. Berlin* **6**, 649 (1964); *Chem. Abstr.* **63**, 665 (1965).
125. M. Schulz and H. Steinmaus, *Z. Naturforsch.* **19b**, 263 (1964).
125a. J. C. P. Schwarz, *J. Chem. Soc.* p. 2644 (1954).
126. B. S. Shasha, W. M. Doane, and R. K. Rohwedder, *Tetrahedron Letters* p. 1479 (1966).
127. B. S. Shasha, W. M. Doane, C. R. Russell, and C. E. Rist, *Nature* **204**, 186 (1964).
128. B. S. Shasha, W. M. Doane, C. R. Russell, and C. E. Rist, *J. Org. Chem.* **30**, 2324 (1965).
129. J. Smrt, *Collection Czech. Chem. Commun.* **29**, 2049 (1964).
129a. J. Smrt, *Collection Czech. Chem. Commun.* **32**, 198 (1967).
129b. J. Smrt and F. Sorm, *Collection Czech. Chem. Commun.* **32**, 3380 (1967).
130. J. Staněk, M. Černý, J. Kocourek, and J. Pacák, "The Monosaccharides," Chapter 10. Academic Press, New York, 1963.
131. J. Staněk, M. Černý, and J. Pacák, "The Oligosaccharides." Academic Press, New York, 1965.
132. J. M. Sugihara, *Advan. Carbohydrate Chem.* **8**, 1 (1953).
133. J. Swiderski and K. Wolko-Samochocka, *Roczniki Chem.* **36**, 1767 (1962); *Chem. Abstr.* **59**, 8963 (1963).
134. E. A. Talley, D. D. Reynolds, and W. L. Evans, *J. Am. Chem. Soc.* **65**, 575 (1943).
135. R. S. Tipson, H. S. Isbell, and J. E. Stewart, *J. Res. Natl. Bur. Std.* **62**, 257 (1959).
136. A. M. Unrau, *Can. J. Chem.* **41**, 2394 (1963).
137. M. G. Vafina, V. A. Derevitskaya, and N. K. Kochetkov, *Izv. Akad. Nauk SSSR, Ser. Khim.* p. 1814 (1965).
138. F. Weygand and H. Ziemann, *Ann. Chem.* **657**, 179 (1962).
139. S. Winstein and R. E. Buckles, *J. Am. Chem. Soc.* **65**, 613 (1943).
140. S. Winstein, E. Grunwald, R. E. Buckles, and C. Hanson, *J. Am. Chem. Soc.* **70**, 816 (1948).

141. S. Winstein, E. Grunwald, and L. L. Ingraham, *J. Am. Chem. Soc.* **70**, 821 (1948).
142. M. L. Wolfrom and R. M. deLederkremer, *J. Org. Chem.* **30**, 1560 (1965).
143. M. L. Wolfrom and D. I. Weisblat, *J. Am. Chem. Soc.* **66**, 805 (1944).
144. J. Zemlicka, *Chem. & Ind. (London)* p. 581 (1964).
144a. J. Zemlicka, *Collection Czech. Chem. Commun.* **33**, 3796 (1968).
145. J. Zemlicka and S. Chladek, *Tetrahedron Letters* p. 3057 (1965).
146. J. Zemlicka and S. Chladek, *Collection Czech. Chem. Commun.* **31**, 3775 (1966).
146a. J. Zemlicka and S. Chladek, *Collection Czech. Chem. Commun.* **33**, 3293 (1968).
147. J. Zemlicka, S. Chladek, A. Holy, and J. Smrt, *Collection Czech. Chem. Commun.* **31**, 3198 (1966).
148. G. Zemplen, L. Mester, and E. Eckhart, *Acta Chim. Acad. Sci. Hung.* **4**, 73 (1954); *Chem. Abstr.* **49**, 12305 (1955).

CHAPTER 6

# Thioorthocarboxylates, Thioorthocarbonates, and Related Compounds

## I. INTRODUCTION

This chapter discusses the preparation, properties, and chemical transformations of trithioorthocarboxylates, tetrathioorthocarbonates, and related substances. Trithioorthocarboxylates and tetrathioorthocarbonates are substances having three and four alkylthio or arylthio groups bonded to a central carbon atom: $RC(SR')_3$ and $C(SR)_4$. The discussion which follows concerns not only these two classes of compounds, but also substances derived from them by oxidation, halogenation, or metalation reactions. Also considered are acylic and heterocyclic substances in which oxygen and nitrogen, as well as at least one sulfur atom, are singly bonded to a tetrahedral carbon atom at the carboxyl or carbonate level of oxidation.

The various methods of synthesizing this diverse group of compounds are considered first. Table I in Section III lists some of their physical properties and supplies references to the original literature. The last section of the chapter discusses some of the more important chemical reactions of these compounds.

The chemistry of thioortho esters is more complex than that of ordinary orthocarboxylates and orthocarbonates. This is due primarily to the fact that sulfur in thioortho esters can be oxidized to the sulfoxide, sulfone, and even the sulfonate level without disrupting carbon–sulfur bonds, and also to the fact that $d$-orbital resonance renders carbanions derived from thioorthoformates sufficiently stable so that they are involved as intermediates in some of the reactions of these compounds.

The chemistry of thioortho esters was reviewed by Post in 1943 (*126a*), and in part by Reid in 1962 (*128*).

## II. SYNTHESES OF TRITHIOORTHOCARBOXYLATES, TETRATHIOORTHOCARBONATES, AND RELATED COMPOUNDS

### A. From Carboxylic Acids or Thiolcarboxylic Acids

Formic acid reacts with mercaptans at room temperature, and with thiophenols at higher temperatures, to form trialkyl and triaryl trithioorthoformates.

$$HCO_2H + 3 RSH \rightarrow HC(SR)_3 + 2 H_2O \tag{1}$$

This reaction was first reported in 1907 by Holmberg, who used hydrogen chloride or zinc chloride as catalysts (*74, 75*). Shortly thereafter Houben and Schultze found that catalysts are unnecessary, and that satisfactory yields of trialkyl and triaryl trithioorthoformates can be prepared by simply mixing the reactants and allowing the mixture to stand at ambient or higher temperatures (*77, 78*). This reaction has been used to prepare trimethyl trithioorthoformate (*78*), triethyl trithioorthoformate (*75–78, 125*), tripropyl trithioorthoformate (*125, 132*), tri-*tert*-butyl trithioorthoformate (*11*), tri-isoamyl trithioorthoformate (*2*), triamyl trithioorthoformate (*2*), triheptyl trithioorthoformate (*2*), trioctadecyl trithioorthoformate (*2*), tricarboxymethyl trithioorthoformate (*74, 77*), tribenzyl trithioorthoformate (*78*), triphenyl trithioorthoformate (*78*), tri-3-thienyl trithioorthoformate (*24*), and a mixture of complex trithioorthoformates from 1,3,4-thiadiazol-2,5-dithiol (*52*).

The only higher aliphatic carboxylic acid which has been converted to trithioortho esters by direct reaction with thiols is trifluoroacetic acid, which reacts with ethane-1,2-dithiol and propane-1,2-dithiol at reflux temperatures to form the cyclic thioortho esters **1** ($n = 0, 1$) (*15a, 30*).

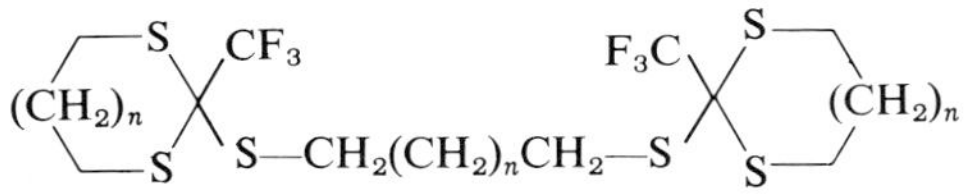

**1**

It is interesting that formic acid can be converted to trithioorthoformates by direct reaction with thiols, but not to orthoformates by direct reaction with alcohols—this in spite of the fact that thiols are less nucleophilic toward acyl carbon than are alcohols (*99a, 122a*). A crude estimate based on bond energies indicates that reaction (1) should be about 20 kcal/mole more exothermic than the analogous reaction between formic acid and an alcohol. This, coupled with the smaller loss of hydrogen bonding association energy in the thiol reaction, apparently accounts for the larger equilibrium constants for the thiol–formic acid reactions than for analogous alcohol–formic acid reactions. Similar arguments apply to conversions of other acyl derivatives to trithioortho esters.

Formic acid reacts with hydrogen sulfide in a nitromethane solution of hydrogen chloride, or in boron trifluoride etherate, to form 2,4,6,8,9,10-hexathiaadamantane (**2**, R = H) (*56, 121*).

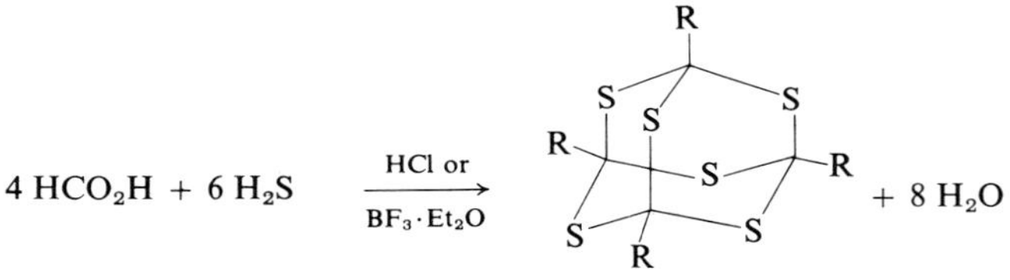

$$4 \, HCO_2H + 6 \, H_2S \xrightarrow[BF_3 \cdot Et_2O]{HCl \text{ or}} \mathbf{2} + 8 \, H_2O \qquad (2)$$

Substituted hexathiaadamantanes are formed by condensation of formic acid with thiolacetic, thiolpropionic, or thiolbutyric acid in the presence of zinc chloride (*119, 121*). These reactions yield mixtures of compounds of structure **2**, where one, two, three, or all four of the R groups are alkyl groups from the thiol acid RCOSH. Chlorothiolacetic acid is converted by treatment with aluminum chloride in nitrobenzene at 0°C into **2**, R = CH$_2$Cl (*122*). Under the same reaction conditions, dichlorothiolacetic acid and trichlorothiolacetic acid formed mixed ortho ester anhydrides, **3**, R = CHCl$_2$ or CCl$_3$ (*122*).

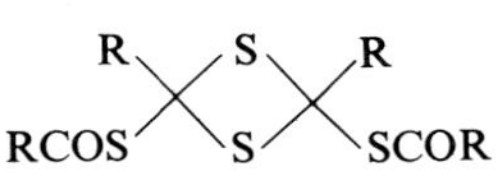

$$RCOS \underset{S}{\overset{R}{\diagup}}\!\!\underset{S}{\overset{S}{\diagdown}}\!\!\underset{S}{\overset{R}{\diagup}} SCOR$$

**3**

Thiolacetic acid, acetic acid, and β-dicarbonyl compounds react in the presence of zinc chloride and hydrochloric acid to form substituted 3,5-dimethyl-2,4,6,9,10-pentathiaadamantanes (*121a*).

## B. From Carboxylate Esters and Thiolcarboxylate Esters

When a mixture of ethyl formate and a mercaptan or thiophenol is saturated with dry hydrogen chloride and allowed to stand in the cold, a two-phase mixture is formed from which a trialkyl or triaryl trithioorthoformate can be isolated.

$$HCO_2C_2H_5 + 3 \ RSH \xrightarrow{HCl} HC(SR)_3 + C_2H_5OH + H_2O \qquad (3)$$

Yields of trithioorthoformates usually exceed 50% of theory and may be nearly quantitative. This procedure, first described by Holmberg in 1907 (74, 75), has been used for preparing trimethyl trithioorthoformate (96), triethyl trithioorthoformate (57, 75), tri-*tert*-butyl trithioorthoformate (11), tridodecyl trithioorthoformate (54), tricarboxymethyl trithioorthoformate (74), tribenzyl trithioorthoformate (96), a series of triperfluoroalkylmethyl trithioorthoformates (138), and triphenyl trithioorthoformate (57). Ethyl formate and ethanethiol undergo an uncatalyzed reaction at 150°C and 8000 atmospheres of pressure which forms triethyl trithioorthoformate in 41% yield (26).

Higher trithioorthocarboxylates have not been prepared by strictly analogous procedures, although a low yield of triethyl trithioorthoacetate was obtained by refluxing a mixture of ethyl acetate and ethanethiol with zinc chloride (129).

A bicyclic bis trithioorthoformate, probably **4**, was obtained when an attempt was made to convert methyl 3α-formyloxy-11,12-diketocholanate to a trimethylenethioketal by treating it with propane-1,3-dithiol and hydrogen chloride (4).

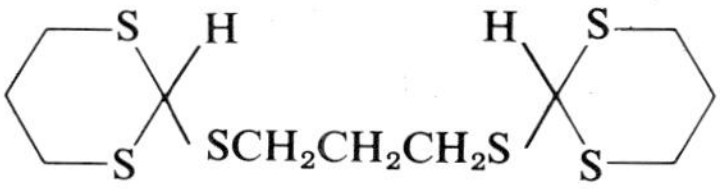

<chemical structure 4>

Triphenyl trithioorthoformate and tri-*p*-tolyl trithioorthoformate have been prepared by refluxing ethyl formate with the arylthiomagnesium bromides in ether (100).

$$3 \ ArSMgBr + HCO_2C_2H_5 \rightarrow HC(SAr)_3 + MgBrOC_2H_5 + MgBr_2 + MgO \qquad (4)$$

Thiolcarboxylate esters undergo acid-catalyzed reactions with thiols and thiophenols to form trithioortho esters.

$$RCOSR' + 2 \ R''SH \xrightarrow{HCl, \ ZnCl_2, \ or \ BF_3} RC(SR'')_2SR' + H_2O \qquad (5)$$

Tribenzyl trithioorthoformate was prepared in 99% yield by the hydrogen chloride-catalyzed reaction of benzyl mercaptan with benzyl thiolformate (*61*). Trialkyl and triaryl trithioorthoacetates can be prepared by means of similar reactions, although in lower yields. Triethyl trithioorthoacetate (*155*), tribenzyl trithioorthoacetate (*153*), triphenyl trithioorthoacetate (*153*), and tri-*m*-tolyl trithioorthoacetate (*153*) were prepared from the thiolacetates and thiols or thiophenols. The reaction can be used to prepare mixed trithioorthoacetates, in which R' and R" of Eq. (5) are different. Ethyl thiolacetate and benzyl mercaptan yield ethyldibenzyl trithioorthoacetate (*91*, *93*), and phenyl thiolacetate and *m*-thiocresol form phenyldi-*m*-tolyl trithioorthoacetate (*153*).

## C. From Carboxamides

Formamide reacts with ethanethiol and with thioglycolic acid in the presence of catalytic amounts of sulfuric acid to form triethyl trithioorthoformate and tricarboxymethyl trithioorthoformate, respectively [Eq. (6)] (*74*, *75*). Neither acetamide nor *N*-methylacetamide reacts with ethanethiol under similar conditions (*129*).

$$HCONH_2 + 3\ RSH \xrightarrow{H^+} HC(SR)_3 + NH_3 + H_2O \quad (6)$$
$$(R = C_2H_5, HO_2CCH_2)$$

Slow addition of phosphorus oxychloride to cold mixtures of *N*-methylformanilide and thiols results in formation of trialkyl trithioorthoformates. Tributyl trithioorthoformate and triisobutyl trithioorthoformate were prepared in 69% and 44% yields by means of this reaction (*49*).

*N*,*N*-Dimethylformamide, on the other hand, reacts with thiols and phosphorus oxychloride at low temperatures to form *N*,*N*-dimethylformamide dithioacetals (*49*).

$$(CH_3)_2NCHO + 2\ RSH \xrightarrow{POCl_3} (CH_3)_2NCH(SR)_2 \quad (7)$$
$$[R = C_4H_9, (CH_3)_2CHCH_2, (CH_3)_3C]$$

## D. From Acid Chlorides and Acid Anhydrides

A generally applicable method of preparing thioorthocarboxylates involves refluxing acid chlorides with thiols and anhydrous zinc chloride.

$$RCOCl + 3\ R'SH \xrightarrow{ZnCl_2} RC(SR')_3 + HCl + H_2O \quad (8)$$

The reaction gives 40% to 70% yields of a variety of triethyl trithioorthocarboxylates (*129*), and has also been used to prepare trimethyl trithioorthoacetate and triphenyl trithioorthobenzoate (153). It gave poor results in attempts to prepare trithioortho esters from higher thiols and acid chlorides having α-hydrogen atoms due to the extreme ease of acid-catalyzed dethiolation of the initially formed trialkyl trithioorthocarboxylates to ketene dithioacetals (129). Triethyl esters of trithioorthoacetic (93, 129, 155), trithioorthopropionic (93, 129), trithioorthoisobutyric (138a), trithioorthooctanoic (129), trithioorthophenylacetic (129), trithioorthoacrylic (93), and trithioorthobenzoic (93, 129) acids were prepared according to Eq. (8), as were hexaethyl hexathioorthoadipate and hexaethyl hexathioorthomalonate (93).

Acetic anhydride reacts with *m*-thiocresol in the presence of boron trifluoride to form tri-*m*-tolyl trithioorthoacetate in 30% yield (153).

Chloroacetyl chloride and hydrogen sulfide undergo an aluminum chloride-catalyzed reaction at $-20°C$ which forms a mixture of **2**, R = $CH_2Cl$ and **3**, R = $CH_2Cl$ (122). β-Chloropropionyl chloride and hydrogen sulfide reacted in pyridine at $-70°C$ to form **3**, R = $CH_2CH_2Cl$ (122).

### E. From Trialkyl Orthocarboxylates and Amide Acetals

Orthocarboxylates react with thiols and thiophenols to form trithioorthocarboxylates [Eq. (9)].

$$RC(OR')_3 + 3\ R''SH \rightarrow RC(SR'')_3 + 3\ R'OH \tag{9}$$

Acid catalysts such as hydrogen chloride (41, 83, 147), zinc chloride (116), and toluenesulfonic acid (41) accelerate these reactions, but transesterification proceeds satisfactorily, at least in the case of orthoformates, at elevated temperatures without an added catalyst (149).

Most examples of this transesterification reaction thus far reported involve orthoformates. The one example of reaction of an orthoacetate with a thiol (41) suggests that the reaction may be a general one for converting orthocarboxylates to trithioorthocarboxylates.

Hydrogen chloride was used to catalyze the conversion of triethyl orthoformate to trimethyl trithioorthoformate, triphenyl trithioorthoformate, and tri-*p*-tolyl trithioorthoformate (147). Triethyl orthodeuteroformate and ethanethiol reacted in the presence of zinc chloride to form triethyl trithioorthodeuteroformate, $DC(SC_2H_5)_3$, in 80% yield (116). Triethyl orthoformate and ethanethiol reacted at 130°C in the absence of added catalysts to form triethyl trithioorthoformate in 87% yield (149). The principal product obtained by allowing the same reactants to stand at room temperature for 1 week was triethyl monothioorthoformate, $HC(OC_2H_5)_2SC_2H_5$ (127).

Dithiols and trithiols react with trialkyl orthoformates and trialkyl orthoacetates in the presence of acids to form cyclic and bicyclic thioortho esters.

Ethanedithiol (*83*) and 1,3-propanedithiol (*147*) yield solid products which are probably the bis dithiane **4** and the bis dithiolane **5**. Triethyl orthoformate

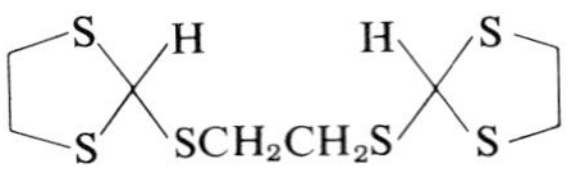

(*41*), triethyl orthodeuteroformate (*116*) and triethyl orthoacetate (*41*) undergo acid-catalyzed reactions with 1,1,1-trimercaptomethylethane [$CH_3C(CH_2SH)_3$] to form trithiabicyclooctane derivatives **6**, R = H, D, and $CH_3$. 1,2,3-Propanetrithiol and triethyl orthoformate form an unstable trithiabicycloheptane derivative, **7** (*41*).

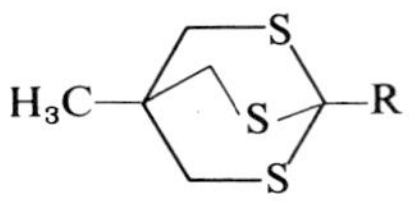

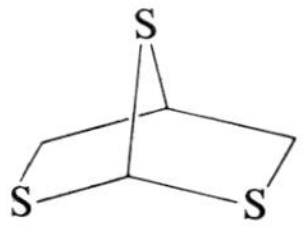

It was reported recently that triethyl orthocarboxylates react with butanethiol in the presence of aluminum chloride to form mixtures of monothio- and dithioortho esters, $RC(OR')_2SC_4H_9$ and $RC(SC_4H_9)_2OR'$ (R = H, $CH_3$, $C_6H_5$; R' = $CH_3$, $C_2H_5$) (*50b, 50c*).

N,N-dimethylamide dimethylacetals react with ethanedithiol to form 2-R-2-dimethylamino-1,3-dithiolanes (R = H, $CH_3$, $C_2H_5$, $C_4H_9$) (*50d*).

### F. From Trihalomethyl Compounds, Halo Ethers, and Halo Thioethers

Gabriel discovered in 1877 that chloroform reacts with salts of thiophenols and thiols in alkaline solutions to form trithioorthoformates (*59*).

$$HCCl_3 + 3\, RS^- \rightarrow HC(SR)_3 + 3\, Cl^- \qquad (10)$$

This reaction, which can be carried out in aqueous or alcoholic solutions, has been used mainly for the synthesis of triaryl trithioorthoformates, $HC(SAr)_3$, where Ar = $C_6H_5$ (*59, 68, 86, 101*), p-$CH_3C_6H_4$ (*6, 82, 86*), p-$ClC_6H_4$ (*7, 82, 86*), p-$CH_3OC_6H_4$ (*86*), p-$FC_6H_4$ (*86*), p-$BrC_6H_4$ (*86*), p-$O_2NC_6H_4$ (*86*), and 3,4-$(O_2N)_2C_6H_3$ (*86*). Yields of triaryl trithioorthoformates usually exceed 70%. Triethyl trithioorthoformate (*59*), tri-*tert*-butyl trithioorthoformate (*11*), and tribenzyl trithioorthoformate (*39*) were prepared from the sodium mercaptides and chloroform, and tri(2-amino-4-methyl-6-pyrimidyl) trithioorthoformate was prepared from 2-amino-4-mercapto-6-methylpyrimidine and chloroform in alcoholic potassium hydroxide solution (*82*).

Other haloforms also yield trithioortho esters. Thus bromoform and p-thiocresol react in aqueous dioxane–sodium hydroxide solution to form tri-p-tolyl trithioorthoformate in 78% yield (70), and chlorodifluoromethane reacts with sodium methylmercaptide in methanol solution to form a mixture of trimethyl trithioorthoformate, trimethyl monothioorthoformate [HC(OCH$_3$)$_2$SCH$_3$], methyl difluoromethyl sulfide, and methyl difluoromethyl ether (71).

The conversion of haloforms to trithioorthoformates probably involves reaction of the trihalomethane with basic species present in the reaction mixture to form a dihalocarbene, which is converted to the trithiorthoformate by a series of fast reactions (68, 70, 71).

In a somewhat different reaction, mercury trifluoromethylmercaptide and iodoform were heated together at 130°C to form tris(trifluoromethyl) trithioorthoformate, HC(SCF$_3$)$_3$ (66, 67).

In 1877 Claesson claimed to have prepared a tetraethyl tetrathioorthocarbonate from carbon tetrachloride and sodium ethylmercaptide (29). Later it was found that the product of this reaction is actually triethyl trithioorthoformate (11, 39). This reaction resembles the conversion of carbon tetrachloride to trialkyl orthoformates by alkoxides, and probably proceeds by a similar mechanism: nucleophilic attack on chlorine by RS$^-$, RO$^-$, or OH$^-$ displaces trichloromethyl carbanion, which either loses chloride ion to form dichlorocarbene, or reacts with the solvent to form chloroform. The trithioorthoformate is formed from dichlorocarbene.

Backer and Stedehouder (11) used the reaction of carbon tetrachloride with sodium alkylmercaptides in ethanol to prepare a number of trialkyl trithioorthoformates, HC(SR)$_3$ [R = CH$_3$, C$_2$H$_5$, C$_3$H$_7$, C$_4$H$_9$, C(CH$_3$)$_3$, and CH$_2$C$_6$H$_5$]. Dialkyl disulfides were by-products of these reactions, and, in addition, the mixed dithioorthoformate C$_2$H$_5$OCH[SC(CH$_3$)$_3$]$_2$ was obtained as a by-product in the synthesis of tri-*tert*-butyl trithioorthoformate.

Under a different set of reaction conditions, carbon tetrabromide reacts with mercury trifluoromethylmercaptide at temperatures above 100°C to form tetrakis(trifluoromethyl) tetrathioorthocarbonate, C(SCF$_3$)$_4$ (66, 67).

The products Laves prepared from 1,1,1-trichloroethane and the sodium salts of thiophenol and ethyl mercaptan, for which he claimed trithioorthoacetate structures (102), were later shown to be 1,2-diethylthio- and 1,2-diphenylthioethanes, RSCH$_2$CH$_2$SR (153). They were probably formed from 1,2-dichloroethane present as an impurity in Laves' methylchloroform. However, two other substituted trichloromethyl compounds have been converted to trithioortho esters. 2-Trichloromethylbenzimidazole is readily converted to trithioortho esters **8** (R = CH$_3$, p-ClC$_6$H$_4$) by reaction with methanethiol and p-chlorothiophenol in ethanolic sodium ethoxide solutions (46). The mixed dithiortho ester **9** (as well as **8**, R = CH$_2$CH$_2$OH) was formed by reaction of 2-trichloromethylbenzimidazole with 3-mercaptoethanol in ethanolic sodium ethoxide solution (47). Ethyl and phenyl 1,2,2,2-tetrachloroethyl sulfones are converted by methanolic sodium methylmercaptide to $RSO_2CHClC(SCH_3)_3$ ($R = C_2H_5, C_6H_5$) (17).

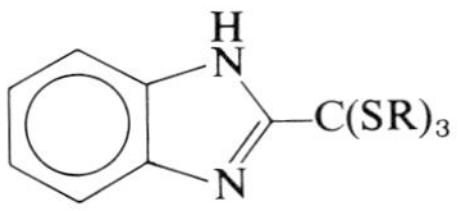

[Structures 8 and 9]

**8** 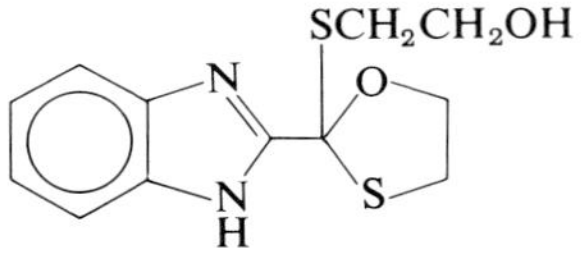  **9**

Chlorodiphenoxymethane reacts with thiophenol in ether–pyridine solution to form triphenyl monothioorthoformate, $(C_6H_5O)_2CHSC_6H_5$, in 42% yield (19). 2,2-Dichlorobenzodioxole undergoes an analogous reaction with thiophenol in tetrahydrofuran–triethylamine solution which yields 2,2-diphenylthiobenzodioxole, **10** (63). Methyl dichloromethyl ether and lead

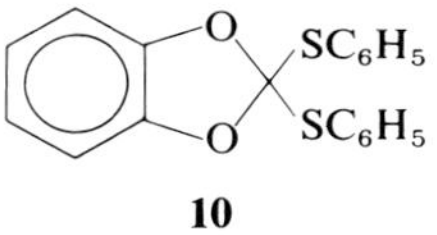

[Structure 10]

**10**

ethylmercaptide yield, instead of the anticipated dithioortho ester, $CH_3OCH(SC_2H_5)_2$, a mixture of trimethyl orthoformate and triethyl trithioorthoformate (62).

Chlorodimethylthiomethane, $CHCl(SCH_3)_2$, reacts with thiols at 0°C to form mixtures of trithioorthoformates (20).

$$ClCH(SCH_3)_2 + RSH \rightarrow HC(SCH_3)_3 + HC(SCH_3)_2SR + HC(SR)_2SCH_3$$
$$+ HC(SR)_3 \quad (11)$$
$$(R = CH_3, C_2H_5, C_6H_5CH_2)$$

The symmetrical trithioortho esters are formed by disproportionation of initially formed alkyldimethyl trithioorthoformates. The same chlorodithioformal reacts with dimethylamine in ether solution to form dimethylformamide dimethylthioacetal, $(CH_3)_2NCH(SCH_3)_2$ (20).

Trichloromethylsulfenyl chloride and sodium alkylmercaptides react to form mixtures of bis(trialkylthiomethyl) disulfides and dialkyl disulfides [Eq. (12)] (14). Thiophenolates, on the other hand, yield bis(triarylthiomethyl) trisulfides, $(ArS)_3CSSSC(SAr)_3$ [Ar = $C_6H_5$, m- (and p-) $CH_3C_6H_4$, p-$(CH_3)_3CC_6H_4$, 2,4,6-$(CH_3)_3C_6H_2$, o- (and p-) $ClC_6H_4$] (14).

$$Cl_3CSCl + RSNa \rightarrow (RS)_3CSSC(SR)_3 + RSSR \quad (12)$$
$$[R = C_2H_5, C_3H_7, C_4H_9, C(CH_3)_3]$$

## G. From Dialkylthio, Trialkylthio, and Related Carbanion Salts

Alkylthiocarbanions are much more stable than analogous alkoxycarbanions, due to resonance interactions between the unshared electron pair on carbon and vacant $d$-orbitals of the sulfur atoms. Di- and trialkylthiomethylide salts can be prepared from dithioacetals, trithioorthoformates, and tetrathioorthocarbonates. Their reactions with a variety of electrophilic reagents, including alkyl halides, carbon dioxide, carbonyl compounds, and dialkyl and diaryl disulfides lead to trithioortho esters. These reactions make it possible to convert trithioorthoformates and tetrathioorthocarbonates to esters of higher trithioortho acids, to convert tetrathioorthocarbonates to trithioorthoformates, and to prepare trithioorthobenzoates from benzaldehyde dithioacetals.

Sodium diethylthiomethylide, $NaCH(SC_2H_5)_2$, reacts with diethyl disulfide in liquid ammonia to form tetraethyl tetrathioorthocarbonate in 43% yield (57). The tetrathioorthocarbonate presumably is formed by reaction of triethylthiomethylide ion, produced from the initially formed trithioorthoformate, with diethyl disulfide.

$$NaCH(SC_2H_5)_2 + C_2H_5SSC_2H_5 \xrightarrow{-C_2H_5SNa} NaC(SC_2H_5)_3 \xrightarrow{NaNH_2} \xrightarrow{C_2H_5SSC_2H_5} C(SC_2H_5)_4 \quad (13)$$

The lithium salt of 1,3-dithiane was similarly converted to 2,2-dimethylthio-1,3-dithiane (11) by reaction with dimethyl disulfide (147).

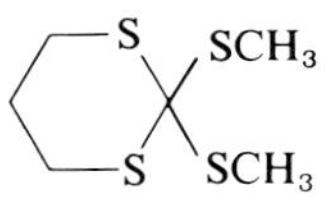

Structure 11: 1,3-dithiane ring with two SCH_3 groups at the 2-position

**11**

Benzaldehyde diethyl dithioacetal, which has only one acidic hydrogen atom, was converted by sodium amide in liquid ammonia to sodium diethylthiophenylmethylide, which reacted with diethyl disulfide to form triethyl trithioorthobenzoate in 76% yield (57).

$$C_6H_5CH(SC_2H_5)_2 \xrightarrow{NaNH_2} C_6H_5\bar{C}(SC_2H_5)_2 \xrightarrow{C_2H_5SSC_2H_5} C_6H_5C(SC_2H_5)_3 \quad (14)$$

Attempts to prepare trithioorthoacetates and trithioorthopropionates by similar reactions failed (57).

Alkali metal trithiomethylide salts can be prepared by reactions of trithioorthoformates with lithium, sodium, or potassium amide in liquid ammonia, or with butyllithium in tetrahydrofuran (5). Trithiomethylides prepared by means of these reactions include $(RS)_3C^-M^+$, $R = C_6H_5$, $M = Li$

(*147, 157*); R = CH$_3$, M = K (*69*); R = C$_2$H$_5$, M = Li (*57*) or K (*15, 69*). The lithium salt of methyltrithiabicyclooctane **6** (R = H) was also prepared (*147*). The same methylides are formed by reaction of tetrathioorthocarbonates with butyllithium in tetrahydrofuran (*147, 157*).

$$C(SR)_4 + C_4H_9Li \rightarrow LiC(SR)_3 + C_4H_9SR \qquad (15)$$

Reactions of tris(alkylthio)carbanions and tris(arylthio)carbanions with alkyl halides yield trialkyl and triaryl trithioorthocarboxylates [Eq. (16)]. Compound **6**, R = (CH$_3$)$_2$CH, was prepared in a similar manner.

$$M^+C(SR')_3^- + RX \rightarrow RC(SR')_3 + MX \qquad (16)$$

| R | R' | References |
|---|---|---|
| CH$_3$ | CH$_3$, C$_6$H$_5$ | *69, 147, 157* |
| C$_2$H$_5$, (CH$_3$)$_2$CH, C$_4$H$_9$, CH$_2$=CHCH$_2$ | C$_2$H$_5$, C$_6$H$_5$ | *57, 157* |
| (C$_2$H$_5$O)$_2$CHCH$_2$, (CH$_3$)$_2$CHCH$_2$, C$_7$H$_{15}$ | C$_2$H$_5$ | *15, 17* |
| C$_3$H$_7$ | C$_6$H$_5$ | *157* |

The same carbanion salts react with carbonyl compounds to give esters of α-hydroxytrithioortho acids (*147*).

$$M^+C(SR)_3^- + R'R''CO \rightarrow R'R''C(OH)C(SR)_3 \qquad (17)$$
$$(R = C_6H_5, R' = H, R'' = C_6H_5, C_3H_7)$$

The lithium salt of **6**, R = H, reacted with benzaldehyde to form **6**, R = CH(OH)C$_6$H$_5$; with benzophenone to form **6**, R = C(OH)(C$_6$H$_5$)$_2$; and with cyclohexanone to form **6**, R = 1-hydroxycyclohexyl (*147*).

Lithium salts of triphenyl trithioorthoformate and **6**, R = Li, react with ethyl chloroformate to form esters of ethyl trithioorthooxalate (*147*).

$$(RS)_3CLi + ClCO_2C_2H_5 \longrightarrow (RS)_3CCO_2C_2H_5 + LiCl$$

$$\left[ (RS)_3C = (C_6H_5S)_3C,\ H_3C\!\!\begin{array}{c}\text{–S}\\\text{–S–}\\\text{–S}\end{array}\!\!\right] \qquad (18)$$

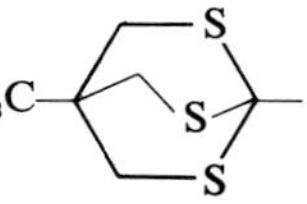

Dialkyl or diaryl disulfides and alkali trialkylthio- or triarylthiomethylides yield tetrathioorthocarbonates, including **6**, R = SCH$_3$ (*147*).

$$MC(SR)_3 + R'SSR' \rightarrow R'SC(SR)_3 + MSR' \qquad (19)$$

| R | R' | References |
|---|---|---|
| C$_2$H$_5$ | C$_2$H$_5$ | 57 |
| C$_6$H$_5$ | C$_6$H$_5$, CH$_3$ | *147, 157* |

Hexaphenyl hexathioorthooxalate was prepared by oxidizing lithium triphenylthiomethylide with iodine or potassium ferricyanide (*147*).

$$2\ \text{LiC(SC}_6\text{H}_5)_3 + \text{I}_2 \rightarrow (\text{C}_6\text{H}_5\text{S})_3\text{CC(SC}_6\text{H}_5)_3 + 2\ \text{LiI} \qquad (20)$$

Lithium triphenylthiomethylide also reacts with carbon dioxide to form triphenyl trithioorthooxalate, (C$_6$H$_5$S)$_3$CCO$_2$H (*157*), with 1,2-epoxypropane to form triphenyl trithio-3-hydroxyorthobutyrate (*147*), and with chlorotrimethylsilane to form (C$_6$H$_5$S)$_3$CSi(CH$_3$)$_3$ (*147*). The reaction of trithiomethylide salts with carbonyl compounds may prove to be a useful route to α-hydroxy acids (*147*).

$$(RS)_3C^-M^+ + R'R''CO \longrightarrow R'R''C(OH)C(SR)_3 \xrightarrow{H^+,H_2O} R'R''C(OH)CO_2H \qquad (21)$$

Bis(arylsulfonyl)methanes, bis(alkylsulfonyl)methanes, and alkylsulfonylarylsulfonylmethanes are weak acids having p$K_a$'s of about 12 (*34a*). In ethanolic sodium ethoxide solutions they form sodium salts which react with alkane and arenesulfonyl chlorides to form trisubstituted trisulfonylmethanes (*18*). Trisulfonylmethanes are, in effect, oxidation products of trithioorthoformates.

$$RSO_2CH_2SO_2R' \xrightarrow{NaOC_2H_5} (RSO_2)(R'SO_2)CH^-Na^+ \xrightarrow{R''SO_2Cl} (RSO_2)(R'SO_2)CHSO_2R'' \qquad (22)$$

| R | R' | R'' |
|---|---|---|
| CH$_3$ | CH$_3$, C$_2$H$_5$ | CH$_3$ |
| C$_2$H$_5$ | C$_2$H$_5$ | CH$_3$, C$_2$H$_5$, C$_6$H$_5$ |
| C$_2$H$_5$ | C$_3$H$_7$, C$_6$H$_5$ | CH$_3$ |
| CH$_3$ | C$_6$H$_5$ | C$_6$H$_5$ |
| Ar | Ar | CH$_3$ |

(Ar = C$_6$H$_5$, *m*- and *p*-CH$_3$C$_6$H$_4$, *p*-FC$_6$H$_4$, *m*- and *p*-ClC$_6$H$_4$, *p*-BrC$_6$H$_4$)

Alkylthioalkyl sulfones (RSSO$_2$R) react with the sodium salts of disulfonylmethanes to form dialkylsulfonalkylthiomethanes (*34, 132*).

$$(RSO_2CHSO_2R')^-Na^+ + R''SSO_2R'' \rightarrow R''SCH(SO_2R)SO_2R' + NaSO_2R'' \quad (23)$$
$$(R = R' = R'' = C_2H_5; R = C_6H_5, R' = R'' = C_2H_5)$$

There is a patent claim that methyloctadecylthiosulfate and dimethylsulfonylmethane react in the presence of alkali to form dimethylsulfonyloctadecylsulfonylmethane (*2*). It seems more likely that octadecyl dimethylsulfonylmethanesulfonate would be formed from these reactants. Phenylthiophenyl sulfone, C$_6$H$_5$SSOC$_6$H$_5$, and salts of bis(sulfonyl)methanes undergo reactions analogous to that of Eq. (23) (*60*).

Trisulfonylmethanes and disulfonylphenylthiomethanes are alkylated by methyl iodide in aqueous alkali to form trisulfonylethanes and disulfonylphenylthioethanes (*102, 103*).

$$HC(SO_2R)(SO_2R')Y + CH_3I \xrightarrow{OH^-} CH_3C(SO_2R)(SO_2R')Y \quad (24)$$

| R | R' | Y |
|---|---|---|
| C$_6$H$_5$ | C$_6$H$_5$ | SO$_2$C$_6$H$_5$ |
| C$_2$H$_5$ | C$_2$H$_5$ | SC$_6$H$_5$, SO$_2$C$_6$H$_5$ |

**H. From Carbonium Salts**

Carbonium ions having sulfur and at least one other electronegative atom such as sulfur, oxygen, or nitrogen bonded to the positively charged carbon atom react with mercaptide ions, alkoxide ions, alcohols, and amines to form thioortho acid derivatives. The carbonium salts are usually prepared by alkylation of thio esters, thio amides, or thiazoles.

2-Ethoxythiolenium tetrafluoroborate reacts with sodium ethoxide to form 2,2-diethoxythiolane (*109*).

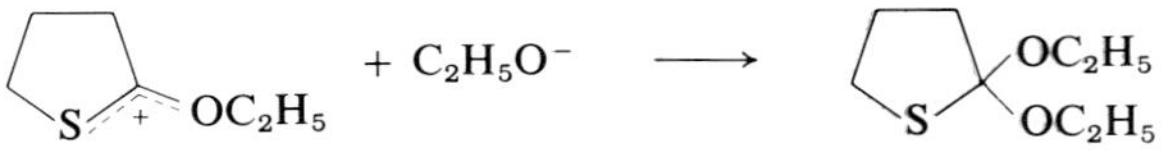

Similarly, 2-methylthio-4,5-tetramethylene-1,3-dithiolium iodide yields 2-alkoxy-2-alkylthio-4,5-tetramethylene-1,3-dithioles in reactions with alcohols (*48*),

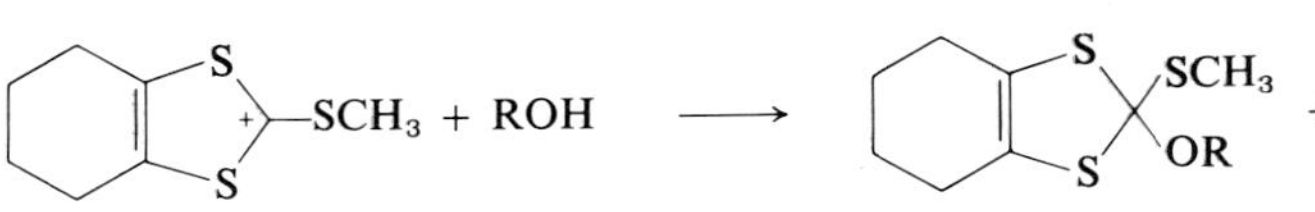

[R = CH$_3$, C$_2$H$_5$, C$_3$H$_7$, (CH$_3$)$_2$CH]

and 2,2-bis(substituted thio)-4,5-tetramethylene-1,3-dithioles in reactions with mercaptides and thiophenolates (48).

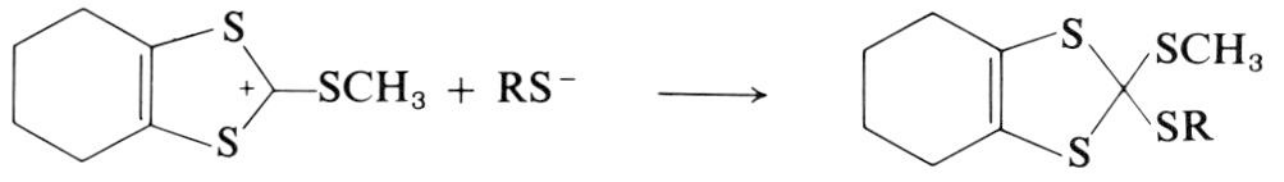

$$\text{[structure]} + \text{RS}^- \longrightarrow \text{[structure 12]} \quad (27)$$

(R = CH$_3$, C$_6$H$_5$, p-ClC$_6$H$_4$, p-O$_2$NC$_6$H$_4$)

The same salt disproportionates in *tert*-butyl alcohol to a mixture of **12** (R = CH$_3$) and the tetracyclic tetrathioorthocarbonate **13** (48).

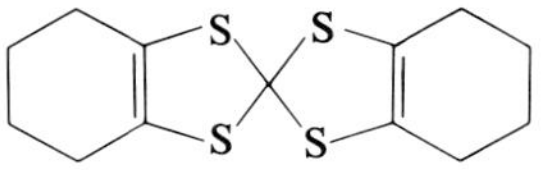

**13**

2-(4-Morpholinyl)-1,3-dithiolenium iodide is converted by sodium cyanide to 2-cyano-2-(4-morpholinyl)-1,3-dithiolane (107).

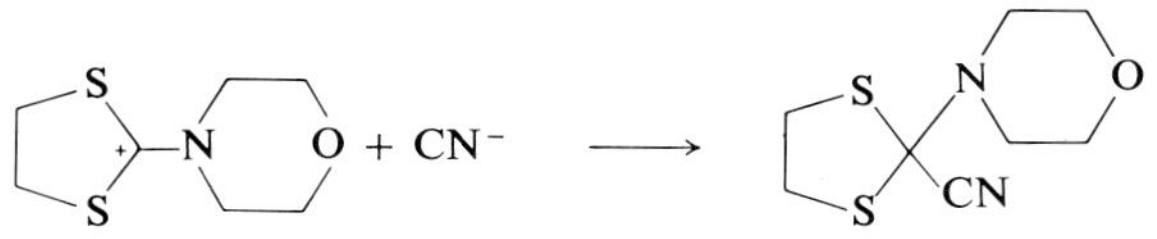

$$\text{[structure]} + \text{CN}^- \longrightarrow \text{[structure]} \quad (28)$$

Dimethylaminoethylthiophenylcarbonium iodide, prepared by ethylation of N,N-dimethylthiobenzamide, reacts with sodium ethylmercaptide to form the diethylthioacetal of N,N-dimethylbenzamide (114, 115).

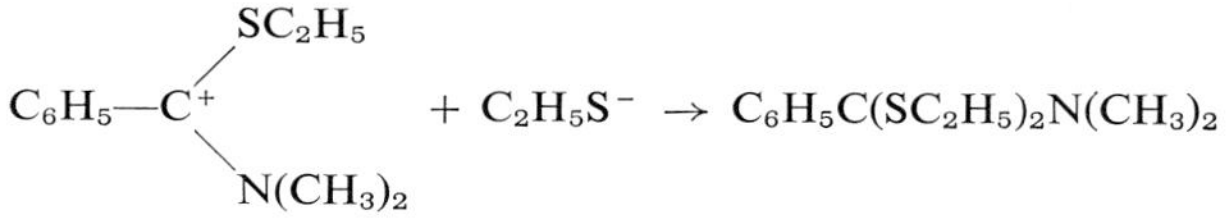

$$\text{C}_6\text{H}_5\text{—C}^+\begin{array}{c}\text{SC}_2\text{H}_5\\ \\ \text{N(CH}_3\text{)}_2\end{array} + \text{C}_2\text{H}_5\text{S}^- \rightarrow \text{C}_6\text{H}_5\text{C(SC}_2\text{H}_5\text{)}_2\text{N(CH}_3\text{)}_2 \quad (29)$$

N-Methylbenzothiazolium salts react with amines (111, 155a) and with alkoxides (110, 112) to form derivatives of monothioortho acids.

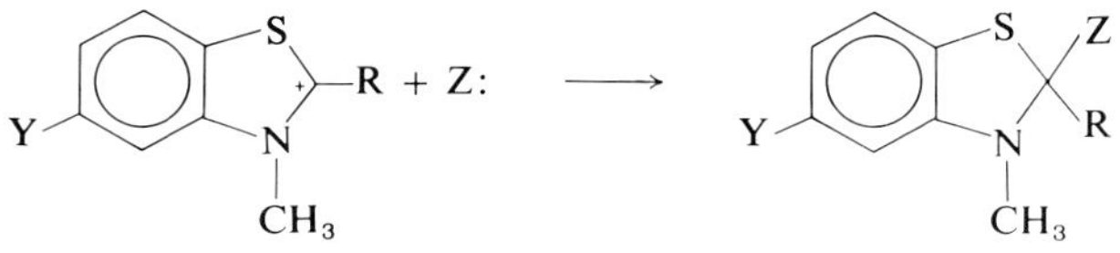

$$\text{[structure]}\text{—R} + \text{Z}: \longrightarrow \text{[structure]} \quad (30)$$

| R | Z | Y | References |
|---|---|---|---|
| H | 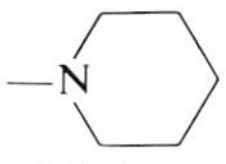 —N(piperidinyl) | H, Cl | 111, 155a |
| C$_6$H$_5$ | —OC$_2$H$_5$ | H | 112 |
| OC$_2$H$_5$ | —OC$_2$H$_5$ | H | 110 |

Trimethylthiocarbonium tetrafluoroborate, $(CH_3S)_3C^+BF_4^-$, reacts with alkanethiols and with thiophenols to form tetrathioorthocarbonates, $C(SR)_4$ [$R = CH_3, C_2H_5, C_3H_7, CH(CH_3)_2, C_6H_5, p\text{-}CH_3C_6H_4, p\text{-}ClC_6H_4, \beta\text{-}C_{10}H_7$], in 85%–97% yields (*129a*).

## I. From Thioimidates, Cyanamide, Nitrile Oxides, Isothioureas, Tosylhydrazones, and other Nitrogen Compounds

A variety of organic nitrogen compounds can be converted to trithioorthocarboxylates, tetrathioorthocarbonates, and related compounds. Several of these reactions yield heterocyclic products which are derivatives of thioorthocarboxylic and thioorthocarbonic acids.

Thioimidic ester hydrochlorides yield trithioorthocarboxylates when treated with thiols or thiophenols (*25, 153*).

$$RC(=NH_2^+Cl^-)SR + 2\,R'SH \rightarrow RC(SR')_3 + NH_4Cl \qquad (31)$$

| R | R' | References |
|---|---|---|
| $CH_3$ | $C_6H_5$ | *153* |
| $C_2F_5, C_3F_7$ | $CH_3, C_2H_5$ | *25* |

This reaction is a counterpart of the Pinner synthesis of orthocarboxylates (p. 2).

Dialkyldithiocarbonate tosylhydrazones decompose, when heated in inert solvents in the presence of strong bases, to salts of toluenesulfonic acid and bis(alkylthio)carbenes (*104, 139, 140*). Among the products formed by secondary reactions of the carbenes are low yields of trialkyl trithioorthoformates.

$$CH_3C_6H_4SO_2NHN{=}C(SR)_2 \xrightarrow{OH^-} [:C(SR)_2] \longrightarrow HC(SR)_3 + \text{other products} \qquad (32)$$
$$(R = CH_3, C_2H_5, C_6H_5CH_2)$$

When the tosylhydrazone is treated with a thiol or thiophenol in butanol–potassium hydroxide solution (*140*), high yields ($\sim 90\%$) of trithioorthoformates are obtained. The reaction is applicable to the synthesis of mixed trithioorthoformates.

$$CH_3C_6H_4SO_2NHN{=}C(SCH_2C_6H_5)_2 + RSH \xrightarrow[C_4H_9OH]{KOH} RSCH(SCH_2C_6H_5)_2 \qquad (33)$$
$$(R = C_6H_5, C_4H_9)$$

A general method of synthesizing tetrathioorthocarbonates involves base-catalyzed reactions of $N,N'$-dinitroso-$S$-alkyl(or aryl)isothioureas with thiols or thiophenols.

$$RS—C(=NH)NH_2 \xrightarrow{HONO} RS—C(=NNO)NHNO \xrightarrow{R'SH} RSC(SR')_3 \quad (34)$$

[R = R' = $C_6H_5$, $p$-$CH_3C_6H_4$, $p$-$ClC_6H_4$, $p$-$FC_6H_4$, $p$-$BrC_6H_4$, $\beta$-$C_{10}H_7$ (6, 7, 13, 88); R = R' = $CH_3$, $C_2H_5$, $CH(CH_3)_2$, $cyclo$-$C_6H_{11}$ (12, 13, 36); R = $C_6H_5$, $o$-$CH_3C_6H_4$, R' = $p$-$ClC_6H_4$ (17)].

The dinitrosoisothioureas are prepared by nitrosation of the corresponding isothioureas, which may be obtained by addition of thiols or thiophenols to cyanamide, or (in the case of $S$-alkylisothioureas) by alkylation of thiourea. The reaction can be used for the synthesis of mixed, as well as symmetrical, tetrathioorthocarbonates. Attempts to prepare tetra-*tert*-butyl tetrathioorthocarbonate by this reaction failed (*12*). Symmetrical tetrathioorthocarbonates are also obtained by heating the equimolar complex formed by an $N,N'$-dinitroso-$S$-alkylisothiourea and the corresponding $S$-alkylisothiourea with a base (*12, 13*).

$$2 \text{ RSC}(=NNO)NHNO \cdot RSC(=NH)NH_2 \rightarrow C(SR)_4 + 6 N_2 + 4 H_2O \quad (35)$$

Spirobicyclic tetrathioorthocarbonates **14** have been prepared from potassium cyanodithioimidocarbonate by the sequence of reactions outlined in Eq. (36) (*38*). When 1,2-dibromoethane and 1,2-ethanedithiol were used, 1,4,6,9-tetrathiaspiro[4.4]nonane (**14**, $n = n' = 2$) was obtained. 1,3-Dibromopropane and 1,3-propanedithiol were converted to 1,5,7,11-tetrathiaspiro[5.5]undecane (**14**, $n = n' = 3$), and dibromomethane plus 1,3-propanedithiol gave 1,4,6,10-tetrathiaspiro[4.5]decane (**14**, $n = 2$, $n' = 3$), together with the spirononane and spiroundecane (*38*).

$$H_2NCN + CS_2 + 2 KOH \xrightarrow{-2 H_2O} (KS)_2C=NCN \xrightarrow{Br(CH_2)_nBr}$$

$$(CH_2)_n \begin{array}{c} S \\ \diagup \\ \diagdown \\ S \end{array} =NCN \xrightarrow{HS(CH_2)_{n'}SH} (CH_2)_n \begin{array}{c} S \\ \diagup \\ \diagdown \\ S \end{array} \begin{array}{c} S \\ \diagdown \\ \diagup \\ S \end{array} (CH_2)_{n'} \quad (36)$$

**14**

2-Substituted-2-thiazolines react with isocyanic acid in ether solution to form 9-substituted 1,2,3,4,6,7-hexahydrothiazolo[3,2-*a*]-*s*-triazine-2,4-diones, **15** (*123*).

$$\begin{array}{c} S \\ \diagup \\ \diagdown \\ N \end{array} -R + 2 \text{ HNCO} \longrightarrow \begin{array}{c} \text{structure of } \mathbf{15} \end{array} \quad (37)$$

**15**

Nitrile oxides react with thione esters and dithio esters to form 3,4,4-trisubstituted 1,4,2-oxathiazoles, **16**, which are derivatives of mono- and dithiocarboxylic acids (80).

$$RC\equiv\overset{+}{N}-O^- + R'CS-YR'' \longrightarrow \underset{\mathbf{16}}{\text{R}-\!\!\begin{array}{c}N-O\\ \phantom{x}\\ S\end{array}\!\!\overset{YR''}{\underset{R'}{\diagdown}}} \qquad (38)$$

| R | R' | YR'' |
|---|---|---|
| $C_6H_5$ | $C_6H_5O$ | $C_6H_5O$ |
| $C_6H_5$ | $\alpha$-$C_{10}H_7$ | $CH_3S$ |
| $C_6H_5$ | $C_6H_5$, $C_6H_5CH_2$ | $HO_2CCH_2S$ |

Condensation of d-penicillamine methyl ester hydrochloride with 2-phenyl (or 2-benzyl-)-4-ethoxymethyleneoxazolone yields methyl phenylpenilloate or methyl benzylpenilloate, **17** (84).

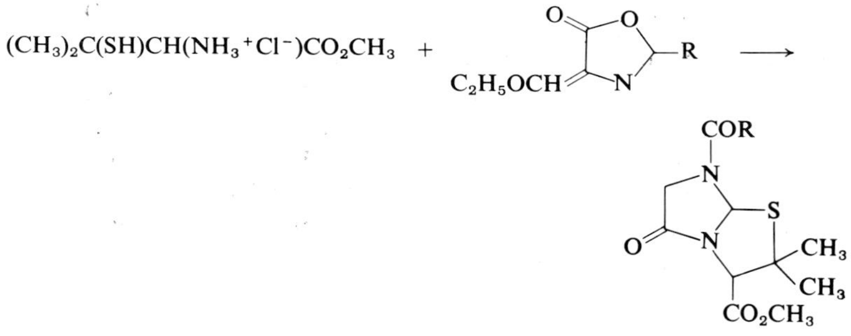

$$(CH_3)_2C(SH)CH(NH_3^+Cl^-)CO_2CH_3 \; + \; \underset{C_2H_5OCH}{\overset{O\diagdown\!\!\!\diagup O}{\diagup\!\!\!\!\diagdown}}\!\!\!-\!R \longrightarrow$$

$$\underset{\mathbf{17}}{\begin{array}{c}\text{COR}\\|\\ \text{N}\\ \text{...ring...}\\ \text{CO}_2\text{CH}_3\end{array}} \qquad (39)$$

$(R = C_6H_5, C_6H_5CH_2)$

Compound **17**, R = $C_6H_5$, is also formed from 4-methoxycarbonyl-5,5-dimethyl-2-thiazoline and 2-phenyloxazolone (84).

### J. From Thioncarbonates

A number of thioorthocarbonates and thioorthocarboxylates are formed by addition, oxidation, and reduction reactions of thioncarbonates.

1,5,7,11-Tetrathiaspiro[5.5]undecane (**14**, $n = n' = 3$) is formed in 33% yield by the reaction of equimolar amounts of trimethylenetrithiocarbonate and diethanolamine in ethanol solution [Eq. (40)] (89). The reaction probably involves aminolysis of part of the cyclic trithiocarbonate followed by reaction of the remainder with the 1,3-propanedithiol thus liberated. The predicted by-product, $N,N,N',N'$-tetra(2-hydroxyethyl)thiourea, was not isolated.

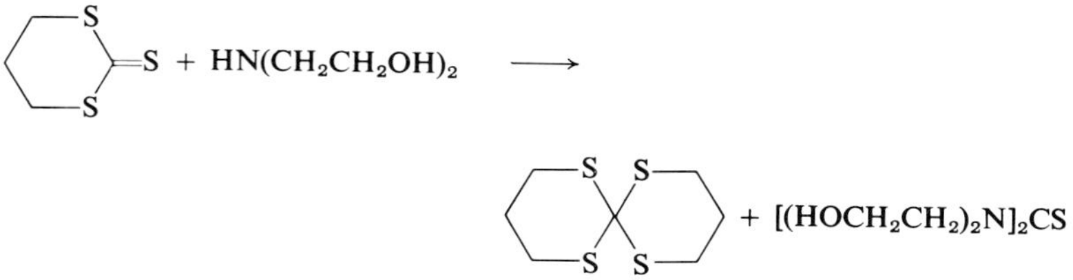

$$\underset{S}{\overset{S}{\bigcirc}}\!\!=\!\!S + HN(CH_2CH_2OH)_2 \longrightarrow$$

$$\underset{S\ \ S}{\overset{S\ \ S}{\bigcirc\!\!\!\!\!\!\!\!\times\!\!\!\!\!\!\!\!\bigcirc}} + [(HOCH_2CH_2)_2N]_2CS \quad (40)$$

Diphenyl trithiocarbonate and diazomethane yield 4,4,5,5-tetraphenyl-thio-1,3-dithiolane, a derivative of hexathioorthooxalic acid (143).

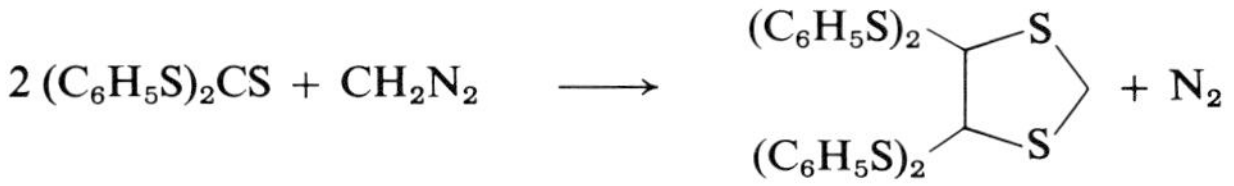

$$2\,(C_6H_5S)_2CS + CH_2N_2 \longrightarrow \underset{(C_6H_5S)_2}{\overset{(C_6H_5S)_2}{\bigg\rangle}}\!\!\!\underset{S}{\overset{S}{\bigg\langle}} + N_2 \quad (41)$$

In an analogous reaction phenyldiazomethane and 3,4-dioxo-1,3-dioxolane-2-thione react to form the hexathioorthooxalate derivative **18** (142).

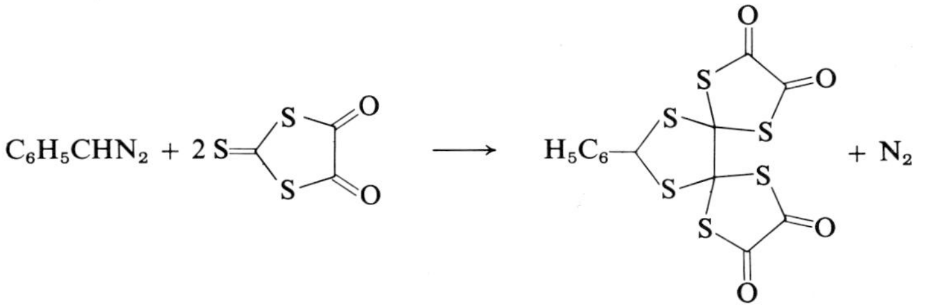

$$C_6H_5CHN_2 + 2\,S\!\!=\!\!\underset{S}{\overset{S}{\bigg\langle}}\!\!\underset{O}{\overset{O}{\bigg\rangle}} \longrightarrow H_5C_6\!\!-\!\!\text{[structure]} + N_2 \quad (42)$$

**18**

Diaryldiazomethanes, 9-diazofluorene, 9-diazoxanthene, and 9-diazo-thiaxanthene react with acyclic and cyclic trithiocarbonates to form derivatives of 1,4,7-trithiaspiro[2.4]heptane [Eqs. (43)–(46)] (141).

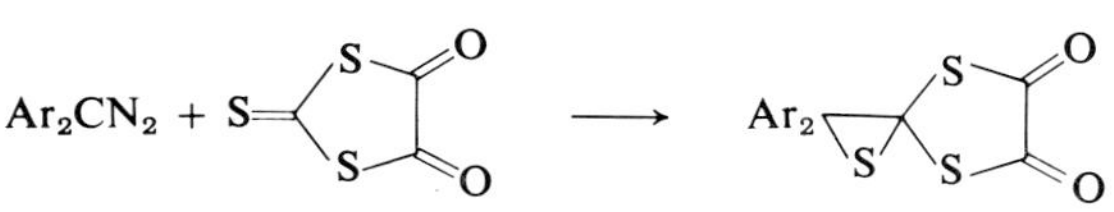

$$Ar_2CN_2 + S\!\!=\!\!\underset{S}{\overset{S}{\bigg\langle}}\!\!\underset{O}{\overset{O}{\bigg\rangle}} \longrightarrow Ar_2\!\!\text{[structure]} \quad (43)$$

$(Ar = C_6H_5, p\text{-}CH_3OC_6H_4)$

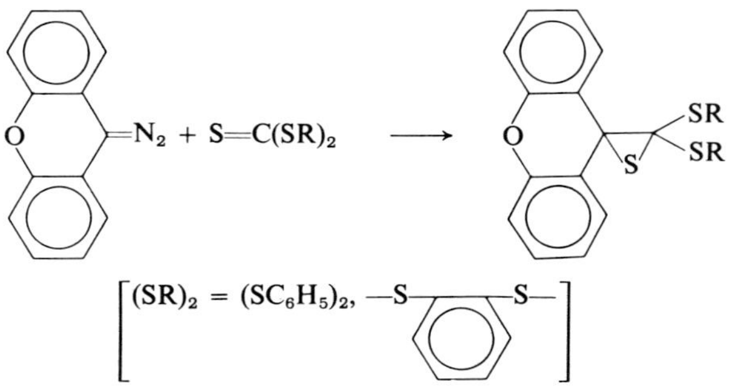

$$\text{structure} \quad =N_2 + S=C(SR)_2 \longrightarrow \text{product} \quad (44)$$

$$[(SR)_2 = (SC_6H_5)_2, \; -S-C_6H_4-S-]$$

$$X\text{-structure}=N_2 + S=\text{(dithiolane-dione)} \longrightarrow \text{product} \quad (45)$$

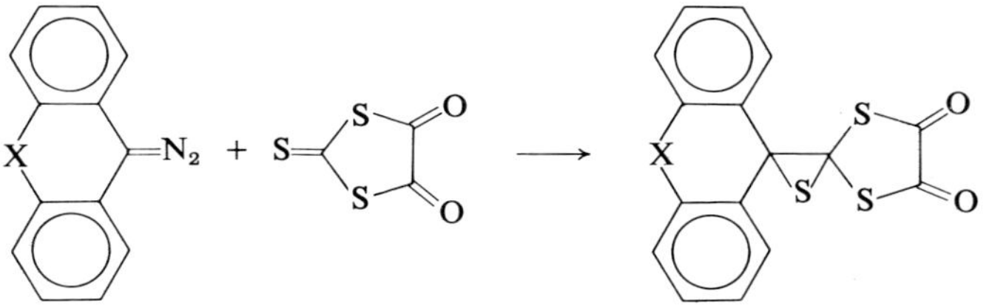

$$(X = O, S, -)$$

Lithium aluminum hydride reduction of D-(1,2,5/3,4,6)-1,2:3,4-di-*O*-isopropylidene-5,6-dimercapto-1,2,3,4-cyclohexanetetrol-5,6-trithiocarbonate (**19**) leads to a cyclic trithioorthoformate acid ester, **20** [Eq. (46)], whose structure was assigned on the basis of its elementary composition and nuclear magnetic resonance spectrum (*106*). This reaction indicates that trithioortho ester acids are more stable than their oxygen analogues, which ordinarily are not isolable.

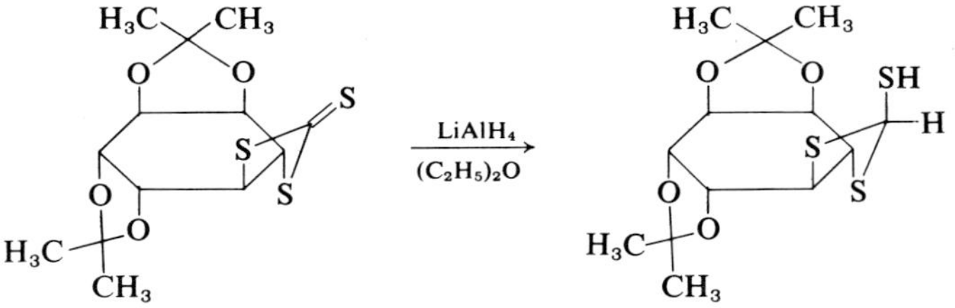

$$\mathbf{19} \xrightarrow{\text{LiAlH}_4} \mathbf{20} \quad (46)$$
$$(C_2H_5)_2O$$

Lead tetraacetate oxidation of 1,3-benzodioxol-2-thione in acetic acid solution produces di(2-acetoxy-1,3-benzodioxolyl) disulfide, **21**, formed from the product of addition of acetic acid to the thione group of the thionocarbonate (*1a*).

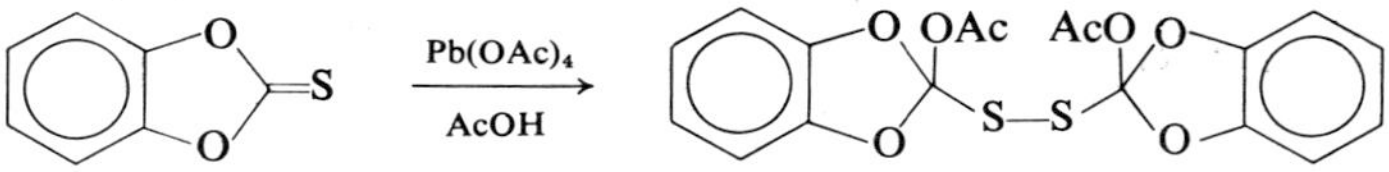

$$\text{benzodioxol-2-thione} \xrightarrow[\text{AcOH}]{\text{Pb(OAc)}_4} \mathbf{21} \quad (47)$$

When 1,2,3-benzothiadiazole is decomposed in the presence of carbon disulfide, a product assumed to be dibenzo-1,4,6,9-tetrathiaspiro[4.4]nonane (**22**) is formed. A possible mechanism for this reaction is indicated in Eq. (48) (*81*).

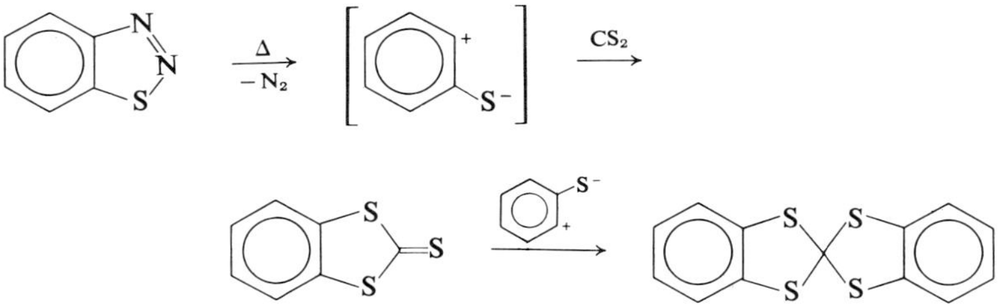

$$\text{(48)}$$

## K. From Other Thioortho Esters

### 1. Oxidation of Thioortho Esters to Sulfoxides and Sulfones

Trithioorthocarboxylates can be oxidized to derivatives having one or more sulfonyl or sulfinyl groups bonded to the ortho ester acyl carbon.

Permanganate was the first oxidizing agent used to convert trithioorthocarboxylates to sulfone derivatives. Yields of products are usually poor when this reagent is used, due to competing reactions such as hydrolysis of the trithioortho ester and sulfonic acid formation.

Laves reported in 1890 that carefully controlled oxidation of triphenyl trithioorthoformate by acidic permanganate produces diphenylsulfonylphenylthiomethane (*101, 102*).

$$HC(SC_6H_5)_3 \xrightarrow{MnO_4^-,\ H^+} C_6H_5SCH(SO_2C_6H_5)_2 \quad (49)$$

Trialkyl trithioorthoformates have been oxidized to trialkylsulfonylmethanes by acidic permanganate (*11, 49*).

$$HC(SR)_3 \xrightarrow{MnO_4^-,\ H^+} HC(SO_2R)_3 \quad (50)$$
$$(R = n\text{-}C_4H_9,\ iso\text{-}C_4H_9)$$

Usually, however, the major oxidation products are dialkylsulfonylmethanes, $CH_2(SO_2R)_2$, and alkanesulfonic acids (*11, 75*).

Triethyl, triphenyl, and tribenzyl trithioorthoacetates were oxidized to trisulfonylethanes by acidic permanganate (*102*).

$$CH_3C(SR)_3 \xrightarrow{MnO_4^-,\ H^+} CH_3C(SO_2R)_3 \quad (51)$$
$$(R = C_2H_5,\ C_6H_5,\ C_6H_5CH_2)$$

Disulfonylalkylthio- and disulfonylarylthiomethanes are oxidized to trisulfones by alkaline permanganate [Eq. (52)]. In a similar reaction, 1,1-diethylsulfonyl-1-phenylthioethane is oxidized to 1,1-diethylsulfonyl-1-phenylsulfonylethane, $CH_3C(SO_2C_2H_5)_2SO_2C_6H_5$ (56, 60, 103).

$$RSCH(SO_2R')(SO_2R'') \xrightarrow{MnO_4^-, OH^-} RSO_2CH(SO_2R')(SO_2R'') \quad (52)$$

| R | R' | R'' | References |
|---|----|-----|------------|
| $C_6H_5$ | $C_6H_5$ | $C_6H_5$ | *101, 102* |
| $CH_3$ | $CH_3$ | $CH_3$ | *8* |
| $C_6H_5$ | $CH_3, C_2H_5$ | $C_2H_5$ | *58, 60* |
| $CH_3$ | $C_6H_5$ | $C_2H_5$ | *60* |

Hydrogen peroxide in acetic acid oxidizes bicyclic trithioorthoformate and trithioorthoacetate **6** (R = H, $CH_3$) to the trisulfones **23** (R = H, $CH_3$) (*41*).

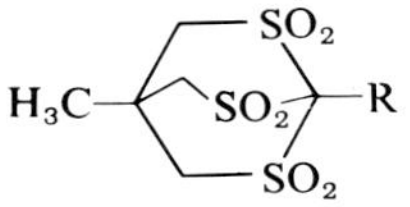

$H_3C$—[ring with $SO_2$, $SO_2$, $SO_2$]—R

**23**

Triarylsulfinylmethanes are oxidized to triarylsulfonylmethanes by the same reagent [Eq. (53)] (*86*). Hydrogen peroxide was used to oxidize several disulfonylmethylsulfides to trisulfonylmethanes [Eq. (54)].

$$HC(SOAr)_3 \xrightarrow[AcOH]{H_2O_2} HC(SO_2Ar)_3 \quad (53)$$

(Ar = $C_6H_5$, $p$-$CH_3OC_6H_4$, $p$-$BrC_6H_4$, $p$-$ClC_6H_4$, $p$-$FC_6H_4$)

$$RSCH(SO_2R')(SO_2R'') \xrightarrow{H_2O_2} RSO_2CH(SO_2R')(SO_2R'') \quad (54)$$

| R | R' | R'' | References |
|---|----|-----|------------|
| $CH_3$ | $C_6H_5$ | $p$-$CH_3C_6H_4$ | *60* |
| $C_2H_5$ | $CH_3, C_2H_5$ | $CH_3, C_2H_5$ | *34, 132* |
| $p$-$CH_3C_6H_4$ | $CH_3$ | $C_6H_5$ | *60* |

An attempt to oxidize tetracyclohexyl tetrathioorthocarbonate to a tetrasulfone with hydrogen peroxide in acetic acid gave cyclohexanesulfonic acid as the only isolable product (*13*).

Peracetic acid in ether solution oxidizes triaryl trithioorthoformates to the corresponding triarylsulfinylmethanes in almost quantitative yields [Eq. (55)] (*86*). This reaction, followed by oxidation of the triarylsulfinylmethanes with hydrogen peroxide [Eq. (53)], provides a superior method for conversion of trithioorthoformates to trisulfonylmethanes.

$$HC(SAr)_3 \xrightarrow[(C_2H_5)_2O]{CH_3CO_3H} HC(SOAr)_3 \qquad (55)$$

(Ar = $C_6H_5$, $p$-$CH_3C_6H_4$, $p$-$CH_3OC_6H_4$, $p$-$ClC_6H_4$, $p$-$FC_6H_4$)

Trithioorthoformates can be oxidized directly to trisulfones by using an excess of peracetic acid (*2, 138*).

$$HC(SR)_3 \xrightarrow{CH_3CO_3H} HC(SO_2R)_3 \qquad (56)$$

(R = $C_5H_{11}$, $C_7H_{15}$, $C_{18}H_{37}$, $C_6H_{13}OCH_2CH_2$, $C_nF_{2n+1}CH_2$)

Monoperphthalic acid also converts trialkyl trithioorthoformates to trialkylsulfonylmethanes (*8, 16, 132*).

$$HC(SR)_3 \xrightarrow[(C_2H_5)_2O]{o\text{-}HO_2CC_6H_4CO_3H} HC(SO_2R)_3 \qquad (57)$$

(R = $CH_3$, $C_2H_5$, $C_3H_7$)

In contrast, monoperphthalic acid oxidation of tetramethyl tetrathioorthocarbonate yielded only trimethylsulfonylmethane, $HC(SO_2CH_3)_3$ (*8*).

Dimethylsulfonylmethylthiomethane, $CH_3SCH(SO_2CH_3)_2$, was oxidized to dimethylsulfonylmethylsulfinylmethane, $CH_3SOCH(SO_2CH_3)_2$, by perbenzoic acid (*9*).

Halogenation followed by hydrolysis of the *S*-halo derivatives is a potential method of selective oxidation of tetrathioorthocarbonates. The only reported examples of such reactions are conversions of tetraaryl tetrathioorthocarbonates to diaryldisulfinylarylthiomethanes (*6, 13*).

$$C(SAr)_4 \xrightarrow{Br_2} (ArSBr_2)_2C(SAr)_2 \xrightarrow{OH^-, H_2O} (ArSO)_2C(SAr)_2 \qquad (58)$$

(Ar = $C_6H_5$, $p$-$CH_3C_6H_4$)

## 2. Reactions of Thioortho Esters with Halogens

Tetrathioorthocarbonates react with bromine in carbon tetrachloride solution to form unstable octabromo derivatives which dissociate on warming to tetrabromides [Eq. (59)] (*6, 12, 13*). Tetramethyl tetrathioorthocarbonate reacts with iodine to form a hexaiodide and a tetraiodide (*12*).

$$C(SR)_4 \xrightarrow{Br_2} C(SBr_2R)_4 \xrightarrow{-Br_2} (RS)_2C(SBr_2R)_2 \qquad (59)$$

(R = $CH_3$, $C_2H_5$, $C_6H_5$, $C_6H_{11}$, $p$-$CH_3C_6H_4$, $\beta$-$C_{10}H_7$)

Trisulfonylmethanes and chlorine or bromine react in aqueous solutions to form halotrisulfonylmethanes [Eq. (60)]. These reactions involve nucleophilic displacement of halide ion from the halogen by trisulfonylmethylcarbanions; the products are oxidizing agents and sources of positive halogen.

$$HC(SO_2R)_2SO_2R' + X_2 \xrightarrow{H_2O} XC(SO_2R)_2SO_2R + H^+ + X^- \quad (60)$$

| R | R' | X | References |
|---|----|---|------------|
| $C_2H_5$, $C_6H_5$ | $C_6H_5$ | Cl, Br | *102, 103* |
| $C_2H_5$ | $C_2H_5$ | Br | *130, 133* |
| $C_3H_7$ | $C_3H_7$ | Cl, Br | *132* |
| $C_4H_9$ | $C_4H_9$ | Cl, Br | *132* |
| $CH_3$ | $C_2H_5$ | Cl, Br | *132* |

Trialkyl esters of methanetrisulfonic acid form similar positive halogen derivatives on treatment with aqueous bromine (*134, 136, 137*).

$$HC(SO_3R)_3 + Br_2 \xrightarrow{H_2O} BrC(SO_3R)_3 + H^+ + Br^- \quad (61)$$
$$(R = CH_3, C_2H_5)$$

### 3. Reduction of Tetrathioorthocarbonates to Trithioorthoformates

Metalation of tetrathioorthocarbonates with butyllithium or phenyllithium in tetrahydrofuran, followed by hydrolysis of the resulting trithiomethyllithium compounds, yields trithioorthoformates [Eq. (62) and (63)] (*147, 157*).

$$C(SR)_4 + R'Li \xrightarrow{-R'SR} (RS)_3CLi \xrightarrow{H_2O} HC(SR)_3 \quad (62)$$
$$(R = C_2H_5, R' = C_6H_5; R = C_6H_5, R' = C_4H_9)$$

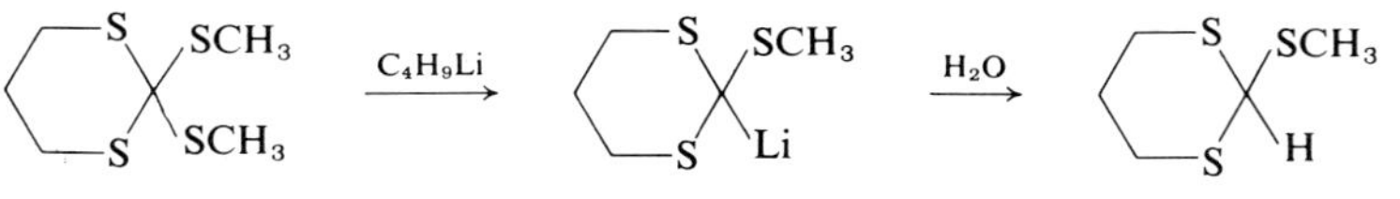

(63)

The reaction is of little synthetic utility, since trithioorthoformates are usually easier to prepare than tetrathioorthocarbonates. It does provide a means of synthesizing isotopically labeled trithioorthoformates, however (*147*).

$$C(SR)_4 \xrightarrow{C_4H_9Li} \xrightarrow{D_2O} DC(SR)_3 \quad (64)$$
$$(R = CH_3, C_6H_5, p\text{-}CH_3C_6H_4)$$

Di-*p*-chlorophenyl-di-*p*-tolyl tetrathioorthocarbonate was reduced to di-*p*-chlorophenyl-*p*-tolyl trithioorthoformate by zinc and acetic acid (*7*).

## 4. Substitution and Elimination Reactions of Trithioorthocarboxylates

Trithioorthoformate 20 yields mixed ortho ester acid anhydride 24 when subjected to mild acid hydrolysis and then acetylated with acetic anhydride (106).

$$20 \xrightarrow[\text{NaOAc}]{H^+, H_2O \quad Ac_2O} \mathbf{24} \qquad (65)$$

where 24 is a ring structure with substituents AcO, OAc, SAc, S, H, AcO, OAc.

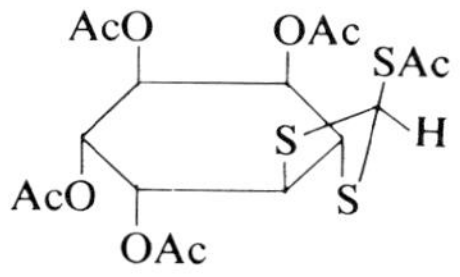

Diethylsulfonylphenylsulfonylmethane is converted by diazomethane to 1,1-diethylsulfonyl-1-phenylsulfonylethane (18).

$$C_6H_5SO_2CH(SO_2C_2H_5)_2 + CH_2N_2 \rightarrow CH_3C(SO_2C_2H_5)_2SO_2C_6H_5 + N_2 \qquad (66)$$

Triethyl trithioorthoacrylate was prepared by dehydrochlorination of triethyl trithioortho-3-chlropropionate with sodium ethoxide (93).

$$CH_2ClCH_2C(SC_2H_5)_3 \xrightarrow{NaOC_2H_5} CH_2=CHC(SC_2H_5)_3 \qquad (67)$$

## 5. Disproportionation Reactions of Mixed Trithioortho Esters

Thioortho esters having two different alkylthio or arylthio groups bonded to the acyl carbon atom undergo acid-catalyzed disproportionation reactions which yield mixtures of symmetrical and unsymmetrical thioortho esters. Disproportionation of initially formed alkyldimethyl trithioorthoformates accounts for the formation of mixtures of trithioorthoformates by reactions of ethyl and benzyl mercaptans with chlorodimethylthiomethane [Eq. (11)] (20).

Arndt reported that recrystallization of tri-*p*-chlorophenylphenyl tetrathioorthocarbonate and tri-*p*-chlorophenyl-*o*-tolyl tetrathioorthocarbonate from acetic acid results in disproportionations to mixtures of tetrathioorthocarbonates (7).

## 6. Alkylation of Carbanions Derived from Trithioorthoformates and Tetrathioorthocarbonates

Esters of higher trithioorthocarboxylates have been synthesized by alkylation of trithiomethylides derived from trithioorthoformates and tetrathioorthocarbonates. These reactions are discussed in Section G.

## L. From Ketene Dithioacetals and Thioketenes

Trithioorthocarboxylate esters having hydrogen situated *alpha* to the acyl carbon exist, in the presence of acids, in equilibrium with ketene dithioacetals. The reverse reaction can be used to convert ketene dithioacetals to thioorthocarboxylates. For example, ketene diethyl dithioacetal and ethanethiol react in the presence of zinc chloride to give a 68% yield of triethyl trithioorthoacetate (*155*).

$$CH_2=C(SC_2H_5)_2 + C_2H_5SH \xrightarrow{ZnCl_2} CH_3C(SC_2H_5)_3 \qquad (68)$$

Bis(trifluoromethyl)thioketene reacts with cyclic trithiocarbonates to form spirocyclic tetrathioorthocarbonates (*127b*).

## M. From Aldose Dithioacetals

Lead tetraacetate oxidation of diethyl dithioacetals of D-galactose, L-arabinose, and glyceraldehyde yields triethyl trithioorthoglyoxalate (isolated as the semicarbazone or thiosemicarbazone) rather than the expected diethyl dithioacetal of glyoxal [Eq. (69)] (*158*). The thioortho acid derivative was probably formed by a bimolecular oxidation–reduction reaction of glyoxal and diethyl dithioacetal.

$$\begin{array}{c} HC(SC_2H_5)_2 \\ | \\ (CHOH)_n \\ | \\ CH_2OH \end{array} \xrightarrow{Pb(OAc)_4} \xrightarrow{H_2NNHCYNH_2} H_2NCYNHN=CHC(SC_2H_5)_3 \qquad (69)$$

$$(Y = O, S)$$

## N. Methanetrisulfonic Acid and Its Esters

Methanetrisulfonic acid, $HC(SO_3H)_3$, is formally a derivative of thioorthoformic acid. It has been prepared by reaction of calcium methylsulfate with pyrosulfuric acid, by oxidation of mercaptomethanetrisulfonic acid, by reaction of potassium sulfite with nitromethane–disulfonic acid, by reaction of diazomethanedisulfonic acid with water, and by sulfonation of methanedisulfonic acid. The various methods of preparing this compound are critically reviewed by Backer and Klaasens (*10*). The trisulfonic acid is a strong acid and is stable in aqueous solutions. Trialkyl esters of this acid were prepared by Samen, who allowed the silver salt to react with alkyl iodides in benzene solution (*134–136*).

$$HC(SO_3^-Ag^+)_3 + 3\ RI \xrightarrow{C_6H_6} HC(SO_3R)_3 + 3\ AgI \qquad (70)$$

$$(R = CH_3, C_2H_5)$$

## III. PROPERTIES AND SOURCES OF THIOORTHOCARBOXYLIC AND THIOORTHOCARBONIC ACID DERIVATIVES

Table I lists the derivatives of thioortho acids which have been described in the literature, together with their principal physical properties and references to the original literature.

## IV. REACTIONS OF THIOORTHOCARBOXYLATES, THIOORTHO-CARBONATES, AND RELATED COMPOUNDS

### A. Reactions Which Form Carbon–Hydrogen Bonds

1. Isotope Exchange Reactions of Trithioorthoformates

Trithioorthoformates are sufficiently acidic to undergo orthoformyl proton exchange in alcoholic or aqueous alcoholic alkoxide or hydroxide solutions (*116, 117, 150*), and to form carbanion salts with alkali metal amides in liquid ammonia (see below). They are such weak acids, however, that their thermodynamic dissociation constants have not been measured.

If rates of proton exchange roughly parallel acidities, triethyl trithioorthoformate is at least seven powers of ten more acidic than triethyl orthoformate. It is about four powers of ten less acidic than chloroform and six powers of ten less acidic than bromoform or iodoform (*150*).

The acidity sequence $HC(OC_2H_5)_3 \ll HC(SC_2H_5)_3 < HCCl_3 < HCBr_3$, $HCI_3$ obviously does not correlate with the electronegativities of the atoms bonded to the orthoformyl carbon. While the electronegativity and polarizibility of $\alpha$-substituents in this series of compounds undoubtedly influence their acidities and rates of proton exchange, the most important factor is probably stabilization of the conjugate bases by $d$-orbital resonance between the carbanion carbon and the $\alpha$-substituent, which in the case of trialkylthiomethyl carbanions involves canonical structures such as

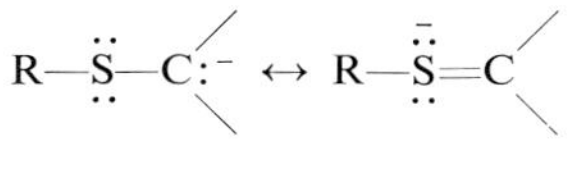

$$R-\ddot{S}-\overset{/}{\underset{\backslash}{C}}:^- \leftrightarrow R-\bar{\ddot{S}}=\overset{/}{\underset{\backslash}{C}}$$

Such resonance is unimportant in stabilizing alkoxycarbanions, which probably accounts for the negligible basicity of trialkyl orthoformates compared to trialkyl trithioorthoformates.

The smaller acidity of triethyl trithioorthoformate than chloroform may be due in part to steric shielding of the formyl protons by the three ethylthio groups. The fact that the bicyclic trithioorthoformate **6**, R = H, undergoes

374

6. THIOORTHOCARBOXYLATES AND THIOORTHOCARBONATES

TABLE I
DERIVATIVES OF THIOORTHO ACIDS

| Compound | M.p., °C (b.p., °C/torr) | $n_D (T, °C)$ | $d (T, °C)$ | References |
|---|---|---|---|---|
| $HC(SO_3H)_3$ | $C_1$ compounds<br>156° (trihydrate) | — | — | 10 |
| $HC(SCF_3)_3$ | $C_4$ compounds<br>(106.5°) | 1.3650 (25°) | — | 66, 67 |
| 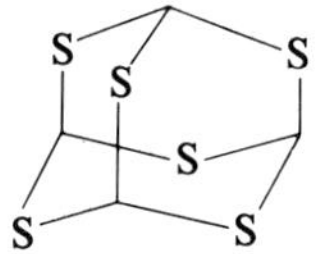 (cage structure) | 330° (dec.) | — | — | 3, 56, 121, 152 |
| 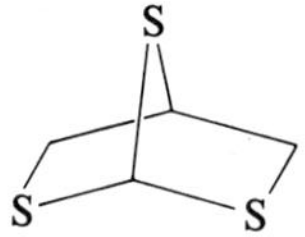 (bicyclic structure) | 198° | — | — | 41 |
| $BrC(SO_2CH_3)_3$ | 140° | — | — | 134 |
| $CH_3SCH(OCH_3)_2$ | (40°/12) | 1.4510 (24°) | 1.0234 (24°) | 71 |
| $CH_3SCH(SO_2CH_3)_2$ | 199° | — | — | 9 |
| $CH_3SOCH(SO_2CH_3)_2$ | 180° | — | — | 9 |
| $HC(SO_2CH_3)_3$ | dec. 350° | — | — | 8, 18 |
| $HC(SO_3CH_3)_3$ | 111° | — | — | 134 |
| $HC(SCH_3)_3$ | (103°–104°/12) | 1.5749 (25°) | 1.1309 (28°) | 11, 20, 71, 78, 96, 145, 147 |

| | | | | |
|---|---|---|---|---|
| C(SCF$_3$)$_4$ | C$_5$ compounds (82°/70) | 1.3980 (25°) | — | 66, 67 |
| [structure: bicyclic S-CH$_3$] | dec. | — | — | 119 |
| 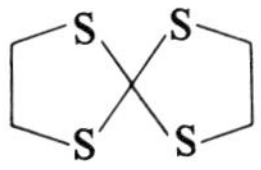 [structure: spiro S] | (142°–143°) | — | — | 38 |
| 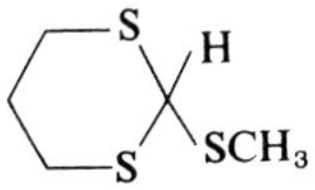 [structure: S-CH, SCH$_3$ ring] | (74°/0.35) | 1.6180 (20°) | — | 147 |
| BrC(SO$_2$CH$_3$)$_2$SO$_2$C$_2$H$_5$ | 137° | — | — | 130, 132 |
| ClC(SO$_2$CH$_3$)$_2$SO$_2$C$_2$H$_5$ | 150° | — | — | 130, 132 |
| C(SBr$_2$CH$_3$)$_2$(SCH$_3$)$_2$ | 122° (dec.) | — | — | 12 |
| (CH$_3$SO$_2$)$_2$CHSC$_2$H$_5$ | 104° | — | — | 132 |
| (CH$_3$S)$_2$C(SO$_2$CH$_3$)$_2$ | 122° | — | — | 9 |
| HC(SO$_2$CH$_3$)$_2$SO$_2$C$_2$H$_5$ | 276° | — | — | 18, 132 |
| CH$_3$C(SCH$_3$)$_3$ | (100°/11) | 1.5680 (25°) | 1.1197 (25°) | 35–37, 69 |
| C$_2$H$_5$SCH(SCH$_3$)$_2$ | (103°/10) | 1.5628 (20°) | — | 20 |
| C(SCH$_3$)$_4$ | (127°/12) 66° | — | — | 12, 35–37, 124, 129a |
| (CH$_3$)$_2$NCH(SCH$_3$)$_2$ | (82°/10) | 1.5258 (20°) | — | 20 |

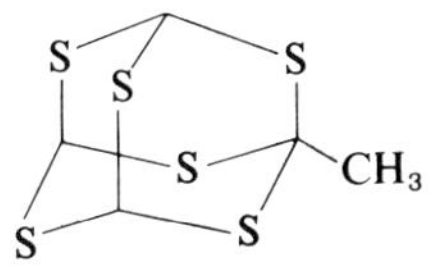

376  6. THIOORTHOCARBOXYLATES AND THIOORTHOCARBONATES

TABLE 1–*continued*

| Compound | M.p., °C (b.p., °C/torr) | $n_D (T, °C)$ | $d (T, °C)$ | References |
|---|---|---|---|---|
| C$_6$ compounds | | | | |
| 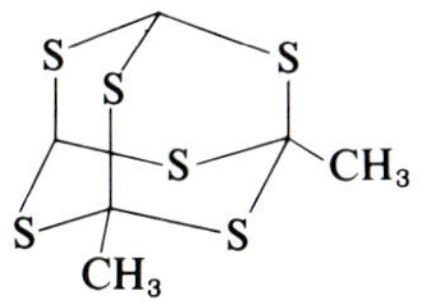 (bicyclic trithioorthoester with CH$_3$) | dec. | — | — | *119* |
| 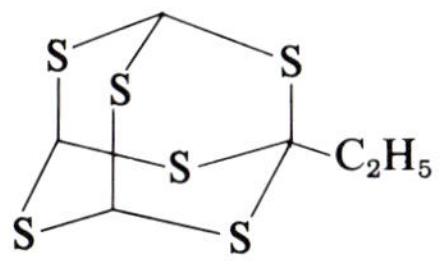 (bicyclic trithioorthoester with C$_2$H$_5$) | dec. | — | — | *120* |
| C$_2$F$_5$C(SCH$_3$)$_3$ | (86°/0.7) | 1.4630 (25°) | 1.360 (25°) | *25* |
| 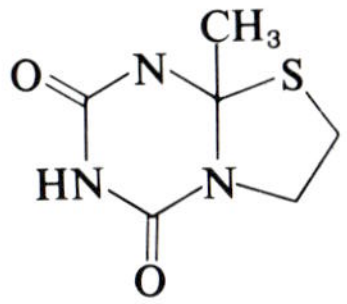 (cyclic thiourea structure with CH$_3$) | 191° | — | — | *123* |
| 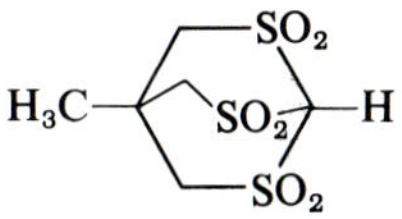 (trisulfone bicyclic structure with H$_3$C and H) | 360° | — | — | *41* |

| Compound | mp/bp | $n_D$ | Ref. |
|---|---|---|---|
| 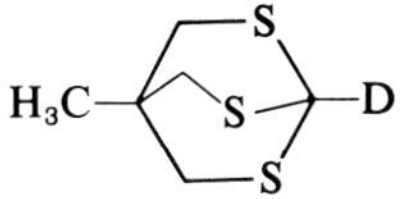 2-D-1-methyl-2,6,7-trithiabicyclo structure | 131° | — | 116–118 |
| 2-H-1-methyl-2,6,7-trithiabicyclo structure | 131° | — | 41 |
| 1,4,6,9-tetrathiaspiro structure | 130° | — | 38 |
| 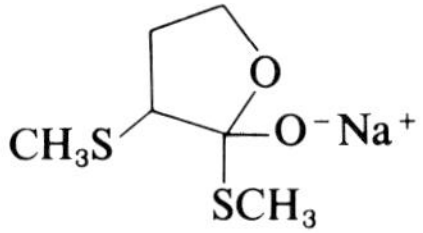 2-methylthio-2-O⁻Na⁺-tetrahydrofuran (and homologues) | — | — | 155c |
| 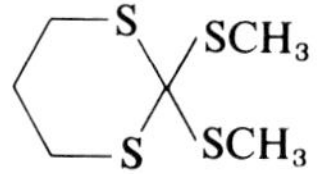 2,2-bis(methylthio)-tetrahydropyran | 32.7° | — | 147 |
| $CH_3SO_2CH(SO_2C_2H_5)_2$ | 225° (105°/11) | 1.5543 (15°) | 16, 18 |
| $CH_3SCH(SC_2H_5)_2$ | $C_7$ compounds (111°/0.7) | 1.4528 (25°) | 16, 20 |
| $C_3F_7-C(SCH_3)_3$ | 171° | 1.483 (25°) | 25 |
| $HC(SCH_2CO_2H)_3$ | >300° | — | 74, 77 |
| methyl-trithiaadamantane structure | — | — | 121a |

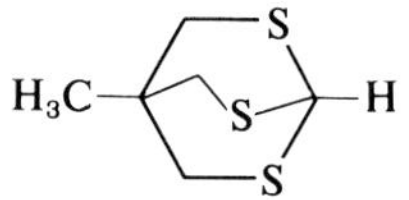

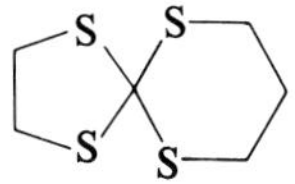

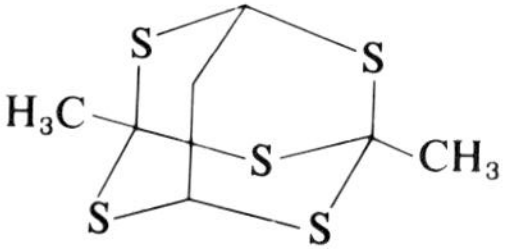

TABLE 1–*continued*

| Compound | M.p., °C (b.p., °C/torr) | $n_D\,(T,\,°C)$ | $d\,(T,\,°C)$ | References |
|---|---|---|---|---|
| 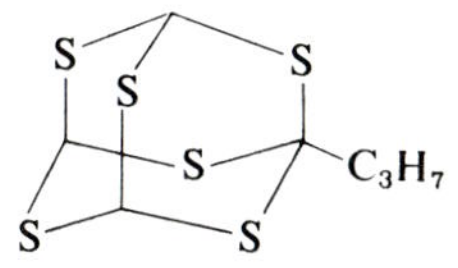 (structure with C₃H₇ and S atoms) | dec. | — | — | *120* |
| 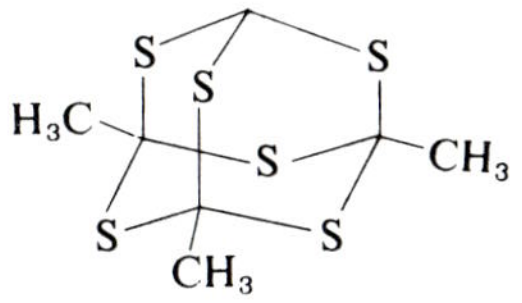 (structure with CH₃ groups and S atoms) | dec. | — | — | *119* |
| 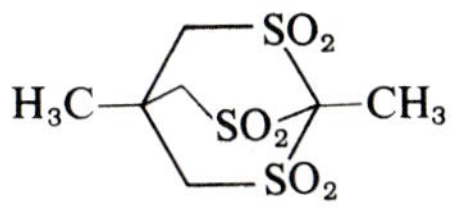 (structure with CH₃, SO₂ groups) | 370° | — | — | *41* |
| 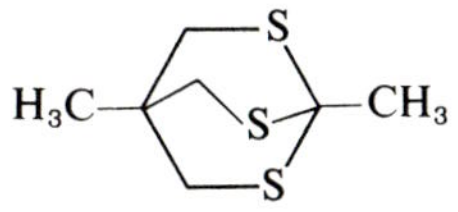 (structure with CH₃ groups) | 135° | — | — | *41* |
| 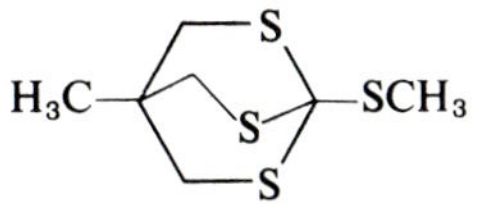 (structure with SCH₃ group) | 124° | — | — | *147* |

| Compound | mp/bp | n | d | Refs |
|---|---|---|---|---|
| 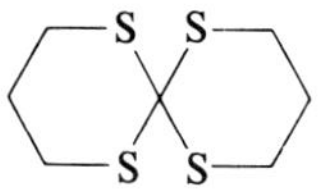 [1,4,5,8-tetrathia-spiro structure] | 120° | — | — | 38, 89 |
| BrC(SO₃C₂H₅)₃ | 60°–65° | — | — | 137 |
| C₂H₅SO₂CHClC(SCH₃)₃ | 73° | — | — | 17 |
| C₂H₅SCH(OC₂H₅)₂ | (78°/17) | 1.4397 (20°) | 0.945 (20°) | 127 |
| CH₃SO₂CH(SO₂C₂H₅)SO₂C₃H₇ | 178° | — | — | 18 |
| HC(SO₂C₂H₅)₃ | 216° | — | — | 18, 34, 131 |
| HC(SO₃C₂H₅)₃ | 75° | — | — | 136 |
| DC(SC₂H₅)₃ | (136°/16) | — | — | 116–118 |
| HC(SC₂H₅)₃ | (127°/12) | 1.5416 (20°) | 1.056 (20°) | 11, 26, 57, 59, 62, 75, 77, 78, 104a, 125, 149, 150, 157 |

C₈ compounds

| Compound | mp/bp | | | Refs |
|---|---|---|---|---|
| 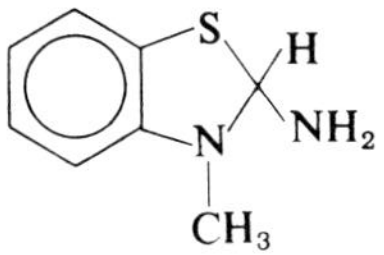 [benzothiazoline with H, NH₂, CH₃] | 100°–110° (dec.) | — | — | 155b |
| [thiazoline with SCH₃, N₃] | — | — | — | 47a |
| [morpholino-dithiolane with CN] | — | — | — | 107 |
| Cl₃CCOS–C(CCl₃)(SCOCCl₃)–CCl₃ | — | — | — | 122 |

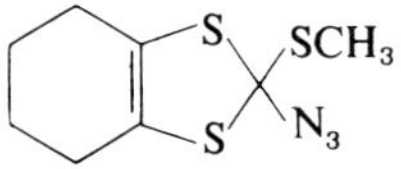

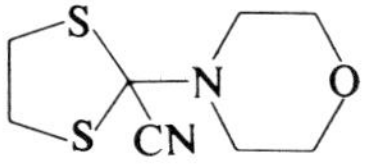

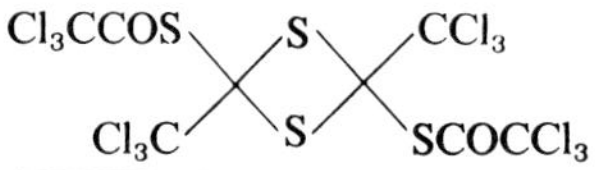

TABLE 1–*continued*

| Compound | M.p., °C (b.p., °C/torr) | $n_D$ (T, °C) | d (T, °C) | References |
|---|---|---|---|---|
| 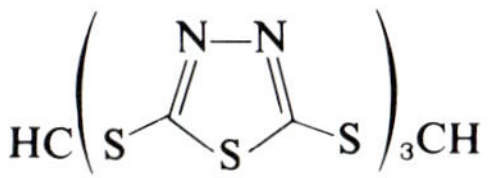 HC(S—N=N—S)$_3$CH | 177° | — | — | *52* |
| 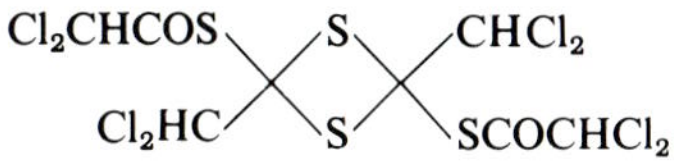 Cl$_2$CHCOS-S-C(CHCl$_2$)(SCOCHCl$_2$) | 168° | — | — | *122* |
| 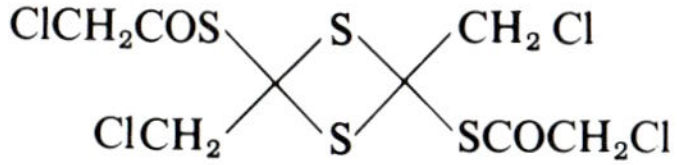 ClCH$_2$COS-S-C(CH$_2$Cl)(SCOCH$_2$Cl) | — | — | — | *122* |
| 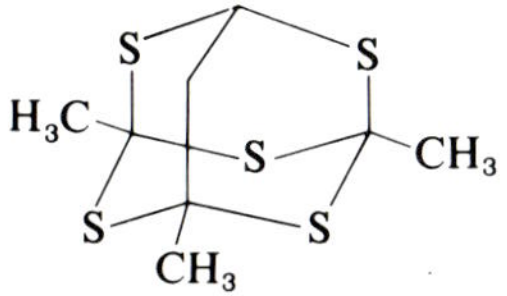 (CH$_3$ adamantane-like cage with S and CH$_3$ groups) | 244°–245° | — | — | *121a* |
| 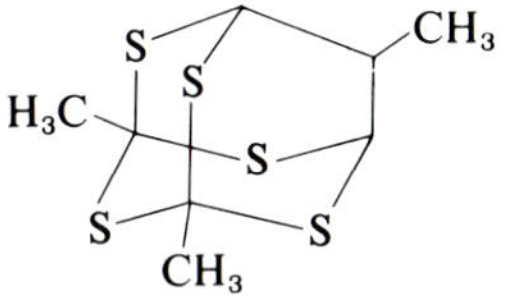 (CH$_3$ adamantane-like cage with S and CH$_3$ groups) | 227°–229° | — | — | *121a* |

| Structure | mp | — | Ref. |
|---|---|---|---|
| 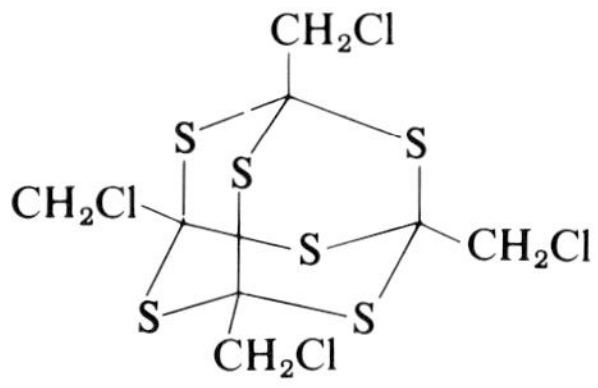 (CH₂Cl-substituted tetrathiaadamantane) | 290° (dec.) | — | 122 |
| 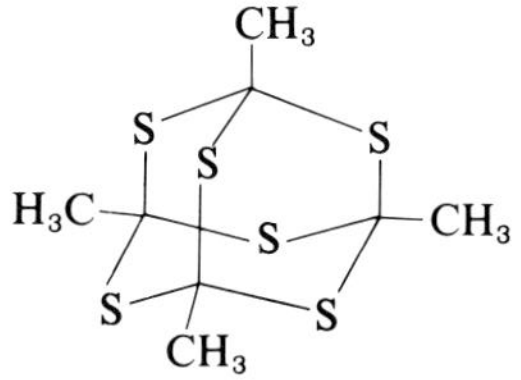 (CH₃-substituted tetrathiaadamantane) | 225° | — | 21, 55, 108, 119 |
| (C₂H₅-substituted tetrathiaadamantane) | 250° | — | 120 |
| (CH(CH₃)₂ bicyclic thiohydantoin) | 218° | — | 123 |
| (bis-dithiolane) | 105° | — | 83 |

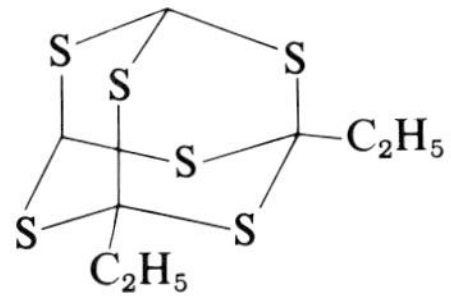

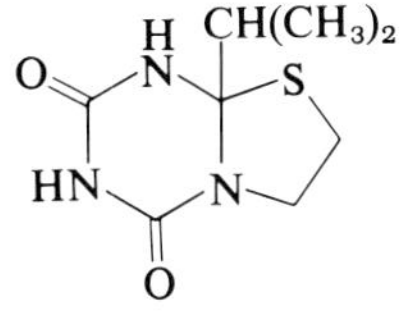

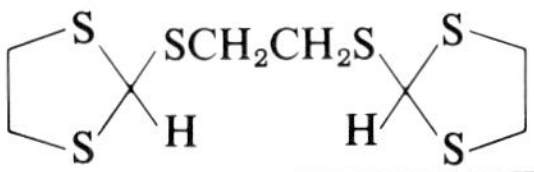

TABLE 1–continued

| Compound | M.p., °C (b.p., °C/torr) | $n_D$ (T, °C) | d (T, °C) | References |
|---|---|---|---|---|
| 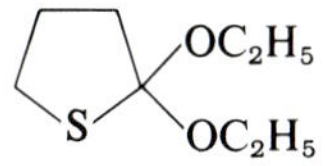 ![structure: 2,2-diethoxy thiolane] | (96°/14) | 1.4765 (20°) | 1.037 (20°) | 109 |
| 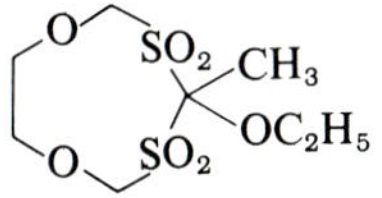 ![structure: cyclic disulfone with CH3, OC2H5] | — | — | — | 1 |
| CH₃C(SC₂H₅)₂OC₂H₅ | (102°/10) | 1.4948 (20°) | — | 148a |
| CH₃C(SC₂H₅)₃ | (127°/13) | 1.5370 (20°) | — | 93, 129, 155 |
| C₉ compounds | | | | |
| 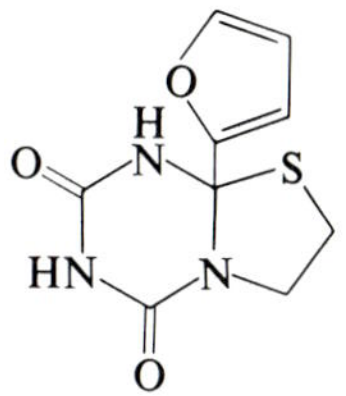 ![structure: furyl fused bicyclic urea] | 197° | — | — | 123 |
| 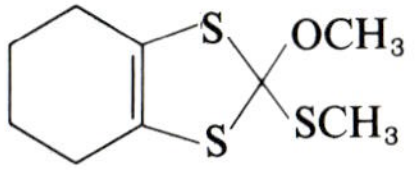 ![structure: benzodithiole with OCH3 and SCH3] | 67° | — | — | 48 |

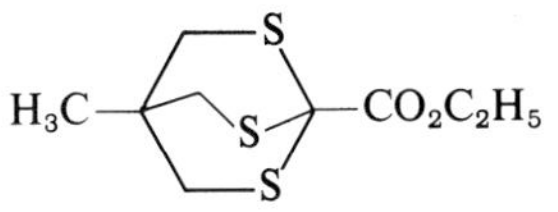

| Structure | Temp | — | — | Ref |
|---|---|---|---|---|
| (2-CO₂C₂H₅, 4-CH₃ trithioorthoester bicyclic) | 105° | — | — | 147 |
| (dimethyl-substituted trithio bicyclic, 4-CH₃) | 264°–266° | — | — | 121a |
| (trimethyl-substituted trithio bicyclic, 4-CH₃) | 185° | — | — | 121a |
| (cyclohexene bis-SCH₃) | 38° | — | — | 48 |
| (C₅H₁₁ trithio bicyclic) | 300° | — | — | 122 |
| (2-CH(CH₃)₂, 4-CH₃ trithio bicyclic) | 66° | — | — | 147 |

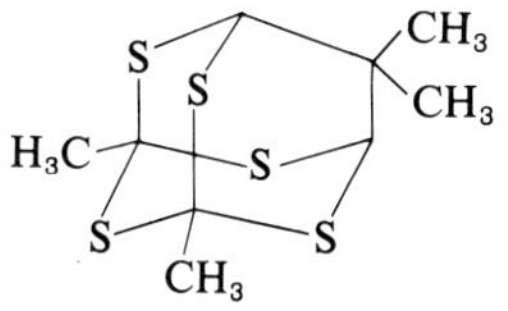

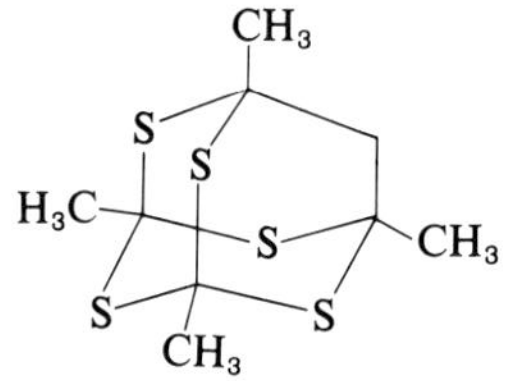

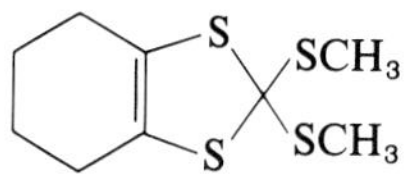

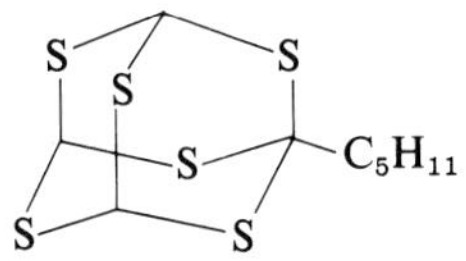

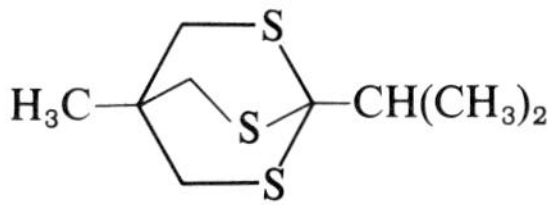

TABLE 1—continued

| Compound | M.p., °C (b.p., °C/torr) | $n_D$ (T, °C) | d (T, °C) | References |
|---|---|---|---|---|
| $CH_2=CHC(SC_2H_5)_3$ | (149°/10) | — | — | 93 |
| $(C_2H_5S)_3CCH=NNHCONH_2$ | 195° | — | — | 158 |
| $(C_2H_5S)_3CCH=NNHCSNH_2$ | 141° | — | — | 158 |
| $C_2H_5C(SC_2H_5)_3$ | (139°/17) | 1.5364 (20°) | — | 57, 93, 129 |
| $C(SC_2H_5)_4$ | 35.5° | — | — | 12, 57, 129a |

$C_{10}$ compounds

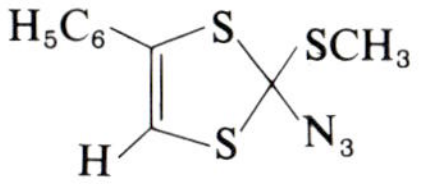

| Compound | M.p., °C (b.p., °C/torr) | $n_D$ (T, °C) | d (T, °C) | References |
|---|---|---|---|---|
| $H_5C_6\diagup S\diagdown C(SCH_3)(N_3)\diagup S\diagdown_H$ | — | — | — | 47a |
| 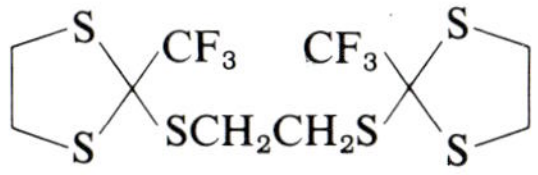 $\diagup S\diagdown C(CF_3)(SCH_2CH_2S)\diagup S\diagdown$ (with CF_3) | 85° | — | — | 15a, 30 |
| $(C_6H_5SO_2)(C_2H_5SO_2)CHSCH_3$ | 98° | — | — | 60 |
| $(CH_3SO_2)(C_2H_5SO_2)CHSC_6H_5$ | 126° | — | — | 60 |
| $(C_6H_5SO_2)(C_2H_5SO_2)CHSO_2CH_3$ | 226° | — | — | 18, 60 |
| $C_6H_5CH_2SCH(SCH_3)_2$ | — | 1.6119 (20°) | — | 20 |
| $C_6H_5N(CH_3)CH(SCH_3)_2$ | 47°–50° | — | — | 29a |
| $C_3F_7C(SC_2H_5)_3$ | (100°/0.7) | 1.4530 (25°) | 1.346 (25°) | 25 |
| 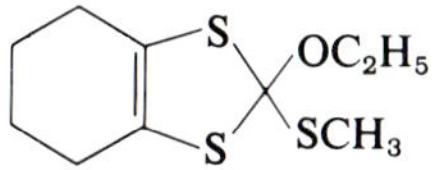 $\diagup S\diagdown C(OC_2H_5)(SCH_3)$ (cyclohexene-fused) | 48° | — | — | 48 |
| $HC(SCH_2CH=CH_2)_3$ | (122°/19) | — | — | 78 |

| Structure | bp/mp | n | Ref. |
|---|---|---|---|
| 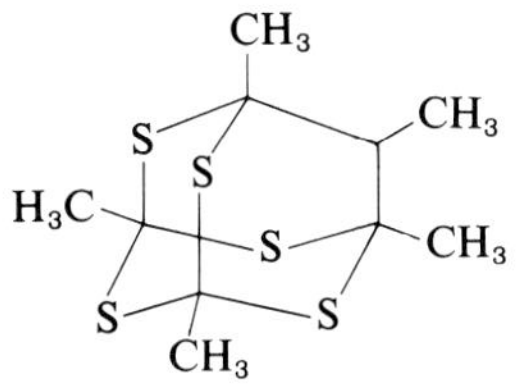 (adamantane-like cage with CH₃ groups) | 92°–94° | — | 121a |
| (adamantane-like cage with C₂H₅ groups) | 210° | — | 120 |
| (adamantane-like cage with C₃H₇ groups) | 182° | — | 120 |
| CH₂=CHCH₂C(SC₂H₅)₃ | (90°/0.45) | 1.5440 (20°) | 157 |
| BrC(SO₂C₃H₇)₃ | 143° | — | 132 |
| ClC(SO₂C₃H₇)₃ | 124° | — | 132 |
| HC(SO₂C₃H₇)₃ | 237° | — | 132 |
| HC(SC₃H₇)₃ | (160°/12) | 1.5248 (15°) | 11, 125, 132 |
| (CH₃)₃CHC(SC₂H₅)₃ | (148°/17) | 1.5376 (20°) | 57, 138a |

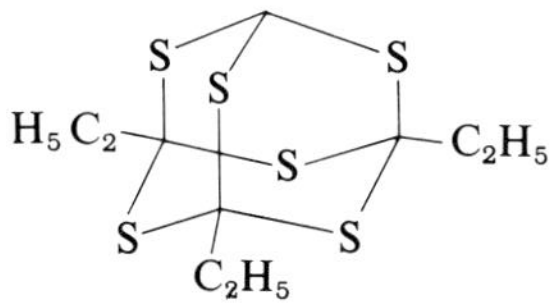

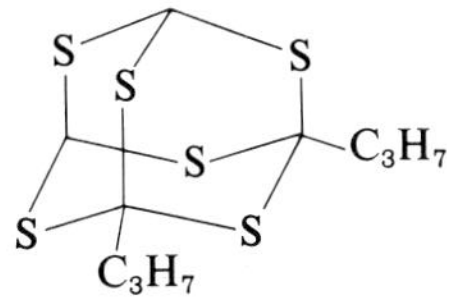

C₁₁ compounds

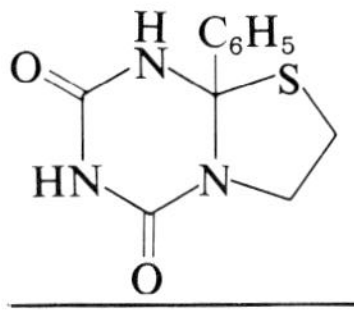

| Structure | bp/mp | n | Ref. |
|---|---|---|---|
| (bicyclic structure with C₆H₅, NH, N, O) | 206° | — | 123 |

386  6. THIOORTHOCARBOXYLATES AND THIOORTHOCARBONATES

TABLE 1–continued

| Compound | M.p., °C (b.p., °C/torr) | $n_D$ (T, °C) | d (T, °C) | References |
|---|---|---|---|---|
| 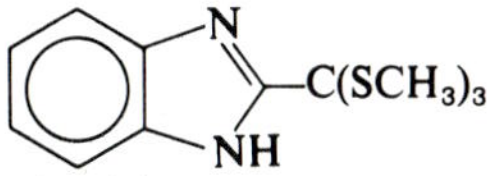 benzimidazole-C(SCH₃)₃ (with NH) | 270° (dec.) | — | — | 46 |
| C₆H₅SO₂CHClC(SCH₃)₃ | 62° | — | — | 17 |
| C₆H₅SO₂CH(SO₂C₂H₅)SC₂H₅ | 108° | — | — | 34 |
| C₆H₅SCH(SO₂C₂H₅)₂ | 86° | — | — | 58, 60, 103 |
| C₆H₅SO₂CH(SO₂C₂H₅)₂ | 165°, 179° | — | — | 18, 58, 60, 103 |
| 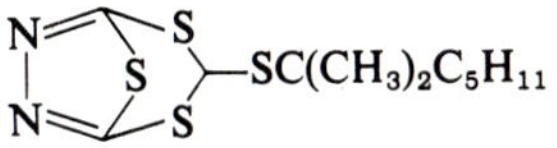 thiazoline–SC(CH₃)₂C₅H₁₁ | 75° | — | — | 52 |
| 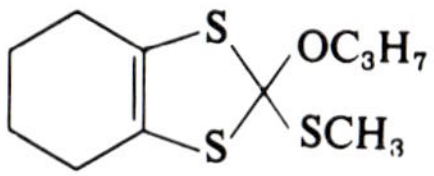 cyclohexene-fused dithiolane with OC₃H₇, SCH₃ | 64° | — | — | 48 |
| 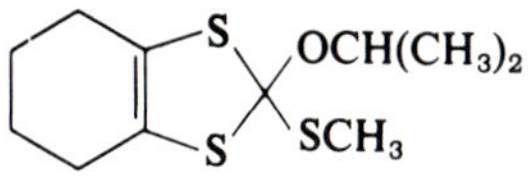 cyclohexene-fused dithiolane with OCH(CH₃)₂, SCH₃ | 65° | — | — | 48 |
| 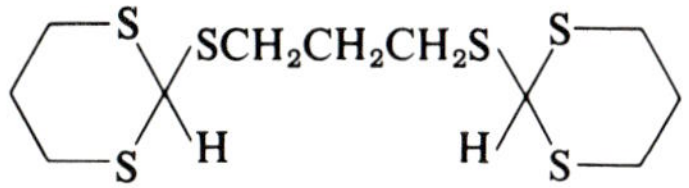 1,3-dithiane with SCH₂CH₂CH₂S, H | 83° | — | — | 4, 147 |
| C₂H₅OCH[SC(CH₃)₃]₂ | (120°/12) | 1.4875 (15°) | — | 11 |

# THIO-o-CARBOXYLATES AND RELATED COMPOUNDS

| Compound | b.p./m.p. | $n_D$ | $d$ | Ref. |
|---|---|---|---|---|
| $C_4H_9C(SC_2H_5)_3$ | (160°/20) | 1.5275 (20°) | — | 57 |
| $(CH_3)_2CHCH_2C(SC_2H_5)_3$ | (164°/17) | 1.5369 (20°) | — | 57 |
| $(CH_3)_2NCH[SC(CH_3)_3]_2$ | (104°/6) | 1.4940 (20°) | — | 49 |
| $(CH_3)_2NCH(SC_4H_9)_2$ | (135°/5) | 1.4979 (20°) | 0.9490 (20°) | 49 |
| $(CH_3)_2NCH[SCH_2CH(CH_3)_2]_2$ | (123°/5) | 1.4960 (20°) | 0.9399 (20°) | 49 |

$C_{12}$ compounds

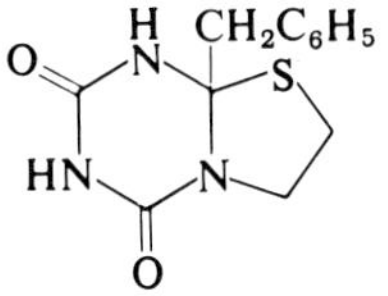

| Compound | m.p./b.p. | Ref. |
|---|---|---|
| [structure: bicyclic with H, CH₂C₆H₅, S, N, HN, O] | 222° | 123 |
| 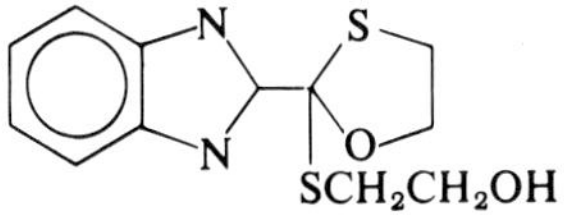 [structure: benzothiazoline with SCH₂CH₂OH] | 252° (dec.) | 47 |
| $ClCH_2CH_2COS-C(CH_2CH_2Cl)(CH_2CH_2Cl)-SCOCH_2CH_2Cl$ / $ClCH_2CH_2$ | — | 122 |
| 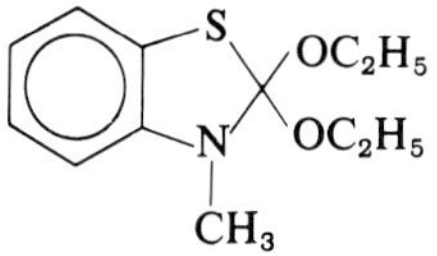 [benzothiazoline with $OC_2H_5$, $OC_2H_5$, $CH_3$] | (126°/1), 28° | 110 |
| [benzothiazoline with H, piperidino N] | 64° | 155a |

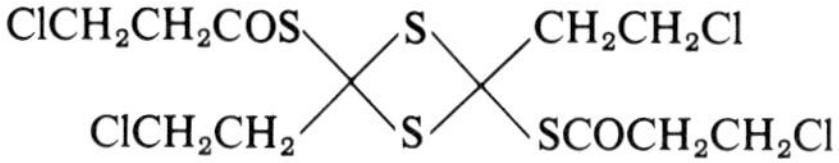

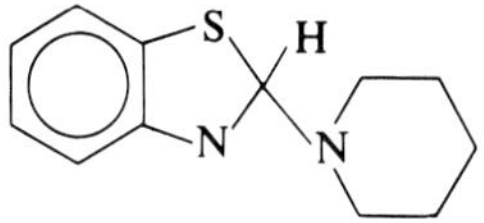

## TABLE 1 – continued

| Compound | M.p., °C (b.p., °C/torr) | $n_D$ (T, °C) | d (T, °C) | References |
|---|---|---|---|---|
| CH$_3$C(SO$_2$C$_2$H$_5$)$_2$SC$_6$H$_5$ | 113° | — | — | 103 |
| CH$_3$C(SO$_2$C$_2$H$_5$)$_2$SO$_2$C$_6$H$_5$ | 109° | — | — | 18, 103 |
| C$_6$H$_5$C[N(CH$_3$)$_2$](SCH$_3$)SC$_2$H$_5$ | (97°/0.08) | — | — | 115 |
| 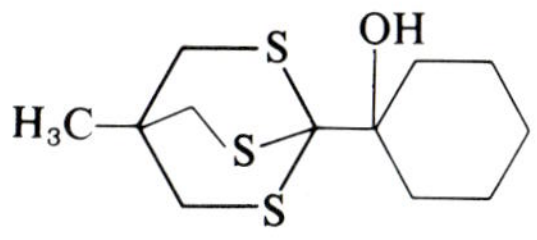 [cyclohexanol-substituted trithiane structure] | 175° | — | — | 147 |
| **C$_{13}$ compounds** | | | | |
| 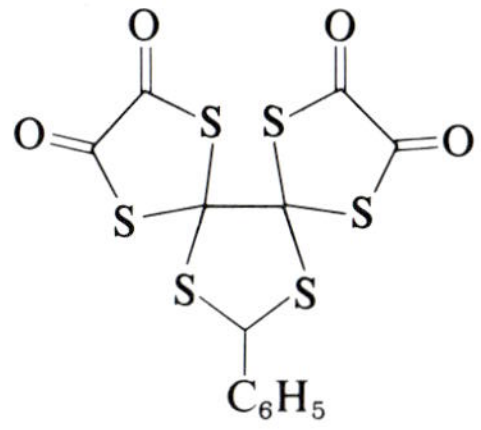 [dispiro tetrathio bis(dione) with C$_6$H$_5$ structure] | 161° (dec.) | — | — | 142 |
| 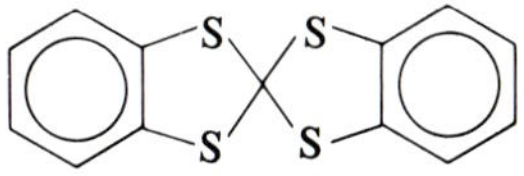 [bis(benzodithiole) spiro structure] | — | — | — | 81 |
| 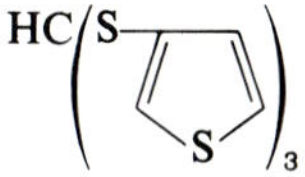 HC(S–thienyl–S)$_3$ structure | — | — | — | 24 |

| | | | | | | |
|---|---|---|---|---|---|---|
| 175° | | 71° | 54°–55° | 143° | 152° | |
| — | — | — | — | — | — | |
| — | — | — | — | — | — | |
| *147* | *48* | *111* | *15a* | *111* | *106* | |

390   6. THIOORTHOCARBOXYLATES AND THIOORTHOCARBONATES

TABLE 1–*continued*

| Compound | M.p., °C (b.p., °C/torr) | $n_D$ (T, °C) | d (T, °C) | References |
|---|---|---|---|---|
| $C_6H_5C(SC_2H_5)_3$ | (178°/12) | 1.5895 (20°) | — | 57, 93, 129 |
| $C_6H_5C(SC_2H_5)_2N(CH_3)_2$ | (109°/0.2) | 1.5732 (16°) | — | 114 |
| 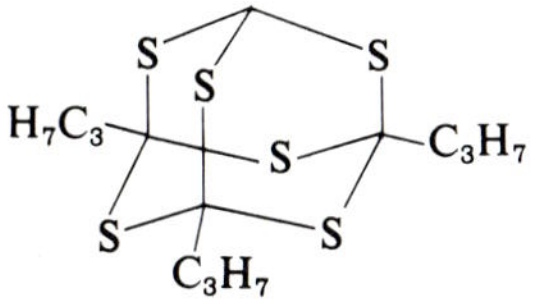 (bicyclic structure with S, C₃H₇ groups) | 199° | — | — | 120 |
| $BrC(SO_2C_4H_9)_3$ | 84° | — | — | 132 |
| $ClC(SO_2C_4H_9)_3$ | 124° | — | — | 132 |
| $(C_2H_5O)_2CHCH_2C(SC_2H_5)_3$ | (118°/0.2) | 1.5128 (15°) | — | 15 |
| $HC(SO_2C_4H_9)_3$ | 230° | — | — | 11, 49, 132 |
| $HC[SO_2CH_2CH(CH_3)_2]_3$ | — | — | — | 49 |
| $HC(SC_4H_9)_3$ | (148°/1) | 1.5180 (20°) | 0.9831 (20°) | 49 |
| $HC[SCH_2CH(CH_3)_2]_3$ | (135°/1) | 1.5102 (20°) | 0.9701 (20°) | 49 |
| $HC[SC(CH_3)_3]_3$ | (117°/4) | — | — | 11 |
|  | 64.5° |  |  |  |
| $HC(SC_4H_9)_3$ | (189°/12) | 1.5158 (15°) | — | 11 |
| $ClSCH(CH_3)_2]_4$ | 61.4° | — | — | 12, 129a |
| $C(SC_3H_7)_4$ | (123°/0.16) | — | — | 129a |

$C_{14}$ compounds

| 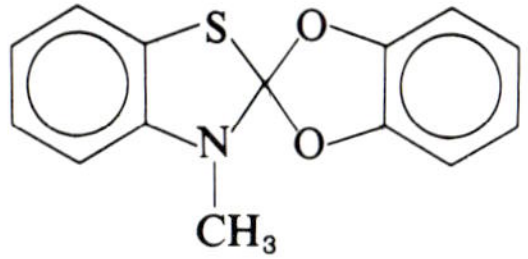 (spiro benzoxazole structure with N–CH₃) | 117°–119° | — | — | 127a |

# THIO-*o*-CARBOXYLATES AND RELATED COMPOUNDS 391

| Compound | mp/bp | n | Ref. |
|---|---|---|---|
| (*p*-BrC$_6$H$_4$SO$_2$)$_2$CHSO$_2$CH$_3$ | 268° | — | 85 |
| 3,5-Cl$_2$C$_6$H$_3$SCH(SO$_2$CH$_3$)SO$_2$C$_6$H$_5$ | 147° | — | 60 |
| (*m*-ClC$_6$H$_4$SO$_2$)$_2$CHSO$_2$CH$_3$ | 238° | — | 85 |
| (*p*-ClC$_6$H$_4$SO$_2$)$_2$CHSO$_2$CH$_3$ | 263° | — | 85 |
| (*p*-FC$_6$H$_4$SO$_2$)$_2$CHSO$_2$CH$_3$ | 204° | — | 85 |
| α-C$_{10}$H$_7$—C(=N—O)(CH$_3$S)(SCH$_3$)... (oxathiazoline with CH$_3$) | 86° | — | 80 |
| (C$_6$H$_5$SO$_2$)$_2$CHSO$_2$CH$_3$ | 231° | — | 18, 85 |
| [benzodithiole, 4-ClC$_6$H$_4$S, SCH$_3$] | 88° | — | 48 |
| [benzodithiole, 4-O$_2$NC$_6$H$_4$S, SCH$_3$] | 93° | — | 48 |
| [benzodithiole, SC$_6$H$_5$, SCH$_3$] | 63° | — | 48 |
| benzimidazole-2-C(SCH$_2$CH$_2$OH)$_3$ | 160° | — | 47 |
| C$_6$H$_5$CH$_2$C(SC$_2$H$_5$)$_3$ | (174°/1.5) | 1.5835 (20°) | 129 |

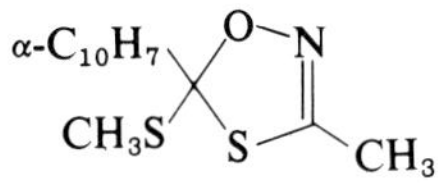

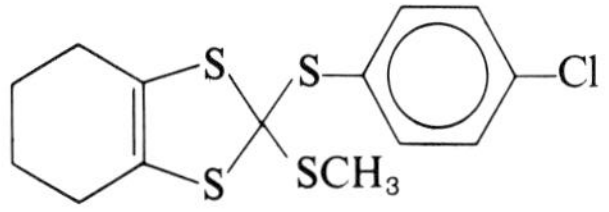

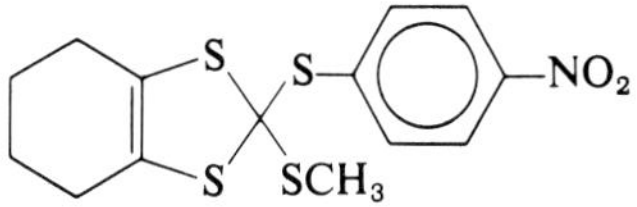

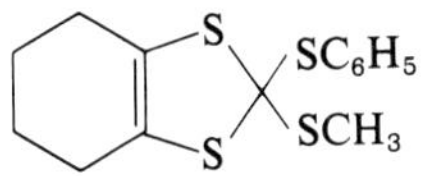

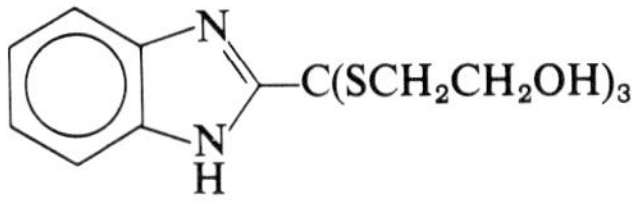

TABLE 1–*continued*

| Compound | M.p., °C (b.p., °C/torr) | $n_D$ (T, °C) | d (T, °C) | References |
|---|---|---|---|---|
| 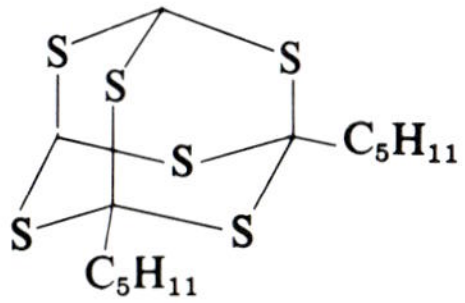 [structure with $C_5H_{11}$ groups] | 218° | — | — | 122 |
| $C_7H_{15}C(SC_2H_5)_3$ | (155°/1) | 1.5168 (20°) | — | 57, 129 |
| $(C_2H_5S)_3CC(SC_2H_5)_3$ | — | — | — | 29 |
| $(C_2H_5S)_3CSSC(SC_2H_5)_3$ | — | 1.594 (25°) | — | 14 |
| **$C_{15}$ compounds** | | | | |
| 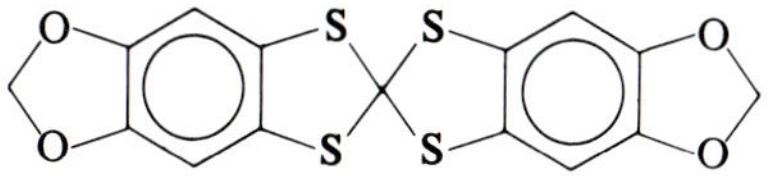 [fused bicyclic dithiolane structure] | 237°–238° | — | — | 37a |
| 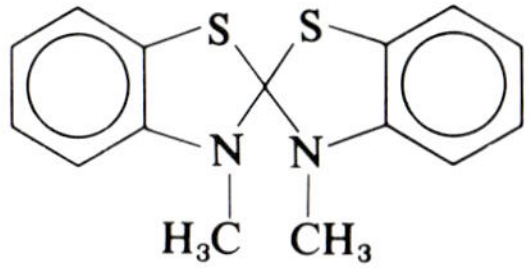 [benzothiazoline spiro structure with $H_3C$, $CH_3$, N] | 204° | — | — | 112a, 127a, 155a |
| 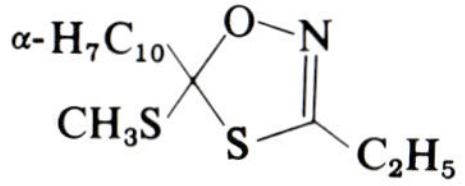 [isoxazoline structure with $\alpha$-$H_7C_{10}$, $CH_3S$, $C_2H_5$] | 68° | — | — | 80 |
| $(C_6H_5SO_2)(p\text{-}CH_3C_6H_4SO_2)CHSCH_3$ | 105° | — | — | 60 |

| | | | |
|---|---|---|---|
| $(p\text{-}CH_3C_6H_4S)(C_6H_5SO_2)CHSO_2CH_3$ | 169° | — | — | 60 |
| $(p\text{-}CH_3C_6H_4SO_2)(C_6H_5SO_2)CHSO_2CH_3$ | 174° | — | — | 60 |
| $(C_2H_5S)_3CCH_2C(SC_2H_5)_3$ | (150°/5) | — | — | 93 |

$C_{16}$ compounds

| Structure | mp | | | Ref |
|---|---|---|---|---|
| 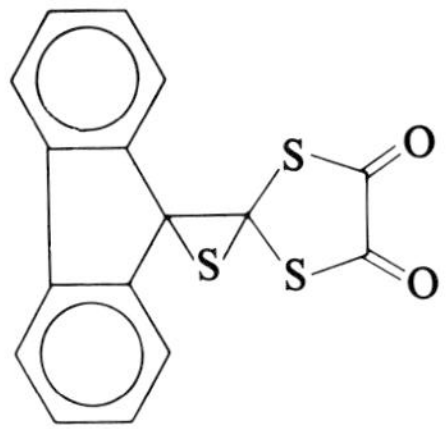 (fluorene-dithiolane-dione) | dec. 100° | — | — | 141 |
| 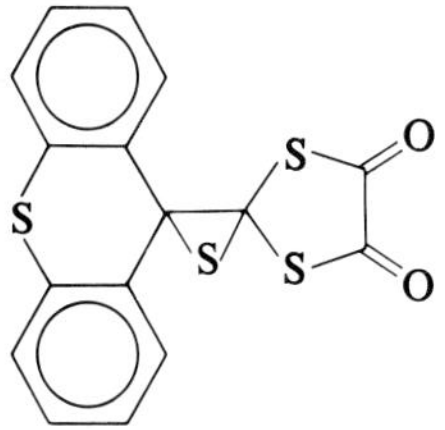 (thioxanthene-dithiolane-dione) | dec. 180° | — | — | 141 |
| 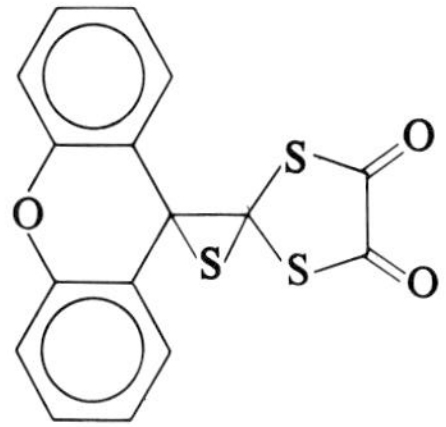 (xanthene-dithiolane-dione) | dec. 140° | — | — | 141 |
| $[CHF_2(CF_2)_3CH_2SO_2]_3CH$ | — | — | — | 138 |

TABLE 1–continued

| Compound | M.p., °C (b.p., °C/torr) | $n_D (T, °C)$ | $d (T, °C)$ | References |
|---|---|---|---|---|
| 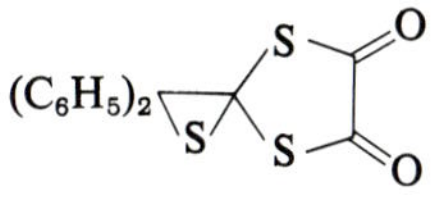 (C₆H₅)₂C(S)₂C(=O)C(=O) [dithiolane dione] | dec. | — | — | 141 |
| 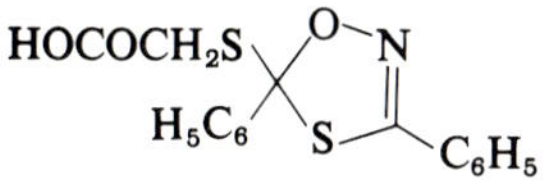 HOCOCH₂S–C(C₆H₅)(N=)–S–C₆H₅ [isoxazoline] | 116° | — | — | 80 |
| 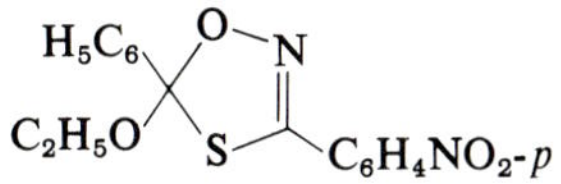 C₂H₅O–C(C₆H₅)(N=)–S–C₆H₄NO₂-p [isoxazoline] | 97° | — | — | 80 |
| 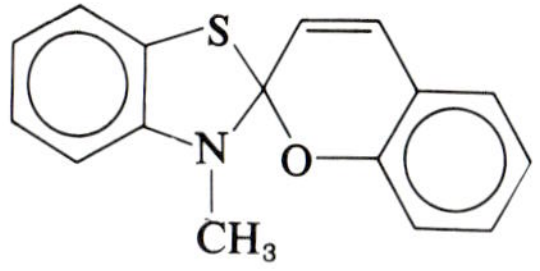 spiro benzothiazoline-chromene (and derivatives) | — | — | — | 63a |
| 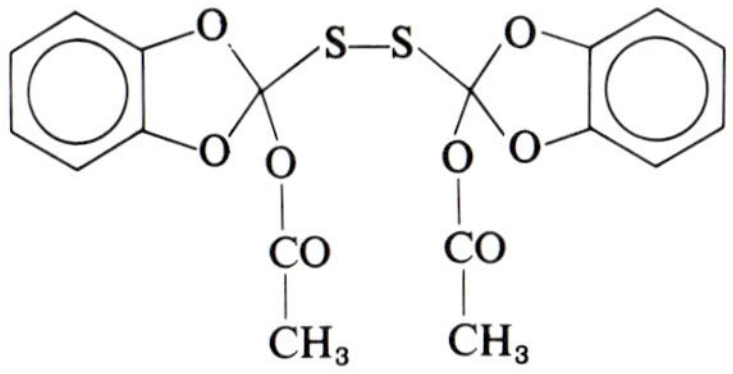 bis(benzodioxole)–S–S– with CO–CH₃ groups | 156° | — | — | 1a |

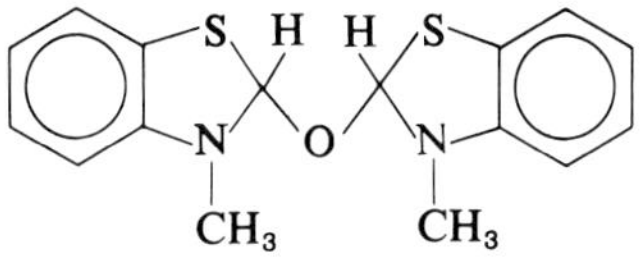

| Structure | mp | | | Ref |
|---|---|---|---|---|
| | 128° | — | — | 155a |
| | 223°–224° | — | — | 127a |
| | 95° (dec.) | — | — | 155b |
| | 78° | — | — | 112 |
| | 130° | — | — | 84 |
| (p-CH₃C₆H₄SO₂)₂CHSO₂CH₃ | 201° | — | — | 85 |
| (m-CH₃C₆H₄SO₂)₂CHSO₂CH₃ | 180° | — | — | 85 |

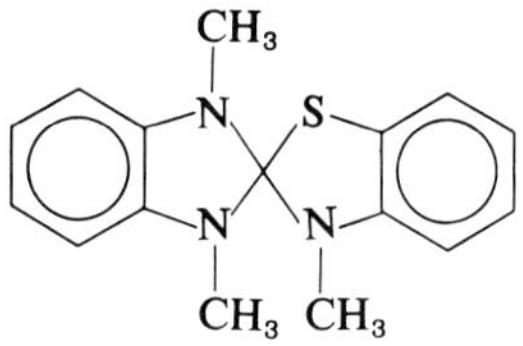

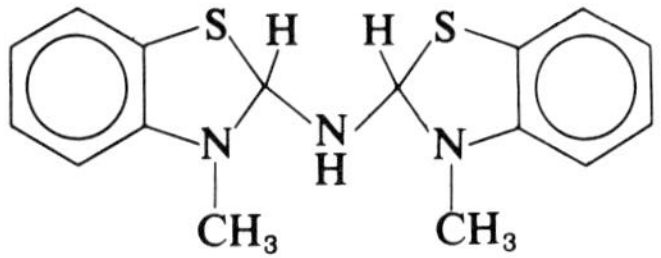

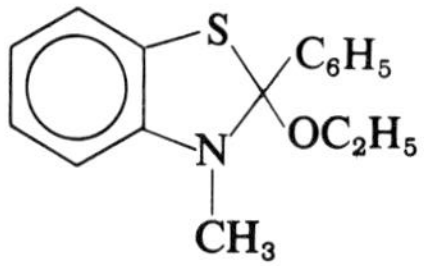

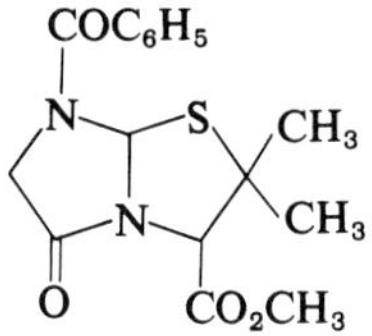

TABLE 1–continued

| Compound | M.p., °C (b.p., °C/torr) | $n_D$ (T, °C) | $d$ (T, °C) | References |
|---|---|---|---|---|
| [structure: HC(S–pyrimidine with NH₂, N, CH₃)₃] | 218° | — | — | 82 |
| 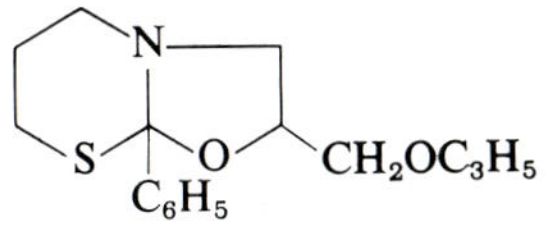 [structure: bicyclic N,S,O ring with CH₂OC₆H₅ and C₆H₅ substituents] | (134°/0.02) | 1.5786 (20°) | — | 50a |
| 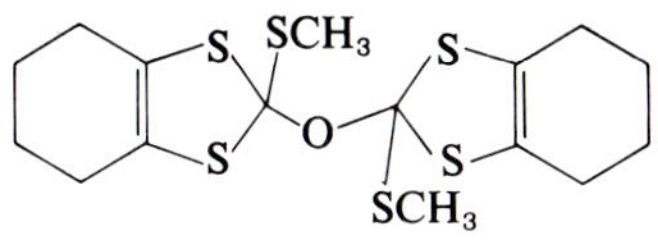 [structure: bis(tetrahydrobenzothiolane) with SCH₃ groups and central O] | 98° | — | — | 48 |
| HC(SO₂C₅H₁₁)₃ | 208° | — | — | 2 |
| HC[SO₂CH₂CH₂CH(CH₃)₂]₃ | 275° | — | — | 2 |

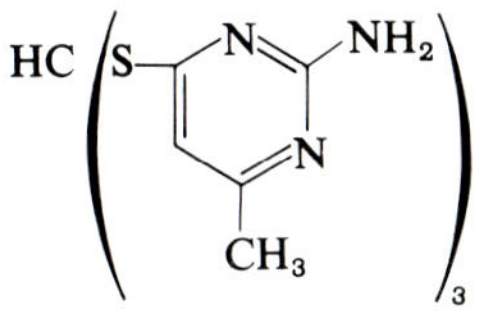

C₁₇ compounds

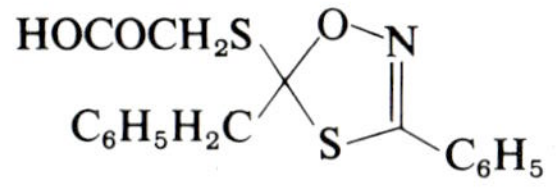

| Compound | M.p., °C (b.p., °C/torr) | $n_D$ (T, °C) | $d$ (T, °C) | References |
|---|---|---|---|---|
| [structure: HOCOCH₂S–C(C₆H₅CH₂)–O–N=C(C₆H₅)–S ring] | 96° | — | — | 80 |

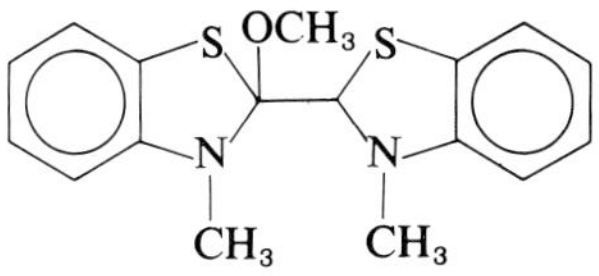

| Structure | mp | — | Ref |
|---|---|---|---|
| (benzothiazoline-OCH₃ structure) | 170° | — | 111 |
| (benzothiazoline-C₂H₅ structure) | 136° | — | 112a |
| (COCH₂C₆H₅ / CH₃ / CO₂CH₃ pyrrolidinone) | 150° | — | 84 |
| 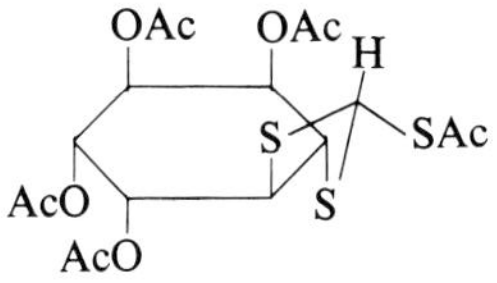 (sugar acetate with SAc) | 132° | — | 106 |
| $C_{18}$ compounds |  |  |  |
| 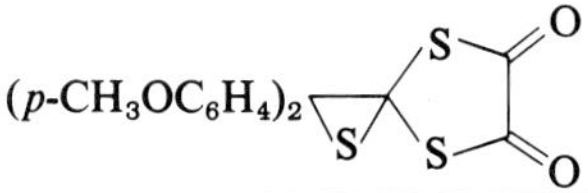 $(p\text{-}CH_3OC_6H_4)_2$ (dithiolane dione) | dec. 90° | — | 141 |

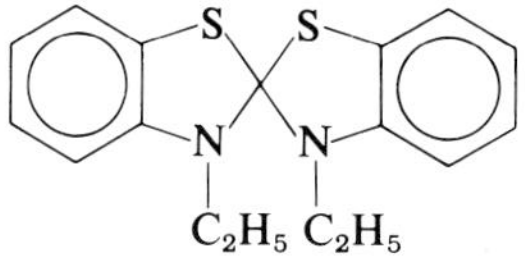

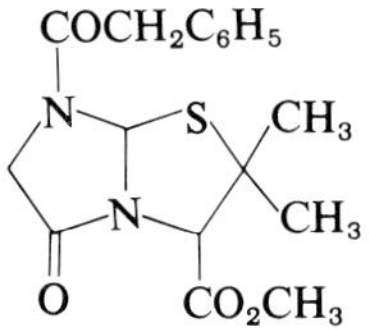

TABLE 1 –continued

| Compound | M.p., °C (b.p., °C/torr) | $n_D(T, °C)$ | $d(T, °C)$ | References |
|---|---|---|---|---|
| [structure with OC₂H₅, S, N–CH₃, benzothiazole groups, CH₃] | 136° | — | — | 155b |
| CH₃C(SCH₂C₆H₅)₂SC₂H₅ | (215°/15) | — | — | 93 |
| (C₂H₅)₃C(CH₂)₄C(SC₂H₅)₃ | (260°/12) | — | — | 93 |
| C₁₉ compounds | | | | |
| (p-BrC₆H₄S)₃CH | 124° | — | — | 86 |
| (p-BrC₆H₄SO₂)₃CH | 257° | — | — | 86 |
| (p-ClC₆H₄SO)₂CHSO₂C₆H₄Cl-p | 210° | — | — | 86 |
| (p-ClC₆H₄SO₂)₃CH | 246° | — | — | 86 |
| (p-ClC₆H₄S)₃CH | 112° | — | — | 7, 82, 86 |
| (p-FC₆H₄SO)₂CHSO₂C₆H₄F-p | 161° | — | — | 86 |
| (p-FC₆H₄SO₂)₃CH | 246° | — | — | 86 |
| (p-FC₆H₄S)₃CH | 40° | — | — | 86 |
| (p-O₂NC₆H₄S)₃CH | 176° | — | — | 86 |
| [structure: α-H₇C₁₀, CH₃S, O–N, S, C₆H₄NO₂-p] | 144° | — | — | 80 |
| [structure: benzodioxole with SC₆H₅, SC₆H₅] | 99° | — | — | 63 |

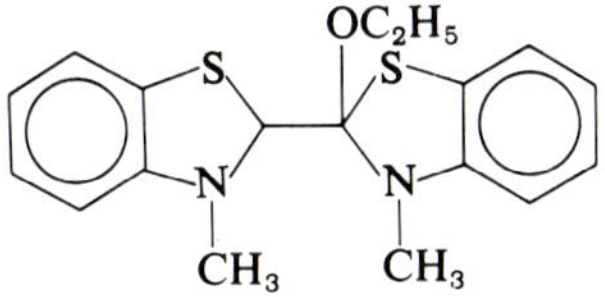

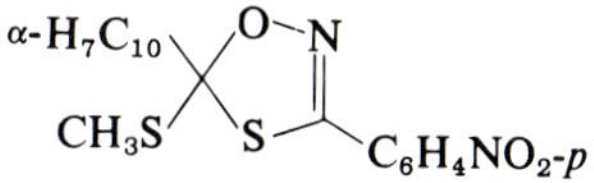

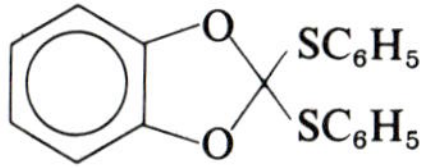

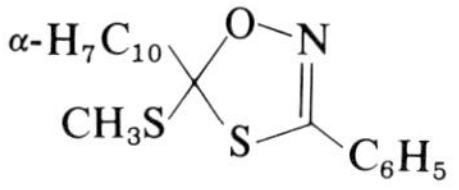

## THIO-o-CARBOXYLATES AND RELATED COMPOUNDS

| Compound | | mp | | Ref. |
|---|---|---|---|---|
| α-H$_7$C$_{10}$—C(CH$_3$S)—O—N=C(C$_6$H$_5$)—S (ring) | | 108° | — | — | 80 |
| (p-ClC$_6$H$_4$S)$_2$CHSC$_6$H$_4$CH$_3$-m | | 97° | — | — | 7 |
| C$_6$H$_5$SCH(OC$_6$H$_5$)$_2$ | | 50° | — | — | 19 |
| C$_6$H$_5$SO$_2$CH(SOC$_6$H$_5$)$_2$ | | 176° | — | — | 86, 101 |
| HC(SO$_2$C$_6$H$_5$)$_3$ | | 215° | — | — | 86, 101, 102 |
| HC(SC$_6$H$_5$)$_3$ | | 40° | — | — | 57, 59, 68, 75, 86, 100, 101, 105, 145, 147, 157 |
| H$_3$C—C(bicyclic with 3 S)—C(OH)(C$_6$H$_5$)$_2$ | | 173° | — | — | 147 |
| C$_4$H$_9$SCH(SCH$_2$C$_6$H$_5$)$_2$ | | — | — | — | 140 |
| N,S,O-heterocycle with CH$_2$OC$_6$H$_5$ and C$_6$H$_5$ | | 133°–134° | — | — | 50a |

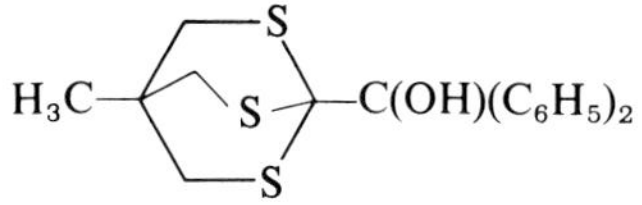

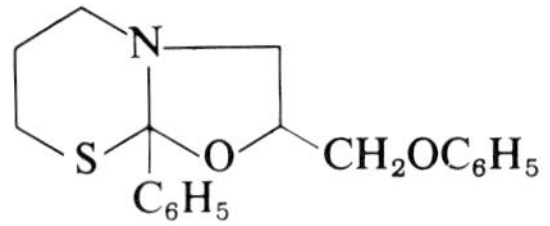

### C$_{20}$ compounds

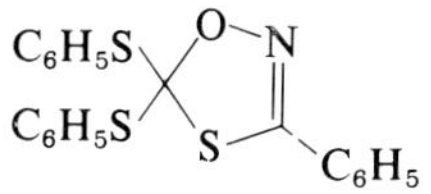

| Compound | | mp | | Ref. |
|---|---|---|---|---|
| C$_6$H$_5$S—C(C$_6$H$_5$)=N—O—C(C$_6$H$_5$S)— (ring) | | 83° | — | — | 80 |
| C$_6$H$_5$O—C(C$_6$H$_5$)=N—O—C(C$_6$H$_5$O)(S)— (ring) | | 79° | — | — | 80 |

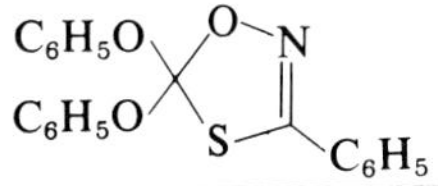

TABLE 1—continued

| Compound | M.p., °C (b.p., °C/torr) | $n_D (T, °C)$ | $d (T, °C)$ | References |
|---|---|---|---|---|
| (3,4-Cl$_2$C$_6$H$_3$S)(p-CH$_3$C$_6$H$_4$SO$_2$)CHSO$_2$C$_6$H$_5$ | 145° | — | — | *60* |
| (C$_6$H$_5$S)$_3$CCO$_2$H | 100° | — | — | *157* |
| 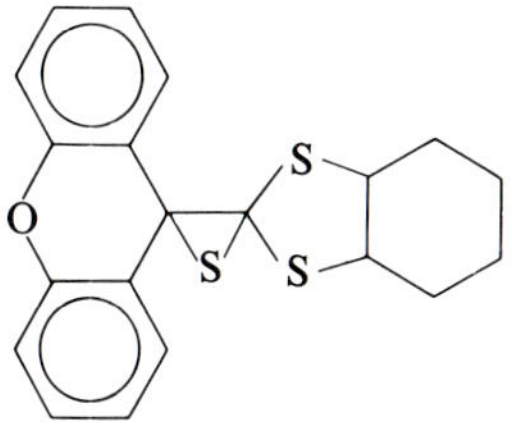 [cyclohexane-fused dithiolane spiro xanthene structure] | dec. 140° | — | — | *141* |
| CH$_3$C(SO$_2$C$_6$H$_5$)$_2$SC$_6$H$_5$ | 194° | — | — | *101* |
| CH$_3$C(SO$_2$C$_6$H$_5$)$_3$ | 182° | — | — | *102* |
| CH$_3$C(SC$_6$H$_5$)$_3$ | 146° | — | — | *147, 153, 157* |
| CH$_3$SC(SC$_6$H$_5$)$_3$ | 96° | — | — | *147* |
| (C$_3$H$_7$S)$_3$CSSC(SC$_3$H$_7$)$_3$ | — | 1.552 (25°) | — | *14* |
| (cyclo-C$_6$H$_{11}$SO$_2$)$_2$CHSO$_2$CH$_2$C$_6$F$_{13}$ | — | — | — | *138* |
| C$_{21}$ compounds | | | | |
| C$_6$H$_5$SCH(SCH$_2$C$_6$H$_5$)$_2$ | 162° | — | — | *140* |
| CF$_3$(CF$_2$)$_{16}$CH$_2$SO$_2$CH(SO$_2$CH$_3$)$_2$ | — | — | — | *138* |
| CF$_3$(CF$_2$)$_{10}$CH$_2$SO$_2$CH[SO$_2$CH$_2$(CF$_2$)$_2$CF$_2$H]$_2$ | — | — | — | *138* |
| C$_2$H$_5$C(SC$_6$H$_5$)$_3$ | 67° | — | — | *157* |
| (CH$_3$SO$_2$)$_2$CHSO$_2$C$_{18}$H$_{37}$ | — | — | — | *2* |
| C$_{22}$ compounds | | | | |
| HC[SO$_2$(CF$_2$)$_6$CF$_3$]$_3$ | — | — | — | *138* |
| HC[SO$_2$CH$_2$(CF$_2$)$_5$CF$_3$]$_3$ | — | — | — | *138* |

# THIO-o-CARBOXYLATES AND RELATED COMPOUNDS

| Compound | | | Ref. |
|---|---|---|---|
| HC[SCH$_2$(CF$_2$)$_5$CF$_3$]$_3$ | — | — | *138* |
| (p-CH$_3$C$_6$H$_4$SO$_2$)$_2$CHSO$_2$CH$_2$C$_6$F$_{13}$ | — | — | *138* |
| 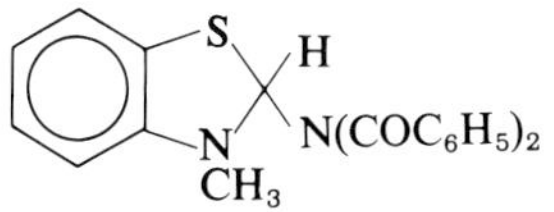 benzothiazoline with N(COC$_6$H$_5$)$_2$ and CH$_3$ substituents | 135° | — | *155b* |
| (C$_6$H$_5$S)$_3$CCO$_2$C$_2$H$_5$ | 86° | — | *147* |
| CH$_2$=CHCH$_2$C(SC$_6$H$_5$)$_3$ | 81° | — | *157* |
| (C$_6$H$_5$S)$_3$CCH$_2$CH(OH)CH$_3$ | 91° | — | *147* |
| p-CH$_3$OC$_6$H$_4$S)$_3$CH | 62° | — | *86* |
| (C$_6$H$_5$CH$_2$SO$_2$)$_2$CHSCH$_2$C$_6$H$_5$ | 214° | — | *103* |
| (p-CH$_3$C$_6$H$_4$SO)$_2$CHSO$_2$C$_6$H$_4$CH$_3$-p | 146° | — | *86* |
| (p-CH$_3$C$_6$H$_4$SO$_2$)$_3$CH | 226° | — | *86* |
| (p-CH$_3$OC$_6$H$_4$SO)$_2$CHSO$_2$C$_6$H$_4$OCH$_3$-p | 162° | — | *86* |
| (p-CH$_3$OC$_6$H$_4$SO$_2$)$_3$CH | 168° | — | *86* |
| C$_3$H$_7$-C(SC$_6$H$_5$)$_3$ | 73° | — | *157* |
| (m-CH$_3$C$_6$H$_4$S)$_2$C(CH$_3$)SC$_6$H$_5$ | 104° | — | *153* |
| (CH$_3$)$_2$CHC(SC$_6$H$_5$)$_3$ | 120° | — | *147* |
| HC(SCH$_2$C$_6$H$_5$)$_3$ | 103° | — | *11, 20, 39, 40, 61, 78, 96, 145, 147, 151* |
| (p-CH$_3$C$_6$H$_4$S)$_3$CH | 109° | — | *6, 70, 78, 82, 86, 100* |
| (C$_6$H$_5$S)$_3$CSi(CH$_3$)$_3$ | 102° | — | *147* |
| HC(SO$_2$C$_7$H$_{15}$)$_3$ | 171° | — | *2* |
| HC(SC$_7$H$_{15}$)$_3$ | — | — | *2* |
| C$_{23}$ compounds | | | |
| (C$_6$H$_5$S)$_3$CCH(OH)C$_3$H$_7$ | 63° | — | *147* |
| C$_4$H$_9$C(SC$_6$H$_5$)$_3$ | 77° | — | *157* |
| (m-CH$_3$C$_6$H$_4$S)$_3$CCH$_3$ | 118° | — | *153* |
| CH$_3$C(SCH$_2$C$_6$H$_5$)$_3$ | 67° | — | *153* |

TABLE 1—continued

| Compound | M.p., °C (b.p., °C/torr) | $n_D (T, °C)$ | $d(T, °C)$ | References |
|---|---|---|---|---|
| **C$_{25}$ compounds** | | | | |
| $(p\text{-BrC}_6\text{H}_4\text{S})_4\text{C}$ | 211° | — | — | 88 |
| $(p\text{-ClC}_6\text{H}_4\text{S})_4\text{C}$ | 219° | — | — | 7, 88, 129a |
| $(p\text{-FC}_6\text{H}_4\text{S})_4\text{C}$ | 169° | — | — | 88 |
| $(p\text{-ClC}_6\text{H}_4\text{S})_3\text{CSC}_6\text{H}_5$ | 191° | — | — | 7 |
| $\text{CH[SO}_2(\text{CH}_2)_2\text{OCH}_2(\text{CF}_2)_4\text{CF}_3]_3$ | — | — | — | 138 |
| $(\text{C}_6\text{H}_5\text{S})_2\text{C(SBr}_2\text{C}_6\text{H}_5)_2$ | 179° | — | — | 13 |
| $(\text{C}_6\text{H}_5\text{S})_2\text{C(SOC}_6\text{H}_5)_2$ | 42° | — | — | 13 |
| $\text{C}_6\text{H}_5\text{C(SC}_6\text{H}_5)_3$ | 87° | — | — | 153 |
| $\text{C(SC}_6\text{H}_5)_4$ | 159° | — | — | 7, 13, 88, 129a, 147, 157 |
| $[3,5\text{-}(\text{CH}_3\text{O})_2\text{C}_6\text{H}_3\text{S}]_3\text{CH}$ | 125° | — | — | 86 |
| $(\text{cyclo-C}_6\text{H}_{11}\text{S})_4\text{C}$ | 169° | — | — | 13 |
| $\text{HC[SO}_2(\text{CH}_2)_2\text{OC}_6\text{H}_{13}]_3$ | 97° | — | — | 2 |
| **C$_{26}$ compounds** | | | | |
| 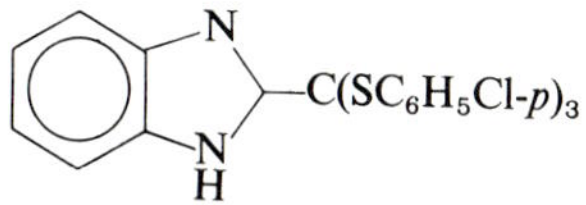 ![structure] benzimidazoline–C(SC$_6$H$_5$Cl-$p$)$_3$ | 193° | — | — | 46 |
| 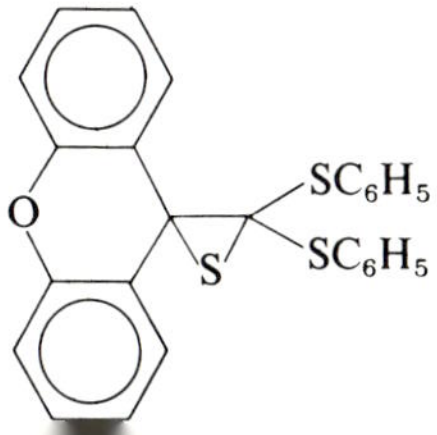 xanthene with SC$_6$H$_5$, SC$_6$H$_5$, S substituents | 114° | — | — | 141 |

| Compound | m.p. | | | Ref. |
|---|---|---|---|---|
| $(o\text{-}CH_3C_6H_4S)C(SC_6H_4Cl\text{-}p)_3$ | 193° | — | — | 7 |
| $C_6H_5CH(OH)C(SC_6H_5)_3$ | 141° | — | — | 147 |
| $(C_4H_9S)_3CSSC(SC_4H_9)_3$ | — | — | — | 14 |
| $[(CH_3)_3CS]_3CSSC[SC(CH_3)_3]_3$ | 60° | — | — | 14 |
| | $C_{27}$ compound | | | |
| 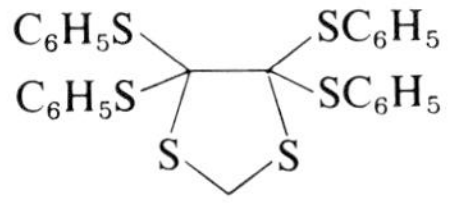 $C_6H_5S$    $SC_6H_5$<br>       \\__/<br>       /  \\<br>$C_6H_5S$    $SC_6H_5$<br>     S      S | dec. 140° | — | — | 143 |
| | $C_{29}$ compounds | | | |
| $C(SBr_2C_6H_4CH_3\text{-}p)_4$ | — | — | — | 6 |
| $(p\text{-}CH_3OC_6H_4S)_4C$ | 156° | — | — | 88 |
| $(p\text{-}CH_3C_6H_4S)_4C$ | 147° | — | — | 6, 88, 129a |
| | $C_{31}$ compound | | | |
| $(\alpha\text{-}C_{10}H_7S)_3CH$ | 134° | — | — | 78 |
| | $C_{37}$ compound | | | |
| $HC(SC_{12}H_{25})_3$ | 37° | — | — | 54 |
| | $C_{38}$ compounds | | | |
| $(o\text{-}ClC_6H_4S)_3CSSSC(SC_6H_4Cl\text{-}o)_3$ | 76° | — | — | 14 |
| $(p\text{-}ClC_6H_4S)_3CSSSC(SC_6H_4Cl\text{-}p)_3$ | 141° | — | — | 14 |
| $(C_6H_5S)_3CC(SC_6H_5)_3$ | 197° | — | — | 147 |
| $(C_6H_5S)_3CSSSC(SC_6H_5)_3$ | 99° | — | — | 14 |
| | $C_{41}$ compounds | | | |
| $(\beta\text{-}C_{10}H_7S)_2C(SBr_2C_{10}H_7\text{-}\beta)_2$ | 130° | — | — | 13 |
| $(\beta\text{-}C_{10}H_7S)_4C$ | 136° | — | — | 13, 129a |

TABLE 1—continued

| Compound | M.p., °C (b.p., °C/torr) | $n_D$ (T, °C) | d (T, °C) | References |
|---|---|---|---|---|
| $CH(SO_2CH_2C_6H_4CH_2CH_2C_5H_{11})_3$ | $C_{43}$ compound — | — | — | 138 |
| $(m\text{-}CH_3C_6H_4S)_3CSSSC(SC_6H_4CH_3\text{-}m)_3$ | $C_{44}$ compounds 82° | — | — | 14 |
| $(p\text{-}CH_3C_6H_4S)_3CSSSC(SC_6H_4CH_3\text{-}p)_3$ | 119° | — | — | 14 |
| $HC(SO_2C_{18}H_{37})_3$ | $C_{55}$ compound 135° | — | — | 2 |
| $[2,4,6\text{-}(CH_3)_3C_6H_2S]_3CSSSC[SC_6H_2(CH_3)_3\text{-}2,4,6]_3$ | $C_{56}$ compounds 170° | — | — | 14 |
| $[p\text{-}(CH_3)_3CC_6H_4S]_3CSSSC[SC_6H_4C(CH_3)_3\text{-}p]_3$ | 114° | — | — | 14 |

orthoformyl proton exchange about 1000 times faster than triethyl trithioorthoformate (*116, 117*) may be due in part to diminished steric hindrance in the bicyclic compound.

2. CONVERSION OF TETRATHIOORTHOCARBONATES TO TRITHIOORTHOFORMATES AND TRITHIOORTHOFORMATE DERIVATIVES TO DIALKYLSULFONYLMETHANES

Tetrathioorthocarbonates undergo electrophilic replacement reactions with organolithium compounds to form trithiomethylide derivatives which react with water to form trithioorthoformates (*147, 157*). These reactions are discussed in Section II, K, 3 (p. 370).

Methylsulfonylethylsulfonylpropylsulfonylmethane,

$$CH_3SO_2CH(SO_2C_2H_5)SO_2C_3H_7,$$

reacts with aqueous alkali to form ethylsulfonylpropylsulfonylmethane (*18*). Oxidation of trialkyltrithioorthoformates with aqueous permanganate yields dialkylsulfonylmethanes rather than trialkylsulfonylmethanes in most instances (*11, 75*).

$$HC(SR)_3 \xrightarrow[H_2O]{MnO_4^-} CH_2(SO_2R)_2 \qquad (71)$$

$$[R = CH_3, C_2H_5, C_3H_7, (CH_3)_3C]$$

These reactions probably involve nucleophilic attack by hydroxide or water on sulfonyl sulfur of trialkylsulfonylmethanes to form dialkylsulfonylmethylides, which yield the products on protonation. In a similar reaction, 2-[tris(*p*-chlorophenylthio)]methylbenzimidazole was converted to 2-[bis-(*p*-chlorophenylthio)methyl]benzimidazole by refluxing with ethanolic sodium ethoxide (*46*).

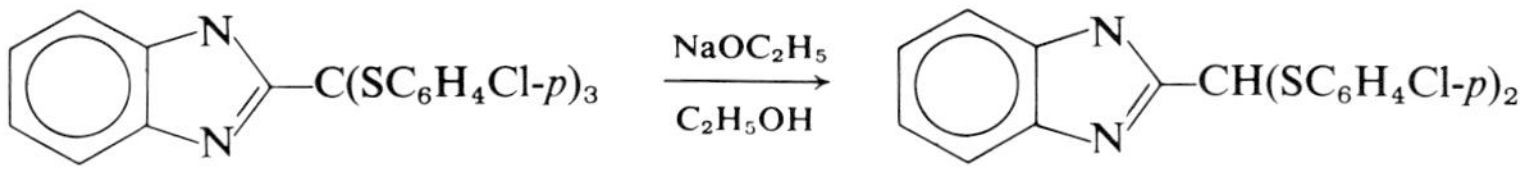

$$\text{benzimidazole-}C(SC_6H_4Cl\text{-}p)_3 \xrightarrow[C_2H_5OH]{NaOC_2H_5} \text{benzimidazole-}CH(SC_6H_4Cl\text{-}p)_2 \qquad (72)$$

## B. Reactions Which Form Carbon–Carbon Bonds

1. ALKYLATION OF TRITHIOMETHYLIDE SALTS

Carbanion salts derived from trithioorthoformates, tetrathioorthocarbonates, and related sulfoxides and sulfones react with alkyl halides, carbonyl compounds, and disulfides to form trithioorthocarboxylates and tetrathioorthocarbonates. These reactions are discussed in Section II, G (p. 357).

## 2. Reactions of Trithioorthocarboxylates with Substances Having Activated Methyl or Methylene Groups

Malonic ester and triethyl trithioorthoformate react in the presence of acetic anhydride and zinc chloride to form diethyl ethylthiomethylenemalonate in 50% yield (64).

$$CH_2(CO_2C_2H_5)_2 + HC(SC_2H_5)_3 \xrightarrow[ZnCl_2]{Ac_2O} C_2H_5SCH=C(CO_2C_2H_5)_2 \quad (73)$$

3-Coumaranone and triethyl trithioorthoformate undergo a similar reaction (146).

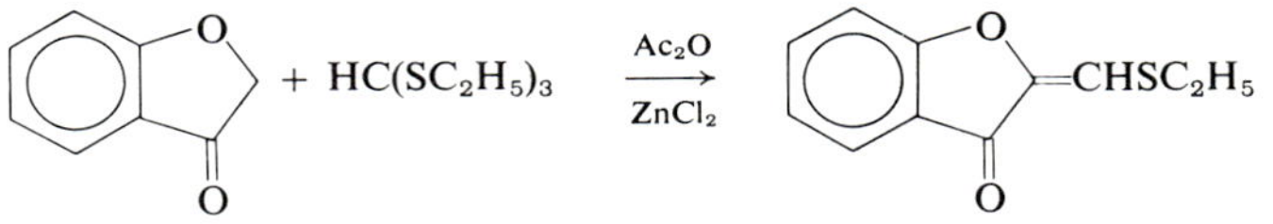

<chemical_structure> + HC(SC_2H_5)_3 →[Ac_2O, ZnCl_2] <chemical_structure>=CHSC_2H_5    (74)

Malononitrile reacts with trialkyl trithioorthoformates in acetic acid solution to form $RSCH=C(CN)_2$ (28).

A number of methyl-substituted heterocyclic quaternary salts react with triethyl trithioorthoformate to form ethylthiomethylene derivatives (**25**) which are useful as cyanine dye intermediates (42, 45, 51).

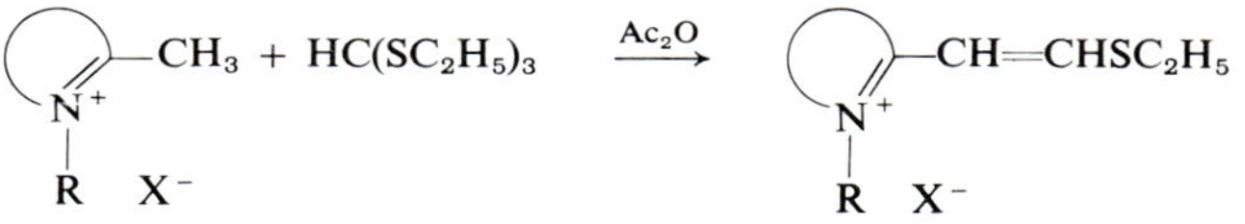

<chemical_structure>−CH_3 + HC(SC_2H_5)_3 →[Ac_2O] <chemical_structure>−CH=CHSC_2H_5    (75)

**25**

Ethylthiomethylene quaternary salts (**25**) were probably unisolated intermediates in syntheses of cyanine dye intermediates from ethyl-substituted quaternary salts, triethyl trithioorthoformate, and alkoxyethylidene or alkylthioethylidene derivatives of cyanoacetic ester and malononitrile.

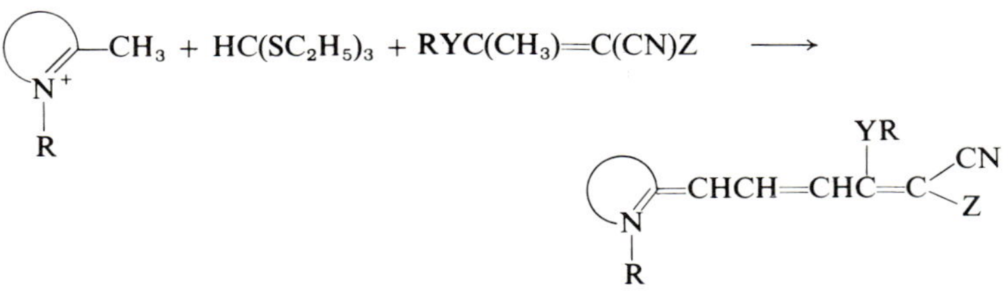

<chemical_structure>−CH_3 + HC(SC_2H_5)_3 + RYC(CH_3)=C(CN)Z ⟶

<chemical_structure>=CHCH=CHC(YR)=C(CN)(Z)    (76)

(Y = O, S; Z = CN, CO_2R)

2,3-Dimethylthiazolium (43, 98), 2-methyl-3-alkylbenzoxazolium (43, 98), 2-methyl-3-alkylbenzothiazolium (43, 44, 98), 2-methyl-3-alkylnaphthothiazolium, and 1,3,3-trimethylindolenium (*43*, *98*) salts were used in these reactions.

$N,N$-Dimethylbenzamide diethylthioacetal reacts with active methylene compounds to form dimethylaminobenzylidene compounds (*114*).

$$C_6H_5C(SC_2H_5)_2N(CH_3)_2 + H_2CXY \rightarrow C_6H_5C[N(CH_3)_2]\!=\!CXY \qquad (77)$$
$$(X = CN, Y = CN, CONH_2; X = H, Y = NO_2)$$

### 3. Addition of Trialkyl Trithioorthoformates to Carbon–Carbon Double Bonds

Triethyl trithioorthoformate undergoes boron trifluoride-catalyzed addition reactions with a variety of unsaturated substances such as methyl vinyl ether, isobutylene, styrene, indene, and cyclopentadiene [Eqs. (78)–(80)] (*31–33*).

$$RR'C\!=\!CH_2 + HC(SC_2H_5)_3 \xrightarrow{BF_3} RR'C(SC_2H_5)CH_2CH(SC_2H_5)_2 \qquad (78)$$
$$(R = H, R' = CH_3O, C_6H_5; R = R' = CH_3)$$

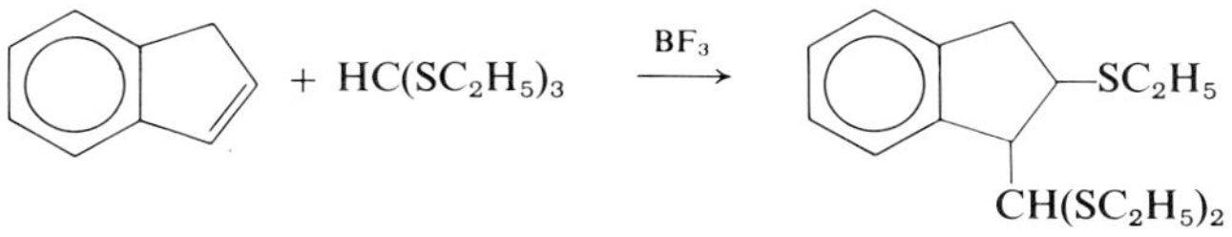

indene + HC(SC₂H₅)₃ →[BF₃] indane with SC₂H₅ and CH(SC₂H₅)₂ substituents (79)

cyclopentadiene + HC(SC₂H₅)₃ →[BF₃] cyclopentene with CH(SC₂H₅)₂ and SC₂H₅ substituents (80)

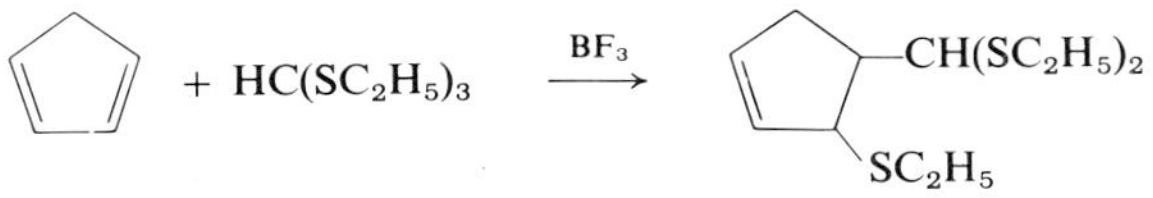

2-Ethylthio-1,3-dioxolane adds to methyl vinyl ether to form 2-(2'-ethylthio-2'-methoxyethyl)-1,3-dioxolane (*31*).

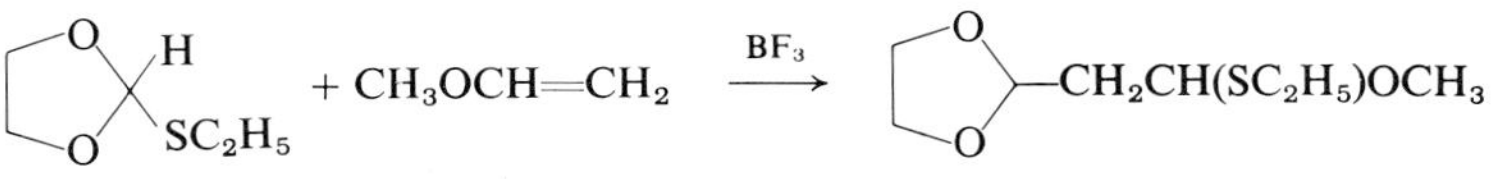

2-ethylthio-1,3-dioxolane + CH₃OCH=CH₂ →[BF₃] 1,3-dioxolane-CH₂CH(SC₂H₅)OCH₃ (81)

These reactions presumably involve electrophilic attack by diethylthiocarbonium ion [or, in the case of Eq. (81), dioxolenium ion] on the ethylenic compound, followed by reaction of the resulting carbonium ion with ethanethiol.

Trimethinecyanine dyes (**26**) react with trialkyl trithioorthoformates to form ethylthiomethylene derivatives (**27**) (*94*) which are hydrolyzed to formyl derivatives (*95*).

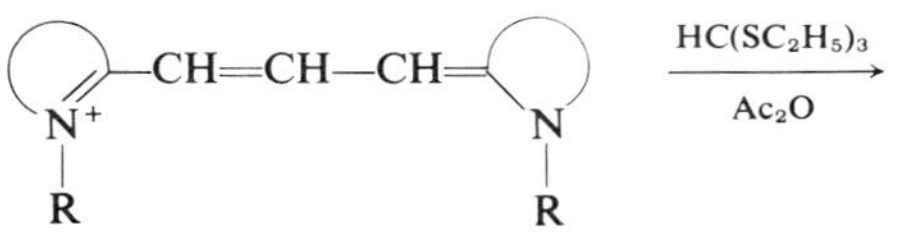

## 6. THIOORTHOCARBOXYLATES AND THIOORTHOCARBONATES

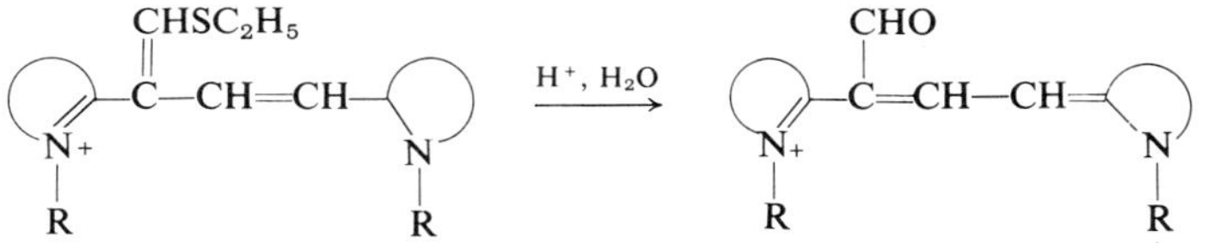

$$\text{26} \xrightarrow{\text{HC}(SC_2H_5)_3, \text{Ac}_2O} \text{27} \xrightarrow{H^+, H_2O} \quad (82)$$

Ethylthiomethylene derivatives (29) are probably transient intermediates in the conversion of trimethinecyanine dyes to trinuclear pentamethinecyanine dyes by reaction with quaternary salts and triethyltrithioorthoformate in the presence of acetic anhydride (72, 73, 97, 99).

$$26 + \text{HC}(SC_2H_5)_3 \xrightarrow{\text{Ac}_2O} 27 \longrightarrow \quad (83)$$

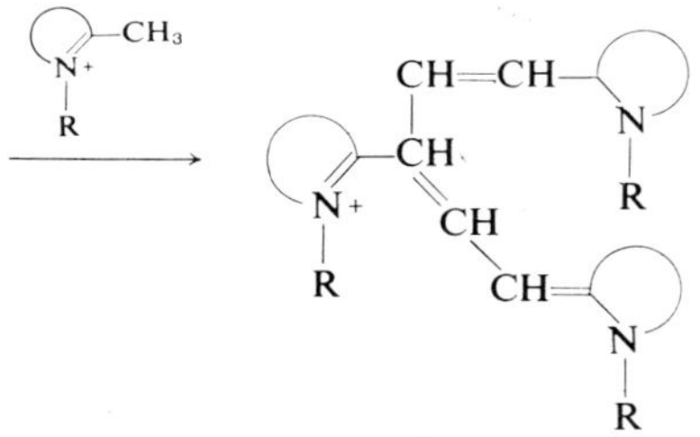

### 4. Reaction of Phenylacetylene with Triethyl Trithioorthoformate

Phenylacetylene undergoes a coupling reaction with triethyl trithioorthoformate in the presence of zinc chloride which yields phenylpropargyl aldehyde diethyl dithioacetal (79). This reaction is probably a generally applicable method of synthesizing substituted propargyl aldehyde dithioacetals.

$$C_6H_5C\equiv CH + HC(SC_2H_5)_3 \xrightarrow{ZnCl_2} C_6H_5C\equiv CCH(SC_6H_5)_2 + C_2H_5SH \quad (84)$$

### 5. Reactions of Trialkyl Trithioorthoformates with Diazoketones and Diazoesters

Trimethyl and triethyl trithioorthoformates in ether–boron trifluoride solutions react with ethyl diazoacetate and with diazoketones to form complex dithioacetals. Ethyl 2,3,3-trialkylthiopropionates are formed when ethyl diazoacetate is added to ether–boron trifluoride solutions of trimethyl or triethyl trithioorthoformate (*144, 145*).

$$HC(SR)_3 + N_2CHCO_2C_2H_5 \xrightarrow{BF_3} (RS)_2CHCH(SR)CO_2C_2H_5 + N_2 \quad (85)$$
$$(R = CH_3, C_2H_5)$$

Attempts to carry out this reaction using triphenyl trithioorthoformate, triethyl trithioorthoacetate, tribenzyl trithioorthoformate, and triethyl trithioorthobenzoate were unsuccessful (*145*).

Diazomethyl ketones undergo similar reactions with trimethyl and triethyl trithioorthoformates (*146*).

$$HC(SR)_3 + R'COCHN_2 \xrightarrow{BF_3} R'COCH(SR)CH(SR)_2 + N_2 \quad (86)$$

($R = CH_3$, $R' = p\text{-}CH_3OC_6H_4$, $p\text{-}O_2NC_6H_4$, $o\text{-}ClC_6H_4$, $CH_3$; $R = C_2H_5$, $R' = p\text{-}CH_3OC_6H_4$, $p\text{-}O_2NC_6H_4$)

Yields of trialkylthio ketones from aryl diazomethyl ketones were almost quantitative. *o*-Anisoyl diazomethyl ketone reacts with triethyl trithioorthoformate in the presence of boron trifluoride to form 2-(diethylthiomethyl)-3-coumaranone (*146*).

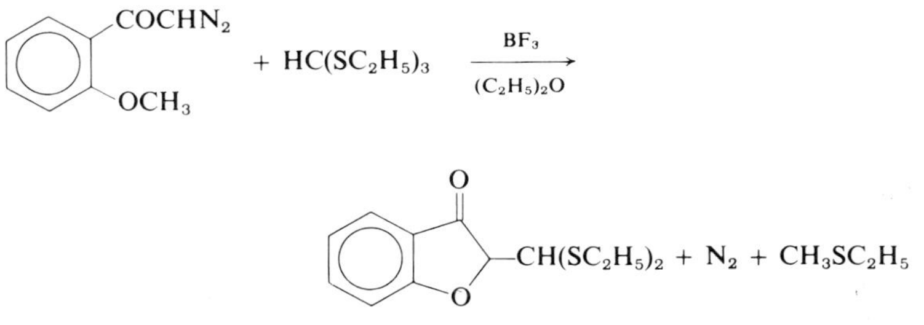

$$\underset{OCH_3}{\underset{|}{C_6H_4}}\text{-COCHN}_2 + HC(SC_2H_5)_3 \xrightarrow[(C_2H_5)_2O]{BF_3}$$

$$\text{(3-coumaranone)-CH(SC_2H_5)_2} + N_2 + CH_3SC_2H_5 \quad (87)$$

The fact that these reactions occur at room temperature in the absence of irradiation suggests that they involve attack by dialkylthiomethyl carbonium ions on the diazo compound to form a carbonium ion intermediate which reacts with the trifluoroalkylthioborate ion to form the products and regenerate the catalyst.

$$HC(SR)_3 + BF_3 \xrightarrow{(C_2H_5)_2O} H\overset{+}{C}(SR)_2 \cdot BF_3SR^-$$

$$H\overset{+}{C}(SR)_2 + R'COCHN_2 \longrightarrow R'CO\overset{+}{C}HCH(SR)_2 + N_2 \quad (88)$$

$$R'CO\overset{+}{C}HCH(SR)_2 + BF_3SR^- \longrightarrow R'COCH(SR)CH(SR)_2 + BF_3$$

## 6. Conversion of Trialkyl Trithioorthocarboxylates to Ketene Dithioacetals

Trialkyl trithioorthocarboxylates having hydrogen atoms *alpha* to the orthoacyl carbon atom undergo facile acid-catalyzed dealkanethiolation to form ketene dithioacetals (*57, 129, 155*).

$$RR'CHC(SC_2H_5)_3 \xrightarrow{H^+} RR'C=C(SC_2H_5)_2 + C_2H_5SH \qquad (89)$$
$$(R = H, R' = H, CH_3, C_3H_7, C_6H_{13}, C_6H_5; R = CH_3, R' = CH_3, C_2H_5)$$

The ease with which this reaction occurs is illustrated by the fact that distillation of triethyl trithioorthoacetate from a flask which was rinsed with hydrochloric acid and dried resulted in formation of ketene diethyl dithioacetal in 82% yield (*155*). When the trithioortho esters are distilled from catalysts such as potassium bisulfate, yields of ketene dithioketals ranging from about 60% to 90% are obtained.

Triethyl trithioorthoacetate reacts with acetyl chloride to form ketene diethyl dithioacetal and ethyl thiolacetate (*20*). Treatment of triethyl 3,3-diethoxytrithioorthopropionate with 2,4-dinitrophenylhydrazine hydrochloride yields $2,4-(O_2N)_2C_6H_3NHN=CHCH=C(SC_2H_5)_2$ (*15*).

Dethiolation of trithioorthocarboxylates forms the basis of a procedure for converting carboxylic acids to aldehydes having one less carbon atom (*129*).

$$RCH_2CO_2H \longrightarrow RCH_2COCl \xrightarrow{C_2H_5SH} RCH_2C(SC_2H_5)_3 \xrightarrow{H^+}$$
$$RCH=C(SC_2H_5)_2 \xrightarrow{[O]} RCH=C(SO_2C_2H_5)_2 \xrightarrow{H_2O}$$
$$RCHO + CH_2C(SO_2C_2H_5)_2 \qquad (90)$$

## 7. Reactions of Trithiomethylides and Dialkylthiocarbonium Salts

Alkali metal trithiomethylide salts, prepared either by reaction of tetrathioorthocarbonates with organolithium compounds or by reaction of trithioorthoformates with alkali metal amides in liquid ammonia, decompose on standing to form mixtures of products containing tetraalkylthioethylenes or tetraarylthioethylenes (*57, 69, 104, 148*).

$$2 (RS)_3C^-M^+ \rightarrow (RS)_2C=C(SR)_2 + 2 M^+SR^- \qquad (91)$$
$$(R = C_6H_5, M = Li, Na; R = C_2H_5, M = Na, K; R = CH_3, M = K)$$

These reactions probably involve decompositions of the trithiomethylides to dithiocarbenes, which couple to form the ethylenes. This hypothesis is supported by the fact that decompositions of dialkylthiocarbonate tosylhydrazones (*104, 140*) or dialkylthiodiazomethanes (*139*), which should yield dialkylthiocarbenes, also form product mixtures containing tetraalkylthioethylenes.

Diphenyldithiocarbene, formed from lithium triphenylthiomethylide, reacts with activated ethylenes to form diphenylthiocyclopropanes *(148)*.

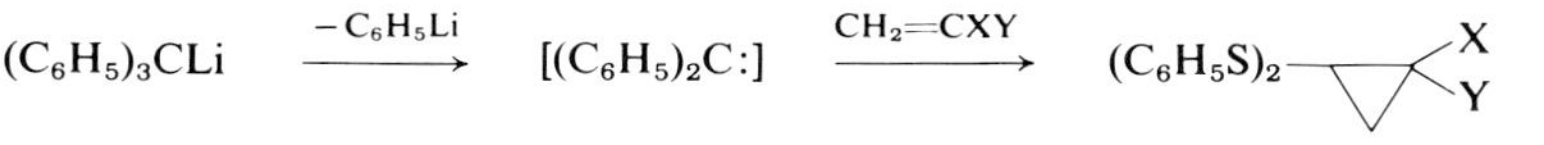

$$(C_6H_5)_3CLi \xrightarrow{-C_6H_5Li} [(C_6H_5)_2C:] \xrightarrow{CH_2=CXY} (C_6H_5S)_2\!\!\triangle\!\!{}^X_Y \quad (92)$$

| X | Y |
|---|---|
| [–S(CH$_2$)$_3$—S–] | |
| C$_6$H$_5$S | C$_6$H$_5$S |
| CH$_3$O | CH$_3$O |

Dimethylthiocarbonium tetrafluoroborate [(CH$_3$S)$_2$CH$^+$BF$_4^-$] is converted by tertiary amines to tetramethylthioethylene. This reaction also probably involves a carbene intermediate *(118a)*.

## C. Reactions Which Form Carbon–Oxygen Bonds

### 1. Hydrolysis of Trithioorthocarboxylates

In contrast to hydrolysis reactions of orthocarboxylate esters, which have been studied in considerable detail (see Chap. 2), only qualitative observations concerning hydrolysis reactions of trithioorthocarboxylates have been published. Trithioorthocarboxylates undergo hydrolysis to thiolcarboxylate esters, thiols, and carboxylic acids.

$$RC(SR')_3 + H_2O \xrightarrow{H^+} 2\ R'SH + RCOSHR' \xrightarrow[H_2O]{H^+} R'SH + RCO_2H \quad (93)$$

These reactions are catalyzed by acids *(39, 74, 75, 151, 155, 158)*, although no quantitative data on reactivity or catalytic effectiveness of hydrogen ion or other acids have been published.

Triethyl trithioorthoformate hydrolyzes rapidly in aqueous hydrochloric acid *(75)*, and tricarboxymethyl trithioorthoformate, HC(SCH$_2$CO$_2$H)$_3$, hydrolyzes when heated in acidic aqueous solution *(74)*. Triethyl trithioorthoacetate hydrolyzes on standing in dilute aqueous ethanolic sulfuric acid at room temperature *(155)*. Triphenyl trithioorthoformate is hydrolyzed by concentrated hydrochloric acid at 100°C *(59)*. Triethyl trithioorthoglyoxalate semicarbazone is hydrolyzed to glyoxalic acid semicarbazone by refluxing aqueous ethanolic hydrochloric acid *(158)*. Heating tribenzyl trithioorthoformate with concentrated hydrochloric acid in a sealed tube at 150°C converts it to formic acid and benzyl mercaptan *(39)*. A procedure for estimating trithioorthocarboxylates by hydrolyzing them in aqueous hydrochloric acid and determining the liberated thiol iodometrically has been described (*151*).

Electronegative substituents in either the acyl or thiol portions of trithioorthocarboxylates diminish acid-catalyzed hydrolytic reactivity. 2-Tris-(*p*-chlorophenylthio)methyl benzimidazole is unaffected by heating with concentrated hydrochloric acid (in which the imidazole ring is protonated) for 2 hours (*46*). Trialkylsulfonylmethanes can be recrystallized from aqueous acid (*8, 18*).

Triethyl trithioorthoformate (*75*), triethyl trithioorthoacetate (*129, 155*), and triphenyl trithioorthoacetate (*153*) are completely inert in alkaline solution. The fact that triethyl trithioorthoacetate is unaffected by boiling with 2.5 N sodium hydroxide for 1 hour (*129*) suggests that its hydrolysis in acid solution is specific hydrogen ion-catalyzed. If the reaction were general acid-catalyzed, water at 100°C should be sufficiently effective as a general acid to cause some hydrolysis.

The mechanism of hydrolysis of trithioorthocarboxylates is probably similar to that of ordinary orthocarboxylates. Trithioorthocarboxylates should be less basic than orthocarboxylates, and dialkylthiocarbonium ions should be less stable than dialkoxycarbonium ions. Both of these factors should act to make trithioorthocarboxylates less reactive toward acid hydrolysis than their oxygen analogues. In the absence of experimental data, further speculation on these reactions is unwarranted.

## 2. Alcoholysis of Thioorthocarboxylates

Triethyl trithioorthoformate reacts with ethyl and with butyl alcohol in the presence of zinc chloride to form triethyl orthoformate and tributyl orthoformate in 46% and 66% yields (*65, 113*).

$$HC(SC_2H_5)_3 + 3\ ROH \xrightarrow{ZnCl_2} HC(OR)_3 + 3\ C_2H_5SH \qquad (94)$$
$$(R = C_2H_5, C_4H_9)$$

If this reaction is general for trialkyl trithioorthocarboxylates, it would provide a means of synthesizing ortho esters from acyl halides and acid anhydrides, since these can be converted to trithioorthocarboxylates.

2,2-Diethoxythiolane reacts with methanol in the presence of mercuric chloride to form a product assumed to be a mercuric chloride complex of diethylmethyl 3-mercaptoorthobutyrate (*109*).

$$\underset{S}{\overset{\phantom{S}}{\bigsqcup}}\!\!\begin{matrix}OC_2H_5\\OC_2H_5\end{matrix} + CH_3OH + HgCl_2 \longrightarrow$$

$$ClHgSCH_2CH_2CH_2C(OC_2H_5)_2OCH_3 \qquad (95)$$

## D. Reactions Which Form Carbon–Sulfur Bonds

Triethyl and tributyl trithioorthoformates undergo acid-catalyzed reactions with acetaldehyde which produce low yields of acetaldehyde dithioacetals (*125*).

$$HC(SR)_3 + CH_3CHO \xrightarrow{H^+} CH_3CH(SR)_2 + H_2O \qquad (96)$$
$$(R = C_2H_5, C_4H_9)$$

Refluxing triethyl trithioorthoformate with chloral hydrate resulted in formation of a 5% yield of the hemithioacetal, $CCl_3CH(OH)SC_2H_5$ (*126*). Acrolein reacts with triethyl trithioorthoformate in ethanol containing ammonium chloride to form 1,3-diethylthiopropene (*129b*).

The patent literature describes syntheses of thienol ethyl ethers from cholestenone and from an $\alpha,\beta$- unsaturated ketone by reaction with triethyl trithioorthoformate (*53, 90*).

Trithioorthoformates react with acyl chlorides to form thiol esters and chlorodi(substituted thio)methanes (*20*).

$$HC(SR)_3 + R'COCl \rightarrow R'COSR + HCCl(SR)_2 \qquad (97)$$
$$(R = CH_3, R' = CH_3, C_6H_5; R = C_6H_5, R' = CH_3)$$

Triethyl trithioorthobenzoate and acetyl chloride formed ethyl thiolacetate and ethyl dithiobenzoate (*20*).

$$C_6H_5C(SC_2H_5)_3 + CH_3COCl \rightarrow CH_3COSC_2H_5 + C_6H_5CS_2C_2H_5 + C_2H_5Cl \quad (98)$$

Trimethyl trithioorthoformate reacts with acetic anhydride to form methyl thiolacetate and methyl thiolformate (*20*).

Pyrolysis of tetratrifluoromethyl tetrathioorthocarbonate at 400°C yields bis(trifluoromethyl) trithioorthocarbonate at low pressures and tetrakis-(trifluoromethylthio)ethylene at atmospheric pressure (*67*).

$$C(SCF_3)_4 \xrightarrow{400°C} \begin{array}{l} \xrightarrow{1\ mm} (CF_3S)_2C{=}S \\ \xrightarrow{760\ mm} (F_3CS)_2C{=}C(SCF_3)_2 \end{array} \qquad (99)$$

Presumably the initial pyrolysis product is tris(trifluoromethylthio)methyl radical, which either loses trifluoromethyl radical to form the trithiocarbonate, or couples to form an unstable hexathioorthooxalate, which decomposes to form the tetraalkylthioethylene.

Tetramethyl tetrathioorthocarbonate is converted by trityl tetrafluoroborate in methylene chloride solution to trityl methyl sulfide and trimethylthiocarbonium tetrafluoroborate (*154*).

$$(CH_3S)_4C + (C_6H_5)_3C^+BF_4^- \rightarrow (CH_3S)_3C^+BF_4^- + (C_6H_5)_3CSCH_3 \quad (100)$$

## E. Reactions Which Form Carbon–Nitrogen Bonds

Triethyl trithioorthoacetate and triethyl trithioorthoisobutyrate react with aniline to form $N,N'$-diphenylamidines *(138a, 155)*.

$$RC(SC_2H_5)_3 + 2\ C_6H_5NH_2 \rightarrow C_6H_5N{=}CR{-}NHC_6H_5 + 3\ C_2H_5SH \quad (101)$$
$$[R{=}CH_3, (CH_3)_2CH]$$

$N,N$-dimethylaminodibutylthiomethane and $p$-toluidine formed $N,N$-dimethyl-$N'$-$p$-tolylformamidine *(50)*.

$$(CH_3)_2NCH(SC_4H_9)_2 + p\text{-}CH_3C_6H_4NH_2 \rightarrow (CH_3)_2NCH{=}NC_6H_4CH_3\text{-}p$$
$$+ 2\ C_4H_9SH \quad (102)$$

## F. Miscellaneous Reactions of Trithioorthocarboxylates

Triethyl trithio-3,3-diethoxyorthopropionate reacts with phosphorus pentasulfide to form 1,2-dithiole-3-thione *(15)*.

$$(C_2H_5O)_2CHCH_2C(SC_2H_5)_3 \xrightarrow{P_2S_5} \underset{S}{\overset{S}{\underset{|}{\bigg\langle}}}{=}S \quad (103)$$

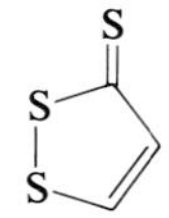

Triethyl trithioorthoformate and phosphorus pentasulfide yield ethyl dithioformate and triethyl tetrathiophosphate *(22, 23)*.

$$3\ HC(SC_2H_5)_3 + P_2S_5 \xrightarrow{110°C} 3\ HCS_2C_2H_5 + (C_2H_5)_3PS_4 \quad (104)$$

Trialkyl trithioorthoformates react with sulfur in the presence of zinc chloride, ferric bromide, or aluminum chloride to form complex products useful as vulcanization catalysts and lubricants *(156)*.

Addition of trialkyl trithioorthocarboxylates to hydrocarbon rocket fuels makes them hypergolic with nitric acid oxidizers *(27)*.

## REFERENCES

1. A. I. Abdullaev, I. G. Ali-Zadi, and M. A. Dadasheva, U.S.S.R. Patent 191,545 (1967); *Chem. Abstr.* **68**, 69060 (1968).
1a. T. J. Adley, A. K. M. Anissizzaman, and L. N. Owen, *J. Chem. Soc.*, C p. 807 (1967).
2. American Cyanamid Co., Netherlands Patent Appl. 6,413,362 (1965); *Chem. Abstr.* **63**, 497 (1965).
3. E. K. Anderson and I. Lindqvist, *Arkiv Kemi* **9**, 169 (1956).
4. S. Archer, T. R. Lewis, C. M. Martini, and M. Jackman, *J. Am. Chem. Soc.* **76**, 4915 (1954).
5. J. F. Arens, *in* "Organic Sulfur Compounds" (N. Kharasch, ed.), Chapter 23, Pergamon Press, Oxford, 1961.

6. F. Arndt, *Ann. Chem.* **384**, 322 (1911).
7. F. Arndt, *Ann. Chem.* **396**, 1 (1913).
8. H. J. Backer, *Rec. Trav. Chim.* **65**, 53 (1946).
9. H. J. Backer, *Rec. Trav. Chim.* **67**, 884 (1948).
10. H. J. Backer and K. H. Klaasens, *Rec. Trav. Chim.* **49**, 1107 (1931).
11. H. J. Backer and P. L. Stedehouder, *Rec. Trav. Chim.* **52**, 437 (1933).
12. H. J. Backer and P. L. Stedehouder, *Rec. Trav. Chim.* **52**, 923 (1933).
13. H. J. Backer and P. L. Stedehouder, *Rec. Trav. Chim.* **52**, 1039 (1933).
14. H. J. Backer and E. Westerhuis, *Rec. Trav. Chim.* **71**, 1071 (1952).
15. R. J. S. Beer and R. A. Slater, *J. Chem. Soc.* p. 4069 (1964).
15a. D. R. Pender and D. L. Coffen, *J. Org. Chem.* **33**, 2504 (1968).
16. H. Boehne and H. Gran, *Ann. Chem.* **581**, 133 (1953).
17. H. Boehme, H. Lohmeyer, and J. Wickop, *Ann. Chem.* **587**, 51 (1954).
18. H. Boehme and R. Marx, *Chem. Ber.* **74**, 1667 (1941).
19. H. Boehme and R. Neiderlein, *Chem. Ber.* **95**, 1859 (1962).
20. H. Boehme and J. Roehr, *Ann. Chem.* **648**, 21 (1961).
21. J. Bongartz, *Chem. Ber.* **19**, 2182 (1886).
22. K. C. Brannock, *J. Am. Chem. Soc.* **73**, 4953 (1951).
23. K. C. Brannock, U.S. Patent 2,622,095 (1952); *Chem. Abstr.* **47**, 9343 (1953).
24. J. W. Brookes, U.S. Patent 2,522,489 (1950); *Chem. Abstr.* **44**, 11083 (1950).
25. H. C. Brown and R. Pater, *J. Org. Chem.* **27**, 2858 (1962).
26. T. L. Cairns, A. W. Larcher, and B. C. McKusick, *J. Org. Chem.* **18**, 748 (1953).
27. D. R. Carmody and A. Fletz, U.S. Patent 2,998,699 (1961); *Chem. Abstr.* **55**, 27892 (1961).
28. Chinoin Gyogys zer es Vegyeszíti Termekek Gyara, Hungarian Patent 128,404 (1941); *Chem. Abstr.* **46**, 2570 (1952).
29. P. Claesson, *J. Prakt. Chem.* [2] **15**, 212 (1877).
29a. D. H. Clemens, E. Y. Shropshire, and W. D. Emmons, *J. Org. Chem.* **27**, 3664 (1962).
30. D. L. Coffen, *Chem. Commun.* p. 1089 (1967).
31. J. W. Copenhaver, U.S. Patent 2,500,486 (1950); *Chem. Abstr.* **44**, 5379 (1950).
32. J. W. Copenhaver, U.S. Patent 2,677,707 (1954); *Chem. Abstr.* **49**, 1812 (1955).
33. J. W. Copenhaver, U.S. Patent 2,677,708 (1954); *Chem. Abstr.* **49**, 1812 (1955).
34. D. W. Cowie and D. T. Gibson, *J. Chem. Soc.* p. 306 (1933).
34a. D. J. Cram, "Fundamentals of Carbanion Chemistry," p. 80. Academic Press, New York, 1965.
35. C. W. N. Cumper, A. Melnikoff, E. F. Mooney, and A. I. Vogel, *J. Chem. Soc., B* p. 874 (1966).
36. C. W. N. Cumper, A. Melnikoff, and A. I. Vogel, *J. Chem. Soc., A* p. 242 (1966).
37. C. W. N. Cumper, A. Melnikoff, and A. I. Vogel, *J. Chem. Soc., A* p. 323 (1966).
37a. F. Dallacker, E. Kaiser, and P. Uddrich, *Ann. Chem.* **689**, 179 (1965).
38. J. J. D'Amico and R. H. Campbell, *J. Org. Chem.* **32**, 2567 (1967).
39. M. Dennstedt, *Chem. Ber.* **11**, 2265 (1878).
40. M. Dennstedt, *Chem. Ber.* **13**, 238 (1880).
41. W. von E. Doering and L. K. Levy, *J. Am. Chem. Soc.* **77**, 509 (1955).
42. F. P. Doyle and J. D. Kendall, British Patent 620,801 (1949); *Chem. Abstr.* **43**, 7842 (1949).
43. H. D. Edwards, British Patent 642,517 (1950); *Chem. Abstr.* **45**, 6105 (1951).
44. H. D. Edwards and F. P. Doyle, U.S. Patent 2,721,799 (1955); *Chem. Abstr.* **50**, 1505 (1956).

45. H. D. Edwards, F. P. Doyle, and S. J. Palling, U.S. Patent 2,839,402 (1958); *Chem. Abstr.* **53**, 943 (1959).
46. B. C. Ennis, G. Holan, and E. L. Samuel, *J. Chem. Soc.*, C p. 30 (1967).
47. B. C. Ennis, G. Holan, and E. L. Samuel, *J. Chem. Soc.*, C p. 33 (1967).
47a. E. Fanghaenel, *Z. Chem.* **5**, 386 (1965).
48. E. Fanghaenel and R. Mayer, *Z. Chem.* **4**, 384 (1964).
49. B. P. Federov and F. M. Stoyanovich, *Izv. Akad. Nauk SSSR, Otd. Khim. Nauk* p. 1828 (1960).
50. B. P. Federov and F. M. Stoyanovich, U.S.S.R. Patent 148,804 (1962); *Chem. Abstr.* **58**, 8968 (1963).
50a. R. Feinauer, *Angew. Chem. Intern. Ed. Engl.* **5**, 894 (1966).
50b. C. Feugeas and D. Olschwang, *Compt. Rend.* **C266**, 1506 (1968).
50c. C. Feugeas and D. Olschwang, *Bull. Soc. Chim. France* p. 325 (1969).
50d. C. Feugeas and D. Olschwang, *Bull. Soc. Chim. France* p. 332 (1969).
51. G. E. Ficken and J. D. Kendall, *J. Chem. Soc.* p. 3202 (1959).
52. E. K. Fields, *Ind. Eng. Chem.* **49**, 1361 (1957).
53. Z. Foldi, Hungarian Patent 135,687 (1949); *Chem. Abstr.* **44**, 4047 (1950).
54. R. L. Frank, S. S. Drake, P. V. Smith, and C. Stevens, *J. Polymer Sci.* **3**, 50 (1948).
55. A. Fredga, *Arkiv Kemi, Mineral. Geol.* **25B**, No. 8, 1–5 (1947); *Chem. Abstr.* **42**, 5918 (1948).
56. A. Fredga and K. Olsson, *Arkiv Kemi* **9**, 163 (1956).
57. A. Frohling and J. F. Arens, *Rec. Trav. Chim.* **81**, 1009 (1962).
58. E. Fromm, *Ann. Chem.* **253**, 166 (1889).
59. S. Gabriel, *Chem. Ber.* **10**, 185 (1877).
60. D. T. Gibson, *J. Chem. Soc.* p. 2637 (1931).
61. Y. A. Goldfarb and E. N. Karaulova, *Izv. Akad. Nauk SSSR, Otd. Khim. Nauk* p. 1102 (1959).
62. H. Gross and A. Rieche, *Chem. Ber.* **94**, 538 (1961).
63. H. Gross, J. Rusche, and H. Bornowski, *Ann. Chem.* **675**, 142 (1964).
63a. R. Gugliemetti and J. Metzger, *Bull. Soc. Chim. France* p. 2824 (1967).
64. K. D. Gundermann, *Ann. Chem.* **578**, 48 (1952).
65. W. E. Hanford and W. E. Mochel, U.S. Patent 2,229,651 (1941); *Chem. Abstr.* **35**, 2904 (1941).
66. J. F. Harris, U.S. Patent 3,062,894 (1962); *Chem. Abstr.* **58**, 8907 (1963).
67. J. F. Harris, *J. Org. Chem.* **32**, 2063 (1967).
68. J. Hine, *J. Am. Chem. Soc.* **72**, 2438 (1950).
69. J. Hine, R. P. Bayer, and G. G. Hammer, *J. Am. Chem. Soc.* **84**, 1751 (1962).
70. J. Hine, A. M. Dowell, and J. E. Singley, *J. Am. Chem. Soc.* **78**, 479 (1956).
71. J. Hine and J. J. Porter, *J. Am. Chem. Soc.* **82**, 6118 (1960).
72. Y. Hishiki, *J. Sci. Res. Inst.* (*Tokyo*) **48**, 130; *Chem. Abstr.* **49**, 4295 (1955).
73. Y. Hishiki, *Kagaku Kenkyusho Hokoku* **30**, 279 (1954); *Chem. Abstr.* **52**, 8807 (1958).
74. B. Holmberg, *Ann. Chem.* **353**, 131 (1907).
75. B. Holmberg, *Chem. Ber.* **40**, 1740 (1907).
76. B. Holmberg, *Chem. Ber.* **45**, 364 (1912).
77. J. Houben, *Chem. Ber.* **45**, 2942 (1912).
78. J. Houben and K. M. L. Schultze, *Chem. Ber.* **44**, 3235 (1911).
79. B. W. Howk and J. C. Sauer, U.S. Patent 2,840,613 (1958); *Chem. Abstr.* **52**, 17186 (1958).
80. R. Huisgen, W. Mack, and E. Anneser, *Angew. Chem.* **73**, 656 (1961).

81. R. Huisgen and V. Weberndorfer, *Experientia* **17**, 566 (1961).
82. R. Hull, *J. Chem. Soc.* p. 4845 (1957).
83. W. H. Hurtley and S. Smiles, *J. Chem. Soc.* p. 2263 (1926).
84. A. B. A. Jansen and R. Robinson, *Monatsh. Chem.* **98**, 1017 (1967).
85. G. Jeminet and A. Kergomard, *Compt. Rend.* **259**, 2248 (1964).
86. G. Jeminet and A. Kergomard, *Compt. Rend.* **259**, 4051 (1964).
88. G. Jeminet and A. Kergomard, *Bull. Soc. Chim. France* p. 3233 (1967).
89. T. P. Johnston, C. R. Stringfellow, and A. Gallagher, *J. Org. Chem.* **27**, 4068 (1962).
90. J. D. Kendall, British Patent 555,560 (1943); *Chem. Abstr.* **39**, 3437 (1945).
91. J. D. Kendall, British Patent 556,815 (1943); *Chem. Abstr.* **39**, 1880 (1945).
92. J. D. Kendall, British Patent 555,935 (1943); *Chem. Abstr.* **39**, 1546 (1943).
93. J. D. Kendall, U.S. Patent 2,389,153 (1945); *Chem. Abstr.* **40**, 1540 (1946).
94. J. D. Kendall and F. P. Doyle, U.S. Patent 2,533,816 (1950); *Chem. Abstr.* **45**, 2803 (1951).
95. J. D. Kendall and F. P. Doyle, U.S. Patent 2,534,913 (1950); *Chem. Abstr.* **45**, 3273 (1951).
96. J. D. Kendall and J. R. Majer, *J. Chem. Soc.* p. 687 (1948).
97. J. D. Kendall and J. R. Majer, *J. Chem. Soc.* p. 690 (1948).
98. J. D. Kendall and J. H. Mayor, British Patent 674,003 (1952); *Chem. Abstr.* **47**, 987 (1953).
99. T. Kimura, *Proc. Imp. Acad. (Tokyo)* **13**, 261 (1937); *Chem. Abstr.* **32**, 171 (1938).
99a. G. Klopman, *J. Am. Chem. Soc.* **90**, 223 (1968).
100. I. I. Lapkin and N. I. Panova, *Zh. Obshch. Khim.* **32**, 745 (1962).
101. E. Laves, *Chem. Ber.* **23**, 1414 (1890).
102. E. Laves, *Chem. Ber.* **25**, 347 (1892).
103. E. Laves, *Chem. Ber.* **25**, 361 (1892).
104. D. Lemal and E. H. Banitt, *Tetrahedron Letters* p. 245 (1964).
104a. L. Lunazzi and F. Taddei, *Spectrochim. Acta* **23A**, 841 (1967).
105. G. Maccagnani, *Boll. Sci. Fac. Chim. Ind. Bologna* **21**, 131 (1963); *Chem. Abstr.* **60**, 3640 (1964).
106. G. E. McCasland, S. Furuta, A. Furst, L. F. Johnson, and J. N. Schoolery, *J. Org. Chem.* **28**, 456 (1963).
107. R. Mayer and K. Schafer, *J. Prakt. Chem.* [4] **26**, 279 (1964).
108. R. Mecke and H. Spiesecke, *Chem. Ber.* **88**, 1997 (1955).
109. H. Meerwein, P. Borner, O. Fuchs, J. Sasse, H. Schrodt, and J. Spille, *Chem. Ber.* **89**, 2060 (1956).
110. H. Meerwein, W. Florian, N. Schon, and G. Stopp, *Ann. Chem.* **641**, 37 (1961).
111. J. Metzger, H. Larive, R. Dennilauler, R. Baralle, and C. Gaurat, *Bull. Soc. Chim. France* p. 2857 (1964).
112. J. Metzger, H. Larive, R. Dennilauler, R. Baralle, and C. Gaurat, *Bull. Soc. Chim. France* p. 2888 (1964).
112a. W. H. Mills, L. M. Clark, and J. A. Aeschlimann, *J. Chem. Soc.* **123**, 2362 (1923).
113. W. E. Mochel, C. L. Agre, and W. E. Hanford, *J. Am. Chem. Soc.* **70**, 2268 (1948).
114. T. Mukaiyama and T. Yamaguchi, *Bull. Chem. Soc. Japan* **39**, 2005 (1966).
115. T. Mukaiyama, T. Yamaguchi, and H. Nohira, *Bull. Chem. Soc. Japan* **38**, 2107 (1965).
116. S. Oae, W. Tagaki, and A. Ohno, *Tetrahedron* **20**, 417 (1964).
117. S. Oae, W. Tagaki, and A. Ohno, *J. Am. Chem. Soc.* **83**, 5036 (1961).

118. S. Oae, W. Tagaki, and A. Ohno, *Proc. Intern. Symp. Mol. Struct. Spectry.*, *Tokyo*, *1962* B319; *Chem. Abstr.* **61**, 9057 (1964).
118a. R. A. Olofson, S. W. Walinsky, and J. L. Jernow, *J. Am. Chem. Soc.* **90**, 6554 (1968).
119. K. Olsson, *Acta Chem. Scand.* **12**, 366 (1958).
120. K. Olsson, *Arkiv Kemi* **14**, 371 (1959).
121. K. Olsson, *Arkiv Kemi* **26**, 435 (1967).
121a. K. Olsson, *Arkiv Kemi* **28**, 53 (1967).
122. K. Olsson, H. Baekstrom, and R. Engwall, *Arkiv Kemi* **26**, 219 (1966).
122a. R. G. Pearson, *J. Am. Chem. Soc.* **85**, 3533 (1963).
123. H. P. Penner and A. R. Conklin, *J. Heterocyclic Chem.* **4**, 93 (1967).
124. W. G. Perdok and G. Terpstra, *Rec. Trav. Chim.* **65**, 493 (1946).
125. H. W. Post, *J. Org. Chem.* **5**, 244 (1940).
126. H. W. Post, *J. Org. Chem.* **6**, 830 (1941).
126a. H. W. Post, "Chemistry of the Aliphatic Ortho Esters." Reinhold, New York, 1943.
127. H. W. Post and E. R. Erickson, *J. Am. Chem. Soc.* **55**, 3851 (1933).
127a. H. Quast and E. Schmitt, *Chem. Ber.* **101**, 1137 (1968).
127b. M. S. Raasch, U.S. Patent 3,336,334 (1967); *Chem. Abstr.* **69**, 19138 (1968).
128. E. E. Reid, "Organic Chemistry of Sulfur," Vol. IV, p. 42. Chem. Publ. Co., New York, 1962.
129. L. C. Rinzema, J. Stoffelsma, and J. F. Arens, *Rec. Trav. Chim.* **78**, 354 (1959).
129a. G. L. Roof and W. P. Tucker, *J. Org. Chem.* **33**, 3333 (1968).
129b. E. Rothstein, D. J. Stanbank, and R. Whitely, *J. Chem. Soc.*, C p. 746 (1968).
130. E. Samen, *Arkiv Kemi, Mineral. Geol.* **12B**, No. 51 (1938); *Chem. Abstr.* **32**, 4520 (1938).
131. E. Samen, *Arkiv Kemi, Mineral. Geol.* **14B**, No. 28 (1941); *Chem. Abstr.* **35**, 3876 (1941).
132. E. Samen, *Arkiv Kemi, Mineral. Geol.* **15B**, No. 15 (1942); *Chem. Abstr.* **37**, 5015 (1943).
133. E. Samen, *Arkiv Kemi, Mineral. Geol.* **24B**, No. 6 (1947); *Chem. Abstr.* **42**, 6313 (1948).
134. E. Samen, *Arkiv Kemi* **1**, 231 (1949).
135. E. Samen, *Acta Chem. Scand.* **4**, 397 (1950).
136. E. Samen, *Arkiv Kemi* **3**, 303 (1951).
137. E. Samen, *Svensk Kem. Tidskr.* **63**, 31 (1951).
138. C. S. Scanley, U.S. Patent 3,333,007 (1967); *Chem. Abstr.* **67**, 90393 (1967).
138a. R. Scarpati, C. Santacroce, and D. Sica, *Gazz. Chim. Ital.* **95**, 302 (1965).
139. U. Schoellkopf and E. Wiskott, *Angew. Chem. Intern. Ed. Engl.* **2**, 485 (1963).
140. U. Schoellkopf and E. Wiskott, *Ann. Chem.* **694**, 44 (1966).
141. A. Schoenberg, B. Koenig, and E. Frese, *Chem. Ber.* **98**, 3303 (1965).
142. A. Schoenberg, B. Koenig, and E. Singer, *Chem. Ber.* **100**, 767 (1967).
143. A. Schoenberg, S. Nickel, and D. Cernik, *Chem. Ber.* **65**, 289 (1932).
144. A. Schoenberg and K. Praefke, *Tetrahedron Letters* p. 2043 (1964).
145. A. Schoenberg and K. Praefke, *Chem. Ber.* **99**, 2371 (1966).
146. A. Schoenberg, K. Praefke, and J. Kohtz, *Chem. Ber.* **99**, 3076 (1966).
147. D. Seebach, *Angew. Chem. Intern. Ed. Engl.* **6**, 442 (1967).
148. D. Seebach, *Angew. Chem. Intern. Ed. Engl.* **6**, 443 (1967).
148a. Y. A. Sinnema and J. F. Arens, *Rec. Trav. Chim.* **76**, 949 (1957).

149. M. F. Shostakovskii, A. V. Bogdanova, G. I. Plotkinova, and A. N. Dolgikh, *Izv. Akad. Nauk SSSR, Otd. Khim. Nauk* p. 1901 (1960).
150. L. H. Slaugh and E. Bergman, *J. Org. Chem.* **26**, 3158 (1962).
151. J. A. Smythe, *Proc. Univ. Durham Phil. Soc.* **4**, 75 (1911).
152. H. Stetter, *Angew. Chem. Intern. Ed. Engl.* **1**, 286 (1962).
153. D. S. Tarbell and A. H. Herz, *J. Am. Chem. Soc.* **75**, 1668 (1953).
154. W. P. Tucker and G. L. Roof, *Tetrahedron Letters* p. 2747 (1967).
155. H. C. Volger and J. F. Arens, *Rec. Trav. Chim.* **76**, 847 (1957).
155a. J. J. Vorsanger, *Bull. Soc. Chim. France* p. 119 (1964).
155b. J. J. Vorsanger, *Bull. Soc. Chim. France* p. 971 (1968).
155c. A. F. Wagner, U.S. Patent 2,872,458 (1959); *Chem. Abstr.* **53**, 12184 (1959).
156. I. D. Webb, U.S. Patent 2,966,522 (1960); *Chem. Abstr.* **55**, 8294 (1961).
157. G. A. Wildschut, H. J. T. Bos, L. Brandsma, and J. F. Arens, *Monatsh. Chem.* **98**, 1043 (1967).
158. M. L. Wolfram and E. Usdin, *J. Am. Chem. Soc.* **75**, 4619 (1953).

CHAPTER 7

# Amide Acetals, Ester Aminals, and Ortho Amides

## I. INTRODUCTION

This chapter discusses the synthesis and chemical properties of substances having nitrogen or nitrogen and oxygen bonded to an orthoacyl carbon atom. These include the amide acetals (**1**), ester aminals (**2**), the ortho amides (**3**), and heterocyclic compounds related to them. The corresponding derivatives of orthocarbonic acid (**1**, **2**, and **3**, R = R'O or R'R''N) are also discussed.

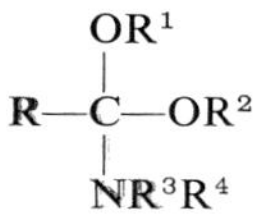

$$\text{R}-\underset{\underset{\text{NR}^3\text{R}^4}{|}}{\overset{\overset{\text{OR}^1}{|}}{\text{C}}}-\text{OR}^2 \qquad \text{R}-\underset{\underset{\text{NR}^4\text{R}^5}{|}}{\overset{\overset{\text{OR}^1}{|}}{\text{C}}}-\text{NR}^2\text{R}^3 \qquad \text{R}-\underset{\underset{\text{NR}^5\text{R}^6}{|}}{\overset{\overset{\text{NR}^1\text{R}^2}{|}}{\text{C}}}-\text{NR}^3\text{R}^4$$

$$\textbf{1} \qquad\qquad\qquad \textbf{2} \qquad\qquad\qquad \textbf{3}$$

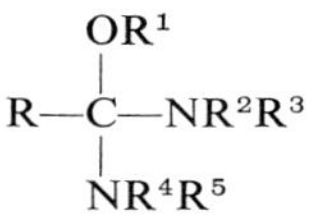

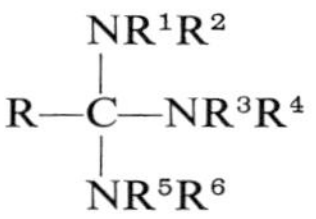

Until 1956, when Meerwein reported the synthesis of a number of amide acetals, ortho amide derivatives were only postulated intermediates in substitution and isomerization reactions of amides and other carboxylic acid derivatives (see, for example, refs *3, 12, 63*). In the sixties, Meerwein, Bredereck, Arnold, and their co-workers made particularly important contributions to the synthesis and applications of ortho amide derivatives.

Some of these compounds—the amide acetals in particular—are useful synthetic intermediates which are more reactive even than the corresponding ortho esters. Dialkylformamide acetals react with primary amino groups to form easily hydrolyzed dialkylaminomethylene derivatives in which the original amino group is protected during some kinds of transformations of other parts of the molecule. Other members of this diverse group of substances yield unusually stable carbonium ions, some of which do not react rapidly with water.

Methods of synthesizing the various types of ortho acid amides are presented first, followed by a compilation of ortho amide derivatives which have been reported in the literature and some of their physical properties. Finally, their chemical properties are considered.

Trinitromethyl compounds $RC(NO_2)_3$ and tetranitromethane are not included, since they cannot be prepared by oxidation of ortho amides, and also because their chemical properties are entirely different from those of the ortho acid derivatives.

## II. SYNTHESES OF AMIDE ACETALS, ESTER AMINALS, AND ORTHO AMIDES

### A. Amide Acetals

Amide acetals can be divided into two groups: heterocyclic amide acetals, in which the orthoacyl carbon atom is part of a hetero ring; and nonheterocyclic amide acetals, in which the orthoacyl carbon is not part of a ring. In the discussion which follows, amide acetals whose orthoacyl carbon is not part of a hetero ring are classed as nonheterocyclic, even if they have a hetero ring which does not include the ortho amide function.

Systematically, the nonheterocyclic amide acetals are named as substituted amines. Dimethylformamide diethyl acetal, for example, is indexed by Chemical Abstracts as 1,1-diethoxytrimethylamine. In order to emphasize the structures of these compounds, they are named as amide acetals in the discussion which follows. The heterocyclic amide acetals are named systematically.

1. NONHETEROCYCLIC AMIDE ACETALS

a. *Carboxamide Acetals.* The largest and most useful group of amide acetals are those derived from $N,N$-dialkylformamides. The first such compound, $N,N$-dimethylformamide diethyl acetal, was reported by Meerwein and co-workers in 1956 (*114*). Since then a large number of $N,N$-dialkylcarboxamide dialkyl acetals have been described in the literature.

i. *Preparation of amide acetals from alkoxydialkylaminocarbonium salts.* A number of $N,N$-dialkylaminoalkoxycarbonium salts react with alkoxides in alcohol or with alcohol–tertiary amine mixtures to form carboxamide acetals. Meerwein's original formamide acetal synthesis involved the reaction of $N,N$-dimethylaminoethoxycarbonium tetrafluoroborate (prepared by alkylation of dimethylformamide with triethyloxonium tetrafluoroborate) with cold ethanolic sodium ethoxide [Eq. (1), R = H] (*115*). The overall yield of dimethylformamide diethyl acetal was 72%. $N,N$-Dimethylacetamide diethyl acetal was prepared in 64% yield by the same procedure [Eq. (1), R = CH$_3$] (*115*). The preparation of amide diethyl acetals from aminoethoxycarbonium tetrafluoroborates should be a generally applicable reaction, since the carbonium salts can be prepared in high yields from most carboxamides (*158*).

$$(CH_3)_2NCRO \xrightarrow{(C_2H_5)_3O^+BF_4^-} (CH_3)_2\overset{+}{N}CROC_2H_5 \xrightarrow{NaOC_2H_5}$$
$$(CH_3)_2NCR(OC_2H_5)_2 \quad (1)$$

A closely related synthesis was reported by Bredereck and co-workers in 1961 (*37*). They treated the product formed by warming an equimolar mixture of dimethyl sulfate and dimethylformamide (essentially $N,N$-dimethylaminomethoxycarbonium methylsulfate) with alcoholic alkoxide to obtain the dimethylformamide dialkyl acetal. This reaction, which has most often been used for the preparation of $N,N$-dimethylformamide dimethyl acetal [Eq. (2), R = R″ = CH$_3$, R′ = H] (*5, 19, 53, 84*), is a fairly general one.

$$R_2NCR'O \xrightarrow{R''_2SO_4} R_2\overset{+}{N}CR'OR''R''SO_4^- \xrightarrow{NaOR''} R_2NCR'(OR'')_2 \quad (2)$$

Most $N,N$-dialkylformamides and $N$-formyl heterocyclic amines can be converted to formamide acetals in fair to good yields by successive treatment with a dialkyl sulfate and the corresponding alkoxide. Bredereck and coworkers have synthesized a number of formamide acetals in yields ranging from 42% to 89% by this procedure (*53*). The carbonium salts were prepared by mixing equimolar amounts of the $N,N$-dialkylformamide and the dialkyl sulfate and allowing the mixtures to stand (at elevated temperatures in the case of amides having bulky $N$-alkyl groups) until the reactions were complete (*39*). The carbonium alkylsulfate salts were then added slowly to the alcoholic alkoxide at a temperature below 20°C. Higher temperatures increase the extent of side reactions, principally ether and amide formation due to attack by alkoxide on alkyl rather than acyl carbon of the carbonium salt. The reaction between the salt and alkoxide is practically instantaneous.

The same general procedure has been used to prepare acetals of $N,N$-dimethylacetamide, $N,N$-dimethylpropionamide, and $N,N$-dimethylbenzamide

[Eq. (2), R = CH$_3$, R' = CH$_3$, C$_6$H$_5$, R" = C$_2$H$_5$; R = CH$_3$, R' = C$_2$H$_5$, R" = CH$_3$] (84).

N,N-Dimethylformamide di-*tert*-butyl acetal was prepared by reaction of N,N-dimethylamino-*tert*-butoxycarbonium perchlorate with potassium *tert*-butoxide in *tert*-butyl alcohol (5).

$$(CH_3)_2NCHO \xrightarrow{COCl_2} (CH_3)_2\overset{+}{N}CHClCl^- \xrightarrow{(CH_3)_3COH}$$
$$(CH_3)_2\overset{+}{N}CHOC(CH_3)_3Cl^- \xrightarrow{NaClO_4} (CH_3)_2\overset{+}{N}CHOC(CH_3)_3ClO_4^- \xrightarrow{KOC(CH_3)_3}$$
$$(CH_3)_2NCH[OC(CH_3)_3]_2 \quad (3)$$

Eilingsfeld, Seefelder, and Weidinger (75–77) and Bosshard, Jenny, and Zollinger (16, 62) discovered independently and almost simultaneously that N,N-dimethylaminochlorocarbonium chloride (N,N-dimethylformamide chloride) can be converted to N,N-dimethylformamide acetals by reaction with alkoxides. The amide chlorides, which are best described as carbonium salts (16, 75), are prepared by reaction of N,N-dialkylamides with phosgene in an inert solvent at or slightly below room temperature. The formamide acetal is obtained by adding a solution of the carbonium salt (in chloroform, for example) to an alcoholic alkoxide solution (76) or by adding the alkoxide solution, with stirring, to the freshly prepared dialkylaminochlorocarbonium chloride [Eq. (4), R = H, R' = CH$_3$, C$_2$H$_5$, C$_3$H$_7$, cyclo-C$_6$H$_{11}$, C$_6$H$_5$CH$_2$] (21).

$$RCON(CH_3)_2 \xrightarrow{COCl_2} R\overset{+}{C}ClN(CH_3)_2Cl^- \xrightarrow{NaOR'} RC(OR')_2N(CH_3)_2 \quad (4)$$

Reported yields of N,N-dimethylformamide dialkyl acetals range from about 30% to 70% overall (16, 21, 76). Bredereck obtained a somewhat higher yield of the diethyl acetal using a benzene solution of ethanol and triethylamine instead of ethanolic sodium ethoxide (21).

N,N-Dimethylacetamide dimethyl acetal (56), N,N-dimethylacetamide diethyl acetal (16, 62), and N,N-dimethylbenzamide diethyl acetal (16, 62, 76, 77) have also been prepared from amide chlorides and alkoxides [Eq. (4), R = CH$_3$, R' = CH$_3$, C$_2$H$_5$; R = C$_6$H$_5$, R' = C$_2$H$_5$].

N,N,N',N'-Tetramethylformamidinium chloride is converted by potassium *tert*-butoxide in *tert*-butyl alcohol to N,N-dimethylformamide di-*tert*-butyl acetal in 70% yield (5).

$$[(CH_3)_2N]CH^+Cl^- + (CH_3)_3CO^-K^+ \xrightarrow[-KCl]{}$$
$$(CH_3)_2NCH[OC(CH_3)_3]_2 + (CH_3)_2NH \quad (5)$$

The complex amidinium salt **4**, obtained by reaction of N,N-dimethylformamide with cyanuric chloride, reacts with sodium methoxide to form a mixture of the $N,N$-dimethylaminomethylene derivative of formamide dimethyl acetal and $N,N$-dimethylformamide dimethyl acetal (*89*).

$(CH_3)_2NCHO$ + [cyanuric chloride] $\longrightarrow$

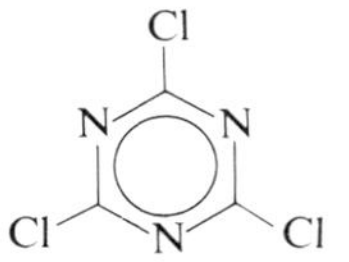

$[(CH_3)_2N\cdots CH\cdots N\cdots CH\cdots N(CH_3)_2]^+ Cl^-$  $\xrightarrow{NaOCH_3}$
**4**

$$(CH_3)_2NCH=N-CH(OCH_3)_2 + (CH_3)_2NCH(OCH_3)_2 \quad (6)$$

ii. *Transacetalization and transamination reaction of formamide acetals.* Meerwein and co-workers found that $N,N$-dimethylformamide diethyl acetal reacts smoothly with butanol to form $N,N$-dimethylformamide dibutyl acetal and ethanol (*115*).

$$(CH_3)_2NCH(OC_2H_5)_2 + 2\ C_4H_9OH \rightarrow (CH_3)_2NCH(OC_4H_9)_2 + 2\ C_2H_5OH \quad (7)$$

The reaction requires no catalyst, and is driven to completion by distilling off the ethanol. This procedure has been used to prepare a number of higher $N,N$-dimethylformamide dialkyl acetals, using $N,N$-dimethylformamide dimethyl acetal as the starting material (*5, 19, 53*). Typically, $N,N$-dimethylformamide dimethyl acetal is refluxed with an excess of the desired alcohol, and methanol is removed at the still head as it is formed. When no more methanol is obtained, the residue is fractionally distilled to collect the product. Yields are usually high (50%–90%), but are lower for secondary alcohols than for primary alcohols. Arnold and Kornilov were unable to obtain $N,N$-dimethylformamide di-*tert*-butyl acetal by the transacetalization procedure, presumably because of steric hindrance (*5*).

Dimethylformamide acetals also undergo transamination reactions. Meerwein prepared 1-(diethoxymethyl)piperidine and 1-(diethoxymethyl)-morpholine in 52% and 64% yields by heating $N,N$-dimethylformamide diethyl acetal with piperidine and with morpholine (*115*).

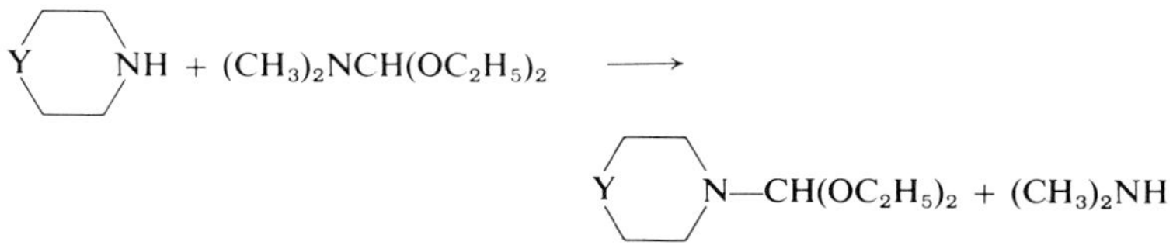

$$Y\underset{\diagdown}{\diagup}NH + (CH_3)_2NCH(OC_2H_5)_2 \longrightarrow$$

$$Y\underset{\diagdown}{\diagup}N-CH(OC_2H_5)_2 + (CH_3)_2NH$$

$$(Y = CH_2, O) \quad (8)$$

A patent describes the synthesis of 1-(dimethoxymethyl)pyrrolidine, $N,N$-dipropylformamide dimethyl acetal, and 1-dimethoxymethyl-2,5-dimethylpyrrolidine by similar reactions of $N,N$-dimethylformamide dimethyl acetal (*161*).

iii. *Reactions of trialkyl orthoformates with nitrogen compounds.* Formamide acetals should be the kinetically controlled products of reactions of dialkoxycarbonium ions with secondary amines (*102*). It is surprising that no simple amide acetals have been prepared by acid-catalyzed reactions of trialkyl orthocarboxylates with secondary amines. Reaction conditions could probably be found which would permit the isolation of formamide acetals from reactions of dialkoxycarbonium tetrafluoroborates [which can be prepared from trialkyl orthocarboxylates and boron trifluoride etherate (*113*)] with secondary amines.

$N$-Alkylarenesulfonamides react with trialkyl orthoformates to yield $N$-alkyl-$N$-arenesulfonylformamide dialkyl acetals (*68, 165*),

$$\text{ArSO}_2\text{NHR} + \text{HC(OR')}_3 \rightarrow \text{ArSO}_2\text{NR—CH(OR')}_2 + \text{R'OH} \quad (9)$$
$$(\text{Ar} = \text{C}_6\text{H}_5, \text{R} = \text{R'} = \text{C}_2\text{H}_5; \text{Ar} = p\text{-CH}_3\text{C}_6\text{H}_4, \text{R} = \text{CH}_3, \text{R'} = \text{CH}_3, \text{C}_2\text{H}_5;$$
$$\text{Ar} = p\text{-CH}_3\text{C}_6\text{H}_4, \text{R} = \text{R'} = \text{C}_2\text{H}_5)$$

and $N,N$-difluoroformamide and $N,N$-difluoroacetamide dimethyl acetals were synthesized from the trimethyl orthocarboxylates and difluoroamine (*90*).

$$\text{RC(OCH}_3)_3 + \text{HNF}_2 \xrightarrow{25°C} \text{RC(OCH}_3)_2\text{NF}_2 + \text{CH}_3\text{OH} \quad (10)$$
$$(\text{R} = \text{H, CH}_3)$$

$N$-Alkyl- and $N,N$-dialkylureas are converted by triethyl orthoformate to mixtures of products which contain $N$-alkyl- or $N,N$-dialkyl-$N'$-diethoxymethylureas, $\text{RR'NCONHCH(OC}_2\text{H}_5)_2$ (*158b, 158c*).

Triethyl orthoformate reacts with isocyanates in the presence of Lewis acids ($\text{BF}_3$, $\text{ZnCl}_2$, $\text{AlCl}_3$) to form ethyl $N$-diethoxymethylcarbamates [Eq. (11)] (*145a, 145b*). This reaction probably involves attack of diethoxycarbonium ion on isocyanate nitrogen, followed by reaction of the resulting oxocarbonium ion with ethanol or an ethoxy complex ion derived from the catalyst.

$$\text{HC(OC}_2\text{H}_5)_3 + \text{RNCO} \xrightarrow{\text{BF}_3} \text{RN(CO}_2\text{C}_2\text{H}_5)\text{CH(OC}_2\text{H}_5)_2 \quad (11)$$
$$(\text{R} = \text{CH}_3, n\text{-C}_3\text{H}_7, \text{C}_6\text{H}_5)$$

iv. *Reaction of 1,1-difluoroalkyldimethylamines with alkoxides.* A method of synthesizing $N,N$-dimethylformamide and $N,N$-dimethylbenzamide dialkyl acetals, which would probably be suitable for synthesizing other $N,N$-dialkylcarboxamide acetals, involves conversion of the $N,N$-dialkylamide to a

1,1-difluoroalkyl dialkylamine by reaction with carbonyl fluoride, followed by reaction of the difluoroalkyl tertiary amine with an alkoxide in ether or tetrahydrofuran (*56, 159*). The difluoro compounds differ from the analogous products obtained from amides and phosgene in being covalent rather than salt-like.

$$RCON(CH_3)_2 \xrightarrow{COF_2} RCF_2N(CH_3)_2 \xrightarrow{NaOR'} RC(OR')_2N(CH_3)_2 \quad (12)$$
$$(R = H, R' = CH_3, C_2H_5; R = C_6H_5, R' = CH_3)$$

v. *Miscellaneous reactions yielding amide acetals.* Chlorodiphenoxymethane reacts with triethylamine in benzene solution to form diphenoxymethyltriethylammonium chloride, an unstable, hygroscopic salt (*127*).

$$(C_6H_5O)_2CHCl + (C_2H_5)_3N \xrightarrow{C_6H_6} (C_6H_5O)_2CH\overset{+}{N}(C_2H_5)_3Cl^- \quad (13)$$

The reaction of dichlorocarbene (generated *in situ* from ethyl trichloroacetate or chloroform) with sodium methoxide and diethylamine in hexane yields *N,N*-diethylformamide dimethyl acetal [Eq. (14)] (*142*).

$$:CCl_2 + 2\,NaOCH_3 + (C_2H_5)_2NH \xrightarrow{C_6H_{12}} (C_2H_5)_2NCH(OCH_3)_2 + 2\,NaCl \quad (14)$$

A yield of 51% of the amide acetal was obtained using ethyl trichloroacetate as the carbene source, while a 27% yield resulted from the reaction using chloroform.

*N,N*-Dimethylacetamide diethyl acetal is formed by the addition of ethanol to 1-ethoxy-1-dimethylaminoethylene (*115*). The reaction is of no synthetic utility, since the starting material was prepared by dealcoholation of the amide acetal.

b. *Carbamide Acetals.* Carbamide acetals are orthocarbonic acid derivatives having one or two secondary amino and three or two alkoxy groups bonded to the same carbon atom. The two classes of compounds, **5** and **6**, are acetals of alkyl carbamates and ureas, respectively.

$$R_2NC(OR')_3 \qquad (R_2N)_2C(OR')_2$$
$$\mathbf{5} \qquad\qquad\quad \mathbf{6}$$

Tetramethylurea diethyl acetal (**7**) was first prepared by the reaction of bis(dimethylamino)ethoxycarbonium tetrafluoroborate with sodium ethoxide (*115*).

$$[(CH_3)_2N]_2C{=}O \xrightarrow{(C_2H_5)_3O^+BF_4^-} [(CH_3)_2N]_2\overset{+}{C}OC_2H_5BF_4^- \xrightarrow{NaOC_2H_5}$$
$$[(CH_3)_2N]_2C(OC_2H_5)_2 \quad (15)$$
$$\mathbf{7}$$

The same product is obtained by treating bis(dimethylamino)chlorocarbonium chloride (tetramethylchloroformamidinium chloride) with alcohol-free sodium ethoxide.

$$[(CH_3)_2N]_2C=O \xrightarrow{COCl_2} [(CH_3)_2N]_2\overset{+}{C}ClCl^- \xrightarrow{NaOC_2H_5} 7 \quad (16)$$

Bis(dimethylamino)difluoromethane, prepared from tetramethylurea and carbonyl fluoride, is coverted to tetramethylurea dimethyl acetal (**6**, R = R' = CH$_3$) by reaction with sodium methoxide (*56*).

Alkyl carbamate acetals (**5**) have been prepared from alkyl $N,N$-dialkyl-carbamates, tetraalkyl ureas, and urea acetals.

Meerwein obtained $N,N$-dimethylaminotriethoxymethane (**8**) in high yield by ethanolysis of tetramethylurea diethyl acetal [Eq. (17)], and also by reaction of dimethylaminodiethoxycarbonium tetrafluoroborate with sodium ethoxide [Eq. (18)] (*115*).

$$[(CH_3)_2N]_2C(OC_2H_5)_2 + C_2H_5OH \rightarrow (CH_3)_2NC(OC_2H_5)_3 + (CH_3)_2NH \quad (17)$$
$$\mathbf{8}$$

$$(CH_3)_2NCO_2C_2H_5 \xrightarrow{(C_2H_5)_3O^+BF_4^-} (CH_3)_2N\overset{+}{C}(OC_2H_5)_2BF_4^- \xrightarrow{NaOC_2H_5} \mathbf{8} \quad (18)$$

The facile transformation of urea acetals to alkyl carbamate acetals [Eq. (17)] provides an explanation for the observation that alcoholysis of $N,N,N',N'$-tetraalkyl urea–dimethyl sulfate complexes [Eq. (19)] (*22*) or $N,N,N',N'$-tetraalkylchloroformamidinium chlorides [Eq. (20)] (*74, 149*) leads to dialkylaminotrialkoxymethanes (**5**) rather than to urea acetals (**6**).

$$(R_2N)_2CO \xrightarrow{R'_2SO_4} (R_2N)_2\overset{+}{C}OR'R'SO_4^- \xrightarrow[R'OH]{NaOR'} R_2NC(OR')_3 \quad (19)$$
$$(R = CH_3, R' = CH_3, C_2H_5; R = C_2H_5, R' = CH_3)$$

$$(R_2N)_2CO \xrightarrow{COCl_2} (R_2N)_2\overset{+}{C}ClCl^- \xrightarrow[R'OH]{NaOR'} R_2NC(OR')_3 \quad (20)$$
$$[R_2 = (CH_3)_2, -(CH_2)_5-, R' = C_2H_5; R = R' = CH_3]$$

## 2. Heterocyclic Amide Acetals

There are many amide acetals whose orthoacyl carbon atom is part of one or more hetero rings. For discussion, these are divided into three groups: those which are derived from simple (i.e., unsubstituted) orthocarboxylic acids; those which are derived from substituted orthocarboxylic acids; and those which are derived from orthocarbonic acid.

a. *Heterocyclic Amide Acetals Derived from Unsubstituted Orthocarboxylic Acids.* Acyclic $N,N$-dialkylamide acetals react with 1,2- and 1,3-diols to form 2-($N,N$-dialkylamino)-1,3-dioxolanes and 2-($N,N$-dialkylamino)-1,3-dioxanes in high yields [Eq. (21)]. 2-Dimethylamino-5,5-dimethyl-1,3-dioxane was prepared by a similar reaction of $N,N$-dimethylformamide dimethyl acetal with 2,2-dimethyl-1,3-propanediol (5).

$$(CH_3)_2N-CR(OR^1)_2 + HOCR^2{}_2(CH_2)_nCR^2{}_2OH \longrightarrow$$

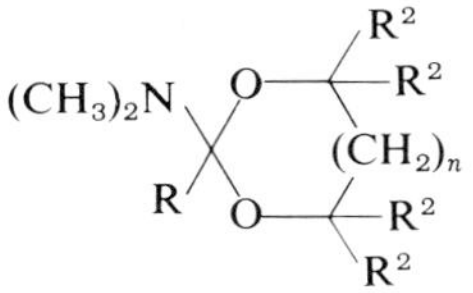

$$(CH_3)_2N \begin{array}{c} O \\ \diagup \\ R \end{array} \begin{array}{c} R^2 \\ R^2 \\ (CH_2)_n \\ R^2 \\ R^2 \end{array} \quad (21)$$

| R | $R^1$ | $R^2$ | $n$ | References |
|---|---|---|---|---|
| H, CH$_3$ | C$_2$H$_5$ | H | 0 | *115* |
| H | CH$_3$ | H, CH$_3$ | 0 | *53* |
| CH$_3$, C$_2$H$_5$, C$_3$H$_7$ | CH$_3$ | H | 0 | *84* |
| H | CH$_3$ | H | 1 | *53* |

$N,N$-Dimethylformamide diethyl acetal underwent simultaneous aminolysis and alcoholysis when heated with diethanolamine to form 1-aza-4,6-dioxabicyclo[3.3.0]octane (5).

$$(CH_3)_2NCH(OCH_3)_2 + HN(CH_2CH_2OH)_2 \longrightarrow$$

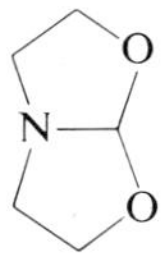

$$\begin{array}{c} \text{[bicyclic structure]} \end{array} + 2\,CH_3OH + (CH_3)_2NH \quad (22)$$

Adenosine and cytidine react with $N,N$-dimethylformamide dimethyl acetal to form 6-$N$-2′,3′-$O$-bis(dimethylaminomethylene)adenosine and 4-$N$-2′,3′-$O$-bis(dimethylaminomethylene)cytidine (**9** and **10**), respectively (*172*).

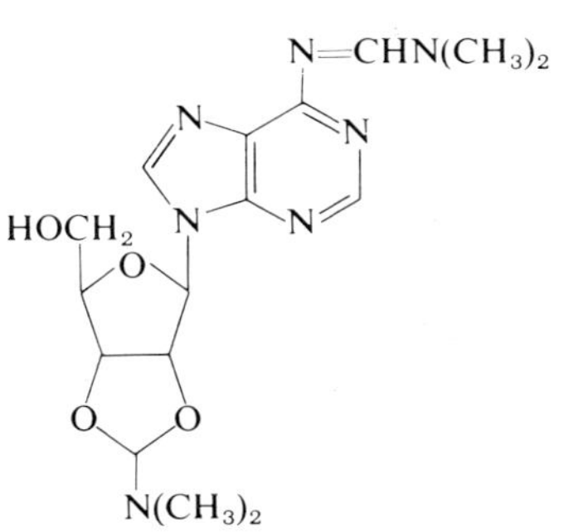

**9**

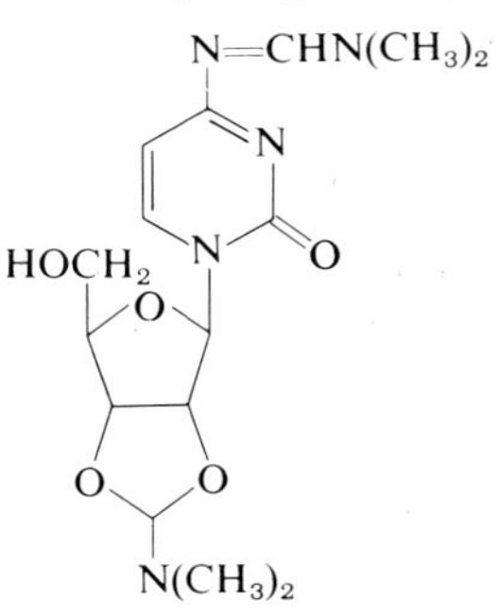

**10**

Ethyl *N*-propyl-*N*-diethoxymethyl carbamate reacts with ethylene glycol to form ethyl *N*-propyl-*N*-2-dioxolanylcarbamate (*145a, 145b*).

$$(CH_2H_5O)_2CHN(C_3H_7)CO_2C_2H_5 + HOCH_2CH_2OH \xrightarrow{C_2H_5OH}$$

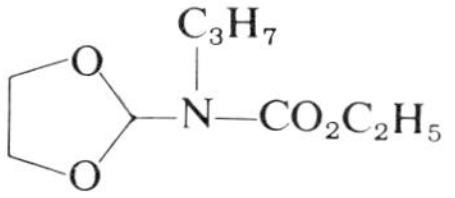

[dioxolane structure with N(C$_3$H$_7$)—CO$_2$C$_2$H$_5$ substituent] (23)

1,1-Difluoroalkyl tertiary amines react with dihydroxy compounds and their sodium derivatives to form heterocyclic amide acetals. The disodio derivative of 2,2-dimethyl-1,3-propanediol and 1,1-difluorotrimethylamine yield 2-dimethylamino-5,5-dimethyl-1,3-dioxane (*56*).

$$HCF_2N(CH_3)_2 + (CH_3)_2C(CH_2ONa)_2 \longrightarrow$$

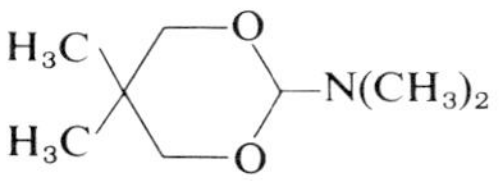

[1,3-dioxane with 5,5-dimethyl and 2-N(CH$_3$)$_2$ substituents] (24)

1,2-*O*-Isopropylidene-α-D-xylofuranose and diethyl(2-chloro-1,1,2-trifluoroethyl)amine react in methylene chloride to form a mixture of diastereomeric amide acetals, **11** (*93*).

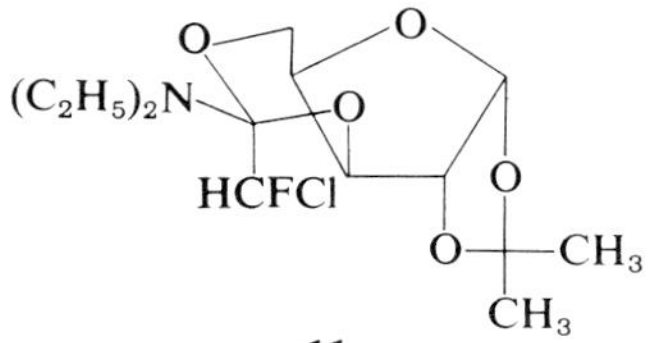

[structure of compound **11** with (C$_2$H$_5$)$_2$N, HCFCl, and isopropylidene-xylofuranose moieties]

**11**

Ethyl benzimidate hydrochloride and bis(2-hydroxycyclohexyl)amine react in ethanol solution to form the benzamide acetal **13** (*144*).

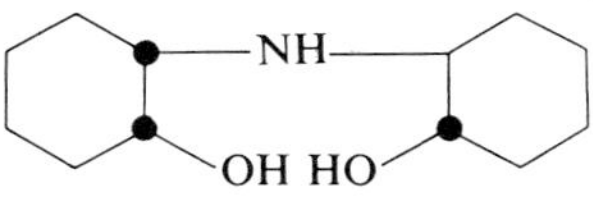

[structure of compound **12**: bis(2-hydroxycyclohexyl)amine with NH, OH HO] + $C_6H_5C(=NH_2{}^+)OC_2H_5Cl^-$ $\longrightarrow$

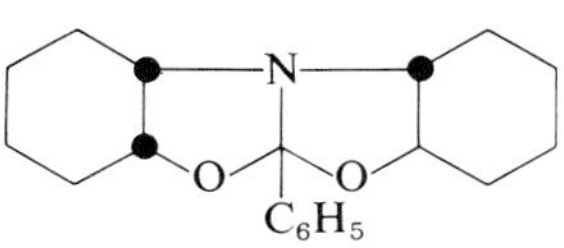

**12**

[structure of compound **13**: fused bicyclic benzoxazoline with C$_6$H$_5$] (25)

**13**

Compound **13** is also the product of successive treatment of the *N*-benzoyl derivative of **12** with thionyl chloride and sodium hydroxide, and of reaction of cyclohexene oxide with 2-phenylhexahydrobenzoxazole (*144*).

Ethyl benzimidate adds to diketene to form 2-ethoxy-6-methyl-2-phenyl-3,4-dihydro-2$H$-1,3-oxazine-4-one (*105*).

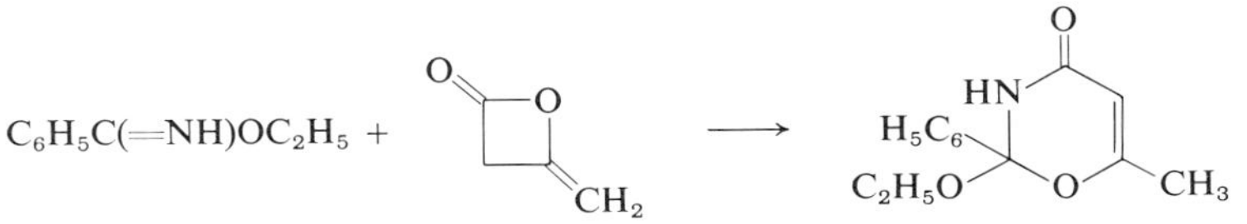

$$C_6H_5C(=NH)OC_2H_5 + \text{[diketene]} \longrightarrow \text{[oxazinone product]} \quad (26)$$

2-Oxazolines and 5,6-dihydro-4$H$-1,3-oxazines add to epoxides to form bicyclic amide acetals (*80, 82*).

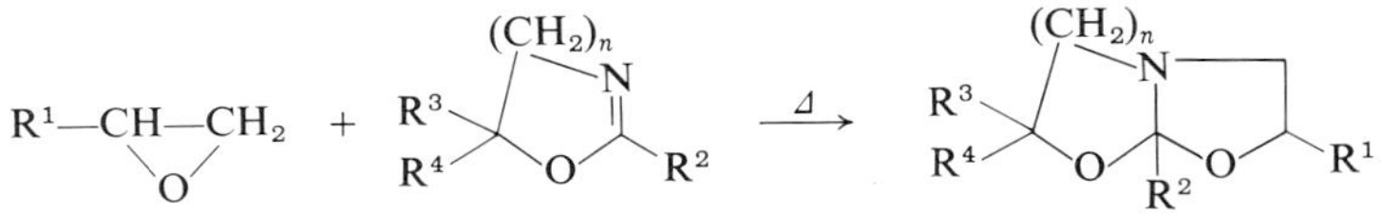

$$R^1-CH-CH_2 + \text{[oxazoline]} \xrightarrow{\Delta} \text{[bicyclic product]} \quad (27)$$

| $R^1$ | $R^2$ | $R^3$ | $R^4$ | $n$ |
|---|---|---|---|---|
| $C_6H_5OCH_2$ | $CH_3, C_2H_5,$ $C_3H_7, C_6H_5$ | H | H | 1 |
| $C_6H_5, C_3H_5OCH_2,$ $C_{12}H_{25}OCH_2$ | $C_2H_5$ | H | H | 1 |
| H, $CH_3$ | $C_6H_5$ | H | H | 1 |
| $C_6H_5OCH_2$ | $C_6H_5$ | H, $C_2H_5$ | $CH_3$ | 2 |
| $C_6H_5$ | $CH_3$ | H | H | 2 |
| $C_3H_5OCH_2$ | $C_{11}H_{23}$ | H | $C_6H_5$ | 2 |

Benzoylhydrazones are oxidized at low temperatures by lead tetraacetate in methylene chloride to 2-acetoxy-2-phenyl-5,5-disubstituted-$\Delta^3$-1,3,4-oxadiazines [Eq. (28)] (*94*). The reaction involves initial formation of $R'R''C(OAc)N=NCOC_6H_5$, which cyclizes when the temperature is raised from $-40°$ to $-20°C$ (*95, 95a*).

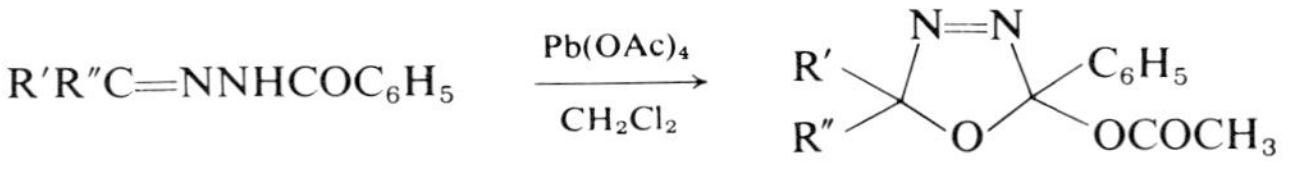

$$R'R''C=NNHCOC_6H_5 \xrightarrow[CH_2Cl_2]{Pb(OAc)_4} \text{[oxadiazine product]} \quad (28)$$

$[R'R'' = (CH_2)_5; R' = CH_3, R'' = C_6H_5]$

1,3-Dioxolane and 2- and 4-methyl-1,3-dioxolane are converted to ethyl $N$-(2-dioxolanyl)carbamates by photochemical reactions with ethyl azidoformate. The reaction presumably involves insertion of carbethoxynitrene into the carbon–hydrogen bond of the dioxolane (*119*).

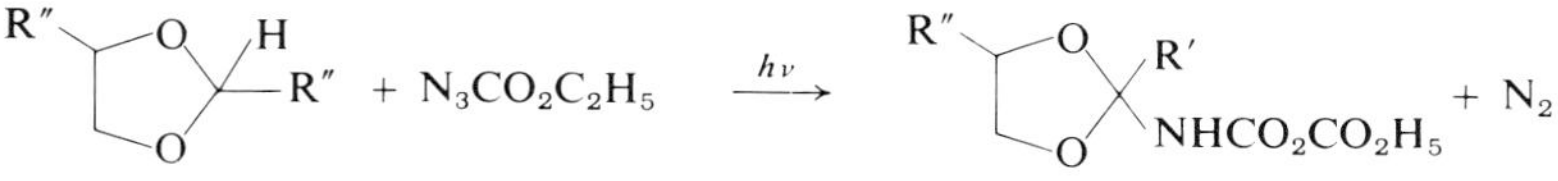

$$R''\text{-}\underset{O}{\overset{O}{\diagdown}}\overset{H}{\underset{}{\diagup}}R'' + N_3CO_2C_2H_5 \xrightarrow{h\nu} R''\text{-}\underset{O}{\overset{O}{\diagdown}}\overset{R'}{\underset{NHCO_2CO_2H_5}{\diagup}} + N_2 \quad (29)$$

(R' = R" = H; R' = H, R" = CH$_3$; R' = CH$_3$, R" = H)

o-Benzoylacetanilide undergoes electrochemical reduction at −0.2 volts in aqueous sulfuric acid to give a product whose composition and properties suggest the orthoamide hemiacetal structure **14** (*108*).

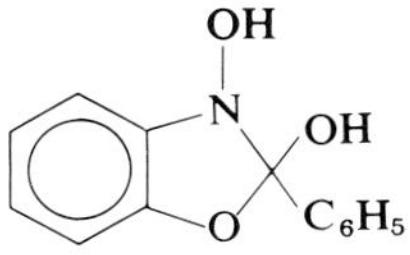

**14**

b. *Heterocyclic Amide Acetals Derived from Substituted Orthocarboxylic Acids.* N,N-Dimethylacetamide dimethyl acetal reacts with methyl vinyl ketone to form the bicyclic amide acetal **15**. An eneamine derived from the amide acetal by loss of methanol probably undergoes electrophilic addition of the vinyl ketone to form the product (*120*). **15** is also formed by reaction of methyl vinyl ketone with 1-dimethylamino-1-ethylthioethylene (*117a*).

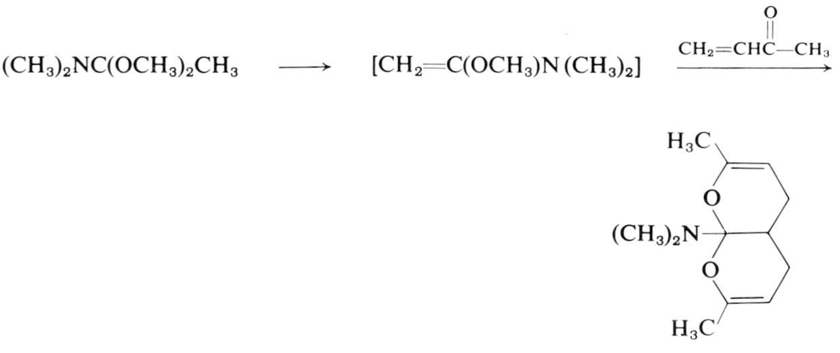

$$(CH_3)_2NC(OCH_3)_2CH_3 \longrightarrow [CH_2=C(OCH_3)N(CH_3)_2] \xrightarrow{CH_2=CHC-CH_3\overset{O}{\parallel}}$$

$$\text{(CH}_3)_2\text{N}\text{—structure 15—} \quad (30)$$

**15**

Carbonium salts derived from lactams, like their open-chain analogues, react with alkoxides in alcohol to form heterocyclic amide acetals. N-Methylpyrrolidone diethyl acetal (**17**), was prepared by ethylation of the lactam with triethyloxonium tetrafluoroborate followed by reaction of the resulting salt with sodium ethoxide (*114*).

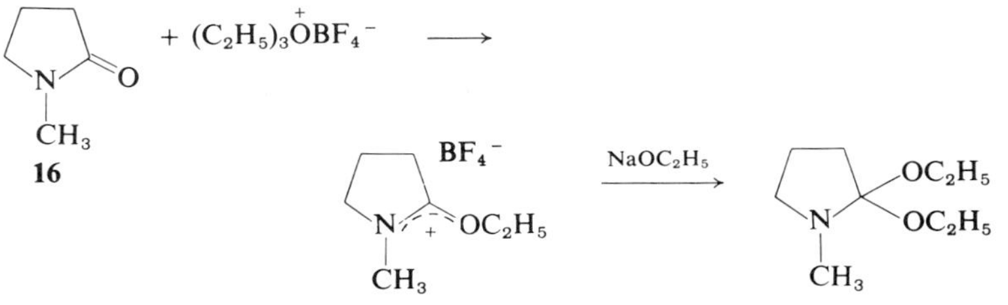

[Structure 16: N-methylpyrrolidone] + $(C_2H_5)_3\overset{+}{O}BF_4^-$ ⟶

[Structure: pyrrolidinium with $OC_2H_5$, $BF_4^-$, N-CH₃] $\xrightarrow{NaOC_2H_5}$ [Structure 17: N-methylpyrrolidine with two $OC_2H_5$ groups] (31)

The same compound is obtained by successive treatment of N-methyl-pyrrolidone with phosgene and sodium ethoxide (21, 76, 77).

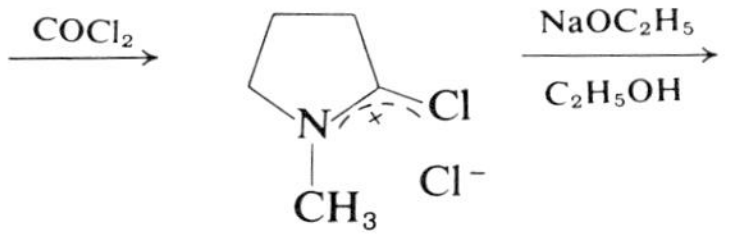

**16** $\xrightarrow{COCl_2}$ [Structure 18: chloride salt] $\xrightarrow[C_2H_5OH]{NaOC_2H_5}$ **17** (32)

N-Methylpyrrolidone dimethylacetal has been prepared from the amide chloride **18** and sodium methoxide (77), and also from the N-methylpyrroli-done–dimethyl sulfate addition complex and sodium methoxide (53). 1-Methylquinolone diethyl acetal, **19**, was prepared by a reaction sequence analogous to that of Eq. (31) (115).

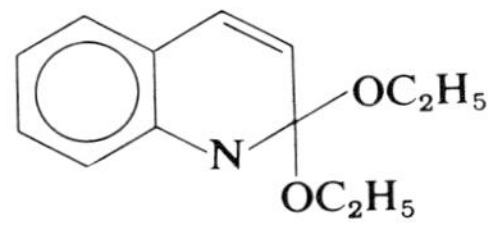

[Structure 19: quinoline with two $OC_2H_5$ groups]

The carbonium salt prepared by reaction of N-(3-bromopropyl)phthalimide with silver perchlorate reacts with potassium ethoxide and with potassium p-nitrophenolate to form polycyclic amide acetals **20** (103).

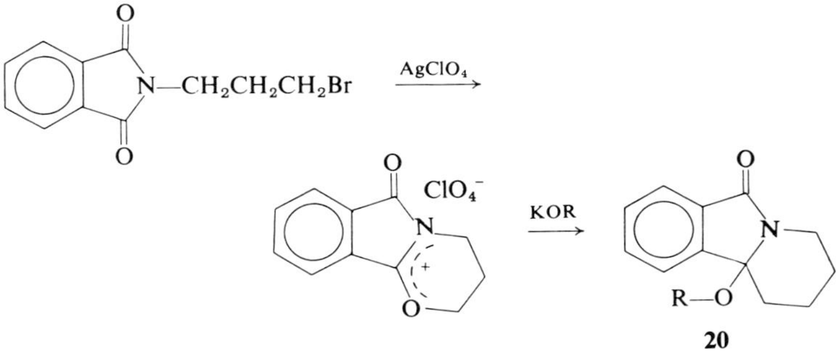

[Structure: phthalimide-N-CH₂CH₂CH₂Br] $\xrightarrow{AgClO_4}$

[Structure: cyclic cation with $ClO_4^-$] $\xrightarrow{KOR}$ [Structure **20**: polycyclic amide acetal with R—O] (33)

(R = $C_2H_5$, p-$O_2NC_6H_4$)

3-(2-Bromoethyl)-3-phenyl-2-benzofuranone is converted to 6a-(substituted amino)-3a-phenyl-2,3,3a,6a-tetrahydro-4,5-benzofuro[2,3-b]furans (**24**) by warming with secondary amines. Ortho amide hemiacetals (**22**) or their conjugate bases (**23**) may be transient intermediates in these reactions (*166*).

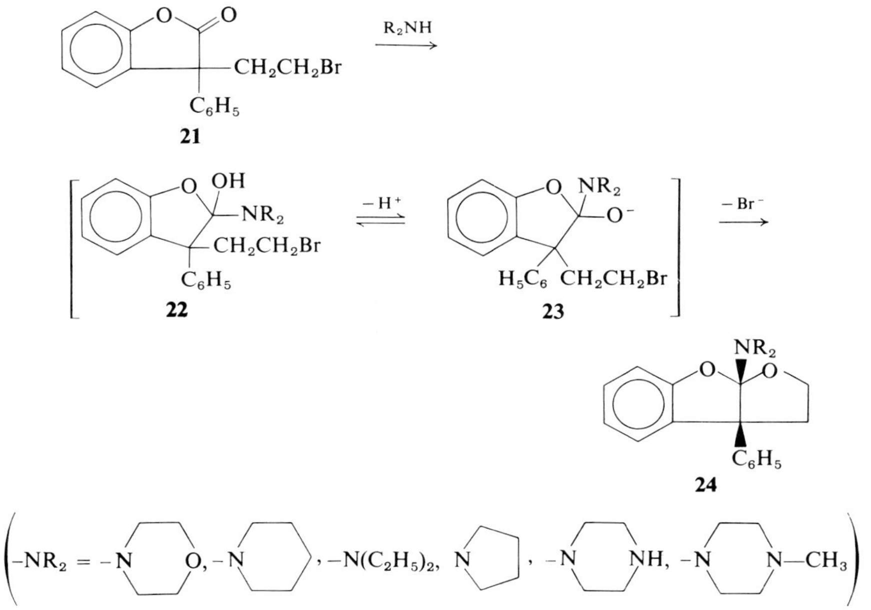

$$\left(-NR_2 = -N\bigcirc O, -N\bigcirc, -N(C_2H_5)_2, N\bigcirc, -N\bigcirc NH, -N\bigcirc N-CH_3\right)$$

(34)

In addition to the tricyclic orthoamides **24**, varying amounts of rearranged amides **25** (formed by aminolysis of the lactone followed by cyclization of the resulting substituted phenol) were produced in these reactions. The 3-bromomethyl and 3-(3-bromopropyl) homologues of **21** yielded products other than ortho amides in reactions with secondary amines. The 6a-pyrrolidinyl ortho amide $\left(\textbf{24}, NR_2 = -N\bigcirc\right)$ yields an N-methyl quaternary salt when methylated with methyl iodide in chloroform.

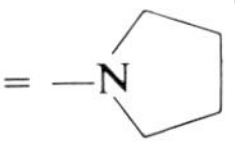

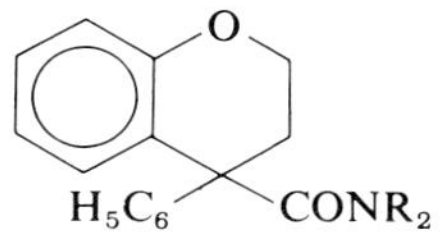

434                  7. AMIDE ACETALS, ESTER AMINALS, AND ORTHO AMIDES

Reaction of cyclopropylamine, a primary amine, with **21** yielded a product whose NMR spectrum indicated that it was an equilibrium mixture of the ortho amide **24**, $NR_2 = -NH\text{-cyclo-}C_3H_5$ and the isomeric imidate **25** (*169*).

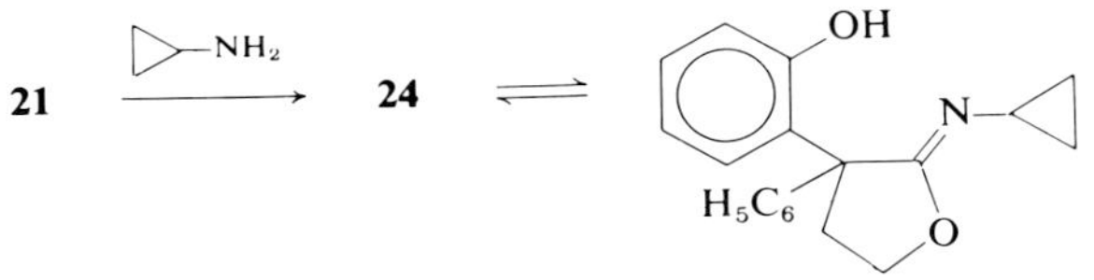

$$21 \xrightarrow{\triangleright-NH_2} 24 \rightleftharpoons 26$$

$$\left( NR_2 = NH-\triangleleft \right) \qquad (35)$$

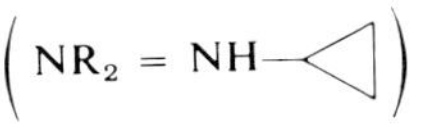

When liquid ammonia in acetonitrile was used, a mixture consisting mostly of **25**, R = H, and some **24**, R = H, was obtained. Fractional crystallization afforded the pure cyclic ortho amide **24**, R = H (*170*). Compound **24**, R = H, was also obtained by neutralization of the imidate hydrochloride which it formed on reaction with hydrochloric acid (*170*).

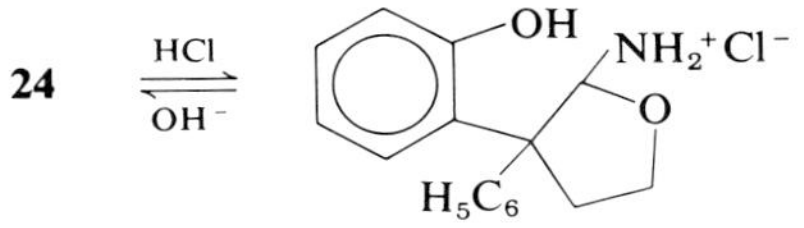

$$24 \underset{OH^-}{\overset{HCl}{\rightleftharpoons}} \text{[structure]} \qquad (36)$$

(R = H)

Refluxing 3-(2-haloethyl)-3-phenyloxindole with sodium methoxide yields a halogen-free product having no carbonyl absorption in the infrared. The NMR spectrum of this product in deuterochloroform, together with its elementary composition and molecular weight, suggest that in solution it is an equilibrium mixture of tricyclic ortho amide **27** and its open-chain isomer [Eq. (37)] (*167*). The mechanism of formation of the ortho amide is probably analogous to that of the closely related reaction of Eq. (34).

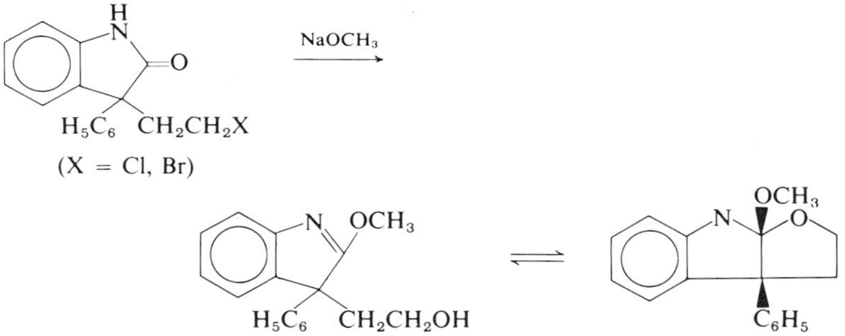

$$\text{[oxindole structure, } H_5C_6, CH_2CH_2X\text{]} \xrightarrow{NaOCH_3} \text{[imidate structure]} \rightleftharpoons 27 \qquad (37)$$

(X = Cl, Br)

Scarpati, Del Re, and Maone reported in 1959 that ketene diethyl acetal and phenyl isocyanate react in a 1:2 ratio to form 1,3-diphenylbarbituric acid 4-diethyl acetal (*125*). Effenberger, Gleiter, and Kiefer later showed that β,β-diethoxyacrylamides are intermediates in this reaction, and that by carrying out the reaction in two steps it is possible to prepare barbituric acid acetals with different substituents at the 1- and 3-positions [Eq. (38)] (*73*). The reactions were carried out in inert solvents or without solvents, and require no catalyst. The first step in reaction (38) presumably involves electrophilic addition of the isocyanate to the ketene acetal. The N—H group of the resulting diethoxyacrylamide then adds to a second isocyanate molecule to form an *N*-acryloylurea derivative, which cyclizes to the barbituric acid acetal.

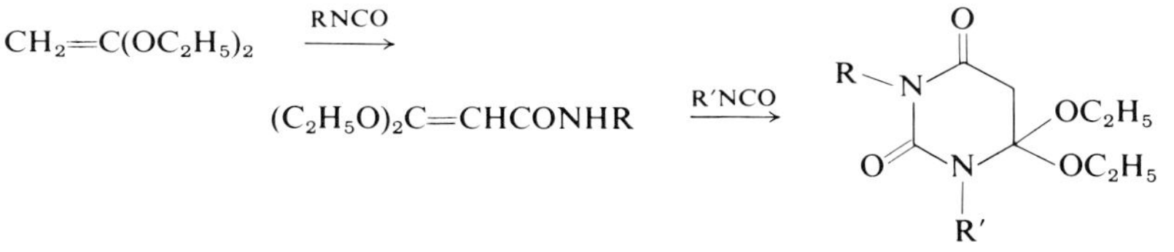

$$CH_2=C(OC_2H_5)_2 \xrightarrow{RNCO}$$

$$(C_2H_5O)_2C=CHCONHR \xrightarrow{R'NCO} \tag{38}$$

(R = R' = $C_6H_5$, $p$-$O_2NC_6H_4$, $p$-$ClC_6H_4$, $p$-$C_2H_5OC_6H_4$, $p$-$CH_3C_6H_4SO_2^-$; R = $C_6H_5$, R' = $p$-$ClC_6H_4$, $o$-$CH_3C_6H_4$; R = $p$-$ClC_6H_4$, $o$-$CH_3C_6H_4$, R' = $C_6H_5$)

Substituted ketene acetals also react with phenyl isocyanate, but the products are substituted 4,4-dialkoxyazetidine-2-ones rather than barbituric acid derivatives (*125*).

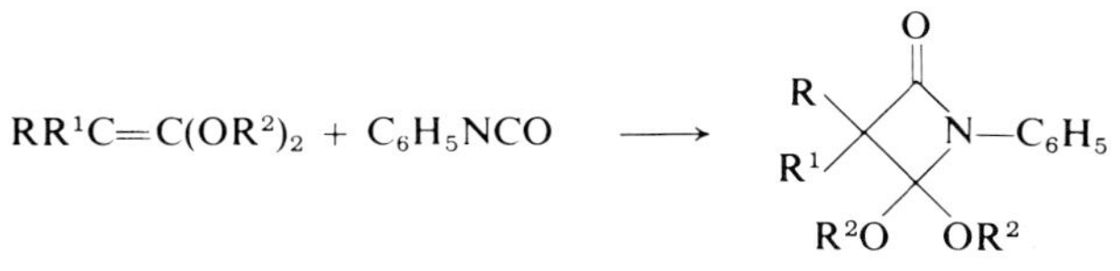

$$RR^1C=C(OR^2)_2 + C_6H_5NCO \longrightarrow \tag{39}$$

(R = $C_6H_5$, $R^1$ = H, $R^2$ = $C_2H_5$; R = $R^1$ = $R^2$ = $CH_3$)

These are the products to be expected from cyclization of an intermediate α-substituted β,β-dialkoxyacrylamide (in the case of a monosubstituted ketene acetal) or the zwitterion $(R^2O)_2\overset{+}{C}CRR^1$—CO—$\overset{-}{N}$—$C_6H_5$ (in the case of a disubstituted ketene acetal). Isothiocyanates and dimethylketene dimethyl acetal undergo an analogous reaction (*126*).

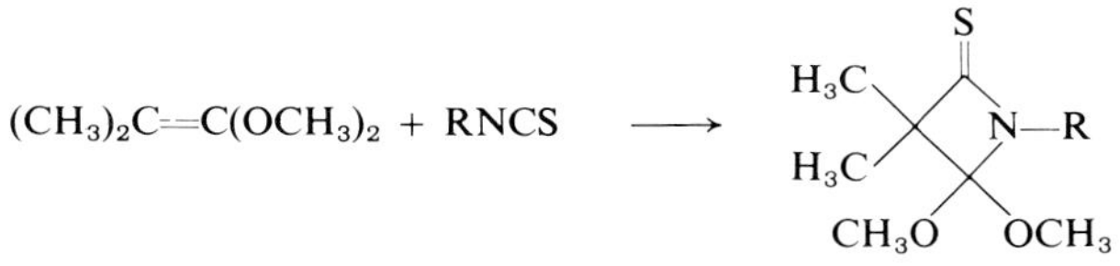

$$(CH_3)_2C=C(OCH_3)_2 + RNCS \longrightarrow \tag{40}$$

(R = $C_4H_9$, $C_6H_5$)

436   7. AMIDE ACETALS, ESTER AMINALS, AND ORTHO AMIDES

Ketene acetals react with phenyl azide to form 5,5-dialkoxy-1-phenyl-1,2,3-triazolines, and with aroyl azides to form 2-aryl-4,4-dialkoxyoxazolines (*125a*).

A substituted 4,4-diethoxyazetidine-2-one is also the product of the reaction of diethyl *N*-phenylimidocarbonate with diphenylacetyl chloride in refluxing carbon tetrachloride–triethylamine solution (*117*).

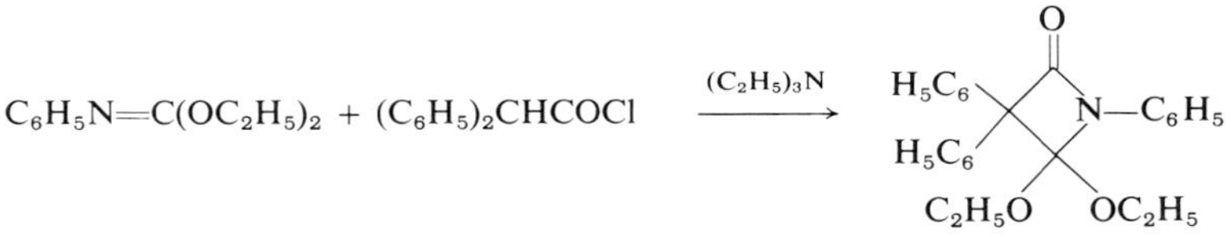

$$C_6H_5N=C(OC_2H_5)_2 + (C_6H_5)_2CHCOCl \xrightarrow{(C_2H_5)_3N} \text{[azetidinone product]} \quad (41)$$

The recently discovered *N*-sulfonylamines also undergo cycloaddition reactions with ketene acetals (*59*).

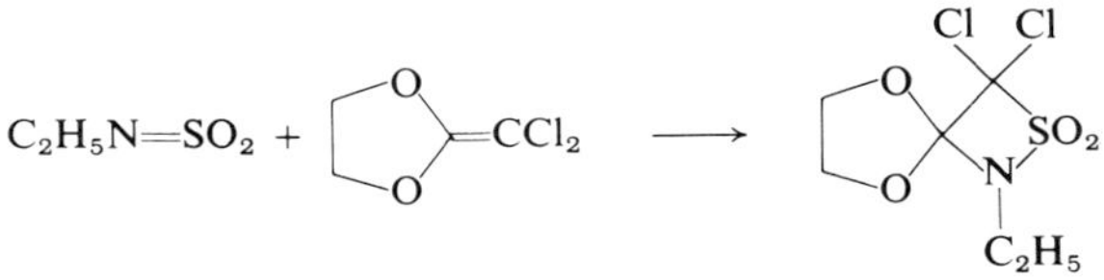

$$C_2H_5N=SO_2 + \text{[dioxolane=CCl}_2\text{]} \longrightarrow \text{[cycloadduct]} \quad (42)$$

Among the products of electrochemical methoxylation of 2,6-dimethoxypyridine is the cyclic ortho amide **28**. Similar methoxylation of *N*-methylpyrrole affords **29a**, which was converted to **29b** by acid hydrolysis in 0.05 *N* acetic acid (*150*).

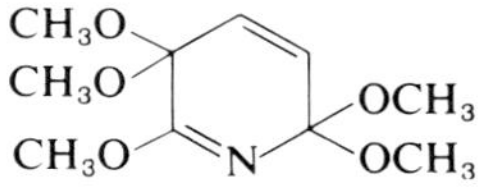

**28** 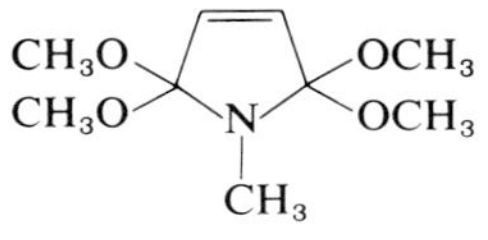     **29a** 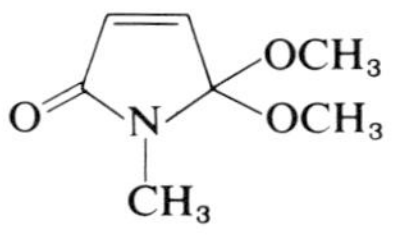     **29b**

A number of heterocyclic amide acetals have been synthesized by methylating various heterocyclic amide hemiacetals (also called cyclols—see below) with methyl iodide and silver oxide (*91, 121, 129, 130*). A representative example of this reaction is given in Eq. (43) (*130*). Other amide acetals prepared in this manner are listed in Table I, Section III.

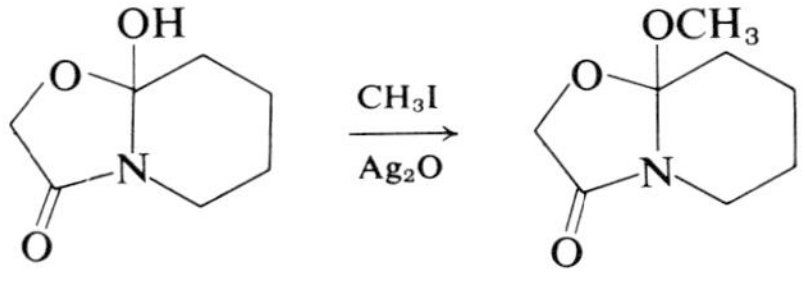

$$\text{[hemiacetal]} \xrightarrow[\text{Ag}_2\text{O}]{\text{CH}_3\text{I}} \text{[methyl ether]} \quad (43)$$

Dimethylketenimines, o-quinonediazides and alcohols react to form 2-amino-2-alkoxy-3,3-dimethyl-2,3-dihydrobenzoxadiazepines, **29c** (*123d*).

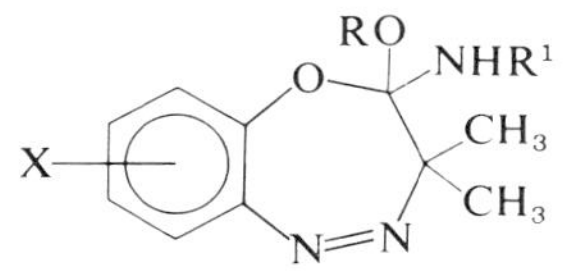

**29c**

c. *Heterocyclic Amide Hemiacetals—The Cyclols.* N-Hydroxyalkyl carboxamides should exist in equilibrium with isomeric substances formed by intramolecular addition of the hydroxyl group to the amide carbonyl group.

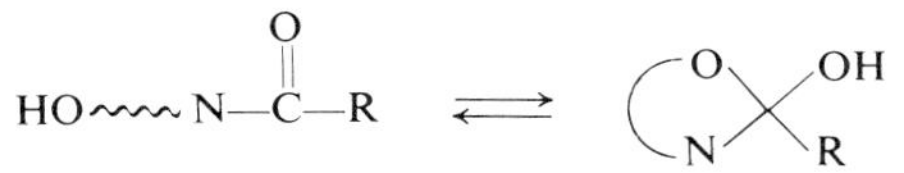

$$\text{HO}\sim\sim\text{N--C--R} \rightleftarrows \text{cyclic form} \tag{44}$$

The cyclic addition products may be regarded both as ortho acid amides and as amide hemiacetals. In most cases they are so unstable relative to the hydroxyamide (due to loss of the resonance energy of the amide group accompanying their formation) that the ortho amide isomers cannot be detected experimentally.

Compounds having heterocyclic amide hemiacetal structures were named cyclols by Wrinch [(*164*) and references therein], and the term was adopted by others. Cyclols have frequently been proposed as transient intermediates in amide isomerizations and other reactions (*2, 10, 13, 85, 104, 131, 168*), but the first examples of stable compounds possessing cyclol structures were the ergot alkaloids and peptides derived from them. Stoll and co-workers demonstrated that these alkaloids have the general structures **30** (5R:8R:2'R:5'S:11'S:12'S) (*98, 143*).

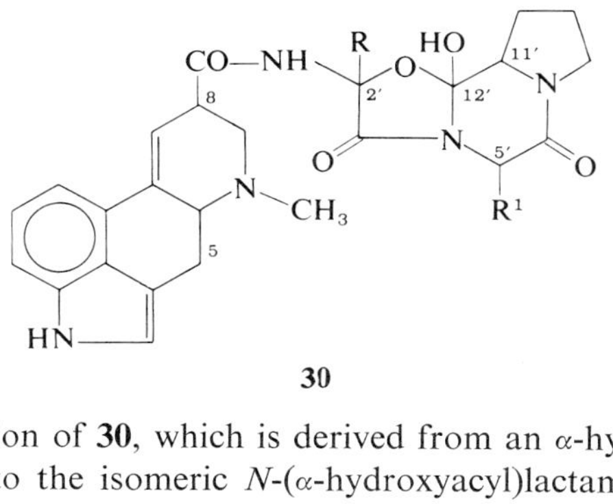

**30**

The cyclol portion of **30**, which is derived from an α-hydroxyacyl lactam, is stable relative to the isomeric *N*-(α-hydroxyacyl)lactam and macrocyclic peptide, whereas simpler cyclols are unstable relative to their isomerization products. What structural features render the cyclol the thermodynamically most stable product in equilibria of the type generalized by Eq. (45)?

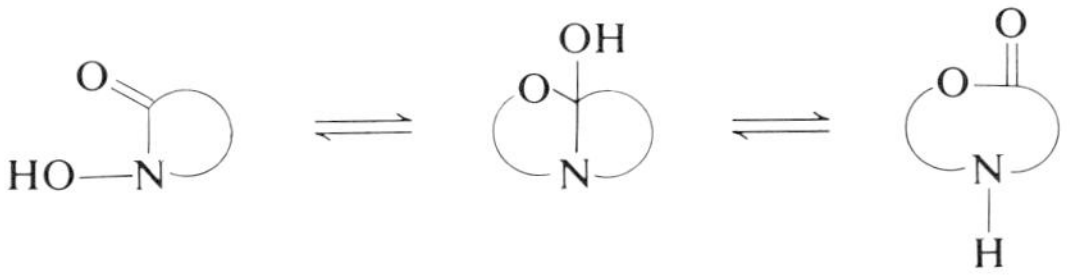

$$\underset{HO-N}{\overset{O}{\bigcirc}} \rightleftharpoons \underset{N}{\overset{OH}{\bigcirc}} \rightleftharpoons \underset{\underset{H}{N}}{\overset{O}{\bigcirc}} \quad (45)$$

This question was answered by Hoffman, Frey, and co-workers, and by Shemyakin, Antonov, and co-workers. Both these groups undertook to study systematically, using simple model compounds, the factors affecting the position of equilibria between cyclols and substances isomeric with them (91, 130). Their findings are summarized below.

Cyclol formation is strongly influenced by the electrophilicity and steric accessibility of the lactam carbonyl group. N-Acylation activates amide carbonyl groups toward nucleophilic attack, and it is observed that $N$-($\alpha$-hydroxyacyl)lactams form cyclols readily, whereas $N$-($\beta$-hydroxyalkyl)-lactams do not (91).

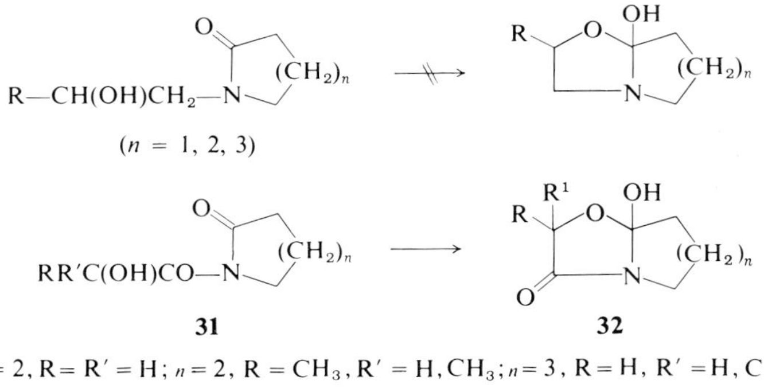

$$\text{R—CH(OH)CH}_2\text{—N}\underset{(n = 1, 2, 3)}{\overset{O}{\bigcirc}}(CH_2)_n \nrightarrow \underset{N}{\overset{R\quad O\quad OH}{\bigcirc}}(CH_2)_n$$

$$\text{RR'C(OH)CO—N}\overset{O}{\underset{31}{\bigcirc}}(CH_2)_n \longrightarrow \underset{\underset{O}{N}}{\overset{R^1\quad OH}{\underset{32}{\bigcirc}}}(CH_2)_n \quad (46)$$

($n = 2, R = R' = H; n = 2, R = CH_3, R' = H, CH_3; n = 3, R = H, R' = H, CH_3$)

The activating carbonyl must be exocyclic to the lactam ring: $N$-($\beta$-hydroxyalkyl)glutarimides (33) do not form cyclols (91).

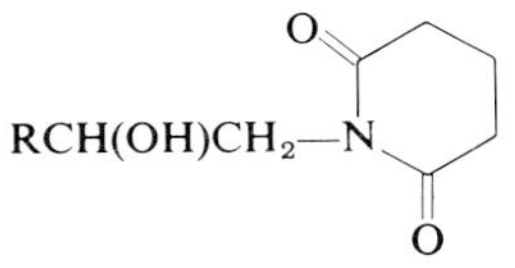

$$\text{RCH(OH)CH}_2\text{—N}\underset{\underset{O}{33}}{\overset{O}{\bigcirc}}$$

Carbonyl groups in five-membered rings show reduced electrophilicity, and $N$-($\alpha$-hydroxyacyl)butyrolactams (**31**, $n = 1$) do not isomerize to cyclols.

Alkyl substituents in the oxazolidone ring stabilize the cyclol relative to the hydroxyacyl lactam. For example, in chloroform **32**, $n = 2$, $R = R' = H$, isomerizes to a mixture of about one part **31** and two parts **32**. Compound **32**, $n = 2$, $R = H$, $R' = CH_3$, however, contained no detectable amount of **31** under the same conditions (*91*). The stabilizing effect of alkyl substituents on the cyclic compound in equilibria between cyclic and open-chain isomers is a well-known phenomenon (*57*).

Cyclols of the type **32** are in equilibrium not only with hydroxyacyl lactams **31**, but also with macrocyclic ester-amides of type **34**.

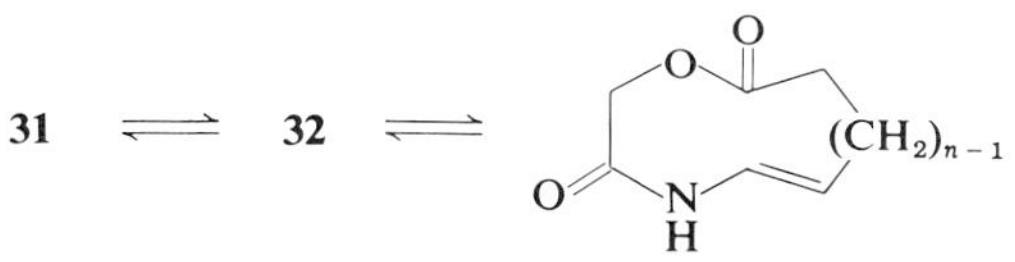

$$31 \rightleftharpoons 32 \rightleftharpoons \underset{\mathbf{34}}{\text{[structure]}} \quad (CH_2)_{n-1} \qquad (47)$$

In order for the equilibrium to favor the macrocyclic compound, it is necessary that the ring contain 11 or more atoms. Hydroxyacyllactams which would yield 10-membered macrocycles do not do so, but exist in equilibrium with the cyclol. Those which would form 9-membered macrocyclic rings tend to isomerize completely to the cyclol (*130*).

Most of the cyclols which have been synthesized (see Table I) were prepared from $N$-($\alpha$-hydroxyacyl)lactams. These in turn are usually synthesized by treating the appropriate lactam with a 2-benzyloxy-substituted acid chloride and then removing the protecting benzyl group by hydrogenolysis [Eq. (48)]. Diketopiperazines and other substances having lactam functions have also been converted to cyclols by similar procedures, as have $N$-($\beta$-hydroxyacyl)lactams and salicyllactams.

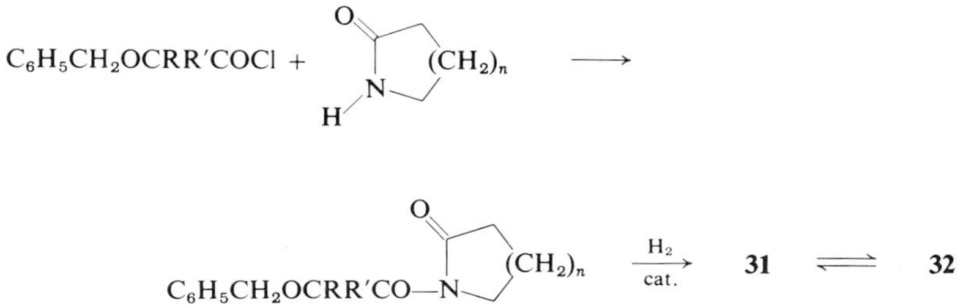

$$C_6H_5CH_2OCRR'COCl + \underset{H}{\text{[lactam]}}(CH_2)_n \longrightarrow$$

$$C_6H_5CH_2OCRR'CO-N\text{[structure]}(CH_2)_n \xrightarrow[\text{cat.}]{H_2} 31 \rightleftharpoons 32 \quad (48)$$

In one instance the synthesis of a cyclol was approached from the macrocyclic ester-amide rather than from the hydroxyacyllactam (*132*).

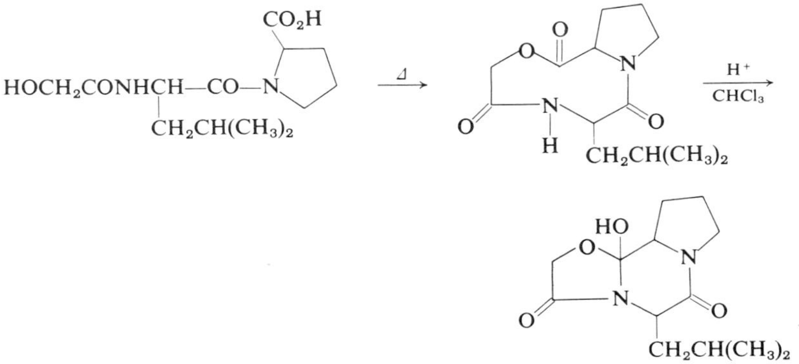

$$\text{HOCH}_2\text{CONHCH—CO—N} \begin{array}{c} \text{CO}_2\text{H} \\ | \\ \end{array} \xrightarrow{\Delta} \quad \xrightarrow[\text{CHCl}_3]{\text{H}^+} \quad \quad (49)$$

Among natural products containing the cyclol ortho amide hemiacetal system are the ergot alkaloids (**30**), and the antibiotic griseolutein B (**35**) (*118*).

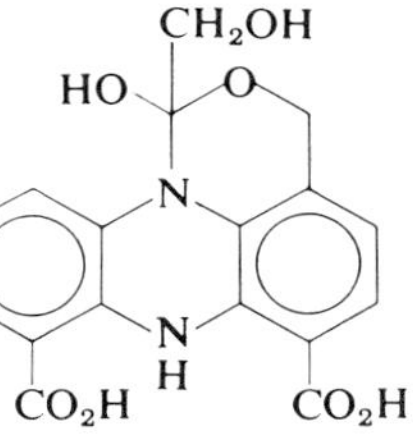

**35**

d. *Heterocyclic Carbamide Acetals.* Few heterocyclic amide acetals derived from carbonic acid have been reported. Meerwein and co-workers synthesized 1,3-dimethyl-2,2-diethoxyimidazolidine (**36**), 2,2-diethoxy-3-methylbenzoxazoline (**37**), and 1,3-dimethyl-2,2-diethoxybenzimidazoline (**38**) from 1,3-dimethylimidazolidone, 3-methylbenzoxazolone, and 1,3-dimethylbenzimidazolone by ethylating the heterocyclic carbonyl compounds with triethyloxonium tetrafluoroborate and treating the resulting carbonium salts with sodium ethoxide (*115*).

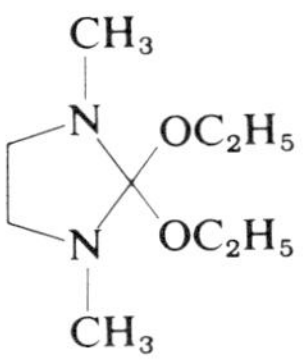

**36** 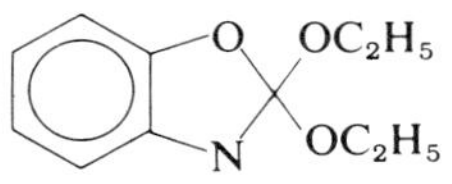  **37** 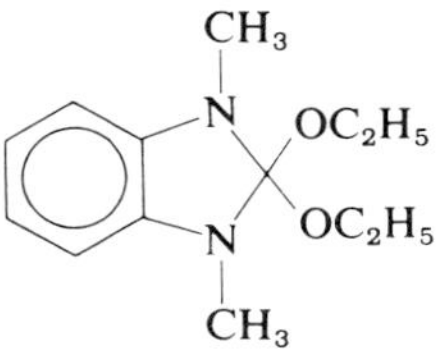  **38**

2,2-Dichloro-1,3-benzodioxole reacts with piperidine to form 2,2-dipiperidino-1,3-benzodioxole, **39** (*92*).

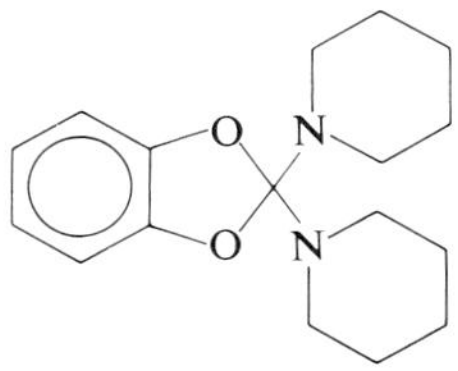

**39**

The carbonium salt obtained by ethylation of 1-methyl-4,6-diphenyl-1,2-dihydropyrimidin-2-one with triethyloxonium tetrafluoroborate reacts with sodium ethoxide to form the diethyl acetal of the starting material, which is a carbamide acetal (*165a*).

## B. Ester Aminals and Related Compounds

Ester aminals are substances having two nitrogen atoms and an oxygen atom bonded to an orthoacyl carbon. They have received less attention than the amide acetals, and only a limited number of them have been synthesized. This is in part due to their high reactivity and their tendency to disproportionate to amide acetals and ortho amides (*53*).

### 1. Aminals of Carboxylate Esters

All of the ester aminals which have been described in the literature are derivatives of orthoformic acid. The formate aminals are most conveniently prepared by reaction of N,N,N',N'-tetrasubstituted formamidinium salts with sodium or potassium alkoxides. In the procedure developed in Bredereck's laboratories, $N,N$-dimethylformamide is converted to $N,N$-dimethylaminomethoxycarbonium methyl sulfate by reaction with dimethyl sulfate. This salt is treated with dimethylamine to obtain the tetramethylformamidinium methyl sulfate, and the amidinium salt is added to the alkoxide. The synthesis of bis(dimethylamino)alkoxymethanes by this method is outlined in Eq. (54) (*38–42, 50, 53, 160*). The amidinium salt is obtained in greater than 90% yield, and is converted to the ester aminals in yields ranging from 45% to 70%. As indicated, the reaction produces aminal esters of primary, secondary, and tertiary alcohols.

$$HCON(CH_3)_2 \xrightarrow{(CH_3)_2SO_4} (CH_3)_2\overset{+}{N}CHOCH_3CH_3SO_4^- \xrightarrow[-CH_3OH]{+(CH_3)_2NH}$$
$$[(CH_3)_2N]_2\overset{+}{C}HCH_3SO_4^- \xrightarrow{RONa} [(CH_3)_2N]_2CHOR \quad (50)$$
$$[R = CH_3, C_2H_5, C_3H_7, (CH_3)_2CH, (CH_3)_3C]$$

Formate aminals disproportionate if heated or allowed to stand for long periods of time (*50, 53, 67*).

$$[RN(CH_3)]_2CHOR' \rightleftarrows RN(CH_3)CH(OR')_2 + HC[N(CH_3)R]_3 \quad (51)$$
$$[R = CH_3, C_6H_5; R' = CH_3, C_2H_5, CH(CH_3)_2, C(CH_3)_3]$$

Equilibrium mixtures containing the ester aminal, the amide acetal, and the ortho amide are obtained. This fact makes it difficult to prepare and store pure samples of formate aminals. Purer products are obtained if the esters are prepared by adding the tetramethylformamidinium salt to an alcohol-free suspension of the alkoxide in ether or cyclohexane rather than to an alcohol solution of the alkoxide (*50, 53*).

$N,N$-Dimethyl-$N,N'$-diphenylformamidinium tetrafluoroborate is also converted to formate aminals by alcoholic alkoxide solutions [Eq. (52)] (*66, 67*). These substances are stable under refrigeration, but on standing at room temperature or being heated are converted to a mixture of $N,N',N''$-trimethyl-$N,N',N''$-triphenylorthoformamide and unidentified more volatile substances.

$$[C_6H_5N(CH_3)]_2\overset{+}{C}HBF_4^- \xrightarrow[ROH]{Na\overset{+}{O}R^-} [C_6H_5N(CH_3)]_2CHOR \quad (52)$$
$$[R = CH_3, CH(CH_3)_2, C(CH_3)_3]$$

According to a patent claim, formate aminals may be prepared by aminolysis of formamide acetals. $N,N$-Dimethylformamide dimethyl acetal was heated with several heterocyclic and one aliphatic secondary amine to produce product mixtures containing formamide acetals (from transamination) and formate aminals (from aminolysis) (*161*).

$$(CH_3)_2NCH(OCH_3)_2 \xrightarrow{R_2NH} R_2NCH(OCH_3)_2 + (R_2N)_2CHOCH_3 \quad (53)$$

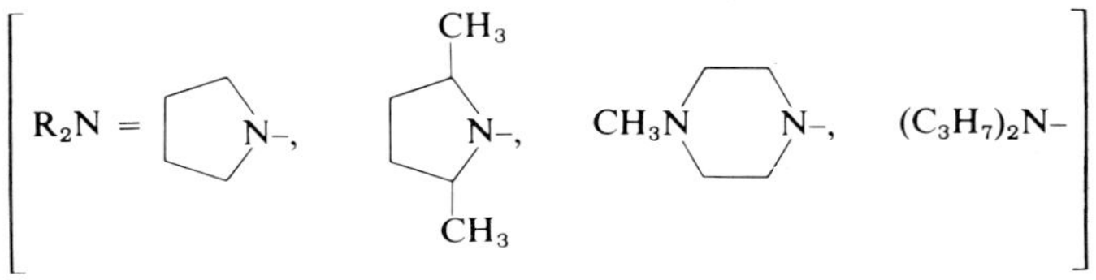

$$\left[ R_2N = \underset{}{\underset{N-,}{\bigcirc}} \quad \underset{CH_3}{\underset{\underset{CH_3}{|}}{\overset{\overset{CH_3}{|}}{\bigcirc}}} N-, \quad CH_3N\underset{}{\bigcirc}N-, \quad (C_3H_7)_2N- \right]$$

## 2. Heterocyclic Ester Aminals and Acid Aminals

There are a few heterocyclic compounds which have, as part of a hetero ring, an orthoacyl carbon bonded to one oxygen atom and two or three nitrogen atoms. These substances are regarded as heterocyclic ester aminals, or, when the oxygen atom is in a hydroxyl group, acid aminals.

2-Ethyl-1-phenylimidazoline adds to epoxides at elevated temperatures to form 4-oxa-1,6-diazabicyclo[3.3.0]octanes which are derivatives of orthopropionic acid (79).

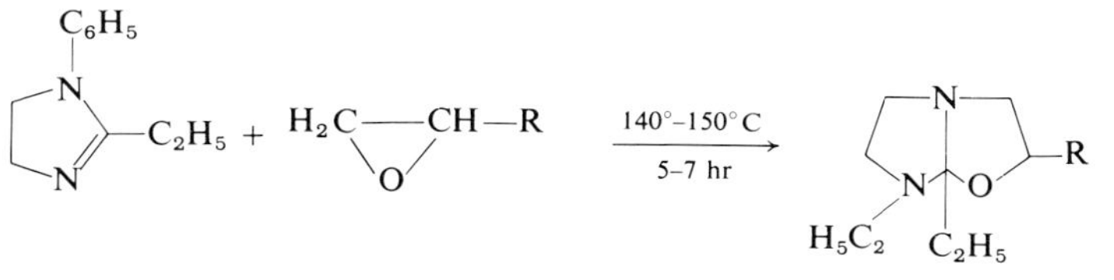

$$\text{2-Ethyl-1-phenylimidazoline} + H_2C\text{—}CHR\text{(epoxide)} \xrightarrow[5-7\ hr]{140°-150°C} \text{product} \quad (54)$$

(R = H, CH$_3$, C$_6$H$_5$, C$_6$H$_5$OCH$_2$, C$_3$H$_5$OCH$_2$, C$_{12}$H$_{22}$OCH$_2$)

Dimethylformamide diethyl acetal reacts with isocyanates at elevated temperatures to form $O,N$-acetals of 1,3-disubstituted parabanic acids (49).

$$(CH_3)_2NCH(OC_2H_5)_2 + 3\ RN=C=O \longrightarrow$$

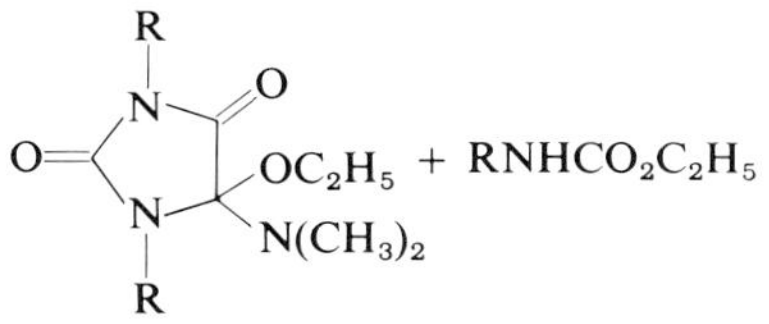

$$\text{parabanic acid derivative} + RNHCO_2C_2H_5 \quad (55)$$

(R = C$_6$H$_5$, cyclo-C$_6$H$_{11}$)

When methanol is added to a benzene solution of bis(1,3-diphenyl-2-imidazolinylidene), a purple color develops which disappears on warming. The product of this reaction is assumed to be 2-methoxy-1,3-diphenylimidazolidine, formed by reaction of methanol with the carbene produced by dissociation of the starting material (148).

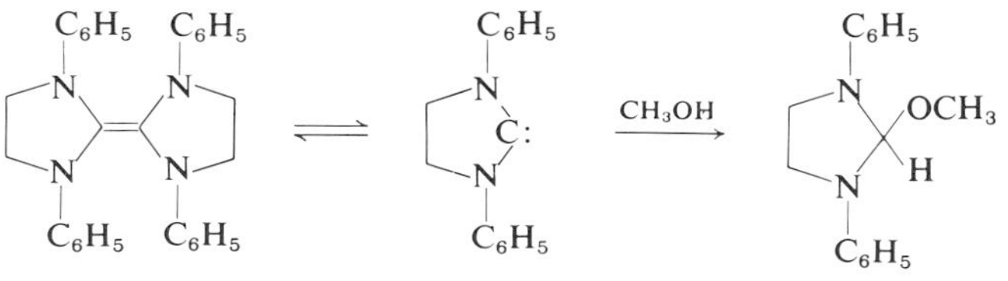

$$\text{bis(imidazolinylidene)} \rightleftharpoons \text{carbene} \xrightarrow{CH_3OH} \text{2-methoxy-1,3-diphenylimidazolidine} \quad (56)$$

The carbonium salt formed by reaction of $N$-(3-bromopropyl)phthalimide with silver perchlorate reacts with piperidine in acetonitrile to form the complex aminal **40** (103).

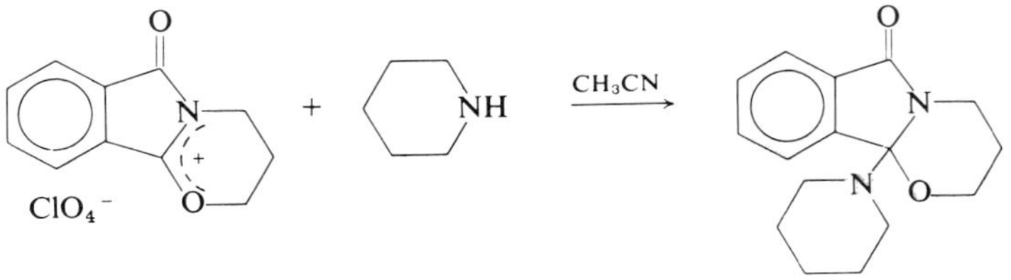

444                    7. AMIDE ACETALS, ESTER AMINALS, AND ORTHO AMIDES

[Reaction scheme showing isoindolinone-oxazinium perchlorate + piperidine → product **40**]  (57)

A structurally analogous quaternary salt is formed by reaction of the same carbonium ion with pyridine.

Nitration of 4-arylpyrimidines with nitric acid–acetic anhydride was claimed to yield the aminal acid anhydrides **41** (*110*). The assigned structures are consistent with the infrared and NMR spectra and elementary compositions of the reaction products, but are surprising in view of their resistance to hydrolysis and thermal decomposition.

The acid aminal (cyclol) structure **42** has been assigned to the alkaloid rhetsinene (*61*). Similar cyclol structures have been proposed for transient intermediates in isomerizations of *N*-(ω-aminoacyl) lactams to cyclic dipeptides (*1*) and the reverse of this reaction (*88*).

2-Chloro-1,3-dimethylbenzimidazolium tetrafluoroborate and *N*-methyl-

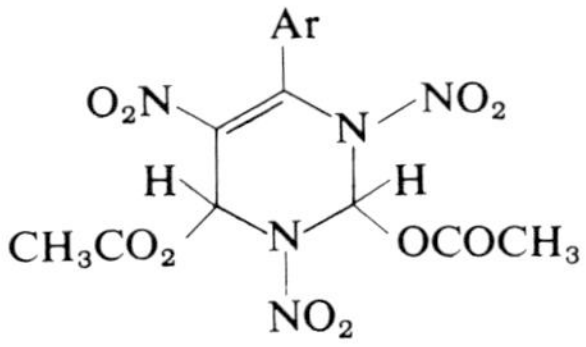

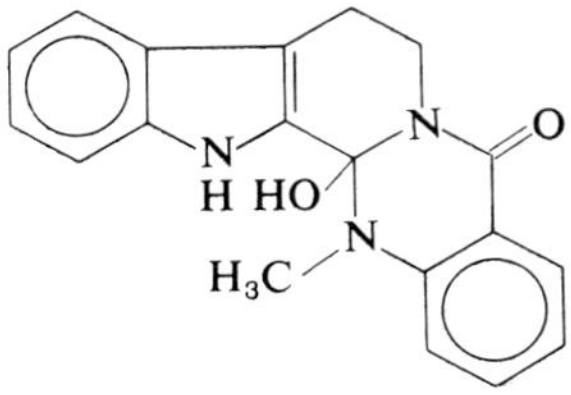

[Structures **41**, **42**, and **43**]

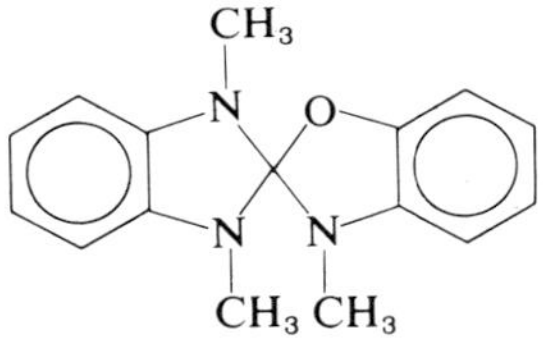

*o*-aminophenol react to form the tetracyclic orthocarbonic acid derivative **43** (*123*).

## C. Ortho Amides and Related Compounds

Ortho amides are substances having three nitrogen atoms bonded to an orthoacyl carbon or four nitrogens bonded to an orthocarbonyl carbon. All of the acyclic orthocarboxamides which have been reported are derivatives of orthoformic acid. Attempts to prepare ortho amides of higher orthocarboxylic acids thus far have yielded other products instead.

1. ORTHOFORMAMIDES

One obvious approach to the synthesis of ortho amides is the reaction of secondary amines with trialkyl orthocarboxylates. The first authentic orthocarboxamide, $N,N',N''$-trimethyl-$N,N',N''$-triphenylorthoformamide was synthesized from $N$-methylaniline and triethyl orthoformate. When a mixture of the theoretical ratio of the reactants was heated for a few hours or allowed to stand at room temperature for several days, the ortho amide was formed in about 20% yield (66, 67). While the reaction yields a very pure product, all efforts to improve the yield were unsuccessful.

$$3\ C_6H_5NHCH_3 + HC(OC_2H_5)_3 \rightarrow HC[N(CH_3)C_6H_5]_3 + 3\ C_2H_5OH \quad (58)$$

There is currently some confusion concerning the structure of the products obtained from $N$-alkylanilines and triethyl orthoformate. Hagedorn and Lichtel (92a, 92b), who apparently were unaware of Clemens' work, claimed to obtain tetrakis($N$-methyl-$N$-phenylamino)ethylene,

$$[C_6H_5N(CH_3)]_2C=C[N(CH_3)C_6H_5]_2,$$

from the same starting materials. The reported decomposition temperatures of the two products are similar, as are the melting points of the homologous $N$-ethyl-$N$-phenyl compounds. The orthoamides and tetraaminoethylenes have almost identical elementary compositions, and Hagedorn's spectroscopic data are compatible with either structure. The orthoformamide structure is probably correct. Clemens obtained the same product from several different starting materials, and also found that refluxing the ortho amide with methyl iodide gave a 73% yield of phenyltrimethylammonium iodide and a 77% yield of $N,N'$-diphenyl-$N,N'$-dimethylformamidinium iodide—a result difficult to account for if the starting material was the tetraaminoethylene.

Giacolone (87), and later Post and co-workers (109) claimed to have obtained $N,N',N''$-triphenylorthoformamide by the reaction of aniline with triethyl orthoformate. The product of this reaction is actually $N,N'$-diphenylformamidine (7, 8, 107, 124). It is doubtful if ortho amides can be prepared from trialkyl orthocarboxylates and primary amines, due to the ease with which the ortho amides decompose to amidines. Orthoformamides are formed by disproportionation of formate aminals [Eq. (51)] (50, 53, 67).

The disproportionation reactions yield mixtures of ortho acid derivatives which are difficult to separate into pure components, and hence are not useful for the synthesis of ortho amides.

Dropwise addition of dimethylformamide to an ether solution of tetrakis-(dimethylamino)titanium results in precipitation of titanium dioxide and formation of an 83% yield of tris(dimethylamino)methane *(153, 154, 157)*.

$$\text{Ti}[\text{N}(\text{CH}_3)_2]_4 + 2\ \text{HCON}(\text{CH}_3)_2 \rightarrow \text{TiO}_2 + 2\ \text{HC}[\text{N}(\text{CH}_3)_2]_3 \quad (59)$$

Since many secondary amines can be converted into tetrakis(disubstituted amino)titanium derivatives *(17)*, this reaction should be generally applicable to the synthesis of orthoformamides. $N,N$-dialkylamides of higher orthocarboxylic acids are converted to vinylidene-bis(dialkylamines) under the same reaction conditions.

Treatment of $N,N,N',N'$-tetramethylformamidinium salts with an alkali metal $N,N$-dimethyl amide at $-20°C$ affords tris(dimethylamino)methane in about 70% yield *(27, 28, 28a)*.

$$[(\text{CH}_3)_2\text{N}]_2\overset{+}{\text{C}}\text{HX}^- + \text{M}^+\text{N}(\text{CH}_3)_2^- \rightarrow \text{HC}[\text{N}(\text{CH}_3)_2]_3 + \text{M}^+\text{X}^- \quad (60)$$
$$(\text{X} = \text{Cl, ClO}_4, \text{CH}_3\text{SO}_4; \text{M} = \text{Li, Na, K})$$

In a similar reaction $N,N',N''$-trimethyl-$N,N',N''$-triphenylorthoformamide was prepared by reaction of the sodium salt of $N$-methylaniline with $N,N'$-dimethyl-$N,N'$-diphenylformamidinium iodide *(67)*.

Aminolysis of *tert*-butoxybis(dialkylamino)-methanes yields orthoformamides *(53a)*.

Several $N,N',N''$-trialkyl-$N,N',N''$-triarylorthoformamides were prepared by reactions of the sodium salts of $N$-alkylanilines with sources of dihalocarbenes. Highest yields were obtained by addition of chlorodifluoromethane to a refluxing suspension of the $N$-alkylaniline salt in 1,2-dimethoxyethane *(65–67)*. The same products were obtained in lower yields when sodium trichloroacetate or chloroform was substituted for chlorodifluoromethane.

$$\text{ArNHR} \xrightarrow[(\text{CH}_3\text{OCH}_2)_2]{\text{NaH}} \text{Ar}\overset{-}{\text{N}}\text{RNa}^+ \xrightarrow{\text{CHF}_2\text{Cl}} \text{HC}(\text{NRAr})_3 + \text{NaCl} + 2\ \text{NaF} \quad (61)$$
$(\text{Ar} = \text{C}_6\text{H}_5, \text{R} = \text{CH}_3, \text{C}_2\text{H}_5; \text{Ar} = p\text{-O}_2\text{NC}_6\text{H}_4, \text{R} = \text{CH}_3, \text{C}_3\text{H}_7; \text{R} = \text{CH}_3,$
$\text{Ar} = o\text{-NO}_2\text{C}_6\text{H}_4, o\text{-BrC}_6\text{H}_4, m\text{-C}_2\text{H}_5\text{OC}_6\text{H}_4, \alpha\text{-C}_{10}\text{H}_7, 2,3\text{-xylyl}, p\text{-CH}_3\text{C}_6\text{H}_4)$

Tris(acylamino)methanes, $\text{HC}(\text{NHCOR})_3$, which may be regarded as tris acyl derivatives of orthoformamide, were discovered by Bredereck and co-workers in 1958 *(46, 47)*. These substances, which are all rather high-melting solids, are useful reagents for the synthesis of pyrimidines and other heterocyclic compounds.

Bredereck's group developed a number of methods for synthesizing tris(acylamino)methanes. Tris(formylamino)methane, the simplest member of the series, was obtained by reaction of formamide with dialkyl sulfates, alkyl tosylates, triethyl orthoformate, ethyl formimidate hydrochloride, formamidinium methyl sulfate, alkyl halides, triethyloxonium tetrafluoroborate, or acid halides (20, 46, 47). The highest yields were obtained using dialkyl sulfates. All of these reactions probably involve formation of a series of carbonium ion intermediates and reaction of these with formamide.

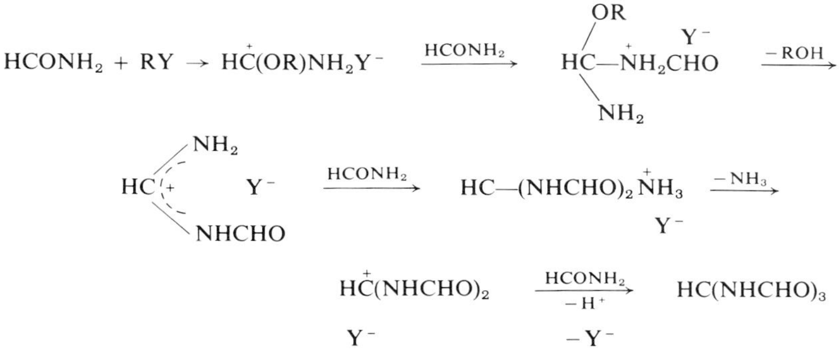

$$HCONH_2 + RY \rightarrow H\overset{+}{C}(OR)NH_2Y^- \xrightarrow{HCONH_2} HC\underset{NH_2}{\overset{OR}{\underset{|}{\overset{|}{-}}}}\overset{+}{N}H_2CHO \; Y^- \xrightarrow{-ROH}$$

$$HC\underset{NHCHO}{\overset{NH_2}{\underset{-}{\overset{+}{\diagdown}}}} Y^- \xrightarrow{HCONH_2} HC-(NHCHO)_2 \overset{+}{N}H_3 \xrightarrow{-NH_3}$$
$$Y^-$$

$$\overset{+}{HC}(NHCHO)_2 \xrightarrow[-H^+]{HCONH_2} HC(NHCHO)_3 \quad (62)$$
$$Y^- \qquad\qquad -Y^-$$

Higher acyl derivatives of orthoformamide are best prepared from the carboxamides and triethyl orthoformate (43, 44). The best procedure involves adding the ortho ester dropwise to a stirred, refluxing solution of the amide and an acid catalyst in toluene (43).

$$HC(OC_2H_5)_3 + 3\, RCONH_2 \xrightarrow[C_6H_4CH_3]{H^+} HC(NHCOR)_3 + 3\, C_2H_5OH \quad (63)$$
$$(R = CH_3, C_2H_5, C_3H_7, C_6H_5, p\text{-}CH_3OC_6H_4, p\text{-}O_2NC_6H_4, C_6H_5CH_2)$$

Tris(acylamino)methanes can also be prepared by transacylation reactions, in which tris(formylamino)methane is heated with an acid anhydride. The fact that gas is evolved during the reaction suggests that one of the products is a mixed formic anhydride, which decomposes to carbon monoxide and a carboxylic acid (44).

$$HC(NHCOR)_3 + (R'CO)_2O \rightarrow HC(NHCOR')_3 \quad (64)$$
$$(R = H, R' = CH_3, C_2H_5, C_3H_7, C_6H_5;\; R = C_2H_5, R' = C_3H_7;$$
$$R = CH_3, R' = C_2H_5, C_3H_7;\; R = C_4H_9, R' = CH_3)$$

As indicated, higher tris(acylamino)methanes also undergo transacylation reactions, but yields are generally low. Tris(ethylcarbamyl)methane undergoes transacylation when heated in toluene with carboxamides (141).

$$(C_2H_5OCONH)_3CH \xrightarrow{RCONH_2} HC(NHCOR)_3 \quad (65)$$
$$(R = H, CH_3, C_2H_5)$$

Potassium phthalimide and the sodium salt of saccharine both react with $N,N$-dimethylaminochlorocarbonium chloride to form $N,N$-dimethylaminobis(imido)methanes (14).

$$\overset{+}{\text{ClCHN(CH}_3)_2\text{Cl}^-} + \underset{\text{(Y = CO, SO}_2\text{)}}{\text{[phthalimide-N}^-\text{M}^+\text{]}} \longrightarrow \text{[phthalimide-N]}_2\text{CHN(CH}_3)_2 \quad (66)$$

Ethyl or methyl carbamate and triethyl orthoformate undergo an acid-catalyzed reaction which forms a product melting at 133.5°C. Heating this product in ethanol yields an isomeric substance which melts at 188°C. It was suggested (141) that these two substances are amide–imido tautomers of the expected tris(carbamyl)methane; they appear not to be polymorphic modifications of the same compound [Eq. (67)]. The same products are obtained from reactions of ethyl carbamate with either ethyl $N$-carbethoxyformimidate or $N,N'$-dicarbethoxyformamidine.

$$3\ \text{ROCONH}_2 + \text{HC(OC}_2\text{H}_5)_3 \xrightarrow[\text{C}_6\text{H}_5\text{CH}_3]{\text{H}^+} \text{HC[N=C(OH)OR]}_3 \xrightarrow[\Delta]{\text{C}_2\text{H}_5\text{OH}}$$
$$\text{HC(NHCO}_2\text{R)}_3 \quad (67)$$
$$(\text{R = CH}_3,\ \text{C}_2\text{H}_5)$$

As mentioned previously, ortho amides of higher orthocarboxylic acids are unknown. Busz and Kekulé claimed to have prepared 1,1,1-tripiperidinoethane by reaction of 1,1,1-trichloroethane with piperidine (60), and Barnes, Kundiger, and McElvain later obtained a compound of different properties, for which they claimed the same structure, by the reaction of ketene diethylacetal with piperidine (11). Boehme and Soldan later showed that Kekulé's product was 1,2-dipiperidinoethane, and that McElvain's was 1,1-dipiperidinoethylene (14). Baganz and Domaschke attempted to prepare an orthobenzamide by reaction of benzotrichloride with piperidine, but obtained 1,2-dichloro-1,2-diphenylethylene instead (9). Apparently higher orthocarboxamides are inherently unstable: they either decompose to ketene aminals (153, 154), or yield carbenoid $\alpha$-elimination products which dimerize.

## 2. Orthocarbamides

Tetrakis(dimethylamino)methane, the only acyclic derivative of orthocarbamide known, was prepared by reaction of lithium dimethylamide with $N,N,N',N'$-tetramethylchloroformamidinium chloride in benzene (155).

$$[(\text{CH}_3)_2\text{N}]_2\overset{+}{\text{CClCl}^-} + 2\ \text{LiN(CH}_3)_2 \rightarrow \text{C[N(CH}_3)_2]_4 + 2\ \text{LiCl} \quad (68)$$

## 3. Heterocyclic Ortho Amides

A few ortho amides are known whose orthoacyl carbons are incorporated into heterocyclic rings. $N,N',N'',N'''$-Tetraethyl-$\Delta^{2,2'}$-bis(imidazolidine) (**44**) is converted by phenyl isothiocyanate to the mercapto-$N$-arylformimidoylimidazolium inner salt **45**, which undergoes cycloaddition reactions with phenyl isocyanate, ethyl isocyanate, and methyl isothiocyanate to form the heterocyclic ortho amide derivatives **46** [Eq. (69)] (*163*).

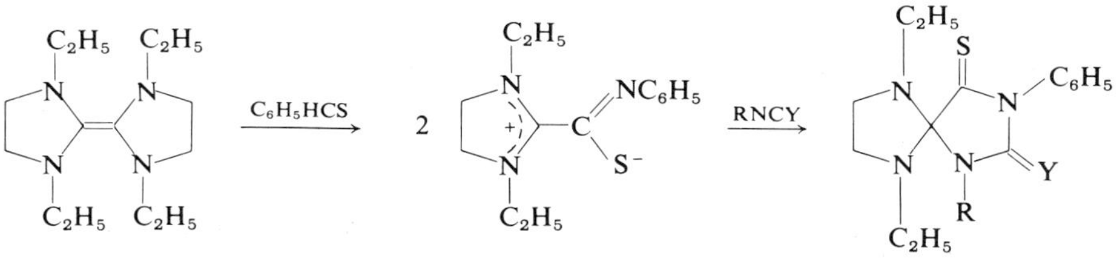

(Y = O, R = $C_2H_5$, $C_6H_5$; Y = S, R = $CH_3$)

(69)

Heating aryl isocyanates with dialkylformamides results in formation of pentaaryl-1,3,6,8,10-pentaazaspiro[4,5]decane-2,4,7,9-tetrones, **46a** (*72a, 145aa*). $N,N$-dimethyl-$N'$-phenylformamidine reacts with aryl isocyanates to form 2-dimethylamino-1,3,5-triarylhexahydro-1,3,5-triazin-4,6-diones, **46b**, which undergo further reactions with aryl isocyanates (*72a, 123c*). Heterocyclic amidines undergo analogous reactions (*123c*).

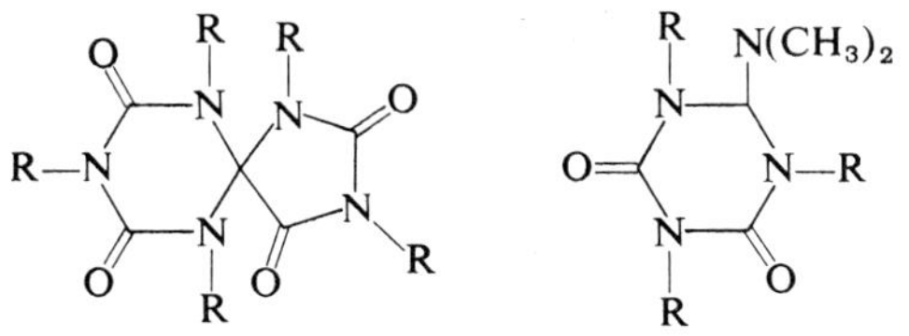

Three reactions leading to heterocyclic orthocarbamides have been reported. 2-Chloro-4,5-dioxoimidazolinium chloride and its 1,3-diisopropyl derivative react with heterocyclic amines to form orthocarbamide derivatives **47** (*140*). In a similar reaction, 2-chloro-1,3-dimethylbenzimidazolium chloride and $N,N'$-dimethyl-$o$-phenylenediamine formed the spirocyclic orthocarbamide **48** (*123*).

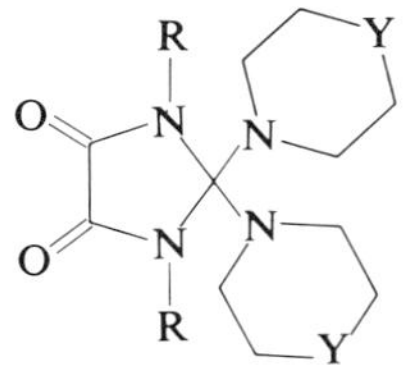

**47** 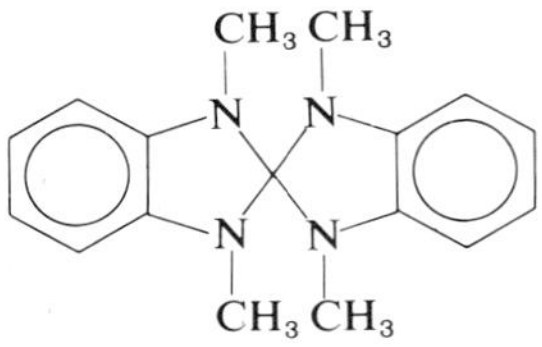 **48**

[R = H, Y = CH₂; R = CH(CH₃)₂, Y = O]

$N,N,N',N',N''$-penta-substituted guanidines react with isocyanates to form 1,2-dihydro-1,3,5-trisubstituted-1,1-bis(dialkylamino)-($3H,5H$)-s-triazin-4,6-diones (*123b*).

## III. PROPERTIES AND SOURCES OF AMIDE ACETALS, ESTER AMINALS, AND ORTHO AMIDES

Table I lists the ortho amide derivatives which have been described in the literature, together with their principal physical properties and references to the original literature.

### TABLE I
### Derivatives of Ortho Amides

| Compound | M.p., °C (B.p., °C/torr) | $n_D$ ($T$, °C) | References |
|---|---|---|---|
| | C₃ compounds | | |
| HC(OCH₃)₂NF₂ | — | — | 90 |
| | C₄ compounds | | |
| HC(NHCHO)₃ | 164° | — | 20, 46, 47, *141* |
| CH₃C(N₃)₂OC₂H₅ | (46°/10) | 1.466 (20°) | *136, 137* |
| CH₃C(OCH₃)₂NF₂ | — | — | 90 |
| 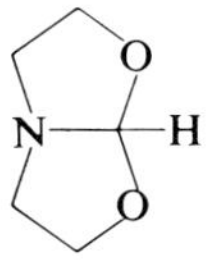 ![structure]N—H | (66°/5) | 1.4673 (20°) | 5 |
| | C₅ compounds | | |
| 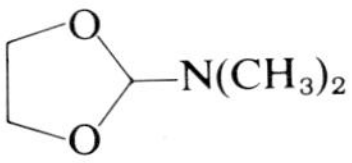 ![structure]—N(CH₃)₂ | (144°/760) | 1.4280 (20°) | *53, 115* |

# TABLE 1—continued

| Compound | M.p., °C (B.p., °C/torr) | $n_D$ (T, °C) | References |
|---|---|---|---|
| $(CH_3)_2NCH(OCH_3)_2$ | (102°/740) | 1.3963 (20°) | 5, 19, 21, 23, 53, 56, 76, 84, 89, 160 |
| **$C_6$ compounds** | | | |
| 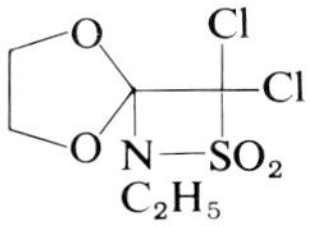 2-(dichloromethyl)-3-ethylsulfonyl-1,3-dioxolane [N-SO$_2$C$_2$H$_5$, C(Cl)$_2$Cl on dioxolane] | — | — | 59 |
| 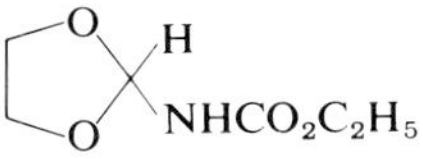 2-(ethoxycarbonylamino)-1,3-dioxolane [NHCO$_2$C$_2$H$_5$, H on dioxolane] | 55.5° | — | 119 |
|  2-methyl-2-(dimethylamino)-1,3-dioxolane [CH$_3$, N(CH$_3$)$_2$ on dioxolane] | (50°/28) | 1.4260 (20°) | 84, 115 |
| 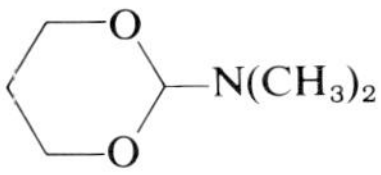 2-(dimethylamino)-1,3-dioxane [N(CH$_3$)$_2$ on dioxane] | (54°/10) | 1.4387 (20°) | 53 |
| 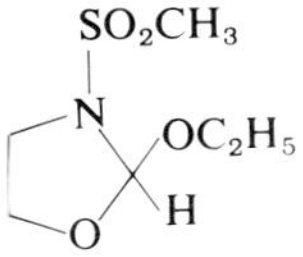 3-methylsulfonyl-2-ethoxy-1,3-oxazolidine [SO$_2$CH$_3$, OC$_2$H$_5$, H on oxazolidine] | (124°/0.5) | 1.462 (25°) | 68 |
| $(CH_3)_2NCH=NCH(OCH_3)_2$ | (98°/14) | 1.4649 (20°) | 89 |
| $CH_3C(OCH_3)_2N(CH_3)_2$ | (119°) | 1.4047 (25°) | 56, 120 |
| $(CH_3)_2NC(OCH_3)_3$ | (131°) | 1.4098 (20°) | 22, 74, 149 |
| $[(CH_3)_2N]_2CHOCH_3$ | (128°/740) | 1.4158 (20°) | 38, 40, 42, 50, 53, 160, 161a |
| **$C_7$ compounds** | | | |
| 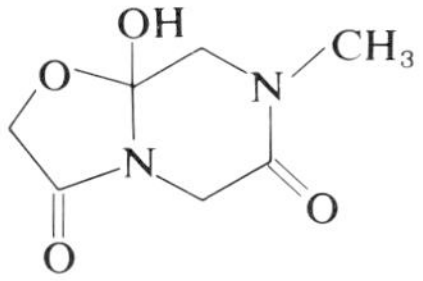 bicyclic OH, N-CH$_3$ compound (oxazolo-piperazine-dione) | Oil | — | 130 |
| 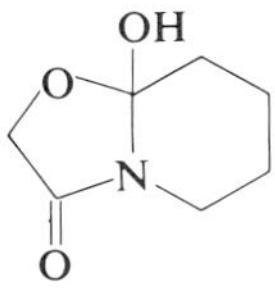 bicyclic OH compound (oxazolo-piperidinone) | 89°, 97° | — | 91, 130 |

TABLE 1–continued

| Compound | M.p., °C (B.p., °C/torr) | $n_D$ (T, °C) | References |
|---|---|---|---|
|  2,2-dimethoxy-3-pyrrolin-5-one (O=C−N, ring with OCH₃, OCH₃) | (103°/5) 43°–45° | — | 150 |
| 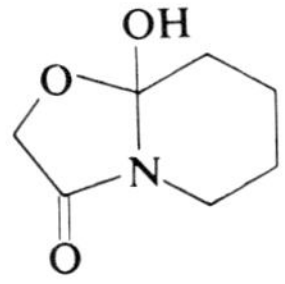 Bicyclic hemiketal with OH, O, N, C=O | 97° | — | 129 |
| 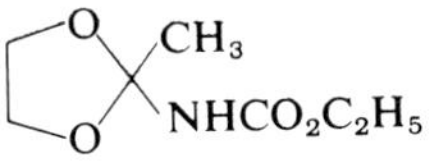 2-methyl-2-(ethoxycarbonylamino)-1,3-dioxolane | (107°/0.25) | — | 119 |
| 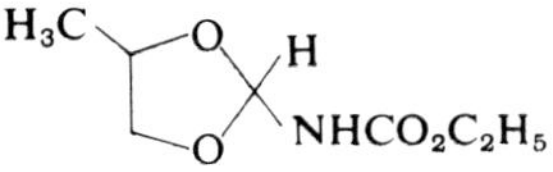 4-methyl-2-(ethoxycarbonylamino)-1,3-dioxolane | 47° | — | 119 |
| HC(NHCOCH₃)₃ | 262° | — | 43, 44, 141 |
| HC[N=C(OH)OCH₃]₃ | 127° | — | 141 |
| HC(NHCO₂CH₃)₃ | 161° | — | 141 |
| 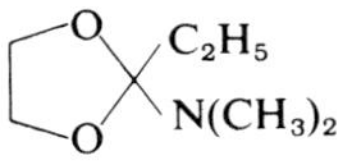 2-ethyl-2-(dimethylamino)-1,3-dioxolane | (42°/12) | 1.4316 (22°) | 84 |
| 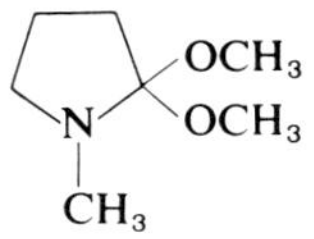 1-methyl-2,2-dimethoxypyrrolidine | (52°/18) | — | 77 |
| 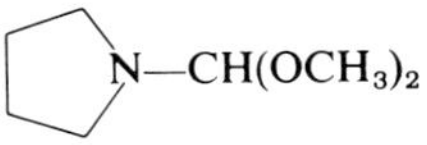 1-(dimethoxymethyl)pyrrolidine (N—CH(OCH₃)₂) | (63°/25) | 1.4320 (20°) | 53, 56, 161 |
| 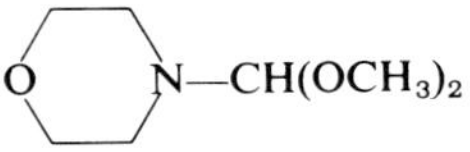 4-(dimethoxymethyl)morpholine | (87°/15) | 1.4811 (20°) | 53 |
| 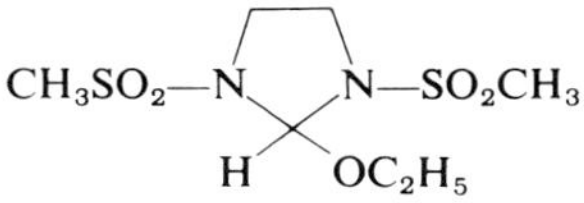 CH₃SO₂—N−N—SO₂CH₃ with H, OC₂H₅ (imidazolidine) | 79° | — | 68 |
| C₂H₅C(OCH₃)₂N(CH₃)₂ | (30°/12) | 1.4178 (21°) | 84 |
| (C₂H₅)₂NCH(OCH₃)₂ | (144°/760) | 1.4070 (20°) | 53, 142 |
| (CH₃)₂NCH(OC₂H₅)₂ | (136°/760) | 1.4075 (20°) | 5, 16, 19, 21, 37, 42, 53, 56, 62, 76, 84, 114, 115 |

TABLE 1—continued

| Compound | M.p., °C (B.p., °C/torr) | $n_D$ (T, °C) | References |
|---|---|---|---|
| $CH_3SO_2N(CH_3)CH(OC_2H_5)_2$ | (101°/2) | 1.433 (25°) | 68 |
| $C_2H_5OCH[N(CH_3)_2]_2$ | (143°/736) | 1.4175 (20°) | 50, 53 |
| $[(CH_3)_2N]_2C(OCH_3)_2$ | (70°/44) | 1.4242 (25°) | 56 |
| $HC[N(CH_3)_2]_3$ | (146°/760) | 1.4348 (25°) | 27, 28, 53a. 153, 154, 157, 158 |

$C_8$ compounds

| 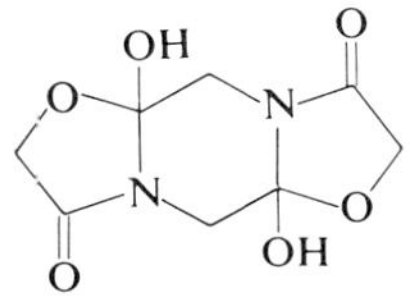 | 186° (dec.) | — | 130 |
| 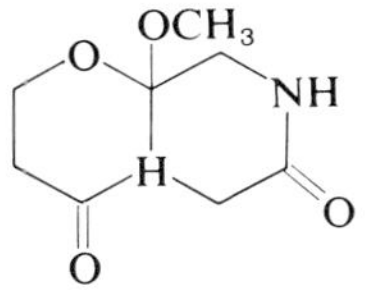 | 118° | — | 130 |
| 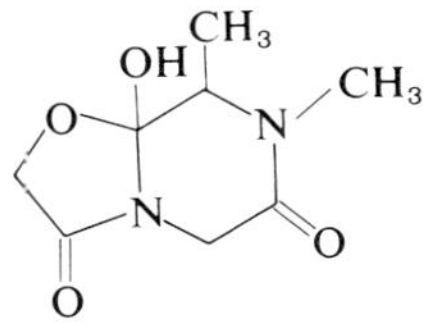 | 148° | — | 130 |
| 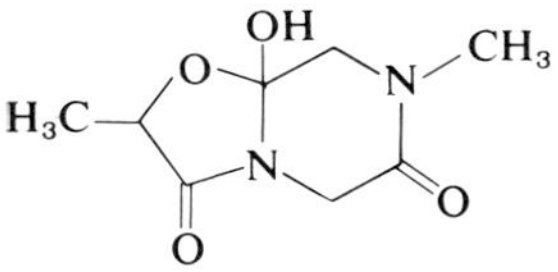 | 150° | — | 130 |
| 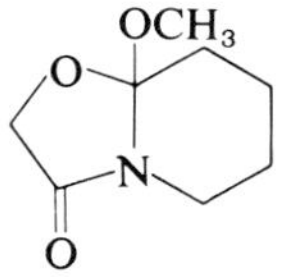 | 37° (45°–48°/0.002) | — | 130 |
| 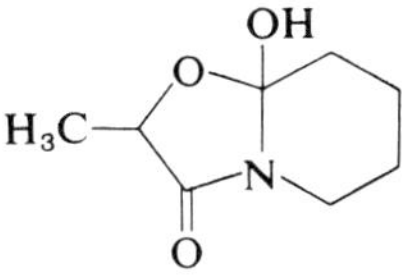 | 103°, 92° | — | 91, 130 |

454   7. AMIDE ACETALS, ESTER AMINALS, AND ORTHO AMIDES

TABLE 1 – *continued*

| Compound | M.p., °C (B.p., °C/torr) | $n_D$ ($T$, °C) | References |
|---|---|---|---|
| 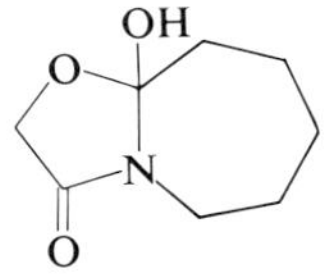 [bicyclic structure with OH, O, N, C=O] | 78° | — | 91 |
| 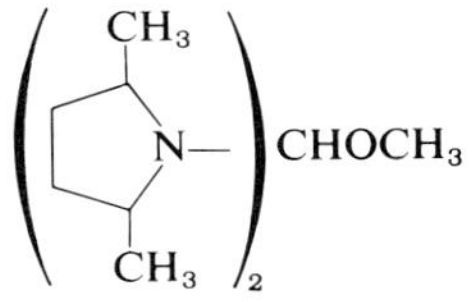 [pyrrolidine with 2,5-dimethyl]₂ N—CHOCH₃ | (135°/24) | — | 161 |
| 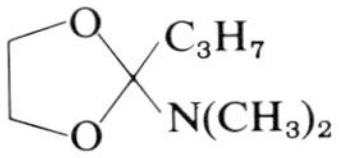 [1,3-dioxolane with C₃H₇ and N(CH₃)₂] | (55°/12) | 1.4338 (21°) | 84 |
| 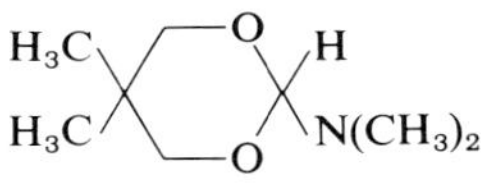 [1,3-dioxane with H₃C, H₃C, H, N(CH₃)₂] | (178°/760) | 1.4348 (25°) | 5, 52 |
| 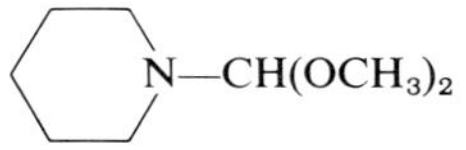 [piperidine]N—CH(OCH₃)₂ | (83°/15) | 1.4411 (20°) | 53, 161 |
| (CH₃)₂NC(CH₃)(OC₂H₅)₂ | (58°/13) | 1.4112 (18°) | 16, 62, 84, 115 |
| (C₂H₅)₂NC(OCH₃)₃ | (48°/12) | 1.4212 (20°) | 22 |
| [(CH₃)₂N]₂CHOC₃H₇ | (52°/12) | 1.4232 (20°) | 53 |
| (CH₃)₂CHO—CH[N(CH₃)₂]₂ | (40°/10) | 1.4160 (20°) | 42, 50, 53 |
| C₉ compounds | | | |
| 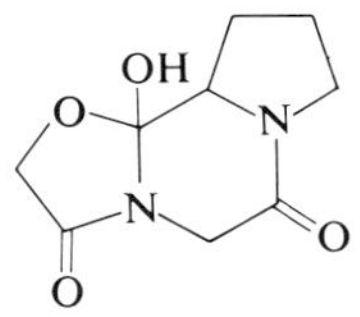 [tricyclic structure with OH, O, N, C=O, pyrrolidine ring] | 148° | — | 129 |
| 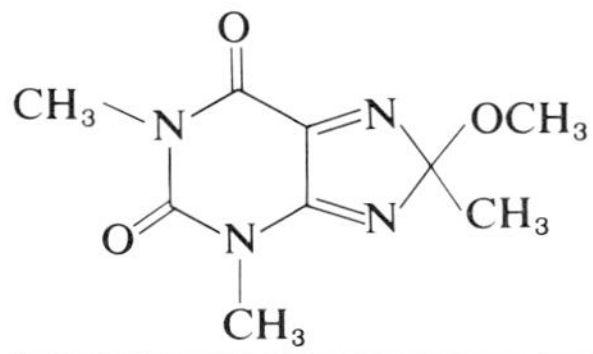 [purine/imidazole fused structure with CH₃-N, OCH₃, CH₃, N-CH₃] | 232° | — | 89a |

AMIDE ACETALS, ESTER AMINALS, AND ORTHO AMIDES 455

TABLE 1–*continued*

| Compound | M.p., °C (B.p., °C/torr) | $n_D$ (T, °C) | References |
|---|---|---|---|
| 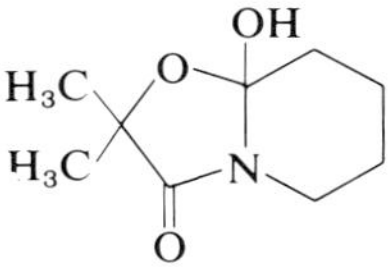 [structure: bicyclic with OH, gem-dimethyl, O, N, C=O] | 123° | — | *91* |
| 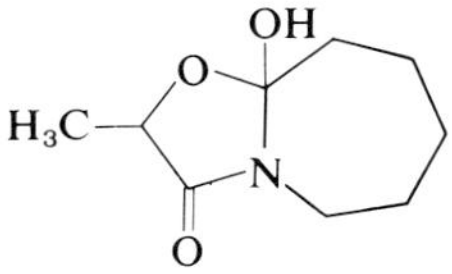 [structure: bicyclic with OH, CH₃, O, N, C=O, 7-membered ring] | 138° | — | *91* |
| 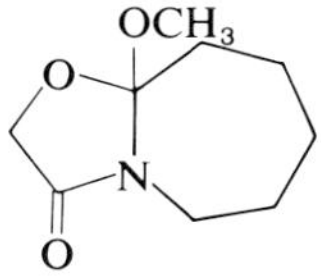 [structure: bicyclic with OCH₃, O, N, C=O, 7-membered ring] | (90°/0.05) | 1.4860 (20°) | *91, 130* |
| 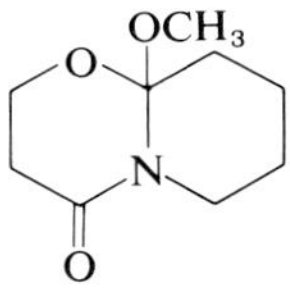 [structure: bicyclic with OCH₃, O, N, C=O, 6-membered ring] | 112° | — | *130* |
| (CH₃)₂NCH(OCH₂CH=CH₂)₂ | (60°/8) | 1.4340 (20°) | *5* |
| 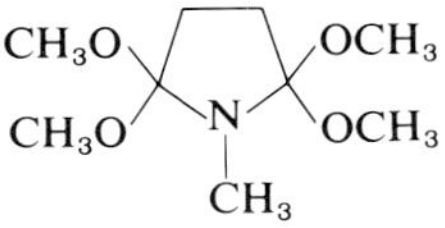 [structure: pyrrolidine with CH₃O, OCH₃ groups and N-CH₃] | (95°–105°/7) 52°–53° | — | *150* |
| 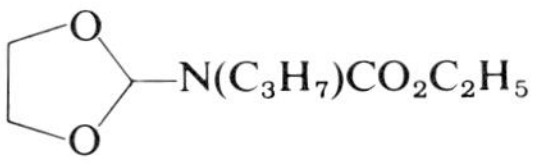 [1,3-dioxolane]—N(C₃H₇)CO₂C₂H₅ | (131°/14) | 1.440 (20°) | *145a, 145b* |
| 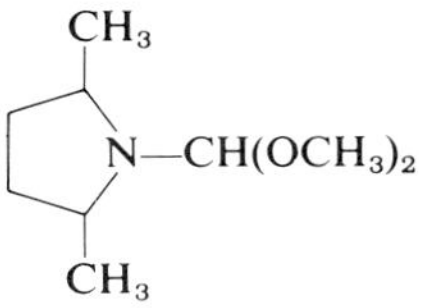 [2,5-dimethylpyrrolidine]N—CH(OCH₃)₂ | (91°/49) | — | *161* |
| 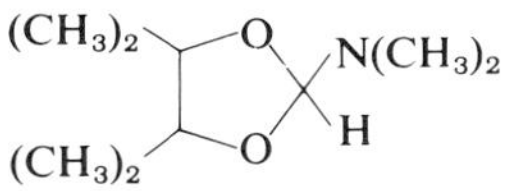 [structure: dioxolane with (CH₃)₂ groups, N(CH₃)₂, H] | (69°/10) | 1.4340 (20°) | *53* |

TABLE 1–continued

| Compound | M.p., °C (B.p., °C/torr) | $n_D$ ($T$, °C) | References |
|---|---|---|---|
| 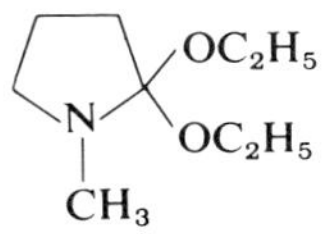 *N-methylpyrrolidine with 2,2-diOC₂H₅* (CH₃ on N, OC₂H₅, OC₂H₅) | (62°/14) | 1.4365 20° ($d_{20°} = 0.963$) | 21, 76, 114 |
| 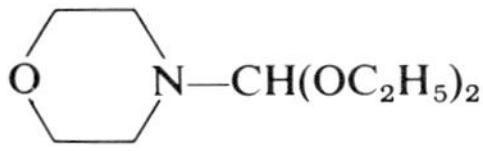 *morpholine*–N–CH(OC₂H₅)₂ | (95°/13) | — | 115 |
| CH₃N(CO₂C₂H₅)CH(OC₂H₅)₂ | (95°/13) | 1.4202 (20°) | 145a, 145b |
| 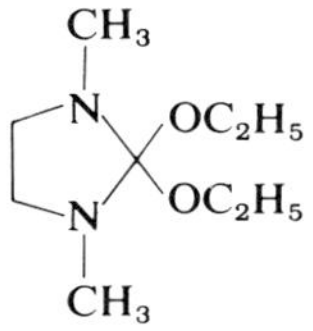 *N,N′-dimethylimidazolidine-2,2-diOC₂H₅* | (30°/0.01) | — | 115 |
| (CH₃O)₂CHN(C₃H₇)₂ | (70°/13) | — | 161 |
| [(CH₃)₂CH]₂NCH(OCH₃)₂ | (61°/12) | 1.4160 (20°) | 53 |
| (CH₃)₂NCH(OC₃H₇)₂ | (96°/50) | 1.4083 (20°) | 53, 76 |
| (C₂H₅)₂NCH(OC₂H₅)₂ | (162°/760) | — | 77 |
| (CH₃)₂NCH[OCH(CH₃)₂]₂ | (69°/50) | 1.4000 (20°) | 5, 53 |
| (CH₃)₂NC(OC₂H₅)₃ | (76°/23) | 1.4089 (20°) | 22, 74, 115, 149 |
| (CH₃)₃COCH[N(CH₃)₂]₂ | (60°/19) | 1.4260 (20°) | 42, 44, 49 |
| [(CH₃)₂N]₂C(OC₂H₅)₂ | (61°–65°/18) | — | 74, 115 |
| C[N(CH₃)₂]₄ | 122°–124.5° | — | 155 |

C₁₀ compounds

| Compound | M.p., °C (B.p., °C/torr) | $n_D$ ($T$, °C) | References |
|---|---|---|---|
| 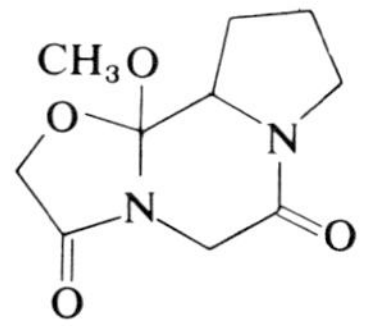 *bicyclic structure with CH₃O* | 172°–174° | — | 129, 130 |
| *bicyclic structure with OCH₃, OCH₃* | 222° | — | 130 |

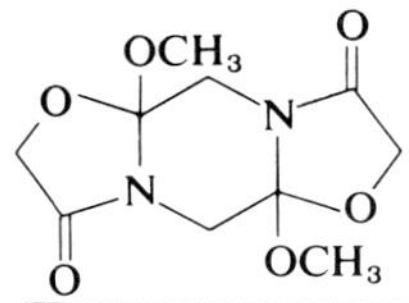

AMIDE ACETALS, ESTER AMINALS, AND ORTHO AMIDES 457

TABLE 1–continued

| Compound | M.p., °C (B.p., °C/torr) | $n_D$ (T, °C) | References |
|---|---|---|---|
| 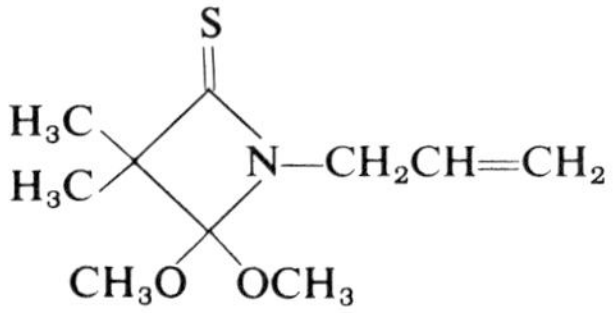 (H₃C)₂C—N(CH₂CH=CH₂)—C(=S)—C(OCH₃)₂ (β-thiolactam with gem-dimethyl and gem-dimethoxy) | 24°–25° | — | 126 |
| 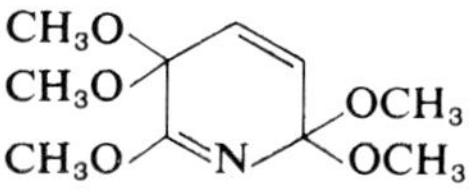 3,3,6-tri(methoxy)-2-methoxy-... dihydropyridine derivative | (115°/8) | 1.4722 (20°) | 150 |
| HC[—N=C(OH)OC₂H₅]₃ | 133.5° | — | 141 |
| HC(NHCO₂C₂H₅)₃ | 188°, 242° | — | 43, 44, 141 |
|  (pyrrolidino)₂CHOCH₃ | (74°/0.85) | — | 161 |
| 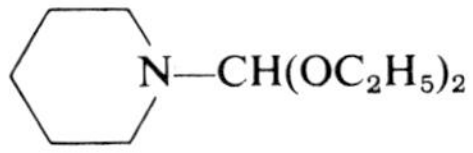 (piperidino)—CH(OC₂H₅)₂ | (84°/12) | — | 115 |
| (CH₃O)₂CHN(CH₃)C₆H₁₁ | (116°/24) | 1.4378 (25°) | 56 |
| C₁₁ compounds | | | |
| 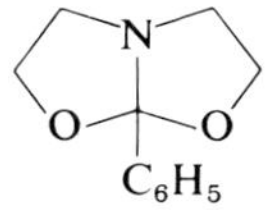 2-phenyl-oxazolidine bicyclic (C₆H₅ substituent) | (80°/0.03) | 1.5379 (20°) | 82 |
| 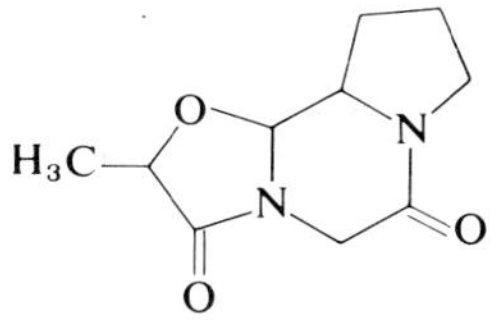 H₃C-substituted bicyclic oxazolidinone-pyrrolidine | 2R: m.p. 108°, [α]_D + 45°; 2S: m.p. 116°, [α]_D + 43° | — | 121 |
| p-CH₃C₆H₄SO₂N(CH₃)CH(OCH₃)₂ | 123° | — | 68 |
| C₆H₅C(OCH₃)₂N(CH₃)₂ | (67°/5) | 1.5045 (25°) | 56, 159 |
| 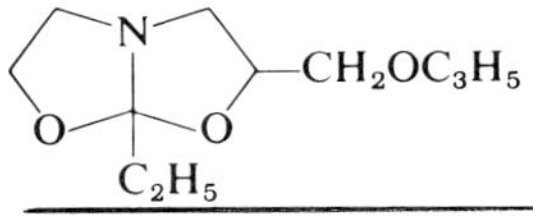 bicyclic oxazolidine—CH₂OC₃H₅ (C₂H₅ substituent) | (67°/0.03) | 1.4669 (20°) | 82 |

TABLE 1 –continued

| Compound | M.p., °C (B.p., °C/torr) | $n_D$ (T, °C) | References |
|---|---|---|---|
| (spiro imidazolidine with H₅C₂-N, N-CH₃, S, =S, H₅C₂, CH₃ substituents) | 88° | — | 163 |
| 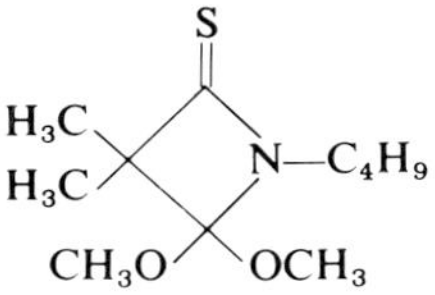 (azetidinethione with H₃C, H₃C, CH₃O, OCH₃, N–C₄H₉, S) | (103°/0.6) | — | 126 |
| $C_3H_7N(CO_2C_2H_5)CH(OC_2H_5)_2$ | (110°/19) | 1.4224 (20°) | 145a, 145b |
| $(CH_3)_2NCH(OC_4H_9)_2$ | (93°/12) | 1.4169 (20°) | 53, 115 |
| $(CH_3)_2NCH[OCH_2CH(CH_3)_2]_2$ | (129°/100) | 1.4123 (21°) | 19 |
| $(CH_3)_2NCH[OCH(CH_3)C_2H_5]_2$ | (120°/100) | 1.4134 (21°) | 19 |
| $(C_4H_9)_2NCH(OCH_3)_2$ | (75°/12) | 1.4248 (20°) | 53 |
| $[(CH_3)_2CH]_2NCH(OC_2H_5)_2$ | (76°/16) | 1.4195 (20°) | 53 |
| $(CH_3)_2NCH[OC(CH_3)_3]_2$ | (57°/8) | 1.4140 (20°) | 5 |

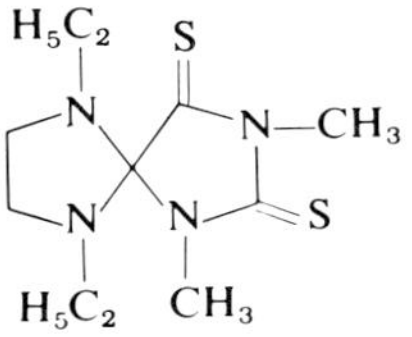

$C_{12}$ compounds

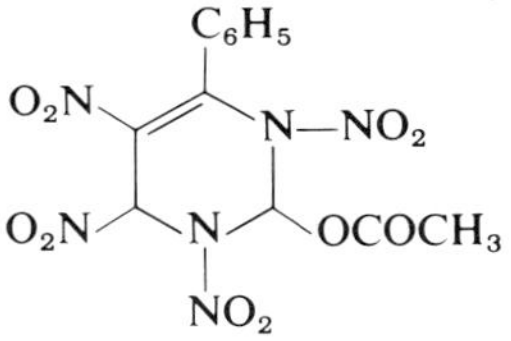

| Compound | M.p., °C (B.p., °C/torr) | $n_D$ (T, °C) | References |
|---|---|---|---|
| (dihydropyrimidine with $C_6H_5$, $O_2N$, $O_2N$, N–$NO_2$, N–$NO_2$, OCOCH₃) | 147° | — | 110 |
| (benzoxazine fused with cyclohexanone, OH) | 164° | — | 130 |
| 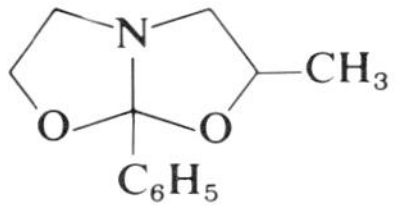 (bicyclic oxazolidine with CH₃, $C_6H_5$) | (87°/0.05) | 1.5321 (20°) | 82 |
| 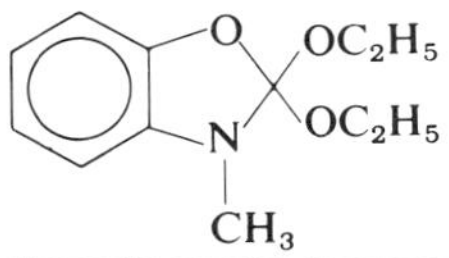 (benzoxazoline with $OC_2H_5$, $OC_2H_5$, N–CH₃) | (97°/1) | — | 115 |

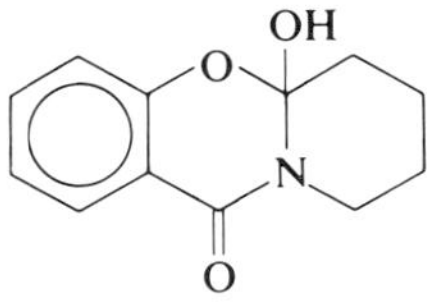

TABLE 1–continued

| Compound | M.p., °C (B.p., °C/torr) | $n_D$ (T, °C) | References |
|---|---|---|---|
| 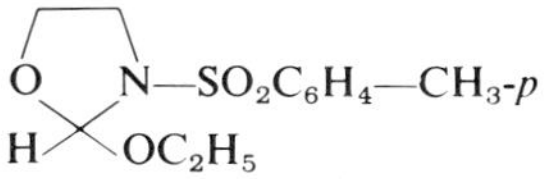 (oxazolidine with N—SO$_2$C$_6$H$_4$—CH$_3$-p and OC$_2$H$_5$, H substituents) | 63° | — | 68 |
| 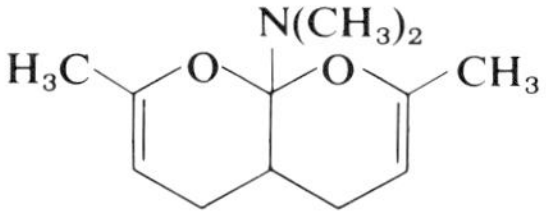 H$_3$C—(pyran fused system)—CH$_3$ with N(CH$_3$)$_2$ | (96°/7) | — | 117a, 120 |
| 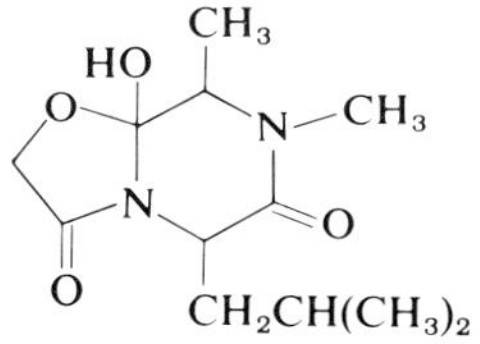 bicyclic structure with HO, CH$_3$, N—CH$_3$, O, CH$_2$CH(CH$_3$)$_2$ | 117° | — | 130 |
| 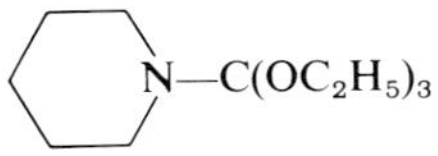 piperidine-N—C(OC$_2$H$_5$)$_3$ | (100°–105°/13) | — | 75, 149 |
| 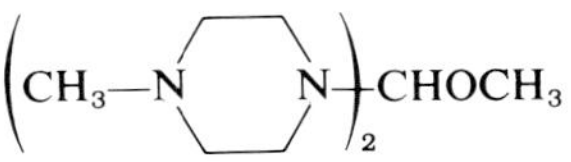 (CH$_3$—N⟨piperazine⟩N—CHOCH$_3$)$_2$ | (106°–110°/0.9) | — | 161 |
| [(CH$_3$)$_2$N]$_2$CHN(CH$_3$)C$_6$H$_5$ | (72°/0.005) | 1.5363 (20°) | 53a |
| C$_{13}$ compounds | | | |
| 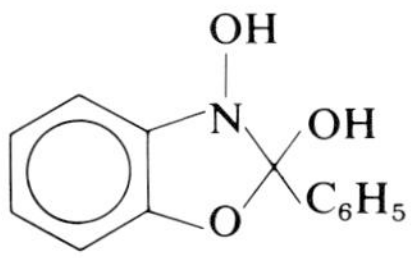 benzoxazoline with N—OH, OH, C$_6$H$_5$ | 135° | — | 108 |
| 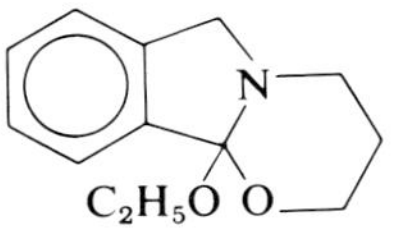 isoindoline fused with C$_2$H$_5$O, O | (110°/0.001) | — | 103 |
| 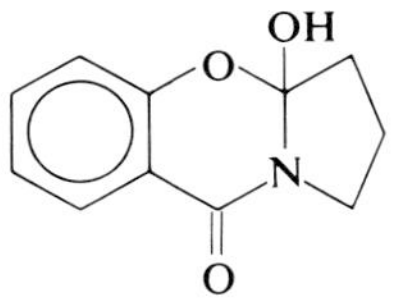 benzoxazine fused with OH, O, N, C=O | 132° | — | 130 |

TABLE 1–continued

| Compound | M.p., °C (B.p., °C/torr) | $n_D$ (T, °C) | References |
|---|---|---|---|
| 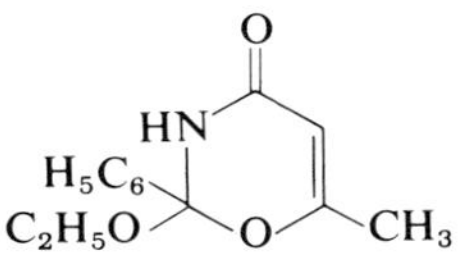 H₅C₆, C₂H₅O–C(O)–CH=C(CH₃)–NH (cyclic) | 88° | — | 105 |
| 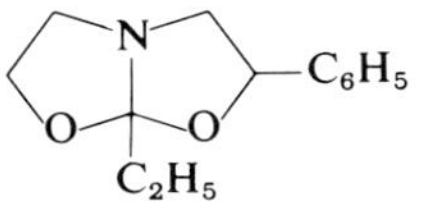 Bicyclic oxazolidine with C₂H₅ and C₆H₅ substituents | (92°/0.05) | 1.5278 (20°) | 82 |
| 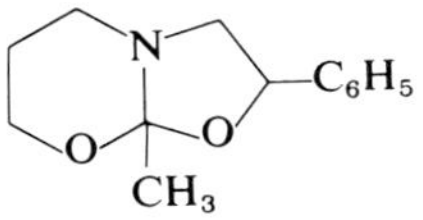 Bicyclic oxazine with CH₃ and C₆H₅ substituents | (107°/0.05) | 1.5365 (20°) | 82 |
| 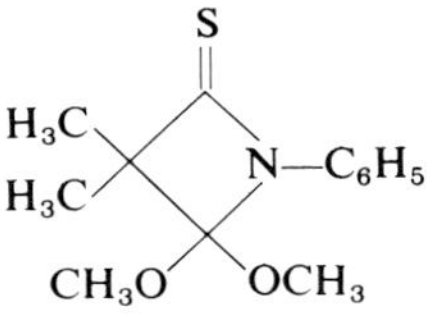 (H₃C)₂C–C(=S)–N(C₆H₅) with CH₃O, OCH₃ (azetidine) | 76° | — | 126 |
| 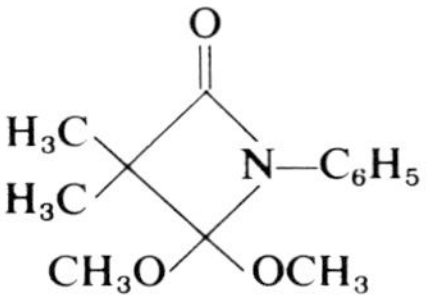 (H₃C)₂C–C(=O)–N(C₆H₅) with CH₃O, OCH₃ (azetidine) | 45° | — | 125 |
| 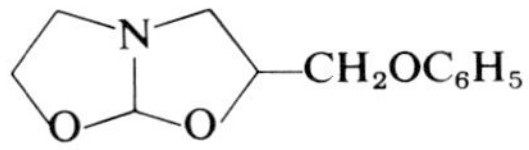 Bicyclic oxazolidine with –CH₂OC₆H₅ | 75° (116°–120°/0.1) | — | 82 |
| 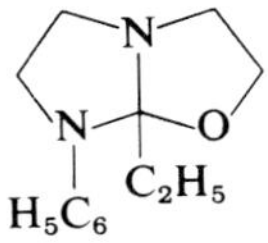 Bicyclic imidazolidine with C₂H₅ and H₅C₆ on N | (108°/0.2) | 1.5615 (20°) | 79 |
| 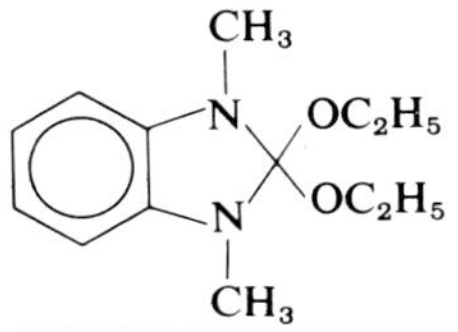 Benzimidazoline with N–CH₃, N–CH₃, OC₂H₅, OC₂H₅ | 68° (90°/0.005) | — | 115 |

## AMIDE ACETALS, ESTER AMINALS, AND ORTHO AMIDES

TABLE 1—continued

| Compound | M.p., °C (B.p., °C/torr) | $n_D$ (T, °C) | References |
|---|---|---|---|
| p-$CH_3C_6H_4SO_2N(CH_3)CH(OC_2H_5)_2$ | 57° | — | 68 |
| $C_6H_5SO_2N(C_2H_5)CH(OC_2H_5)_2$ | 47° | — | 165 |
| $(CH_3)_2NC(C_6H_5)(OC_2H_5)_2$ | (118°/13) | 1.4896 (22°) | 16, 62, 76, 84 |
| 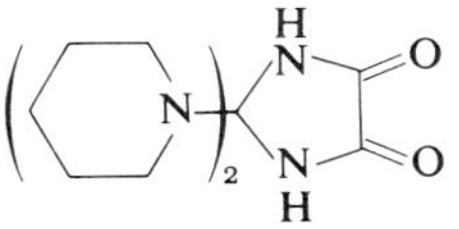 (piperidine-hydantoin structure)[2] | 149° | — | 140 |
| $[(CH_3)_2N]_2CHN(C_2H_5)C_6H_5$ | (72°/0.0005) | 1.5235 (25°) | 53a |
| $HC(NHCOC_3H_7)_3$ | 227° | — | 43, 44 |
| $(CH_3)_2NCH[OCH_2C(CH_3)_3]_2$ | (75.5°/5) | 1.4122 (20°) | 5, 19 |
| $HC[N(C_2H_5)_2]_3$ | (100°/13) | 1.4545 (25°) | 53a |

$C_{14}$ compounds

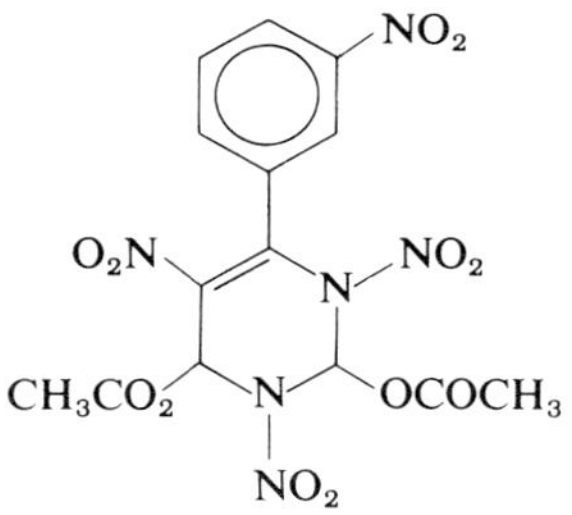

(structure with m-$NO_2$-phenyl, $O_2N$, $CH_3CO_2$, $N-NO_2$, $OCOCH_3$, $NO_2$) — 162° — 111

(structure with o-$NO_2$-phenyl, $O_2N$, $CH_3CO_2$, $N-NO_2$, $OCOCH_3$, $NO_2$) — 147° — 110

(structure with $C_6H_5$, $O_2N$, $CH_3CO_2$, $N-NO_2$, $OCOCH_3$, $NO_2$) — 147° — 110

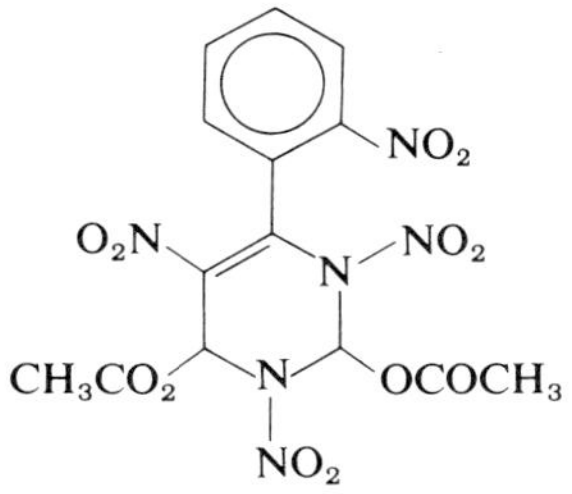

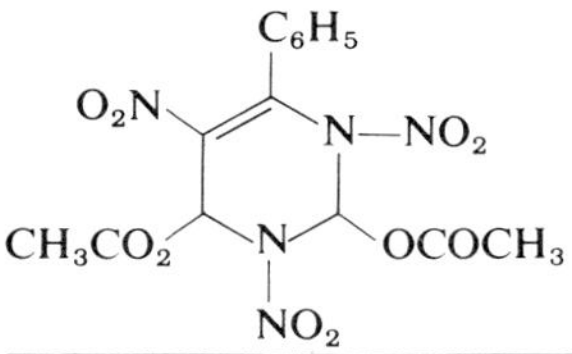

TABLE 1—continued

| Compound | M.p., °C (B.p., °C/torr) | $n_D$ (T, °C) | References |
|---|---|---|---|
| 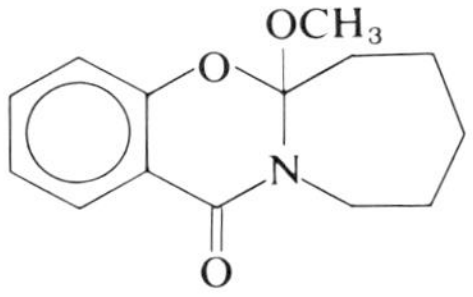 [benzoxazine fused with azepanone, OCH₃ substituent] | 117° | — | 130 |
| 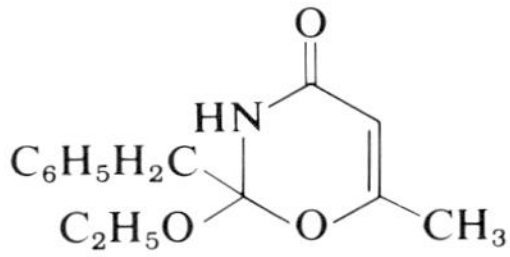 $C_6H_5H_2C$, $C_2H_5O$ — [6-methyl-4H-1,3-oxazin-4-one with HN] —$CH_3$ | 115° | — | 105 |
| 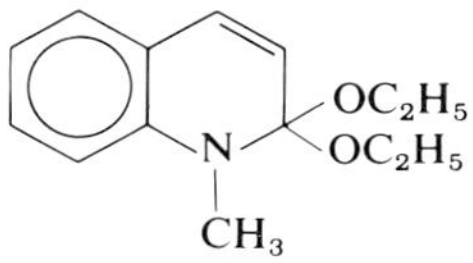 [1-methyl-2,2-diethoxy-1,2-dihydroquinoline] | (93°/0.01) | — | 115 |
| 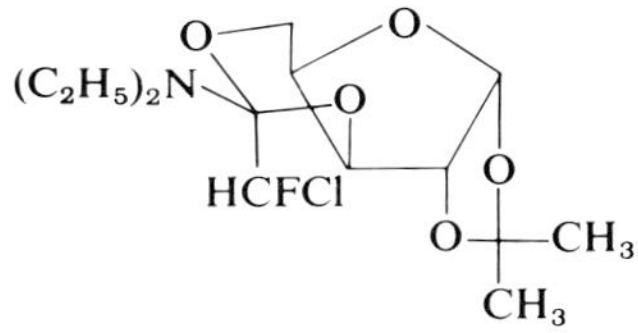 $(C_2H_5)_2N$—[bicyclic sugar derivative with HCFCl and two CH₃] | — | — | 93 |
| 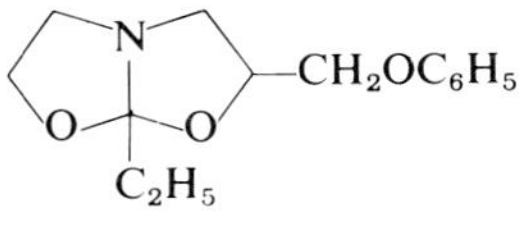 [bicyclic N,O,O-acetal with $C_2H_5$]—$CH_2OC_6H_5$ | (121°/0.1) 64.5° | — | 82 |
| 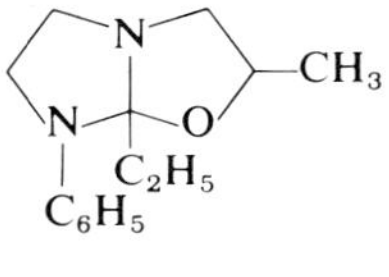 [bicyclic N,O,O-acetal with $C_2H_5$, $C_6H_5$]—$CH_3$ | (115°/0.4) | 1.5519 (20°) | 79 |
| $C_6H_5N(CO_2C_2H_5)CH(OC_2H_5)_2$ | (96°/0.02) | 1.4917 (20°) | 145a, 145b |
| 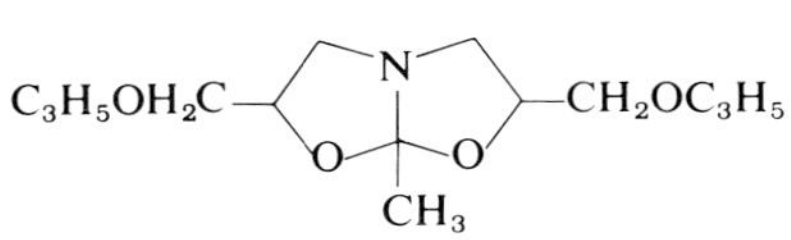 $C_3H_5OH_2C$—[bicyclic N,O,O-acetal with $CH_3$]—$CH_2OC_3H_5$ | (126°/0.3) | 1.4678 (20°) | 80 |
| p-$CH_3C_6H_4SO_2N(C_2H_5)CH(OC_2H_5)_2$ | 59° | — | 68 |
| $CH_3OCH[N(C_3H_7)_2]_2$ | (105°/7) | — | 161 |

TABLE 1–*continued*

| Compound | M.p., °C (B.p., °C/torr) | $n_D$ (T, °C) | References |
|---|---|---|---|
| 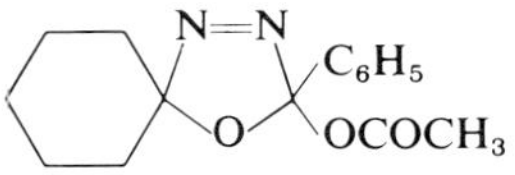 [cyclohexane spiro 1,3-dioxa-4,5-diazole with C₆H₅ and OCOCH₃] | 67° (dec.) | — | 94 |
| **C₁₅ compounds** | | | |
| 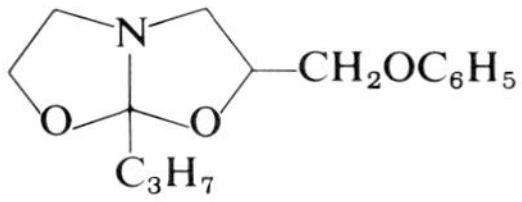 [bicyclic N,O,O-acetal with C₃H₇ and –CH₂OC₆H₅] | (146°/0.6) 54° | — | 82 |
| 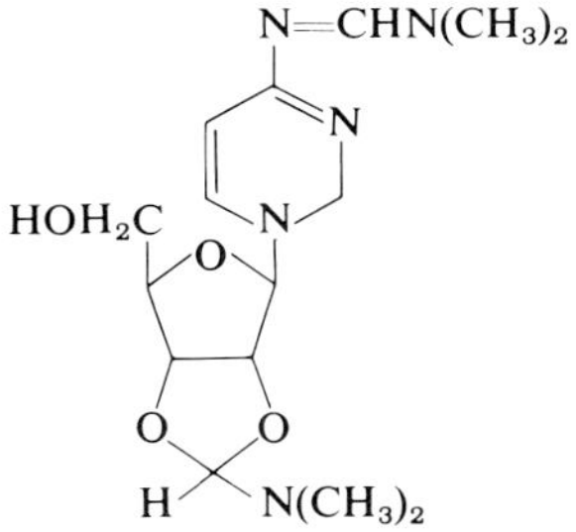 [nucleoside with HOH₂C, pyrimidine N=CHN(CH₃)₂, and dioxolane H/N(CH₃)₂] | 166° | — | 172 |
| (CH₃)₂NCH(OC₆H₁₁)₂ | (77°/0.25) | 1.4685 (20°) | 5, 15, 62 |
| **C₁₆ compounds** | | | |
| 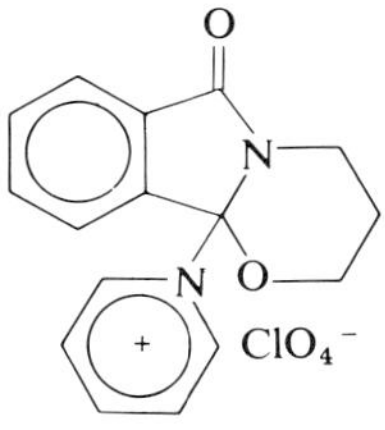 [tricyclic isoindolinone fused with pyridinium ClO₄⁻] | 240° | — | 103 |
| 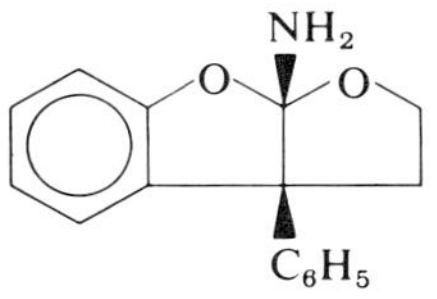 [benzofuran-fused dioxolane with NH₂ and C₆H₅] | 95° | — | 170 |
| 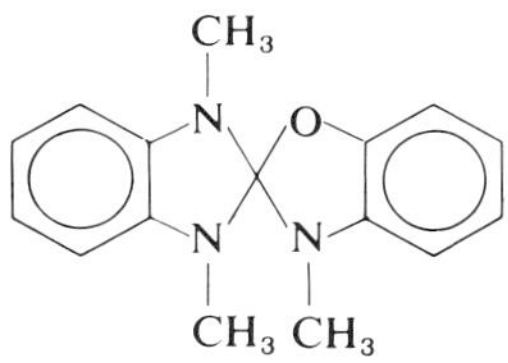 [spiro bis-benzoxazole/benzimidazoline with three CH₃ groups] | — | — | 123 |

TABLE 1–*continued*

| Compound | M.p., °C (B.p., °C/torr) | $n_D$ (T, °C) | References |
|---|---|---|---|
| 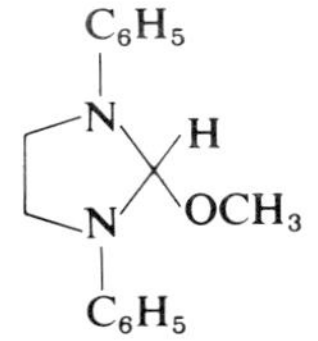 [structure: 2-phenyl-N,N'-diphenyl imidazolidine with H and OCH₃] | — | — | 128 |
| 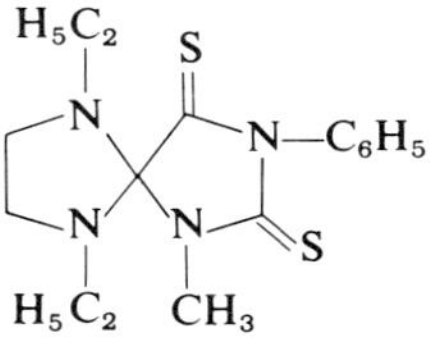 [structure: spirocyclic thiohydantoin with H₅C₂, N-C₆H₅, CH₃] | 131° | — | 163 |
| [C₆H₅N(CH₃)]₂CHOCH₃ | (145°/0.5) | 1.5909 (25°) | 78 |
| 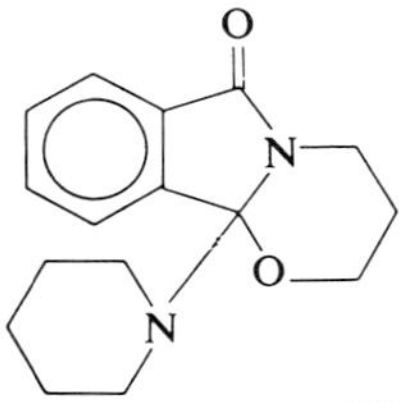 [structure: isoindolinone fused with piperidine and oxazine] | 99° | — | 103 |
| 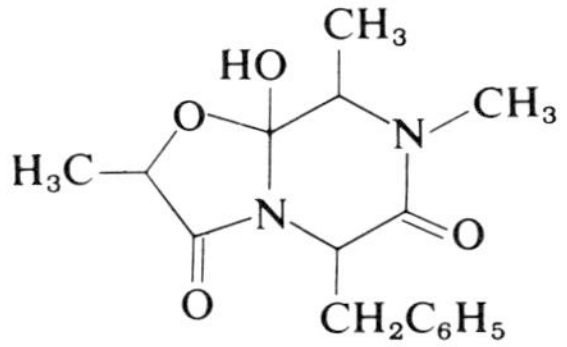 [structure: bicyclic morpholine-dione with HO, CH₃, CH₂C₆H₅] | 197° | — | 130 |
| 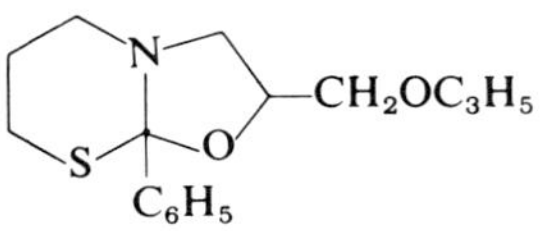 [structure: thiazine-oxazolidine with C₆H₅ and CH₂OC₃H₅] | (134°/0.02) | 1.5786 (20°) | 80 |
| 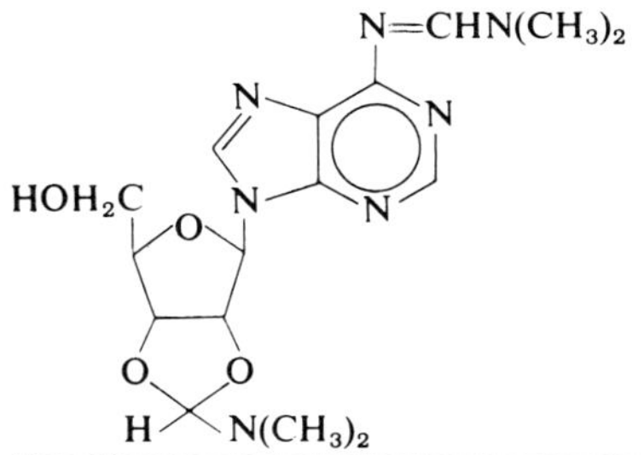 [structure: adenosine derivative with N=CHN(CH₃)₂ and N(CH₃)₂ acetal] | Amorphous | — | 172 |

## TABLE 1—continued

| Compound | M.p., °C (B.p., °C/torr) | $n_D$ (T, °C) | References |
|---|---|---|---|
| $C_{17}$ compounds | | | |
| 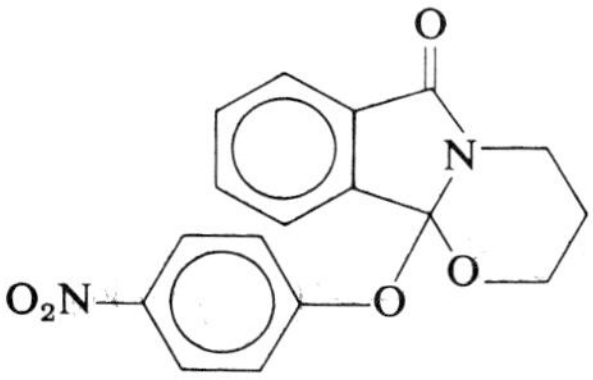 [structure: phthalimide-type fused bicyclic with O-aryl-$NO_2$ substituent] | 110° | — | 103 |
| 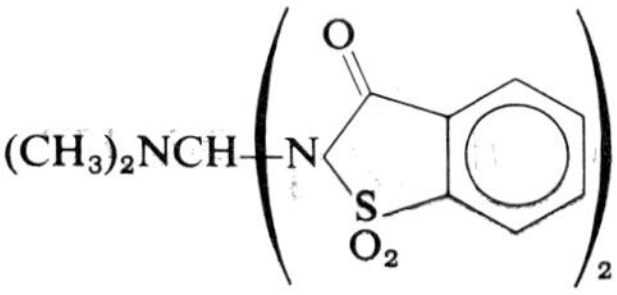 $(CH_3)_2NCH$–[N-benzisothiazolone-$S,S$-dioxide]$_2$ | 170° (dec.) | — | 14 |
| 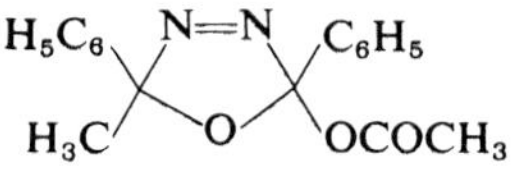 $H_5C_6$, $H_3C$ $\diagdown$ $N=N$ $\diagup$ $C_6H_5$, $OCOCH_3$ with O bridge | 58° (dec.) | — | 94 |
| 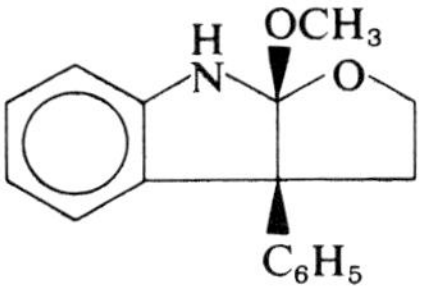 [structure: indoline with $OCH_3$, $C_6H_5$, fused O-containing ring] | — | — | 167 |
| 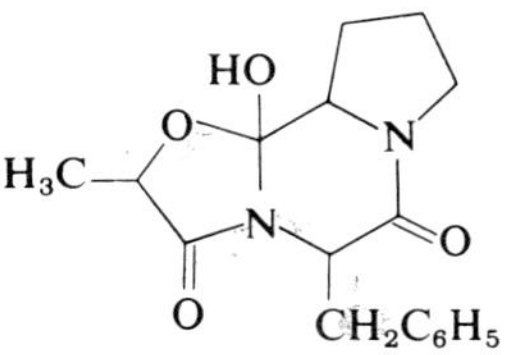 [structure: bicyclic diketopiperazine with HO, $CH_3$, $CH_2C_6H_5$ substituents] | 193°–196°; $[\alpha]_D$ −22.5° (ethanol) | — | 98, 121 |
| 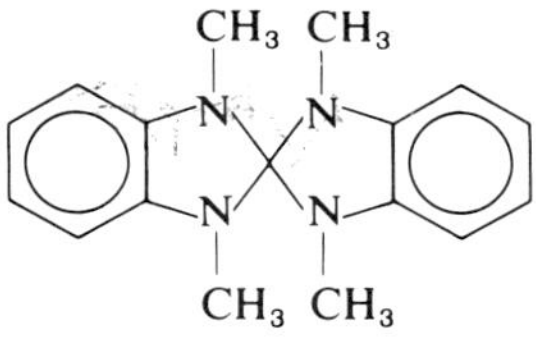 [structure: spiro bis-benzimidazoline with four $CH_3$ groups on N] | — | — | 123 |
| $(CH_3)_2NCH(OCH_2C_6H_5)_2$ | (118°–120°/0.25) | 1.5353 (20°) | 5, 16, 19, 62 |

TABLE 1–*continued*

| Compound | M.p., °C (B.p., °C/torr) | $n_D$ (T, °C) | References |
|---|---|---|---|
| 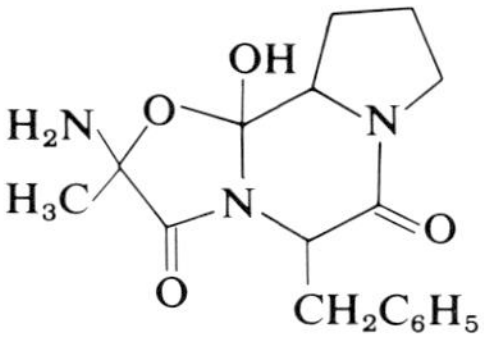 [structure: H₂N, H₃C, OH, pyrrolidine-fused oxazolidinone with CH₂C₆H₅] | 183° | — | 96, 98 |
| [structure: benzodioxole with two piperidines] | 85° | — | 92 |
| 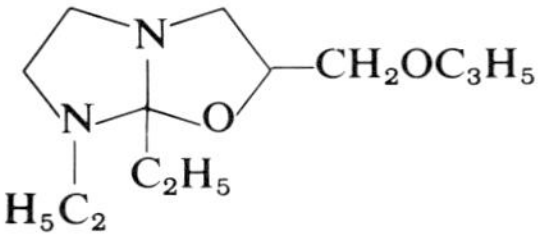 [structure: bicyclic with N, H₅C₂, C₂H₅, —CH₂OC₃H₅] | (136°/0.05) | 1.5428 (20°) | 79 |
| [structure: spiro compound with H₅C₂, S, N, O, C₂H₅] | 106° | — | 163 |
| 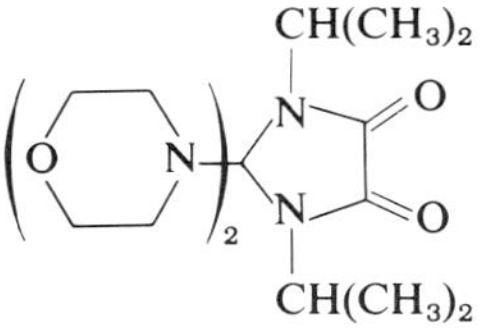 [structure: bis-morpholino with CH(CH₃)₂ groups] | 142° | — | 140 |

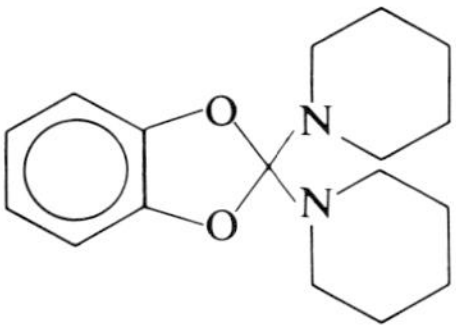

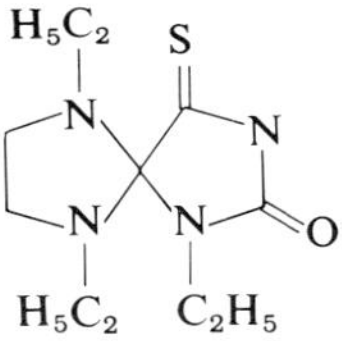

$C_{18}$ compounds

| | | | |
|---|---|---|---|
| 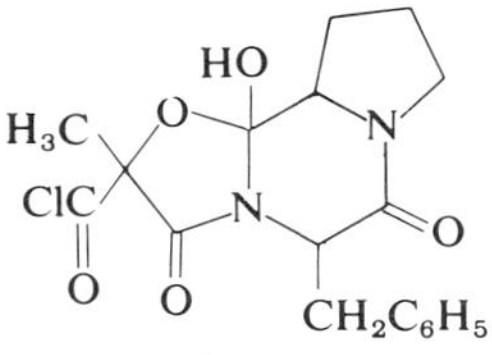 [structure: H₃C, ClC, HO, pyrrolidine-fused with CH₂C₆H₅] | 126° | — | 96, 98 |
| 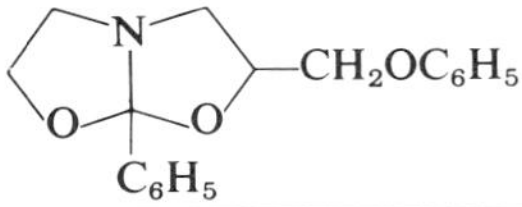 [structure: bicyclic N,O with —CH₂OC₆H₅ and C₆H₅] | (196°/0.6) 82.5° | — | 82 |

## TABLE 1—continued

| Compound | M.p., °C (B.p., °C/torr) | $n_D$ (T, °C) | References |
|---|---|---|---|
| 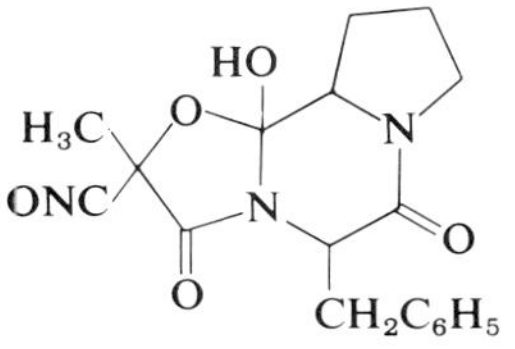 [structure: bicyclic pyrrolidine-oxazolidine with HO, H₃C, ONC, O, CH₂C₆H₅] | Two diastereomers: m.p. 225° m.p. 154° | — | 98 |
| 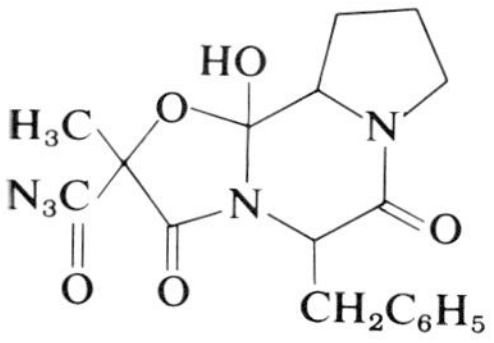 [structure: bicyclic pyrrolidine-oxazolidine with HO, H₃C, N₃C, O, CH₂C₆H₅] | 100° (dec.) | — | 96, 98 |
| 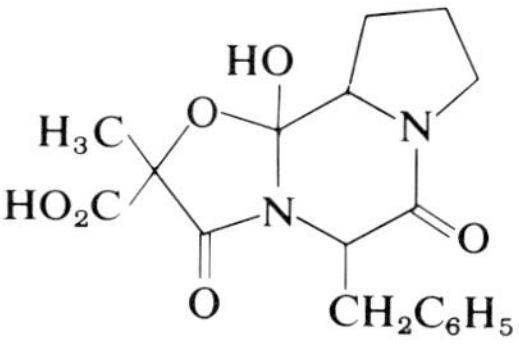 [structure: bicyclic pyrrolidine-oxazolidine with HO, H₃C, HO₂C, O, CH₂C₆H₅] | 122° | — | 96, 98 |
| [C₆H₅N(CH₃)]₂CHOCH(CH₃)₂ | (140°/0.25) | 1.5696 (25°) | 78 |
| | C₁₉ compounds | | |
| 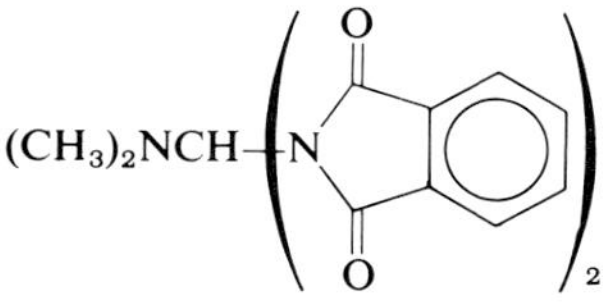 (CH₃)₂NCH—(N-phthalimide)₂ | 197° | — | 14 |
| 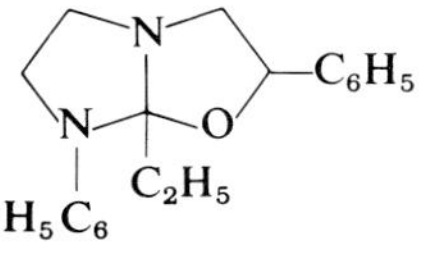 [structure: imidazolidine-oxazolidine with C₆H₅, C₂H₅, H₅C₆] | (172°/0.2) | 1.5920 (20°) | 79 |
| 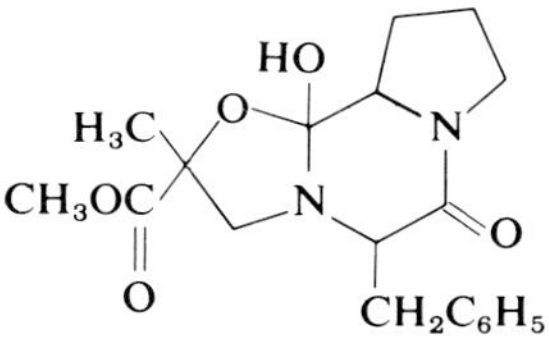 [structure: bicyclic pyrrolidine-oxazolidine with HO, H₃C, CH₃OC, O, CH₂C₆H₅] | Two diastereomers: m.p. 174°, m.p. 211° | — | 98 |

## TABLE 1–continued

| Compound | M.p., °C (B.p., °C/torr) | $n_D$ (T, °C) | References |
|---|---|---|---|
| *p*-CH₃C₆H₄—N⌐⌐N—C₆H₄CH₃-*p*, H, OC₂H₅ (imidazolidine) | 120° | — | 68 |
| 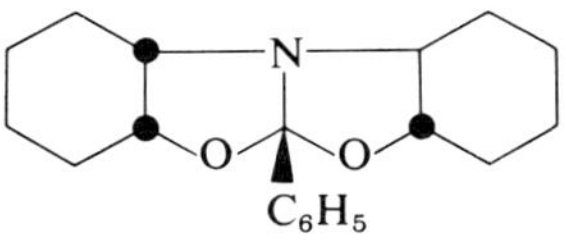 (bis-cyclohexyl fused bicyclic with N, O, O, C₆H₅) | 151° | — | 144 |
| $(C_6H_5O)_2CH\overset{+}{N}(C_2H_5)_3Cl^-$ | Hygroscopic solid | — | 127 |
| $[C_6H_5N(CH_3)]_2CHOC(CH_3)_3$ | (145°/0.7) | 1.5609 (25°) | 78 |
| $HC[N(C_3H_7)_2]_3$ | (82°/0.0005) | 1.4550 (20°) | 53a |

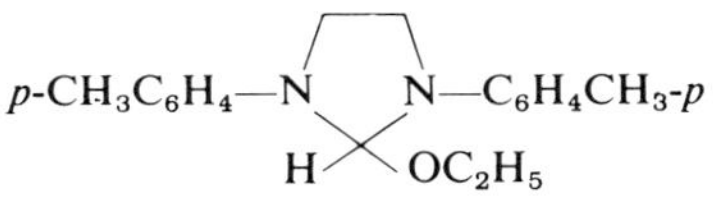

C₂₀ compounds

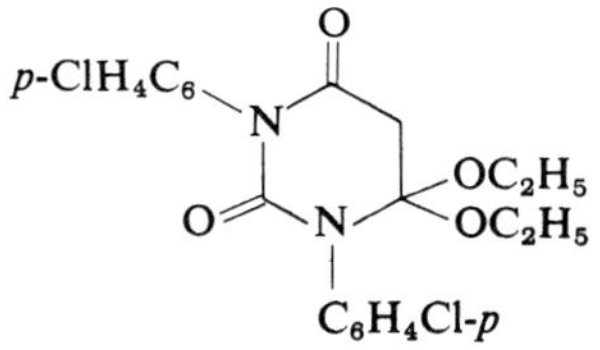

| Compound | M.p., °C | $n_D$ | References |
|---|---|---|---|
| *p*-ClH₄C₆–N, O, OC₂H₅, OC₂H₅, N–C₆H₄Cl-*p* (dihydrouracil) | 143° | — | 73 |
| *p*-O₂NH₄C₆–N, O, OC₂H₅, OC₂H₅, N–C₆H₄NO₂-*p* | 169° | — | 73 |
| H₅C₆–N, O, OC₂H₅, OC₂H₅, N–C₆H₄Cl-*p* | 138° | — | 73 |
| *p*-ClH₄C₆–N, O, OC₂H₅, OC₂H₅, N–C₆H₅ | 138° | — | 73 |

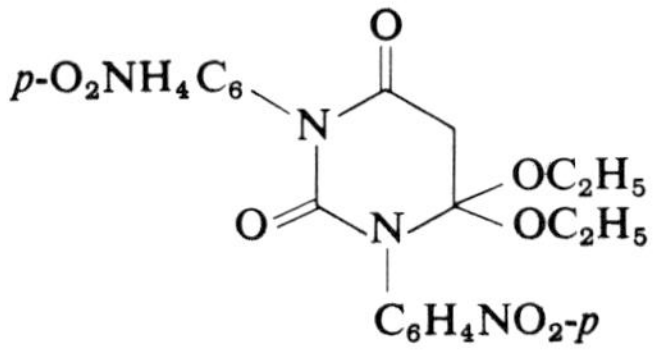

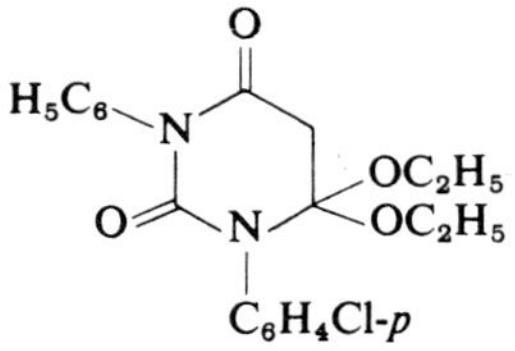

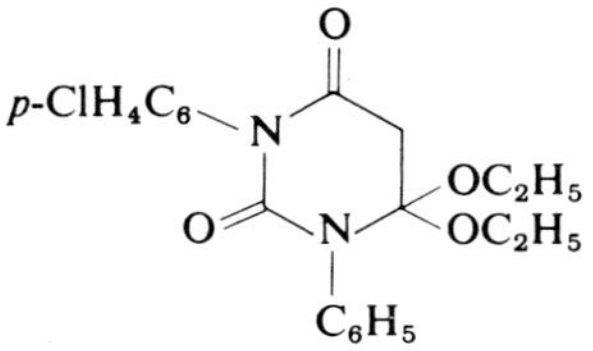

# AMIDE ACETALS, ESTER AMINALS, AND ORTHO AMIDES

TABLE 1–*continued*

| Compound | M.p., °C (B.p., °C/torr) | $n_D$ ($T$, °C) | References |
|---|---|---|---|
| 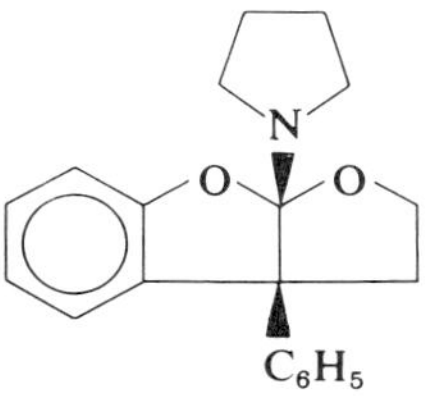 (pyrrolidine benzofuran phenyl compound) | 124° | — | 166 |
| 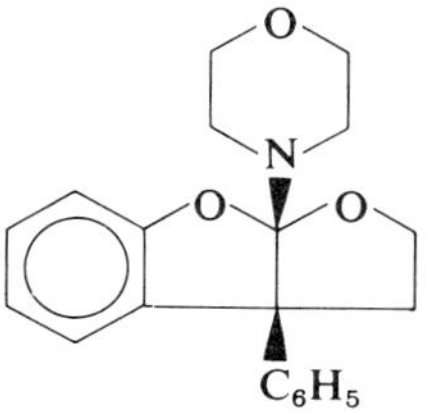 (morpholine benzofuran phenyl compound) | 114° | — | 166 |
| 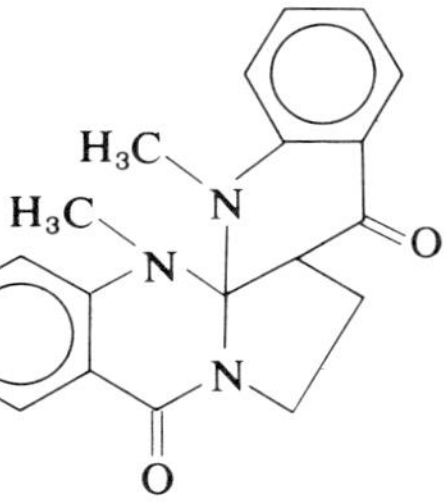 (dimethyl diamino pyrrolidinone compound) | 198° | — | 133 |
| 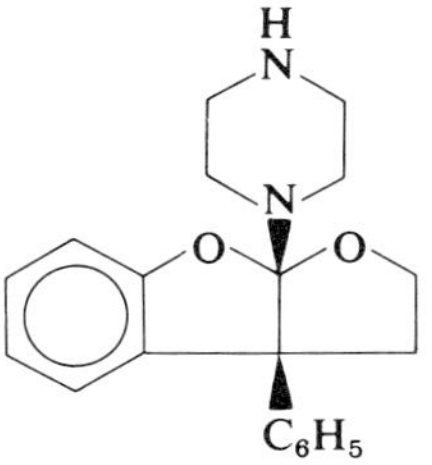 (piperazine benzofuran phenyl compound) | 153° | — | 166 |
| 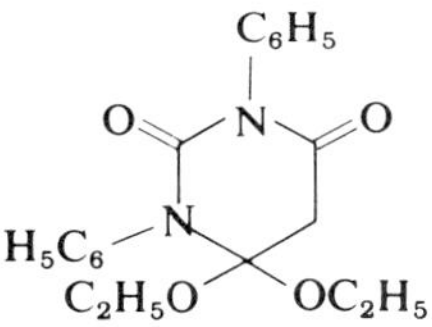 (diphenyl diethoxy dioxo pyrimidine compound) | 149° | — | 125 |

TABLE 1–*continued*

| Compound | M.p., °C (B.p., °C/torr) | $n_D$ ($T$, °C) | References |
|---|---|---|---|
| 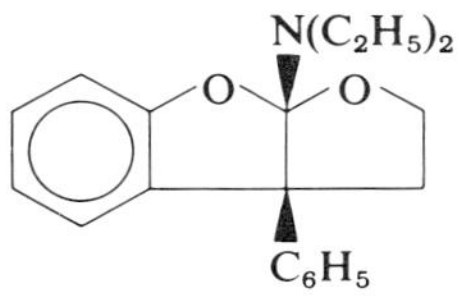 (structure with N(C$_2$H$_5$)$_2$ and C$_6$H$_5$) | 72° | — | 166 |
| 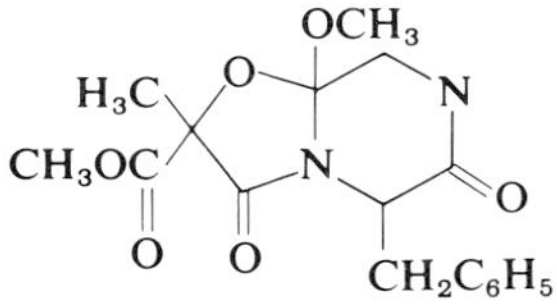 (structure with OCH$_3$, H$_3$C, CH$_3$OC(=O), CH$_2$C$_6$H$_5$) | Two diastereomers: m.p. 122°; m.p. 142° | — | 98 |
| 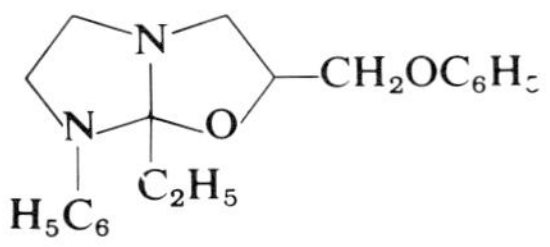 (structure with –CH$_2$OC$_6$H$_5$, H$_5$C$_6$, C$_2$H$_5$) | (176°/0.05) | 1.5800 (20°) | 79 |
| 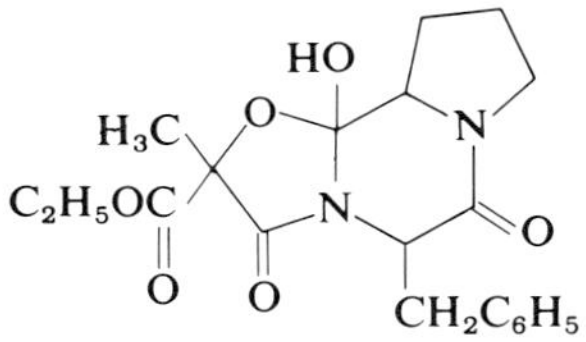 (structure with HO, H$_3$C, C$_2$H$_5$OC(=O), CH$_2$C$_6$H$_5$) | Two diastereomers: m.p. 136°, m.p. 204° | — | 94, 96, 98 |
| 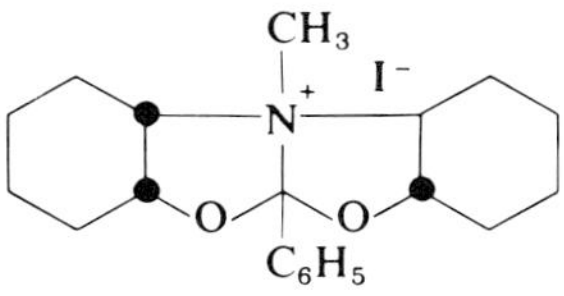 (structure with CH$_3$, N$^+$, I$^-$, C$_6$H$_5$) | 204° (dec.) | — | 144 |
| 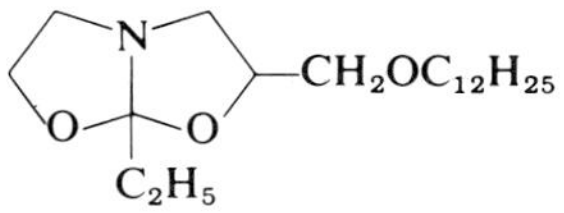 (structure with –CH$_2$OC$_{12}$H$_{25}$, C$_2$H$_5$) | (148°/0.05) | 1.4586 (20°) | 82 |
| 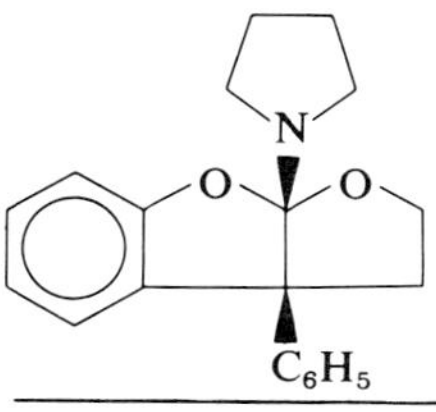 (structure with pyrrolidine N, C$_6$H$_5$) | 139° | — | 166 |

## TABLE 1—continued

| Compound | M.p., °C (B.p., °C/torr) | $n_D$ (T, °C) | References |
|---|---|---|---|
| $C_{21}$ compounds | | | |
| [structure: benzofuran fused with dioxolane, N-piperidinyl, C₆H₅] | 138° | — | 166 |
| [structure: benzofuran fused with dioxolane, N-(4-methylpiperazinyl), C₆H₅] | 107° | — | 166 |
| [structure: barbiturate-type, H₅C₆–N, N–C₆H₄CH₃-p, OC₂H₅, OC₂H₅] | 167° | — | 73 |
| [structure: barbiturate-type, p-CH₃H₄C₆–N, N–C₆H₅, OC₂H₅, OC₂H₅] | 160° | — | 73 |
| [structure: spiro imidazolidine, H₅C₂–N, N–C₆H₅, S, H₅C₂, C₆H₅, O] | 144° | — | 163 |

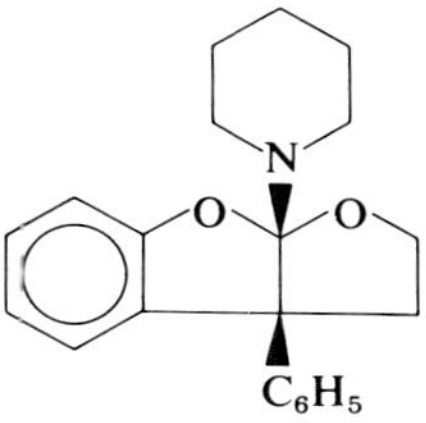

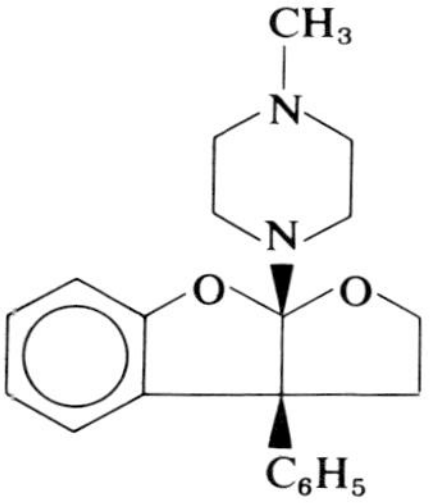

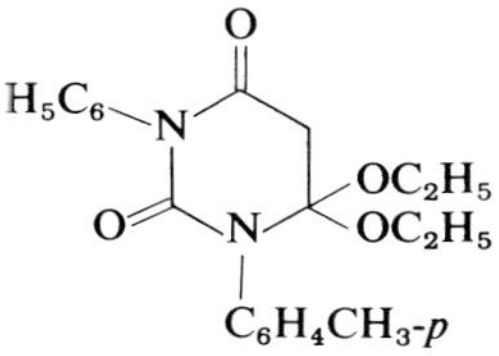

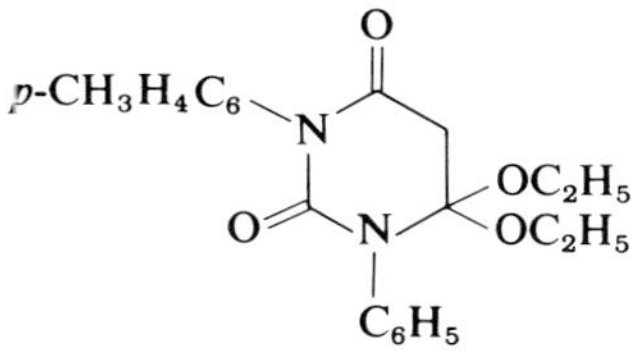

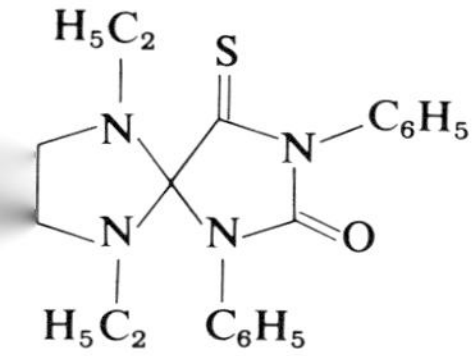

TABLE 1–continued

| Compound | M.p., °C (B.p., °C/torr) | $n_D$ (T, °C) | References |
|---|---|---|---|
| 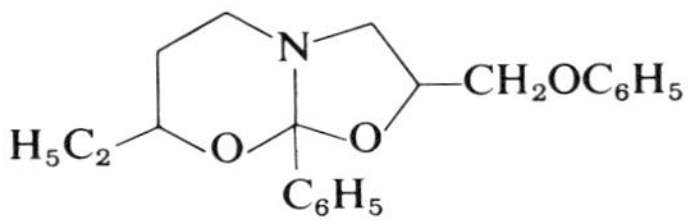 [bicyclic structure with $H_5C_2$, $C_6H_5$]—$CH_2OC_6H_5$ | (164°/0.05) | 1.5560 (20°) | 82 |
| 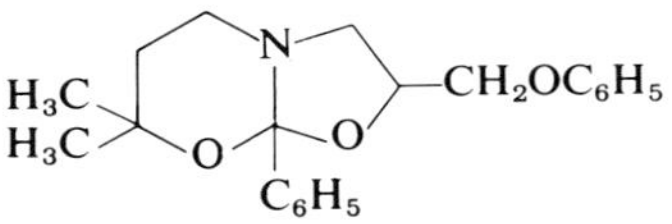 [bicyclic structure with $H_3C$, $H_3C$, $C_6H_5$]—$CH_2OC_6H_5$ | (175°/0.05) 106.5° | — | 82 |

$C_{22}$ compounds

| Compound | M.p., °C | $n_D$ | References |
|---|---|---|---|
| $(p\text{-}O_2NC_6H_4CONH)_3CH$ | 228° | — | 43 |
| $(C_6H_5CONH)_3CH$ | 231° | — | 43, 44, 141 |
| $[2,4\text{-}(O_2N)_2C_6H_3N(CH_3)]_3CH$ | — | — | 65 |
| $[o\text{-}BrC_6H_4N(CH_3)]_3CH$ | — | — | 65 |
| $[p\text{-}O_2NC_6H_4N(CH_3)]_3CH$ | 260°–265° | — | 65–67 |
| $[o\text{-}O_2NC_6H_4N(CH_3)]_3CH$ | — | — | 65 |
| $[C_6H_5N(CH_3)]_3CH$ | 263° | — | 65–67 |
| 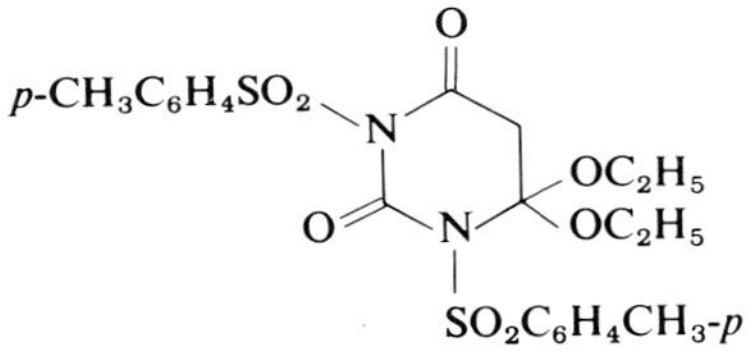 [cyclic urea with $p\text{-}CH_3C_6H_4SO_2$, $OC_2H_5$, $OC_2H_5$, $SO_2C_6H_4CH_3\text{-}p$] | 116° | — | 73 |

$C_{24}$ compound

| Compound | M.p., °C | $n_D$ | References |
|---|---|---|---|
| 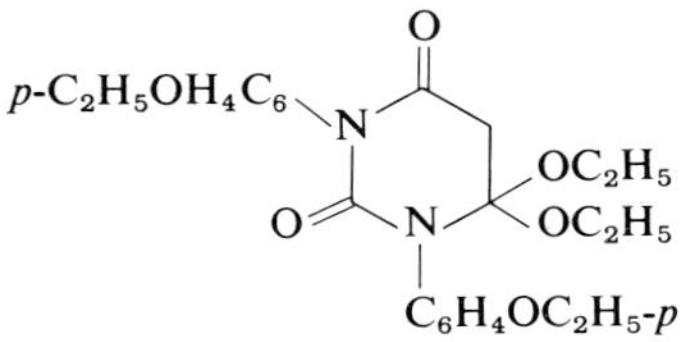 [cyclic urea with $p\text{-}C_2H_5OH_4C_6$, $OC_2H_5$, $OC_2H_5$, $C_6H_4OC_2H_5\text{-}p$] | 149° | — | 73 |

$C_{25}$ compounds

| Compound | M.p., °C | $n_D$ | References |
|---|---|---|---|
| 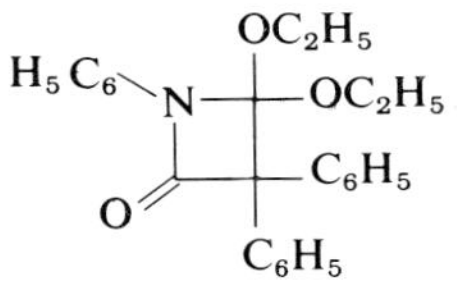 [β-lactam with $H_5C_6$, $OC_2H_5$, $OC_2H_5$, $C_6H_5$, $C_6H_5$] | 139° | — | 117 |
| $(C_6H_5CH_2CONH)_3CH$ | 222° | — | 43 |
| $(p\text{-}CH_3OC_6H_4CONH)_3CH$ | 225° | — | 43 |

TABLE 1—*continued*

| Compound | M.p., °C (B.p., °C/torr) | $n_D$ ($T$, °C) | References |
|---|---|---|---|
| 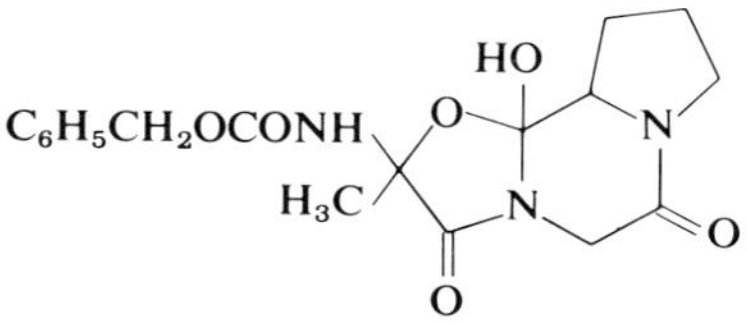 $C_6H_5CH_2OCONH$—[bicyclic structure with HO, O, N, CH_3, O] | 173° | — | 96, 98 |
| $[C_6H_5N(C_2H_5)]_3CH$ | 180°–186° | — | 65–67 |
| $[p\text{-}CH_3C_6H_4N(CH_3)]_3CH$ | — | — | 65 |
| $C_{26}$ compound | | | |
| 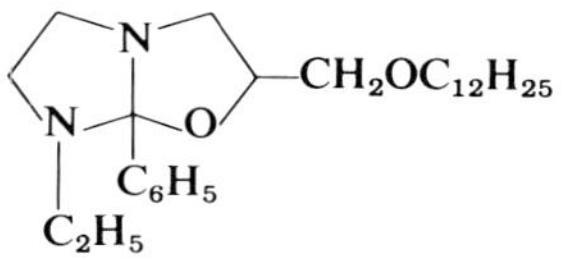 [bicyclic N,N,O structure with $C_6H_5$, $C_2H_5$]—$CH_2OC_{12}H_{25}$ | (213°/0.1) | 1.5071 (20°) | 79 |
| $C_{27}$ compounds | | | |
| 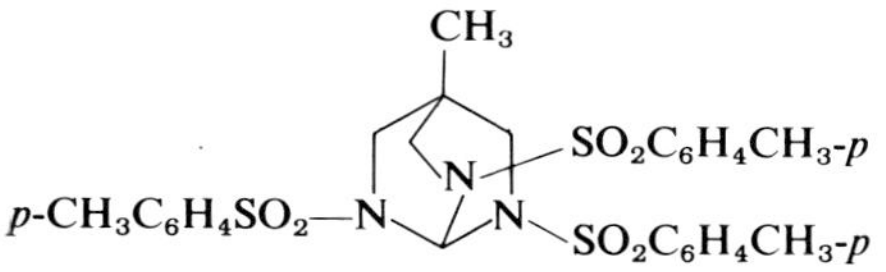 [bicyclic structure with CH_3, two $SO_2C_6H_4CH_3$-$p$ groups], $p\text{-}CH_3C_6H_4SO_2$—N | — | — | 68 |
| $[p\text{-}C_2H_5OC_6H_4N(CH_3)]_3CH$ | — | — | 65 |
| 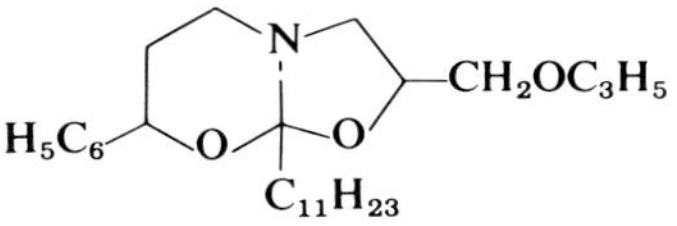 [bicyclic structure with $H_5C_6$, $C_{11}H_{23}$]—$CH_2OC_3H_5$ | (194°/0.05) | 1.5030 (20°) | 82 |
| $C_{28}$ compounds | | | |
| $[p\text{-}O_2NC_6H_4N(C_3H_7)]_3CH$ | — | — | 65 |
| $[2,5\text{-}(CH_3)_2C_6H_3N(CH_3)]_3CH$ | — | — | 65 |
| $C_{29}$ compounds | | | |
| [phenanthrene-fused dioxolane-imidazolidinone spiro structure with two $C_6H_5$ groups] | 200° | — | 77a |

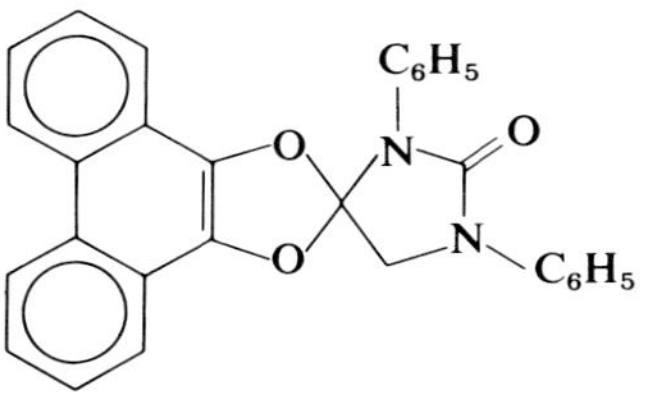

TABLE 1—*continued*

| Compound | M.p., °C (B.p., °C/torr) | $n_D$ ($T$, °C) | References |
|---|---|---|---|
| [Structure: diphenyl-substituted spiro hydantoin with piperazine, C=O groups] | 255° | — | 163 |
| 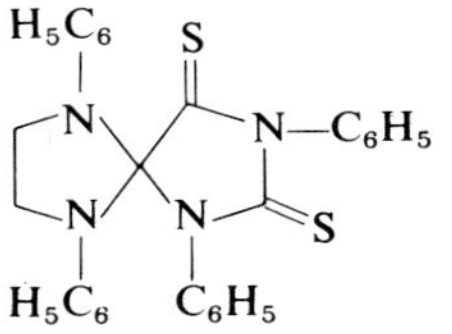 [Structure: diphenyl-substituted spiro thiohydantoin with piperazine, C=S groups] | 208° | — | 163 |
| [2-CH$_3$-5-CH(CH$_3$)$_2$—C$_6$H$_3$N(CH$_3$)]$_3$CH | — | — | 65 |

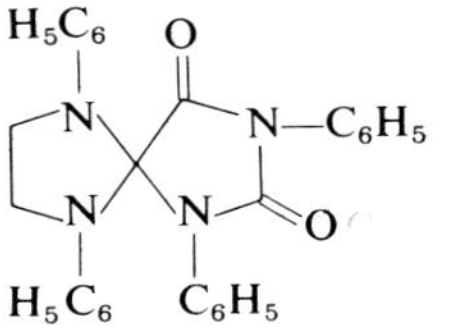

C$_{33}$ compounds

| Compound | M.p., °C | $n_D$ | References |
|---|---|---|---|
| 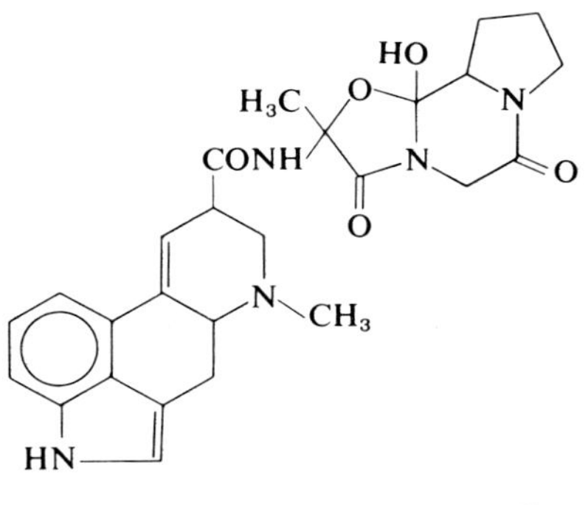 [Ergot alkaloid structure with HO, H$_3$C, CONH, O, N—CH$_3$, HN groups] | 236° $[\alpha]_D^{2°}$ + 385° (pyridine) | — | 96, 98 |
| 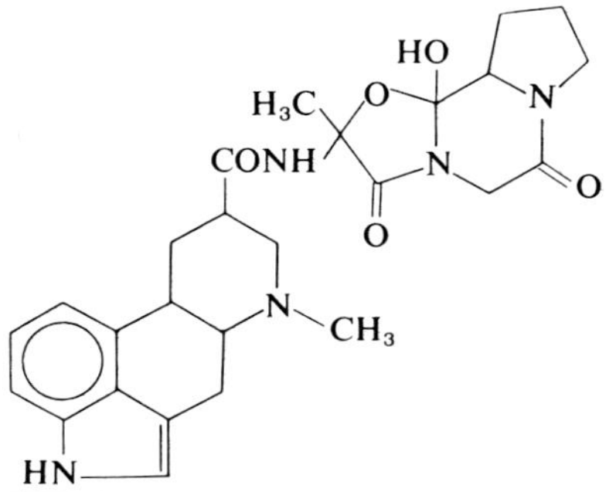 [Ergot alkaloid structure with HO, H$_3$C, CONH, O, N—CH$_3$, HN groups] | 242° $[\alpha]_D^{2°}$ − 63° (pyridine) | — | 98 |

TABLE 1—continued

| Compound | M.p., °C (B.p., °C/torr) | $n_D$ (T, °C) | References |
|---|---|---|---|
| [α-$C_{10}H_7N(CH_3)$]$_3$CH | $C_{34}$ compound — | — | 65 |
| [$C_6H_5N(C_8H_{17})$]$_3$CH | $C_{43}$ compound — | — | 65 |
| [$C_6H_5N(C_{12}H_{25})$]$_3$CH | $C_{55}$ compound — | — | 65 |
| [$C_6H_5N(C_{18}H_{37})$]$_3$CH | $C_{72}$ compound — | — | 65 |

## IV. REACTIONS OF AMIDE ACETALS, ESTER AMINALS, AND ORTHO AMIDES

This section discusses those reactions of amide acetals, ester aminals, and ortho amides which do not yield other ortho amide derivatives. Transacetalization, transamination, and disproportionation reactions are discussed in Section II of this chapter.

Many of the reactions of amide acetals, ester aminals, and ortho amides are analogous to those of ortho esters. A few are unique to the ortho amide derivatives. The ortho amide derivatives tend to be substantially more reactive than ortho esters, presumably due to their greater ease of conversion to resonance-stabilized carbonium ions. The ease of ionization of the ortho amide derivatives is demonstrated by the fact that solutions of them in aprotic solvents exhibit small but measurable electrical conductivity (*115, 134*).

### A. Reactions Which Form Carbon–Oxygen Bonds

1. HYDROLYSIS REACTIONS

Since the kinetics of hydrolysis reactions of ortho amide derivatives have not been investigated, only qualitative observations pertinent to their mechanisms appear in the literature. Most of these compounds hydrolyze rapidly under acidic conditions, and several are known to hydrolyze under neutral or alkaline conditions.

Dimethylformamide acetals are weak bases (5, 115), and it is reasonable to assume that the conjugate acid present in largest concentration in aqueous solutions is the N-protonated species $(CH_3)_2\overset{+}{N}HCH(OR)_2$. This thermodynamically most stable conjugate acid does not give rise to hydrolysis products under neutral or alkaline conditions. Hydrolysis of the hydrochloride salts of dimethylformamide acetals yields dimethylammonium chloride and formate esters [Eq. (70)], while the free bases hydrolyze to dimethylformamide and alcohols under neutral or alkaline conditions [Eq. (71)] (52a, 115).

$$(RO)_2CH\overset{+}{N}H(CH_3)_2 + H_2O \rightarrow (CH_3)_2\overset{+}{N}H_2 + HCO_2R + ROH \quad (70)$$

$$(RO_2)CHN(CH_3)_2 + H_2O \rightarrow (CH_3)_2NCHO + 2\,ROH \quad (71)$$

A similar dependence of hydrolysis products on the pH of the reaction medium is exhibited by the heterocyclic benzamide acetal **49**, which yields a benzamide derivative when refluxed in aqueous ethanol and a benzoate ester when allowed to stand in dilute hydrochloride acid [Eq. (72)] (144).

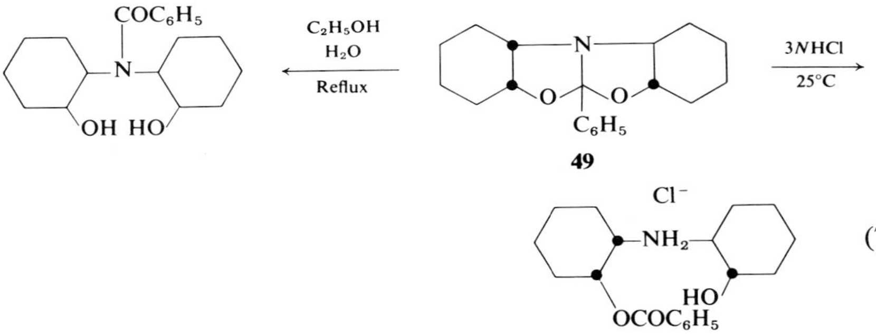

Hydrolysis of amide acetals in acidic solutions presumably involves dissociation of the N-protonated conjugate acid to a dialkoxycarbonium ion, which reacts with water to form the ester [Eq. (73)].

$$R_2NCR^1(OR^2)_2 + H^+ \rightleftarrows R_2\overset{+}{N}HCR^1(OR^2)_2 \rightleftarrows$$

$$R_2NH + R^1C\overset{OR^2}{\underset{OR^2}{{\diagup}\atop{\diagdown}}} \xrightarrow{H_2O} R_2NH_2^+ + R^1CO_2R^2 + R^2OH \quad (73)$$

In less acidic or neutral solutions the preferred reaction pathway is via the more stable dialkylaminoalkoxycarbonium ion, possibly formed by a rate-limiting proton transfer from water or some other general acid to alkoxy oxygen [Eq. (74)]. Kinetic studies would be required to ascertain whether an

$S_E2$ mechanism analogous to that of ortho ester hydrolysis is involved in this reaction.

$$R_2NCR^1(OR^2)_2 + HA \longrightarrow R^2OH + A^- + R^1C\begin{smallmatrix}NR_2\\+\\OR^2\end{smallmatrix} \xrightarrow{H_2O}$$

$$R^1CONR_2 + R^2OH \quad (74)$$

Unlike triethyl orthoformate, which is stable in neutral and alkaline solutions, dimethylformamide diethyl acetal undergoes rapid hydrolysis in water or aqueous alkali to dimethylformamide and ethanol (115). It was not established whether hydroxide ion accelerates the hydrolysis reaction, or whether it occurs at the same rate in neutral and alkaline solutions. In either event, the reaction under these conditions may occur by a different mechanism than that of the acid-catalyzed reaction. The observation that dimethylformamide acetals undergo rapid formyl proton exchange with deuteriomethanol (135) suggests that neutral and alkaline hydrolysis may proceed through a carbenoid intermediate.

$$B: + HC(OR)_2N(CH_3)_2 \longrightarrow \overset{+}{B}H + RO^- + (CH_3)_2\overset{..}{N}COR \xrightarrow{H_2O}$$
$$(CH_3)_2NCH(OR)OH \longrightarrow ROH + (CH_3)_2NCHO \quad (75)$$

The carbene mechanism could not operate in the case of ortho amide derivatives of higher orthocarboxylic acids. The indications are that such amide acetals are considerably less reactive than formamide acetals toward alkaline hydrolysis. For example, the cyclol methyl ethers **50** and **51** are unaffected by aqueous alkali, although they hydrolyze to the cyclols under acidic conditions (91, 121).

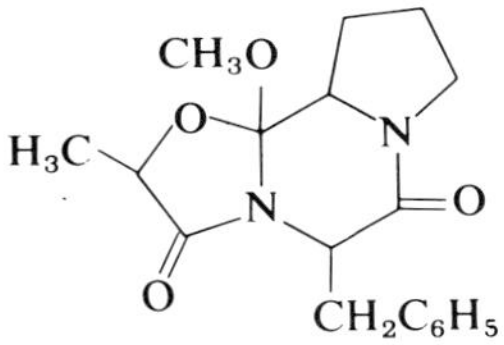

**50** 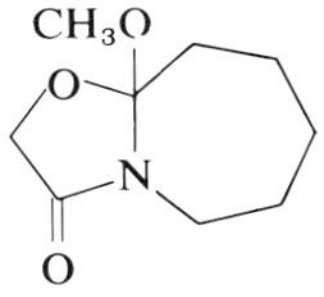          **51**

The heterocyclic ortho amide derivatives **24** are unaffected by neutral water, but hydrolyze slowly in dilute acid (166).

Not all ortho amide derivatives of higher orthocarboxylic acids are inert in neutral and alkaline aqueous solutions. The heterocyclic orthobenzoic acid derivative **49** is completely hydrolyzed by prolonged heating with water or alkali (144). The heterocyclic ester aminals obtained by addition of 2-ethyl-1-phenylimidazoline to epoxides [Eq. (54)] hydrolyze in alkaline as well as in acidic solutions (79). 1-Phenyl-4,4-dimethoxy-3,3-dimethylazetidine-2-one and its thione analogue hydrolyze in refluxing aqueous dioxane (125, 126). Diazidoethoxyethane, $CH_3(N_3)_2OC_2H_5$, hydrolyzes to ethyl acetate and hydrazoic acid under neutral or alkaline, as well as acidic, conditions (136). These reactions may be due, at least in part, to general acid catalysis by water.

The effects of structure on hydrolytic reactivities of ortho amide derivatives have not been investigated. The qualitative information available indicates that most amide acetals and ester aminals hydrolyze rapidly in acidic solutions. Orthoformamide derivatives—at least the formamide acetals—also hydrolyze rapidly in neutral and alkaline solutions. Many other ortho amide derivatives hydrolyze slowly in neutral and alkaline solutions. A few ortho amide derivatives, such as 2,2-diethoxy-3-methylbenzoxazoline and the $N,N',N''$-trialkyl-$N,N',N''$-triarylorthoformamides hydrolyze slowly even in acidic solutions (89a, 115).

## 2. ALCOHOLYSIS REACTIONS

Amide acetals react with alcohols under weakly acidic conditions to form ortho esters. This fact is the basis of a synthesis of ortho esters from $N,N$-dialkyl amides. The dialkyl amide is converted to an amide acetal as described previously, and the amide acetal is alcoholized to the ortho ester by adding a slight excess of acetic acid to the alcoholic solution of the acetal and allowing the solution to stand for a few hours at room temperature (76). Triethyl orthopropionate was obtained in 62% overall yield, and triethyl orthobenzoate in 45% overall yield, from propionyl and benzoyl piperidides.

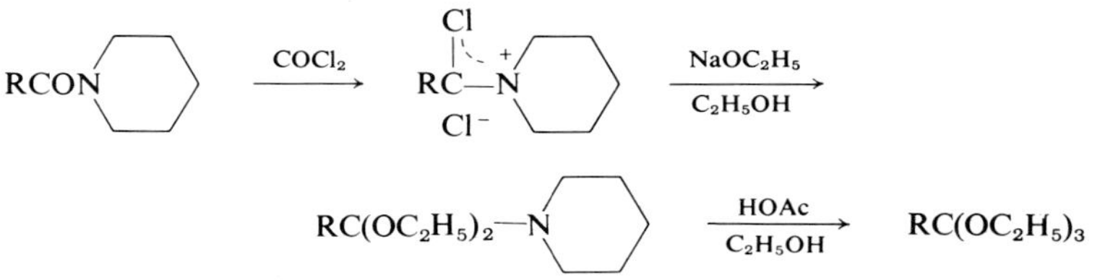

$$RCON\langle\rangle \xrightarrow{COCl_2} RC(Cl)\text{—}N^+\langle\rangle\ Cl^- \xrightarrow[C_2H_5OH]{NaOC_2H_5} RC(OC_2H_5)_2\text{—}N\langle\rangle \xrightarrow[C_2H_5OH]{HOAc} RC(OC_2H_5)_3 \qquad (76)$$

$$(R = C_2H_5,\ C_6H_5)$$

Similarly, boiling tris(ethylcarbamyl)methane with ethanol containing a little sulfuric acid gave triethyl orthoformate (141).

$$HC(NHCO_2C_2H_5)_3 + 3\ C_2H_5OH \xrightarrow{H^+} HC(OC_2H_5)_3 + 3\ H_2NCO_2C_2H_5 \qquad (77)$$

1,1-Diazido-1-ethoxyethane is converted to triethyl orthoacetate in 49% yield by reaction with ethanolic sodium ethoxide (136).

$$CH_3C(N_3)_2OC_2H_5 + 2\ C_2H_5ONa \rightarrow 2\ NaN_3 + CH_3C(OC_2H_5)_3 \qquad (78)$$

## 3. ACETALIZATION REACTIONS

It is reasonable to expect amide acetals to undergo transacetalization reactions with carbonyl compounds, since the resonance stabilization of the amide formed by the reaction should provide a considerable driving force for formation of the acetal or ketal. The only examples of this method of acetalization, which takes place under very mild reaction conditions, are conversions of 3-cholestanone, 5α-androstan-3,17-dione, and 5α-pregnan-3,20-dione to 3-ethylene ketals by reaction with 2-dimethylamino-1,3-dioxolane in methylene chloride or benzene containing a little acetic acid (*146*).

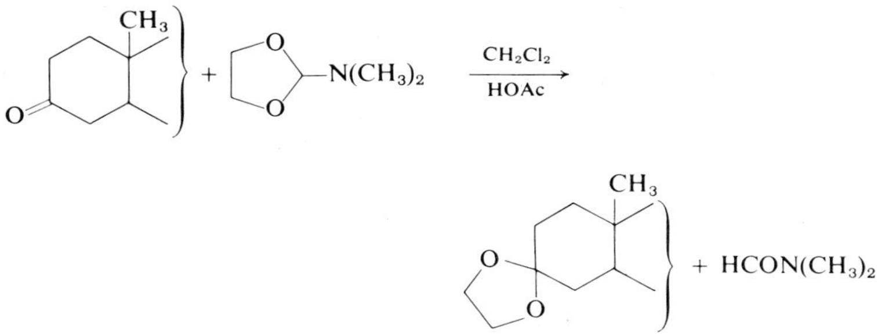

$$\text{[steroid ketone]} + \text{[dioxolane]}-N(CH_3)_2 \xrightarrow[\text{HOAc}]{CH_2Cl_2} \text{[ethylene ketal]} + HCON(CH_3)_2 \qquad (79)$$

## 4. ESTERIFICATION OF CARBOXYLIC ACIDS BY AMIDE ACETALS

Brechbuehler and co-workers and Vorbruggen reported simultaneously in 1963 that $N,N$-dimethylformamide dialkyl acetals react with carboxylic acids in inert solvents such as methylene chloride, chloroform, or benzene to form carboxylate esters in high yields (*18, 19, 147*). The reaction probably involves bimolecular attack by the carboxylate ion of an alkoxydialkylaminocarbonium carboxylate ion pair on the alkyl group of the alkoxycarbonium ion.

$$RCO_2H + (CH_3)_2NCH(OR')_2 \xrightarrow{-R'OH}$$

$$HC\overset{+}{\underset{OR'}{\overset{N(CH_3)_2}{\diagup}}} RCO_2^- \longrightarrow RCO_2R' + CHON(CH_3)_2 \qquad (80)$$

This view is supported by the fact that the (S)-2-butyl acetal of dimethylformamide reacts with benzoic acid in boiling benzene to form (R)-2-butyl benzoate of high optical purity, and by the fact that dimethylformamide acetals of highly hindered alcohols such as neopentyl react with carboxylic acids only very slowly (*19*). The reaction is insensitive to steric hindrance in the carboxylic acid: methyl, ethyl, and benzyl mesitoates were prepared in

82%, 93%, and 94% yields from mesitoic acid and the appropriate dimethylformamide acetals (*19*).

Reaction (80) gives esters of primary and secondary alcohols in yields usually exceeding 80% and sometimes approaching 100%. It has been used to prepare methyl esters of benzoic and mesitoic acids; ethyl esters of benzoic, mesitoic, pyridine-3-carboxylic, adipic, *o*-hydroxybenzoic, and 3β-hydroxy-16,17-secoandrost-5-ene-16,17-dioic acids; benzyl esters of benzoic, mesitoic, fumaric, and β-naphthoic acids; and isobutyl, secondary butyl, and neopentyl esters of benzoic acid (*18, 19, 106, 147*). The reaction also affords high yields of benzyl esters of N-protected amino acids and peptides.

Reaction (80) utilizes only half of the alcohol used in synthesizing the dimethylformamide acetal to esterify the carboxylic acid. This may prove disadvantageous if the alcohol is expensive or is optically active. Complete utilization of the alcohol can be achieved by heating a solution of the acid, the desired alcohol, and *N,N*-dimethylformamide dineopentyl acetal. The dineopentyl acetal apparently undergoes transacetalization with the alcohol to form a formamide mixed acetal, which then alkylates the carboxylic acid (*19, 58*).

$$RCO_2H + R'OH + (CH_3)_2NCH[OCH_2C(CH_3)_3]_2 \rightarrow$$
$$RCO_2R' + (CH_3)_2NCHO + 2\,(CH_3)_3CCH_2OH \quad (81)$$

The bicyclic amide acetals from reactions of 2-substituted oxazolines with epoxides (*82*) react with carboxylic acids to form ester amides of diethanolamine (*81*).

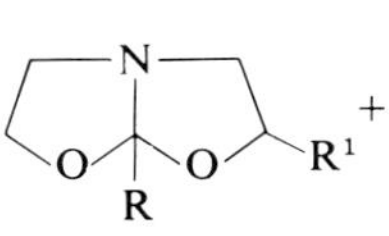

$$\text{[bicyclic oxazolidine structure]} + 2\,R^2CO_2H \longrightarrow R^2CO_2CH_2CH_2N(COR)CH_2CHR^1OCOR^2 \quad (82)$$

$(R = C_2H_5, R^1 = C_6H_5, CH_2OC_6H_5, R^2 = H, CH_3, C_2H_5;$
$R = R^1 = C_6H_5, R^2 = CH_3)$

## 5. Alkylation of Phenols by Amide Acetals

Phenol and *N,N*-dimethylformamide dibenzyl acetal react in methylene dichloride to form benzyl phenyl ether in 64% yield. Ethyl 1,3,5-trichlorophenyl ether was prepared similarly in 87% yield (*147*).

$$(CH_3)_2NCH(OR)_2 + ArOH \xrightarrow{CH_2Cl_2} ArOR + ROH + HCON(CH_3)_2 \quad (83)$$

$(Ar = C_6H_5, R = C_6H_5CH_2; Ar = 1,3,5-Cl_3C_6H_2, R = C_2H_5)$

## B. Reactions Which Form Carbon–Nitrogen Bonds

1. SYNTHESES OF AMIDINES FROM ORTHO AMIDE DERIVATIVES

Dimethylformamide dimethyl and diethyl acetals react with a variety of substances having primary amino groups to form $N$-dimethylaminomethylene derivatives ($N,N$-dimethyl-$N'$-substituted formamidines).

$$(CH_3)_2NCH(OR)_2 + R'NH_2 \rightarrow R'N{=}CHN(CH_3)_2 + 2\ ROH \quad (84)$$

[R = $CH_3$ or $C_2H_5$, R' = $p$-$CH_3C_6H_4$, $o$-$(m,p)$-$O_2NC_6H_4$, 2,4-$(O_2N)_2C_6H_3$-, 2,4,6-$(O_2N)_3C_6H_2$, 3,4-$Cl_2C_6H_3$, 2-[$C(C_6H_5)$=$NCH_2CONHCH_3$]-4-$O_2NC_6H_3$, 2-(3,4)-pyridyl, 3-(4-)-(1,2,4-triazolyl), 5-(1,2,3,4-tetrazolyl), $C_3H_7CO$, $C_6H_5CO$, $C_2H_5OCO$, $H_2NCO$, $H_2NCS$, $H_2NCONHCO$-, $(CH_3)_2NCH{=}NCONHCO$, $(CH_3)_2N$, $p$-$CH_3C_6H_4NH$, $p$-$(O_2N)C_6H_4NH$, 2,4-$(O_2N)_2C_6H_3NH$, $p$-$CH_3C_6H_4SO_2$, $(CH_3)_2NCH{=}NSO_2$]

Amino compounds which have been converted to $N$-dimethylaminomethylene derivatives in this way include aromatic amines (32, 86, 115, 159); heterocyclic amines (32) including nucleoside derivatives (99a, 100, 101, 138, 139, 171, 172); carboxamides and carbamates (32, 159); urea, thiourea, and biuret (32, 99); nitroguanidine (29a); hydrazine derivatives (159); and sulfonamides (159). Yields of amidines usually exceed 75%. The $N$-dimethylaminomethylene group is hydrolyzed to a formyl group at pH 5–8. A very mild method of formylating primary amino groups is to treat the amino compound with a dimethylformamide acetal and then hydrolyze the resulting $N,N$-dimethylformamidine derivative (32).

Dimethylaminomethylene groups, introduced by reactions with dimethylformamide acetals, have been used to protect the amino groups of adenine and cytosine residues of nucleoside and nucleotide derivatives during acylation and other reactions of the carbohydrate moeity (100, 101, 138, 139, 171, 172). The protecting group is completely removed by allowing the amidine to stand in 10% acetic acid for 5 hours at room temperature.

While acetals of higher carboxamides should also react with amino compounds to form substituted amidines, few such reactions have been reported. $N,N$-Dimethylbenzamide dimethyl acetal reacts with amides and sulfonamides to form substituted benzamidines (159).

$$C_6H_5C(OCH_3)_2N(CH_3)_2 + RNH_2 \rightarrow C_6H_5C(=NR)N(CH_3)_2 + 2\ CH_3OH \quad (85)$$
$$(R = CCl_3CO,\ CF_3CO,\ p\text{-}CH_3C_6H_4SO_2)$$

$N$-Methylpyrrolidone diethyl acetal and semicarbazide yield $N$-methylpyrrolidone semicarbazone (115).

$$\underset{\underset{CH_3}{|}}{\left\langle\!\!\!\bigcirc\!\!\!\right\rangle_{\!N}}\!\!(OC_2H_5)_2 + H_2NCONHNH_2 \longrightarrow \underset{\underset{CH_3}{|}}{\left\langle\!\!\!\bigcirc\!\!\!\right\rangle_{\!N}}\!\!{=}NNHCONH_2 + 2\ C_2H_5OH \quad (86)$$

Reactions of ester aminals and ortho amides with amino compounds have received little attention. The only such reactions thus far reported are conversions of *p*-nitroaniline, urea, and *p*-toluenesulfonamide to dimethylaminomethylene derivatives by reaction with tris(dimethylamino)methane (27).

$$RNH_2 + HC[N(CH_3)_2]_3 \rightarrow RN=CHN(CH_3)_2 + 2(CH_3)_2NH \quad (87)$$
$$(R = p\text{-}O_2NC_6H_4, H_2NCO, p\text{-}CH_3C_6H_4SO_2)$$

Fluorinolysis of tris(formylamino)methane yields $N,N,N'$-trifluoroformamidine, $F_2NCH=NF$ (*11a*).

## 2. SYNTHESES OF NITROGEN HETEROCYCLES FROM ORTHO AMIDE DERIVATIVES

Formamide acetals, formate aminals, and tris(formylamino)methanes have been used as reagents in syntheses of a variety of heterocyclic compounds, including triazines, pyrimidines, purines, triazoles, and condensed polycyclic compounds.

Ureas, amidines, and guanidines react with dimethylformamide dialkyl acetals to form substituted 1,3,5-triazines. Benzamidine and dimethylformamide diethyl acetal yield 2,5-diphenyl-1,3,5-triazine (29).

$$2 C_6H_5-C(NH_2)=NH + (CH_3)_2NCH(OC_2H_5)_2 \longrightarrow$$

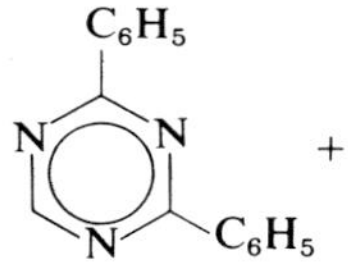

[2,5-diphenyl-1,3,5-triazine structure] $+ (CH_3)_2NH + 2 C_2H_5OH + NH_3$ (88)

*S*-Methylisothiourea reacts with the same acetal to form 2-amino-4-methylthio-1,3,5-triazine (29).

$$2 H_2NC(SCH_3)=NH + (CH_3)_2NCH(OC_2H_5)_2 \longrightarrow$$

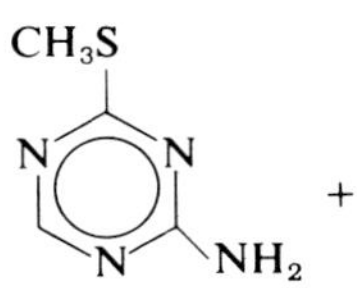

[2-amino-4-methylthio-1,3,5-triazine structure] $+ (CH_3)_2NH + 2 C_2H_5OH + CH_3SH$ (89)

2,4-Diamino-1,3,5-triazines are formed by reactions of guanidine, N-substituted guanidines, and N,N-disubstituted guanidines with dimethylformamide diethyl acetal (29, 31, 32). The substituted guanidines give either 2-amino-4-(substituted amino)-1,3,5-triazines, 2,4-bis(substituted amino)-

1,3,5-triazines, or mixtures of the two; yields range from poor to fair. These reactions are summarized by Eq. (90).

$$RR'N-C(NH_2)=NH + (CH_3)_2NCH(OC_2H_5)_2 \longrightarrow$$

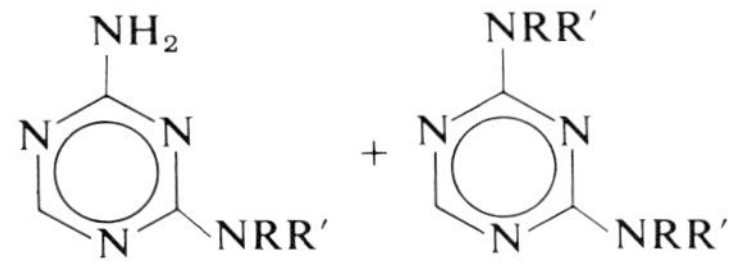

[diagram of triazines I and II with NH$_2$/NRR' and NRR'/NRR' substituents] (90)

[I: R = H, R' = H, CH$_3$, C$_2$H$_5$, C$_6$H$_{11}$; R = CH$_3$, R' = C$_6$H$_5$; RR' = (—CH$_2$—)$_5$
 II: R = H, R' = C$_4$H$_8$, C$_6$H$_5$, $m$-CH$_3$C$_6$H$_4$, $p$-CH$_3$C$_6$H$_4$, C$_6$H$_5$CH$_2$, $p$-CH$_3$OC$_6$H$_4$]

The mechanisms of these reactions have not been investigated. It seems likely that the amide acetal condenses with 2 moles of the amidine, guanidine, or urea to form a complex formamidine of the type

$$HN=CY-N=CH-N=CY-NH_2,$$

which then undergoes cyclization and elimination of either NH$_3$ or YH.

Guanylurea and substituted guanidines react with dimethylformamide dimethyl acetal to form 5-azacytosine and $N$-methylated 5-azacytosines in high yields (122).

$$RNHCON=C(NH_2)NR'R'' + (CH_3)_2NCH(OCH_3)_2 \longrightarrow$$

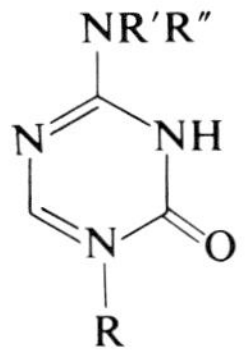

[diagram of triazinone with NR'R'', NH, O, and N-R substituents] + (CH$_3$)$_2$NH + 2 CH$_3$OH (91)

(R = R' = R'' = H; R = R' = CH$_3$, R'' = H; R = R' = R'' = CH$_3$)

A similar reaction of $O$-methylguanylisourea yields 2-amino-4-methoxy-1,3,5-triazine (122).

$$\overset{OCH_3}{\underset{|}{HN=C}}-N=C(NH_2)_2 + (CH_3)_2NCH(OCH_3)_2 \longrightarrow$$

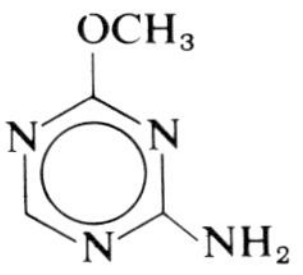

[diagram of triazine with OCH$_3$ and NH$_2$ substituents] + (CH$_3$)$_2$NH + 2 CH$_3$OH (92)

Dimethylformamide diethyl acetal and chloromethanesulfonamide react to form 5-(N,N-dimethylamino)-1,3,4-oxathiazol-3,3-dioxide (103a).

2-Amino-4-methylpyrimidine is formed when bis(dimethylamino)methoxymethane is heated successively with acetone and guanidine carbonate. 1-Dimethylamino-1-butene-3-one was probably an unisolated intermediate in this reaction (25).

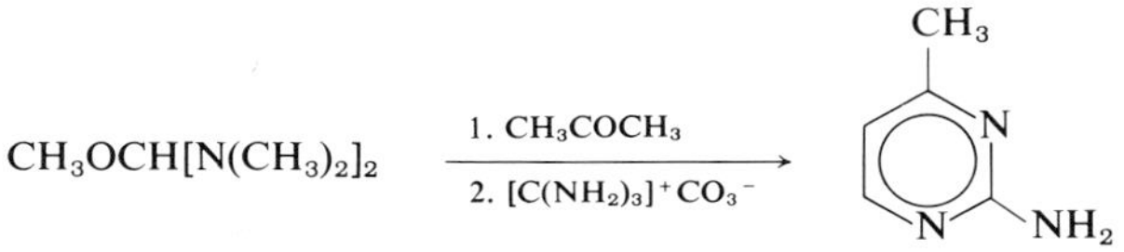

$$CH_3OCH[N(CH_3)_2]_2 \xrightarrow[\text{2. }[C(NH_2)_3]^+CO_3^-]{\text{1. }CH_3COCH_3} \text{4-methyl-2-aminopyrimidine} \quad (93)$$

The ortho amide derivative which has been most extensively used in syntheses of nitrogen heterocycles is tris(formylamino)methane, whose preparation and reactions were thoroughly studied by Bredereck and co-workers.

Anilinoguanidinium sulfate and tris(formylamino)methane form 1-phenyl-3-formamido-1,2,4-triazole in 90% yield (55).

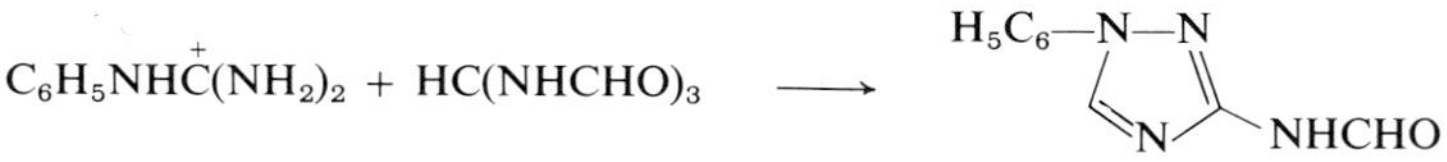

$$C_6H_5N\overset{+}{H}\overset{.}{C}(NH_2)_2 + HC(NHCHO)_3 \longrightarrow \text{1-phenyl-3-formamido-1,2,4-triazole} \quad (94)$$

Substances having active methylene groups condense with tris(formylamino)methane to form 4,5-disubstituted pyrimidines. 4-Amino-5-arylpyrimidines are obtained from arylacetonitriles (33, 145), 4-hydroxy-5-arylpyrimidines from arylacetamides (33, 145), 4-amino-5-cyanopyrimidine from malononitrile (33), and 4-hydroxy-5-substituted pyrimidines from cyanoacetamide, other derivatives of the monoamide of malonic acid, and dialkyl malonates [Eqs. (95) and (96)] (36, 54).

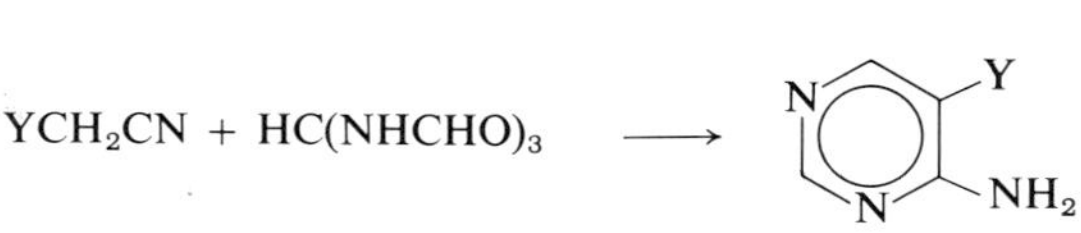

$$YCH_2CN + HC(NHCHO)_3 \longrightarrow \text{4-amino-5-Y-pyrimidine} \quad (95)$$

[Y = $C_6H_5$, m (and p)-$O_2NC_6H_4$, p-$HCONHC_6H_4$, CN]

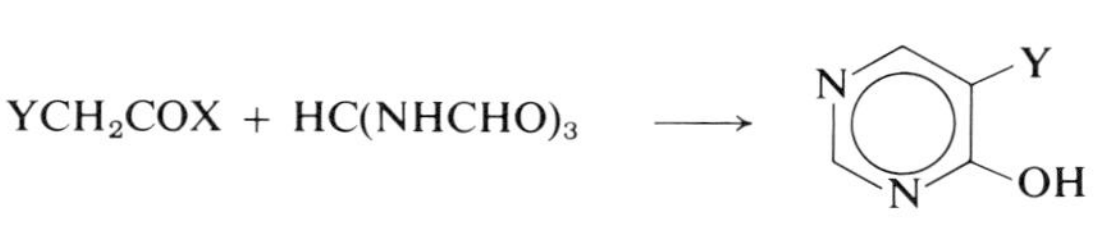

$$YCH_2COX + HC(NHCHO)_3 \longrightarrow \text{4-hydroxy-5-Y-pyrimidine} \quad (96)$$

[X = $NH_2$, Y = CN, $CO_2C_2H_5$, $CONH_2$, p-$O_2NC_6H_4$; X = OR, Y = $CO_2R$ (R = $CH_3$, $C_2H_5$, $C_4H_9$)]

Ketones condense with tris(formylamino)methane to form 4-substituted and 4,5-disubstituted pyrimidines [Eq. (97)] (*45, 158a*). All of these reactions probably involve initial condensation of the active methylene compound with tris(formylamino)methane to form formylaminomethylene derivatives, which cyclize to the pyrimidines.

$$RCH_2COR' + HC(NHCHO)_3 \longrightarrow \text{[pyrimidine with R, R' substituents]} \quad (97)$$

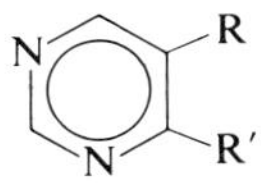

[R = H, R' = H, $CH_3$, $CH_2CH(CH_3)_2$, $C(CH_3)_3$, $C_6H_5$,
6-methyl-2-pyridyl; R = $CH_3$, R' = H, $CH_3$, $C_6H_5$;
R = $C_2H_5$, R' = $CH_3$, $C_6H_5$; R, R' = $-(CH_2)_n-$
(n = 3, 4), R = $CH_2CO_2C_2H_5$, R' = $CH_3$]

When tris(formylamino)methane is heated above 150°C, a complex series of reactions occur which produce 1,3,5-triazine, carbon monoxide, formamide, ammonia, and water (*30, 32a, 55*). A mechanism involving successive decarbonylation, formamide elimination, and cyclization of the resulting N-formylformamidine, has been proposed for this reaction (*30*). 1,3,5-Triazine is also obtained by heating tris(formylamino)methane with formamide (*55*).

Urea and tris(formylamino)methane condense to form 2,4-dihydroxy-1,3,5-triazine, which is difficult to prepare by other means (*29*).

$$2\ H_2NCONH_2 + HC(NHCHO)_3 \longrightarrow \text{[2,4-dihydroxy-1,3,5-triazine]} + 3\ HCONH_2 + NH_3 \quad (98)$$

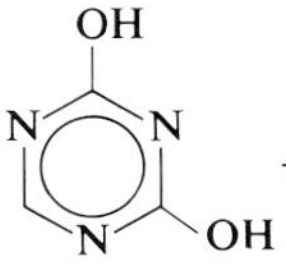

Guanidine and N-substituted guanidines are converted to 2-amino- and 2-(substituted amino)-1,3,5-triazines by heating with tris(formylamino)-methane in dimethylformamide at 160°–170°C (*55, 123a*).

$$RR'NC(NH_2)=NH + HC(NHCHO)_3 \xrightarrow[160°C]{(CH_3)_2NCHO}$$

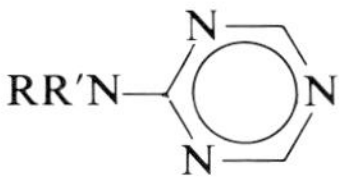

$$RR'N-\text{[1,3,5-triazine ring]} + 2\ HNONH_2 + H_2O \quad (99)$$

[R = H, R' = H, $CH_3$, $C_2H_5$, $C_3H_7$, $CH(CH_3)_2$, $C_4H_9$, $C_6H_5$, $C_6H_5CH_2$,
p-$CH_3C_6H_4$; R = R' = $CH_3$; RR' = $-(CH_2)_5-$]

Under the same reaction conditions, guanylurea is converted to a mixture of approximately equal amounts of 2-amino-1,3,5-triazine and 2-amino-4-hydroxy-1,3,5-triazine (122).

Purine is obtained in yields of up to about 60% by heating various derivatives of aminoacetonitrile (aminoacetonitrile hydrogen sulfate, phthalimidoacetonitrile, hippuronitrile, or methyleneaminoacetonitrile) with a mixture of tris(formylamino)methane and formamide (33). The reaction of the aminoacetonitrile hydrogen sulfate with tris(formylamino)methane is described by Eq. (100).

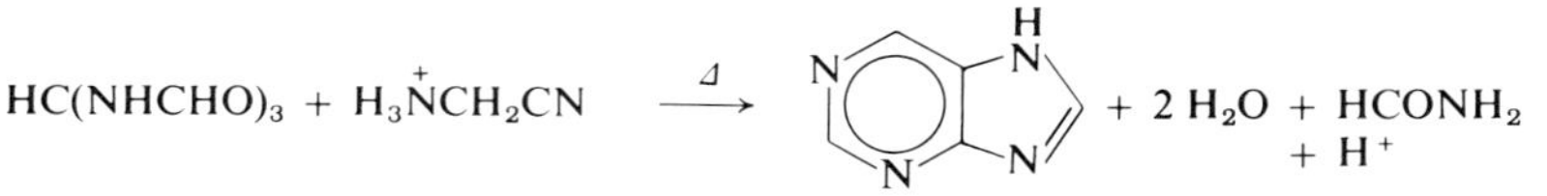

$$HC(NHCHO)_3 + H_3\overset{+}{N}CH_2CN \xrightarrow{\Delta} \text{[purine]} + 2\,H_2O + HCONH_2 + H^+ \quad (100)$$

2-Phenyl-5-amino-4$H$-pyrazol-(2$H$)-3-one and 1-phenyl-5-amino-(2$H$)-pyrazolone react with tris(formylamino)methane at elevated temperatures to form 1- and 2-phenyl-3-hydroxypyrazolo[3,4-$d$]pyrimidines [Eqs. (101) and (102)] (32).

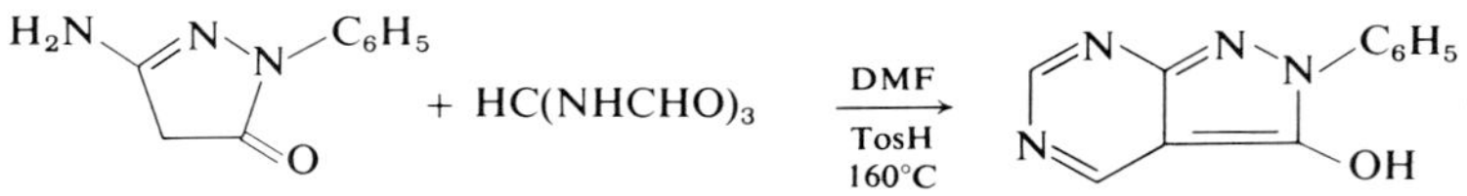

$$\text{[pyrazolone]} + HC(NHCHO)_3 \xrightarrow[\text{TosH}]{\text{DMF}}_{160°C} \text{[pyrazolopyrimidine]} \quad (101)$$

$$\text{[pyrazolone]} + HC(NHCHO)_3 \xrightarrow[\text{TosH}]{\text{DMF}}_{150°C} \text{[pyrazolopyrimidine]} \quad (102)$$

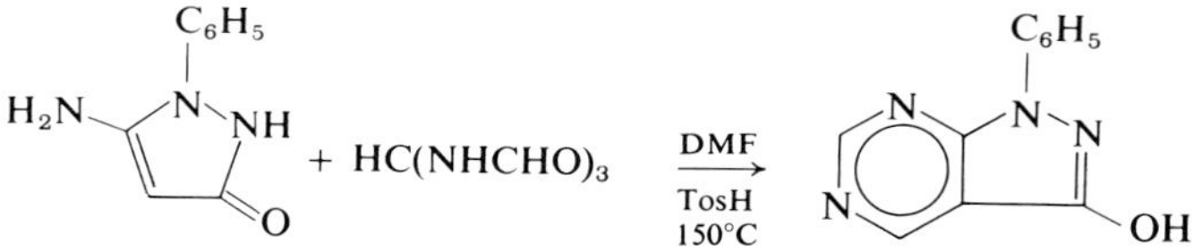

Pyrimido[4,5-$d$]pyrimidine derivatives are obtained by condensing substituted 3-aminopyrimidines with tris(formylamino)methane (35).

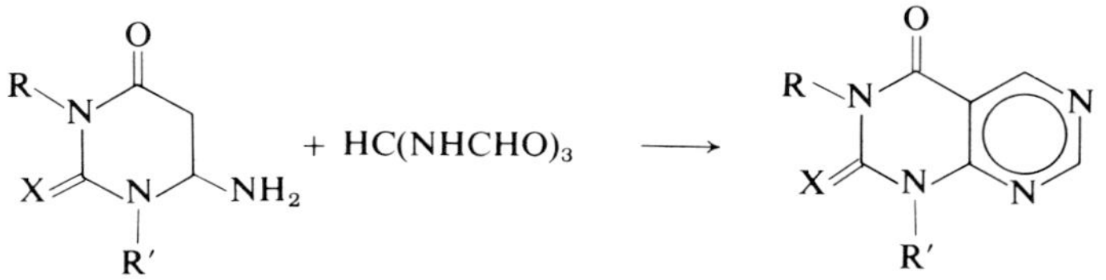

$$\text{[aminopyrimidine]} + HC(NHCHO)_3 \longrightarrow \text{[pyrimidopyrimidine]} \quad (103)$$

(X = O, R = H, R' = H, CH$_3$; X = O, R = CH$_3$, R' = H, CH$_3$;
X = O, R = R' = C$_6$H$_5$CH$_2$; X = S, R = R' = H)

Other condensed polycyclic nitrogen heterocycles synthesized using tris(formylamino)methane as a reagent include 1,2,6,7-tetraphenyl-3,5-

dioxodipyrazolo[3,4-*b*:4′,3′-*e*]pyridine (**52**), prepared from the ortho amide and 1,2-diphenyl-3-amino-5-pyrazolone (*34*), and 1,3,7,9-tetramethyl-2,4,6,8-tetraoxooctahydrodipyrimido[4,5-*b*:5′,4′-*e*]pyridine (**53**), obtained as a minor product of the reaction of 1,3-dimethyl-4-methylaminouracil and the ortho amide (*35*).

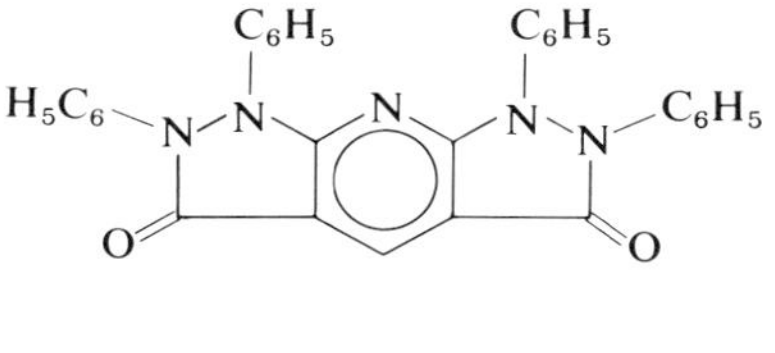

**52** 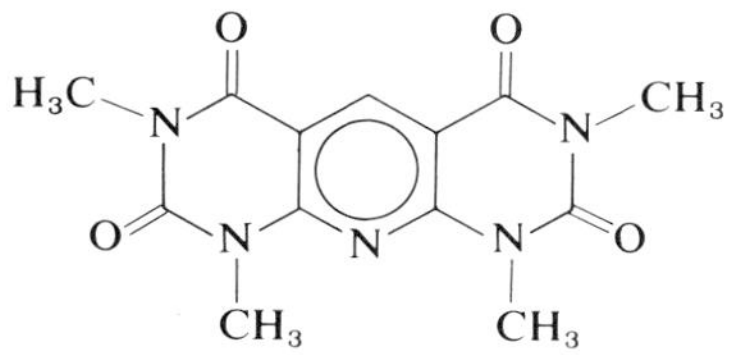          **53**

DeRuggieri, Gandolfi, and Guzzi prepared a number of steroid derivatives having pyrimidine rings fused to the 2,3- and 16,17-positions of the steroid systems by condensing steroidal formyl ketones, aminonitriles, ketonitriles, and aminocarboxamides with tris(formylamino)methane (*69–72*). A representative example of these reactions is the synthesis of [17,16-*d*]pyrimidino-5α-androstan[3,2-*c*]pyrazole (*69*).

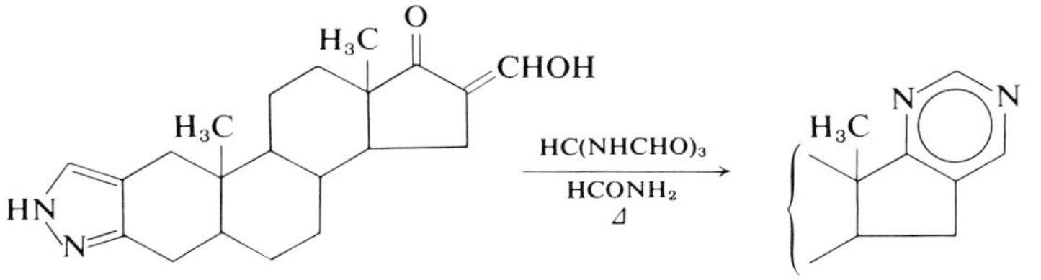

(104)

1,1-Diazido-1-ethoxyethane, an orthoacetamide derivative, reacts with aniline to form 1-phenyl-5-methyltetrazole in 83% yield (*136*).

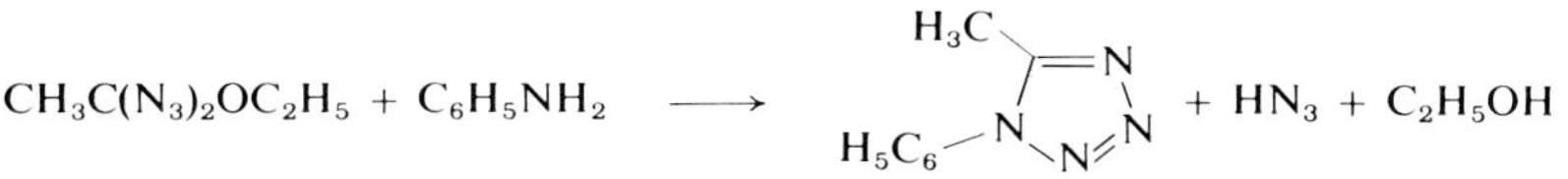

$CH_3C(N_3)_2OC_2H_5 + C_6H_5NH_2 \longrightarrow$  + $HN_3$ + $C_2H_5OH$   (105)

2 - Ethoxy - 6 - methyl - 2 - phenyl - 3,4 - dihydro - 2*H* - 1,3 - oxazine - 4 - one, formed by addition of ethyl benzimidate to diketene, loses ethanol when refluxed in benzene to form 6-methyl-2-phenyl-4*H*-oxazine-4-one (*105*).

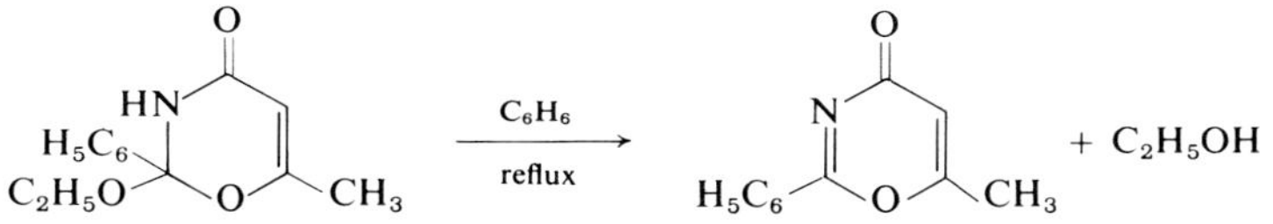

$\xrightarrow[\text{reflux}]{C_6H_6}$  + $C_2H_5OH$   (106)

Dimethylformamide diethyl acetal reacts with phenyl and cyclohexyl isocyanates to form 1,3-disubstituted 5-ethoxy-5-dimethylaminoimidazolidine-2,4-diones [Eq. (107), Y = O] (49).

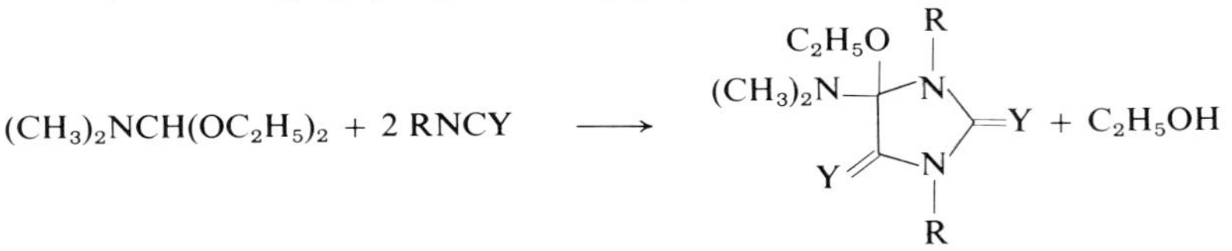

$$(CH_3)_2NCH(OC_2H_5)_2 + 2\ RNCY \longrightarrow \underset{\substack{|\\R}}{\overset{\substack{R\\|}}{\underset{Y}{\overset{C_2H_5O}{\underset{(CH_3)_2N}{\diagup\!\!\!\diagdown}}}}}\!\!=Y + C_2H_5OH \quad (107)$$

[Y = O, R = $C_6H_5$, $C_6H_{11}$; Y = S, R = $CH_3$, $C_2H_5$, $(CH_3)_2CH$, $C_4H_9$, $C_6H_{11}$]

Alkyl isothiocyanates undergo analogous reactions with the amide acetal [Eq. (107), Y = S] (52a–52c). It seems likely that the isothiocyanate reactions involve initial formation of a zwitterionic adduct by attack of amide acetal nitrogen on isothiocyanate carbon, followed by isomerization of the adduct and reaction of the resulting thiooxamide monoacetal with a second mole of the isothiocyanate (52a). Aryl isothiocyanates yield S-alkylation products [ArN=C(SC$_2$H$_5$)CON(CH$_3$)$_2$] of the initial adducts (52a, 52b).

3,5-Disubstituted 1-aza-4,6-dioxabicyclo-[3.3.0]octanes react with aldehydes to form substituted oxazolidines (81a).

## 3. Alkylation of Nitrogen by Formamide Acetals

Heterocyclic compounds containing the –CONH– grouping, such as 1-methyluracil, 6-azauracil, and 6-azauridine, react with dimethylformamide dialkyl acetals to form N-alkyl derivatives [Eq. (108)] (171). Whether dimethylformamide acetals would be generally useful reagents for alkylating secondary amines and N-substituted carboxamides remains to be established.

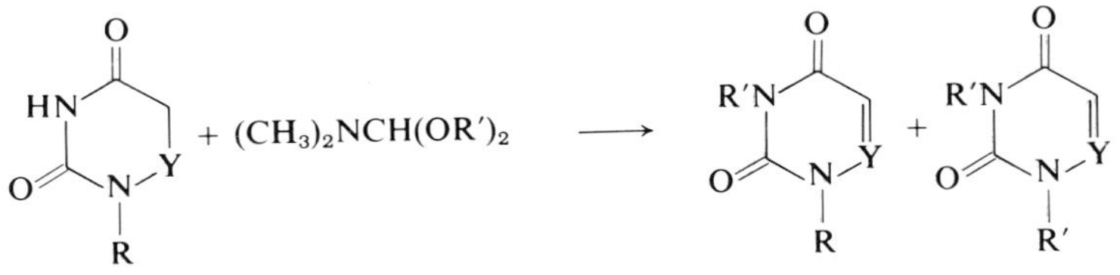

(Y = N, R = H, R' = $C_2H_5$; Y = N, R = 1-ribofuranosyl, R' = $C_2H_5$; Y = CH, R = R' = $CH_3$)

## 4. Miscellaneous Reactions of Ortho Amide Derivatives Which Form Carbon–Nitrogen Bonds

6-Amino-4-hydroxypyrimidine is converted to 6-formylamino-4-hydroxypyrimidine by heating with tris(formylamino)methane in quinoline solution (35).

$O,N$- and $O,O$-Acetals of acyl isocyanates are obtained by treating $N$-halocarboxamides with dimethylformamide diethyl acetal at 10°C and 80°C, respectively. These reactions probably involve $\alpha$-elimination of hydrogen halide from the halo amide to form an acyl nitrene, followed by insertion of the nitrene into the formyl C—H bond of the amide acetal and loss of ethanol or dimethylamine from the resulting acylurea acetal (52).

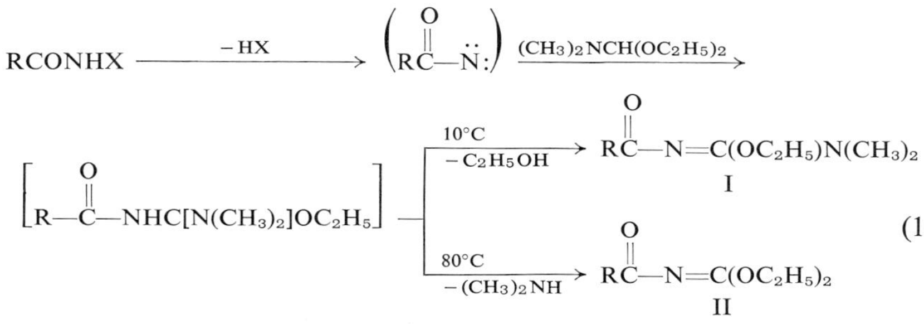

$$\text{RCONHX} \xrightarrow{-\text{HX}} \left( \underset{\text{RC}-\ddot{\text{N}}:}{\overset{\text{O}}{\|}} \right) \xrightarrow{(CH_3)_2NCH(OC_2H_5)_2}$$

$$\left[ \underset{\text{R}-\overset{\text{O}}{\overset{\|}{\text{C}}}-\text{NHC[N(CH}_3)_2]\text{OC}_2\text{H}_5}{} \right] \begin{array}{c} \xrightarrow[-\text{C}_2\text{H}_5\text{OH}]{10°C} \quad \text{RC}-\text{N}=\text{C(OC}_2\text{H}_5)\text{N(CH}_3)_2 \\ \text{I} \\ \\ \xrightarrow[-(\text{CH}_3)_2\text{NH}]{80°C} \quad \text{RC}-\text{N}=\text{C(OC}_2\text{H}_5)_2 \\ \text{II} \end{array} \quad (109)$$

[I: X = Cl, R = CH$_3$, C$_2$H$_5$, C$_3$H$_7$, CH(CH$_3$)$_2$; X = Br, R = CH$_3$, C$_2$H$_5$, CH(CH$_3$)$_2$.
II: X = Cl, R = C$_2$H$_5$, C$_3$H$_7$, CH(CH$_3$)$_2$]

Two moles of dimethylformamide acetal must be used per mole of halo amide, since the hydrogen halide produced by the elimination reaction converts the amide acetal to ethyl halide, ethanol, and dimethylformamide.

### C. Reactions Which Form Carbon–Carbon Bonds

1. REACTIONS OF ORTHO AMIDE DERIVATIVES WITH SUBSTANCES HAVING ACTIVATED METHYL OR METHYLENE GROUPS

Amide acetals, ester aminals, ortho amides, and tris(acylamino)methanes react with compounds having activated methylene groups to form dialkylaminoalkylene or acylaminomethylene derivatives. These reactions are analogous, mechanistically and otherwise, to reactions of trialkyl orthocarboxylates with active methylene compounds. The ortho amide derivatives are considerably more reactive, however, and the condensations usually occur smoothly without a catalyst.

a. *Condensation Reactions of Amide Acetals.* Dimethylformamide dialkyl acetals undergo condensation reactions with compounds having activated methyl or methylene groups to form dialkylaminomethylene derivatives of the carbon acid. The reactions occur at ordinary or slightly elevated temperatures, and require no catalyst. The amide acetal itself apparently is sufficiently basic to deprotonate the carbon acid, which thereby serves as the source of protons for converting the amide acetal to a dimethylaminoalkoxycarbonium ion and alcohol. Combination of the carbonium and carbanions yields an $O,N$-acetal, which is converted to the dimethylaminomethylene compound by loss of alcohol. Dimethylformamide acetal reacts not only with methylene compounds having two activating substituents, but also with ketones and other substances having a single activating group. These reactions are summarized by Eq. (110).

$$XYCH_2 + (CH_3)_2NCH(OR)_2 \rightarrow XYC=CHN(CH_3)_2 + 2\,ROH \quad (110)$$

| X | Y | R | References |
|---|---|---|---|
| $C_6H_5CO$ | $C_6H_5$ | $CH_3, C_2H_5, CH(CH_3)_2,$ $C(CH_3)_3, CH_2C(CH_3)_3,$ $CH_2C_6H_5$ | 5 |
| CN | $CN, CO_2CH_3$ | $C_2H_5$ | 24, 115 |
| $CH_3CO$ | $H, CH_3, CO_2C_2H_5$ | $C_2H_5$ | 24, 115 |
| H | $NO_2, C_6H_5CO, C_2H_5CO,$ $C_6H_5-CH=CHCO,$ 4-pyrimidyl, $2,4\text{-}(NO_2)_2C_6H_3$ | $C_2H_5$ | 24, 48, 115 |
| $o\text{-}O_2NC_6H_4$ | $CO_2C_2H_5$ | $C_2H_5$ | 86a |
| $NO_2$ | $C_2H_5, C_6H_5, CO_2C_2H_5,$ $COC_6H_5$ | $CH_3$ | 128a |
| $XY = -[CH=CHCH=CH]-$ (cyclopentadiene) | | $C_2H_5$ | 115 |
| $XY = -[COCH_2C(CH_3)_2CH_2CO]-$ (dimedone) | | $CH_3$ | 5 |

Arnold and Kornilov studied the effect of structural changes in the alkoxy groups of dimethylformamide dialkyl acetals on their rates of condensation with deoxybenzoin at 50°C [Eq. (110), X = $C_6H_5$, Y = $C_6H_5CO$, R = alkyl, allyl, benzyl] (5). The reactions were carried out in a mixture of triethylamine and the alcohol corresponding to the alkoxy group of the amide acetal. For this reason, the different reactions occurred in solvents of different polarity, and are not strictly comparable. The qualitative results [relative reactivities of $(CH_3)_2NCH(OR)_2$ decrease in the order R =

$$C(CH_3)_3 > CH(CH_3)_2 > C_2H_5 \sim CH_2CH=CH_2 > CH_2C(CH_3)_3$$
$$\sim CH_2C_6H_5 > CH_3]$$

suggest that reactivity is insensitive to electronic effects of alkoxy alkyl groups, but is determined largely by their size. Apparently, release of steric strain as the amide acetal is converted to a carbonium ion is a significant driving force for the reaction.

N-Carbethoxy-N-propylformamide diethyl acetal condenses with ethyl cyanoacetate to form the unsaturated ester **54**. Dimethylformamide dineopentyl acetal and cyanoacetic acid, when heated in benzene solution, undergo

$$C_2H_5OCON(C_3H_7)CH=C(CN)CO_2C_2H_5$$
**54**

consecutive condensation and decarboxylation reactions which produce 3-dimethylaminoacrylonitrile in 96% yield (*19*).

$$(CH_3)_2NCH[OCH_2C(CH_3)_3]_2 + NCCH_2CO_2H \xrightarrow[\Delta]{C_6H_6}$$
$$(CH_3)_2NCH=CHCN + CO_2 + 2\,(CH_3)_3CCH_2OH \quad (111)$$

Heterocyclic quaternary salts with activated methyl or methylene groups react with dimethylformamide dialkyl acetals to form symmetrical trimethine cyanine dyes (*6, 62, 112*). The thiazolium salts of Eq. (112) yielded dyes with dimethylformamide diethyl acetal, but did not react with triethyl orthoformate.

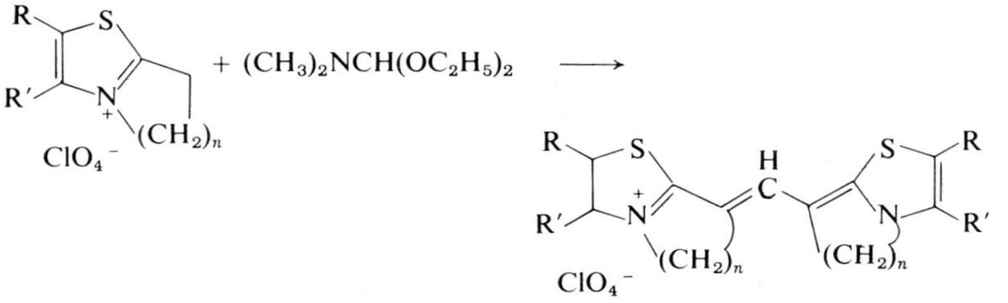

$(R = R' = C_6H_5, n = 4;\ R = H,\ R' = CH_3,\ n = 2;\ R = H,\ R' = C_6H_5,\ n = 2)$

Substituted trimethinium perchlorates **55** are converted to pentamethinium perchlorates by heating with dimethylformamide dimethyl or diethyl acetal in dimethylformamide solution (*4*).

$$[(CH_3)_2NC(CH_3)=CHCR=N(CH_3)_2]^+ClO_4^- + (CH_3)_2NCH(OR)_2 \xrightarrow[\Delta]{DMF}$$
$$[(CH_3)_2N=CH-CH=C[N(CH_3)_2]-CH=CR-N(CH_3)_2]^+ClO_4^- \quad (113)$$
**55**
$(R = H,\ CH_3,\ C_6H_5)$

Amide acetals of higher carboxamides, as well as acetals of N-methylpyrrolidone 1-methylquinolone, 1,3-dimethylimidazolidineone, 1,3-dimethylbenzimidazolone, and 3-methylbenoxazolone undergo condensation reactions with active methylene compounds which are analogous to those of formamide acetals (115). Condensation reactions of N,N-dimethylacetamide diethyl acetal are described by Eq. (114) (115).

$$CH_3C(OC_2H_5)_2N(CH_3)_2 + XYCH_2 \rightarrow XYC=C(CH_3)N(CH_3)_2 \quad (114)$$

| X | Y |
|---|---|
| H | $COC_6H_5$, $NO_2$, 2,4-$(O_2N)_2C_6H_3$ |
| CN | CN, $CO_2CH_3$ |
| XY = [–CH=CH—CH=CH–] | |

These reactions are spontaneous and exothermic, and were carried out without a solvent. N-Methylpyrrolidone diethyl acetal undergoes similar reactions (115).

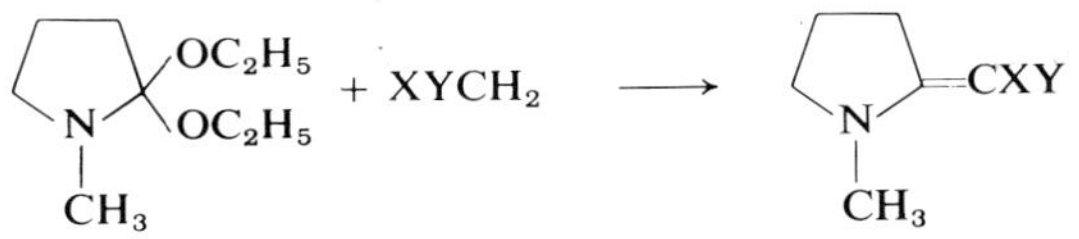

$$\underset{\underset{CH_3}{|}}{\text{pyrrolidine}}\!\!<\!\!\overset{OC_2H_5}{\underset{OC_2H_5}{}} + XYCH_2 \longrightarrow \underset{\underset{CH_3}{|}}{\text{pyrrolidine}}\!=\!CXY \quad (115)$$

[X = CN, Y = CN, $CO_2CH_3$; X = H, Y = $NO_2$; XY = –CH=CHCH=CH–; X = H, Y = $COC_6H_5$, 2,4-$(O_2N)_2C_6H_3$]

1-Methyl-2-quinolone diethyl acetal and N-methylpyrrolidone diethyl acetal yield monomethinecyanine dyes when condensed with 1,2-dimethylquinolinium iodide (115). The reaction of the quinolone acetal, which forms a symmetrical dye, is given in Eq. (116).

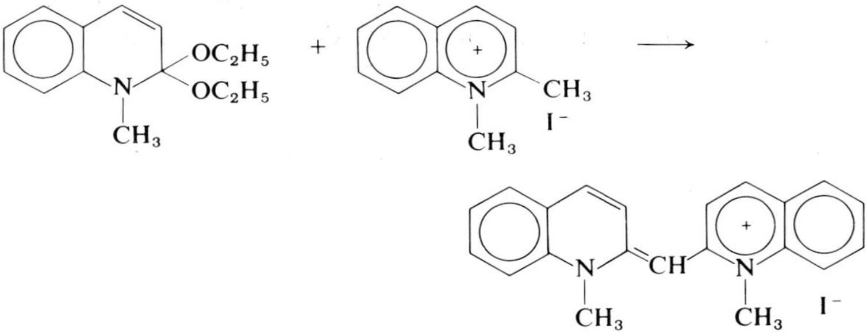

$$\text{(quinoline-}OC_2H_5\text{)}_2\text{N-CH}_3 + \text{quinolinium-CH}_3 \; I^- \longrightarrow \text{symmetrical dye} \; I^- \quad (116)$$

b. *Condensation Reactions of Ester Aminals.* Ethyl formate bis(dimethylamino)aminal is even more reactive than dimethylformamide acetals in condensations with active methyl or active methylene compounds, and affords higher yields of dimethylaminomethylene derivatives with the less reactive carbon acids *(24)*.

$$HC(OC_2H_5)[N(CH_3)_2]_2 + XYCH_2 \rightarrow$$
$$XYC=CHN(CH_3)_2 + C_2H_5OH + (CH_3)_2NH \quad (117)$$
$$(X = CN, Y = CO_2CH_3; X = H, Y = C_6H_5CO, CH_3COC_2H_5CO;$$
$$X = CH_3, Y = CH_3CO)$$

*Tert*-butyl formate bis(dimethylamino)aminal was used to convert a number of methyl-substituted heterocyclic and aromatic compounds to dimethylaminomethylene derivatives *(54a)*.

c. *Condensation Reactions of Ortho Amides and Tris(acylamino)methanes.* Tris(dimethylamino)methane reacts smoothly with active methylene compounds to form dimethylaminomethylene derivatives *(27, 152)*.

$$HC[N(CH_3)_2]_3 + XYCH_2 \longrightarrow XYC=CHN(CH_3)_2 + 2(CH_3)_2NH$$
$$[X = CN, Y = CN, CO_2CH_3; X = C_6H_5CO, Y = H; \quad (118)$$

$$XY = \text{(fluorene)}, \text{(xanthene)}]$$

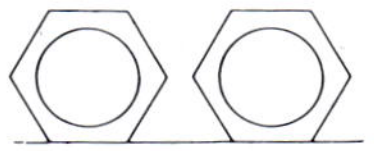

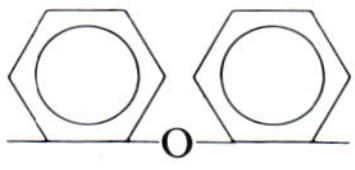

The course of reaction of tetrakis(dimethylamino)methane with carbon acids depends on the strength of the acid. The orthocarbamide and diethyl malonate react in ether at room temperature to form hexamethylguanidinium bis(carbomethoxy)methylide in high yield, rather than the expected aminomethylene compound *(152)*.

$$C[N(CH_3)_2]_4 + CH_2(CO_2CH_3)_2 \xrightarrow{(C_2H_5)_2O}$$
$$[(CH_3)_2N]_3\overset{+}{C}\overset{-}{C}H(CO_2CH_3)_2 + (CH_3)_2NH \quad (119)$$

The weaker carbon acid, fluorene, yields the expected bis(dimethylamino)methylene derivative.

$$C[N(CH_3)_2]_4 + \text{(fluorene)} \xrightarrow[6\ hr]{140°C} \text{(fluorenyl)}-C[N(CH_3)_2]_2 + 2(CH_3)_2NH \quad (120)$$

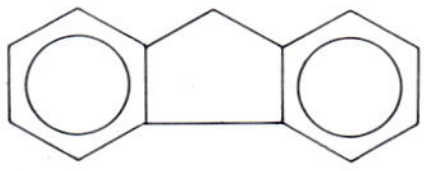

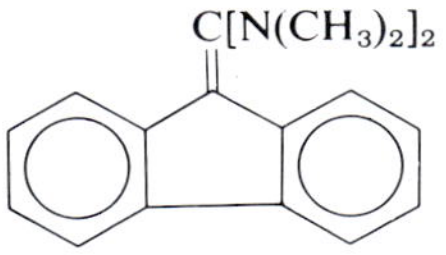

Tris(acylamino)methanes also react with active methylene compounds. Dimethyl malonate and methyl cyanoacetate condense with tris(formylamino)methane to form formylaminomethylene derivatives [Eq. (121)] (54). Higher tris(acylamino)methanes and ketones yield acylaminomethylene ketones [Eq. (122)] (43).

$$YCH_2CO_2CH_3 + HC(NHCHO)_3 \rightarrow HCONHCH{=}C(Y)CO_2CH_3 \quad (121)$$
$$(Y = CO_2CH_3, CN)$$

$$RCOCH_2R^1 + HC(NHCOR^2)_3 \rightarrow RCOC(R^1){=}CHNHCOR^2 \quad (122)$$

| R | $R^1$ | $R^2$ |
|---|---|---|
| $C_6H_5$ | H, $C_2H_5$, $C_3H_7$ | H, $C_2H_5$, $C_3H_7$ |
| $C_6H_5$ | H | $C_6H_5$, $p$-$CH_3OC_6H_4$ |
| ($-CH_2CH_2CH_2CH_2-$) | | $CH_3$ |

The monomethine dyes **56** were obtained by heating 1-substituted-5-amino-(2$H$)-pyrazolones with tris(formylamino)methane in dimethylformamide [Eq. (123)] (34). These products presumably were formed by reaction of an intermediate formylaminomethylene compound (derived from the 3-imido tautomer of the aminopyrazolone) with a second mole of the pyrazolone.

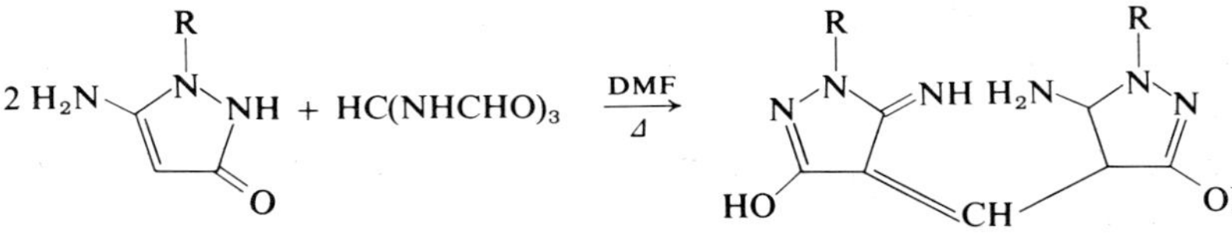

$$2\,H_2N{-}\underset{\text{pyrazolone}}{\diagup}{-}NH + HC(NHCHO)_3 \xrightarrow[\Delta]{DMF} \text{products} \quad (123)$$

(R = $CH_3$, $C_6H_5$)  **56**

### 2. Formation of Tetrakis(dialkylamino)ethylene Derivatives by Thermal Decomposition of Ortho Amide Derivatives

Thermal decomposition of formate aminals yields tetrakis(dialkylamino)-ethylenes [Eq. (124)] (148, 160).

$$2\,(R_2N)CHOR' \xrightarrow[-2R'OH]{\Delta} [2\,(R_2N)_2C\text{:}] \longrightarrow (R_2N)_2C{=}C(NR_2)_2 \quad (124)$$

The reaction probably involves $\alpha$-elimination of an alcohol from the ester aminal, followed by coupling of the resulting carbenoid intermediate. The tetraaminoethylene derivatives are conveniently prepared by heating dimethylformamide dimethyl acetal with acyclic or heterocyclic amines [Eq. (125)]. $N,N'$-Dialkyl-1,2-diamines and $N,N'$-dialkyl-1,3-diamines react with the amide acetal to form bis(imidazolidines) and bis(hexahydropyrimidines) [Eq. (126)] (*160–162*).

$$2\,(CH_3)_2NCH(OCH_3)_2 + 4\,R_2NH \rightarrow$$
$$(R_2N)_2C=C(NR_2)_2 + 2\,(CH_3)_2NH + 4\,CH_3OH \quad (125)$$

($R_2NH$ = pyrrolidine, piperidine, morpholine, $N'$-methylpiperazine)

$$2\,(CH_3)_2NCH(OCH_3)_2 + RNH(CH_2)_nNHR \longrightarrow$$

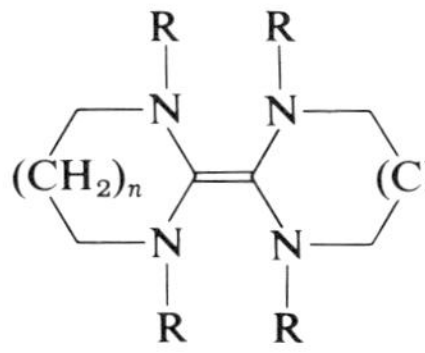

$$\underset{\substack{R\quad R\\|\quad\;|\\N\quad N\\|\quad\;|\\R\quad R}}{(CH_2)_n\!\!\bigg\rangle\!\!=\!\!\bigg\langle\!\!(CH_2)_n} + 4\,CH_3OH + 2\,(CH_3)_2NH \quad (126)$$

($n = 0$, R = $CH_3$, $C_2H_5$, $C_3H_7$, $C_4H_9$, $C_6H_{13}$, $C_6H_5CH_2$; $n = 1$, R = $CH_3$)

The initial steps in these reactions probably involve aminolysis and transamination of the dimethylformamide acetal to a formate aminal or an orthoformamide. These reactions must be carried out with exclusion of air, since the tetraaminoethylenes react rapidly with oxygen to form ureas, imidazolidones, or hexahydropyrimidones (*162*).

Tetrakis(dimethylamino)ethylene is obtained in about 80% yield by heating tris(dimethylamino)methane to 150°–190°C in the absence of oxygen (*26, 156*).

$$2\,HC[N(CH_3)_2]_3 \rightarrow 2\,[(CH_3)_2N]_2C=C[N(CH_3)_2]_2 + 2\,(CH_3)_2NH \quad (127)$$

Pyrolysis of tetrakis(dimethylamino)methane yields dimethylamine, tar, and tris(dimethylamino)ethylene [Eq. (128)] (*151*). A free radical mechanism was proposed for this reaction.

$$C[N(CH_3)_2]_4 \xrightarrow{150°C} [(CH_3)_2N]_2C=CHN(CH_3)_2 + (CH_3)_2NH \quad (128)$$

### 3. Formation of Ketene Aminals from Ortho Amide Derivatives

Meerwein and co-workers found that refluxing impure $N,N$-dimethylacetamide diethyl acetal with metallic sodium yields pure 1-dimethylamino-1-ethoxyethene, a ketene $O,N$-acetal (*115*).

$$2\,CH_3C(OC_2H_5)_2N(CH_3)_2 + 2\,Na \rightarrow$$
$$2\,CH_2=C(OC_2H_5)N(CH_3)_2 + 2\,NaOC_2H_5 + H_2 \quad (129)$$

Bredereck showed that this reaction is generally applicable to amide acetals, and prepared a number of substituted ketene $O,N$-acetals in 50%–75% yields by refluxing amide acetals with calcium metal (23).

$$2\ RCH_2C(OR^2)_2NR^1{}_2 + Ca \rightarrow 2\ RCH{=}C(OR^2)NR^1{}_2 + Ca(OR^2)_2 + H_2 \quad (130)$$

(R = $CH_3$, $R^1$ = $CH_3$, $C_2H_5$, $R^2$ = $CH_3$, $C_2H_5$;
R = $C_2H_5$, $R^1$ = $CH_3$, $R^2$ = $CH_3$, $C_2H_5$; R = $CH_3$, $R^1$ = $CH_3$, $R^2$ = $C_2H_5$)

The polycyclic ortho amides **57** undergo intramolecular elimination to form the isomeric ketene aminals **58** (*133*).

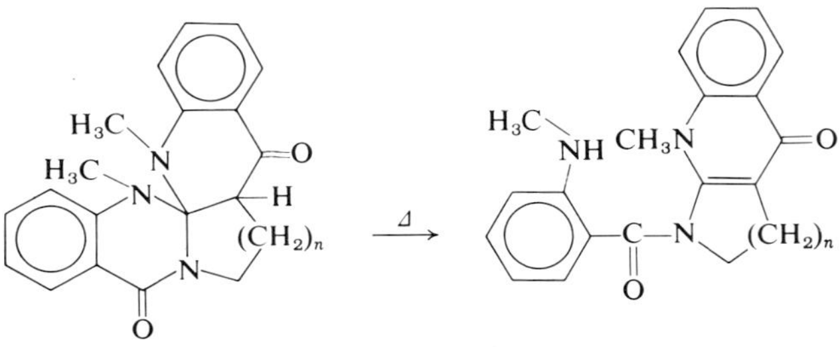

(131)

**57**   **58**

($n$ = 1, 2)

### 4. Reactions of Acetamide Acetals with Activated Alkenes, Alkyl Halides, and Acyl Halides

When dimethylacetamide dimethyl acetal is treated with acrylonitrile or with methyl acrylate in benzene solution, substituted ketene $O,N$-acetals are formed which may react with a second mole of the $\alpha,\beta$-unsaturated compound to form cyclobutene derivatives [Eq. (132)] (*120*). The initial product is

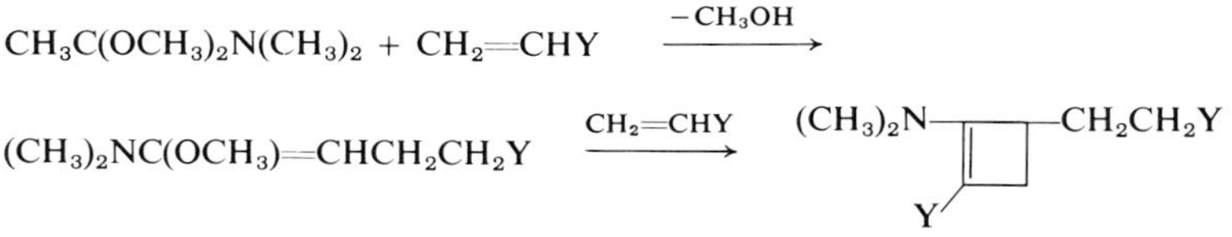

$$CH_3C(OCH_3)_2N(CH_3)_2 + CH_2{=}CHY \xrightarrow{-CH_3OH}$$

$$(CH_3)_2NC(OCH_3){=}CHCH_2CH_2Y \xrightarrow{CH_2{=}CHY} \underset{Y}{(CH_3)_2N{-}\square{-}CH_2CH_2Y} \quad (132)$$

(Y = CN, $CO_2CH_3$)

probably formed by electrophilic addition of the activated vinyl compound to the ketene $O,N$-acetal resulting from loss of methanol from the amide acetal.

Methyl vinyl ketone undergoes a more complex reaction with the same amide acetal to form the bicyclic ortho amide **59** (*120*). This result can be rationalized by assuming that an initial electrophilic addition is followed by enolization, cyclization, a second electrophilic addition, and a second enolization–cyclization.

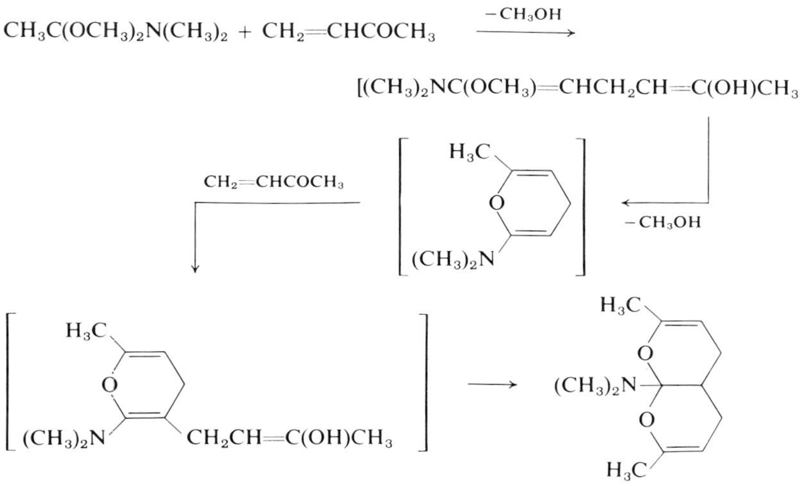

$$CH_3C(OCH_3)_2N(CH_3)_2 + CH_2=CHCOCH_3 \xrightarrow{-CH_3OH}$$

$$[(CH_3)_2NC(OCH_3)=CHCH_2CH=C(OH)CH_3]$$

(133)

**59**

Analogous electrophilic additions of benzyl chloride and benzoyl chloride to the acetamide acetal, followed by alkaline hydrolysis of the initial products, yield phenylpropionamide derivatives (120).

$$CH_3C(OCH_3)_2N(CH_3)_2 + RCl \longrightarrow$$
$$(CH_3)_2NC(OCH_3)=CHR \xrightarrow{H_2O} (CH_3)_2NCOCH_2R \quad (134)$$
$$(R = C_6H_5CH_2, C_6H_5CO)$$

## 5. OTHER REACTIONS OF ORTHO AMIDE DERIVATIVES WHICH FORM CARBON–CARBON BONDS

Amide acetals react with hydrogen cyanide at room temperature to form $O,N$-acetals of $\alpha$-ketonitriles in good yields (51).

$$RC(OC_2H_5)_2N(CH_3)_2 + HCN \rightarrow RC(OC_2H_5)[N(CH_3)_2]CN + C_2H_5OH \quad (135)$$
$$(R = H, CH_3, C_3H_7)$$

Bis(dimethylamino)-*tert*-butoxymethane undergoes a similar reaction which yields bis(dimethylamino)acetonitrile (51).

$$[(CH_3)_2N]_2CHOC(CH_3)_3 + HCN \rightarrow [(CH_3)_2N]_2CHCN + (CH_3)_3COH \quad (136)$$

Bis(dimethylamino)-*tert*-butoxymethane and benzaldehydes react at elevated temperatures to form low yields of 1,2-bis(dimethylamino)vinyl aryl ketones (*50*).

$$p\text{-XC}_6\text{H}_4\text{CHO} + [(\text{CH}_3)_2\text{N}]_2\text{CHOC}(\text{CH}_3)_3 \xrightarrow{110°-140°\text{C}}$$
$$p\text{-XC}_6\text{H}_4\text{COC}[\text{N}(\text{CH}_3)_2]{=}\text{CHN}(\text{CH}_3)_2 \quad (137)$$
$$(X = p\text{-CH}_3\text{C}_6\text{H}_4, p\text{-CH}_3\text{OC}_6\text{H}_4)$$

2-Acetoxy-2-phenyl-5,5-disubstituted-$\Delta^3$-1,3,4-oxadiazolines decompose on warming with evolution of nitrogen and formation of substituted acetoxyepoxides (*94*).

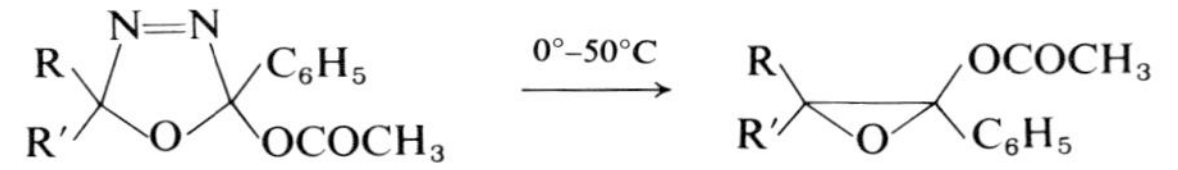

$$\underset{\text{R}'}{\overset{\text{R}}{\diagdown}}\!\!\underset{\text{O}}{\overset{\text{N=N}}{\diagup\!\!\diagdown}}\!\!\underset{\text{OCOCH}_3}{\overset{\text{C}_6\text{H}_5}{\diagup}} \xrightarrow{0°-50°\text{C}} \underset{\text{R}'}{\overset{\text{R}}{\diagdown}}\!\!\underset{\text{O}}{\diagup\!\!\diagdown}\!\!\underset{\text{C}_6\text{H}_5}{\overset{\text{OCOCH}_3}{\diagup}} + \text{N}_2 \quad (138)$$

$$[\text{R, R}' = (\text{CH}_2)_5; \text{R} = \text{C}_6\text{H}_5, \text{R}' = \text{CH}_3, \text{C}_6\text{H}_5]$$

The tetracyclic ortho amide **60** reacts with methylmagnesium iodide to form an adduct which was hydrolyzed to acetophenone and 2,2'-dihydroxydicyclohexylamine (*144*). This appears to be the only reported example of a reaction between an ortho amide derivative and a Grignard reagent.

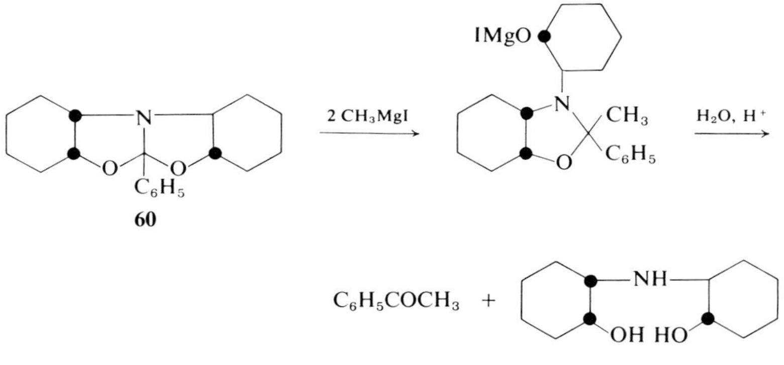

$$\text{C}_6\text{H}_5\text{COCH}_3 + \quad \text{(cyclohexyl-NH-cyclohexyl with OH, HO)} \quad (139)$$

Gross and Costisella (*91a*) discovered a reaction of dimethylformamide acetals whose indirect result is the conversion of an aldehyde to a carboxylic acid having one more carbon atom than the original aldehyde. The amide acetal is heated with diethyl phosphite to form tetraethyl dimethylaminomethylenediphosphonate, which is converted to a carbanion salt by treatment with sodium hydride. Reactions of the carbanion salt with aldehydes yield

1-dimethylaminoalkenyl phosphonates, which are hydrolyzed by aqueous acid to diethyl phosphite, dimethylamine, and the carboxylic acid.

$$(C_2H_5O)_2PHO + (CH_3)_2NCH(OCH_3)_2 \xrightarrow{60°-80°C}$$
$$[(C_2H_5O)_2P(O)]_2CHN(CH_3)_2 \xrightarrow[\text{2. RCHO}]{\text{1. NaH}} RCH=C[N(CH_3)_2]P(O)(OC_2H_5)_2 \xrightarrow{H^+, H_2O}$$
$$RCH_2CO_2H + (C_2H_5O)_2PHO + (CH_3)_2NH_2^+ \quad (140)$$
$$(R = C_6H_5, \; p\text{-}O_2NC_6H_4, \; 3,4\text{-methylenedioxyphenyl})$$

### D. Reactions of Ortho Amide Derivatives Which Form Carbon–Hydrogen Bonds

Hydrogenolysis of the tetracyclic ortho amide derivative **56** yields benzyl-bis(2-hydroxycyclohexyl)amine *(144)*.

$$60 \xrightarrow[\text{PtO}_2]{H_2} \underset{\text{OH HO}}{\underset{|}{\text{structure}}} \overset{CH_2C_6H_5}{\underset{|}{N}} \quad (141)$$

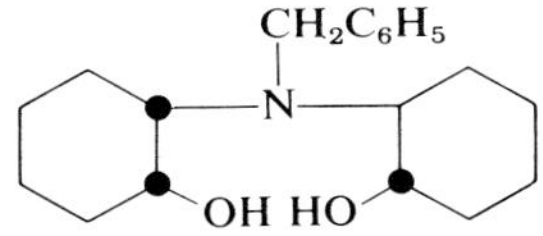

In contrast to trialkyl orthoformates, which undergo formyl proton exchange only under acidic conditions, dimethylformamide acetals undergo rapid formyl proton exchange with deutero alcohols in the absence of catalysts [Eq. (142)] *(135)*. The facility with which these reactions take place suggests that they occur by a mechanism not available to the orthoformates. One possible mechanism involves formation of an ylide intermediate.

$$(CH_3)_2NCH(OR)_2 \longrightarrow [(CH_3)_2\overset{+}{N}H-\overset{..}{C}(OR)_2] \xrightarrow[-ROH]{+ROD} (CH_3)_2NCD(OR)_2 \quad (142)$$

### E. Conversion of Ortho Amide Derivatives to Carbonium Salts

Since nitrogen is better able to support positive charge than oxygen, mesomeric alkoxyaminocarbonium ions and diaminocarbonium ions derived from amide acetals, ester aminals, and ortho amides should be more stable than analogous dialkoxycarbonium ions derived from ortho esters. This prediction is borne out by the fact that dimethylformamide diethyl acetal and diethoxycarbonium ion react to form dimethylaminoethoxycarbonium ion and triethyl orthoformate *(116)*.

$$(CH_3)_2NCH(OC_2H_5)_2 + (C_2H_5O)_2CH^+Y^- \rightarrow$$
$$(CH_3)_2N(C_2H_5O)CH^+Y^- + HC(OC_2H_5)_3 \quad (143)$$
$$(Y^- = BF_4^-, \; SbCl_6^-)$$

Carbonium ions derived from amide acetals, ester aminals, and ortho amides are so stable that these compounds are partly dissociated, as demonstrated by their electrical conductivity (*115, 134*). $N,N',N''$-Trialkyl-$N,N',N''$-triaryl orthoformamides react with strong acids to form bis(substituted amino)-carbonium ions ($N,N'$-dialkyl-$N,N'$-diarylformamidinium ions) [Eq. (144)] (*64, 66, 67*). These amidinium ions are so stable that they react only slowly with water (*61a*).

$$(ArNR)_3CH + 2\,HY \rightarrow (ArNR)_2CH^+Y^- + ArNH_2R^+Y^- \quad (144)$$

Amide acetals are conveniently converted to alkoxyaminocarbonium tetrafluoroborates by reaction with boron trifluoride etherate. Meerwein and co-workers prepared several carbonium salts by this means [Eqs. (145), (146)] (*114, 115*).

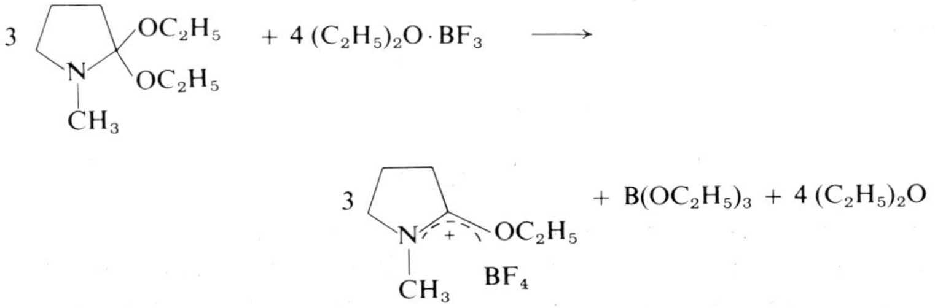

3 [pyrrolidine with OC$_2$H$_5$, OC$_2$H$_5$, N-CH$_3$] + 4 (C$_2$H$_5$)$_2$O · BF$_3$ ⟶

3 [pyrrolidinium with OC$_2$H$_5$, N-CH$_3$, BF$_4$] + B(OC$_2$H$_5$)$_3$ + 4 (C$_2$H$_5$)$_2$O  (145)

$3\,RC(OC_2H_5)_2N(CH_3)_2 + 4\,(C_2H_5)_2O \cdot BF_3 \rightarrow$
$\quad 3\,(CH_3)_2N\text{···}\overset{+}{C}R\text{···}OC_2H_5\,BF_4^- + B(OC_2H_5)_3 + 4\,(C_2H_5)_2O \quad (146)$
$[R = H, CH_3, (CH_3)_2N]$

Dimethylformamide diethyl acetal is converted to a carbonium hexachloroantimonate salt by reaction with triphenylmethyl hexachloroantimonate (*116*).

$(CH_3)_2NCH(OC_2H_5)_2 + (C_6H_5)_3C^+SbCl_6^- \rightarrow$
$\quad (CH_3)_2N\text{—}\overset{+}{C}H\text{—}OC_2H_5 + (C_6H_5)_3COC_2H_5 \quad (147)$

The heterocyclic benzamide acetal 5-phenyl-2,3:7,8-dicyclohexano-1-aza-4,6-dioxabicyclo[3.3.0]octane reacts reversibly with strong acids to form a heterocyclic carbonium ion (*144*).

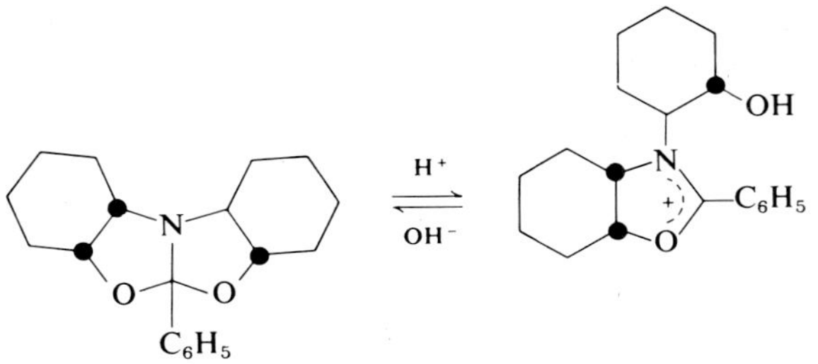

$$\underset{C_6H_5}{\text{[bicyclic structure]}} \underset{OH^-}{\overset{H^+}{\rightleftharpoons}} \text{[ring-opened cation with OH, C}_6\text{H}_5\text{]} \quad (148)$$

## REFERENCES

1. V. K. Antonov, T. E. Agadzhanyan, T. R. Telesnina, and M. M. Shemyakin, *Tetrahedron Letters* p. 727 (1964).
2. V. K. Antonov, A. M. Shkrob, and M. M. Shemyakin, *Tetrahedron Letters* p. 439 (1963).
3. J. F. Arens and T. R. Rix, *Koninkl. Ned. Akad. Wetenschap. Proc.*, **B57**, 270 (1954).
4. Z. Arnold and A. Holy, *Collection Czech. Chem. Commun.* **28**, 2040 (1963).
5. Z. Arnold and M. Kornilov, *Collection Czech. Chem. Commun.* **29**, 645 (1964).
6. F. S. Babichev, F. A. Mikhailenko, V. K. Kibirev, and V. A. Bogolyubskii, *Ukr. Khim. Zh.* **32**, 204 (1966); *Chem. Abstr.* **64**, 17749 (1966).
7. H. J. Backer and W. L. Wanmaker, *Rec. Trav. Chim.* **68**, 247 (1949).
8. H. J. Backer and W. L. Wanmaker, *Rec. Trav. Chim.* **67**, 257 (1948).
9. H. Baganz and L. Domaschke, *Chem. Ber.* **95**, 2095 (1962).
10. G. Barnath, K. Kovacs, and K. L. Lang, *Tetrahedron Letters* p. 2713 (1968).
11. H. D. Barnes, D. Kundiger, and S. M. McElvain, *J. Am. Chem. Soc.* **62**, 1281 (1940).
11a. W. C. Behnke and H. E. Doorenbos, U.S. Patent 3,393,238 (1968); *Chem. Abstr.* **69**, 76673 (1968).
12. H. Beiner, *Ann. Chem.* **686**, 102 (1965).
13. M. L. Bender, *Chem. Rev.* **60**, 54 (1960).
14. H. Boehme and F. Soldan, *Chem. Ber.* **94**, 3109 (1961).
15. F. G. Borgardt, A. K. Seeler, and P. Noble, *J. Org. Chem.* **31**, 2806 (1966).
16. H. H. Bosshard, E. Jenny, and H. Zollinger, *Helv. Chim. Acta* **44**, 1203 (1961).
17. D. C. Bradley and T. M. Thomas, *J. Chem. Soc.* p. 3857 (1960).
18. H. Brechbuehler, H. Buechi, E. Hatz, J. Schreiber, and A. Eschenmoser, *Angew. Chem. Intern. Ed. Engl.* **2**, 212 (1963).
19. H. Brechbuehler, H. Buechi, E. Hatz, J. Schreiber, and A. Eschenmoster, *Helv. Chim. Acta* **48**, 1746 (1965).
20. H. Bredereck, German Patent 1,046,604 (1958); *Chem. Abstr.* **55**, 2487 (1961).
21. H. Bredereck and K. Bredereck, *Chem. Ber.* **94**, 2278 (1961).
22. H. Bredereck, F. Effenberger, and H. P. Beyerlin, *Chem. Ber.* **97**, 1834 (1964).
23. H. Bredereck, F. Effenberger, and H. P. Beyerlin, *Chem. Ber.* **97**, 3081 (1964).
24. H. Bredereck, F. Effenberger, and H. Botsch, *Chem. Ber.* **97**, 3397 (1964).
25. H. Bredereck, F. Effenberger, H. Botsch, and H. Rehn, *Chem. Ber.* **98**, 1081 (1965).
26. H. Bredereck, F. Effenberger, and H. J. Bredereck, *Angew. Chem. Intern. Ed. Engl.* **5**, 971 (1966).
27. H. Bredereck, F. Effenberger, and T. Brendle, *Angew. Chem. Intern. Ed. Engl.* **5**, 132 (1966).
28. H. Bredereck, F. Effenberger, and T. Brendle, German Patent 1,217,391 (1966); *Chem. Abstr.* **65**, 5366 (1966).
28a. H. Bredereck, F. Effenberger, T. Brendle, and H. Muffler, *Chem. Ber.* **101**, 1885 (1968).
29. H. Bredereck, F. Effenberger, and A. W. Hofmann, *Angew. Chem. Intern. Ed Engl.* **1**, 331 (1962).
29a. H. Bredereck, F. Effenberger, and M. Hajek, *Chem. Ber.* **101**, 3062 (1968)
30. H. Bredereck, F. Effenberger, and A. Hofmann, *Chem. Ber.* **96**, 3260 (1963).
31. H. Bredereck, F. Effenberger, and A. Hofmann, *Chem. Ber.* **96**, 3265 (1963).

32. H. Bredereck, F. Effenberger, and A. Hofmann, *Chem. Ber.* **97**, 61 (1964).
32a. H. Bredereck, F. Effenberger, A. Hofmann, and M. Hajek, *Angew. Chem. Intern. Ed. Engl.* **2**, 655 (1963).
33. H. Bredereck, F. Effenberger, G. Rainer, and H. P. Schosser, *Ann. Chem.* **659**, 133 (1962).
34. H. Bredereck, F. Effenberger, and W. Resemann, *Chem. Ber.* **95**, 2796 (1962).
35. H. Bredereck, F. Effenberger, and R. Sauter, *Chem. Ber.* **95**, 2049 (1962).
36. H. Bredereck, F. Effenberger, and E. Schweizer, *Chem. Ber.* **95**, 803 (1962).
37. H. Bredereck, F. Effenberger, and G. Simchen, *Angew. Chem.* **73**, 493 (1961).
38. H. Bredereck, F. Effenberger, and G. Simchen, *Angew. Chem. Intern. Ed. Engl.* **1**, 331 (1962).
39. H. Bredereck, F. Effenberger, and G. Simchen, *Chem. Ber.* **96**, 1350 (1963).
40. H. Bredereck, F. Effenberger, and G. Simchen, *Chem. Ber.* **98**, 1078 (1965).
41. H. Bredereck, F. Effenberger, and G. Simchen, German Patent 1,205,528 (1965); *Chem. Abstr.* **64**, 6504 (1966).
42. H. Bredereck, F. Effenberger, and G. Simchen, German Patent 1,205,548 (1965); *Chem. Abstr.* **64**, 9593 (1966).
43. H. Bredereck, F. Effenberger, and H. J. Treiber, *Chem. Ber.* **96**, 1505 (1963).
44. H. Bredereck, R. Gompper, F. Effenberger, H. Keck, and H. Heise, *Chem. Ber.* **93**, 1398 (1960).
45. H. Bredereck, R. Gompper, and B. Geiger, *Chem. Ber.* **93**, 1402 (1960).
46. H. Bredereck, R. Gompper, H. Rempfer, K. Klemm, and H. Keck, *Angew. Chem.* **70**, 269 (1958).
47. H. Bredereck, R. Gompper, H. Rempfer, K. Klemm, and H. Keck, *Chem. Ber.* **92**, 329 (1959).
48. H. Bredereck and G. Simchen, *Angew. Chem. Intern. Ed. Engl.* **2**, 738 (1963).
49. H. Bredereck, G. Simchen, and E. Goeknel, *Angew. Chem. Intern. Ed. Engl.* **3**, 704 (1964).
50. H. Bredereck, G. Simchen, H. Hofmann, P. Horn, and R. Wahl, *Angew. Chem. Intern. Ed. Engl.* **6**, 356 (1967).
51. H. Bredereck, G. Simchen, and P. Horn, *Angew. Chem. Intern. Ed. Engl.* **4**, 523 (1965).
52. H. Bredereck, G. Simchen, and H. Porkert, *Angew. Chem. Intern. Ed. Engl.* **5**, 841 (1966).
52a. H. Bredereck, G. Simchen, and S. Rebsdat, *Chem. Ber.* **101**, 1872 (1968).
52b. H. Bredereck, G. Simchen, and S. Rebsdat, *Chem. Ber.* **101**, 1863 (1968).
52c. H. Bredereck, G. Simchen, and S. Rebsdat, *Angew. Chem.* **77**, 507 (1965).
53. H. Bredereck, G. Simchen, S. Rebsdat, W. Kantlehner, P. Horn, R. Wahl, H. Hofmann, and P. Grieshaber, *Chem. Ber.* **101**, 41 (1968).
53a. H. Bredereck, G. Simchen, and H. U. Schwenk, *Chem. Ber.* **101**, 3058 (1968).
54. H. Bredereck, G. Simchen, and H. Traut, *Chem. Ber.* **98**, 3883 (1965).
54a. H. Bredereck, G. Simchen, and R. Wahl, *Chem. Ber.* **101**, 4048 (1968).
55. H. Bredereck, O. Smerz, and R. Gompper, *Chem. Ber.* **94**, 1883 (1961).
56. M. Brown, U.S. Patent 3,092,637 (1963); *Chem. Abstr.* **59**, 12764 (1963).
57. T. C. Bruice and V. K. Pandit, *J. Am. Chem. Soc.* **82**, 5858 (1960).
58. H. Buchi, K. Steen, and A. Eschenmoser, *Angew. Chem. Intern. Ed. Engl.* **3**, 62 (1964).
59. E. M. Burgess and G. M. Atkins, *J. Am. Chem. Soc.* **89**, 2502 (1967).
60. J. Busz and A. Kekulé, *Chem. Ber.* **20**, 3247 (1887).
61. A. Chatterjee, S. Bose, and C. Ghosh, *Tetrahedron* **7**, 257 (1959).

61a. M. W.-L. Cheng, M.A. thesis, University of California, Santa Barbara, California (1966).
62. CIBA Ltd., British Patent 911,475 (1959); *Chem. Abstr.* **58**, 13852 (1963).
63. L. Claisen, *Ann. Chem.* **287**, 360 (1895).
64. D. H. Clemens, U.S. Patent 3,143,571 (1964); *Chem. Abstr.* **61**, 9435 (1964).
65. D. H. Clemens, U.S. Patent 3,214,471 (1965); *Chem. Abstr.* **64**, 3415 (1966).
66. D. H. Clemens and W. D. Emmons, *J. Am. Chem. Soc.* **83**, 2588 (1961).
67. D. H. Clemens, E. Y. Shropshire, and W. D. Emmons, *J. Org. Chem.* **27**, 3664 (1962).
68. G. Crank and F. W. Eastwood, *Australian J. Chem.* **18**, 1967 (1965).
69. P. De Ruggieri and C. Gandolfi, U.S. Patent 3,114,749 (1963); *Chem. Abstr.* **60**, 5599 (1964).
70. P. De Ruggieri, C. Gandolfi, and U. Guzzi, U.S. Patent 3,198,790 (1965); *Chem. Abstr.* **63**, 18218 (1965).
71. P. De Ruggieri, C. Gandolfi, and U. Guzzi, *Gazz. Chim. Ital.* **96**, 152 (1966).
72. P. De Ruggieri, C. Gandolfi, and U. Guzzi, *Gazz. Chim. Ital.* **96**, 179 (1966).
72a. E. Dyer, T. E. Majewski, and J. D. Travis, *J. Org. Chem.* **33**, 3931 (1968).
73. F. Effenberger, R. Gleiter, and G. Kiefer, *Chem. Ber.* **99**, 3892 (1966).
74. H. Eilingsfeld, G. Neubauer, M. Seefelder, and H. Weidinger, *Chem. Ber.* **97**, 1232 (1964).
75. H. Eilingsfeld, M. Seefelder, and H. Weidinger, *Angew. Chem.* **72**, 836 (1960).
76. H. Eilingsfeld, M. Seefelder, and H. Weidinger, *Chem. Ber.* **96**, 2671 (1963).
77. H. Eilingsfeld, M. Seefelder, and H. Weidinger, German Patent 1,119,872 (1959); *Chem. Zentr.* **133**, 10275 (1962).
77a. S. Farid, D. Hess, G. Pfundt, K. H. Scholz, and G. Steffan, *Chem. Commun.* p. 638 (1968).
78. F. S. Fawcett, C. W. Tullock, and D. D. Coffman, *J. Am. Chem. Soc.* **84**, 4275 (1962).
79. R. Feinauer, *Angew. Chem. Intern. Ed. Engl.* **5**, 894 (1966).
80. R. Feinauer, *Angew. Chem. Intern. Ed. Engl.* **5**, 895 (1966).
81. R. Feinauer, *Angew. Chem. Intern. Ed. Engl.* **6**, 178 (1967).
81a. R. Feinauer and E. Henckel, *Ann. Chem.* **716**, 135 (1968).
82. R. Feinauer and W. Seeliger, *Ann. Chem.* **698**, 174 (1966).
84. C. Feugeas, D. Olschwang, and M. Chatzopoulos, *Compt. Rend.* **C265**, 113 (1967).
85. G. Fodor, *Experientia* **11**, 129 (1955).
86. R. I. Fryer, J. V. Earley, and L. H. Sternbach, *J. Org. Chem.* **32**, 3798 (1967).
86a. E. E. Garcia, J. G. Riley, and R. I. Fryer, *J. Org. Chem.* **33**, 2868 (1968).
87. A. Giacolone, *Gazz. Chim. Ital.* **62**, 577 (1932).
88. G. I. Glover, R. B. Smith, and H. Rapoport, *J. Am. Chem. Soc.* **87**, 2003 (1965).
89. H. Gold, *Angew. Chem.* **72**, 956 (1960).
89a. H. Goldner, G. Dietz, and E. Carstens, German Patent 1,249,871 (1967); *Chem. Abstr.* **68**, 78314 (1968).
90. W. H. Graham and J. P. Freeman, *J. Am. Chem. Soc.* **89**, 716 (1967).
91. R. G. Griot and A. J. Frey, *Tetrahedron* **19**, 1661 (1963).
91a. H. Gross and B. Costisella, *Angew. Chem. Intern. Ed. Engl.* **7**, 391 (1968).
92. H. Gross, J. Rusche, and H. Bornowski, *Ann. Chem.* **675**, 142 (1964).
92a. I. Hagedorn and K. E. Lichtel, *Chem. Ber.* **99**, 524 (1966).
92b. I. Hagedorn, K. E. Lichtel, and H. D. Winkelmann, *Angew. Chem. Intern. Ed. Engl.* **4**, 702 (1965).
93. L. D. Hall and L. Evelyn, *Chem. & Ind. (London)* p. 183 (1968).

94. R. W. Hoffmann and H. J. Luthardt, *Tetrahedron Letters* p. 411 (1966).
95. R. W. Hoffmann and H. J. Luthardt, *Tetrahedron Letters* p. 3501 (1967).
95a. R. W. Hoffmann and H. J. Luthardt, *Chem. Ber.* **101**, 3851 (1968).
96. A. Hofmann, A. J. Frey, and H. Ott, *Experientia* **17**, 206 (1961).
97. A. Hofmann, A. Frey, and H. Ott, French Patent M3208 (1965); *Chem. Abstr.* **64**, 5164 (1966).
98. A. Hofmann, H. Ott, R. Griot, P. A. Sadler, and A. J. Frey, *Helv. Chim. Acta* **46**, 2306 (1963).
99. A. Holy, *Collection Czech. Chem. Commun.* **31**, 2973 (1966).
99a. A. Holy, S. Chladek, and J. Zemlicka, *Collection Czech. Chem. Commun.* **34**, 253 (1969).
100. A. Holy and J. Smrt, *Collection Czech. Chem. Commun.* **31**, 3800 (1966).
101. A. Holy, J. Smrt, and F. Sorm, *Collection Czech. Chem. Commun.* **32**, 2980 (1967).
102. S. Huenig, *Angew. Chem. Intern. Ed. Engl.* **3**, 548 (1964).
103. S. Huenig and L. Geldern, *J. Prakt. Chem.* [4] **24**, 246 (1964).
103a. H. A. Jacobson and A. Senning, *Chem. Commun.* p. 1245 (1968).
104. D. S. Jones, G. W. Kenner, and R. C. Sheppard, *Experientia* **19**, 126 (1963).
105. T. Kato and Y. Yamamoto, *Chem. & Pharm. Bull. (Tokyo)* **15**, 1334 (1967).
106. R. W. Kierstead, A. Farone, and A. Boris, *J. Med. Chem.* **10**, 177 (1967).
107. E. B. Knott and R. A. Jeffreys, *J. Org. Chem.* **14**, 879 (1949).
108. M. LeGuyader and D. Peltier, *Compt. Rend.* **261**, 471 (1965).
109. C. D. Lewis, R. G. Krupp, H. Tieckelmann, and H. W. Post, *J. Org. Chem.* **12**, 303 (1947).
110. B. M. Lynch and L. Poon, *Can. J. Chem.* **45**, 1431 (1967).
111. D. Martin and A. Weise, *Chem. Ber.* **99**, 3367 (1966).
112. H. Meerwein, *Vortrag. Basler Chem. Ges.* **29**, 5 (1959).
113. H. Meerwein, K. Bodenbenner, P. Borner, F. Kunert, and K. Wunderlich, *Ann. Chem.* **632**, 38 (1960).
114. H. Meerwein, P. Borner, O. Fuchs, H. J. Sasse, H. Schrodt, and J. Spille, *Chem. Ber.* **89**, 2060 (1956).
115. H. Meerwein, W. Florian, N. Schon, and G. Stopp, *Ann. Chem.* **641**, 1 (1961).
116. H. Meerwein, V. Hederich, H. Morschel, and K. Wunderlich, *Ann. Chem.* **635**, 1 (1960).
117. C. Metzger and R. Wegler, *Chem. Ber.* **101**, 1120 (1968).
117a. T. Mukaiyama, S. Aizawa, and T. Yamaguchi, *Bull. Chem. Soc. Japan* **40**, 264 (1967).
118. S. Nakamura, K. Maeda, and H. Umazawa, *J. Antibiotics (Tokyo)* **A17**, 33 (1964).
119. H. Nozaki, S. Fujita, H. Takaya, and R. Noyari, *Tetrahedron* **23**, 45 (1967).
120. M. Oishi, M. Ochiai, M. Nagai, and Y. Ban, *Tetrahedron Letters* p. 497 (1968).
121. H. Ott, A. J. Frey, and A. Hofmann, *Tetrahedron* **19**, 1675 (1963).
122. A. Piskala, *Collection Czech. Chem. Commun.* **32**, 3966 (1967).
123. H. Quast and E. Schmitt, *Chem. Ber.* **101**, 1137 (1968).
123a. G. L. Reginier, R. J. Canevari, M. J. Laubie, and J. C. LeDourac, *J. Med. Chem.* **11**, 1151 (1968).
123b. R. Richter, *Tetrahedron Letters* p. 5037 (1968).
123c. R. Richter, *Chem. Ber.* **101**, 3002 (1968).
123d. W. Ried and P. Junker, *Ann. Chem.* **713**, 119 (1968).
124. R. M. Roberts, *J. Am. Chem. Soc.* **72**, 3608 (1950).
125. R. Scarpati, G. Del Re, and T. Maone, *Rend. Accad. Sci. Fis. Mat. (Soc. Nazl. Sci., Napoli)* [4] **26**, 1–20 (1959) (reprint); *Chem. Abstr.* **55**, 11423 (1961).

125a. R. Scarpati, M. L. Graziano, and R. A. Nicolaus, *Gazz. Chim. Ital.* **98**, 681 (1968).
126. R. Scarpati, D. Sica, and C. Santacroce, *Gazz. Chim. Ital.* **94**, 1430 (1964).
127. H. Scheibler, U. Faas, and B. Hadji-Walassis, *J. Prakt. Chem.* [4] **7**, 70 (1958).
128. W. Seeliger, E. Aufderhaar, W. Diepers, R. Feinauer, R. Nehrung, and H. Hellmann, *Angew. Chem. Intern. Ed. Engl.* **5**, 875 (1966).
128a. T. Severin and H.-J. Boehme, *Chem. Ber.* **101**, 2925 (1968).
129. M. M. Shemyakin, V. K. Antonov, A. M. Shkrob, and L. B. Senyavina, *Tetrahedron Letters* p. 701 (1962).
130. M. M. Shemyakin, V. K. Antonov, A. M. Shkrob, V. I. Shchelakov, and Z. E. Agadzhanyan, *Tetrahedron* **21**, 3537 (1965).
131. M. M. Shemyakin, Y. A. Ovchinnikov, V. K. Antonov, A. A. Kiryushkin, V. T. Ivanov, V. I. Shchelokov, and A. M. Shkrob, *Tetrahedron Letters* p. 47 (1964).
132. R. C. Sheppard, *Experientia* **19**, 125 (1963).
133. A. M. Shkrob, Y. I. Krylova, V. K. Antonov, and M. M. Shemyakin, *Tetrahedron Letters* p. 2701 (1967).
134. G. Simchen, H. Hofmann, and H. Bredereck, *Chem. Ber.* **101**, 52 (1968).
135. G. Simchen, S. Rebsdat, and W. Kantlehner, *Angew. Chem. Intern. Ed. Engl.* **6**, 875 (1967).
136. Y. A. Sinnema and J. F. Arens, *Rec. Trav. Chim.* **74**, 901 (1955).
137. Y. A. Sinnema and J. F. Arens, *Rec. Trav. Chim.* **75**, 1423 (1956).
138. J. Smrt and F. Sorm, *Collection Czech. Chem. Commun.* **32**, 3169 (1967).
139. J. Smrt and F. Sorm, *Collection Czech. Chem. Commun.* **32**, 3380 (1967).
140. H. D. Stachel, *Angew. Chem.* **73**, 64 (1961).
141. A. V. Stravrovskaya, T. V. Protopopova, and A. P. Skoldinov, *Zh. Organ. Khim.* **3**, 1749 (1967).
142. W. Stilze, German Patent 1,161,285 (1964); *Chem. Abstr.* **60**, 9156 (1964).
143. A. Stoll, *Fortschr. Chem. Org. Naturstoffe* **9**, 114 (1952).
144. T. Taguchi and Y. Kawazoe, *J. Org. Chem.* **26**, 2699 (1961).
145. G. Tsatsaronis and F. Effenberger, *Chem. Ber.* **94**, 2876 (1961).
145aa. H. Ulrich, B. Tucker, F. A. Stuber, and A. A. R. Sayigh, *J. Org. Chem.* **33**, 3928 (1968).
145a. H. von Brachel, German Patent 1,156,780 (1963); *Chem. Abstr.* **60**, 5344 (1964)
145b. H. von Brachel and R. Merten, *Angew. Chem. Intern. Ed. Engl.* **1**, 592 (1962).
146. H. Vorbruggen, *Steroids* **1**, 45 (1963).
147. H. Vorbruggen, *Angew. Chem. Intern. Ed. Engl.* **2**, 211 (1963).
148. H. Wanzlick, *Angew. Chem. Intern. Ed. Engl.* **1**, 75 (1962).
149. H. Weidinger and H. Eilingsfeld, German Patent 1,122,936 (1960); *Chem. Abstr.* **57**, 4552 (1962).
150. N. L. Weinberg and E. A. Brown, *J. Org. Chem.* **31**, 4054 (1966).
151. H. Weingarten, *J. Org. Chem.* **32**, 3713 (1967).
152. H. Weingarten and N. K. Edelman, *J. Org. Chem.* **32**, 3293 (1967).
153. H. Weingarten and W. A. White, *J. Am. Chem. Soc.* **88**, 850 (1966).
154. H. Weingarten and W. A. White, *J. Org. Chem.* **31**, 2874 (1966).
155. H. Weingarten and W. A. White, *J. Am. Chem. Soc.* **88**, 2885 (1966).
156. H. Weingarten and W. A. White, *J. Org. Chem.* **31**, 3427 (1966).
157. H. Weingarten and W. A. White, *Monsanto Tech. Rev.* **12**, 18 (1967); *Chem. Abstr.* **67**, 63650 (1967).
158. L. Weintraub, S. R. Oles, and N. Kalish, *J. Org. Chem.* **33**, 1679 (1968).
158a. G. Westphal and H. Stroh, *Ann. Chem.* **711**, 124 (1968).
158b. C. W. Whitehead, *J. Am. Chem. Soc.* **75**, 671 (1953).

158c. C. W. Whitehead and J. J. Traverso, *J. Am. Chem. Soc.* **77**, 5872 (1955).
158d. J. D. Wilson and H. Weingarten, *J. Org. Chem.* **33**, 1246 (1968).
159. H. E. Winberg, U.S. Patent 3,121,084 (1964); *Chem. Abstr.* **60**, 13197 (1964).
160. H. E. Winberg, U.S. Patent 3,239,534 (1966); *Chem. Abstr.* **64**, 17425 (1966).
161. H. E. Winberg, U.S. Patent 3,239,519 (1966); *Chem. Abstr.* **64**, 15864 (1966).
161a. H. E. Winberg, U.S. Patent 3,361,757 (1968); *Chem. Abstr.* **69**, 19155 (1968).
162. H. E. Winberg, J. E. Carnahan, D. D. Coffman, and M. Brown, *J. Am. Chem. Soc.* **87**, 2055 (1965).
163. H. E. Winberg and D. D. Coffman, *J. Am. Chem. Soc.* **87**, 2776 (1965).
164. D. Wrinch, *Nature* **199**, 564 (1963).
165. H. L. Yale and J. T. Sheehan, *J. Org. Chem.* **26**, 4315 (1961).
165a. O. A. Zagulyaeva and V. P. Mamaev, *Izv. Sib. Otd. Akad. Nauk SSSR, Ser. Khim. Nauk* p. 55 (No. 5) (1967).
166. H. E. Zaugg, F. E. Chadde, and R. J. Michaels, *J. Am. Chem. Soc.* **84**, 4567 (1962).
167. H. E. Zaugg and R. W. DeNet, *J. Am. Chem. Soc.* **84**, 4574 (1962).
168. H. E. Zaugg and R. W. DeNet, *J. Org. Chem.* **29**, 2769 (1964).
169. H. E. Zaugg, R. W. DeNet, and R. J. Michaels, *J. Org. Chem.* **28**, 1795 (1963).
170. H. E. Zaugg, V. Papendick, and R. J. Michaels, *J. Am. Chem. Soc.* **86**, 1399 (1964).
171. J. Zemlicka, *Collection Czech. Chem. Commun.* **28**, 1060 (1963).
172. J. Zemlicka and A. Holy, *Collection Czech. Chem. Commun.* **32**, 3159 (1967).

# Author Index

Numbers in parentheses are reference numbers and indicate that an author's work is referred to although his name is not cited in the text. Numbers in italics show the page on which the complete reference is listed.

## A

Abdullaev, A. I., 382(1), *414*
Abramenko, P. I., 253(643), 256(1, 1a, 642, 642a, 643, 644), 258(645), *278*, *297*
Abrams, E. L. A., 183(125), 189(125), 190(125), 191(125), 207(125), *219*
Acton, E. M., 312(121a), *346*
Adams, J., 143(229a), 145(229a), *175*
Adkins, H., 13(1), 13(53), 56(1), 57(1), *122*, *123*, 155(59, 220), 156(59, 220), 157, 159(59, 220), 160(59, 220), *170*, *175*
Adlef, W., 13(24), 43(24), *122*
Adley, T. J., 366(1a), 394(1a), *414*
Aeschlimann, J. A., 392(112a), 397(112a), *417*
Agadzhanyan, T. E., 436(130), 438(130), 439(130), 444(1), 451(130), 453(130), 455(130), 456(130), 459(130), 462(130), 464(130), *501*, *505*
Agre, C. L., 21(251), *129*, 412(113), *417*
Ainsworth, C., 181(3), 183(1, 3, 5), 184(3), 187(3), 201(1, 2, 3, 4, 5, 33), 203(3), 212(1), *215*, *216*
Aizawa, S., 431(117a), 459(117a), *504*
Akunian, E., 158(151), 159(151), 160(151), *172*
Albert, A., 211(6), *215*

Aldanova, N. A., 247(540), *294*
Aldridge, C. L., 3(202, 203), 4(202), 6(202), 8(202), *128*, 275(395b), 276(395b), *289*
Al-Dulaimi, K. S., 183(76b), 193(76b), *217*
Alexander, E. R., 19, 20(2), 56(2), *122*, 155(2), 159(3, 4, 5), *168*
Ali-Zadi, I. G., 382(1), *414*
Allbut, A. D., 137(158a), *173*
Allen, C. H. F., 158(6), 160(6), *168*, 212(7, 8, 9), *215*
Allen, G. R., 148(7), *168*
Allen, I., 137(236), *175*
Alperovich, M. A., 253(3, 6, 414), 256(4), 257(3), 258(2, 5, 6, 416e), 263(414), *278*, *290*
Altmark, E. M., 159(163), *173*
Altona, C., 71(2a), *122*
Amakasu, O., 43(340b), *132*
Anderson, B. C., 26(353), *132*, 145(288), *177*
Anderson, C. B., 37(4), 57(4), *122*
Anderson, E. K., 374(3), *414*
Anderson, G. de W., 245(7), *278*
Andrews, K. J. M., 265(7a), *278*
Andrews, T. M., 186(47), *216*
Animali, E., 257(8), *278*
Anish, A. W., 257(9, 12), 258(10, 11, 13), *278*

Anissizzaman, A. K. M., 366(1a), 394(1a), *414*
Annenkova, V. M., *176*
Anneser, E., 364(80), 391(80), 392(80), 394(80), 396(80), 398(80), 399(80), *416*
Anschutz, R., 17(5, 6), 56(5), 57(5, 6), 61(5, 6), *122*, 187(10), 190(10), *215*
Ansell, M. F., 149(8), 151(8), 159(8), 160(8), 161(8), *168*
Antaki, H., 198(11), *215*
Anthes, H. I., 14(204), 29(204), 33(204), 47(204), *128*, 277(395c), *289*
Antonov, V. K., 436(129, 130), 436(2), 437(131), 438(130), 439(130), 444(1), 451(130), 452(129), 453(130), 454(129), 455(130), 456(129, 130), 458(130), 459(130), 462(130), 464(130), 469(133), 496(133), *501*, *505*
Apjok, A., 22(180a), 37(180a), *127*
Arbusov, A., 151(9), 159(9, 11), 160(9, 10), *168*
Arbusov, A. E., 149(12), 151(12), 159(12), *168*
Arbusov, B. A., 21(6a, 9), 54(6a, 6b, 7, 8, 8a, 9, 10), 55(7), 57(6a, 8a), *122*, 165(13), 167(13), *168*
Archer, S., 183(104), 189(104), *218*, 351(4), 386(4), *414*
Arcis, A., 229(223), *284*
Arens, J. F., 16(130, 327), 221(103, 93), 34(294, 345a), *125*, *126*, *130*, *131*, *132*, 185(74), 188(74), *217*, 351(57, 129), 352(129, 155), 353(129, 155), 357(5, 57), 358(57, 157), 359(57, 147), 370(157), 372(155), 379(57, 157), 382(129, 148a, 155), 384(57, 129), 385(57, 157), 387(57), 390(57, 129), 391 (129), 392(57, 129), 399(57, 157), 400(157), 401(157), 402(157), 405(157), 410(57, 129, 155), 411(155), 412(129, 155), 414 (155), *414*, *416*, *418*, *419*, 420(3), 450(136, 137), 478(136), 487(136), *501*, *505*
Arndt, F., 151(14), *168*, 354(6, 7), 363(6, 7), 369(6), 370(7), 371(7), 398(7), 399(7), 401(6), 402(7), 403(6, 7), *415*
Arnhold, M., 13(11), 56(11), 57(11), *122* 165(15), 167(15), *168*
Arni, U., 260(614), *296*
Arnold, W., 226(521), *293*
Arnold, Z., 159(16), *168*, 422(5), 423(5), 424(5), 428(5), 450(5), 451(5), 452(5), 454(5), 455(5), 456(5), 458(5), 461(5), 463(5), 465(5), 476(5), 490(5), 491(4), *501*
Asako, T., 238(412a), *290*
Asnina, F. I., 258(317), *287*
Asou, T., 14(181), *127*
Astle, M. J., 24(12), 56(12), *122*, 158(17), 162(17), *168*
Atavin, A. S., 54(325a), 57(325a), 61(325a), *131*
Atkins, G. M., 436(59), 451(59), *502*
Atkinson, J. D., 212(11a), *215*, 236(13a), *278*
Atkinson, M. R., 204(12), *215*
Attenburrow, J., 196(19), *215*
Aubert, P., 196(12a, 13), *215*, 242(15), 243(14), *278*
Aufderhaar, E., 464(128), *505*
Auterhoff, H., 43(13), 44(13), *122*
Auwers, X., 14(14, 15), *122*, 144(18), *168*

## B

Babayan, A. T., 47(16), *122*
Babers, F. H., 326(30), *343*
Babicher, F. S., 246(17), 253(16), 258(18), 260(16), 264(17), *278*, 491(6), *501*
Babievskii, K. K., 233(560b), 235(560b), 237(560b), *295*
Babin, D. R., 43(16a), *122*
Bach, G., 198(31), *216*, 253(19, 20), 255(21, 22), 257(20, 99), 258(99), 259(19, 20), *278*, *280*
Bachman, G. B., 225(23, 413), 226(23), 227(413), *278*, *290*
Bachurina, M. P., 253(648), 259(647, 648), *297*
Backer, H. J., 186(14, 15, 16), *215*, 349(11), 351(11), 354(11), 355(11), 356(14), 363(12, 13), 367(11), 368(8, 13), 369(8, 9, 12, 13), 372, 374(8, 9, 10, 11), 375(9, 12), 379(11), 384(12), 385(11), 386(11), 390(11, 12), 392(14), 400(14), 401(11), 402(13), 403(13, 14), 404(14), 405(11), 412(8), *415*, 445(7, 8), *501*
Baekstrom, H., 350(122), 379(122), 380(122), 381(122), 383(122), 387(122), 392(122), *418*
Baer, E., 159(110, 111), *171*

Baganz, H., 16, 17(18), 24(17), 55(17), 56 (17), 57(17), 59(17), 60(17), *122*, 144(21), 144(19, 21), 153(20), 162(19), 167(20), *168*, 189, 197(17), 204(17), *215*, 268(24, 25), *278*, 448(9), *501*
Bailey, J., 246(26), 247(27), 265(28), *278*
Baines, H., 14(21), *122*
Baker, B. R., 137(22), 151(21a, 22a), *169*, 205(18d), 207(18a), 208(18b, 18c), *215*
Baldwin, J. E., 275(29), *278*
Ballester, M., *132*, 233(589), 236(589), *295*
Ballou, C. E., 158(23), 159(23a), *169*
Bamfield, P., 260(30), 261(30), *278*
Ban, Y., 431(120), 451(120), 459(120), 496(120), 497(120), *504*
Banitt, E. H., 362(104), 410(104), *417*
Bank, H. M., 197(159), *220*
Banks, P., 183(129b), *219*
Banus, J., 13(22), *122*
Baralle, R., 254(405, 406), *290*, 361(111, 112), 389(111), 395(112), 397(111), *417*
Baran, J. S., 23(23), *122*
Barany, H. C., 257(31), 258(31), *278*
Barber, H. J., 196(19), *215*
Barbulescu, N., 13(24), 43(24), *122*
Baril, O. L., 149(282), *177*
Barnath, G., 437(10), *501*
Barnes, H. D., 448(11), *501*
Barnes, R. A., 30(25), 61(25), 62(25), 63(25), *122*, 135(24), 145(24), *169*
Barni, E., 256(32), 258(32, 155, 156), *278*, *282*
Barre, R., 45(26), *122*, 228(33), *278*
Barton, D. H. R., 43(27), *122*
Barvyn, N. S., 253(629), 258(385, 392), 259(629), *289*, *297*
Basaglia, W., 257(35), 260(34), *279*
Basignana, P., 246(36), 257(36), 258(36), *279*
Bassett, H., 12(28), 13(27a), 15, *122*, 164(25, 26), 165(25), 167(25), *169*
Bauer, H., 246(488), *292*
Bauriedel, W., 258(211a), *284*
Bayer, R. P., 258(69), 375(69), 410(69), *416*
Bayliss, M., 227(529), *294*
Beauregard, Y., 166(170a), *173*
Beck, F., 339(1a), *342*
Beckett, A. H., 212(20), *215*
Bedell, S. F., 160(41), *169*
Bedoukian, P. Z., 225(37), 227(37), *279*

Beer, R. J. S., 358(15), 390(15), 410(15), 414(15), *415*
Behme, M. T. A., 138(27), *169*
Behnke, W. C., 482(11a), *501*
Behringer, H., 239(38), *279*
Behun, J. D., 63(211), *128*, 145(184), *174*
Beilenson, B., 258(39), *279*
Beilstein, F., 13(29), *122*
Beier, G., 16(20), *122*
Beilfuss, H. R., 212(7, 8), *215*
Beiner, H., 194(21), *215*, 420(12), *501*
Beiser, W., *131*
Belenkaya, I. A., 197(131a), *219*
Belikova, A. M., 160(27a), *169*
Belikov, V. M., 233(560b), 235(560b), 237(560b), *295*
Bell, R. P., 139(28), 143(29, 303), 154(28), 155(28, 29), 156(29), *169*
Bell, S. C., 183(125), 189(125), 190(125), 191(125), 207(125, 213), *219*, *222*
Belleau, B., 47(30), *122*
Benard, M., *132*, 201(110, 205), 203(110), 205(205), *218*, *221*
Benary, E., 237(40), *279*
Bender, D. R., 349(15a), 384(15a), 389(15a), *415*
Bender, M. L., 437(13), *501*
Bender, R., 258(502), *293*
Bendich, A., 15(63), 15(63a), *123*
Bengler, I., 135(35), 144(35), *169*
Benko, P., 183(22), 208(129a), *215*, *219*
Benson, R. E., 35(357), 36(129), *125*, *132*
Bercot, P., 159(231), *175*
Berestecki, M., 234(538), *294*
Berezhnaya, M. I., 159(192a), *174*
Berger, J. G., 189(181), *21*
Bergman, E., 373(150), 379(150), *419*
Bergmann, E., 227(41, 600), 258(41), *279*, *296*
Bergmann, E. D., 48(31), *122*, 158(160), 159(160), 160(160), *173*
Bergmann, M., 159(33, 34), 163(32, 32a), *169*, 300, 303(17), 323(17), 339(1a), *342*, *343*
Beringer, F. M., 37(31a), *122*
Berlin, K. D., 230(42), *279*
Berlin, L., 259(527, 528), 260(526), *294*
Berlin, P., 313(122), 327(122), *346*
Berlin, T., 227(600), *296*
Bernemann, P., *130*

Bernreuther, E., 225(616), 226(616), *296*
Bernstein, S., 27(32, 83, 84, 85, 86), *122*, 135 (35, 89, 90, 91), 137(90), 144(35), 147(93), 151(90, 92, 262), 152(89, 262), *169, 170, 176*, 268(172), *282*
Bert, L., 225(45), 227(43, 44, 45, 46), *279*
Bertin, D., 28(33), *123*
Beving, H. F. G., 325(1b), 337(1b), *342*
Beyer, D. L., 5(271), *130*
Beyer, E., 51(291), 25(289), 50(287, 289, 290, 319), 51(288, 289), *130, 131*
Beyer, H., 203(23), 203(23, 96), 212(206), *215, 218, 221*
Beyer, O., 39(41), 62(41), *123*
Beyerlin, H. P., 61(49), *123*, 427(22), 451(22, 23), 454(22), 456(22), 496(23), *501*
Beyerstedt, F., 3(35), 29(35), 32(35), 33(34), 33(36), *123*, 277(47), *279*
Bianchi, M., 20(273a), *130*
Bihlmayer, G. A., 238(48), *279*
Bilhuber, E., 160(300), *177*
Biocchi, L., 30(264), *129*
Biograchov, E., 48(31), *122*
Bladon, P., 42(37), *123*
Blaise, E. E., 17(38), *123*, 228(49), *279*
Blohm, H., 233(433), 236(433), 237(433), *291*
Bochkov, A. F., 303(66c, 69, 71, 71a, 73), 304(64, 68, 71), 308(65, 66b, 68, 71), 309(73), 320(71, 73, 73a), 321(1d, 64, 66d, 68, 71, 73, 73a), 322(1d, 64, 68, 71), 323(66e), 325(64, 68, 69, 71), 327(71a), 328(66c), 336(65, 66, 66c, 67, 68, 69, 70, 71, 71a, 72, 73a), 337(66c, 66d, 71, 71a, 73a), 338(1c, 71), *342, 344*
Bodenbenner, H., 145(36), *169*
Bodenbenner, K., 37(241), 38(39, 41), 39(40), 58(40), 59(39), 62(40, 41), 63(40), *123, 129*, 138(199), 164(199), *174*, 225(402), *290*, 425(113), *504*
Bodroux, F., 44(42), 57(42), *123*
Bodrowski, W., 49(264a), *129*
Boeden H., *346*
Boehme, H., 16(43), 45(43b), 83(43a), *123*, 144(37), 153(38), 167(38), *169*, 303(123), 313(123), 323(123), 338(123), 342(123), *346*, 356(17, 19, 20), 358(17), 359(18), 363(17), 369(16), 371(20), 374(18, 20), 375(18, 20), 377(16, 18, 20), 379(17, 18), 384(18, 20), 386(17, 18), 388(18), 391(18), 399

(19), 401(20), 405(18), 410(20), 412(18), 413(20), *415*, 448(14), 465(14), 467(14), 490(128a), *501, 505*
Boekelheide, V., 191(135), *219*
Bogdanova, A. V., 353(149), 379(149), *419*
Bogert, M. T., 260(296), 269(296), *286*
Bogolyubskaya, L. T., 253(3), 257(3), *278*
Bogolyubskii, V. A., 491(6), *501*
Bohler, F., 34(247), *129*
Bohlmann, F., 23(44), 30(44), 30(45), 31(44), 31(45), *123*, 135(39, 40), *169*, 226(281), 227(50, 51, 52), *279, 286*
Bolstad, A. N., 32(205), 33(205), 34(205), *128*
Bondarev, G. P., 226(108), *281*
Bongartz, J., 381(21), *415*
Bonitsch, G., 186(137), 187(137), *219*
Bonner, W. A., 314(2), 315(2), *342*
Bootsma, H., 34(345a), *132*
Boren, H. B., 325(1b), 337(1b), *342*
Borgardt, F. G., 463(15), *501*
Boris, A., 480(106), *504*
Borner, P., 37(241, 242), 60(242), *129*, 138(199, 200), 145(200), 164(199, 200), *174*, 225(402), *290*, 360(109), 382(109), 412(109), *417*, 421(114), 425(113), 431(114), 452(114), 456(114), 500(114), *504*
Bornowski, H., 18(119), 25(119), *125*, 398(63), *416*, 441(92), 466(92), *503*
Bornstein, J., 160(41), *169*
Bos, H. J. T., 358(157), 359(157), 370(157), 379(157), 385(157), 399(157), 400(157), 401(157), 402(157), 405(157), *419*
Boschan, R., 33(297), *130*, 301(119), *346*
Bose, A. K., 3(46), *123*
Bose, S., 444(61), *502*
Bosshard, H. H., 423, 452(16), 454(16), 461(16), 465(16), *501*
Bost, R. W., 205(109), *218*
Botsch, H., 484(25), 490(24), 493(24), *501*
Bott, H. G., 300, 301(3), 316(3), 319(3), 323(3), 325(3), *342*
Bottenbruch, L., 304(43), 309(43), 325(43), 326(43), *343*
Bottomley, A. C., 159(42), *169*
Boudroux, F., 224(53, 54, 55), 227(55), *279*
Bowles, P. E., 203(174), *220*
Bradley, D. C., 446(17), *501*
Bradshaw, R., 137(157), *173*

Brandsma, L., 358(157), 359(157), 370(157), 379(157), 385(157), 399(157), 400(157), 401(159), 402(157), 405(157), *419*

Brannock, K. C., 151(240), 165(43), *169, 175*, 414(22, 23), *415*

Braun, E., 300(4), 301(4), 323(4, 21), 327(21), *342, 343*

Braun, R. A., 22(47), 24(47), 31(47, 48), 51(153a), 58(153a), 63(153a), *123, 126*, 163(44, 45, 46, 47), 164(46), *169*

Brechbuehler, H., 422(19), 423(19), 451(19), 452(19), 458(19), 461(19), 464(19), 479(18, 19), 480(18, 19), 491(19), *501*

Bredereck, F., 188(27), *216*

Bredereck, H., 61(49), *123*, 188(25, 28, 29), 207(24, 25), 212(26), *216*, 246(56), *279*, 422(39, 53), 423(21), 424(53), 427(22), 428(53), 432(21, 53), 441(38, 39, 40, 41, 42, 50, 53), 442(50, 53), 443(49), 445(50, 53), 446(27, 28, 28a, 46, 47, 53a), 447(20, 43, 44, 46, 47), 450(20, 46, 47, 53), 451(21, 22, 23, 38, 40, 42, 50, 53), 452(21, 37, 42, 43, 44, 53), 453(27, 28, 50, 53, 53a), 454(22, 42, 50, 52, 53), 455(53), 456(21, 22, 42, 44, 49, 53), 457(43, 44), 458(53), 459(53a), 461(43, 44, 53a), 468(53a), 472(43, 44), 475(134), 476(52a), 481(29a, 32), 482(27, 29, 31, 32), 484(25, 33, 36, 54, 55), 485(29, 30, 32a, 45, 55), 486(32, 33, 35), 487(34, 35), 488(35, 49, 52a, 52b, 52c), 489(52), 490(24, 48), 493(24, 27, 54a), 494(34, 43, 54), 495(26), 496(23), 497(51), 498(50), 500(134), *501, 502, 505*

Bredereck, H. J., 495(26), *501*

Bredereck, K., 423(21), 432(21), 451(21), 452(21), 456(21), *501*

Bredt-Savelsberg, M., 149(48), 151(48), 160(48), *169*

Breitfous, V. L., 226(283), *286*

Bregant, N., 258(57, 58), *279*

Brendle, T., 446(27, 28, 28a), 453(27, 28), 482(27), 493(27), *501*

Brescia, F., 138(49, 50, 51), *169*

Bressel, U., *126*

Brice, C., 321(79), 333(79), *345*

Briggs, R. M., 262(116a), *282*

Brimacombe, J. S., 307(5), *342*

Brockmann, H., 260(59), *279*

Bronsted, J. N., 138(53), 141, 143(53), 144(53), 145(53), *169*

Brooker, L. G. S., 3, 4(50), 5(50), *123*, 194(30), *216*, 244(143), 245(87, 98), 246(86, 143, 288, 585), 249(81, 82, 91, 92, 559, 586), 250(263, 264), 253(77, 78, 79, 80, 83, 84, 90, 144, 287, 585), 254(76a), 255(75a), 256(65, 77, 78, 79, 80, 81, 85, 89, 97, 144, 606, 607), 257(63, 66, 72, 73, 74, 75, 83, 86, 93, 94, 96, 144, 585), 258(60, 61, 62, 64, 65, 67, 68, 69, 70, 76, 80, 84, 85, 88, 91, 92, 93, 94, 95, 96, 144, 314, 585, 586), 259(60, 61, 90, 94, 96, 287, 288a), 260(88, 314), 263(71, 144), 264(144), *279, 280, 282, 285, 286, 287, 294, 295, 296*

Brookes, J. W., 349(24), 388(24), *415*

Brooks, C. J. W., 43(27), *122*

Broquet, C., 151(53a), *169*

Brown, D. J., 210(30a), 211(6), *215, 216*

Brown, E. A., 436(150), 452(150), 455(150), 457(150), *505*

Brown, E. W., 186(48), 187(48), *216*

Brown, H. C., 16(342), *132*, 140(54), *169*, 362(25), 376(25), 377(25), 384(25)

Brown, M., 35(357), *132*, 423(56), 426(56), 427(56), 429(56), 451(56), 452(56), 453(56), 457(56), 495(162), *502, 506*

Brownfield, R. B., 27(32, 51), *122, 123*

Brownfield, R. P., 135(35), 144(35), *169*

Bruhl, J. W., 65(52), *123*

Bruice, T. C., 439(57), *502*

Brunken, J., 198(31), *216*, 253(100), 257(99, 100), 258(99), *280*

Brunner, H., 226(101), *280*

Buchanan, J. G., 231(102), *280*, 302(6), 307(6), 321(6), 340(6), *342*

Buchi, H., 480(58), *502*

Buchler, E., 210(32), *216*

Buckles, R. E., 22(364), 40(364), *132*, 140(298), *177*, 301(140), 306(139), 316(139), *346*

Buddrus, J., 49(264a), *129*

Budesinsky, B., 43(349), *132*

Büchi, H., 422(19), 423(19), 51(19), 452(19), 458(19), 461(19), 464(19), 479(18, 19), 480(18, 19), 491(19), *501*

Bunton, C. A., 141, 142(55), 143(56), 145(55), *169*

Burgess, E. M., 436(59), 451(59), *502*

Burn, D., 152(57), 158(57), *169*

Burness, D. M., 151(119), *171*, 212(7, 8), *215*

Busch, H. M., 19, 20(2), 56(2), *122*, 159(3), *168*
Busz, J., 448(60), *502*
Buting, W. E., 201(33), *216*
Butler, D. N., 183(165), *220*
Butler, H. W., 212(34), *216*
Butler, J. A. V., 138(141), *172*
Butler, J. C., 13(133), 16(133), *126*
Butler, K., 206(35), 212(35), *216*

C

Cagniant, P., 227(103), *280*
Cairns, T. L., 351(26), 379(26), *415*
Campbell, R. H., 363(38), 375(38), 377(38), 379(38), *415*
Canevari, R. J., 485(123a), *504*
Carbon, J. A., 211(182), *221*
Cardillo, B., 270(104), *280*
Carmody, D. R., 414(27), *415*
Carnahan, J. E., 495(162), *506*
Carothers, W. H., 160(58), 161(58), *170*
Carpel, W. J., 151(154), 152(154), *172*
Carroll, B. H., 250(105), *280*
Carroll, P. M., 161(133), *172*
Carstens, E., 210(84a), *217*, 454(89a), 478(89a), *503*
Carswell, H. E., 13(53), *123*, 155(59), 156(59), 157, 159(59), 160(59), *170*
Carty, D. T., 13(133), 16(133), *126*
Caserio, M. C., 154(237), *175*
Casnati, G., 270(104, 106), *280*, *281*
Castle, R. N., 212(89a, 98, 123a), *218*, *219*
Cavallo, J. F., 212(36), *216*
Cernik, D., 365(143), 403(143), *418*
Černý, M., 299(130, 131), 301(130), 315(130, 131), *346*
Chadde, F. E., 433(166), 469(166), 470(166), 471(166), 477(166), *506*
Chaikin, S. W., 64(284), *130*
Chamberlain, E. M., 151(242), 152(242), *175*
Chapman, A. W., 193(37), *216*
Chatterjee, A., 444(61), *502*
Chatzopoulas, M., 192(56), *216*, 422(84), 423(84), 428(84), 451(84), 452(84), 454(84), 461(84), *503*
Chauvelier, J., 227(107), *281*
Chaux, R., 227(166), *282*
Checchi, S., 246(501), *293*
Cheeseman, G., 211(6), *215*
Chelpanova, L. P., 226(108), *281*

Chemische Fabriken A.-G., 11(199a), 55(199a), *128*
Cheng, C. C., 210(183a), 211(183, 184), *221*
Cheng, M. W.-L., 500(61a), *503*
Chernyuk, I. N., 256(109, 110, 111, 112, 464, 465), *281*, *292*
Chichibabin, A. E., 44, *123*, 224(113, 114, 115), 226(115), 227(114, 115, 116, 119, 120), 228(117), 229(118), *281*
Chifu, Y., 264(268), *285*
Childress, S. J., 207(213), *222*
Chiodoni, U., 258(121), *281*
Chladek, S., *133*, 135(62, 307), 137(61a, 307, 308), 145(307), 161(61), *170*, *177*, 306(7a, 9), 307(7, 7a, 7b, 145, 146, 146a, 147), 331(7, 9, 145, 146, 147), 334(8, 145, 146), *342*, *347*, 481(99a), *504*
Chou, T. C., 210(71a, 169a), *217*, *220*
Chu, E. J. H., 13(58), *123*
Cignarella, G., 186(196), 209(38, 196), *216*, *221*
Cioffi, M., *131*, 137(245), *176*
Cipera, J. D. T., 303(80), 321(80), 332(80), 340(80), *345*
Claesson, P., 355(29), 392(29), *415*
Claisen, L., 5(59), *123*, 149(65), 150(64), 151(63, 64, 65, 67), 155, 158(67), 159(64, 67, 68, 69), 160(66, 67, 65), *170*, 180, 183(39), 186(39), 188(39), *216*, 232(122), 236(122), 237(122, 123), 244(123), *281*, 420(63), *503*
Clancey, D. J., 137(70), *170*
Clark, H. G., 165(229), 166(229), *175*
Clark, J., 211(39a), *216*
Clark, L. M., 259(124), *281*, 392(112a), 397(112a), *417*
Clark, R. D., 227(204), *284*
Clark, R. H., 205(39b), *216*
Clarke, R. L., 3(206), 4(206), 7(206), 29(207), *128*, 225(396), 275(395d), 377(396), *289*
Claus, C. J., 81(60, 61), *123*, 231(125, 126, 127), *281*
Cleaver, C. S., 13, *123*
Clemens, D. H., 3(208), 4(208), 6(208), *128*, 189(40, 41), *216*, 275(396a), *289*, 384(29a), *415*, 442(66, 67), 445(66, 67), 446(65, 66, 67), 472(65, 66, 67), 473(65, 66, 67), 474(65), 475(65), 500(64, 66, 67), *503*
Clement, H., 227(128), *281*

Cloke, J. B., 5(122), *125*
Clunie, J. C., 143(30), *169*
Cocker, W., 159(42), *169*
Coenen, M., 257(129), *281*
Coffen, D. L., 349(15a, 30), 384(15a, 30), 389(15a), *415*
Coffman, D. D., 449(163), 458(163), 464(78, 163), 466(163), 467(78), 468(78), 471(163), 474(163), 495(162), *503, 506*
Cogrossi, C., 260(130), *281*
Cohen, H., 152(71), *170*
Cohen, S., 15(63), 15(63a), *123*
Cole, W., 151(154), 152(154), *172*
Coleman, G. H., 227(159), *282*
Colles, W., 1(63b), *123*
Collins, E., 28(323e), *131*, 135(254), 136(254), 144(254), *176*
Collins, M., 257(370), 258(370), 259(370), *288*
Comley, W. A., 13(367), *133*, 226(618), 227(618), *296*
Comoy, P., 42(268), *129*
Coni, C., 273(461b), *292*
Conia, J.-M., 149(175), 151(175), 160(175), *173*
Conklin, A. R., 363(123), 376(123), 381(123), 382(123), 385(123), 387(123), *418*
Connolly, J. M., 15(64), *123*
Cook, A. H., 212(42), *216*, 237(133), 240(131), 242(131), 243(132), 246(132), 260(134, 135), *281*
Cooley, G., 152(57), 158(57), *169*
Cope, A. C., 15(65), *123*
Copenhaver, J. W., 55(66), *123*, 267(136a), 268(136, 137), 269(136), *281*, 407(31, 32, 33), *415*
Corbett, W. M., 338(9a), *342*
Corder, E. H., 143(229a), 145(229a), *175*
Cordes, E. H., 138(27, 117, 118), 139(292, 293), 140(72), 142(222), *169, 170, 171, 175, 177*
Corey, E. J., 49(66a), *123*
Cornforth, J. W., 5(67), *123*, 159(73), *170*, 196(43), *216*, 240(138), 242(138), 246(138), *281*
Cornforth, R. H., 5(67), *123*
Cornu, P., 42(268), *129*
Corse, J., 33(297), *130*, 301(119), *346*
Costisella, B., 17(116a), 55(116a), *125*, 167(128b), *172*, 213(67), *217*, 498, *503*

Courduvalis, C., 53(263), *129*
Coxon, B., 323(10), *342*
Coxon, J. M., 231(139), *281*
Cowie, D. W., 360(34), 368(34), 379(34), 386(34), *415*
Cram, D. J., 359(34a), *415*
Cramer, F., 135(95), 137(95), 144(95), *171*, 306(15), 319(15), 334(15), *343*
Cramer, F. B., 339(107), *345*
Crank, G., 22(68), 22(69), 23(68), 24(68), 25(69), 55(68), 58(68), 60(68), *123*, 144(74), *170*, 197(44), 200(44), 212(44), *216*, 277(139a), *281*, 306(11), *342*, 425(68), 451(68), 452(68), 453(68), 457(68), 459(68), 461(68), 462(68), 468(68), 473(68), *503*
Crawford, R. J., 46(70), 61(70), *123*, 149(75), 151(75), 160(75), *170*
Creitz, E. C., 323(60), *344*
Cressman, H. W., 246(86), 257(86), *280*
Cressman, H. W. J., 257(76, 88), 260(88), *279*
Cresswell, W. T., 54(352a), *132*
Criegee, R., 51(71), *123*
Croxall, W. J., 241(393), *289*
Csepreghy, G., 190(101), 191(101), *218*
Cummings, J. R., 184(60a), *217*
Cumper, C. W. N., 363(36), 375(35, 36, 37), *415*
Curry, M. J., 34(209), 55(209), 56(209), 60(209), *128*
Curtius, T., 203(45, 46), *216*
Cuvigny, T., 159(76), *170*, 266(140), *281*
Czerminski, J., 159(213), *175*

**D**

Dadasheva, M. A., 382(1), *414*
Daily, A. E., 186(49), *216*
Dains, F. B., 186(47, 48, 49, 50), 187(48), *216*
Dale, J. K., 300(12), 323(12), 334(12), *343*
Dallacker, F., 392(37a), *415*
DalMonte, D., 256(140a), 257(8), 258(140a), 259(140a), *278, 281*
D'Amico, J. J., 363(38), 375(38), 377(38), 379(38), *415*
d'Angelo, J., 151(53a), 159(230, 231), *169, 175*
Dangyan, M. T., 168(77), *170*

Danielisz, M., 258(249), 260(249), 265(249), *285*
Daniels, T. C., 14(72), *124*
Danno, T., 160(155a), *173*
Darwent, B. de B., 143(31), *169*
Das, B., 258(141), *282*
Dashevskaya, B. I., 258(318), *287*
Dauphin, 225(45), 227(45), *279*
Davie, W. R., *128*, 275(396b), 276(396b), *289*
Davies, M. I., 152(57), 158(57), *169*
David, A., 53(158), *126*
David, S., 159(78), *170*
Davis, C. S., 197(80), 198(51), *216*, *217*
Dawson, G. A., 256(410), 260(411), *290*
Day, A. R., 212(122, 163), 212(162), *219*, *220*
Day, J. N. E., 139(79), *170*
deBollemont, E. G., 236(141a, 141b), *282*
DeCat, A., 192(202), *221*
Degener, E., 14(300), *130*, 231(505), *293*
Degginger, E. R., 63(211), *128*, 145(184), *174*
DeHaas, G. H., 149(260a), 151(260a), 160(260a), *176*
Deichmeister, M. V., 255(383), 258(385), *289*
Delange, R., 114(72a), *124*, 227(417, 418), *290*
del Arco, L. V., 30(190), *127*
Del Bino, G., 20(303a, b), *130*, 273(508b, 508c), *293*
de Lederkremer, R. M., 315(142), *347*
Delfs, D., 237(194), *283*
Delin, S., 15(329), 21(329), *131*
Del Re, G., 435(125), 460(125), 469(125), 478(125), *504*
Deluzarche, A., 227(103), *280*
Demushkina, L. B., 303(71a), 327(71a), 336(71a), 337(71a), *344*
DeNet, R. W., 434(167, 169), 437(168), *506*
Dennilauler, R., 197(99), *218*, 254(380, 405, 406, 407), 255(380), *289*, *290*, 361(111, 112), 389(111), 395(112), 397(111), *417*
Dennstedt, M., 354(39), 355(39), 401(39, 40), 411(39), *415*
Deno, N. C., 226(142), 229(142), *282*
Dent, S. G., 244(143), 246(86, 143), 253(144), 254(76a), 256(144), 257(86, 144), 258(144), 263(144), 264(144), *279*, *280*, *282*
Depner, M., 16(317, 318), *131*

Depoorter, H., 247(145, 146, 438), 257(439), 258(437, 439), *282*, *291*
Derevitskaya, V. A., 304(137), 321(12a, 12b, 137), 322(137), 338(12a), *343*, *346*
Derflinger, G., 238(48), *279*
Derkosch, J., 238(48), *279*
Dermer, O. C., 61(72b, c), *124*
De Ruggieri, P., 183(52), *216*, 487(69, 70, 71, 72), *503*
De Smet, P., 258(522, 523), *293*
DeStevens, G., 191(52a), *216*, 258(147, 148, 149, 150, 151), 263(147), 264(147), *282*
Detert, D. H., 322(80a), 341(80a), *345*
Deutsch, A., 12(73), 13(73), 55(73), *124*
Dewar, M. J. S., 260(152), *282*
Dewidar, A. M., 13(274), *130*, 165(221), 167(221), *175*
De Wolfe, R. H., 138(81, 82), 141(81), 142(55), 143(81), 144(81), 145(55, 81), *169*, *170*, 181(53, 147), 182(53), 183(53), 184(53), 185(53), 186(54), *216*, *219*
Diassi, P. A., 27(74), *124*, 135(84), 144(84), 152(83), *170*
Dickhauser, E., 314(100), *345*
Diedericksen, J., 254(420), *290*
Diels, O., 235(154), 236(153), 237(154), *282*
Diepers, W., 464(128), *505*
Dietsche, W., 70(74a), *124*, 168(84a), *170*, 213(54a), 214(54a), 215(54a), *216*
Dietz, G., 210(84a), *217*, 454(89a), 478(89a), *503*
Di Modica, G., 256(32), 258(32, 155, 156), *278*, *282*
Dimroth, K., 25(75), *124*, 164(85), *170*, 213(54b), *216*
Djerassi, C., 226(157), *282*
Dmitrenko, I. P., 258(575), *295*
Doane, W. M., 310(127, 128), 311(13), 312(14, 126), 323(13, 127, 128), 327(14, 126), 334(127), 338(127), *343*, *346*
Dobner, O., 14(76), *124*
Dodson, R. M., 149(215), 151(215), 159(215), 160(215), *175*
Doering, W. von E., 22(78), 53(77), 58(78), *124*, 145(86), *170*, 353(41), 354(41), 374(41), 376(41), 377(41), 378(41), *415*
Doja, M. A., 245(158), *282*
Dolby, L. C., 22(79), *124*, 137(87), *170*
Dolgikh, A. N., 353(149), 379(149), *419*
Dolliver, M. A., 151(87a), *170*

# AUTHOR INDEX

Domaschkle, L., 17(18), 24(17), 55(17), 56(17), 57(17), 59(17), 60(17), *122*, 144 (19, 21), 153(20), 162(19), 167(20), *168*, 189, 197(17), 204(17), *215*, 448(9), *501*
Dombrovskii, A. V., 253(466), 256(466), *292*
Donovan, J., 135(286a), *177*, 210(204b), *221*
Doorenbos, H. E., 482(11a), *501*
Doppstadt, A., 303(44), 321(44), *343*
Dornfield, C. A., 227(159), *282*
Dornow, A., 212(54c), *216*, 226(160), *282*
Dorries, A., 16(43), *123*
Dost, H., 240(304), 256(303), *287*
D'Osualdo, V., 187(154), 203(154), 204(154), 205(154), *220*
Dougherty, P., 257(370), 258(370), 259(370), *288*
Douglass, I. B., 18(80), *124*
Dowell, A. M., 355(70), 401(70), *416*
Doyle, G., 30(25), 61(25), 62(25), 63(25), *122*, 135(24), 145(24), *169*
Doyle, F. P., 236(173), 249(161), 263(306), *282*, *287*, 406(42, 44, 45), 407(94, 95), *415*, *416*, *417*
Drake, S. S., 351(54), 403(54), *416*
Dreger, E., 13(164), *127*
Dreyfuss, P. D., 257(162), 260(162), *282*
Driver, J., 14(21, 81), *122*, *124*
Drummond, P. E., 160(41), *169*
Ducker, J. W., 152(57), 158(57), *169*
Duffin, J. F., 196(87), *217*, 240(165), 241(163), 242(308), 246(307), 260(164, 309, 310, 597), 261(596), *282*, *287*, *296*
Dufraise, C., 227(166), *282*
Duggleby, P. M., 49(82), *124*, 145(88), *170*
Dughi, M., 191(52a), *216*
Dundon, M. L., 262(166a), *282*
Dunlop, A. P., 47(324), *131*
Durand, M. H., 226(171), 227(171), *282*
Durmashkina, V. V., 256(384), 258(384), *289*
Dusza, J. P., 27(83, 84, 85, 86), *124*, 135(89, 90, 91), 137(90), 144(90, 91), 147(93), 151(90, 92, 262), 152(89, 90, 91, 262), *170*, *176*, 268(172), *282*
Dvolaitzky, M., 161(194), *174*
Dyatkin, B. L., 14(176), 61(176), *127*
Dyer, E., 449(72a), *503*
Dykstra, H. B., 159(94), 160(58, 94), 161(58, 94), *170*

Dynamit-Nobel, A.-G., 20(87a), *124*
Dyson, G. M., 15(64, 88), *123*, *124*

## E

Earley, J. V., 481(86), *503*
Eastwood, F. W., 22(68), 22(69), 23(68), 24(68), 25(69), 55(68), 58(68), 60(68), *123*, 144(74), *170*, 197(44), 200(44), 212(44), *216*, 230(295a), 277(139a, 295a), *281*, *286*, 306(11), 308(63), *342*, *344*, 425(68), 451(68), 452(68), 453(68), 457(68), 459(68), 461(68), 462(68), 468(68), 473(68), *503*
Ebert, F., 237(40), *279*
Eckhart, E., 315(148), *347*
Eckstein, C., 135(95), 137(95), 144(95), *171*
Eckstein, F., 306(15), 319(15), 334(15), *343*
Edelman, N. K., 493(152), *505*
Edens, C. O., 158(6), 160(6), *168*
Edgar, A. R., 231(102), *280*, 302(6), 307(6), 321(6), 340(6), *342*
Edwards, H. D., 236(173), *282*, 406(43, 44, 45), 407(43), *415*, *416*
Effenberger, F., 61(49), *123*, 188(25, 27, 28), 207(24, 25), 212(26), *216*, 246(56), *279*, 422(37, 39), 427(22), 435(73), 441(38, 39, 40, 41, 42), 446(27, 28, 28a), 447(43, 44), 451(22, 23, 38, 40, 42), 452(37, 42, 43, 44), 453(27, 28), 454(22, 42), 456(22, 42, 44), 457(43, 44), 461(43, 44), 468(73), 471(73), 472(43, 44, 73), 481(29a, 32), 482(27, 29, 31, 32), 484(25, 33, 36, 145), 485(29, 30, 32a), 486(32, 33, 35), 487(34, 35), 488(35), 490(24), 493(24, 27), 494(34, 43), 495(26), 496(23), *501*, *502*, *503*, *505*
Eger, H. H., 145(259), *176*
Eggart, F. G., 26(89), *124*
Eglinton, G., 52(90), 53(90), 57(90), *124*
Eilingsfeld, H., 9, 11(92), 61(91, 359), *124*, *132*, 201(55), *216*, 423, 427(74, 149), 432(76, 77), 451(74, 76, 149), 452(76, 77), 456(74, 76, 77, 149), 459(75, 149), 461(76, 77), 478(76), *503*, *505*
Elderfield, R. C., 151(95a), 159(95a), *171*, 227(216), *284*, 301(90), 323(90), 335(90), *345*
Eliazyan, M. A., 20(234), 32(234), 62(234), 63(234), *128*

Eliel, E. L., 15(93a), 56(93a), 60(93a), *124*, 225(173aa), 226(173aa), *282*
Elliott, D. F., 196(19), *215*
Ellis, B., 152(59), 158(57), *169*
Elmanovitch, N. A., 226(283), *286*
Emeleus, H. J. E., 13(22), *122*
Emmons, W. D., 189(40, 41), *216*, 384(29a), *415*, 442(66, 67), 445(66, 67), 446(66, 67), 472(66, 67), 473(66, 67), 500(66, 67), *503*
Engels, H., 300(25), 315(25), *343*
England, B. T., 210(30a), *216*
Engwall, R., 350(122), 379(122), 380(122), 381(122), 383(122), 387(122), 392(122), *418*
Ennis, B. C., 15(143a), 82(93b), *124*, *126* 355(46), 356(47), 386(46), 387(47), 391(47), 402(46), 405(46), 412(46), *416*
Epler, H., 12(195a), 55(195a), *127*
Epsztein, R., 265(173a, 173b, 173c, 400a), 266(173a, 173b), *283*, *290*
Ercoli, A., 27(95, 105, 106, 107, 108), 28(94, 95, 96, 105, 106, 107, 351b), 31(106), *124*, *125*, *132*, 135(97, 98, 121, 122, 123), 136(97, 98, 120), 144(97, 98, 120, 121, 122), 152(96, 99), 160(99), *171*
Erhart, W. A., 181(186), 182(186), 185(185, 186, 187), 186(185, 186), 197(186), *221*
Erickson, A. E., 151(242), 152(242), *175*
Erickson, E. R., 13(281), 28(281), 52(282), *130*, *175*, 232(490), 233(490), 237(490), 253(490), *292*, 353(127), 379(127), *418*
Erickson, J. G., 11, *124*, 266(174), *283*
Erickson, S. P., 43(185), 44(185), *127*
Erlenmeyer, H., 226(174a), *283*
Errera, G., 233(177), 237(175), 238(176, 177), *283*
Eschenmoser, A., 422(19), 423(19), 451(19), 452(19), 458(19), 461(19), 464(19), 479(18, 19), 480(18, 19, 58), 491(19), *501*, *502*
Espinosa, F. G., 302(15a, 109a, 109b, 110, 111, 112), 303(50), 310(15a, 109a, 109b, 111, 112), 316(50), 321(50), 325(15a, 111, 112), 327(50), 340(15a, 109a, 109b, 110, 111, 112), *343*, *344*, *345*, *346*
Esser, F., 200(211), *221*
Etting, N., 189(68), *217*
Evans, M. E., 158(100), 160(100), *171*
Evans, W. L., 303(134), 320(66a), 327(134), 338(66a), *344*, *346*
Evelyn, L., 429(93), 462(93), *503*

Ewlampieff, W. W., 158(101), 159(101, 102), 162(102), *171*
Exner, O., 160(103), *171*

**F**

Faas, U., 34(98), 59(98), *124*, 426(127), 468(127), *505*
Fajardo-Pinzon, B., 3(213), 5(213), 33(212), 34(212), *128*
Fanghaenel, E., 360(48), 361(48), 379(47a), 382(48), 383(48), 384(48), 386(48), 389(48), 391(48), 396(48), *416*
Farbenfabriken Bayer, A.-G., 137(104), *171*
Farid, S., 473(77a), *503*
Farkas, I., 299(36), *343*
Farmer, E. H., 226(101), *280*
Farone, A., 480(106), *504*
Farrand, R., 49(152a), *126*
Fawcett, J. S., 43(27), *122*, 464(78), 467(78), 468(78), *503*
Fawey, E., 159(73), *170*
Fearing, R. B., 64(284), *130*
Feather, P., 152(57), 158(57), *169*
Federov, B. P., 352(49), 366(49), 387(49), 390(49), 414(50), *416*
Fedorova, I. P., 258(319), *287*
Feinauer, R., 396(50a), 399(50a), *416*, 430(80, 82), 443(79), 457(82), 458(82), 460(79, 82), 462(79, 80, 82), 463(82), 464(80, 128), 466(79, 82), 476(79), 470(79, 82), 472(82), 473(79, 82), 478(79), 480(81, 82), 488(81a), *503*, *505*
Feist, F., 237(194), *283*
Feldmeuhle Papier, 12(99), 13(99), *124*
Felgner, W., 255(22), *278*
Feuer, H., 34(100), *125*, 137(105), *171*
Feugeas, C., 192(56), *216*, 354(50b, 50c, 50d), *416*, 422(84), 423(84), 428(84), 451(84), 452(84), 454(84), 461(84), *503*
Fialkov, Y. A., 253(623), 258(623), 259(623), *296*
Ficken, G. E., 249(198, 199, 201, 202, 203, 311), 258(200), 260(200, 203), *283*, *284*, *287*, 406(51), *416*
Field, L., 227(204), *284*
Fields, E. K., 349(52), 380(52), 386(52), *416*
Fieser, L. F., 27(100a), *125*, 136(105a), *171*
Fieser, M., 27(100a), *125*, 136(105a), *171*

Fife, T. H., 140(108), 146(106, 108), 159(107, 108), 160(107, 108), *171*
Filileeva, L. I., 253(542), 259(542), *294*
Filler, R., 242(498), *293*
Filmfabrik Agfa Wolfen, V. E. B., 11(101), *125*
Findlay, J. A., 43(360a), *132*
Finkenor, L., 28(323e), *131*, 135(254), 136(254), 144(254), *176*
Firestine, J. C., 257(205, 206), 258(207, 208), *284*
Fischer, E., 159(109), *171*, 300, 303(17), 315(16), 323(17), *343*
Fischer, H., 191(57), *217*
Fischer, H. O. L., 159(23a, 110, 111, 113), 162(112, 113), *169*, *171*
Fischer, O., 270(209), *284*
Fisher, N. I., 194(58), *217*, 258(210), *284*
Fisscher, G., 253(211), 258(211), *284*
Fleischer, G., 151, 152(251), *176*
Fletcher, H. G., 159(157a), *173*, 303(93, 98), 305(96, 97), 308(94), 309(18, 94, 95), 319(18), 320(18, 93, 96, 97), 323(98), 325(94), 336(18, 97), 339(97), 340(97), *343*, *345*
Fletz, A., 414(27), *415*
Flisik, A. C., 55(165a), *127*
Florian, W., 361(110), 387(110), *417*, 422(115), 424(115), 426(115), 427(115), 428(115), 432(115), 440(115), 450(115), 451(115), 452(115), 454(115), 456(115), 457(115), 458(115), 460(115), 462(115), 475(115), 476(115), 477(115), 478(115), 481(115), 490(115), 492(115), 495(115), 500(115), *504*
Fock, J., 17(18), *122*, 144(21), *168*
Fodor, G., 437(85), *503*
Foldi, Z., 413(53), *416*
Fomine, D. A., 226(283), *286*
Forrest, G. C., 42(37), *123*
Forrest, T. P., 18(160a), 43(16a), 56(160a), *122*, *126*
Frank, R. L., 159(114), 160(114), 161(114), *171*, 351(54), 403(54), *416*
Franke, W., 258(211a), *284*
Franks, N. E., 303(20), 319(19, 20), 324(20), 325(19, 20), 335(19), 338(19), *343*
Frantantoni, J. C., 161(133), *172*
Franz, R., 226(235), *285*
Fravolini, A., 258(241), *285*

Fredga, A., 350(56), 368(56), 374(56), 381(55), *416*
Freeman, J. H., 182(59), 183(59), 207(59), *217*
Freeman, J. P., 425(90), 450(90), *503*
Freiberg, J., 17(116a), 55(116a), *125*, 167(128b), *172*, 213(67), *217*
Frese, E., 365(141), 393(141), 394(141), 397(141), 400(141), 402(141), *418*
Freudenberg, K., 163(32a, 114a), *169*, *171*, 300(21, 25), 308(23), 315(22, 25), 322(22), 323(21, 23), 327(21, 24, 25), *343*
Frey, A. J., 436(91, 121), 437(98), 438(91), 439(91), 451(91), 453(91), 454(91), 455(91), 457(121), 465(98, 121), 466(96, 98), 467(96, 98), 470(96, 98), 473(96, 98), 474(96, 98), 477(91, 121), *503*, *504*
Fridinger, T. L., 48(127a), *125*
Fridman, S. G., 197(60), *217*, 253(215), 258(212, 213, 214, 321), 259(215), *284*, *287*
Fried, J., 27(74), *124*, 135(84), 144(84), *170*, 227(216), *284*
Friedrich, E. C., 37(4), 57(4), *122*
Fritsch, P., 14(102), *125*
Frohling, A., 21(103), *125*, 351(57), 357(57), 358(57), 359(57), 379(57), 384(57), 385(57), 387(57), 390(57), 392(57), 399(57), 410(57), *416*
Fromageot, H. P. H., 306(26, 27), 307(26), *343*
Fromageot, H. P. M., 135(115, 115a), 137(115), 144(115), *171*
Fromm, E., 368(58), 386(58), *416*
Frost, A. A., 139(116), 160(116), *171*
Frush, H. L., 302, 307(28), 314(61), 315(59), 323(60), 327(28), 328(28), 339(28), *343*, *344*
Fry, D. J., 184(88), *217*, 233(313), 239(313), 246(312), 259(218), 260(217), *284*, *287*
Fryer, R. I., 481(86), 490(86a), *503*
Fuchs, O., 37(242), 60(242), *129*, 138(200), 145(200), 164(200), *174*, 360(109), 382(109), 412(109), *417*, 421(114), 431(114), 452(114), 456(114), 500(114), *504*
Fujihara, M., 160(155a), *173*
Fujita, S., 430(119), 451(119), 452(119), *504*
Fullington, J. G., 138(27, 117, 118), *169*, *171*
Furst, A., 366(106), 371(106), 389(106), 397(106), *417*

Furuta, S., 366(106), 371(106), 388(106), 397(106), *417*
Fusco, A. J., 257(370), 258(370), 259(370), *288*
Fuson, R. C., 151(119), *171*, 233(221), 234(221), 237(221), *284*

## G

Gabriel, S., 354(59), 379(59), 399(59), 411(59), *416*
Gadekar, S. M., 184(60a), *217*
Gaitseva, E. A., *176*
Gallagher, A., 364, 379(89), *417*
Galton, S. A., 37(31a), *122*
Gandino, M., 246(36), 257(36), 258(36), *279*
Gandolfi, C., 183(52), *216*, 487(69, 70, 71, 72), *503*
Ganuschak, N. I., 253(466), 256(466), *292*
Gaoni, Y., 47(103a, b), *125*
Garcia, E. E., 183(187a), 212(187a), *221*, 490(86a), *503*
Gardi, R., 27(95, 105, 106, 107, 108), 28(94, 95, 96, 104, 105, 106, 107, 351b), 31(106), *124*, *125*, *132*, 135(97, 98, 121, 122, 123), 136(97, 98, 120, 122), 144(97, 98, 120, 121, 122), *171*
Gardner, W. H., 34(100), *125*
Gardner, W. H., 137(105), *171*
Garcia, E. E., 236(560a), *294*
Garegg, P. J., 325(1b), 337(1b), *342*
Gartner, H., 236(153), *282*
Gasco, A., 256(32), 258(32), *278*
Gaspar, B., 257(162), 260(162), *282*
Gastaldi, C., 253(222), 260(222), *284*
Gassman, P. G., 219
Gaudiano, G., 226(496), *293*
Gaurat, C., 254(405, 406), *290*, 361(111, 112), 389(111), 395(112), 397(111), *417*
Gazzola, A. L., 210(64), *217*
Geiger, B., 485(45), *502*
Geiger, E. M., 249(381, 382), 258(381), 259(382), *289*
Geldern, L., 432(103), 443(103), 459(103), 463(103), 464(103), 465(103), *504*
Gelin, R., 229(223), 274(223a), *284*
Gelin, S., 229(223), 274(223a), *284*
Geller, L. E., 226(157), *282*
Geoigian, V., 135(123b), 152(123b), *171*
Gerencevic, N., 258(224), *284*

German, L. S., 14(176), 61(176), *127*
Gershtein, R. A., 241(225), *284*
Ghelardoni, M., 207(606), 214(606), *217*
Ghigi, E., 253(234), 260(234), *285*
Ghosh, C., 444(61), *502*
Giacolone, A., 180, 186(61), *217*, 445, *503*
Giam, C. S., 305(29), 307(29), 328(29), *343*
Gibbons, R. A., 163(302), *177*
Gibson, D. T., 360(34, 60), 368(34, 60), 379(34), 384(60), 386(34, 60), 391(60), 392(60), 393(60), 400(60), *415*, *416*
Gilman, H., 14(108a), *125*, 226(235), *285*
Ginsberg, S., 198(62), *217*
Giza, C. A., 15(93a), 56(93a), 60(93a), *124*
Gladkaya, V. A., 253(511), 259(511), *293*
Gleiter, R., 435(73), 468(73), 471(73), 472(73), *503*
Glickman, S. A., 180(63), 183(63), *217*
Glockner, H., 257(236), 259(236), *285*
Glover, G. I., 444(88), *503*
Gnevysheva, T. G., 257(558), *294*
Gobe, I., 256(399), 260(399), *289*
Godtfredsen, W. O., 152(179), *173*
Goebel, W. F., 326(30), *343*
Goeknel, E., 443(49), 456(49), 488(49), *502*
Goerdler, J., 253(237), 260(237), *285*
Goetze, J., 253(300), 257(239, 240), 258(238, 300), *285*, *286*
Goguadze, V. P., *125*
Gold, H., 424(89), 451(89), *503*
Goldfarb, Y. A., 352(61), 401(61), *416*
Goldfarb, Y. L., 159(124b), 160(125), *171*
Goldman, L., 210(64), *217*
Goldner, H., 454(89a), 478(89a), *503*
Goldschmid, H. R., 305(29, 31), 307(29), 327(31), 328(29), 332(31), *343*
Goldschmidt, C., 186(65), *217*
Goldsworthy, J., 159(73), *170*
Golub, D. K., 197(60), *217*, 258(214), *284*
Golubushina, G. M., 258(320), *287*
Gomes, L. M., 159(126), *172*
Gompper, R., 183(66), 188(28, 29), *217*, 446(46, 47), 447(44, 46, 47), 450(46, 47), 451(44), 456(44), 457(44), 461(44), 472(44), 484(55), 485(45, 55), *502*
Goodman, J. J., 151(262), 152(262), *176*
Goodman, L., 25(365), 59(365), *132*, 144(299), *177*
Goodman, L. M., 312(121a), *346*
Gorelikov, A. I., 256(109, 110), *281*

Gorgues, A., 159(126a), *172*
Gorichok, S. I., 253(467), 256(467), *292*
Gorichok, Y. O., 253(467), 256(467), 256 (471), *292*
Gorin, P. A. J., 320(31a), 325(32), *343*
Gornostaeva, S. E., 258(576), *295*
Gosselink, E. P., 45(195), *127*
Goto, T., 42(112), *125*, 187(175), *220*, 258 (559a), *294*
Gots, J. S., 3(46), *123*
Gottschlich, A., 303(44), 321(44), *343*
Gougoutas, J. Z., 43(368), *133*
Grace, W. R., *125*
Graf, L., 315(109), 327(109), *345*
Grafen, P., 149(248), 151(248), 158(248), 159(248), 160(248), *176*
Graffin, P., 151(153a), 159(153a), *172*
Graham, W. H., 425(90), 450(90), *503*
Gran, H., 369(16), 377(16), *415*
Grandolini, G., 258(241), *285*
Granger, R., 227(419), *290*
Grard, J., 227(242), *285*
Grard, M., 159(127), *172*
Graziano, M. L., 276(516a), *293*, 436(125a), *505*
Green, D. P. L., 137(127a), *172*, *343*
Greene, F. D., 53(113, 114), *125*
Greer, F., 3(46), *123*
Gresham, T. L., 151(87a), *170*
Gresham, W. F., 20(199), 32(199), *128*, 267 (243), *285*
Grichkevich-Trokhimovsky, E., 226(244, 245), *285*
Grieg, M. E., 42(131), *126*
Grieshaber, P., 422(53), 424(53), 428(53), 432(53), 441(53), 442(53), 445(53), 450 (53), 451(53), 452(53), 453(53), 454(53), 455(53), 456(53), 458(53), *502*
Griffin, B. E., 135(115, 128), 137(115, 128), 144(115, 128), *171*, *172*, 306(26, 33, 34), 307(26), 331(35), *343*
Griffin, G. W., 154(128a), *172*
Grignard, V., 224(246), 227(246), *285*
Grineva, N. I., 160(27a), *169*
Griot, R. G., 436(91), 437(98), 438(91), 439 (91), 451(91), 453(91), 454(91), 455(91), 465(98), 466(98), 467(98), 470(98), 473 (98), 474(98), 477, 477(91), *503*, *504*
Gripenberg, J., 227(247), *285*
Grisafulli, M., 270(106), *281*

Groenenveld, W. L., 137(128aa), *172*
Grohe, K., 35(300a), *130*, 137(237a), *175*
Gross, H., 16(117), 17(116a), 18(116, 118, 119, 160a), 25(119), 29(117), 52(117), 55 (116a, 117), 56(160a), 61(117), 66(118a), *125*, 162(129), 167(128b), *172*, 213(67), *217*, 270(248), *285*, 299(36), *343*, 356(62), 379(62), 398(63), *416*, 441(92), 466(92), 498, *503*
Gruenfeld, N., 43(186), *127*
Grunwald, E., 301(140, 141), *346*, *347*
Gruz, B. E., 253(620, 627, 628), 258(620), 259(627, 628), *296*, *297*
Gudz, P. F., 256(372), *288*
Gugliemetti, R., 394(63a), *416*
Gundermann, K. D., 406(64), *416*
Gussman, P. G., 135(123a), 160(123a), *171*
Gut, J., 188(132, 133), 207(132, 133), 208 (132a), *219*, 237(494), *293*
Gutsulyak, B. M., 253(466, 467), 256(466, 467, 468, 469, 470, 471), *292*
Guttman, H., 160(147), *172*
Guzzi, U., 183(52), *216*, 487(70, 71, 72), *503*

## H

Hadyi-Walassis, B., 426(127), 468(127), *505*
Hafner, K., 258(249), 260(249), 265(249), *285*
Hagedorn, I., 183(70, 71), 189, 193(70), *217*, 445, *503*
Hagopian, L., 159(107), 160(107), *171*
Haines, R. M., 53(77), *124*
Hajek, M., 481(29a), 485(32a), *501*, *502*
Hall, H. K., 142(130), *172*
Hall, L. D., 323(10), *342*, 429(93), 462(93), *503*
Hallot, A., 42(268), *129*
Hamal, H., 260(526), *294*
Hamann, H. C., 210(71a), *217*
Hamer, F. M., 159(131), *172*, 180(72), 183 (72), 194(58), *217*, 245(254, 255), 253(250, 256), 256(250, 251, 256), 257(250), 258 (39, 210, 250, 252, 253, 257, 258), 260(250, 256, 259, 260), 262(250, 251), 263(251, 261), 264(251), *279*, *284*, *285*
Hammen, H. W., 253(237), 260(237), *285*
Hammer, G. G., 358(69), 375(69), 410(69), *416*

Hammett, L. P., 140(132), *172*
Hammick, D. L., 54(120), *125*
Hamprecht, G., 313(120), *346*
Hampton, A., 161(133), *172*
Hanada, I., 183(78), 204(78), *217*
Handy, R. W., 55(165a), *127*
Hanford, W. E., 21(121, 251), *125*, *129*, 412 (65, 113), *416*, *417*
Hanson, C., 301(140), *346*
Harhash, A. H. E., 43(259), *129*
Harned, H. S., 138(134), *172*
Harries, C., 160(135), *172*
Harris, D. A., 158(136), 159(136), *172*
Harris, D. L., 185(72a), *217*
Harris, G., 240(131), 242(131), 243(132), 246(132), *281*
Harris, J. F., 35(121a), 62(121a), *125*, 355 (66, 67), 374(66, 67), 375(66, 67), 413(67), *416*
Hartigan, R. H., 5(122), *125*
Hartke, K., 183(73), *217*
Hartke, K. S., 183(188), 212(188), *221*
Hartmann, I., 34(308), 57(308), *131*
Hartog, J., 152(294), 161(294), *177*
Hartshorn, M. P., 231(139), *281*
Hase, M., 258(238), *285*
Hass, K., 12(195a), 55(195a), *127*
Haszeldine, R. N., 13(22), *122*
Hatz, E., 422(19), 423(19), 451(19), 452(19), 458(19), 461(19), 465(19), 479(18, 19), 480(18, 19), 491(19), *501*
Hauser, C. R., 13(177), *127*
Hauser, H., 34(143), 45(141, 142), *126*
Hawkins, E. G. E., *125*, 158(137), 159(137), *172*
Haworth, W. N., 300, 301(3, 37), 303(37, 38, 39), 314, 315(38, 40), 316(3), 319(3, 39), 323(3, 37, 38), 325(3), 327(39), 332(38), 333(38), *342*, *343*
Haynes, L. J., 299(41), *343*
Hederich, V., 138(202), 140(202), 164(201), *174*, 499(116), 500(116), *504*
Hedgley, E. J., 308(42), 323(42), 334(42), *343*
Heiber, F., 15, *125*
Heidenreich, K., 203(45, 46), *216*
Heijo, K., 269(562c), *295*
Heilbron, I. M., 237(133), 240(131), 242 (131), 253(256), 256(256), 260(256), *281*
Heimke, P., 259(527, 528), *294*

Heinrich, P., 164(85), *170*
Heise, H., 188(28), *216*, 447(44), 452 (44), 456(44), 457(44), 461(44), 472(44), *502*
Heitsch, C. W., 54(349a), *132*
Helberg, J., 212(54c), *216*
Helferich, B., 303(44, 49), 304(43, 49), 308 (47), 309(43, 48), 314(46), 315(46), 319 (48), 321(44, 45), 322(49), 323(47), 325 (43), 326(43, 48), 337(49), *343*
Hellmann, H., 464(128), *505*
Helmic, L. S., 212(131), *219*
Henckel, E., 488(81a), *503*
Hendess, R. W., 183(188a), 212(188a), *221*
Henecka, H., 3(127), 15(126), *125*
Henery-Logan, K. R., 48(127a), *125*
Henkel, K., 226(262), *285*
Hennant-Roland, M., 54(345b), *132*
Hennes, J. H., 31(128, 183, 184), *125*, *127*, 148(138), 153(138), 163(167), 167(138), *172*, *173*
Hennig, H., 183(125a), *219*
Hensley, L. C., 258(13), *278*
Herbrandsen, H. F., 186(136), 190(136), *219*, 236(492), *293*
Herbst, P., 227(52), *279*
Herrman, H. J., 49(233), 57(233), *128*
Herschdorfer, S., 77(326), *131*
Hertler, W. R., 36(129), *125*
Herz, A. H., 352(153), 353(153), 355(153), 362(153), 400(153), 401(153), 412(153), *419*
Herzog, H. L., 28(323e), *131*, 135(254), 136 (254), 144(254), *176*
Heseltine, D. W., 246(266), 249(265), 250 (263, 264), 253(77, 78, 79, 80, 84, 266), 254 (76a), 256(77, 78, 79, 80), 258(80, 84), 262(265), *279*, *280*, *285*
Hess, D., 473(77a), *503*
Hess, E., 2(286), *130*
Hesse, A., 51(292, 323b, c, 323d), 51(323a), 56(323a), 59(292), *130*, *131*
Hessenland, M., 14(15), *122*, 144(18), *168*
Heslinga, L., 185(74), 188(74), *217*
Hesslinga, L., 16(130), *126*
Hewett, C. L., 227(267), *285*
Hewson, K., 210(117), *218*

Heyns, K., 302(109b, 110, 111, 112), 303 (50), 310(109b, 111, 112), 316(50), 321 (50), 325(111, 112), 327(50), 340(109b, 110, 111, 112), *344, 345, 346*
Hibbert, H., 42(131), *126*
Higgins, T. D., 20(298), *130*, 181(148), 183 (148), *219*
Hiho, Y., 264(268), *285*
Hilgert, H., 34(98), 59(98), *124*
Hilgetag, G., 185(100), 186(100), *218*
Hill, A. J., 3(132), *126*, 197(75), *217*
Hill, M. E., 13(133), 16, *126*
Hillers, S., 47(332a), *131*
Hinde, R. W., 15(143a), *126*
Hine, J., 12(134, 136), 13(135, 136, 137), *126*, 354(68), 355(68, 70, 71), 358(69), 374(71), 375(69), 399(68), 401(70), 410(69), *416*
Hinz, G., 42(243), *129*
Hinz, H., 254(420), *290*
Hiraga, K., 238(412a), *290*
Hirata, Y., 42(112), *125*
Hirose, H., 160(206), *174*
Hirst, E. L., 300, 301(3, 37), 303(37, 38, 39), 314, 315(38, 40), 316(3), 319(3, 37), 323 (3, 37, 38), 325(3), 327(39), 332(38), 333 (38), *342, 343*
Hiscock, A. K., 152(57), 158(57), *169*
Hishiki, Y., 256(269), 257(270, 271), 258 (270), 263(270, 272), 264(269, 270, 272), *286*, 408(72, 73), *416*
Hodgson, D., 23(139), *126*
Hoff, D. R., 211(182), *221*
Hoffman, J. A., 30(25), 61(25), 62(25), 63 (25), *122*, 135(24), 145(24), *169*
Hoffmann, R. W., 34(143), 38(142a), 45 (141, 142), 46(142c), 47(140, 142b), 55 (142c), 61(142b), *126*, 430(94, 95, 95a), 463(94), 465(94), 470(94), 498(94), *504*
Hofmann, A. W., 188(25), 189(76), 207(24, 25), *216, 217*, 436(121), 437(98), 457(121), 465(98, 121), 466(96, 98), 467(96, 98), 470 (96, 98), 473(96, 98), 474(96, 98), 477(121), 481(32), 482(31, 32), 482(29), 485(29), 485(30, 32a), 486(32), *501, 502, 504*
Hofmann, H., 422(53), 423(53), 428(53), 432(53), 441(50, 53), 442(50, 53), 445(50, 53), 450(53), 451(50, 53), 452(53), 453(50, 53), 454(50, 53), 455(53), 456(53), 458 (53), 475(134), 498(50), 500(134), *502, 505*

Hoft, E., 18(118), *125*
Hogenkamp, H. P. C., 161(209a), *174*
Holan, G., 15(143a), 82(93b), *124, 126*, 355(46), 356(47), 386(46), 387(47), 391 (47), 402(46), 405(46), 412(46), *416*
Holand, S., 265(173a, 173b), 266(173a, 173b), *283*
Holden, K. G., 135(123b), 152(123b), *171*
Holm, T., 41(144), 47(144), 56(144), 57 (144), 61(144), *126*
Holmberg, B., 21(145), *126*, 349(74, 75, 76), 351(74, 75), 352(74, 75), 367(75), 377(74), 379(75), 399(75), 405(75), 411(74, 75), 412(75), *416*
Holscher, H. A., 236(277), *286*
Holsten, J. R., 227(204), *284*
Holt, G., 49(82), *124*, 145(88), *170*
Holum, J. R., 159(151a), *172*
Holum, L. B., 210(118), *218*
Holy, A., 135, 139(140, 140a, 221a, 221b), *172, 174, 175*, 306(51, 52, 53, 53a, 54, 92a, 115a, 115b), 307(147), 331(52, 53, 53a, 54, 147), *344, 345, 346, 347*, 428(172), 463 (172), 464(172), 481(99, 99a, 100, 101, 172), 491(4), *501, 504, 506*
Horak, J., 15(146), *126*
Horn, P., 422(53), 424(53), 432(53), 441(50, 53), 442(50, 53), 445(50, 53), 450(53), 451 (50, 53), 452(53), 453(50, 53), 454(50, 53), 455(53), 456(53), 458(53), 497(51), 498 (50), *502*
Hornel, J. C., 138(141), *172*
Horsitz, L., 258(509), *293*
Hosgut, R., 227(516), *293*
Houben, J., 5(147, 148), 21(149), *126*, 349 (77, 78), 374(78), 377(77), 379(77, 78), 384(78), 401(78), 403(78), *416*
House, H. O., 149(141a), 151(141a), 159 (141a), *172*
Howk, B. W., 45(150, 151), *126*, 265(273, 274, 275), 266(274), *286*, 408(79), *416*
Huang, P. C., 207(18a), *215*
Huang, T. C., 54(152), *126*
Huber, W., 236(276, 277), *286*
Hudson, C. S., 303(98), 314(118), 323(98), *345, 346*
Huenig, S., 425(102), 432(103), 443(103), 459(103), 463(103), 464(103), 465(103), *504*
Hughes, G. K., 260(278), *286*

Huisgen, R., 364(80), 367(81), 388(81), 391(80), 392(80), 394(80), 396(80), 398(80), 399(80), *416*, *417*
Hull, R., 49(152a), *126*, 210(76a), *217*, 354(82), 396(82), 398(82), 401(82), *417*
Huls, J. P., 226(279), *286*
Huls, R., 226(611), *296*
Hultquist, M. E., 269(280), *286*
Hummel, L., 203(96), *218*
Hunger, K., 165(142), 167(142), *172*
Hunter, H., 19, 20(153), *126*
Hurd, C. H., *172*
Hurtley, W. H., 353(83), 354(83), 381(83), *417*
Hussein, F. A., 181(77), 183(76b, 77), 93(76b, 77), *217*
Hyagarajan, G., 198(112a), *218*

**I**

Ichikawa, Z., 186(180), 187(176), *220*, *221*
Ikeda, C. K., 51(153a), 58(153a), 63(153a), *126*
Ikuna, S., 43(347), *132*
Ingold, C. K., 13, *126*, 139(79), 140(144), 143(144), *170*, *172*
Ingraham, L. L., 40(366), *133*, 301(141), *347*
Inhoffen, H. H., 149(145), 151(145, 146), 152(146), 160(145), *172*, 226(281), 227(52), *279*, *286*
Inoue, S., 212(174a), *220*
Inouye, S., 315(54a), *344*
Iotsitch, J. I., 45(155), *126*, 226(283), 228(284, 285), *286*
Irie, T., 183(78), 204(78), *217*
Iris, R. C., 186(79), *217*
Isak, A. M., 256(478), *292*
Isbell, H. S., 299(57), 302, 307(28), 314(61, 115), 315(59), 323(60), 325(115), 327(28, 55, 56), 328(28), 331, 334(56), 338(55, 56, 115), 339(28, 55, 56), *343*, *344*, *346*
Ische, F., 226(160), *282*
Isler, O., 160(147, 148, 149), *172*, 268(636), *297*
Ito, M., 187(124), *219*
Ito, T., 315(54a), *344*
Ivanov, B. E., 213(97a), *218*
Ivanov, V. T., 437(131), *505*
Ivanova, N. G., 227(416a), 265(416a), *290*
Ivanova, Z. M., 258(321), *287*

Ivanyuk, E. G., 33(370), 55(370), *133*
Ivaschenko, Y. N., 17(156, 157), *126*
Ivers, O., 315(22), 322(22), *343*
Iyengar, D. S., 212(168), *220*

**J**

Jacken, J., 186(203), *221*
Jackman, M., 351(4), 386(4), *414*
Jacob, W., 308(23), 323(23), *343*
Jacobson, H. A., 484(103a), *504*
Jacques, J., 161(194), *174*
Jadot, J., 53(158), *126*
Jakob, W., 163(114a), *171*
James, T. H., 245(404), *290*
Jansen, A. B. A., 364(84), 395(84), 397(84), *417*
Jao, L. K., 140(108), 146(108), 159(108), 160(108), *171*
Jarman, M., 135(128), 136(150), 137(128), 144(128, 150), *172*, 306(33, 34), 334(62), *343*, *344*
Jarque, R. G., 14(351), 57(351), *132*, 236(590), 237(286, 590), *286*, *295*
Jeffery, G. H., 54(352a), *132*
Jeffreys, R. A., 180(95), 183(95), 186(95), *218*, 445(107), *504*
Jelgasin, S. A., 227(120), *281*
Jeminet, G., 354(86), 363(88), 391(85), 395(85), 398(86), 399(86), 401(86), 402(86, 88), 403(88), *417*
Jenkins, G. S., 197(80), *217*
Jenkins, P. W., 246(288), 253(287), 259(287, 288a), *286*
Jennen, J. J., 258(289), *286*
Jenny, E., 423, 452(16), 454(16), 461(16), 464(16), *501*
Jensen, E. V., 149(208), 151(208), 160(208), *174*, 274(416c), *290*
Jensen, J. L., 138(81), 141(81), 143(81), 144(81), 145(81), *170*
Jepson, J. B., 196(81), *217*, 243(290), 246(290), *286*
Jernow, J. L., 269(453a), *291*, 411(118a), *418*
Jewson, F. T., 3(191), *127*
Jochinke, H., 315, 321(45), *343*
Johanassian, A., 158(151), 159(151), 160(151), *172*
Johns, C. O., 237(601), *296*
Johnson, A. W., 260(30), 261(30), *278*

Johnson, B. D., 184(60a), *217*
Johnson, J. H., 151(21a), *169*
Johnson, J. V., 197(75), *217*
Johnson, L. F., 366(106), 371(106), 389(106), 397(106), *417*
Johnson, M. A., 137(22), *169*, 208(18b, 18c), *215*
Johnson, M. C., 212(11a), *215*, 236(13a), *278*
Johnson, O. H., 159(151a), *172*
Johnson, R. N., 210(160a), *220*
Johnson, W. S., 14(369), *133*
Johnston, T. P., 364(89), 379(89), *417*
Joly, R., 27(160), 28(160), *126*, 135(152), 152(152), *172*
Jonas, J., 18(160a), 56(160a), *126*
Jones, C. D., 225(396), 277(396), *289*
Jones, D. S., 437(104), *504*
Jones, E. R. H., 52(90), 53(90), 57(90), *124*
Jones, G. D., 29(207), *128*
Jones, J. E., 250(105), 253(291), 257(291), 258(291), *286*
Jones, R. G., 17(161), *126*, 159(153), *172*, 190(169), 191(169), *220*, 225(295), 227(295), 233(294), 236(293, 294), 237(292, 293, 294), *286*
Jonsson, A., 14(162), *126*
Josan, J. S., 230(295a), 277(295a), *286*, 308(63), *344*
Joseph, J. P., 147(93), 151(92), *170*, 268(172), *282*
Joy, H. V., 260(296), 269(296), *286*
Julia, M., 151(153a), 159(153a), *172*
Julia, S., 151(153a), 159(153a), *172*
Julian, P. L., 151(154), 152(154), *172*
Jung, J. P., 226(174a), *283*
Junge, H., 260(59), *279*
Junker, P., 437(123d), *504*
Justoni, R., 269(296a), *286*

### K

Kaak, R., 236(153), *282*
Kadish, A. F., 205(18d), *215*
Kagal, S. A., 243(297), *286*
Kagan, H. B., 161(194), *174*
Kainrath, P., 245(298), *286*
Kaiser, E., 392(37a), *415*
Kalabin, G. A., 54(325a), 57(325a), 61(325a), *131*
Kalfus, K., 212(81a), *217*
Kalik, M. A., 159(124b), *171*
Kalish, N., 422(158), 453(158), *505*
Kamlet, M. J., 237(299), *286*
Kampfer, H., 253(300), 258(300), *286*
Kanaoka, M., 183(82), 184(83), 201(83), 212(83), *217*
Kanaoke, M., 39(163a), 58(163a), *126*
Kann, E., 159(33), *169*
Kano, H., 189(84), *217*
Kantlehner, W., 422(53), 423(53), 428(53), 432(53), 441(53), 442(53), 445(53), 450(53), 451(53), 452(53), 453(53), 454(53), 455(53), 456(53), 458(53), 477(135), 499(135), *502, 505*
Kao, C. H., 2(313), 5(313), *131*
Karaulova, E. N., 352(61), 401(61), *416*
Kariyone, T., 30(163), *126*, 231(301), *286*
Karmarkar, S. S., 243(297, 302), *286, 287*
Kaski, T. K., 226(456), *291*
Kashiro, J., 39(163a), 58(163a), *126*
Kaslow, C. E., 160(155), *172*
Kasperczyk, K., 53(158), *126*
Katerberg, G. J., 16(130), *126*, 185(74), 188(74), *217*
Kato, T., 430(105), 460(105), 462(105), 487(105), *504*
Katrizky, A. R., 246(488), *292*
Katsui, N., 43(186), *127*
Katz, J. R., 54(163b), *127*
Kaufman, W., 13(164), *127*
Kawanisi, M., 185(187), *221*
Kawasaki, K., 253(452), 256(452), *291*
Kawazoe, Y., 429(144), 468(144), 470(144), 476(144), 477(144), 498(144), 499(144), 500(144), *505*
Kawazu, M., 160(155a), *173*
Kay, G., 12(362, 363), 13(362, 363), *132*
Kazandji, S. Y., 181(77), 183(77), 193(77), *217*
Kazmirowski, H. G., 210(84a), *217*
Keck, H., 188(28, 29), *216*, 446(46, 47), 447(44, 46, 47), 450(46, 47), 452(44), 456(44), 457(44), 460(44), 472(44), *502*
Kehm, B. B., 191(85), *217*
Keiler, A., 253(430), 259(430), *291*
Kekulé, A., 448(60), *502*
Keller, P., 26(89), *124*
Kelley, J. L., 151(22a), *169*
Kelley, R. W., 161(155b), *173*
Kempter, G., 240(304), 256(303), *287*

Kendall, J. D., 21(165), *127*, 184(88), 187 (86, 89), 196(87), *217*, *218*, 233(313), 239 (313), 240(165), 241(163), 242(308), 246 (307, 312), 249(161, 201, 202, 203, 305, 311), 258(200), 259(218), 260(164, 200, 203, 217, 309, 310, 597), 261(596), 262 (305a), 263(306), *282*, *283*, *284*, *287*, *296*, 351(96), 352(91, 93), 353(93), 371(93), 374(96), 382(93), 384(93), 390(93), 393 (93), 398(93), 401(96), 406(42, 51, 98), 407 (94, 95, 98), 408(97), 413(90), *415*, *416*, *417*
Kenner, G. W., 437(104), *504*
Kent, P. W., 159(156), *173*
Kent, R. E., 3(214), 4(214), 6(214), 29(214), *128*
Kergomard, A., 354(86), 363(88), 391(85), 395(85), 398(86), 399(86), 401(86), 402 (86, 88), 403(88), *417*
Kessel, A. Y., 166(210), 168(210), *174*
Kesslin, G., 55(165a), *127*, 137(157), *173*, 277(313a), *287*
Ketley, A. D., 13(135), *126*
Keyes, G. H., 246(86), 248(81, 82), 253(83, 84), 256(81, 85), 256(83, 86), 258(84, 85, 314), 260(314), *280*, *287*
Kharasch, M. S., 227(315), *287*
Kheifets, S. A., 258(316, 385, 388), *287*, *289*
Khorlin, A. Y., 303(69, 71, 71a, 73), 304(64, 68, 71), 308(65, 68, 71), 309(73), 320(71, 73, 73a), 321(64, 68, 71, 73, 73a), 322(64, 68, 71), 325(64, 68, 69, 71), 327(71a), 336 (65, 66, 67, 68, 69, 70, 71, 71a, 72, 73a), 337 (71, 71a, 73a), 338(71), *344*
Khorlina, I. M., 231(635), *297*
Kibirev, V. K., 246(17), 264(17), *278*, 491 (6), *501*
Kidd, J., 338(9a), *342*
Kiefer, G., 435(73), 468(73), 471(73), 472 (73), *503*
Kiely, D. E., 159(157a), *173*
Kierstead, R. W., 480(106), *504*
Kilpatrick, M., 138(158), *173*
Kilpatrick, M. L., 138(158), *173*
Kimura, M., 26(165c), *127*
Kimura, T., 408(99), *417*
Kimura, Y., 30(163), *126*, 231(301), *286*
King, F. E., 4(166), *127*
King, H., *127*
King, J. F., 137(158a), *173*
King, R. W., 54(349a), *132*

Kinoshita, T., 212(89a), *218*
Kiprianov, A., 3(167), 4(168, 169), *127*
Kiprianov, A. I., 194(90), *218*, 246(334), 247 (334), 253(324, 325, 328, 330, 333, 335, 336, 621, 649), 256(322), 257(324), 258 (317, 318, 319, 320, 321, 322, 326, 327, 328, 329, 330, 331, 332, 333, 335, 336, 337, 338, 339, 621, 649), 260(323), *287*, *288*, *296*, *297*
Kirby, E. C., 247(340), 271(340), *288*
Kirby, F. G., 13(177), *127*
Kirk, D. N., 152(57), 158(57), *169*, 231(139), *281*
Kirmalova, M. L., 159(124b), 160(125), *171*
Kirme, W., 230(341), *288*
Kirmse, W., 12(170), 14(170), *127*
Kirpal, A., *127*
Kirrman, A., 225(342), 226(342), *288*
Kirsanov, A. V., 17(172, 173), 57(173), *127*
Kiryushkin, A. A., 437(131), *505*
Kishi, Y., 42(112), *125*
Kistiakowsky, G. B., 151(87a), *170*
Kitaeva, S. K., 149(159), 151(159), 160(159), *173*
Klaasens, K. H., 372, 374(10), *415*
Klamann, D., 49(264a), *129*
Klein, F., 5(279), *130*
Klein, J., 158(160), 159(160), 160(160), *173*
Klein, R. J., 51(247a), *129*
Kleiner, H. J., 200(211, 212), *221*, *222*
Kleinjot, O. J., 55(174), *127*
Klemm, K., 188(29), *216*, 446(46, 47), 447 (46, 47), 450(46, 47), *502*
Klimko, V. T., 268(343), *288*
Klimov, E. M., 321(12a), 338(12a), *343*
Klingensmith, C. W., 320(66a), 338(66a), *344*
Klopman, G., 350(99a), *417*
Klyushnik, G. I., 253(622), 259(622), *296*
Knevel, A. M., 197(80), *217*
Knorr, E., 299, *344*
Knott, E. B., 9, *127*, 180(95), 183(95), 186 (95), 190(92), 191(91), 192(91, 92), 196 (12a, 13, 93, 94, 94a, b, c), *215*, *218*, 233 (346), 238(346), 239(346), 240(346, 348, 390, 350, 353), 242(15, 353), 243(14, 349, 350, 351), 246(345, 347, 352), 251(365, 366, 367, 368, 369), 258(257, 258), 271 (344), *278*, *285*, *288*, 445(107), *504*
Knovenagel, E., 160(161), *173*

Knunyants, I. L., 14(176), 61(176), *127*, 260 (354), *288*
Kobayashi, E., 268(355, 428), *288*, *291*
Koch, G., 149(145), 151(145), 160(145), *172*
Kochelov, F., 45(155), *126*, 228(284, 285), *286*
Kochetkov, N. K., 303(66c, 69, 71, 71a, 73), 304(64, 68, 71, 137), 308(65, 66b, 68, 71), 309(93), 320(71, 73, 73a), 321(1d, 12a, 12b, 64, 66d, 68, 71, 73, 73a), 322(1d, 64, 68, 71, 137), 323(66e), 325(64, 68, 69, 71), 327(71a), 328(66c), 336(65, 66, 66c, 67, 68, 69, 70, 71, 71a, 72, 73a), 337(66c, 66d, 71, 71a, 73a), 338(1c, 12a, 71), *342*, *343*, *344*, *346*
Kochevar, I. H., 159(264), *176*
Kocourek, J., 299(130), 301(130), 315(130), *346*
Köcher, E., 253(503), 260(503), *293*
Koelsch, C. F., 183(104), 189(104), *218*
Koenig, B., 365(141, 142), 388(142), 393(141), 394(141), 397(141), 400(141), 402(141), *418*
Koenigs, W., 299, *344*
Kofron, W. G., 13(177), *127*
Kohler, E. P., 227(362), *288*
Kohtz, J., 241(520), 246(520), 271(519, 520), *293*, 406(146), 409(146), *418*
Koizumi, T., 166(211a), *174*
Kolling, G., 149(145), 151(145, 146), 152(146), 160(145), *172*
Kondratenko, N. V., 253(623, 628), 258(623), 259(623, 628), *296*, *297*
Kondyref, N. V., 226(283), *286*
Konig, W., 247(363), 250(364), 253(366, 369), 256(364), 257(369), 258(368, 368a, 369), 260(365, 366, 367), 262, *288*
Kopp, H., 12(178), *127*
Koral, M., 257(370), 258(370), 259(370), *288*
Kormandy, M. F., 182(103), 183(103), *218*
Korner, G., 270(209), *284*
Kornilov, M., 422(5), 423(5), 424(5), 428(5), 450(5), 451(5), 452(5), 454(5), 455(5), 456(5), 458(5), 461(5), 463(5), 465(5), 476(5), 490(5), *501*
Korte, F., 8(179), *127*, 165(142), 167(142), *172*
Korystov, V. I., 258(578), *295*

Korytnik, W., 299(75), 303(75), 316(75), 319(75), 322(75), 326(75), 332(75), 340(75), *345*
Kosakada, A., 39(163a), 58(163a), *126*
Koskikallio, J., 138(162), *173*
Kosloski, C. L., 160(41), *169*
Koster, H., 149(253), 151(253), 152(253), 160(253), *176*
Kotova, L. I., 253(215), 259(215), *284*
Koutcheroff, L. M., 226(371), 227(371), *288*
Kovacic, A., 183(95a), *218*
Kovacs, K., 437(10), *501*
Kovacs, K. G., 41(320), *131*, 164(249), *176*
Kovacs, O. J., 22(180a), 37(180a), 41(180), *127*
Koval, A. A., 56(324a), *131*
Kovalev, G. B., 159(163), *173*
Koyama, T., 14(181), *127*
Kozyrkin, B. I., 159(304), *177*
Krabbe, W., *131*
Krainer, Z. Y., 256(372, 472, 473, 474), *288*, *292*
Kramar, V., 149(141a), 151(141a), 159(141a), *172*
Kramm, D. E., 137(70), *170*
Kranzfelder, A. L., 226(373), 227(373), *288*
Kratochvil, M., 18(160a), 56(160a), *126*
Krbavcic, A., 183(169b), 212(169b), *220*
Kresge, A. J., 138(165), *173*
Krieger, P. E., 207(199), *221*
Krieger, C., 27(51), *123*
Krievoy, M. M., 141(163a), 141(164), *173*
Kritchevsky, D., 226(373a), *288*
Kroger, C. F., 183(97), 203(23, 96, 97), *215*, *218*
Krokhina, S. S., 213(97a), *218*
Kroutzberger, A., 256(374), *288*
Kruber, O., 227(595a), *296*
Krueger, R. A., 14(201), *128*
Kruger, K. E., 16, *122*
Kruger, R. A., 183(105, 106), 184(105, 106), *218*
Krupp, R. G., 180(102), *218*, 445(109), *504*
Kruptsov, B. P., 267(429), 268(429), *291*
Krylova, Y. I., 469(133), 496(133), *505*
Kucherov, V. F., 159(304), *177*, 267(375, 631, 632), 268(630, 631, 632), 269(632), *288*, *297*
Kueffner, K., 257(236), 259(236), *285*
Kuhn, E., 253(454), 256(454), *291*

Kuhn, R., 8(182), *127*, 139(166), *173*
Kukhtin, V. I., 209(101a), *218*
Kulakov, V. N., 159(211), *174*
Kulik, E. Z., 258(389), *289*
Kulikov, S. G., 158(178), 159(178), *173*
Kunce, K., 149(215), 151(215), 159(215), 160(215), *175*
Kundiger, D., *128*, 448(11), *501*
Kundiger, D. G., 31(183, 184), *127*, 163(167), *173*, 274(375a), *288*
Kunert, F., 37(241), *129*, 138(199), 164(199), *174*, 225(402), *290*, 425(113), *504*
Kunze, H., *127*
Kupchan, S. M., 43(185, 186, 187), 44(185), *127*
Kuraishi, T., 212(98), *218*
Kurasawa, E., 183(78), 204(78), *217*
Kurepina, G. F., 253(643), 256(643, 644), 258(645), *297*
Kurtz, P., 3(127), 15(126), *125*
Kuryla, W. C., 35(188), *127*
Kussner, C. L., 210(193), 212(194a), 212(194), 212(194b), *221*
Kwart, H., 14(189), *127*, 138(168), 140(168), 143(168), 144(186), 145(168), *173*

## L

Lacasa, F., 30(190), *127*
Lacher, J. R., 34(269), 57(269), *129*
Ladenberg, A., 149(170), 153(169), 167(169), 168(170), *173*, 229(377), *288*
Ladouceur, B., 45(26), *122*, 228(33), *278*
Lafyatis, P. G., 24(12), 56(12), *122*, 158(17), 162(17), *168*
Lagenauer, C., 143(229a), 145(229a), *175*
Lal, A. B., 253(378, 379), 260(378, 379), *288*, *289*
Lalancette, J. M., 166(170a), *173*
Laland, S., 163(170b), *173*
La Mer, V. K., 138(49, 50, 51), *169*
Lampe, F. W., 141(195), 164(195), *174*
Lander, G. D., 3(191), *127*
Lang, L. K., 22(180a), 37(180a, 320a), 41(180), *127*, *131*, 437(10), *501*
Langenkamp, B., 237(194), *283*
Lanka, W. A., 277(521a), *293*
Lao, A., 20(192), *127*
Lao, A. Y., 145, *173*

Lapkin, I. I., 351(100), 399(100), 401(100), *417*
Lapporte, S. J., 14(193), 40(193), *127*
Larcher, A. W., 351(26), 379(26), *415*
Lardicci, L., *130*, 273(508b, 508c), *293*
Larive, H., 197(99), *218*, 249(381, 382), 254(380, 405, 406, 407), 255(380), 258(381), 259(382), *289*, *290*, 361(111, 112), 389(111), 395(112), 397(111), *417*
Larsen, R. G., 227(362), *288*
Larsen, R. J., 306(84), *345*
Latham, K. G., 4(166), *127*
Laubie, M. J., 485(123a), *504*
Laurent, H., *173*
Laursen, R. J., 135(174), *173*
Lavalle, J. E., 257(370), 258(370), 259(370), *288*
Laves, E., 355(102), 360(102, 103), 367(101, 102), 368(101, 102, 103), 370(102, 103), 386(103), 388(103), 399(101, 102), 400(101, 102), 401(103), *417*
Lavie, D., 43(187), *127*
Lavrova, K. F., *176*
Lawson, A., 196(81), *217*, 243(290), 246(290), *286*
Lawrence, T. F. W., 249(161), *282*
Lawson, J., 5(336), *131*
Lawton, V. D., 196(81), *217*, 243(290), 246(290), *286*
Lawton, W. R., 160(155), *172*
Lazaris, A. Y., 57(194), *127*
Lea, B. A., 259(218), 260(217), *284*
LeCoq, A., 159(172a), *173*
LeDourac, J. C., 485(123a), *504*
Leffler, J. E., 184(138), 199(138), 210(138), 211(138), *219*
Leftwick, A. P., 152(57), 158(57), *169*
LeGuyader, M., 431(108), 459(108), *504*
Lehmann, C., 26(89), *124*
Lehmann, G., 185(100), 186(100), *218*
Leicester, J., 54(352a), *132*
Leifer, A., 257(370), 258(370), 259(370), *288*
Leis, D. G., 35(188), *127*
Lemal, D., 362(104), 410(104), *417*
Lemal, D. M., 45(195), *127*
Lemieux, R. U., 299(76), 303(78, 80, 81, 82, 83), 304(78, 81, 82, 83), 315(77), 321(78, 79, 80, 81, 82, 83), 322(80a, 82), 332(80), 333(79), 335(78), 340(80), 341(80a), *345*
Lempert, K., 159(173), *173*

Leng, J., 260(30), 261(30), *278*
Lenhard, R. H., 27(32), *122*, 135(35), 144(35), *169*
Lenz, A., 12(195a), 55(195a), *127*
Leonard, N. J., 135(174), *173*, 186(136), 190(136), *219*, 236(492), *293*, 306(84), *345*
Leowe, L., 151(14), *168*
LePerchec, P., 149(175), 151(175), 160(175), *173*
Letch, R. A., 226(382a), *289*
Levai, L., 190(101), 191(101), *218*
Levaillant, R., 13(197), *128*, 153(176, 177), 166(176, 177), 167(176, 177), *173*
Levas, E., 159(172a), *173*
Levene, P. A., 316(85), 320(86), 321(87), 323(85), *345*
Levin, Y. A., 209(101a), 213(97a), *218*
Levina, R. Y., 158(178), 159(178), *173*
Levkoev, I. I., 253(629), 255(383), 256(384), 258(383, 385, 386, 387, 388, 389, 390, 391, 392, 565), 259(629), *289, 295, 297*
Levy, L. K., 22(78), 58(78), *124*, 145(86), *170*, 353(41), 354(41), 374(41), 376(41), 377(41), 378(41), *415*
Levy, P. R., 211(201), *221*
Lewis, C. D., 180, *218*, 445(109), *504*
Lewis, I. C., 141(276), *176*
Lewis, T. R., 351(4), 386(4), *414*
Leyva, R. D., 186(79), *217*
Lezkoev, I. L., 257(558), *294*
Liang, Y. T. S., 43(185), 44(185), *127*
Libeer, J., 253(440, 442, 583), 257(440, 441, 442, 443), 258(443), 259(583), *291, 295*
Lica, S., 256(399), 260(399), *289*
Lichtel, K. E., 183(70), 189(68), 193(70), *217*, 445, *503*
Liddicoet, T. H., 277(521a), *293*
Liddle, A. M., 338(9a), *342*
Lienhard, G. E., 151(178a), *173*
Lieske, C. N., 22(79), *124*, 137(87), *170*
Lifshits, E. B., 241(225), 253(629), 259(629), *284, 297*
Lihkosherstov, V. M., 159(192a), *174*
Liisberg, S., 152(179), *173*
Limpricht, H., 14(198), *128*
Lin, H. H., 210(71a, 152, 169a), *217, 220*
Lincoln, L. L., 246(266), 253(266), *285*
Lindahl, R. G., 227(247), *285*
Lindgren, C. R., 40(366), *133*
Lindlar, H., 160(147, 148), *172*

Lindqvist, I., 374(3), *414*
Lindsay, D. G., 151(179a), *173*
Linstead, R. P., 226(382a), *289*
Lions, F., 260(278), *286*
Lippert, W., 167(180), *173*
Lo, C. P., 241(393), *289*
Loder, D. G., 20(199), 32(199), *128*
Loefler, P. K., 183(189), 211(191), *221*
Loev, B., 182(103), 183(103), *218*
Lohaus, H., 225(394), 227(394), *289*
Lohmeyer, H., 83(43a), *123*, 144(37), *169*, 356(17), 358(17), 363(17), 379(17), 386(17), *415*
Lohwasser, H., 183(140), *219*
Long, F. A., 140(188), *174*
Long, L., 158(100), 160(100), *171*
Lonza Elektrizitatswerke, 11(199a), 55(199a), *128*
Lorenz, H., 260(395), *289*
Lorenz, R. R., 183(104), 189(104), *218*
Lovall, A., 151(242), 152(242), *175*
Lovell, C. H., 151(233), 160(233), *175*
Lovett, W. E., 15(341), *132*, 137(278), *176*
Lown, J. W., 149(8), 151(8), 159(8), 160(8), 161(8), *168*
Lu, S.-C., 145(181), 146(181), *173*
Ludman, C. J., 47(199b), *128*
Lukashina, L. I., 246(541), *294*
Lukes, R. M., 149(244), 151(244), 159(244), 160(244), 161(244), *175*
Luk'yanov, A. V., 308(88), 327(88), *345*
Lunazzi, L., 379(104a), *417*
Lunt, J. C., 159(182), *173*
Luthardt, H. J., 430(94, 95, 95a), 463(94), 465(94), 470(94), 498(94), *504*
Lyall, J. M., 210(30a), *216*
Lynch, B. M., 444(110), 458(110), 461(110), *504*
Lyons, R. E., 14(72), *124*

## M

McBane, M. S., 45, *128*
Maccagnani, G., 399(105), *417*
McCarthy, J. R., 135(182a), *173*, 306(89), 318(89), 331(89, 89a), *345*
McCasland, G. E., 366(106), 371(106), 389(106), 397(106), *417*

McCaully, R. J., 137(183), *174*
McDonald, R. A., 14(201), *128*
McDonald, R. M., 183(105, 106), 184(105, 106), *218*
McElvain, S. M., 3(35, 202, 203, 206, 208, 213, 221, 222, 223, 225), 4(202, 206, 208, 214, 217, 222, 223, 225, 226, 227, 230), 5 (213, 222), 6(202, 208, 214, 223, 225, 228), 7(206, 227, 228), 8(202, 225), 11(228), 13 (1), 14(204, 229), 29, 30(216, 230), 31 (216, 223), 32(35, 205, 223), 33(34, 36, 204, 205, 212, 218, 230), 34(209, 212, 218, 219, 220), 45, 46(216, 222), 47(231), 56 (1, 209), 57(1, 204), 59(223, 224), 60(209), 62(223), 63(211), *122*, *123*, *128*, *132*, 145 (184, 185), 149(186), *174*, 190(107, 108), 192(108), *218*, 225(396), 226(396), 227 (397), 228(398), 274(395a), 275(395a, 395b, 395d, 396a, 396b, 396c, 396d, 397a, 398a, 398b, 398c, 398d), 276(395b, 396b, 396c, 398d), 277(47, 395a, 395c, 396, 398c, 398e, 598a), *279*, *289*, *296*, 448 (11), *501*
McGregor, S. D., 45(195), *127*
McIntyre, D., 140(188), *174*
McKay, G. R., 30(216), 31(216), 46(216), *128*, 145(185), *174*, 275(396c), 276(396c), *289*
McKee, M. M., 205(109), *218*
McKee, R. L., 183(195a), 183(195), 205 (109), 210(195), 212(195a), 212(195), *218*, *221*
McKenna, J. F., 7, *128*, 149(189), *174*
MacKenzie, C. A., 150(190), 151(190), 155(190), 158(190), 159(190), 160(190), *174*
McKillop, A., 183(190, 192), *221*
McKusick, B. C., 351(26), 379(26), *415*
McMahon, R. E., 15(295), *130*
McPhillamy, H. B., 301(90), 323(90), 335 (90), *345*
McShane, H. F., 3(217), 4(217), *128*, 275 (396d), *289*
Ma, S. Y., 2(313), 5(313), *131*
Macek, K., 43(349), *132*
Macierewicz, B., 159(187), *174*
Mack W., 364(80), 391(80), 392(80), 394 (80), 396(80), 398(80), 399(80), *416*
Maeda, K., 440(118), *504*
Magerlein, B. J., 324(91), 327(91), *345*

Mah, T. S., 13(312), 55(312), *131*
Maikova, A. I., 213(121a, b, c), *219*
Maillard, J., *132*, 201(110, 111, 205), 203 (110), 205(205), *218*, *221*
Maines, S., 135(35), 144(35), *169*
Maira, S., 153(191), 167(191), *174*
Maire, M., 17(38), *123*, 158(192), 159(192), *174*, 228(49), *279*
Majer, J. R., 21(165), *127*, 260(134), *281*, 351(96), 374(96), 401(96), 408(97), *417*
Majewski, T. E., 449(72a), *503*
Makin, S. M., 159(192a), *174*, 267(429), 268 (429), *291*
Malleis, O. O., 186(50), *216*
Mamaev, V. P., 441(165a), *506*
Mamalas, P., 183(112), 198(112), *218*
Mamlok, L., 161(194), *174*
Mancini, F., 151(193), 152(193), *174*
Mandacescu, L., 256(399), 260(399), *289*
Mann, M. J., 225(295), 227(295), *286*
Mansour, A. K. E., 43(259), *129*
Mantica, E., 273(486a), *292*
Mantione, R., 226(399a), 227(399a), *289*
Maone, T., 435(125), 460(125), 469(125), 478(125), *504*
Marenets, M. S., 253(624), 258(624), *296*
Mariani, B., 257(35), 258(121), 260(34, 130, 400), *279*, *281*, *290*
Mariano, J. P., 269(453a), *291*
Marisco, J. W., 210(64), *217*
Markl, G., 73(301), *130*
Markovits-Kornis, R., 159(173), *173*
Marquet, A., 161(194), *174*
Marshall, D. R., 159(156), *173*
Marshall, H., 40(366), *133*
Marszak, I., 265(173b, 173c, 400a), 266 (173b), *283*, *290*
Marth, J., 212(206), *221*
Martin, D., 49(232a, 233), 57(232a, 233), *128*, 461(111), *504*
Martin, E. L., 258(401), *290*
Martin, G. J., 226(399a), 227(399a), *289*
Martin, M. L., 226(399a), 227(399a), *289*
Martin, R., 198(112a), *218*
Martin, R. H., 141(195), 164(195), *174*
Martini, C. M., 351(4), 386(4), *414*
Martirosyan, G. T., 47(16), *122*
Marvell, E. N., 159(4, 5), *168*
Marx, M., 27(330, 331), *131*, 135(263), 144 (263), *176*

Marx, R., 359(18), 371(18), 374(18), 375(18), 377(18), 379(18), 384(18), 386(18), 388(18), 391(18), 405(18), 412(18), *415*
Mason, G., 167(196), *174*
Masuda, M., 253(452), 256(452), *291*
Masuya, H., 238(412a), *290*
Mathews, H. R., 198(112a), *218*
Matsoyan, S. G., 20(234), 32(234), 62(234) 63(234), *128*
Matsuko, K., 187(128), *219*
Matthey, G., 270(240), *285*
Mattocks, A. R., 53(235), *128*, 137(197), *174*
Mauer, S., 27(32), *122*
Mautner, H. G., 211(113), *218*
Mayer, R., 360(48), 361(48, 107), 379(107), 382(48), 383(48), 384(47a, 48), 386(48), 388(48), 391(48), 396(48), *416*, *417*
Mayer, W., 35(300a), *130*, 137(237a), *175*
Mayor, J. H., 406(98), 407(98), *417*
Mazurek, M., 303(92), 304(92), 323(92), 324(92), *345*
Meek, E. G., 151(198), *174*
Mecke, R., 381(108), *419*
Meerwein, H., 3(236), 4(236), 5(236), 6(236), 13(237), 25(238, 244, 245), 37(241), 38(239), 39(240), 42(243, 244, 245), 55(245), 60(242), 62(244), *129*, 138(199, 200, 202), 140(202), 145(200), 164(199, 200, 201), *174*, 225(402), *290*, 360(109), 361(110), 382(109), 387(110), 412(109), *417*, 421, 422(115), 424(115), 425(113), 426(115), 427(115), 428(115), 431(114), 432(115), 440(115), 450(115), 451(115), 452(114, 115), 454(115), 456(114, 115), 457(115), 458(115), 460(115), 462(115), 475(115), 476(115), 477(115), 478(115), 481(115), 490(115), 491(112), 492(115), 495(115), 499(116), 500(114, 115, 116), *504*
Mees, C. E. K., 245(403, 404), *290*
Mehrotra, R. C., 20(261), 22(246), 25(246), *129*
Meier, R., 34(247), *129*
Meier, W., 251(369), 253(369), 257(369), 258(369), *288*
Melnikoff, A., 363(36), 375(35, 36, 37), *415*
Mendelsohn, H., 151(262), 152(262), *176*
Meresz, O., 308(42), 323(42), 334(42), *343*

Merlini, L., 226(496), *293*
Merten, R., 200(206b), *221*, 425(145b), 429(145b), 455(145b), 456(145b), 458(145b), 462(145b), *505*
Mertes, M. P., 135(202a, b), *174*, 306(92a, 92b), *345*
Mester, L., 315(148), *347*
Metzger, C., 436(117), 472(117), *504*
Metzger, J., 254(405, 406, 407), *290*, 361(111, 112), 389(111), 394(63a), 395(112), 397(111), *416*, *417*
Meyer, E. W., 151(154), 152(154), *172*
Meyer, J. T., 186(50), *216*
Meyer, R. B., 207(18a), *215*
Meyer, R. F., 183(114), 194(114), *218*
Michael, A., 149(203), 151(203), 159(203), *174*
Michaeldidis, C., 260(408), *290*
Michaelis, H., 258(289), *286*
Michaels, R. J., 8(374), *133*, 433(166), 434(169, 170), 463(170), 469(166), 470(166), 471(166), 477(166), *506*
Middleton, E. B., 256(410), 258(409), 260 411), *290*
Mier, J. D., 152(71), *170*
Miginiac-Groizeleau, L., 230(412), *290*
Mikhailenko, F. A., 194(90), *218*, 256(322), 258(322), *287*, 491(6), *501*
Mikhailov, B. M., 160(204, 205), 165(13), 167(13), *168*, *174*
Mikhailyuchenko, N. K., 260(422), *290*
Miki, T., 28(254), *129*, 238(412a), *290*
Milas, N. A., 51(247a), *129*
Milbrand, H., 162(112), *171*
Miller, E. J., 301(37), 303(37), 319(37), 323(37), *343*
Miller, H. F., 225(413), 227(413), *290*
Miller, T. G., 35(248), *129*
Mills, J. A., 299(75), 303(75), 316(75), 319(75), 322(75), 326(75), 332(75), 340(75), *345*
Mills, W. H., 392(112a), 397(112a), *417*
Minasyan, R. B., 47(16), *122*
Miroshnichenko, Z. I., 253(414), 263(414), *290*
Mirskova, A. N., 54(325a), 57(325a), 61(325a), *131*
Mirviss, S. B., 33(218), 34(218), *128*
Mirzaev, A. M., 149(214), 151(214), 158(214), 159(214), *175*

Mizuno, Y., 184(115), 186(115), 187(115), *218*, 256(416), 263(415), 264(415), *290*
Mkhitaryan, V. G., 19, 20(249), 24(250), 56 (250), *129*
Mochalin, V. B., 227(416a), 265, *290*
Mochalina, E. P., 14(176), 61(176), *127*
Mochel, W. E., 21(121, 251), *125*, *129*, 412 (65, 113), *416*, *417*
Moedritzer, K., 29(252), *129*
Möhrle, H., 43(13), 44(13), *122*
Mokova, A. P., 209(101a), *218*
Molosnova, V. P., 17(172, 173), 57(173), *127*
Montavon, M., 160(147, 148, 149), *172*
Montgomery, J. A., 183(121, 195, 195a, 195b), 184(121), 209(121), 210(116, 117, 118, 119, 120, 121, 195), 212(194, 194a, 194b, 195, 195a), *218*, *221*
Montgomery, R., 303(20), 319(19, 20), 324 (20), 325(19, 20), 335(19), 338(19), *343*
Mooney, E. F., 375(35), *415*
Moore, R. F., 43(253), *129*
Moravec, J., 306(92c), *345*
Morgan, A. R., 303(81, 82, 83), 304(81, 82, 83), 321(81, 82, 83), 322(82), *345*
Morgenthau, J. L., 81(60, 61), *123*, 231 (125, 126, 127), *281*
Morita, K., 28(254), *129*, 149(207, 208), 151 (207, 208), 159(207), 160(207, 208), 161 (207), *174*, 274(416b, 416c, 416d), *290*
Morita, K. I., 149(209), 160(206, 209), *174*
Morley, C. G. D., 161(209a), *174*
Morris, L. R., 34(219, 220), *128*, 227(397), *289*
Morrow, D., 5(336), *131*
Morschel, H., 138(202), 140(202), 164(201), *174*, 499(116), 500(116), *504*
Moshchitskii, S. D., 17(156, 157), *126*
Moskalenko, Z. I., 258(416e), *290*
Moskaleva, R. N., 253(648), 258(646), 259 (647, 648), *297*
Moskva, V. V., 166(234), *175*, 213(121a, b, c, 136b, c), 214(136c), *219*
Moubacher, R., 43(321), *131*
Moulin, F., 6(255a), 11(255a), *129*
Moureu, C., 227(46, 417, 418), *279*, *290*
Mousseron, M., 227(419), *290*
Moyer, J. D., 314(61), 323(60), *344*
Mozzarachio, P., 256(140a), 258(140a), 259 (140a), *281*
Muehlmann, F. L., 212(122), *219*

Mueller, A., 314(46), 315(46), *343*
Muffler, H., 446(28a), *501*
Muhlmann, R., 260(59), *279*
Mukaiyama, T., 182, *219*, 361(114, 115), 390(114), 407(114), *417*, 431(117a), 459 (117a), *504*
Muller, J., 253(100), 257(100), *280*
Mullineaux, R. D., 3(221), *128*, 275(397a), *289*
Mumm, O., 254(420), *290*
Murakami, H., 212(123a), *219*
Mushkalo, L. K., 260(421, 422, 423), *290*
Mustafa, A., 43(256, 257, 258, 259, 322), *129*, *131*
Mutscher, M., 183(97), 203(96, 97), *218*
Mylo, B., 226(617), *296*

N

Nader, F., 225(173aa), 226(173aa), *282*
Nadolski, K., 49(233), 57(233), *128*
Nagai, M., 431(120), 451(120), 459(120), 496(120), 497(120), *504*
Nagasaki, A., 48(260), *129*, 268(428), 272 (424, 425, 426, 427), *290*, *291*
Nakagome, T., 212(123a), *219*
Nakamura, S., 272(424, 425), *290*, 440(118), *504*
Nakano, H., 187(124), *219*
Nanney, P., 159(42), *169*
Naqui, S., 212(168), *220*
Narain, R. P., 20(261), 22(246), 25(246), *129*
Nasimi, G., 256(156), *282*
Naumov, Y. A., 258(2), *278*
Nawa, H., 28(254), *129*
Nazarov, I. N., 267(429), 268(429), *291*
Nealy, D. L., 37(361), *132*, 144(296), *177*
Neath, G., 211(39a), *216*
Nebel, I., 149(145), 151(145), 160(145), *172*
Nedelec, L., 28(33), *123*
Nef, J. U., 13, 36(261a), *129*
Neggiani, P. P., 272(461c), 273(487), *292*
Nehrung, R., 464(128), *505*
Neiderlein, R., 45(43b), *123*, 153(38), 167 (38), *169*, 356(19), 399(19), *415*
Nelson, J. W., 3(222), 4(222), 5(222), 6, 45 (222), 46(222), *128*, 228(398), *289*
Nemirovskii, V. D., 226(108), *281*

Ness, R. K., 303(93, 98), 305(96, 97), 308 (94), 309(18), 314(93, 95), 319(18), 320 (18, 93, 96, 97), 323(98), 325(94), 336(18, 97), 339(97), 340(97), *343, 345*
Nesterov, L. V., 166(210), 168(210), *174*
Neubauer, G., 61(91), *124*, 427(74), 451(74), 456(74), *503*
Neunhoeffer, H., 183(125a), *219*
Neunhoeffer, O., 253(430), 258(431), 259 (430), 260(432), *291*
Newman, M. S., 53(263), *129*, 152(273), *176*
Newth, F. H., 299(41), *343*
Nicholl, L., 233(433), 236(433), 237(433), *291*
Nichols, J., 224(530), 225(530), 227(530), *294*
Nickel, S., 365(143), 403(143), *418*
Nicolaus, R. A., 48(315a), *131*, 137(245), *176*, 276(516a), *293*, 436(125a), *505*
Nicoletti, R., 30(264), *129*
Nida, T., 315(54a), *344*
Nielson, D. G., 5(302), *130*
Nifatev, E. E., 159(211), *174*
Niki, S. N., 184(60a), *217*
Nishida, K., 244(453), 246(453), *291*
Nishimura, M., 149(207), 151(207), 159 (207), 160(206, 207), 161(207), *174*, 184 (115), 186(115), 187(115), *218*, 263(415), 264(415), 274(416b), *290*
Nisi, C., 183(155, 157), 184(155), 188(155, 157), 201(157), 202(155), 206(156), 208 (155), 212(155), *220*
Noble, P., 463(15), *501*
Noell, C. W., 210(134), *219*
Noelting, E., 270(434, 435), *291*
Nohira, H., 361(115), *417*
Nomura, M., 187(127, 128), *219*
Noppel, H. E., 183(66), *217*
Normant, H., 226(399a), 227(436), *289, 291*
Novello, F. C., 183(125), 189(125), 190(125), 191(125), 207(125), *219*
Noyari, R., 430(119), 451(119), 452(119), *504*
Noyes, P. R., 20(298), 20(298), *130*, 181 (148), 183(148), *219*
Nozaki, H., 430(119), 451(119), 452(119), *504*
Nrdel, F., 49(264a), *129*
Nukina, J., 48(260), *129*
Nukina, S., 272(427), *291*
Nurenbach, A., 213(54b), *216*
Nys, J., 186(203), 212(204), *221*, 247(145, 146, 438), 253(440, 442), 257(439, 440, 441, 442, 443), 258(437, 439, 443), *282, 291*

## O

Oae, S., 22(265), *129*, 200(126), *219*, 227 (444), *291*, 353(116), 354(116), 373(116, 117), 377(116, 117, 118), 379(116, 117, 118), 405(116, 117), *417, 418*
Ochiai, M., 431(120), 451(120), 459(120), 496(120), 497(120), *504*
Oda, R., 48(260), *129*, 187(127, 128), *219*, 225(560aa), 272(424, 425, 426, 427), *290, 291, 294*
Odom, H. C., 159(279), *176*
Ogata, T., 192(129), *219*, 244(453), 246 (453), 253(452), 256(445, 450, 452), 261 (448), 262(446, 447, 449, 451), 263(447, 449, 450, 451), 264(446, 447, 449, 450), *291*
Ohle, H., 314, *345*
Ohme, R., 11(265a), *129*
Ohno, A., *129*, 166(211a), *174*, 200(126), *219*, 227(444), *291*, 353(116), 354(116), 373(116, 117), 377(116, 117, 118), 379 (116, 117, 118), 405(116, 117), *417, 418*
Oishi, M., 431(120), 451(120), 459(120), 496 (120), 497(120), *504*
Okamoto, Y., 140(54), *169*
Olds, W. F., 137(183), *174*
Oles, S. R., 422(158), 453(158), *505*
Oliveto, E. P., 28(323e), *131*, 135(254), 136 (254), 144(254), *176*
Olofson, R. A., 269(453a), *291*
Olofson, K., 411(118a), *418*
Olschwang, D., 192(56), *216*, 354(50b, 50c, 50d), *416*, 422(84), 423(84), 428(84), 451 (84), 452(84), 454(84), 461(84), *503*
Olsson, K., 350(56, 119, 121, 121a, 122), 353(122), 368(56), 374(56, 121), 375(119), 376(119, 120), 377(121a), 378(119, 120), 379(122), 380(121a, 122), 381(119, 120, 122), 383(121a, 122), 385(120, 121a), 387 (122), 390(120), 392(122), *416, 418*
Omran, S. M. A. E., 43(259), *129*
Opanasenko, E. P., 254(475), 256(475, 476, 477, 478, 479, 480, 481, 483), 257(477), *292*

Ordogh, F., 183(129b), *219*
Orlando, C. M., *127*, 277(313a), *287*
Osbond, J. M., 23(266), 31(266), *129*, 135(212), *175*
Oscapowicz, J., 159(187, 213), *174*, *175*
Oszczapowics, J., 258(547), *294*
Ott, H., 436(121), 437(98), 457(121), 465(98, 121), 466(96, 98), 467(96, 98), 470(96, 98), 473(96, 98), 474(96, 98), 477(121), *504*
Ouannes, C., 161(194), *174*
Ovchinnikov, Y. A., 437(131), *505*
Overend, W. G., 163(170b), *173*
Owen, L. N., 366(1a), 394(1a), *414*
Ozansoy, M., 151(14), *168*

**P**

Pacák, J., 299(130, 131), 301(130), 315(130, 131), *346*
Pacsu, E., 41(267), *129*, 299(106), 303(104, 108), 314(106), 315(106, 109), 326(104, 105), 327(102, 108, 109), 332(103, 106), 333(103, 106), 339(107), *345*
Pailer, M., 253(454), 256(454), 258(455), *291*
Palling, S. J., 406(45), *416*
Pallos, L., 183(22, 129b), 208(129a), *215*, *219*
Palomaa, M. H., 226(456), *291*
Panouse, J. J., 42(268), *129*
Pandit, V. K., 439(57), *502*
Panizzi, L., 237(457, 458), *291*
Panova, N. I., 351(100), 399(100), 401(100), *419*
Papendick, V., 8(374), *133*, 434(170), 463(170), *506*
Parfentev, L. N., 149(214), 151(214), 158(214), 159(214), *175*
Parham, W. E., 149(215, 216), 151(215, 216), 159(215), 160(215, 216), *175*, 233(221), 234(221), 237(221, 459), *284*, *291*
Park, J. D., 34, 57(269), *129*
Parrish, F. W., 158(100), 160(100), *171*
Parshikov, N. G., 158(178), 159(178), *173*
Partridge, M. W., 4(166), *127*, 206(35), 212(35, 130), *216*, *219*
Pascual, J., 30(190), *127*
Passalacqua, T., 236(460), *292*
Pasternak, Y., 226(461), 227(461), *292*
Pastour, P., 159(217), *175*
Patchett, A. A., 307(121), 311(121), *346*

Pater, R., 362(25), 376(25), 377(25), 384(25), *415*
Patton, D. S., 135(123a), 160(123a), *171*
Paudler, W. W., 212(131), *219*
Pauer, F., 8(270), *130*
Paulmier, C., 159(217), *175*
Paulsen, H., 302(15a, 109a, 109b, 110, 111, 112), 303(50), 310(15a, 109a, 109b, 111, 112), 316(50), 321(50), 325(15a, 111, 112), 327(50), 340(15a), 340(109a, 109b, 110, 111, 112), *343*, *344*, *345*, *346*
Pauly, H., 159(218), *175*
Pearl, I. A., 5(271), *130*
Pearson, R. G., 139(116), 160(116), *171*, 350(122a), *418*
Peltier, D., 431(108), 459(108), *504*
Penner, H. P., 363(123), 376(123), 381(123), 382(123), 385(123), 387(123), *418*
Perdok, W. G., 375(124), *418*
Perkouskaya, E. K., 258(557), *294*
Perlin, A. S., 303(92, 113), 304(92), 305(29, 31), 307(29), 323(92, 113), 324(92, 113, 114), 325(32), 327(31), 332(31), 333(114), 335(114), 336(14), *343*, *345*, *346*
Perry, C. A., 260(135), *281*
Peshkar, L., 183(73), *217*
Pesin, V. G., 197(131a), *219*
Pessina, R., 269(296a), *286*
Petersen, J., 235(154), 237(154), *282*
Petrenko, O. E., 256(479, 480, 481, 482, 483), *292*
Petrenko-Kritchenko, P., 152(219), *175*
Petrov, A. A., 226(108), *281*
Petrow, V., 24(272), 59(272), *130*, 152(57), 158(57), *169*, 183(112), 198(111, 112), *215*, *218*
Pfankuch, E., 5(148), *126*
Pfleiderer, W., 210(32), *216*
Pfeiffer, G. J., 155(220), 156(220), 157, 159(220), 160(220), *175*
Pfeiffer, M., *175*
Pfundt, G., 473(77a), *503*
Philpott, P. G., 23(266), 31(266), *129*, 135(212), *174*
Piacenti, F., 20(273, 273a, 303a, b), 55(273), *130*, 272(461c), 273(461a, 461b, 486a, 487, 508b, 508c), *292*, *293*
Pianka, M., 257(31), 258(31), *278*
Pigman, W. W., 314(115), 325(115), 338(115), *346*

Pike, R. M., 13(274), *130*, 165(221), 167(221), *175*
Pilyugin, G. T., 250(485), 253(466, 467, 475), 256(109, 110, 111, 112, 462, 463, 463a, 464, 465, 466, 467, 468, 469, 470, 471, 472, 473, 474, 475, 476, 477, 478, 479, 480, 481, 482, 483, 484, 507), 257(477), *281*, *292*, *293*
Pinchas, S., 48(31), *122*
Pinner, A., 2(275, 276, 278), 3(277),5(279), 8, 55(276), 57(275), *130*
Pino, P., 20(273a, 303a, b), *130*, 273(461b, 486, 486a, 487, 508b, 508c), *292*, *293*
Pirsch, J., *130*
Pischel, H., 135(221a, 221b), 137(221c), *175*, 306(115a, 115b), 307(115c), *346*
Piskala, A., 188(132, 133), 207(132, 133), 208(131b, 132a), *219*, 483(122), 486(122), *504*
Pleininger, H., 239(489), 246(488, 489), *292*
Pletcher, T., 142(222), *175*
Plotkinova, G. I., 353(149), 379(149), *419*
Plucket, H., 28(323e), 42(268), *129*, *131*, 135(254), 136(254), 144(254), *176*
Pogosyan, G. M., 20(234), 32(234), 62(234), 63(234), *128*
Pokrovskaya, K. I., 257(558), *294*
Polansky, O. E., 238(48), *279*
Poletto, J. F., 148(7), *168*
Pollack, M. A., *172*
Pollitzer, E. L., 12(136), *126*
Polyakov, A. I., *176*
Poon, L., 444(110), 458(110), 461(110), *504*
Poos, G. I., 149(244), 151(244), 159(244), 160(244), 161(244), *175*
Poppe, E. J., 253(211), 258(211), *284*
Porck, A., 308(47), 323(47), *343*
Porkert, H., 454(52), 489(52), *502*
Porter, J. J., 13(137), *126*, 355(71), 374(71), *416*
Portnaya, B. S., 258(385), *289*
Portnyagina, V. A., 253(577), 258(577), *295*
Portsmouth, D., 307(5), *342*
Post, H. W., 13(281), 15(344), 16(344), 28(281), 45(280), 52(282), 57(344), *130*, *132*, 153(224), 154(226), 155(225, 227), 159(223, 225), 160(225), 167(224), *175*, 180(102), 183(197), 189(197), 190(197), *218*, *221*, 232(490), 233(490), 237(490), 253(490), *292*, 349(125), 353(127), 379(125, 127), 385(125), 413(125, 126), *418*, 445(109), *504*
Povarov, L. S., 160(204), *194*
Powell, W. J., 13, *126*
Praefke, K., 241(520), 246(520), 268(24, 25), 271(517, 518, 519, 520), *278*, *293*, 374(145), 399(145), 401(145), 406(146), 409(144, 145, 146), *418*
Prasad, R. N., 210(134), *219*
Preobrazhenskii, N. A., 226(491), *293*
Preston, J., 165(229), 166(229), *175*
Preto, R. J., 138(165), *173*
Price, A. K., 154(128a), *172*
Price, C. C., 3(46), *123*, 186(136), 190(136), 191(135), 191(135), 210(71a, 169a), *217*, *219*, *220*, 236(492), *293*
Price, M., 143(229a), 145(229a), *175*
Price, M. B., 14(189), *127*, 138(168), 140(168), 143(168), 144(168), 145(168), *173*
Princivalle, E., 253(222), 260(222), *284*
Proskurina, T. S., 54(325a), 57(325a), 61(325a), *131*
Prostenik, M., 258(224), *284*
Protopopova, T. V., 268(493), *293*, 447(141), 448(141), 450(141), 452(141), 457(141), 472(141), 478(141), *505*
Prystas, M., 237(494), *293*
Pummerer, R., 159(297), *177*, 262(612a), *296*
Pyrkin, R. Y., 213(97a), *218*

Q

Quast, H., 390(127a), 392(127a), 395(127a), *418*, 444(123), 449(123), 463(123), 465(123), *504*
Quelet, R., 159(230, 231), *175*, 225(495), 227(495), *293*
Querry, M. U., 205(18d), *215*
Quilico, A., 226(496), *293*

R

Raap, R., 46(70), 61(70), *123*, 149(75), 151(75), 160(75), *170*
Raasch, M. S., 372(127b), *418*
Rabe, A., 300, 303(17), 323(17), *343*
Rabe, S., 17(18), *122*, 144(21), *168*
Rabinowitz, I., 3(132), *126*
Rackow, S., 49(233), 57(233), *128*

Rainer, G., 484(33), 486(33), *502*
Rajagopalan, P., 212(136a), *219*
Ralph, R. K., 236(497), *293*
Ramberg, R., 205(207), *221*
Ramirez, C., 186(79), *217*
Ramsey, B. G., 14(283), 57(283), *130*, 138(232), *175*
Rao, Y. S., 242(498), *293*
Raphael, R. A., 227(499), *293*
Rapoport, H., 151(233), 160(233), *175*, 198(112a), *218*, 444(88), *503*
Rath, P. C., 246(500), *293*
Rathbone, R. J., 159(131), *172*, 180(72), 183(72), *217*, 260(259, 260), 262, 263(261), *285*
Rathore, B. S., 230(42), *279*
Rathsam, G., 2(333), *131*, 165(266), *176*, 267(537), 274(537), *294*
Ravindranathan, R. V., 183(192), *221*
Raymond, A. L., 299(116), *346*
Razumov, A. I., 166(234), *175*, 213(121a, b, c, 136b, c), 214(136c), *219*
Razvadovskaya, L. V., 260(354), *288*
Reade, J. H., 253(256), 256(256), 260(256), *285*
Rebsdat, S., 422(53), 423(53), 428(53), 432(53), 441(53), 442(53), 445(53), 450(53), 451(53), 452(53), 453(53), 454(53), 455(53), 456(53), 458(53), 476(52a), 477(135), 488(52a, 52b, 52c), 499(135), *502*, *505*
Redeman, C. R., 64(284), *130*
Reed, L. J., 233(221), 234(221), 237(221, 459), *284*, *291*
Rees, W. W., 53(114), *125*
Reese, C. B., 135(115, 115a, 128, 235), 136(150), 137(115, 127a, 128, 235), 144(115, 128, 150, 235), *171*, *172*, *175*, 306(26, 27, 33, 34, 117), 306(27), 331(35), 334(62, 117), *343*, *344*, *346*
Reese, E. B., 151(179a), *173*
Regan, C. M., 137(236), *175*
Reginier, G. L., 485(123a), *504*
Rehn, H., 484(25), *501*
Reid, D. H., 247(340), 271(340), *288*
Reid, E. E., 349, *418*
Reiman, P. A., 137(183), *174*
Reinfeld, E., 226(281), *286*
Reinmuth, O., 227(315), *287*
Reist, H. R., 151(233), 160(233), *175*
Reitter, H., 2(286), *130*, 274(500a), *293*

Reitzenstein, F., 186(137), 187(137), *219*
Remberger, H., H., 151(95a), 159(95a), *171*
Rempfer, H., 188(29), *216*, 446(46, 47), 447(46, 47), 450(46, 47), *502*
Renner-Kuhn, E., 258(455), *291*
Resemann, W., 246(56), *279*, 487(34), 494(34), *502*
Reynolds, D. D., 303(134), 327(134), *346*
Reynolds, G. A., 212(7, 8, 9), *215*
Ricca, A., 270(104, 106), *280*, *281*
Rich, F. V., 303(108), 327(108), *345*
Richardson, E. E., 274(375a), *288*
Richman, J., 208(200), *221*
Richter, E., 183(139), 184(138), 199(138), 210(138), 211(138), 212(139), *219*
Richter, R., 449(123c), 450(123b), *504*
Richtmeyer, N. K., 314(118), *346*
Ridi, M., 246(501), *293*
Rieche, A., 16(117), 18(118), 25(289), 29(117), 50(287, 289, 290, 319), 51(288, 289, 291, 292, 323a, 323c, 323d), 52(117), 55(117), 56(323a), 59(292, 319), 61(117), *125*, *130*, *131*, 270(240), *285*, 356(62), 379(62), *416*
Ried, W., 183(140), 197(142), 198(142), 209(141), 210(142), *219*, 253(503), 258(502), 260(503), *293*, 437(123d), *504*
Riedel, H. W., 258(249), 260(249), 265(249), *285*
Riester, O., 253(300), 257(129), 258(300, 504), *281*, *286*, *293*
Riley, J. G., 490(86a), *503*
Ringer, O., 138(260), *176*
Ringler, I., 27(32), *122*
Rinzema, L. C., 21(293), *130*, 351(129), 352(129), 353(129), 382(129), 384(129), 390(129), 391(129), 392(129), 410(129), 412(129), *418*
Rist, C. E., 310(127, 128), 311(13), 312(14), 323(13, 127, 128), 327(14), 334(127), 338(127), *343*, *346*
Ritchie, E., 23(139), *126*
Ritvay-Emandity, C., 190(101), 191(101), *218*
Rix, T. R., 34(294), *130*, 420(3), *501*
Roberts, J. D., 15(295), *130*, 137(236), 154(237), *175*, 226(634), *297*
Roberts, M. E., 186(47), *216*
Roberts, R. M., 9(296), 20(298), 33(297), *130*, 138(82), *170*, 180(144, 145), 181(147,

148, 150), 183(144, 148, 149, 150), 186 (143, 146), 188(143), 189(145), 193(149, 151), *219*, *220*, 301(119), *346*, 445(124), *504*
Robins, M. J., 135(182a), *173*, 306(89), 318 (89, 331(89, 89a), *345*
Robins, R. K., 135(182a), *173*, 210(134, 151a, 152), *219*, *220*, 306(89), 318(89), 331 (89, 89a), *345*
Robinson, C. A., 159(255), *176*
Robinson, C. H., 28(323e), *131*, 135(254), 136(254), 144(254), *176*
Robinson, R., 159(73), *170*, 364(84), 395 (84), 397(84), *417*
Rockwell, D. M., 185(153), 197(153), 198 (153), *220*
Rodova, F. Z., 258(578, 580), *295*
Roedig, A., 14(300), 35(300a), 73(301), *130*, 137(237a), *175*, 231(505), *293*
Roehr, J., 356(20), 371(20), 374(20), 375 (20), 277(20), 384(20), 401(20), 410(20), 413(20), *415*
Roger, R., 5(302), *130*
Rogovik, M. I., 256(109), *281*, 256(506, 507), *293*
Rohwedder, R. K., 312(126), 327(126), *346*
Roof, G. L., 362(129a), 375(129a), 384 (129a), 390(129a), 402(129a), 403(129a), 413(154), *418*, *419*
Rose, B., 15(303), *130*
Rosencrantz, D. R., 22(79), *124*, 137(87), *170*
Rosenoff, A. E., 258(508), 259(508a), 273 (508a), *293*
Rossi, R., 20(303a, b), *130*, 273(508b, 508c), *293*
Rotdy, B., 183(129b), *219*
Roth, C. B., 258(509), *293*
Rothstein, E., 413(129b), *418*
Roudolf, R. J., 226(283), *286*
Rouessac, F., 149(175), 151(175), 160(175), *173*
Rouet, J., 274(223a), *284*
Roussel-UCLAF, 27(304), 27(304, 305, 306), *130*, *131*
Rout, M. K., 246(500), 258(141), 260(561), 261(561), *282*, *293*, *295*
Rowe, F. M., 253(510), *293*
Rowland, B. I., 193(218), *222*
Royals, E. E., 151(240), 155(239), *175*

Rozum, Y. S., 253(511), 259(511), *293*
Rudenko, B. A., 267(631), 268(631), *297*
Rudzik, A. D., 208(200), *221*
Ruechardt, C., 313(120), *346*
Ruegg, R., 160(147, 148, 149), *172*
Ruggieri, P., 151(99), 152(99), 160(99), *171*
Ruh, R. P., 34(307), *131*
Rumscheidt, C., 149(48), 151(48), 160(48), *169*
Runti, C., 183(155, 157, 158), 184(155), 187 (154), 188(155, 157), 201(157), 202(155), 203(154), 204(154), 205(154), 206(156), 207(158), 208(155), 212(155), *220*
Rusche, J., 18(119), 25(119), 66(118a), *125*, 162(129), *172*, 398(63), *416*, 441(92), 466 (92), *503*
Ruske, W., 34(308), 57(308), *131*
Russell, C. R., 310(127, 128), 311(13), 312 (14), 323(13, 127, 128), 327(14), 334(127), 338(127), *343*, *346*
Russell, P. B., 150(241), 151(241), *175*, 207 (158a), *220*, 233(512), 236(512), *293*
Russo, F., 207(60b), 214(60b), *217*
Ruyle, W. A., 151(242), 152(242), *175*
Ruyle, W. V., 307(121), 311(121), *346*
Ruzicka, L., 160(243), *175*
Ryan, K. J., 312(121a), *346*
Rynbrandt, D. J., 159(264), *176*
Ryser, G., 160(147), *172*


**S**

Saam, J. C., 197(159), *220*
Sabata, B. K., 246(500), *293*
Sadler, P. A., 436(98), 465(98), 466(98), 467 (98), 470(98), 473(98), 494(98), *504*
Sah, P. P. T., 2(309, 311, 313), 3(309), 5 (309, 311, 313), 8(311), 13(312), 14(310), 55(312), 57(309), 61(309), *131*, 187(159a), *220*, 225(514), 227(514), 228(513), *293*
Saikachi, N., 186(177), 187(177), *220*
Salmon, M. G., 207(60b), 214(60b), *217*
Samaras, N. T., 138(134), *172*
Samen, E., 349(132), 360(132), 368(132), 369(132), 370(130, 132, 133, 134, 136, 137), 372(134, 135, 136), 374(134), 375 (130, 132), 379(131, 136, 137), 385(132), 390(132), *418*

Samuel, E. L., 15(143a), 82(93b), *124*, *126*, 355(46), 356(47), 386(47), 387(47), 391(47), 402(46), 405(46), 412(46), *416*
Samuels, H., 303(38), 315(38), 323(38), 332(38), 333(38), *343*
Sandri, E., 256(140a), 257(8), 258(140a), 259(140a), *278*, *281*
Sandstrom, J., 201(160), *220*
Santacroca, C., 48(315b), *131*, 353(138a), 385(138a), 414(138a), *418*, 435(126), 457(126), 458(156), 460(126), 478(126), *505*
Santos, A. G., 237(286), *286*
Santroch, J., 43(314), *131*
Sarett, L. H., 149(244), 151(244), 159(244), 160(244), 161(244), *175*
Sarkisyants, S. A., 33(370), 55(370), *133*
Sasada, Y., 43(340b), *132*
Sasaki, K., 26(315), *131*, 187(175), *220*
Sasse, H. J., 37(242), 60(242), *129*, 138(200), 145(200), 164(200), *174*, 421(114), 431(114), 452(114), 456(114), 500(114), *504*
Sasse, J., 360(109), 382(109), 412(109), *417*
Sataki, K., 186(178), 187(178), *220*
Sathe, V. R., 243(515), *293*
Sato, K., 182, *219*
Sauer, J. C., 45(150, 151), *126*, 265(273, 274, 275), 266(274), *286*, 408(79), *416*
Saunders, M., 25(339), 56(339), *132*, 231(545), *294*
Sauter, F., 183(160aa), *220*
Sauter, R., 212(26), *216*, 486(35), 487(35), 488(35), *502*
Savard, J., 227(516), *293*
Sayigh, A. A. R., 449(145aa), *505*
Scanley, C. S., 351(138), 369(138), 393(138), 400(138), 401(138), 402(138), 404(138), *418*
Scarpati, R., 48(315a, 315b), *131*, 137(245), *176*, 276(516a), *293*, 353(138a), 385(138a), 414(138a), *418*, 435(125, 126), 436(125a), 457(126), 458(126), 460(125, 126), 469(125), 478(125, 126), *504*, *505*
Schacke, B., 259(528), *294*
Schade, W., 51(291), *130*
Schaeffer, H. J., *131*, 135(245a), *176*, 210(160a, 161), *220*
Schafer, H., 183(66), *217*
Schafer, K., 31(107), 379(107), *417*
Schaffner, K., 204(12), *215*
Schauwecker, O., 160(135), *172*

Scheeren, J. W., 52(315d), 60(315d), *131*
Scheibler, H., 16(317, 318), *131*, 426(127), 468(127), *505*
Scheit, K.-H., 135(245b), *176*, 306(52, 121b), 331(52), *344*, *346*
Scherillo, G., 48(315a), *131*, 137(245), *176*
Schinz, H., 159(287), 160(243, 287), 161(287), *175*, *177*
Schipper, E., 212(163), 212(162), *220*
Schmid, L., 160(246), *176*
Schmidt, E., 197(142), 198(142), 210(142), *219*
Schmidt, R., 159(247), *176*
Schmidt, U., 149(248), 151(248), 158(248), 159(248), 160(248), *176*
Schmidt, W., 240(304), 256(303), *287*
Schmitt, E., 390(127a), 392(127a), 395(127a), *418*, 444(123), 449(123), 463(123), 465(123), *504*
Schmitt, H., 226(539), *294*
Schmitt, J., 42(268), *129*
Schmitz, E., 11(265a), 25(289), 50(287, 289, 290, 319), 51(288, 289, 291), 59(319), *129*, *130*, *131*
Schneider, G., 22(180a), 34(143), 37(180a, 319a, 320a), 38(142a), 41(180, 320), 46(142c), 47(142b), 55(142c), 61(142b), *126*, *127*, *131*, 164(249), *176*
Schoellkopf, U., 362(139, 140), 399(140), 400(140), 410(139, 140), *418*
Schoen, A. L., 262(166a), *282*
Schoenberg, A., 241(520), 246(520), 271(517, 518, 519, 520), *293*, 365(141, 142, 143), 374(145), 388(142), 393(141), 394(141), 397(141), 399(145), 400(141), 401(145), 402(141), 403(143), 406(146), 409(144, 145, 146), *418*
Scholz, K. H., 473(77a), *503*
Scholze, H., 327(24), *343*
Schon, N., 361(110), 387(110), *417*, 422(115), 424(115), 426(115), 427(115), 428(115), 432(115), 440(115), 450(115), 451(115), 452(115), 454(115), 456(115), 457(115), 458(115), 460(115), 462(115), 475(115), 476(115), 477(115), 478(115), 481(115), 490(115), 492(115), 495(115), 500(115), *504*
Schonberg, A., 43(321, 322), *131*
Schoolery, J. N., 366(106), 371(106), 389(106), 397(106), *417*

Schopf, C., 226(521), *293*
Schosser, H. P., 484(33), 486(33), *502*
Schreiber, J., 422(19), 423(19), 451(19), 451(19), 458(19), 461(19), 465(19), 479(18, 19), 480(18, 19), 491(19), *501*
Schrodt, H., 37(242), 60(242), *129*, 138(200), 145(200), 164(200), *174*, 360(109), 382(109), 412(109), *417*, 421 (114), 431 (114), 452(114), 456 (114), 500(114), *504*
Schroeder, J. P., 3(223), 4(223), 6(223), 31 (223), 32(223), 59(223, 224), 62(223), *128*, 275(398a), *289*
Schromm, K., 25(75), *124*
Schubert, W. M., 277(521a), *293*
Schukina, M. N., 159(250), *176*
Schulte-Hurmann, W., 309(48), 319(48), 326(48), *343*
Schultze, K. M. L., 21(149), *126*, 349(78), 374(78), 401(78), 403(78), *416*
Schulz, G., *173*
Schutz, M., 303(123, 124, 125), 304(124, 125), 313(122, 123), 321(125), 322(125), 323(123, 124), 327(122), 338(123), 342(123), *346*
Schulz, W., 13(360), *132*
Schvchenko, V. I., 56(327a), *131*
Schwarcz, M. J., 137(87), *170*
Schwartz, M. J., 22(79), *124*
Schwarz, G., 254(524), 257(524), 258(522, 523), *293*
Schwarz, J. C. P., 308(125a), 319(125a), 323(125a), *346*
Schweizer, E., 484(36), *502*
Schwenk, H. U., 446(53a), 453(53a), 459(53a), 461(53a), 468(53a), *502*
Schwenk, W., 151, 152(251), *176*
Sciaki, R., 151(193), 152(193), *174*
Scudi, J. V., 24(343), 25(343), *132*
Seebach, D., 353(147), 354(147), 357(147), 358(147), 359(147), 370(147), 374(147), 375(147), 377(147), 378(147), 383(147), 386(147), 388(147), 389(147), 399(147), 400(147), 401(147), 402(147), 403(147), 405(147), 410(148), 411(148), *418*
Seebeck, E., 43(338), *132*
Seed, L., 167(252), *176*
Seefelder, M., 9, 11(92), 61(91), *124*, 201(55), *216*, 423, 427(74), 432(76, 77), 451(74), 76), 452(76, 77), 456(74, 76, 77), 459(75), 461(76), 478(76), *503*

Seefluth, H., 185(100), 186(100), *218*
Seeler, A. K., 463(15), *501*
Seeliger, W., 430(82), 457(82), 458(82), 460(82), 462(82), 463(82), 464(128), 466(82), 470(82), 472(82), 473(82), 480(82), *503*, *505*
Seidel, C. F., 160(243), *175*
Seifert, H., 226(524a), *294*
Selditz, P., 183(97), 203(97), *218*
Selman, J., 54(163b), *127*
Semnikova, N. I., 256(1), *278*
Senda, S., 186(179), 187(179), *221*
Senning, A., 484(103a), *504*
Senyavina, L. B., 436(129), 452(129), 454(129), 456(129), *505*
Serini, A., 149(253), 151(253), 152(253), 160(253), *176*
Serratosa, F., 236(591, 592), 258(591), *296*
Seven, R. P., 159(114), 160(114), 161(114), *171*
Severin, T., 490(128a), *505*
Seymour, D., 33(297), 51(292, 323a, b, c, d), 56(323a), 59(292), *130*, *131*, 159(163), *173*, 210(169a), *220*, 301(119), *346*
Sgarbi, R., 260(130, 400), *281*, *290*
Shapiro, E., 28(323e), *131*, 135(254), 136(254), 144(254), *176*
Shapiro, S. H., 14(204), 29(204), 33(204), 57(204), *128*, 277(395c), *289*
Shasha, B. S., 310(127, 128), 311(13), 312(14, 126), 323(13, 127, 128), 327(14, 126), 334(127), 338(127), *343*, *346*
Shavsha, T. G., 21(6a), 54(6a, b), 57(6a), *122*
Shavsha-Tolkacheva, T. G., 54(7), 55(7), *122*
Shaw, B. L., 52(90), 53(90), 57(90), *124*
Shaw, G., 183(165), 191(164), 204(12), *215*, *220*, 236(497, 525), 240(131), 242(131), 243(132), 246(132), *281*, *293*, *294*
Shchelakov, V. I., 436(130), 437(131), 438(130), 439(130), 451(130), 453(130), 455(130), 456(130), 458(130), 459(130), 462(130), 464(130), *505*
Shchetinskaya, E., 258(18), *278*
Shcowenaars, M., 253(524), 257(524), *293*
Sheehan, J. C., 159(255), *176*
Sheehan, J. T., 193(223), 207(223), *222*, 425(165), 461(165), *506*

Shemyakin, M. M., 436(129, 130), 437(2, 131), 438(130), 439(130), 444(1), 451(130), 452(129), 453(130), 454(129), 455(130), 456(129, 130), 458(130), 459(130), 462(130), 464(130), 469(133), 496(133), *501*, *505*

Shen, T. Y., 307(121), 311(121), *346*

Shen, Z. I., 13(58), *123*

Sheppard, R. C., 437(104), 440(132), *504*, *505*

Sheradsky, T., 183(166), *220*

Sherman, E. C., 47(324), *131*

Sherr, A. E., 159(269), *176*

Shimojo, S., 225(560aa), *294*

Shiner, V. J., 143(56), *169*

Shirley, D. A., 6(325), 63(325), *131*

Shkrob, A. M., 436(129, 130), 437(2, 131), 438(130), 439(130), 451(130), 452(129), 435(130), 454(129), 455(130), 456(129, 130), 458(130), 459(130), 462(130), 464(130), 469(133), 496(133), *501*, *505*

Shorr, L. M., 164(256, 257), 168(256, 257), *176*

Shostakovskii, M. F., 54(325a), 57(325a), 61(325a), *131*, *176*, 353(149), 379(149), *419*

Shropshire, E. Y., 189(41), *216*, 384(29a), *415*, 442(67), 445(67), 446(67), 472(67), 473(67), 500(67), *503*

Shusher, R. S., 258(390), *289*

Sica, D., 48(315b), *131*, 353(138a), 385(138a), 414(138a), *418*, 435(126), 457(126), 458(126), 460(126), 478(126), *505*

Siddall, T. H., 193(167), *220*

Sidhu, G. S., 212(168), *220*

Sieglitz, A., 259(527, 528), 260(526), *294*

Siegmann, C. M., 226(612), *296*

Sigmund, F., 77(326), *131*

Simchen, G., 422(37, 39, 53), 424(53), 428(53), 432(53), 441(38, 39, 40, 41, 42, 50, 53), 442(50, 53), 443(49), 445(50, 53), 446(53a), 450(53), 451(38, 40, 42, 50, 53), 452(37, 42, 53), 453(50, 53, 53a), 454(42, 50, 52, 53), 455(53), 456(42, 49, 53), 458(53), 459(53a), 461(53a), 468(53a), 475(134), 476(52a), 477(135), 484(54), 488(49, 52a, 52b, 52c), 489(52), 490(48), 493(54a), 494(54), 497(51), 498(50), 499(135), 500(134), *502*, *505*

Simon-Ormai, K., 159(173), *173*

Simons, D. M., 26(353), *132*

Sindellari, L., 183(157, 158), 188(157), 201(157), 206(156), 207(158), *220*

Singer, E., 365(142), 388(142), *418*

Singley, J. E., 355(70), 401(70), *416*

Sinnema, Y. A., 16(327), *131*, 382(148a), *418*, 450(136, 137), 478(136), 478(136), *505*

Sklar, A. L., 258(88), 260(88), *280*

Skokal, F. I., 260(423), *290*

Skoldinov, A. P., 268(343, 493), *288*, *293*, 447(141), 448(141), 450(141), 452(141), 457(141), 472(141), 478(141), *505*

Skorova, A. E., 160(125), *171*

Skrabal, A., 138(258, 260), 145(259), *176*

Slack, R., 196(19), *215*

Slater, R. A., 358(15), 390(15), 410(15), 414(15), *415*

Slaugh, L. H., 373(150), 379(150), *419*

Slezak, F. B., 61(72b, c), *124*

Slominskii, Y. L., 260(326), *287*

Slomp, G., 149(208), 151(208), 160(208), *174*, 274(416c), *290*

Slorak, S. A., 212(34, 130), *216*, *219*

Slotboom, A. J., 149(260a), 151(260a), 160(260a), *176*

Smaznaya-Ilina, E. D., 253(553), 258(553), *294*

Smerz, O., 484(55), 485(55), *502*

Smiles, S., 353(83), 354(83), 381(83), *417*

Smith, B., 15(329), 20(328), 21(328, 329), 25(328), *131*, 148(261), 149(261), *176*

Smith, B. H., 183(195b), *221*

Smith, E., 212(42), *216*

Smith, E. A., 151(87a), *170*

Smith, F. A., 314(61), 323(60), *344*

Smith, J. M., 211(222), *222*

Smith, L. I., 224(530), 225(530), 227(529, 530), *294*

Smith, L. L., 27(330, 331), *131*, 135(263), 144(263), 151(262), 152(262), *176*

Smith, R. B., 444(88), *503*

Smith, P. V., 351(54), 403(54), *416*

Smrt, J., 135(140, 140a, 202a, b), 161(61), *170*, *172*, *174*, 267(534), *294*, 306(53, 53a, 54, 92a, 92b, 92c, 129, 129a, 129b), 307(147), 331(53, 53a, 54, 129, 149), *344*, *345*, *346*, *347*, 481(100, 101, 138, 139), *504*, *505*

Smythe, J. A., 401(151), 411(151), 412(151), *419*
Snatzke, G., 23(346), *132*
Snyatkova, V. I., 338(1c), *342*
Snyder, H. R., 190(169), 191(169), *220*
Sobeki, W., 226(531), 227(531), *294*
Sobotka, H., 316(85), 323(85), *345*
Soeder, R. W., 149(215), 151(215), 159(215), 160(215), *175*
Sohnke, H., 25(244), 25(245), 42(244, 245), 55(245), 62(244), *129*
Sokolov, G. P., 47(332a), *131*
Sokolskaya, T. A., 321(1d), 322(1d), *342*
Sokolovsky, R., 229(532), *294*
Soldan, F., 448(14), 465(14), 467(14), *501*
Sondheimer, F., 159(182), *173*, 227(499, 533), *293*, *294*
Sorenson, B. E., 51(153a), 58(153a), 63(153a), *126*
Sorm, F., 135(62, 140, 140a), *170*, *172*, 267(534), *294*, 306(9, 53a, 54, 129b), 331(9, 53a, 54), 334(8), *342*, *344*, *346*, 481(101, 138, 139), *504*, *505*
Southwick, P. L., 183(166), *220*
Sowa, F. J., 7, *128*, 149(189), *174*
Spath, H., 227(535), *294*
Spaziano, V. T., 210(71a, 169a), *217*, *220*
Speck, J. C., 159(264), *176*
Sperley, R. J., 149(216), 151(216), 160(216), *175*
Sperry, J. A., 159(265), 164(265), 167(196, 265), *174*, *176*
Spickett, R. G. W., 212(20), *215*
Spiescke, H., 381(108), *417*
Spille, J., 37(242), 60(242), *129*, 138(200), 145(200), 164(200), *174*, 360(109), 382(109), 412(109), *417*, 421(114), 431(114), 452(114), 456(114), 500(114), *504*
Spoars, J. W., 163(302), *177*
Sprague, J. M., 183(125), 189(125), 190(125), 191(125), 207(125), *219*
Sprague, R. H., 194(30), *216*, 246(86), 256(89), 257(86), 258(151, 536), *280*, *282*, *294*
Stacey, M., 163(170b), *173*, 303(39), 327(39), *343*
Stachel, H. D., 449(140), 461(140), 466(140), *505*
Stackmann, W., 14(76), *124*
Stanbank, D. J., 413(129b), *418*

Staněk, J., 299(130, 131), 301(130), 315(130, 131), *346*
Stanovnik, B., 183(95a, 169b, c, d), 212(169b, c, d), *218*, *220*
Starn, R. E., 3(225), 4(225), 6(225), 8(225), *128*, 190(107), *218*, 275(398b), *289*
Stassevitch, N., 226(283), *286*
Staudinger, H., 2(333), *131*, 165(266), *176*, 267(537), 274(537), *294*
Stavely, H. E., 234(538), *294*
Stedehouder, P. L., 349(11), 351(11), 354(11), 355(11), 363(12, 13), 367(11), 368(13), 369(12, 13), 374(11), 375(12), 379(11), 384(12), 385(11), 386(11), 390(11, 12), 401(11), 402(13), 403(13), 405(11), *415*
Steen, K., 480(58), *502*
Stefanescu, M., 256(399), 260(399), *289*
Steffan, G., 473(77a), *503*
Steinacker, K. H., 23(334, 335), 31(334), 62(334), *131*, 135(267, 268), *176*
Steininger, E., 184(170), *220*
Steinkopf, W., 226(539), *294*
Steinmaus, H., 303(124, 125), 304(124, 125), 321(125), 322(125), 323(124), *346*
Stepanov, F. N., 247(540, 541), *294*
Stephens, J. G., 181(171), 183(171), *220*
Stephenson, O., 24(272), 59(272), *130*
Stern, E. S., 167(196), *174*
Sternbach, L. H., 481(86), *503*
Stetsenko, A. V., 253(324, 325, 542), 257(324), 259(542), *287*, *294*
Stetter, H., 23(334, 335), 31(334), 62(334), *131*, 135(267, 268), *176*, 374(152), *419*
Stevens, C., 351(54), 403(54), *416*
Stevens, C. L., 3(214, 226, 227), 4(214, 226, 227), 5(336), 6(214), 7, 29(214, 226), 33(218), 34(218), *128*, *131*, 149(186), 159(269), *174*, *176*, 275(398c), 277(398c), *289*
Stevens, W., 52(315d), 60(315d), *131*, 239(543), *294*
Stewart, J. E., 314(61), 323(60), 331(135), *344*, *346*
Stickings, C. E., 196(19), *215*
Stiepel, 17(6), 57(6), 61(6), *122*, 187(10), 190(10), *215*
Stiller, E. T., *132*, 237(544), *294*
Stilze, W., 426(142), 452(142), *505*
Stocker, J. H., 150(190), 151(190), 155(190), 158(190), 159(190), 160(190), *174*

Stoeck, G., 151(146), 152(146), *172*
Stoeck, U., 151(146), 152(146), *172*
Stoffelsma, J., *130*, 351(129), 352(129), 353(129), 382(129), 384(129), 390(129), 391(129), 392(129), 410(129), 412(129), *418*
Stoicescu-Crivat, L., 256(399), 260(399), *289*
Stoll, A., 43(338), *132*, 437(143), *505*
Stolle, R., 159(270), *176*, 203(174), 208(173), *220*
Stopp, G., 361(110), 387(110), *417*, 422(115), 424(115), 426(115), 427(115), 428(115), 432(115), 440(115), 450(115), 451(115), 452(115), 454(115), 456(115), 457(115), 458(115), 460(115), 462(115), 475(115), 476(115), 477(115), 478(115), 481(115), 490(115), 492(115), 495(115), 500(115), *504*
Storbeck, W., 209(141), *219*
Story, P. R., 25(339), 56(339), *132*, 231(545), *294*
Stoyanovich, F. M., 352(49), 366(49), 387(49), 390(49), 414(50), *416*
Stravrovskaya, A. V., 447(141), 448(141), 450(141), 452(141), 457(141), 472(141), 478(141), *505*
Streitwieser, A., 141(271), *176*
Stringfellow, C. R., 364(89), 379(89), *417*
Strizhakov, O. D., 57(194), *127*
Stroh, H., 485(158a), *505*
Strong, A. F., 13(133), 16(133), *126*
Stuber, F. A., 449(145aa), *505*
Sturgeon, B., 183(112), 198(112), *218*
Sucrow, W., 23(44), 30(44), 30(45), 31(44), 31(45), *123*, 135(39, 40), *169*
Suda, H., 187(128), *219*
Sudo, G., 269(562c), *295*
Sugasawa, S., 159(272), *176*
Suggate, H. G., 187(89), *218*
Sugihara, A., 187(124), *219*
Sugihara, J. M., 315(132), *346*
Sugihara, S., 212(174a), *220*
Sugimoto, K., 256(546), *294*
Suich, E. D., 3(167), 4(168, 169), *127*
Suitnik, Z. P., 3(167), 4(168, 169), *127*, 258(326, 327), *287*
Suleimanova, M. G., 253(328), 258(328), *287*
Sulston, J. B., 135(235), 137(235), 144(235), *175*

Sulston, J. E., 135(115, 115a, 128), 137(115, 128), 144(115, 128), *171*, *172*, 306(26, 27, 34, 117), 307(26), 334(117), *343*, *346*
Sung, K. P., 54(152), *126*
Suzuki, E., 212(174a), *220*
Suzuki, K., 149(281), *177*
Suzuki, T., 269(562b, 562c), *295*
Suzuki, Z., 149(207, 209), 151(207), 159(207), 160(207, 209), 161(207), *174*, 274(416b, 416d), *290*
Sveshnikov, N. N., 255(383), 258(316, 383, 387, 388, 389, 391, 392), *287*, *289*
Swaminathan, S., 152(273), *176*
Swarts, F., 41(340), *132*
Sweeney, W. M., 34(269), 57(269), *129*
Swiderski, J., 159(187), *174*, 258(547), *294*, 314(133), *346*
Swoboda, W., 160(246), *176*
Sych, E. D., 253(330), 548, 551, 552, 553, 555), 257(555), 258(326, 327, 329, 330, 331, 549, 550, 551, 552, 553, 554, 555, 556, 557), *287*, *294*
Sykes, P. J., 161(155b), *173*
Sytnik, Z. P., 241(225), 257(558), 258(390), *284*, *289*, *294*

T

Tabachnik, F. A., 28(323e), *131*, 135(254), 136(254), 144(254), *176*
Taber, R. C., 249(559), *294*
Taddei, F., 379(104a), *417*
Taft, R. W., 14(283), 57(283), *130*, 138(232), 141(163a, 164, 195, 275, 276), 142(275), 145(274, 275), 164(195), *173*, *174*, *175*, *176*
Tagaki, W., 22(265), *129*, 200(126), *219*, 227(444), *291*, 353(116), 354(116), 373(116, 117), 377(116, 117, 118), 379(116, 117, 118), 405(116, 117), *417*, *418*
Taguchi, T., 429(144), 468(144), 470(144), 476(144), 477(144), 498(144), 499(144), 500(144), *505*
Takahashi, S., 42(112), *125*
Takahashi, T., 186(177, 178, 179, 180), 187(175, 176, 177, 178, 179), *220*, *221*, 258(559a), *294*
Takaya, H., 430(119), 451(119), 452(119), *504*
Takebayashi, M., 236(566), *295*
Takebi, S., 269(562c), *295*

Taki, I. K., 258(560), *294*
Talaty, C. N., 212(136a), *219*
Talbot, R. L., 61(340a), *132*, 145(277), *176*
Talley, E. A., 303(134), 327(134), *346*
Tamura, C., 43(340b, 347), *132*
Tanabe, K., 13(135), *126*
Tanabe, Y., 256(416), *290*
Tanaka, Y., 269(562c), *295*
Tanimoto, S., 187(127, 128), *219*, 225(560aa), *294*
Taniyama, H., 187(175), *220*
Tanno, R., 244(453), 246(453), *291*
Tarbell, D. S., 15(341), *132*, 137(278), *176*, 352(153), 353(153), 355(153), 362(153), 400(153), 401(153), 412(153), *419*
Tarrant, P., 16(342), *132*
Tarsio, P. J., 233(433), 236(433), 237(433), *291*
Tate, B. E., 3(228), 6(228), 7(228), 11(228), *128*, 190(108), 192(108), *218*
Taube, C., 159(113), 162(113), *171*
Taylor, E. C., 181(186), 182(186), 183(139, 187a, 188, 188a, 189, 190, 192), 184(138), 185(185, 186, 187), 186(185, 186), 189(181), 196(204a), 197(186), 199(138), 210(138, 183a), 211(138, 182, 183, 184, 191), 212(139, 187a, 188, 188a), *219*, *221*, 236(560a), *294*
Taylor, N. F., 159(156), *173*
Taylor, W. C., 23(139), *126*
Tchen, S. Y., 151(153a), 159(153a), *172*
Teece, E. G., 314, 315(40), *343*
Tegg, D., 13(133), 16(133), *126*
Telesnina, T. R., 444(1), *501*
Temple, C., 183(121, 195a, b), 184(121), 209(121), 210(119, 120, 121, 193, 195), 212(194, 194a, 194b, 195, 195a), *218*, *221*
Tenenbaum, L. E., 24(343), 25(343), *132*
Teotino, U., 209(38), *216*
Teotino, U. M., 186(196), 209(196), *221*
Teremura, K., 187(128), *219*
Terpstra, G., 375(124), *418*
Ter-Sarkisyan, G. S., 160(205), *174*
Thames, S. F., 159(279), *176*
Thom, E., 15(63), 15(63a), *123*
Thomas, T. M., 446(17), *501*
Thanassi, J. W., 35(248), *129*
Throckmorton, J. R., 149(215), 151(215), 159(215), 160(215), *175*
Thuy, V. M., 151(53a), *169*

Tieckelmann, H., 15(344), 16(344), 57(344), *132*, 180(102), 183(197), 189(197), 190(197), *218*, *221*, 445(109), *504*
Tiemann, F., 14(345), *132*, 144(280), *176*
Tigchelaar-Lutjeboer, H. D. A., 34(345a), *132*
Tikhonova, N. A., 233(560b), 235(560b), 237(560b), *295*
Timmermans, J., 54(345b), *132*
Tinker, J. F., 212(7, 8, 9), *215*
Tipson, R. S., 320(86), 331, *345*, *346*
Tisler, M., 183(95a, 169b, c, d), 212(169b, c, d), *218*, *220*
Tiurina, L. N., 253(6), 258(5, 6), *278*
Tohma, M., 26(165c), *127*
Tolkachev, O. N., 308(88), 327(88), *345*
Tong, B. P., 265(7a), *278*
Tosolini, G., 183(198), *221*
Traut, H., 484(54), 494(54), *502*
Trautner, K., 8(179), *127*
Trautwein, W. P., 302(15a, 109a, 109b, 110, 111, 112), 303(50), 310(15a, 109a, 109b, 111, 112), 316(50), 321(50), 325(15a, 111, 112), 327(50), 340(15a, 109a, 109b, 110, 111, 112), *343*, *344*, *345*, *346*
Traverso, J. J., 188(215, 216), 189(216), 191(215), 194(216), 199(217), 201(216), 203(216), 206(217), *222*, 236(609), *296*, 425 158c),
Travis, J. D., 449(72a), *503*
Traynard, J. C., 226(461), 227(461), *292*
Treibe, W., 271(560c), *295*
Treiber, H. J., 188(27), *216*, 447(43), 452(43), 457(43), 461(43), 472(43), 494(43), *502*
Trepanier, D. L., 207(199), 208(200), *221*
Tri, V. V., 201(111), *218*
Tripathy, P. B., 260(561), 261(561), *295*
Troitskaya, V. I., 253(622, 625, 627, 628, 629), 258(625, 626), 259(622, 625a, 625b, 627, 628, 629), *296*, *297*
Troschenko, A. T., 257(562), *295*
Trout, G. E., 211(201), *221*
Tsatsaronis, G., 484(145), *505*
Tschesche, R., 23(346), *132*
Tsuchihashi, G., 166(211a), *174*
Tsuda, K., 43(340b, 347), *132*
Tsukamoto, T., 149(281), *177*, 269(562a, 562b, 562c), *295*
Tsuruoka, T., 315(54a), *344*

Tsuetkova, N. A., 256(484), *292*
Tucker, B., 449(145aa), *505*
Tucker, W. P., 362(129a), 375(129a), 384 (129a), 390(129a), 402(129a), 403(129a), 413(154), *418*, *419*
Tuemmler, W. B., 261(563), *295*
Tullar, B. F., 183(104), 189(104), *218*
Tullock, C. W., 464(78), 467(78), 468(78), *503*
Turchinovich, G. Y., 246(564a), 247(564), *295*
Turitsyna, N. F., 258(565), *295*
Turnbull, J. H., 151(198), *174*
Turner, D. W., 149(8), 151(8), 159(8), 160(8), 161(8), *168*
Turro, N. J., 48(347a), *132*
Twitchett, H. J., 253(510), *293*

## U

Uddrich, P., 392(37a), *415*
Ulian, F., 183(158), 187(154), 203(154), 204(154), 205(154), 207(158), *220*
Ulm, K., 49(264a), *129*
Ulrich, H., 11(347b), *132*, 449(145aa), *505*
Ulrich, P., 248(615), 260(615), *296*
Umanskaya, L. P., 253(555), 257(555), 258(554, 555, 556, 557), *294*
Umazawa, H., 440(118), *504*
Underwood, H. W., 149(282), *177*
Unrau, A. M., 316(136), 323(136), *346*
Urishibara, Y., 236(566), *295*
Usdin, E., 372(158), 384(158), 411(158), *419*
Ushenko, I. K., 253(6, 333, 567, 568, 571, 572, 573, 577), 256(4, 572), 258(2, 5, 6, 331, 332, 333, 567, 568, 569, 570, 571, 572, 573, 574, 575, 576, 577, 578, 579, 580), 259(571), *278*, *287*, *295*
Utyusheva, Z. M., 158(284), 159(284), *177*

## V

Vafina, M. G., 304(137), 321(12b), 321(137), 322(137), *343*, *346*
Valaeva, Z. Z., 54(8), *122*
Valenta, Z., 43(16a, 314, 348, 360a), 44(348), *122*, *131*, *132*
Van Allan, C. A., 158(285), 159(285), *177*, 212(7, 8,), *215*
Van Alphen, J., 149(286), 151(286), *177*

Van Deenen, L. L. M., 149(260a), 151(260a), 160(260a), *176*
Van der Auwera, A. L., 261(581), *295*
Van der Veek, A. P. M., 71(2a), *122*
Van der Voort, H. G. P., 226(279), *286*
Van Dormael, A., 186(203), 192(202), 212(204), *221*, 246(582), 247(145, 146), 253(583), 259(583), *282*, *295*
van Dorp, D. A., 159(290), *177*
Van Dyke, R. H., 246(86), 257(86), *280*
Vangedal, S., 152(179), *173*
Van Hulsenbeek, A. L., 226(612), *296*
Vankataraman, K., 243(515), *293*
Van Lare, E. J., 246(86), 246(585), 253(90, 584, 585), 254(76a), 257(585), 258(585), 259(90, 584), 276(86), *279*, *280*, *295*
van Leeuwen, P. W. N. M., 137(172b), *173*
Van Leusen, A. M., 196(204a), *221*
Van Oot, J. G., 186(16), *215*
Van Wazer, J. R., 29(252), *129*
Van Zandt, G., 246(86), 249(91, 92, 586), 257(86), 258(91, 92, 586), *280*
Vaughan, W. E., 151(87a), *170*
Vejdelek, Z. J., 43(349), *132*
Velo, L., 270(104), *280*
Venerable, J. T., 3(229), 14(229), *128*, 275(398d), 276(398d), *289*
Verbovskaya, T. M., 246(334), 247(334), 253(649), 258(649), *287*, *297*
Verkade, J. G., 54(349a), *132*
Veyrieres, A., 159(78), *170*
Viguer, P., 226(587, 588), *295*
Vila, J. P., 14(351), 57(351), *132*, 233(589), 236(589, 590, 591, 592), 237(590), 251(591), *295*, *296*
Vince, R., *131*, 135(245a, 286a), *176*, *177*, 210(161, 204b), *220*, *221*
Vincent, E. J., 254(407), *290*
Vincent, M., *132*, 201(110, 111, 205), 203(110), 205(205), *218*, *221*
Vinogradova, V. S., 54(8a), 57(8a), *122*
Vinton, W. H., 257(593, 594, 595), 258(594, 595), 259(594), 260(594), *296*
Viout, P., *133*
Vipond, H. J., 212(34, 130), *216*, *219*
Vitali, R., 27(105, 106, 107, 108), 28(105, 106, 107, 351b), 31(106), *132*, 135(121, 122, 123), 136(120), 144(120, 121, 122), *171*

Vogel, A. I., 54(352, 352a), *132*, 363(36), 375(35, 36, 37), *415*
Vogel, E., 159(287), 160(287), 161(287), *177*
Vogl, O., 26(353), *132*, 145(288), *177*, 211(184, 191), *221*
Vogt, P. J., 181(150), 183(149, 150), 193(149), 193(151), *220*
Vogt, R. R., 226(373), *288*
Volger, H. C., 352(155), 353(155), 372(155), 382(155), 410(155), 411(155), 412(155), 414(155), *419*
Volker, C. E., 212(206), *221*
Vompe, A. F., 258(391), *289*
von Brachel, H., 199(206a), 200(206a, b), *221*, 425(145a, 145b), 429(145a, 145b), 455(145a, 145b), 456(145a, 145b), 458(145a, 145b), 462(145a, 145b), *505*
von Braun, J., 227(595a), *296*
von Buttlar, R., 159(218), *175*
von Hartel, H., 15(353a), 54(353a), 57(353a) *132*
von Hochstetter, H., 300(25), 315(25), 327(25), *343*
von Lippmann, E., 159(34), *169*
von Neyman, H., 159(109), *171*
von Walther, R., 205(207), *221*
von Walther, R. F., 147(289), *177*
Vorbruggen, H., 479(146, 147), 480(147), *505*
Vorsanger, J. J., 254(598), 258(598), *296*, 361(155a), 379(155b), 387(155a), 392(155a), 395(155a, 155b), 398(155b), 401(155b), *419*
Vromen, S., 183(190, 192), *221*

# W

Waddington, H. R. J., 240(165), 260(310), 260(597), 261(596), *287, 296*
Waddington, T. C., 47(199b), *128*
Wagner, A. F., 377(155c), *419*
Wagner, E. C., 182(59), 183(59), 205(39b, 208), 207(59), *216, 217, 221*
Wagner, G., 137(221c), *175*, 307(115c), *346*
Wagner, H., 12(136), *126*
Wahl, H., 254(598), 258(598), *296*
Wahl, R., 422(53), 423(53), 428(53), 432(53), 441(50, 53), 442(50, 53), 445(50, 53), 450(53), 451(50, 53), 452(53), 453(50, 53), 454(50, 53), 455(53), 456(53), 458(53), 493(54a), 498(50), *502*
Waite, J. A., 206(35), 212(35), *216*
Walinsky, S. W., 269(453a), *291*, 411(118a), *418*
Walker, L. E., 275(29), *278*
Walls, H. N., 253(256), 256(256), 260(256), *285*
Walter, J., 13(354), *132*
Walters, P. M., 3(230), 4(230), 29(230, 355), 30(230), 33(230), *128*, 277(398e, 598a), *289, 296*
Walther, R., 186(209, 210), 186(209), 188(209, 210), *221*
Wang, S., 161(133), *172*
Wang, T. C., 151(178a), *173*
Wanmaker, W. L., 186(14, 15), *215*, 445(7, 8), *501*
Wanzlick, H., 200(211, 212), *221*, 443(148), 494(148), *505*
Ward, J. P., 159(290), *177*
Warnant, C., 27(160), 28(160), *126*, 135(152), 152(152), *172*
Warner, G. H., 18(80), *124*
Warner, R. N., 204(12), *215*
Warren, M. E., 151(233), 160(233), *175*
Wazfi, A. S., 183(212a), 187(212a), *222*
Wasserman, H., 53(356), *132*
Waters, W. A., 43(253), *129*
Webb, I. D., 414(156), *419*
Webb, N. E., 212(36), *216*
Weber, L., 28(323e), *131*, 135(254), 136(254), 144(254), *176*
Weberndorfer, V., 367(81), 388(81), *417*
Webster, G. L., 159(3), *168*
Webster, O. W., 35(357), *132*
Weddige, A., 14(358), *132*
Wegler, R., 436(117), 472(117), *504*
Wehrli, H., 26(89), *124*
Wei, P. H. L., 207(213), *222*
Weidinger, H., 9, 11(92), 61(91, 359), *124, 132*, 201(55), *216*, 423, 427(74, 149), 432(76, 77), 451(74, 76, 149), 452(76, 77), 456(74, 76, 77, 149), 459(75), 459(149), 461(76), 478(76), *503, 505*
Weidlich, H. A., 13(360), *132*
Weigel, W., 258(431), 260(432), *291*
Weinberg, N. L., 47(30), *122*, 436(150), 452(150), 455(150), 457(150), *505*
Weindel, A., 274(500a), *293*

Weingarten, H., 446(153, 154, 157), 448 (153, 154, 155), 453(153, 154, 157), 456 (155), 493(152), 495(151, 156), *505*, *506*
Weinstock, L. M., 152(291), *177*
Weintraub, L., 422(158), 453(158), *505*
Weisblat, D. I., 325(143), *347*
Weise, A., 461(111), *504*
Weiss, K., 303(49), 304(49), 322(49), 337(49), *343*
Weiss, R., 237(599), *296*
Weissauer, H., 239(38), *279*
Weizmann, C., 227(600), *296*
Wellman, K. M., 185(72a), *217*
Wenthe, A. M., 139(292, 293), *177*
Westerhof, P., 152(294), 161(294), *177*
Westerhuis, E., 356(14), 392(14), 400(14), 403(14), 404(14), *415*
Westphal, G., 485(158a), *505*
Weygand, C., 151(295), *177*
Weygand, F., 226(262), *285*, 303(138), 321(138), 323(138), *346*
Weyna, P. L., 47(231), *128*
Whalley, E., 138(162), *173*
Wharton, P. S., 53(356), *132*
Wheeler, H. L., 237(601), *296*
White, F. L., 3, 4(50), 5(50), *123*, 194(30), *216*, 245(98), 246(86), 256(97, 606, 607), 257(86, 93, 94, 96), 258(93, 94, 95, 96, 602, 604), 259(94, 96, 603, 605), *280*, *296*
White, W. A., 446(153, 154, 157), 448(153, 154, 155), 453(153, 154, 157), 456(155), 495(156), *505*
Whitehead, C. W., 188(215, 216), 189(216), 191(85, 215), 194(214, 216), 199(217), 201(216), 203(216), 206(217), *217*, *222*, 236(609), 237(608), *296*, 425(158b, 158c), *505*, *506*
Whitely, R., 413(129b), *418*
Whiting, M. C., 52(90), 53(90), 57(90), *124*
Whitman, B., 151, 152(251), *176*
Whittaker, N., 150(241), 151(241), *175*, 207(158a), *220*, 233(512), 236(512, 610), *293*, *296*
Wibaut, J. P., 226(611, 612), *296*
Wiberg, K. B., 193(218), *222*
Wichelhaus, H., 149(170), 168(170), *173*, 186(218a), 187(219), *222*
Wichtl, M., 160(246), *176*
Wickens, J. C., 23(266), 31(266), *129*, 135(212), *174*

Wickop, J., 83(43a), *123*, 144(37), *169*, 356(17), 358(17), 386(17), *415*
Wiechert, R., *173*
Wiegand, E., 13(29), *122*
Wieser, D., 8(182), *127*, 139(166), *173*
Wiesner, K., 43(16a, 314, 348, 360a), 44(348), *122*, *131*, *132*
Wilcox, C. F., 37(361), *132*, 144(296), *177*
Wild, A. M., 24(272), 59(272), *130*
Wild, D., 239(489), 246(489), *292*
Wilder, R., 14(108a), *125*
Wildi, B. S., 261(563), *295*
Wildschut, G. A., 358(157), 359(157), 370(157), 379(157), 385(157), 399(157), 400(157), 401(157), 402(157), 405(157), *419*
Williams, J. H., 205(18d), *215*
Williams, J. R., 48(347a), *132*
Williams, L. A., 196(13), 212(9, 220), *215*, *222*, 242(15), *278*
Williams, W. W., 256(85), 258(85), *279*
Williamson, A. W., 12(362, 363), 13(362, 363), *132*
Williamson, D. M., 152(57), 158(57), *169*
Willis, J. A. D., 212(36), *216*
Willstatter, R., 159(297), *177*, 262(612a), *296*
Wilmut, H. E., 54(120), *125*
Wilson, C. D., 258(613), *296*
Wilson, D. A., 149(8), 151(8), 159(8), 160(8), 161(8), *168*
Wilson, E. J., 315(109), 327(109), *345*
Wilson, J. D., *506*
Wilson, W., 151(198), *174*
Winberg, H. E., 425(161), 426(159), 441(160), 442(161), 449(163), 451(160, 161a), 452(161), 454(161), 455(161), 456(161), 457(159, 161), 458(163), 459(161), 462(161), 464(163), 466(163), 471(163), 474(163), 481(159), 494(160), 495(160, 161, 161a, 162), *506*
Windhoff, J., 49(264a), *129*
Winkelmann, H., 183(71), *217*
Winkelmann, H. D., 183(70), 193(70), *217*, 445(92b), *503*
Winstein, S., 22(364), 25(365), 33(297), 37(4), 40(364, 366), 57(4), 59(365), *122*, *130*, *132*, *133*, 140(298), 144(299), *177*, 301(119, 140, 141), 306(139), 316(139), *346*, *347*
Winter, R. A. E., 49(66a), *123*

Winton, B. S., 180(72), 183(72), *217*, 262, 263(261), *285*
Wiskott, E., 362(139, 140), 399(140), 400(140), 410(139, 140), *418*
Wislicenus, W., 160(300), *177*
Wizinger, R., 239(543), 248(615), 260(395, 408, 614, 615), *289, 290, 294, 296*
Wohl, A., 159(301), *177*, 225(616), 226(616, 617), *296*
Woidich, K., 237(599), *296*
Wolfrom, M. L., 163(302), *177*, 315(142), 321(87), 325(143), *345, 347*, 372(158), 384(158), 411(158), *419*
Wolko-Samochocka, K., 314(133), *346*
Wood, C. A., 226(618), 227(618), *296*
Wood, C. E., 13(367), *133*
Woodward, D. W., 183(221), *222*
Woodward, R. B., 43(368), *133*
Wright, E. V., *127*
Wright, S. H. B., 212(20), *215*
Wright, W. B., 211(222), *222*
Wrinch, D., 437, *506*
Wuhrmann, H., 253(619), *296*
Wunderlich, K., 37(241), *129*, 138(199, 202), 140(202), 164(199, 201), *174*, 225(402), *290*, 425(113), 499(116), 500(116), *504*
Wynberg, H., 14(369), *133*
Wynne-Jones, W. F. K., 138(53, 303), 141, 143(53), 144(53), 145(53), *169, 177*

**Y**

Yagupolskii, L. M., 253(335, 336, 620, 621, 622, 623, 624, 625, 627, 628, 629), 258(335, 336, 620, 621, 623, 624, 625, 626), 259(622, 623, 625a, 625b, 627, 628, 629), *287, 296, 297*
Yale, H. L., 193(223), 207(223), *222*, 425(165), 461(165), *506*
Yamaguchi, T., 361(114, 115), 390(114), 407(114), *417*, 431(117a), 459(117a), *504*
Yamamoto, Y., 430(105), 460(105), 462(105), 487(105), *504*
Yamazaki, E., 189(84), *217*
Yanovskaya, L. A., 159(304), *177*, 267(631, 632), 268(630, 631, 632), 269(632), *297*
Yasnitskii, B. G., 33(370), 55(370), *133*
Yasuda, K., 151(305), 152(305), *177*

Yatsuka, T., 187(176), *220*
Yazlovetskii, I. G., see Yazlovetsky, I. G.
Yazlovetsky, I. G., 303(66c), 303(73), 309(73), 320(73), 321(66d), 321(73), 327(66c), 336(66c), 336(72), 337(66c, 66d), *344*
Yoder, C. M. S., 49(370a), *133*
Yoshizawa, I., 26(165c), *127*
Yoshikawa, Y., 186(180), *221*
Young, F. G., 268(633), *297*
Young, W. G., 226(634), *297*
Yufa, P. A., 253(629), 259(629), *297*
Yufit, S. S., 267(375, 632), 268(632), 269(632), *288, 297*
Yuldasheva, E. K., 21(9), 54(9), *122*
Yuldasheva, L. K., 54(10), *122*

**Z**

Zagulyaeva, O. A., 441(165a), *506*
Zakharkin, L. I., 35(371), *133*, 231(635), *297*
Zaretskii, V. I., 11(372), 61(373), *133*
Zaslowski, J. A., 24(12), 56(12), *122*
Zaslowsky, J. A., 158(17), 162(17), *168*
Zaugg, H. E., 8(374), *133*, 433(166), 434(167, 169, 170), 437(168), 463(170), 469(166), 470(166), 471(166), 477(166), *506*
Zeller, P., 160(147, 148, 149), *172*, 268(636), *297*
Zellstoffwerke, A.-G., 12(99), 13(99), *124*
Zemlicka, J., *133*, 135(62, 306, 306a, 307), 137(61a, 307, 308), 144(306), 145(307), *168, 170, 177*, 306(7a, 9, 144, 144a), 307(7, 7a, 7b, 145, 146, 146a, 147), 331(7, 9, 145, 146, 147), 334(8, 144, 144a, 145, 146), *342, 347*, 428(172), 463(172), 464(172), 481(99a, 171, 172), 488(171), *504, 506*
Zemplen, G., 315(148), *347*
Zenno, H., 186(179), 187(179), *221*, 249(639), 256(637, 638), 258(638), *297*
Zhdanowitz, M., 229(640), *297*
Zhikerava, K. D., 258(579, 580), *295*
Zhiryakov, V. G., 253(641, 643), 256(1a, 642, 642a, 643, 644), 258(392, 641, 645), *278, 289, 297*
Zhmurova, I. N., 258(337, 338, 339), *288*
Zhviglazova, L. E., 250(485), *292*
Zida, P., *133*
Ziegler, C., 183(125), 189(125), 190(125), 191(125), 207(125), *219*

Ziemann, H., 303(138), 321(138), 323(138), *346*
Zilberman, E. N., 57(194), *127*
Zlochevskaya, A. V., 256(111, 112), *281*
Zollinger, H., 423, 451(16), 454(16), 461(16), 465(16), *501*
Zolotukhin, I. O., 303(71a), 327(71a), 336(71a), 337(71a), *344*
Zubarovskii, V. M., 253(648, 649), 258(646, 649), 259(647, 648), *297*
Zuckerman, J. J., 49(370a), *133*

# Subject Index

## A

Acetals
  carbohydrate, 340
  ethylidene, 231
  ketene, 274–277
  synthesis, 154, 224–227, 229, 231, 265–267, 271, 272
Acetamide acetals
  from acetamide, 422
  from acetamide chloride, 423
  alkoxyaminocarbonium salts from, 499
  cyclobutene derivatives from, 496
  $N,N$-difluoro, 425
  from ketene $O,N$-acetals, 426
  ketene $O,N$-acetals from, 495
  reaction with active methylene compounds, 492
    with benzoyl and benzyl chlorides, 497
    with acrylonitrile and acrylates, 496
    with ethanedithiol, 354
    with hydrogen cyanide, 497
    with vinyl ketones, 431, 496
Acetamidines, $N,N'$-disubstituted, 185
Acetic anhydride, trithioorthoacetates from, 353
Acetimidic esters, $N$-aryl, 182
Acetyl cyanide, 153
Acetylenes, reaction with ortho esters, 265

Acid anhydrides, reaction with ortho esters, 52, 413
Active methylene compounds
  reaction with amide acetals, 489
    with ortho esters, 231–244, 406
Acyl halides
  reaction with ortho esters, 153, 413
  trithioortho esters from, 352
Acylation of diols, with ortho esters, 136
Acylglycosyl halides, 299–306, 313, 338
Addition reactions, electrophilic, of ortho esters, 266, 407
Alcohols, alkylation by ortho esters, 146
Aldehydes, synthesis, 224–227, 229, 272
Aldonyl chlorides, 313
Alkenes, reaction with ortho esters, 266, 407
Alkylation, of carbohydrate ortho esters, 319
Alkyllithium reagents, reaction with orthoformates, 230
Amide acetals, see also Acetamide acetals, Benzamide acetals, Formamide acetals, Carbamide acetals, Urea acetals
  alkoxyaminocarbonium salts from, 499
  from carbonium salts, 422, 423
  2-dimethylamino-1,3-dithiolanes from, 354
  dissociation, 499
  heterocyclic, 427–436

Amide acetals—*continued*
　hydrolysis, 475
　ketene *O,N*-acetals from, 495
　from ortho esters, 179
　ortho esters from, 478
　properties (table), 450
　reaction with hydrogen cyanide, 497
Amide chlorides, 11, 423
Amides
　reaction with ortho esters, 183, 187
　synthesis, 190
Amidines, 184, 481
Amines
　reaction with amide acetals, 424, 442, 481
　　with ortho esters, 179, 183, 189, 190, 192, 270, 414
Aminomethylene compounds, 190
Anhydroryanodol orthoacetate, 43, 44
Azacarboxyanine dyes, 194
Azulenes, reaction with orthoformates, 271

## B

Barbituric acids, reaction with ortho esters, 265
Barbituric acid acetals, 435
Benzamide acetals
　from benzamides, 423
　from benzamide chlorides, 423
　benzamidines from, 481
　from benzimidates, 429
　from dimethylaminodifluorotoluene, 426
　hydrolysis, 475
　orthobenzoates from, 478
Benzamide dithioacetals, 361, 407
Benzamidines, 481
Benzimidazoles, 198
Benzimidic esters, 184
Benzodioxole, 2,2-diphenylthio, 356
Benzopyran, 2,2-dialkoxy, 37
Benzothiadiazepines, 209
Benzothiadiazoles, 197
Benzothiazoles, 197
Benzotriazines, 208
Benzotrichloride, orthobenzoates from, 14
Benzoxazoles, 197
Bis(sulfonyl)methanes, 359
Bis(trialkylthiomethyl)disulfides, 356
Bis(triarylthiomethyl)trisulfides, 356
Borate esters, 164

Boudroux–Chichibabin reaction, 44, 224–227
Bromination, of ortho esters, 29
α-Bromo esters, reaction with orthoformates, 229
Bromoform, trithioorthoformates from, 355
α-Bromoortho esters, 277
*tert*-Butylhydroperoxide, 313

## C

Carbamates, 194, 425
Carbamide acetals, 426, 427, 440
Carbenes
　dialkoxy, 46
　dihalo
　　orthoformamides from, 446
　　orthoformates from, 12
Carbocyanine dyes, 250–260
Carbohydrate ortho esters, *see* Ortho esters, carbohydrate
Carbon tetrabromide, tetrathioorthocarbonates from, 355
Carbon tetrachloride, orthoformates from, 13, 355
Carboxylic acids, esterification, 152, 479
Carboxylic anhydrides, reaction with ortho esters, 154
Cevadine orthoacetate, 44
Cevine alkaloids, ortho esters, 43
Chlorodifluoromethane
　orthoformamides from, 446
　trithioorthoformates from, 355
Chloroform
　orthoformamides from, 446
　orthoformates from, 12, 14, 354
Chloropicrin, orthocarbonates from, 15
2,4,6-Collidine, 303
Convallagenin A, orthoacetate, 26
Cyanine dyes, 244–264, 406, 494
Cyclols, 437–440, 444

## D

Deuteroethanol, 137
Dialkoxycarbonium salts, 138, 212, 499
Dialkylsulfonylalkylthiomethanes, 360
Dialkylsulfonylmethanes, 367
Diamines, reaction with ortho esters, 197, 198, 204

## SUBJECT INDEX

Diazepines, dibenzo, 209
Diazines, 204
Diazo ketones, 48, 271
Dichlorocarbene, 12, 426
Diethoxyalkyl acetates, 209, 232
Diethoxychlorophosphine, 165
Diethoxymethyl acetate, 167, 232, 252, 253
Diethoxymethyl carboxylates, synthesis, 52
Diethoxymethyl compounds, 233, 234
2,2-Diethoxythiolane, 412
Diethyl maleate, 272
Diethylthiomethylide salts, 357
Dihydrofuran, 2,2-dialkoxy, 48, 137
2,3-Dihydropyrans, 2,2-dialkoxy, 46
Diisobutylaluminum hydride, 231
Diketene, 267
Dimethine dyes, 248–250
Dimethylaminotriethoxymethane, 427
Dimethylformamide
    formamide acetals from, 422
    formamide dithioacetals from, 352
Diols, acylation with ortho esters, 136
1,3-Dioxanes
    2-alkoxy, 22, 37, 40, 135, 214, 225
    2-dialkylamino, 428
1,3-Dioxenium salts, 36, 309
1,3-Dioxolanes
    2-acyloxy, 315
    2-alkoxy
        reactions, 135, 137, 153, 214, 225, 230, 231, 268, 277
        synthesis, 8, 17, 18, 22, 36, 38–40
    2-alkylperoxy, 50
    2-dialkylamino, 428, 479
    2-ethylthio, 407
    2-hydroperoxy, 51
    2-hydroxy, 42, 43, 314
1,3-Dioxolan-4-one, 2-alkylperoxy, 313
1,3-Dioxolenium salts, 36, 51, 231, 302, 340
Diphenyl methylphosphonate, 166
1,3-Diphenylparabanic acid, 199
Diphenylsulfonylphenylthiomethane, 367
Diphenylthiocarbene, 411
Disulfonylmethylides, 360
Disulfonylmethyl sulfides, 368
Disulfonylphenylthiomethanes, alkylation, 360
1,1-Disulfonyl-1-phenylthioethane, 360
1,3-Dithiane
    2,2-dimethylthio, 357

1,3-Dithiane—*continued*
    lithium salt, 357
    2-substituted thio, 354
Dithioacetals, 413
1,3-Dithiolanes
    2-substituted amino, 354, 364
    2-substituted thio, 354
    4,4,5,5-tetraphenylthio, 365
1,3-Dithioles
    2,2-bis(substituted thio), 361
    2-alkoxy-2-alkylthio, 360
Dithioorthocaetates, trialkyl, 354
Dithioorthobenzoates, trialkyl, 354
Dithioorthoformates, trialkyl, 354, 355
Dyes, *see* Azacarbocyanine dyes, Carbocyanine dyes, Cyanine dyes, Dimethine dyes, Merocyanine dyes, Monomethine dyes, Pentamethine dyes, Trimethine dyes

## E

Enol acetates, 269
Enol ethers
    from orthoformates, 149
    reaction with ortho esters, 267
Ergot alkaloids, 440
Ester aminals, *see also* formate aminals
    dissociation, 499
    heterocyclic, 442
    properties, 450
    synthesis, 441
Ethers, from orthoformates and Grignard reagents, 227
Ethoxydichlorophosphine, 165
Ethoxymethylene compounds, 229, 231–244
Ethyl acetate, trithioorthoacetates from, 351
Ethyl diazoacetate, 271
Ethylene diamine, imidazolines from, 197
Ethylene ketals, synthesis, 161
Ethyl fluoride, 167
Ethyl formate, trithioorthoformates from, 351
Ethyl nitrate, 165
Ethylphosphonic acid, ethyl ester, 165
Ethylthiomethylene compounds, 406

## F

Formamide
    orthoformates from, 11
    trithioorthoformates from, 352

Formamide acetals
  alkoxycarbonium salts from, 499
  N-alkylation reactions of, 488
  N-arenesulfonyl, 425
  arylacetic acids from, 499
  5-azacytosines from, 483
  from carbonium salts, 422
  from dichlorocarbene, 426
  N,N-difluoro, 425
  from dimethylaminodifluoromethane, 425
  from formamide chlorides, 423
  formate aminals from, 442
  formyl proton exchange, 447, 499
  formylation of amino compounds with, 481
  hydrolysis, 475
  reaction with active methylene compounds, 489
    with alcohols, 424
    with amino compounds, 424, 442, 481
    with carboxamides, 481
    with carboxylic acids, 479
    with diethoxycarbonium salts, 499
    with diethyl phosphite, 498
    with diols, 428
    with ethanedithiol, 354
    with N-halocarboxamides, 489
    with hydrazines, 481
    with hydrogen cyanide, 497
    with isocyanates, 488
    with isothiocyanates, 488
    with nucleosides, 428, 481
    with phenols, 480
    with sulfonamides, 481
    with ureas, 481
  transacetalization reactions, 424
  transamination reactions, 424
  1,3,5-triazines from 482
Formamide dithioacetals, 352, 356, 414
Formamidines, 180, 184, 187–190, 194, 481
Formamidinium salts, tetrasubstituted, 189, 423, 441, 445, 446
Formanilides, N-alkyl
  from anilines and orthoformates, 193
  trithioorthoformates from, 352
Formate aminals, 441
  disproportionation, 442
  from formamide acetals, 442
  orthoformamides from, 445
  pyrimidines from, 484

Formate aminals—*continued*
  reaction with active methylene compounds, 492
    with benzaldehyde, 498
    with hydrogen cyanide, 497
  tetraaminoethylenes from, 494
Formic acid
  hexathiaadamantanes from, 350
  trithioorthoformates from, 349
Formimidic esters, N-substituted, 180

G

Germanium tetrachloride, 165
Glycosides, from ortho esters, 163, 334, 337
Grignard reagents, reaction with ortho esters, 44, 224–228
Griseolutein B, 440
Guanidines, from orthocarbonates, 189

H

$\alpha$-Haloalkyl ethers, ortho esters from, 16
Hemiacetals, 154
Heptamethinecyanine dyes, 262
Heterocyclic compounds, nitrogen, 195–212
Hexahydrosyncarpic acid, orthoformate, 23
Hexaoxadimantane, 26
Hexathiaadamantanes, 350, 353
Hexathioorthooxalate, hexaphenyl, 359
Hydrazides, 183, 188
Hydrazidines, 188
Hydrazines, 183
Hydrazones, 183
Hydroformylation of ortho esters, 272
Hydrogen cyanide, 5, 11, 269
Hydrogen sulfide, 166
Hydrolysis
  of ortho amides and amide acetals, 475
  of ortho esters, 134–146, 232–234
  of trithioortho esters, 411
Hydroperoxides, 50
$\alpha$-Hydroxy acids, 359
Hypoxanthine, 210

I

Imidazoles, 197
Imidazolediones, 199
Imidazolidines, 200
Imidazolines, 197
Imidazolones, 198

SUBJECT INDEX

Imidic ester hydrochlorides
  ortho esters from, 2, 8
  synthesis, 3
Imidic esters, *N*-substituted, 9, 179, 201
Imidocarbonates, *N*-tosyl, 193
2-Iminofuran hydrochlorides, 8
2-Iminopyran hydrochlorides, 8
Indocarbocyanine dyes, 251
Inositol, ortho esterification, 308
Iodoform, 355
Isocyanates
  aryl, 199
  *N*-substituted sulfonyl, 194
Isothiocyanates, 206
Isoxazolidines, 5,5-dialkoxy, 48

**K**

Ketals, 149, 154, 228, 265, 479
Ketene acetals, 32, 46, 48, 50, 274–277
Ketene *O*,*N*-acetals, 426, 496
Ketene aminals, 189
Ketene dithioacetals, 353, 372, 410
Ketenes, 267
Koenigs–Knorr reaction, 299

**L**

Lactones, spirocyclic ortho esters from, 39
Lactonium salts, *O*-alkyl, 37
Lincomycin, ortho ester, 307
Lithium aluminum hydride, 231
Lithium borohydride, 231
Lutidine, 303

**M**

Merocyanine dyes, 248–250, 262
Methanetrisulfonic acid, and esters, 370, 372
Methine bridge formation, 253 ff.
Methylene bases, 254
*N*-Methylpyrrolidone acetals, 481, 492
1-Methyl-2-quinolone acetals, 492
Monomethinecyanine dyes, 245–248, 494
Monomethineoxonol dyes, 245
Monothioortho esters, 353–356

**N**

Neocyanine dyes, 262
Neosabadine orthoacetate, 44
Nitriles
  2,2-dialkoxy, 266
  ortho esters from 2–11

Nitrones, 48
Nucleosides
  acetals and ketals, 161
  ortho esters, 135, 306, 331

**O**

Oligonucleotides, ortho esterification, 306
Oligosaccharides, from ortho esters, 337
Organoaluminum compounds, 230
Organozinc compounds, 229
Orthoacetates, *see also* Ortho esters
Orthoacetates
  dialkylacyl, 52
  dialkyl alkylperoxy, 50
  trialkyl
    acetamidines from, 185
    acetimidic esters from, 182
    from alkoxyacetylenes, 34
    from diazidoethoxyethane, 478
    diazines from 204, 206
    from dihaloacetylenes, 36
    from dihaloethyl ethers, 16
    imidazoles and imidazolines from, 197
    ketals from, 228
    from ketene acetals, 33
    ketene aminals from, 189
    oxadiazoles from 201
    oxazoles from, 196
    pteridines from, 211
    purines from 210
    reactions with acetylenes, 265
      with active methylene compounds, 231
      with acylhydrazides, 201
      with amino compounds, 182, 183, 185, 192
      with barbituric acids, 265
      with carboxylic anhydrides, 52
      with chloral, 163
      with chloroacetone, 274
      with dialkyl phosphochloridites, 213
      with diazoesters, 271
      with difluoramine, 425
      with enol ethers, 268
      with Grignard reagents, 228
      with hydrazines and hydrazides, 183
      with hydrogen cyanide, 266
      with hydrogen sulfide, 166
      with lithium aluminum hydride, 231
      with phenols, 148

Orthoacetates—*continued*
  trialkyl—*continued*
    reactions—*continued*
      with phosphorus compounds, 166, 213
      with semicarbazides, 183
      with sulfonyl isocyanates, 194
      with thiols, 353
      with ureas, 194
    thiadiazoles from, 202
    thiazines from, 204
    thiazoles from, 196
    triazines from, 208
    triazoles from 203, 204
Ortho amides, 179, 420, 444
  bicyclic, 496
  dissociation, 499
  heterocyclic, 449
  properties, 450
Orthobenzoates, *see also* Ortho esters
  dialkyl alkylperoxy, 50
  trialkyl
    from benzamide acetals, 478
    benzimidates from, 184
    from benzotrichlorides, 14
    diazines from, 204
    imidazoles and imidazolines from, 197
    oxadiazoles from, 201
    reactions with active methylene compounds, 231
      with acylhydrazides, 201
      with amino compounds, 184
      with hydrogen cyanide, 266
      with hydrogen sulfide, 166
      with sulfonyl isocyanates, 194
    thiadiazoles from, 202
Orthocarbamides, 448
Orthocarbonates
  tetraalkyl
    from chloropicrin, 14
    from halomethyl ethers, 18
    ketals from, 228
    from other orthocarbonates, 21
    orthocarboxylates from, 44, 228
    from phenyl cyanate, 49
    reaction with acetylenes, 45, 265
      with ammonia and amines, 189, 190
      with enol ethers, 268
      with Grignard reagents, 44, 228
      with phenols, 148

Orthocarbonates—*continued*
  tetraalkyl—*continued*
    reaction—*continued*
      with phosphorus compounds, 213
      with sulfonamides, 193
      from trichloromethanesulfenyl chloride, 14
    tetraethyl, deuteroethanol from, 137
Ortho esters, *see also* Orthoacetates, Orthobenzoates, Orthoformates, Orthopropionates
  acyloxy, 52, 53
  bicyclic, 8, 18, 22, 23, 33, 36–39, 43, 46, 49, 51, 53, 135, 164
  α-bromo, 29, 277
  from carbonium salts, 38
  carbohydrate, 298–342
    acetals from 340
    acylation of, 318, 339
    acylglycosyl halides from, 338
    from acylglycosyl halides, 299–306
    alkylation reactions, 319
    *tert*-butylperoxy, 342
    deacylation reactions, 316
    from 1,3-dioxenium salts, 309
    dioxolenium salts from, 340
    elimination reactions, 331
    glycosides from, 334
    hydrolysis reactions, 332
    isomerization to glycosides, 337
    methanesulfonation of, 318
    from ortho esters, 306
    peroxy, 313
    properties, 320–328
    reaction with alcohols, 334
      with antimony pentachloride, 340
      with boron trifluoride, 340
      with carboxylic acids, 339
      with carboxylic anhydrides, 339
      with hydrogen bromide, 339
      with hydrogen chloride, 338
      with titanium tetrachloride, 339
    substitution reactions, 319
    tetracyclic, 308
    from thiono esters, 310
    toluenesulfonation, 318
    transesterification, 308
    tricyclic, 304
    from xanthates, 310
  cyanine dyes from 245–260

Ortho esters—*continued*
  from dihaloalkenes, 34
  function as protecting group, 135
  from α-haloalkyl ethers, 16
  *N*-heterocycles from, 195–212
  hydroformylation of, 272
  hydrogen, 41
  hydrolysis, 134–146
  from imidic ester hydrochlorides, 7, 8
  imidic esters from, 179
  from ketene acetals, 32
  ketene acetals from, 274–277
  methine bridge formation by, 235
  from nitriles, 2, 11
  from orthocarbonates, 44, 228, 265
  pentacyclic, 26, 145
  properties, 54 ff.
  reaction with acetyl cyanide, 153
    with acetylenes, 265
    with active methylene compounds, 231–244
    with acyl halides, 153
    with alcohols (alkylation), 146, 147
    with boron trifluoride, 164
    with carbohydrates, 306
    with Grignard reagents, 224–228
    with metal hydrides, 231
    with thiols, 353
  spirocyclic, 39
  tricyclic, 8, 9, 23, 26, 28, 33, 43, 44, 145
  trithioortho esters from, 353
  from trihalomethyl compounds, 12–15
  transesterification, 18–29
  trithiabicyclooctanes from 354
Orthoformamides, 445
  acyl derivatives, 446
  from dihalocarbenes, 446
  formamidinium cations from, 500
  from formate aminals, 445
  hydrolysis, 478
  from orthoformates, 445
  reaction with active methylene compounds, 494
    with amino compounds, 482
  tetraaminoethylenes from, 495
  from tetrasubstituted amidinium salts, 446
Orthoformates, *see also* 1,3-Dioxanes, 2-alkoxy; 1,3-Dioxolanes, 2-alkoxy; Ortho esters

Orthoformates—*continued*
  alkyl diacyl, 52
  alkyl diaryl, 16
  dialkyl acyl, 52
  dialkyl alkylperoxy, 50
  mixed, 16, 28
  steroid, 26
  trialkyl
    acetals from, 155, 224–227
    *trans*-alkenes from, 230
    *N*-alkylformanilides from, 193
    aminomethylene compounds from, 190
    from carbon tetrachloride, 13
    from chlorodifluoromethane, 13
    from chloroform, 12, 13
    cyanine dyes from, 244–264
    diazines from, 205, 206
    from dichlorodifluoromethane, 13
    from dichlorofluoromethane, 13
    from formamide, 11
    from formamide acetals, 499
    formamidines from, 184
    from formimidate hydrochlorides, 8
    formidic esters from, 179–183
    glycosides from, 163
    from halomethyl ethers, 16
    hydroformylation, 272
    from hydrogen cyanide, 2, 11
    imidazoles from, 198
    from orthothioformates, 20
    oxadiazoles from, 201
    oxazoles from, 196
    pteridines from, 211
    purines from, 209
    quinolines from, 196
    reactions with actylenes, 265
      with active methylene compounds, 231
      with acyl hydrazides, 201
      with alkenes, 266
      with N-alkylanilines, 445
      with N-alkylarenesulfonamides, 425
      with alkyllithium reagents, 230
      with amides and sulfonamides, 182, 183, 187, 190
      with amines, 179–183, 190, 270, 445
      with aromatic compounds (substitution), 269
      with azulenes, 271
      with boron compounds, 164

Orthoformates—*continued*
  trialkyl—*continued*
    reactions—*continued*
      with carbamates, 448
      with carbonyl compounds, 149–151, 273, 274
      with carboxylic acids, 152
      with carboxylic anhydrides, 52
      with dialkyl phosphites, 166
      with diazo compounds, 271
      with diethyl maleate, 272
      with difluoroamine, 425
      with diketene, 267
      with enol acetates, 269
      with enol ethers, 267
      with germanium tetrachloride, 165
      with Grignard reagents, 224–227
      with halosilanes, 164
      with hexafluoroacetone, 163
      with hydrazides, 188
      with hydrazines, 183, 188
      with hydrogen cyanide, 266
      with hydrogen halides, 167
      with $3\beta$-hydroxy-$\Delta^5$-steroids, 147
      with isocyanates, 199, 425
      with ketene, 267
      with lithium aluminum hydride, 231
      with nitric acid, 165
      with organoaluminum compounds, 230
      with organozinc compounds, 229
      with phenols, 148, 270
      with phosphorus compounds, 165, 166, 212–214
      with semicarbazones, 194
      with silicon tetrafluoride, 167
      with sulfonamides, 183, 188
      with sulfonic acids (alkylation), 166
      with sulfonylisocyanates, 194
      with sulfuryl chloride, 166
      with thiols, 353
      with thionyl chloride, 166
      with trihalophospholines, 165
      with ureas, 188, 194, 435
    tetrazines from, 208
    thiadiazines from, 207
    thiadiazoles from, 202
    thiazines from, 204
    thiazoles from, 196
    transesterification, 19

Orthoformates—*continued*
  trialkyl—*continued*
    transition metal alcohol complexes from, 137
    triazines from, 207, 208
    triazoles from, 202–204
    from trihalomethanes, 12–14
    trinuclear cyanine dyes from, 262–264
    tris(acylamino)methanes from, 447
    from tris(alkylcarbamyl)methanes, 478
    from trithioorthoformates, 20, 412
  triaryl, 14
Orthooxalates
  hemi, 17
  hexaalkyl, 16
Orthopropionates, *see also* Ortho esters
  trialkyl
    diazines from, 204
    imidazoles and imidazolines from, 197
    ketene aminals from, 189
    oxadiazoles from, 201
    oxazoles from, 196
    from propionamide acetals, 478
    propionimidic esters from, 184
    purines from, 210
    reaction with active methylene compounds, 231
    with acylhydrazides, 201
    with amino compounds, 182, 184
    with chloral, 163
    with diazo esters, 271
    with Grignard reagents, 228
    with hydrogen cyanide, 266
    thiazoles from, 196
    triazines from, 208
    triazoles from, 204
Ouabagenin orthoformate, 23
Oxacarboxyanine dyes, 251
Oxadiazines, 430
Oxadiazoles, 201
Oxathiazoles, 364
Oxazines, 204, 430
Oxazoles, 196
Oxazolidines, 197
Oxazolones, 196
Oxonol dyes, 196, 245

**P**

Pentamethine dyes, 261, 408
Pentathiaadamantanes, 350

SUBJECT INDEX 555

Phenetoles, 3,5-dialkyl, 274
Phenols
　alkylation, 147, 480
　formylation, 270
Phenyl cyanate, orthocarbonates from, 49
Phenylphosphonic acid, ethyl ester, 165
Phosphinine dyes, 247
Phosphites, dialkyl, 166
Phospholines, trihalo, 165
Phosphorus pentoxide, 165
Phosphorus trichloride, 165, 213
Pinner synthesis of ortho esters, 2–9
Propionamide acetals
　orthopropionates from, 478
　from propionamides, 422
Propionimidic esters, N-aryl, 182
Pteridines, 211
Purines, 209
Pyridazines, 212
Pyrimidines, 205, 212
Pyrrolidone acetals, 431

## Q

Quinazolines, 205, 212
Quinazolones, 205
Quinolines, 196
Quinolone acetals, 432

## R

Reduction, of ortho esters, 30, 231
Reimer–Tiemann reaction, 14
Reformatsky reaction, 229
Rhetsinene, 444
Ryanodol, 43

## S

Scyllitol, 26
Semicarbazones, 194
Silicon tetrafluoride, 167
Steroid ortho esters, 26, 135
Sulfonamides, 183
Sulfones, alkyl thioalky, 360
Sulfonic acids, alkylation, 166
Sulfuryl chloride, 166

## T

Tetrabutylammonium bromide, epimerization catalyst, 304
Tetraethyl pyrophosphate, 165

Tetrahydrofurans, 2,2-dialkoxy, 8, 37
Tetrahydropyrimidines, 204
Tetrakis(dialkylamino)ethylenes, 494, 495
Tetrakis(dimethylamino)methane, 448
　pyrolysis, 495
　reactions with active methylene compounds, 493
Tetrathioorthocarbonates
　properties, 374 ff.
　reaction with organolithium reagents, 370
　spirocyclic, 361, 364, 372
　tetraalkyl, 357
　　from dinitroso-$S$-alkylisothioureas, 363
　　halogenation, 369
　　oxidation, 369
　　reaction with trityl tetrafluoroborate, 413
　　trialkylthiomethylides from, 370
　　from trimethylthiocarbonium salts, 362
　tetraaryl
　　from dinitroso-$S$-arylisothioureas, 363
　　disproportionation reactions, 371
　　halogenation, 369
　　oxidation, 369
　　trithioorthoformates from, 370
　tetrakis(trifluoromethyl), 355
　from trithiomethylides, 359
　trithiomethylides from, 358
Tetrazines, 208
Tetrodotoxin, 42
Theophyllin, 210
Thiacarbocyanine dyes, 251
Thiadiazines, 207
Thiadiazoles, 201
Thiazines, 204
Thiazoles, 196
Thiazolones, 196
Thiazolotriazinediones, 364
Thioimidate salts, 362
Thiolacetic acid, 350
Thiolcarboxylic acids, 350
Thiolane, 2,2-diethoxy, 360
Thiol esters, 351
Thion esters, 166
Thionyl chloride, 166
Thioorthocarbonates, carbohydrate, 384
Transesterification reactions of ortho esters, 18–29
Trialkoxycarbonium salts, 138, 422
Trialkyl orthoacetates, *see* Orthoacetates

Trialkyl orthobenzoates, *see* Orthobenzoates
Trialkyl orthocarboxylates, *see* Ortho esters
Trialkyl orthoformates, *see* Orthoformates
Trialkyl orthopropionates, *see* Orthopropionates
Trialkylthiomethylides, in H-exchange reactions, 373
Triarylsulfinylmethanes, 368
Triazabicyclooctanes, 212
Triazines, 207, 208, 212
Triazoles, 202
1,2,3-Triazolines, 5,5-dialkoxy, 436
Triazolones, 202
Trichloromethanesulfenyl chloride, 15
Triethyl phosphite, 165
Trifluoroacetic acid, 349
Trihalomethyl compounds, ortho esters from, 12, 354
Trimethinecyanine dyes, 250–260, 407
Trimethylthiocarbonium salts, 362, 413
2,4,10-Trioxaadamantanes, 23, 135
2,6,7-Trioxabicyclo[2.2.2]octanes, 230
Tris(acylamino)methanes, 188, 446
Tris(carbamyl)methanes, 448
Tris(dimethylamino)methane, 446
  reaction with active methylene compounds, 493
  tetrakis(dimethylamino)ethylene from, 495
Tris(formylamino)methane
  *N*-heterocycles from, 484–487
  reaction with active methylene compounds, 494
    with 6-amino-4-hydroxypyrimidine, 488
    with fluorine, 482
1,1,1,-Tris(sulfonyl)ethanes, 360
Tris(sulfonyl)methanes, 359, 360, 367, 368, 371, 380
Trithiabicycloheptane, 354
Trithiabicyclooctanes, 354, 358
Trithiaspiraheptane, 365
Trithiomethylides, 357–359, 370, 410, 411
Trithioorthoacetates
  trialkyl
    from acetyl chloride, 352
    from ethyl acetate, 351
    from ketene dithioacetals, 372
    oxidation reactions, 367
    from thioacetates, 352
    from trithiomethylides, 358

Trithioorthoacetates—*continued*
  triaryl
    from acetic anhydride, 353
    oxidation reactions, 367
    from thioacetimidates, 362
    from thiolacetates, 352
Trithioorthoacrylate, triethyl, 371
Trithioorthobenzoates, trialkyl, 353, 357
Trithioortho esters, *see also* Trithioorthoacetates, Trithioorthobenzoates, etc.
  from acid chlorides, 352
  aldehydes from, 410
  bicyclic, 353, 354
  from carboxylic acids, 349
  disproportionation reactions, 371
  ketene dithioacetals from 353, 410
  hydrolysis reactions, 411
  $\alpha$-hydroxy, 358
  from ortho esters, 353
  properties, 374 ff.
  reaction with aniline, 414
  spirocyclic, 365
  from thioimidates, 362
  from thiolcarboxylic acids, 349
  from thiol esters, 351
  from trithiomethylides, 358
Trithioorthoglyoxalate, triethyl, 372
Trithioorthoformates
  reaction with acid chlorides, 413
    with phosphorus pentasulfide, 414
  from thiolformates, 352
  trithiomethylide salts from, 357
  trialkyl
    addition to alkenes, 407
    from bis(alkylthio)carbenes, 362
    from carbon tetrachloride, 355
    from chlorodimethylthiomethane, 356
    from chloroform, 354
    cyanine dyes from, 406
    disproportionation reactions, 356
    from ethyl formate, 351
    from formamide, 352
    form formic acid, 349
    formyl proton exchange, 373
    from methyl dichloromethyl ether, 356
    from *N*-methylformanilide, 352
    from orthoformates, 353
    orthoformates from, 21, 412
    oxidation reactions, 367, 369, 405

Trithioorthoformates—*continued*
  trialkyl—*continued*
    reactions with active methylene compounds, 406
      with acetic anhydride, 413
      with acrolein, 413
      with aldehydes, 413
      with diazo compounds, 408
      with phenylacetylene, 408
      with trimethinecyanines, 407
    thienol ethers from, 413
  triaryl
    from chloroform, 354
    from ethyl formate, 351
    from formic acid, 349
    from orthoformates, 353
    oxidation reactions, 367, 369
    from tetrahioorthocarbonates, 370

Trithioorthoformate, tris(trifluoromethyl), 355
Trithioorthopropionate, trialkyl, 353
Trithioorthopropionates, trialkyl and triaryl, 358

## U

Urea, $N,N'$-diphenyl, 190
Ureas, reaction with ortho esters, 188, 190, 194
Urea acetals, 425, 426
Urethanes, 914

## V

Vinyl ethers, reaction with orthoformates, 267, 268

# Carboxylic Ortho Acid Derivatives

*Preparation and Synthetic Applications*

# ORGANIC CHEMISTRY
## A SERIES OF MONOGRAPHS

*Edited by*
ALFRED T. BLOMQUIST
*Department of Chemistry, Cornell University, Ithaca, New York*

1. Wolfgang Kirmse. CARBENE CHEMISTRY, 1964
2. Brandes H. Smith. BRIDGED AROMATIC COMPOUNDS, 1964
3. Michael Hanack. CONFORMATION THEORY, 1965
4. Donald J. Cram. FUNDAMENTAL OF CARBANION CHEMISTRY, 1965
5. Kenneth B. Wiberg (Editor). OXIDATION IN ORGANIC CHEMISTRY, PART A, 1965; PART B, *In preparation*
6. R. F. Hudson. STRUCTURE AND MECHANISM IN ORGANO-PHOSPHORUS CHEMISTRY, 1965
7. A. William Johnson. YLID CHEMISTRY, 1966
8. Jan Hamer (Editor). 1,4-CYCLOADDITION REACTIONS, 1967
9. Henri Ulrich. CYCLOADDITION REACTIONS OF HETEROCUMULENES, 1967
10. M. P. Cava and M. J. Mitchell. CYCLOBUTADIENE AND RELATED COMPOUNDS, 1967
11. Reinhard W. Hoffman. DEHYDROBENZENE AND CYCLOALKYNES, 1967
12. Stanley R. Sandler and Wolf Karo. ORGANIC FUNCTIONAL GROUP PREPARATIONS, 1968
13. Robert J. Cotter and Markus Matzner. RING-FORMING POLYMERIZATIONS, PART A, 1969; PART B, *In preparation*
14. R. H. DeWolfe. CARBOXYLIC ORTHO ACID DERIVATIVES, 1970
15. R. Foster. ORGANIC CHARGE-TRANSFER COMPLEXES, 1969
16. James P. Snyder (Editor). NONBENZENOID AROMATICS, I, 1969

*In preparation*

C. H. Rochester. ACIDITY FUNCTIONS

QD 305 .A2 D44

DEWOLFE  ROBERT H.

CARBOXYLIC ORTHO
ACID DERIVATIVES

IN BIP 4/85  7000